DATE DUE

Grzimek's ANIMAL LIFE ENCYCLOPEDIA

Volume 1

LOWER ANIMALS

Volume 2

INSECTS

Volume 3

MOLLUSKS AND ECHINODERMS

Volume 4

FISHES I

Volume 5

FISHES II AND AMPHIBIA

Volume 6

REPTILES

Volume 7

BIRDS I

Volume 8

BIRDS II

Volume 9

BIRDS III

Volume 10

MAMMALS I

Volume 11

MAMMALS II

Volume 12

MAMMALS III

Volume 13

MAMMALS IV

Grzimek's
ANIMAL LIFE
ENCYCLOPEDIA

Editor-in-Chief

Dr. Dr. h.c. Bernhard Grzimek

Professor, Justus Liebig University of Giessen
Director, Frankfurt Zoological Garden, Germany
Trustee, Tanzania and Uganda National Parks, East Africa

VAN NOSTRAND REINHOLD COMPANY
New York Cincinnati Toronto London Melbourne

First published in paperback in 1984

Copyright © 1968 Kindler Verlag A.G. Zurich

Library of Congress Catalog Card Number 79-183178

ISBN 0-442-23035-4

Printed in Federal Republic of Germany

Van Nostrand Reinhold Company Inc.
135 West 50th Street
New York, New York 10020

Van Nostrand Reinhold Company Limited
Molly Millars Lane
Wokingham, Berkshire RG11 2PY, England

Van Nostrand Reinhold
480 Latrobe Street
Melbourne, Victoria 3000, Australia

Macmillan of Canada
Division of Gage Publishing Limited
164 Commander Boulevard
Agincourt, Ontario M1S 3C7 Canada

16 15 14 13 12 11 10 9 8 7 6 5 4 3 2 1

EDITORS AND CONTRIBUTORS

DR. SERGE DAAN
Laboratory of Animal Physiology, University of Amsterdam   AMSTERDAM, THE NETHERLANDS

DR. HEINRICH DATHE
Professor and Director, Animal Park and Zoological Research Station, German Academy of Sciences   BERLIN, GERMANY

DR. WOLFGANG DIERL
Zoological Collection of the State of Bavaria   MUNICH, GERMANY

DR. FRITZ DIETERLEN
Zoological Research Institute, A. Koenig Museum   BONN, GERMANY

DR. ROLF DIRCKSEN
Professor, Pedagogical Institute   BIELEFELD, GERMANY

JOSEF DONNER
Instructor of Biology   KATZELSDORF, AUSTRIA

DR. JEAN DORST
Professor, National Museum of Natural History   PARIS, FRANCE

DR. GERTI DÜCKER
Professor and Chief Curator, Zoological Institute, University of Münster   MÜNSTER, GERMANY

DR. MICHAEL DZWILLO
Zoological Institute and Museum, University of Hamburg   HAMBURG, GERMANY

DR. IRENÄUS EIBL-EIBESFELDT
Professor and Director, Institute of Human Ethology, Max Planck Institute for Behavioral Physiology   PERCHA/STARNBERG, GERMANY

DR. MARTIN EISENTRAUT
Professor and Director, Zoological Research Institute and A. Koenig Museum   BONN, GERMANY

DR. EBERHARD ERNST
Swiss Tropical Institute   BASEL, SWITZERLAND

R. D. ETCHECOPAR
Director, National Museum of Natural History   PARIS, FRANCE

DR. R. A. FALLA
Director, Dominion Museum   WELLINGTON, NEW ZEALAND

DR. HUBERT FECHTER
Curator, Lower Animals, Zoological Collection of the State of Bavaria   MUNICH, GERMANY

DR. WALTER FIEDLER
Docent, University of Vienna, and Director, Schönbrunn Zoo   VIENNA, AUSTRIA

WOLFGANG FISCHER
Inspector of Animals, Animal Park   BERLIN, GERMANY

DR. C. A. FLEMING
Geological Survey Department of Scientific and Industrial Research   LOWER HUTT, NEW ZEALAND

DR. HANS FRÄDRICH
Zoological Garden   BERLIN, GERMANY

DR. HANS-ALBRECHT FREYE
Professor and Director, Biological Institute of the Medical School   HALLE A.D.S., GERMANY

GÜNTHER E. FREYTAG
Former Director, Reptile and Amphibian Collection, Museum of Cultural History in Magdeburg   BERLIN, GERMANY

DR. HERBERT FRIEDMANN
Director, Los Angeles County Museum of Natural History   LOS ANGELES, CALIFORNIA, U.S.A.

DR. H. FRIEDRICH
Professor, Overseas Museum   BREMEN, GERMANY

DR. JAN FRIJLINK
Zoological Laboratory, University of Amsterdam   AMSTERDAM, THE NETHERLANDS

DR. DR. H.C. KARL VON FRISCH
Professor Emeritus and former Director, Zoological Institute, University of Munich   MUNICH, GERMANY

DR. H. J. FRITH
C.S.I.R.O. Research Institute   CANBERRA, AUSTRALIA

DR. ION E. FUHN
Academy of the Roumanian Socialist Republic, Trajan Savulescu Institute of Biology   BUCHAREST, RUMANIA

DR. CARL GANS
Professor, Department of Biology, State University of New York at Buffalo   BUFFALO, NEW YORK, U.S.A.

DR. RUDOLF GEIGY
Professor and Director, Swiss Tropical Institute   BASEL, SWITZERLAND

DR. HELMUT O. WAGNER
Director (retired), Overseas Museum, Bremen   MEXICO CITY, MEXICO

DR. FRITZ WALTHER
Professor, Texas A & M University   COLLEGE STATION, TEXAS, U.S.A.

JOHN WARHAM
Zoology Department, Canterbury University   CHRISTCHURCH, NEW ZEALAND

DR. SHERWOOD L. WASHBURN
University of California at Berkeley   BERKELEY, CALIFORNIA, U.S.A.

EBERHARD WAWRA
First Zoological Institute, University of Vienna   VIENNA, AUSTRIA

DR. INGRID WEIGEL
Zoological Collection of the State of Bavaria   MUNICH, GERMANY

DR. B. WEISCHER
Institute of Nematode Research, Federal   MÜNSTER/WESTFALEN,
Biological Institute   GERMANY

HERBERT WENDT
Author, Natural History   BADEN-BADEN, GERMANY

DR. HEINZ WERMUTH
Chief Curator, State Nature Museum, Stuttgart   LUDWIGSBURG, GERMANY

DR. WOLFGANG VON WESTERNHAGEN   PREETZ/HOLSTEIN, GERMANY

DR. ALEXANDER WETMORE
United States National Museum, Smithsonian Institution   WASHINGTON, D.C., U.S.A.

DR. DIETRICH E. WILCKE   RÖTTGEN, GERMANY

DR. HELMUT WILKENS
Professor and Director, Institute of Anatomy, School of Veterinary Medicine   HANNOVER, GERMANY

DR. MICHAEL L. WOLFE
Utah State University   UTAH, U.S.A.

HANS EDMUND WOLTERS
Zoological Research Institute and A. Koenig Museum   BONN, GERMANY

DR. ARNFRID WÜNSCHMANN
Research Associate, Zoological Garden   BERLIN, GERMANY

DR. WALTER WÜST
Instructor, Wilhelms Gymnasium   MUNICH, GERMANY

DR. HEINZ WUNDT
Zoological Collection of the State of Bavaria   MUNICH, GERMANY

DR. CLAUS-DIETER ZANDER
Zoological Institute and Museum, University of Hamburg   HAMBURG, GERMANY

DR. DR. FRITZ ZUMPT
Director, Entomology and Parasitology, South African Institute for Medical Research   JOHANNESBURG, SOUTH AFRICA

DR. RICHARD L. ZUSI
Curator of Birds, United States National Museum, Smithsonian Institution   WASHINGTON, D.C., U.S.A.

Volume I

LOWER ANIMALS

Edited by:

IRENÄUS EIBL-EIBESFELDT

BERNHARD GRZIMEK

OTTO KOEHLER

OTTO KRAUS

BERNHARD RENSCH

PETER RIETSCHEL

ERICH THENIUS

———————

ENGLISH EDITION

GENERAL EDITOR:
George M. Narita

SCIENTIFIC EDITOR:
Erich Klinghammer

TRANSLATOR:
Anne Rasa

SCIENTIFIC CONSULTANT:
Horace W. Stunkard

ASSISTANT EDITORS:
Peter W. Mehren
John B. Brown

PRODUCTION DIRECTOR:
James V. Leone

EDITORIAL ASSISTANT:
Karen Boikess

ART DIRECTOR:
Lorraine K. Hohman

INDEX:
Suzanne C. Klinghammer

CONTENTS

For a more complete listing
of animal names, see systematic classification or the index.

1. **THE FORMS AND LIFE PROCESSES OF THE ANIMAL WORLD** 25

   Introduction to the animal world, by B. Rensch 25

2. **ANIMAL BEHAVIOR** 56

   Animal behavior, by I. Eibl-Eibesfeldt 56

3. **INVERTEBRATE ANIMALS** 80

   Evolution, by E. Thenius 80
   Animal systematics, by H. Wendt 82
   Body forms with the phyla, by P. Rietschel 85

4. **THE UNICELLULAR ANIMALS** 89

   The subkingdom of protozoa, by P. Rietschel and K. Rohde 89

5. **MESOZOANS AND SPONGES** 138

   Subkingdom and phylum: Mesozoa, by P. Rietschel 138
   Subkingdom and phylum: Spongia, by E. F. Kilian 138
   Present-day sponges, by E. F. Kilian 139
   Prehistoric reef communities, by S. Rietschel 166

6. **THE COELENTERATES** 176

   Subkingdom: Coelenterata, by H. R. Haefelfinger 176
   Evolutionary development, by E. Thenius 177
   Phylum: Cnidaria, by H. R. Haefelfinger 178
   Phylum: Acnidaria, by H. R. Haefelfinger 255

7. **THE BILATERALLY SYMMETRICAL ANIMALS** 269

   Subdivision: Bilateralia, by P. Rietschel 269

8. **FLATWORMS AND GNATHOSTOMALIDS** 273

   Phylum: Platyhelminthes, by P. Rietschel 273
   Evolutionary history of the flatworms, by E. Thenius 274
   Class: Turbellaria, by P. Rietschel and P. Röben 275
   Class: Trematoda, by W. Hohorst 288
   Class: Cestoda, by P. Rietschel 298
   Class: Gnathostomulida, by P. Ax 309

9. **KAMPTOZOANS AND BOOTLACE WORMS** 312

   Phylum: Kamptozoa, by P. Rietschel 312
   Class: Nemertini, by H. Friedrich 313

10. **THE ROUNDWORMS** 323

    Phylum: Asc-helminthes, by P. Rietschel 323
    Class: Gastrotricha, by P. Röben 324
    Class: Rotatoria, by J. Donner 328
    Class: Nematoda, by B. Weischer 333
    Class: Nematomorpha, by P. Reitschel 349
    Class: Kinorhyncha, by P. Rietschel 354

11. **PRIAPULIDS, SIPUNCULIDS, AND ECHIURIDS** 356

    Phylum: Priapulida, by P. Rietschel 356
    Phylum: Sipunculida, by P. Rietschel 357
    Phylum: Echiurida, by P. Rietschel 358

12.	THE ANNELIDA	360

The segmented animals, by P. Rietschel — 360
Phylum: Annelida, by P. Rietschel — 362

13.	ONYCHOPHORIDS, WATER BEARS, AND LINGUATULIDS	387

Evolution, by E. Thenius — 387
Phylum: Onychophora, by O. Kraus — 387
Phylum: Tardigrada, by O. Kraus — 390
Phylum: Linguatulida, by O. Kraus — 391

14.	THE ARTHROPODS	397

Phylum: Arthropoda, by P. Rietschel — 397

15.	THE ARACHNIDS AND THEIR RELATIVES	403

Subphylum: Chelicerata, by O. Kraus — 403
Evolutionary development, by E. Thenius — 404

16.	THE CRUSTACEA	433

Subphylum: Diantennata, by P. Rietschel — 433
The evolution of the Crustacea, by E. Thenius — 433
The Lower Crustacea, by P. Rietschel — 435
Subclass: Higher Crustacea, by R. Altevogt — 467

17.	THE TRACHEATES	505

Subphylum: Tracheata, by P. Rietschel — 505
Class: Myriapoda, by O. Kraus — 505
Subclasses: Diplopoda, Pauropoda, Chilopoda, and Symphyla, by O. Kraus — 506

Appendix Systematic Classification — 515
On Zoological Classification and Names — 549
Animal Dictionary:
   English-German-French-Russian — 551
   German-English-French-Russian — 555
   French-German-English-Russian — 566
   Russian-German-English-French — 572
Conversion Tables of Metric to U.S. and British Systems — 578
Supplementary Readings — 583
Picture Credits — 588
Index — 589
Abbreviations and Symbols — 599

Foreword

Man's interest in animal life has perhaps never been greater than it is today. The latest studies in comparative behavior have shown that essential components of human behavior can be understood genetically as something we inherited from our animal ancestors. Therefore, an understanding of animals has become a prerequisite for an understanding of man. We are presently the last member of the many immensely complex ancestral lines in the world of living things.

What would be more understandable than the wish of a great many people to improve their knowledge of our fellow members of the animal world, to learn more about their behavior, and, with the aid of qualified scholars, to obtain a thorough and reliable overview of the entire animal kingdom?

For decades I had attempted to persuade publishers in Germany to produce a comprehensive encyclopedia of animal life based on the model of *Brehms Tierleben* (*Brehm's Animal Life Encyclopedia*) or on *Royal Natural History* by Richard Lydekker in the U.S.A. and England. But no one wanted to assume the great financial risk in such a venture, publishers claiming that people, especially young people, are no longer interested in animals and plants, but in the technology of automobiles, aircraft, and astronautics.

When *Brehms Tierleben*—in German—and Lydekker's *Royal Natural History*—in English—appeared over a century ago, a general interest in animal life was widespread because presentation of Charles Darwin's revolutionary doctrine of evolution had excited the whole world. From university seminars to tavern tables, the proposition was argued: did life on earth, in all its multiplicity, really develop through natural selection in the struggle for existence; and did man also arise in this way? Every thoughtful man who occupied himself with Darwin's teachings no longer looked upon animals as objects of scientific investigation, but as living beings which feel and act—as "our elder brothers," as Goethe's friend Herder had expressed it.

The enormous success of the German edition of *Grzimek's Animal Life Encyclopedia* has proven that the publishers' fears were unfounded. To the contrary, even though each purchaser of the first volume was obliged to buy every successive volume, the first edition was sold out before the first volume was even in print. Three additional editions followed in a short time, as well as translations into French, Dutch, and Italian.

Today, however, man is attracted to animals for other reasons than he was a century ago. The theory of evolution is no longer argued; all findings which have appeared since then have repeatedly confirmed Darwin's theory, and have made its essential features certainty. Still, it was not until 1967 that the legislature of the State of Tennessee, U.S.A., voted to permit instruction of Darwin's theory of evolution in its schools, a fact which is more humorous now than it would have been earlier. In spite of all the evidence, some laymen in the natural sciences and biology still resist the idea that "man evolved from the apes." However, those who know animals well cannot possibly consider Darwin's findings as degrading; the exact opposite is the case. One acquires respect for this greatest of all natural processes, to which we also owe our existence, respect for the fact that we have the same roots and the same structural plan and obey the same laws as the other mammals. In short, we, too, are a part of nature. Nevertheless, we have become more estranged from nature in the past few decades than ever before. For the first time in the history of mankind, which is over one million years old, a preponderance of people in highly developed nations live in cities. All of our ancestors, in many cases right up to our grandparents, lived in villages or at least in the immediate vicinity of the country and the forest. They had daily contact with horses, cattle, pigs, and poultry; they saw the animals of the field and forest, and hunted them. Suddenly, people in industrialized nations find themselves living together only with other people, in space which becomes increasingly crowded. In this kind of world we are less apt to observe animals and to find joy in the varied forms of nature. We forget all too easily that we are also children of nature; most of us can sense very clearly the danger of an inner homelessness.

Thus, what we have lost will become all the more valuable to many of us; the need for an acquaintance with animals and nature is increasing. This deep need is evident on weekends and during holidays, when people in great masses stream out of the cities into the country or visit zoological parks in ever greater numbers. In many cities, two to three times as many people visit zoological parks as attend all sports events combined. Never have dogs, cats, birds, turtles, fishes, and other pets been kept in such large numbers in city apartments and houses as today. People travel by the millions every year to national parks and wildlife refuges where plants and animals are forever protected from human disturbance and competition, and where the wild creatures magically lose their shyness of people. Only by gaining an appreciation for undisturbed nature and a respect for the existence of other creatures does man regain a proper relationship to the surrounding world and to himself.

Hundreds of thousands of people show their anger about the mistreatment of animals, in letters to newspapers and radio and television stations. Untold numbers of young people write asking us to name occupations in which they can be around animals all the time; but such occupations hardly exist anymore. Many people, who in their daily business life and in their free time have little contact with animals, want to learn more about the animal kingdom, and not in the form of sentimental stories, movies, or novels, but from specialists who are closely acquainted with the living animal. Even the biologist, including the zoologist, has become so specialized today that he would be glad to have a collective work to refer to for the basic findings in other areas.

This work has been created for all these people. This is more difficult to do than it was in earlier times, when knowledge of animal life could be brought together in just four volumes. Our work will comprise thirteen volumes in the first edition. Current knowledge of animal life and behavior has become so encompassing and diverse that even these thirteen volumes would by no means suffice if modern means of reproduction did not enable us to illustrate in color practically all the animals shown. Extensive descriptions of the appearance, which were necessary in earlier, similar works, need not be included. Color drawings tell us much more about the animals than even the best photographs can. A good color painting often replaces five to ten photographs, because all the important characteristics of an animal, never simultaneously in view in a single snapshot, can be incorporated into a single picture. Of course, it was not easy, in our era when true-to-life depiction is considered old-fashioned in art, to find enough illustrators and artists who can draw an animal as it actually is in all details. We had to engage people in all parts of the world for one cooperative effort. The illustrations involve much more time and are far more costly than color photographs. Furthermore, there are no color photographs of the majority of the rarer species and subspecies in their natural surroundings. However, we have enhanced the volumes as much as possible with color photographs that document the animal in its natural world, depict interesting behavior, or are effective portrait shots. The distribution of individual species can be surveyed on the maps along the inner edge of the text pages. Hence, lengthy descriptions of habitat and distribution can also be excluded so that more room is available for a more thorough description of the animal's behavior.

Only when one looks at the written copy of such a work does it become completely clear how much our knowledge of animals has increased in the last half century. Almost all of what we know about their life in the wild, as well as the newer information about their behavior in captivity, comes from the last decades. Studies in far-off lands no longer serve the purpose of discovering new species and bringing their bones or skins back to museums. Instead, researchers patiently study the living species and establish how certain forms interact with the other forms of life, both plants and animals, in their natural surroundings. In most distant countries, which were once associated with exciting and dangerous expeditions, institutes have been established in which scientists can do research. These comparative studies can also bear on the roots of our own behavior. Thus, we have quite consciously included man and his evolution in our work; for only when we know and recognize the presence of the animal heritage within us, instead of refuting it in blind arrogance, do we have the possibility of controlling it and making the best of what we have.

With the scope of present-day knowledge of all aspects of animal life, it is impossible for a single zoologist to survey the entire animal kingdom and compile it into a single work. Today such a work can only be written by a group of specialists from various disciplines. The individual sections on orders and families of the animal world had to be put under the scientific direction of specialists here, or even be written by them. We have attempted to engage a number of the best people in this work, from other countries as well as Germany.

For this reason we have avoided all foreign words and technical expressions which are not at once understandable to non-zoologists and non-biologists. Zoological concepts which are too significant to be removed from the text have been clarified as much as possible, so that readers can understand and become familiar with them. It was not only our love for our own terminology which caused us to do this, but also the recognition that today even educated people from other disciplines do not understand the precise meaning of certain technical expressions.

The reader will find the animals in this work described quite differently than in earlier works of this kind. This is due in part to ethological research, which has yielded a wealth of surprising information. One hundred, and even as recently as fifty, years ago one could only surmise such things. For example, the zoological encyclopedist Brehm, writing in his time, describes the cat mother as talking to its young, and states that the camel is a very stupid creature and is stubborn and stinking, with an ear-splitting roar and an "indescribably dumb-looking head on the long ostrich like neck." No one would dispute Brehm's otherwise well-deserved reputation because of this, but to us today it is scarcely conceivable that respected zoologists at that time could anthropomorphize animals with such a lack of restraint, and make judgements according to our own system of values. While some species were described as "stupid," "dumb," "ugly," or "mean," others were characterized as "clever," "gentle," "cute," or "gallant." Today we know that each species obeys innate adaptive behavioral patterns molded according to its way of life; we also know how much our own judgements about animals are subconsciously influenced and distorted by our own likewise innate releasers and key stimuli. With the recognition that we, too, are a very vulnerable part of nature, we approach our fellow creatures much more literally, and no longer make judgements that are based on our own conceptual world. Where comparisons are drawn, we do so with the objectivity of the modern behavioral scientist, whose words we read in the various selections.

In this way we hope that we can present, in spite of our deficiencies, an excerpt from the wealth of knowledge which does justice to the theme and to our readers. We have put the investigations of the last fifty years in the foreground, and thus also the recent findings on the relationships between animal and man. Man's particular position is not at all diminished by this. On the contrary, man's role can be understood in its entirety only by first having an exact knowledge of the animal world from which we have developed. I have to thank the many thousands who made advance orders (including orders for the English edition) on such a comprehensive work, based only on their trust in the editor. They have, in fact, made its publication possible. I would also like to thank all those who have written, drawn, photographed, or otherwise worked on it together with me. The work which bears my name has been a cooperative effort. I am fortunate that such esteemed scholars in all countries were prepared to develop this new *Animal Life Encyclopedia* together with me. The names of these co-editors and authors are presented in the list of contributors in the individual volumes.

This work is based on the classification of animals into naturally related groups. We all hope that it will disseminate knowledge of the animals and love for them, and that it will help to preserve room for the animals on this planet now so overpopulated by man. Future generations of man, our own descendants, should also have the opportunity to live together with the great diversity of these magnificent creatures.

Arusha, Tanzania, East Africa, Winter 1971.

B. Grzimek

1   The Forms and Life Processes
of the Animal World

Introduction to the
animal world, by
B. Rensch

Living creatures with extremely different types of form are characterized as animals. We include in these the microscopically small unicellular species such as amoebae in the same way as we do the highly complicated mammals and birds. To understand nature and the peculiarities of animals, it is necessary to establish first what the numerous species have in common and how they differ from other living creatures, from man on the one hand—who, of course, belongs zoologically to the animal kingdom—and from plants on the other. In this way, it will also be made clear which characteristics all living beings have in common.

Man and animal

It is known that man is extremely similar to the higher animals in his body form and the function of his organs. We have the same types and number of bones as have chimpanzees and gorillas, some differing only in shape and length relationships. Under the microscope, the cells of our muscles, bones, lungs, kidneys, etc., show the same type of construction as in the higher apes. Even the nerve cells of the brain can hardly be distinguished from the corresponding cells of higher animals. This indicates that the functions of the organs and structures are also similar in principle.

This conformity is based on the fact that man has arisen from the animal kingdom, actually from the monkeys. The phylogenetic series leading to man separated from that of the great apes during Tertiary times, probably ten to fifteen million years ago (see Vol. XI, Chapter 2). The oldest types included in the family of man (Hominidae), the Australopithecines (see Vol. X, Chapter 14), numerous fossil remains of which have been found in southeastern and eastern Africa, lived from about three and a half to one million years ago. It walked upright but had, however, a brain no larger than that of the apes living today. The fore-brain, especially, which is very important in complicated thought processes, was very much less developed than in present-day man. Nevertheless, these ancient prehumans were already capable of constructing the simplest stone tools.

The main characteristic which distinguishes present-day man so basically from the other animals is his much greater mental ability, which has enabled him to establish a material as well as an intellectual culture. This has only developed extensively, however, during the last 600,000 years, with the coming into existence of the genus *Homo* (first at the *Homo erectus* level or *Pithecanthropus* level), in which the brain, especially the forebrain, was already substantially larger than those of the prehumans. Together with this, the ability to speak was also present. Speech enabled man to form higher general conceptions, to express basic and logical connections in words, and to communicate personal experiences to other members of the group, especially to the growing youngsters. In this way, an intellectual tradition and a culture growing from generation to generation could be developed which, in present-day man, *Homo sapiens*, have progressed so far that, despite his phylogenetic associations, he is separated from his nearest animal relatives by a deep gulf.

The prehumans and humans therefore belong to the family Hominidae, order Primates, to which the monkeys and marmosets also belong. The crucial differences between us and our nearest relatives within the animal kingdom probably emerged gradually with the development of the genus *Homo*, these being much greater mental abilities, speech, and, through this, a definite culture.

The plant kingdom

Plants differ from animals and man in their forms and life processes to a much greater extent than man and the other animals differ from each other. The green plants are capable of autotrophic feeding, manufacturing an organic substance, sugar, from carbon dioxide and water with the aid of sunlight—that is, from inorganic substances—in their chlorophyll bodies (chloroplasts). It is therefore advantageous for them to intercept as many as possible of the sun's rays. The higher plants have thus, during the course of evolution, developed the largest possible surface area by branching and leafing. Since carbon dioxide, water, and mineral salts were present almost everywhere, they did not need mobility when seeking their food, but could anchor themselves with roots in a permanent position.

The animal kingdom

Animals, in contrast, are not capable of building organic material by the combination of inorganic substances. To live, they are thus dependent on plant and animal material (heterotrophic feeding). Since they are only capable of breaking down the main foodstuffs required—carbohydrates, fats, and proteins—with the aid of enyzmes, and of synthesizing them into characteristic substances, it was of advantage for them to develop, during the course of evolution, internal cavities in which these processes could be conducted (the stomach and intestines). On the other hand, animals must be able to move in order to find the required nourishment, and have thus developed a more or less compact body with limbs and sense organs.

The differences mentioned do not apply, however, without exception. There are also plants which live from the organic substances of other living creatures; on the other hand, among the aquatic animals we also find attached forms, among these being branched "trunks" consisting of numerous single animals, such as sponges, corals, bryozoa, and many others. In addition, among the most primitive living creatures (see Chapter 4) a clear distinction between plants and animals is quite impossible. Bacteria, slime molds, fungi, and flagellates could be included in either kingdom. The flagellates include plantlike species which contain chlorophyll as well as those which have no chlorophyll and feed like animals.

In their detailed construction and in many life processes, plants and animals are very similar to one another. All multicellular creatures are constructed mostly of microscopically small cells which, apart from a few exceptions, contain the same cell organs (organelles): nucleus, mitochondria, ribosomes, and enclosing cell membrane. Nuclear and cell division (mitosis) proceeds in the same way in both plants and animals. Above all, the genetic material, deoxyribose nucleic acid (DNA), consists, in all living creatures, of a few constant and similar basic components. Furthermore, the same laws of inheritance and mutation apply to plants and animals.

Every individual being—plant or animal—develops from germ cells; some plants and animals also develop from bud cells and spores formed from time to time by the parent generation. All these cells are capable of directing the development of the various structures comprising an individual (they are "totipotent"). They originate from the corresponding ancestral undifferentiated cells, these germ cells developing from the existing immature germ cells which were already present in the embryo. Thus, along the chain of generations, a "thread of life" connects the separating totipotential cells with one another. All these threads of life originate in an unbroken series from the ancestral species which, for their part, have diverged from other series of species during the course of evolution. Thus, finally, all plants and animals prove to be divisions of one single, many-branched stream of life, a communal pedigree which divided early into the two main branches of plants and animals.

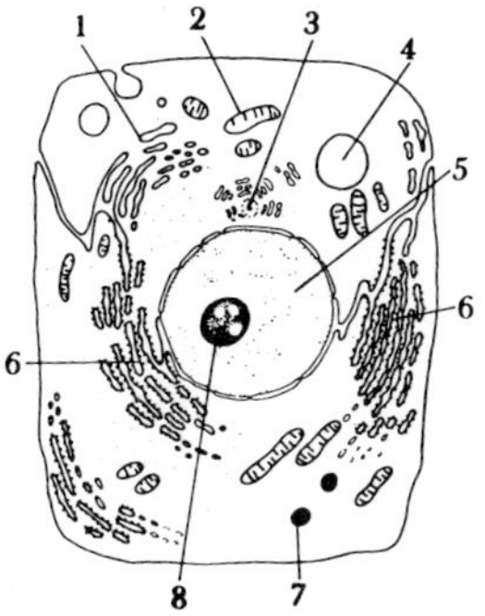

All living beings, therefore, stand in true ancestral relationship to one another. Plants and animals only demonstrate different expressions and developmental levels of a single life principle. The knowledge of this relationship should be crucial for a humane and understanding attitude toward our fellow creatures. Through the fascinating study of the animal species existing with us, we should be better able to understand ourselves, our body form, the function of our organs, and even our behavior.

As mentioned, the cells which comprise living beings have almost exactly the same basic components. They are enclosed by a fine membrane whose form and chemical composition determine which substances are to be taken up and which given out. The interior is filled with protoplasm consisting mainly of proteins and water but also containing many other compounds necessary for life. It is penetrated by many tiny channels (the endoplasmic reticulum) which are important in the transportation of substances and which, at the same time, divide the protoplasm into

Fig. 1-1. Diagram of the structures of an animal cell visible through the electron microscope. 1. Golgi apparatus. 2. Mitochondria. 3. Centrosome. 4. Vacuole. 5. Nucleus. 6. Endoplasmic reticulum. 7. Lipoid droplets. 8. Nucleolus.

numerous areas so that multitudes of highly complicated chemical processes can be carried out simultaneously without interfering with one another. Every active cell contains a large number of mitochondria, rounded or elongated structures mostly less than a thousandth of a millimeter thick; within these, with the aid of respiratory enzymes, energy-rich phosphate substances are synthesized from reduced foodstuffs, these being utilized in all parts of the body where energy for life processes is required. They comprise, to a certain extent, the cell's "power plant." The even smaller ribosomes play a part in the construction of species-specific proteins. Other cell organelles are involved in breaking down substances (lysosomes), forming secretions (dictyosomes), and cell division (nuclei).

All animal cells possess a nucleus enclosed by a membrane, which controls the processes of cell metabolism. It is lacking in only a few relatively short-lived cells such as the red blood corpuscles of mammals, including man. Some cells contain several nuclei (for example, certain protozoa, as well as muscle fibers). The nucleus contains the threadlike chromosomes which can usually be made visible only during cell division. They contain the genetic material, DNA, which ensures that the succeeding generations will develop the same structures. The individual information-carrying structures (genes) are arranged in series on the chromosomes. The number of chromosomes varies greatly among the different animal species. In the mature sex cells it varies between 4 and 127 (23 in man). In the somatic cells of almost all animals, double the number of chromosomes are present, this coming about through the fusion of male and female sex cells at fertilization. Some somatic cells also contain, as a result of doubling, a multiple number of chromosomes.

Many cells of higher animals and man have a diameter of about three-hundredths of a millimeter. Some cell processes, for example the nerve fibers, are, however, several centimeters long, and in the larger animals, even a few meters long, such as the nerve processes in the legs of the larger hooved animals. The largest living cell is the egg yolk of the ostrich, which contains relatively little active protoplasm but an extremely large food reserve (yolk).

Cell reproduction normally occurs through bisection first of the nucleus, then of the cell (mitosis). During this, the nuclear membrane is dissolved, the previously divided centrosome moves to opposing poles of the cell, and a spindlelike structure consisting of a slightly tougher gellike plasma substance is formed between them. The longitudinally divided chromosomes contract into a spiral, attach themselves by two centromeres to the spindle, and arrange themselves in a relatively even row about its center. The two halves separate and move to the opposing poles of the spindle, where they uncoil again, tangle together, and surround themselves with a nuclear membrane. At this point, the cell body bisects—usually exactly in the middle—so that two cells are formed which contain

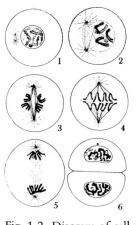

Fig. 1-2. Diagram of cell division. 1. Appearance of the chromosomes in the nucleus, the nuclear spindle between the daughter centrosomes and polar rays. 2. Extension of the nuclear spindle, formation of migration fibrils, and movement of the chromosomes to arrange themselves across the spindle equator. 3. The longitudinally divided chromosomes across the spindle equator. 4 and 5. Separation of the daughter chromosomes, and migration toward the spindle poles. 6. Uncoiling of the chromosomes, formation of a nuclear membrane, division of the cytoplasm.

the same genetic material, since, as a result of the longitudinal division of the chromosomes, all genes have been bisected. (This is discussed later in this chapter, as inheritance.)

Tissue

In multicellular animals, the majority of cells are joined to form tissues. During embryonic development they specialize to take different roles, such as skin, gland, muscle, bone tissue, etc. This division of labor has increased more and more during the evolutionary development of the higher animals, since, as a result of this, increased life processes are made possible. The danger is also present, however, that by severe injury or destruction of only one such tissue, such as the heart musculature or lung tissue, the whole organism could die.

The cells of most tissues are attached by their cell membranes to one another by special processes. In some cases, they are also bound to one another by protoplasmic connections, such as in the loosened germ layer of the mammalian skin. In the case of the invertebrate skin, the glandular epithelium of the intestine, the blood vessel walls, lung tissues, kidney tubules, etc., the cells are attached to one another only laterally, so that they form flat cell layers (epithelia). Skin glands, scales, feathers, and the ceratinized portion of hair are special structures of the uppermost skin cell layer (the skin epithelium). In contrast, connective and supporting tissues grow three-dimensionally. The many-branched bone-building cells are loosely scattered but arranged regularly between the bone substance they secrete, which consists mainly of phosphate and carbonate salts.

In addition to the elongated smooth muscle cells which are predominant in lower animals and present in the intestines, blood vessels, urinary tract, and uterus of the vertebrates, almost all animal groups from the coelomates "upward" also have striped muscle cells, which serve mainly in general movement but also in the movements of other body parts. In vertebrates these striped muscle cells fuse into long, multinucleate muscle fibers capable of contracting especially strongly.

The majority of nerve cells are characterized by long processses which connect them not only to each other but also to sense, muscle, and gland cells. The long processes of many nerve cells, which conduct the impulses from the cell body, combine to form nerves. Free, non-tissue-forming cells include the blood cells and mature sperm cells.

Organs

Animal organs usually consist of different types of tissues; the kidney, for example, consists of epithelial tubulae, connective tissue, blood vessels, and nerve fibers. In addition, organs are often combined to form an organ system which runs a working partnership. The mammalian kidney thus belongs to the excretory system, which also includes the ureter, bladder, and urethra. Although the life processes of all animals are similar, the basics are already present in the protozoa. They have, however, in adapting to extremely varied types of life, developed very different forms and, during the course of earth's history, become modified in various ways.

The activities of all organs and structures are based on metabolism and energy exchange. All animals take up nourishment and convert it within their bodies to species-specific compounds (assimilation). They also break down compounds and use the energy thus released for life processes including muscle movement, nerve excitation, glandular activity, and many others.

Metabolism and energy exchange

The most important food substances of animals are proteins, carbohydrates, fats, mineral salts, and water. The nitrogen-containing proteins are essential for building tissue. Animals such as the butterfly, which feed on nectar—that is, carbohydrates—as fully developed forms (imagoes), must, for protein generation, as in egg reproduction, depend on the protein reserves laid down in their larval stage; they also have a very limited life span. Many species, during the course of evolution, have specialized on very definite types of food, only on meat, insects, blood, or plants, and sometimes even on special plant genera or families—for example the vine aphid, pine weevil, and spurge hawk moth caterpillar. Other species live on a mixed diet. Many parasites and species which form associations to the benefit of both partners (symbionts) live only on or within very definite species or genera of host animals or host plants.

Food substances

Apart from the main food substances, many animals require small amounts of vitamins which are necessary for the growth of various structures (e.g., bones) and cell types (e.g., blood corpuscles and retinal cells), and for the formation of hormones and respiration enzymes, as well as other processes. Lower animals are capable of synthesizing many vitamins themselves. In the higher animals this ability has been lost to some extent since they already take up vitamins or their precursors with their food. Traces of a few rare elements are all essential for many species; these, however, are usually present in the normal food. Vertebrates, for instance, require iodine for the formation of the thyroid gland and fluorine for the formation of tooth enamel; land snails need copper compounds to synthesize the blood-coloring agent hemocyanin.

Water is absolutely essential for metabolic processes. It serves as a transport medium for dissolved substances in the body, and enables many biochemical reactions to occur at all, since these can only be brought about in solution. Water also controls the turgidity of many structures, which is intrinsic for life processes such as the establishment of organ form, normal muscle work, and similar things. Water is usually taken up through the mouth by drinking or eating water-containing food from animals or plants; land snails, earthworms, and many lower animals take up their necessary water through the skin. Apart from this, water is also formed during certain compound conversions within the body, especially in the breakdown of foodstuffs to produce energy (biological oxidation). Mammals consist of 60 to 70 percent of water, tadpoles from 90 to 93 percent, and comb jellyfishes from 98–99 percent.

Most animals take up solid food through the mouth, in many cases chopping it smaller with chewing structures within the mouth. Shell animacules and the higher crustacea also have chewing teeth in the stomach. Seed-eating birds such as chickens, pigeons, and sparrows have a grinding stomach with strong musculature, in which swallowed pebbles help in grinding up the food. Sessile water animals usually circulate fine food particles past themselves by means of ciliary action, while barnacles and goose barnacles do so by movements of their legs. Animals lacking intestines, such as the tapeworm (see Chapter 8), spiny-headed worms (see Chapter 10), the marine pognophores (see Vol. III, Chapter 16), or the female of the barnacle *Sacculina*, which parasitizes higher crustaceans, take up nourishment through the skin.

The majority of animals, however, have an intestine, which in coelenterates, flatworms, and flukes is baglike and often branched. In all higher animals it proceeds as a stretched or coiled tube through the body, terminating in the anus. This allows the food stream to flow through continuously, and enables a division of labor: vertebrates, for example, cut up the food and moisten it with saliva in the mouth cavity, knead it with hydrochloric acid and digestive enzymes in the stomach, and allow other enzymes in the connected portion of the intestine to act so that proteins, carbohydrates, and fats are broken down into simple substances; finally these substances are taken up by the intestine wall (resorption), and the indigestible remains are pushed out through the lower intestine. The food mash is moved slowly through the intestines by progressive contractions of the intestinal musculature from front to back (peristalsis).

Proteins are broken down to amino acids by special enzymes (peptides; in vertebrates these are pepsin, trypsin, and erepsin); carbohydrates are converted to simple sugars (monosaccharides, i.e., glucose and fructose) by carbohydrates, and fat by lipases into glycerine and fatty acids. The cells of the intestinal wall then take up these substances. In vertebrates the amino acids and sugars are carried through the portal vein system to the liver, where some of them are converted and stored; the rest are returned to the bloodstream.

Respiration

Animals obtain the energy necessary for life processes mainly by the "burning" of foodstuffs with the aid of oxygen (biological oxidation). Since oxygen cannot be stored, it must be continually taken up by breathing. According to the physical law of the reciprocal permeation of gases (diffusion), oxygen has the tendency to permeate all thin tissues in which the oxygen content is lower than that of the surrounding air (21 percent) or water (0.5 to 1.5 percent). In nearly all microscopically small species and many larger, naked-skinned ones, no special breathing organ has developed. Many other aquatic animals, especially those with a hard body covering, have thin-skinned, many-branched—that is, of large surface area—gills with a good blood circulation which oxygen can penetrate easily. Air-breathing vertebrates and land snails provide themselves with oxygen principally through their lungs. Insects, millipedes, and many spiders use a system of delicate air channels (tracheae). Even vertebrates, however, occasionally take up oxygen through the skin; frogs, for instance, use this system to take up almost half of the amount they require.

Within the gills and lungs, the oxygen penetrates the extremely finely branched blood system, where it becomes loosely attached to the blood pigment (adsorption) so that the tissue is easily able to give it up again. In vertebrates the red blood corpuscles have the task, with the aid of the blood pigment they contain (hemoglobin), to carry oxygen to all the tissues.

In the finely branched blood vessel system present in most animals, the heart or contractile blood vessel section drives the blood through the body. It thus provides the tissues with nourishment, carries away excretion products and carbon dioxide, and, apart from this, transports hormones and substances protective against poisons (antitoxins). Blood frequently serves to close wounds, especially in vertebrates. Blood vessels are lacking in sponges, coelenterates, lower worms, and in some very small species of vertebrates.

The energy necessary for animals, activity is released from the food compounds by so-called "cellular respiration." The respiratory enzymes in the mitochondria play a decisive role in this. Enzymes are protein compounds which, in small quantities, bring about reactions, especially the breakdown of substances, without being consumed themselves. Carbon dioxide and water are the end products of the complete "burning" of carbohydrates and fats; in the case of proteins, other compounds are also formed. The free energy released by the resulting breakdown processes is utilized in the synthesis of energy-rich compounds (especially adenosine triphosphate = ATP). By the breakdown of this substance in the muscles and other tissues, chemical energy is transformed mainly into mechanical energy for the carrying out of movements.

Heat is also produced in most compound conversions which, in most cold-blooded animals (poikilotherms), is lost to the periodically cooler environment. In warm-blooded animals (homiotherms), that is, birds and mammals, the feathers and covering hair greatly reduce the loss of heat, and special regulatory mechanisms (e.g., contraction of the skin tissue when cold, the regulation of metabolic rate in summer and winter mainly by the thyroid hormones, and its regulation under conditions of extreme exertion by the adrenals) maintain the body temperature at approximately the same level. This raised body temperature is of great advantage since, because of it, almost all biochemical reactions in the body are carried out more rapidly, thus making more intensive life processes possible. In mammals the body temperature reaches 35 to 40°C, in songbirds, up to 43.5°C. Moreover, during flight many large insects have a body temperature from 34 to 35°C, and this temperature is maintained within a bee hive.

Body temperature

Excretory organs

During the processes preceding energy exchange, substances are also produced for which the body has no use or that are even damaging. These elimination products (excretions) are released through the cell membrane in unicellular animals and coelenterates, and, in small quantities, also by certain higher animals. However, most of the higher animals have developed excretory organs through which the dissolved components are released to the environment. In worms, tubelike nephridia are present (see Chapter 12); in vertebrates, kidneys; in insects, malphigian tubules (see Vol. II, Chapters 1 and 10). The nitrogen compounds arising from the breakdown of proteins are converted to chemically inert substances in higher animals so that they can cause no damage on their path to the body surface. Insects, reptiles, and birds form uric acid, which, crystallized, is recognizable as the white portion of the feces. Mammals generally produce urine.

Hormones

To guarantee an ordered course life, all metabolic processes must be continually synchronized with one another. This synchronization is brought about through the nervous system and through "messenger compounds," hormones, minute quantities of which are able to influence organ and structure functions. Most hormones are carried to the organs by the bloodstream. In vertebrates a hormone produced by the thyroid gland, which is in the neck area, (thyroxin) causes a general increase in metabolic activity. It also regulates the metamorphosis of tadpoles and other larval amphibians into adult animals and elicits activity in birds during migration. Adrenaline produced by the adrenal glands acts on the sympathetic nervous system, increasing metabolism very rapidly, mainly by stimulating the liver to release reserve compounds which are very important during short-term exertion. Hormones of the pancreas (insulin and glucagon) primarily control carbohydrate metabolism; other hormones influence calcium-compound metabolism, mineral-salt metabolism, and water-content regulation. All these hormone-secreting glands are, for their part, stimulated and controlled by the hormones of the pituitary gland (hypophysis). Hormones promoting growth and the development of the sex cells, especially the regulation of the female ovum-ripening cycle, are also formed here. Other sex hormones are produced in the interstitial cells of the testes and the cell groups surrounding the ripening ovum (follicles); they control the origination of the differences between the sexes. The male sex hormone (testosterone) is also capable of releasing the fighting instinct and display behavior. The corpus luteum, formed from the residue of the egg follicle, secretes prolactin, which stimulates milk production and also awakens the "mother instinct."

Hormones which are produced by the nerves themselves (acetylcholine, adrenaline, noradrenaline) also play a part in nerve stimulation. These hormones probably represent, evolutionarily, the first hormones to come into existence, since they are found in invertebrates and even in unicellular animals. In arthropods, hormones created in transformed cells (neurosecretory cells) control molting, pupation, the development of the mature form, rapid color change, and the functions of the heart, intestines, and excretory organs.

Since almost all animal species are able to move, because they must search for special food and find a sexual partner and protection from enemies or damaging weather conditions, they have, during the course of evolution, developed sense organs through which they are able to inform themselves about their environment. Such sense organs have mainly been adapted to mechanical stimuli (contact with the environment), but also to chemical stimuli, especially those emitted by food or the sexual partner, and to temperature and light stimuli, which allow responses to be made to objects at a greater distance. Various groups of animals also have sense organs which respond to sound waves, gravity, or electric fields (e.g., in electric fish).

Sense organs

The light rays perceptible by humans represent only a small section of the electromagnetic spectrum. Animals have not developed sense organs for the determination of shorter wave lengths, such as X-rays, or longer ones, such as radio waves, since these sections of the wave-length spectrum are of no significance for their ways of life. Actually, animals respond only to a definite, biologically significant range. Many higher mammals, including man, can perceive sound waves only between sixteen oscillations per second and twenty thousand oscillations per second. Most bats, however, can perceive sound waves up to seventy thousand (the maximum is 175,000). Grasshoppers also have a similar upper-perception limit. Animal sense organs are thus constructed to perceive only a selection of the many environmental factors and to respond only to those which are actually useful to them. Many mammals, including man, cannot hear ultrasonics, which bats and grasshoppers can hear, since ultrasonics have just as little special significance for most mammals as does ultraviolet light, which, however, can be perceived by bees and ants.

Even unicellular animals react in a definite way to different stimuli; almost all respond to touch and chemical stimuli emitted by food or the sexual partner; many react to light, and some even to electric current. Multicellular animals perceive touch or temperature stimuli either through finely branched nerve endings or special sense organs. In arthropods these are sensory hairs, in vertebrates, mechanoreceptors containing lamella-like elastic cells. The invertebrate sense organs end in a more or less elongate nerve process. In vertebrates, apart from such "primary sense organs," (odor receptors, visual receptors), "secondary sense organs" which have no nerve process of their own but are surrounded by finely branched nerve endings to which they transmit their impulses, are also present (mechanoreceptors, taste buds, auditory receptors, gravity receptors of the labyrinth).

During the progressive evolutionary development of animals, the individual sense organs were perfected by including continually more and differently responding sense cells. Thus it became possible to distinguish more details of a stimulus group, and to respond more variably to them. Earthworms, for example, are only capable of distinguishing between light and dark with their scattered light-sensitive cells. Marine annelids can perceive the light direction with their eyepits, and the camera eyes of cephalopods (see Vol. III) and vertebrates, with many more visual cells, are finally capable of seeing a picture, thus allowing response to details within it.

Within the various groups of animals, sense organs with the same capabilities can be present in very different parts of the body. In scaleless fishes, such as the catfish, taste cells are scattered all over the body, even as far back as the tail. Some types of flies and butterflies have taste cells beneath the foot joints (tarsi). In insects, odor receptors combine to form pit organs or pore plates which are present in large numbers on the antennae. The hearing organs of insects are either present in the thorax, as in short-horned grasshoppers and moths, the abdomen (in cicadas), the tibia of the fore or middle leg (in long-horned grasshoppers), or in the antennal base (in mosquitoes).

Some sense organs do not perceive the stimuli from the environment, but inform the animal about processes taking place within its own body or about the immediate position of parts of the body relative to one another. For example, many insects have fine sensory hairs between the head and thorax or thorax and abdomen, serving as such "proprioceptors"; in mammals, these are the tendon organs and muscle spindles.

Man is sensitive to the majority of excitations arising from sense cells. We see, hear, taste, smell, and are sensitive to touch and temperature. It is very probable that other animals, especially the more highly developed ones, can also perceive the same sensations. Many day-active mammals are capable of distinguishing between colors in the same way as ourselves, and it is probable that they also have the same sensibilities, since the excitation processes in the eye and brain occur in the same way as in man. We cannot determine, however, what color experiences a bee, which is also capable of perceiving ultraviolet light, has, or what an electric fish capable of perceiving changes in an electric field, feels. Training experiments have proven, however, that in fishes and bees, a white or gray patch on the edge of a colored area is seen as its complementary color (thus a white or gray section around an orange center is seen as a blue edge).

Nerve cells

Nerve cells (neurons) to which the excitation from the sense cells is transmitted, are found in all multicellular animals with the exception of the unorganized animals (mesozoans and sponges). In coelenterates these cells have only a few threadlike processes which can conduct the impulses in two directions. In all other animal groups, numerous richly branched processes (dendrites) conduct the impulses to the cell body while a longer process (axon or neurite), also highly branched at its tip, conducts the impulses away. In many cases, the axons or dendrites of numerous nerve cells combine to form a nerve leading from a sense organ to a neural center in the brain (sensory, afferent fibers) or from there to muscles and glands (motor or general afferent fibers). If a nerve is severed, the processes detached from the cell body die, but the fiber stump attached to the cell is capable of regrowth.

During the course of evolution, after nerve cells came into being at the level of the lower multicellular animals, they were retained in all developmentally higher forms, increased in number, and perfected, since they were of advantage in many respects. Owing to their long processes, impulses could be conducted from one part of the body to another, enabling the activities of individual organs to be synchronized with one another. This was one of the important factors leading to the creation of larger animals. Certain special connections between nerve trunks which were then retained proved themselves valuable during the further course of evolution, since they enabled sensory stimuli of special meaning for an animal species to be transmitted through a neural center, releasing either muscular or glandular activity advantageous in responding to the sensory stimulus. This type of reflex therefore allows the animal to respond Reflexes automatically in a meaningful way. When a frog, for example, is stimulated by an insect sitting on its back, the corresponding excitation is transmitted through a connection in the spinal cord, and leads to a movement of one hind leg, with which it scratches away the insect. All animals have a large number of such hereditary reflex-connections by which feeding, glandular activity, the course of species-specific reproductive behavior, and other life processes are determined. At the same time, reflexes form the basis of complicated instinctive activities in which sensory stimuli **Instinctive behavior** from the environment also exert a controlling influence. **patterns**

The development of nerve cells has also yielded a further, very important advantage: expired impulses are able to leave traces behind them, these memory traces (engrams) being capable of influencing the **Memory traces** course of later impulses. Learning ability and memory are based on this, capabilities which were of decisive importance in the behavior of all higher animals and for the evolutionary development of man.

In the instigation and transmission of the impulse along the nerve fiber, **Speed of transmission** minute electric currents are formed which allow us to determine the **of an impulse along a** route of the impulse and its cooperative function in the brain; this can **nerve** be done by diversion of the current through the finest electrodes, and its amplification many thousands of times with a cathode ray oscilloscope. The speed of impulse transmission in the nerve is faster in higher animals than in lower ones. In the nerves of the vine snail's creeping foot this is 5 to 20 cm per second; in the leg nerve of the cat, in comparison, 80 to 119 cm per second.

Although in almost all coelenterates the nerve cells are scattered net-like over the whole body, they have, in other animal groups, formed

Neural centers

neural centers by cell aggregation. These have the advantages that many impulses can be connected to one another and that a coherent regulation of the activities of all organs connected to the center can be brought about. In annelids and arthropods such centers are joined to form a ladderlike nervous system, the ventral nerve cord. The thoracic centers, for example, are capable of determining the coordination of the leg movements. These movements can, however, be inhibited or promoted by the controlling brain.

In many branches of the vertebrates the brain has become increasingly rich in nerve cells, and, at the same time, a division of labor has come about within the various areas of the brain. In addition, the forebrain (cerebrum) has divided into many areas with different functions. All complicated behavior patterns are initiated from here, especially those directed by will. Above all, the forebrain also acts as a storage organ for many memory traces.

Sympathetic nervous system

The spinal cord conducts sensory impulses to the brain, and, from there, the motor impulses to parts of the body. At the same time it is, however, an independent neural center which controls the coordinated movements of the body parts, and directs various reflexes. A sympathetic nervous system controls the activity of the internal organs of the body cavity, completely independently of the brain, but it is also involved in the excitation or inhibition of other life processes.

Reproduction

During the course of life, the body tissues change irreversibly. This leads to aging and finally to death. Animal species can therefore only continue to exist if they reproduce themselves. It was thus necessary for special cells or cell groups to remain undifferentiated so that they still retained the property, though their multiplication, of passing all structures and organs on to the following generations (totipotential cells).

Unicellular animals are capable of reproducing by binary fission, in which, as in every normal cell division, first the nucleus and then the cell body divides. The two daughter cells are at first small, and therefore have a greater surface-to-volume ratio. Since this is favorable for metabolism, it represents a rejuvenation. Such asexual reproduction is also present in

Asexual reproduction

sporozoans; these form spores, from every one of which a new individual develops. Some multicellular animals—especially sponges, coelenterates, certain flatworms and annelids, bryozoans, and salps—are able to reproduce by budding, in which juvenile nondifferentiated or hardly differentiated tissue sections are constricted off. A type of asexual reproduction is also involved when the germ cells of mammals divide into two or four, so that uniovular or identical twins or quadruplets develop. In armadillos, for example, uniovular quadruplets or octuplets occur regularly.

Sexual reproduction

Even in the unicellular animals, however, sexual reproduction has also developed: two reproductive cells, almost always ovum and sperm cells originating from two different individuals, unite with one another. In this fertilization, however, the chromosomes of the partners remains separate, so that a double chromosome complement (diploid) results. By far the majority of animals thus have, within the body, cells originating from the division of the fertilized germ cell and, in addition, in the immature sex cells, a double chromosome complement (in certain body cells, even a multiple number). Simultaneously with the evolutionary development of sexuality, therefore, a mechanism had to develop by which the double chromosome complement could be halved again (haploid), otherwise it would, in subsequent generations, become fourfold, eightfold, sixteenfold, and always further increased during fertilization. This reduction to half the chromosome number occurs during maturation division (reduction division or meiosis), in which the corresponding paternal and maternal chromosomes controlling the same inherited characteristics come together. In cell division, entire chromosomes are divided between the two daughter cells rather than, as in normal cell division, split halves of single chromosomes.

The meaning of the separation of the sexes in most animals is, however, less concerned with associated reproductive activity itself than with the fact that the varying inherited abilities of two individuals are brought into a new association with one another. This is important, since, in every generation, severe natural selection processes are at work, and, as a result, only a relatively small percentage of individuals reach maturity. Unusual environmental factors, such as a very cold winter, the appearance of new disease organisms, new enemies, or new, stronger food competitors, may allow only a few animals from a population to survive, those with the correct combination of inherited characteristics. And, in general, the appearance of new contagious diseases results in only a few immune variants surviving.

Many invertebrates, such as tapeworms, flukes, earthworms, land snails, and others, are bisexual (hermaphroditic); they have both male and female germ cells. These types of animal are usually first sexually mature as males, and copulate with one another, the sperm cells of the reproductive partner being retained in a special sperm sac until the ova ripen (protandry). In other cases, a generation change is present, asexually and sexually reproducing generations alternating with one another, as, for example, in malaria parasites, coelenterates, bryozoa, and tunicates; or bisexual generations follow those in which individuals develop from non-fertilized eggs, that is, through parthenogenesis. These parthenogenic types are found among the flukes, rotifers, waterfleas, aphids, bees, and other groups.

The reproduction of many individuals is best ensured in unisexual animals when the same number of males and females are present. Sexual determination in numerous cases is brought about by one of the sexes—usually the male—producing two different types of mature sex cell; one pair of chromosomes of the double complement present in the original sex cells differing from the other. Half of the sperm cells in mammals, for

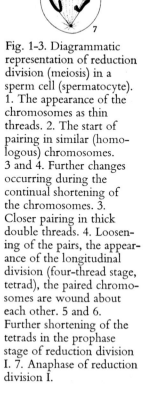

Fig. 1-3. Diagrammatic representation of reduction division (meiosis) in a sperm cell (spermatocyte). 1. The appearance of the chromosomes as thin threads. 2. The start of pairing in similar (homologous) chromosomes. 3 and 4. Further changes occurring during the continual shortening of the chromosomes. 3. Closer pairing in thick double threads. 4. Loosening of the pairs, the appearance of the longitudinal division (four-thread stage, tetrad), the paired chromosomes are wound about each other. 5 and 6. Further shortening of the tetrads in the prophase stage of reduction division I. 7. Anaphase of reduction division I.

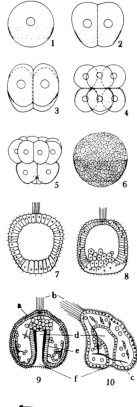

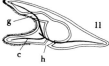

Fig. 1-4. Egg cleavage, germ-cell layer formation (gastrulation), and larva formation in the sea urchin. 1. Fertilized egg. 2-5. Egg cleavage. 5. Sixteen-cell stage. 6. Blastula. 7. Section through the blastula. 8. The moving in of cells of the middle germ-cell layer (mesenchyme cells). 9. Section through the gastrula. 10 and 11. Longitudinal section through a young larva (a) ectoderm, (b) apical cilia, (c) skeletal peg, (d) archenteron, (e) endoderm, (f) stomodaeum, (g) true mouth, (h) anus.

instance, contain a so-called X chromosome, the other half a Y chromosome, while all mature ova have only an X chromosome. Therefore, during fertilization, XY types (males) are produced equally frequently as XX types (females). In other cases, the X chromosome of the double-chromosome complement has no partner (XO= male, XX= female). Apart from these inherited (genotypic) sexual determinations, it is not yet clear how non-inherited (phenotypic) sexual determination comes about, the sex being determined here at a later stage of development through environmental factors. In some unicellular animals, coelenterates, annelids, and other animals, non-inherited sexual determination can come about through the cooperation of many inherited factors determining sexuality, and through an imbalance, at the beginning, between the expression of male and female characteristics. In all hermaphrodites the neutral, original sex cells first divide during the course of embryological development into male and female germ cells.

In many cases of definite unisexuality, the sexes can, however, be transformed by certain influences. In the marine annelid *Ophryotrocha puerilis* the developing animals are first male and become female only in the fully mature form. If one cuts off the posterior two-thirds of a female, however, the remaining anterior part changes to male again. Even in vertebrates the sex determined by inheritance is capable of changing when the sexual hormone of the opposite sex is injected several times during the course of development. Male chicks can, in this way, develop into hens, and female rodents, into males.

During the course of evolutionary advancement in animals, a definite difference between the sexes has developed in most cases. In many arthropods, especially the insects, but predominately in vertebrates, the reciprocal recognition of the sexual partner is facilitated by different colorations, vocalizations, odors, or deviations in the instincts of male and female. Male insects are often more active and equipped with better sense organs than are the females; male vertebrates have varied and special courtship and pairing instincts, and defend the breeding area. Maternal instincts, in so far as they have appeared in animals, are usually more strongly, or exclusively, developed in the female.

The development of individual multicellular animals dependent on fertilization or budding for individual development (ontogeny) takes place by the cells dividing several times and, owing to this, becoming generally differentiated; in this way various tissue types, organ rudiments, and finally complete organs come into being. In the majority of animals the fertilized egg is first divided into increasingly smaller cells through a series of cell divisions. This "egg cleavage" ends in many cases with the formation of a unicellular layered blastocyst (for example, in the coelenterates, echinoderms, and mammals) or in a blastodisc (as in cephalopods, fishes, and birds). Finally, cell multiplication leads to folding and extrusions, and usually to the development of an internal and external cell layer (ectoderm and endoderm). In sponges and coelenterates all tissue and organ structures are formed from these two cell layers, which, incidentally, are not always clearly differentiated in sponges. All higher animals develop, in addition, a middle cell layer (mesoderm) or a comparable cell arrangement.

In numerous animal groups, such as land snails, reptiles, birds, and mammals, further development leads directly to the basic form of the mature animal. In many aquatic animal species and the majority of insects, on the other hand, a completely differently shaped larva is formed, which can be distinguished by special organs and a way of life completely different from that of the adult. Butterfly caterpillars have chewing mouthparts and feed on leaves; the butterfly, in contrast, feeds on water and nectar, using completely differently shaped mouthparts. The metamorphosis to the mature form therefore requires relatively extreme and often rapid structural change. In insects with complete metamorphosis (beetles, butterflies, sawflies, wasps, flies, and others), the nervous system and air channels (trachea) are usually transformed while mouthparts, eyes, antennae, legs, and wings are recreated during the pupal stage from undifferentiated cell complexes (imaginal discs).

During cell division in earlier embryonic stages it becomes clear that in many groups of animals the first cells to be cleaved have all the capabilities of building a complete individual. When one divides the cells of a sea urchin embryo at the four-cell stage, four sea urchin larvae develop. In the same way, the occasional appearance in mammals, including man, of "uniovular" twins or quadruplets is made clear, these being multiple births with the same inherited abilities. In such cases, the first two or four cleavage cells have not adhered to one another sufficiently.

The development of the different tissues and organs is mainly determined by the inherited factors of the cells in question, but also in part by environmental influences around the developing tissue. This "neighbor influence" (induction), discovered by the German Nobel Prizewinner Hans Spemann (1869–1941), was first proven by experiments with newt embryos. Spemann and his co-workers removed the upper lip of the blastula (a tissue section which should develop into a section of the middle cell layer) from newt germ cells at a developmental stage when both embryonic cell layers were already completely developed, and transplanted it in an embryonic cell layer destined to be the future abdomen. Owing to the influence of this tissue, the future development of the abdominal tissue overlying it was induced to form a second spinal cord. An embryonic eye inserted under the abdominal tissue of a newt embryo induces the abdominal tissue to form a lens for the developing eye. Since then, such "neighbor influences" (inductions) have been proven during organ development in many animal groups.

Many animals are capable of forming new tissue sections and even entire organs after loss. Crabs, insect larvae, and newt larvae are able to

Development

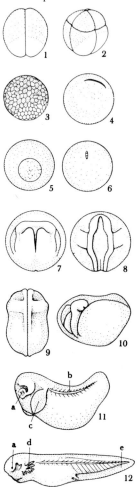

Fig. 1-5. Stages of development in the common frog. 1 and 2. The two-cell and eight-cell stages (stippling: pigmented areas). 3. Morula. 4. Early gastrula with stomodaeum (black). 5 and 6. Gastrula with yolk sac. 7. Formation of the neural crest (precursor of the brain and spinal cord. 8 and 9. Neurula (dorsal view). 10. Neurula (side view). 11. Embryo with primitive segmentation (b) and adhesive-pad precursor (c) together with the eyes (a). 12. Early larva with external gills (d) and tail fin (e).

Regeneration

regenerate lost legs. In fresh-water polyps and flatworms such regeneration abilities go so far that entire individuals can develop from very small portions of the body, since some of the cells scattered throughout the tissues still retain the ability to form all organs, being "totipotent," similar to the germ cells. Sponges, annelids, and starfishes are also capable of extensive organ regrowth (regeneration). In birds and mammals this regeneration is restricted to the healing of wounds and broken bones.

Growth rates

During their development, individual organs and structures show different rates of growth. Some grow at the same rate as the body in general; others, however, grow faster or slower. This has the result that young and old animals or closely related small and large subspecies almost always exhibit different body ratios. In rodents, carnivores, monkeys, and man, the head and brain, for example, grow faster during the embryonic period than does the rump, and are, therefore, relatively larger at birth. They later grow more slowly, and are, in the mature animal, small in comparison with the body. In contrast, the legs are comparatively short at birth but later grow faster than the rump for a certain time.

Life span

Toward the end of life, signs of age appear in all animals, these being usually associated with changes in connective tissue and with the destruction of nerve cells. The average length of life for an individual is, to some extent, hereditarily determined for every animal species. The males of *Hydatina senta*, for example, live for about three days. Many May flies live for a year, but as mature insects for only a few hours in most cases. House mice reach an average age of one and a half years; elephants, sixty to seventy.

Inheritance

To be capable of fulfilling the demands of their species-specific life, the individuals of a species must, in each generation, retain more or less the same form. This is ensured through hereditary factors passed on by the parents. Molecules of the most important hereditary substance, deoxyribonucleic acid, usually abbreviated as DNA, are found in the sexual cells from which a new generation develops. These extremely long, threadlike molecules are able to reproduce each other exactly. Thus, germ cells with the same hereditary factors come into being and carry the same hereditary characteristics on to the next generation. Surprisingly, DNA is formed from the same components in all animals. It consists of two spiraling, intertwining cords in which one sugar and one phosphate group alternate in series. The cords are joined to one another by four identical bases, cytosine and guanine linking together as do adenine and thymine. The differences between the hereditary substances in all other animals and man are mainly based on differences in the serial arrangement of the base pairs and on the length and number of the DNA molecules. Owing to this, the relationships of all species to one another and the unity of life is to a great extent elucidated. Recent investigations have already shown how differences in the arrangement of the base pairs relates to the synthesis of different species-specific proteins, this leading ultimately to the development of a species-specific body form.

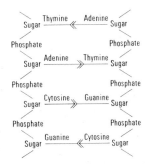

Fig. 1-6. **Base arrangement in a DNA double molecule.**

The unity of all living beings is also illustrated by the fact that the hereditary characteristics of plants and animals follow the same genetic laws. The three laws discovered in 1865 by the Augustine monk and teacher Gregor Mendel (1822–1884) are, briefly, the following:

1. If two individuals of the same species, differing from one another in only one hereditary characteristic, are crossed (as, for example, a black cock and a white hen of the same breed), all the young look alike; they are, in respect to the characteristics in question, in the middle (gray color, mixed inheritance) or exhibit the characteristics of only one of the parents (e.g., color black, covering or dominant inheritance).

2. The descendants of this generation show, in the case of mixed inheritance, fifty percent mixed color (gray) and twenty-five percent each of the pure inherited characteristics (black and white). The characteristics of the grandparents thus appear again. In dominant inheritance, seventy-five percent show the character concerned (black), of which only twenty-five percent are pure, and twenty-five percent the repressed (recessive) character (white).

3. On crossing two individuals which differ from one another in more than one character, the hereditary processes progress independently of one another, and can be combined as required. This rule only applies, however, when the characteristics in question are not present on the same chromosome, that is, when they are not linked.

The way in which the numerical relationships in Mendel's laws come about has only been explicable since the beginning of this century, through cellular investigations. As mentioned already, the double chromosome complement is reduced by half on the maturation of the ova and sperm cells. In the first mixed-inheritance generation, during maturation division, fifty percent of the ova and sperm cells each contain the hereditary tendency for one character (black) and fifty percent for the other (white). On crossing individuals of this first mixed-inheritance generation, four different character combinations could thus appear with the same frequency.

In our example with chickens, the following can thus occur:

1. An individual with the tendency for black can be fertilized by a sperm cell with the same tendency. A genetically pure black chicken thus results.

2. Ova and sperm cells with the character for white can meet each other, on an average, with the same frequency, by which a genetically pure white animal develops.

3. On the other hand, an ovum with the character for black can meet a sperm cell with the character for white, or,

4. Conversely, an ovum with the character for white can meet a sperm cell carrying the character for black. Thus, twenty-five percent in each case—or together fifty percent—mixed color (gray) animals result.

In the case of many characteristics, the hereditary events are difficult to estimate, since several hereditary factors are involved in their occurrence. A single hereditary factor (gene) often controls several character-

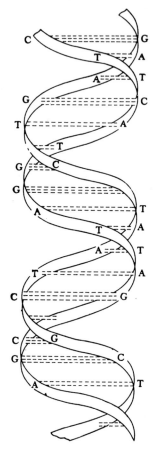

Fig. 1-7. Model of the DNA double helix according to Watson and Crick (1953) A Adenine, C cytosine, G guanine, T thymine. The two molecular chains of the double helix are joined together by hydrogen links (broken horizontal lines).

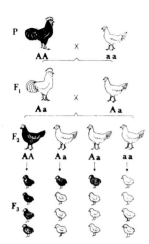

Fig. 1-8. A cross between a white and a black strain of chickens. Differences in one pair of character- istics (= monohybrid). Intermediary crossed (descendants with mixed inheritance of the individual characteristics of both parent strains) A: gene for black, a: gene for white, AA: genetic formula of pure blacks, aa: that of pure whites, Aa: chickens of mixed inheritance (heterozygotes), intermediary crosses, mainly black with white spots which, on further crossing, always give 1/4 black, 1/2 speckled, and 1/4 white offspring. P: parental generation, F_{123}: first, second, and third filial generations.

Evolutionary theory

istics. Apart from this, environmental factors sometimes play a part during development. Because of this, genetic characteristics (genotype) cannot always be inferred from the visible form (phenotype).

Each special hereditary factor (each gene) usually remains unchanged throughout hundreds of thousands of generations; then a sudden change (mutation) may appear. The arrangement of genes on the chromosome can change through an exchange of pieces between two chromosomes, and even the number of chromosomes can change, mainly by multiplica- tion. These mutations, which form the basis of all organisms and their hereditary characteristics, lead aimlessly to changes in form type and abilities. Since they disturb a long harmony developed during evolution, the majority of them are disadvantageous or damaging, and often even fatal. They only prove to be of advantage in a few cases. As a result of such mutations and the constant recombination of hereditary characters through fertilization, all species exhibit variability to a greater or lesser degree, and they are thus able, during the course of time, to change gradually.

Mutations were first discovered at the beginning of this century. It had been known for a long time previously, however, that species were not fixed. This knowledge led to the development of the evolutionary theory, which was mainly initiated by Jean Baptiste Lamarck (1744–1829), in 1809, and later consolidated by Charles Darwin (1809–1882), in 1859. Darwin's colleagues and followers brought forward further evidence in support of this. Fossil remains of animals from stone deposits lying above one another led to the acceptance of the idea that species had gradually changed into other species. Often several new species developed from a single species, and, owing to the continuation of these transformations, new genera, families, orders, and finally new structural forms thus came into being. This was especially evident in animal groups leaving plentiful fossil remains, the Foramenifera, corals, mollusks, and mammals, for example. In many cases a higher developmental level could be established in a long evolutionary series, this usually being characterized by the formation of more advantageous structures.

The evolutionary tree of the horse species is a classical example of such a developmental progression. This group of animals can be traced back to the genus *Eohippus*, which lived during early Tertiary (Eozoic) times. These small animals, about the size of an average dog, had four toes on the fore foot and three on the hind. Development went along several lines during the course of the Tertiary, leading to larger forms, first with three-toed feet, and finally to the present-day one-toed genus *Equus*. At the same time the cushion teeth, adapted to feeding on a mixed diet, developed into crushing teeth with a flat, highly convoluted crown, which were well adapted for chewing grass. The original, almost smooth, cere- brum became larger and more convoluted during the course of evolution.

On the coming into existence of new species, many organs also changed their capabilities. For example, the original five-toed leg of the reptiles became transformed into wings in the case of the birds, which arose during Jurassic times; in the mammals, arising from another branch of the reptiles, they developed into running legs with a reduced number of toes (hoofed animals), or flippers (seals and whales), or wings (bats). Despite their different functions, however, all these limbs have the same basic bone structure; they have a humerus, elbow, radius, carpals, and finger bones. In other cases, the existing organs have been reduced during the course of evolution either completely or into functionless remains (rudiments). Many flightless insects, such as the large ground beetles of the genus *Carabus*, some species of grasshopper, female mealybugs, and others, have only minute rudiments of the wings or hindwings, indicating that these forms must have developed from fully winged ones. Tiny, useless rudiments of the fore and hind legs are found in some reptiles, for example the slowworm and legless skink among the lizards and, among the snakes, the constrictors of the genus *Python*. These species have, therefore, obviously arisen from four-legged ancestors.

Even embryological and juvenile development often show exceptions which can only be explained by evolutionary processes (phylogeny). Small mammalian embryos, including that of man, first develop gill pouches like their oldest evolutionary vertebrate ancestor: the fish. Tooth rudiments first develop within the jaws of the toothless baleen whale and in the toothless upper jaw of Artiodactyls, but these soon disappear. All these "diversions" in individual development are relics of evolutionary processes which proceeded earlier in another direction, and are thus important evidence for the evolutionary development of the species in question.

The evolutionary origins of new species are now explicable to a great extent on the basis of their fundamental causative principles. Usually several causative factors effect these at the same time but in different ways. Through mutations and the manifold recombinations of hereditary factors coming about through fertilization processes, new characteristics and new variants come into being. Since all animals produce variable descendants, it was necessary for the continuation of the species that ninety to ninety-nine percent of these should die out prematurely. Generally, the disadvantageous variants are eliminated rather than the normal types and those which are somehow advantageous. This natural choice (selection) eliminates, for example, variants which are unable to withstand abnormal cold, heat, or dryness, those which are unable to protect themselves sufficiently against enemies, those which are easily susceptible to diseases, those which are not strongly enough developed to compete with members of their own species for food, living space, or reproductive partners, or those which produce too few descendants.

Such weeding out of the inappropriate, and the corresponding favoring of new, suitable variants, leads, in general, to the development of new species. On occasion, even inappropriate characteristics are retained for

▷
The growth on a shadowed rocky wall in the Mediterranean Sea. Below right, the soft coral *Alcyonium palmatum*; next to this, left and above it, the brilliant yellow colonies of the encrusting anemone *Parazoanthus axinellae*; and between these, various sponges.
▷▷ and ▷▷▷
A coral reef in the Red Sea as seen by a diver under normal light conditions (left). Artificial light sources, such as photographic flash or floodlights, are required to reveal the hidden wealth of color.

Above and below: Coral reefs are mainly formed of stony corals (Madreporaria). These can, depending on environmental circumstances, develop extremely variable growth forms, giving the underwater world a multi-formed appearance.

long periods of time—especially when their development is connected with suitable characteristics that are of great biological advantage. It was in this way, for example, that the antlers of the great elk of the ice ages became too heavy, since they were associated with large body size, of especial advantage in the cold climate. Separation (isolation) of large groups of individuals is also beneficial for speciation, since they develop differences after being separated and thereby prevented from mixing with the main population.

Animal species are usually distinguished by their inability to breed with closely related species or their production with them of sterile offspring. As a result of further mutation, gene recombinations, selection, and isolation processes, new genera, families, and other higher orders can arise from these. The formation or disappearance of organs previously present often comes about, and finally, owing to this, a completely different body plan. This development from species (transspeciation) usually leads next to a change at the same organizational level; for example, the development of the catlike body form first led to the development of the wildcat, lynx, lion, tiger, leopard, and the other members of the cat family. In certain cases, evolution also brought degeneration with it. For instance, the adaptation of animals to a parasitic mode of life on other animals or plants was usually associated with the loss of eyes and limbs.

On the other hand, a higher level of development (anagenesis) occurred in many evolutionary series through the creation of adaptive forms and behavior patterns, this also being distinguished by an improvement in the brain. Since these developments were advantageous, they pervaded the variants in question during the course of evolution. For example, the brain of members of today's carnivore families is larger in relation to the body, and, in addition, the cerebrum is much more convoluted, thus containing more nerve cells—according to brain volume analyses—than those of the primitive ancient carnivores from which they developed during the Tertiary. Man also owes his creation to such a higher development from the apes of the Middle and Upper Tertiary (see Vol. X).

Since each species of animal is associated with a particular environment where it finds its food, reproductive partner, and protection, it must, during the course of evolution, develop suitable bodily modifications. The science of these associations with the environment, ecology, first had the task of investigating the adaptation of species to the inanimate: to temperature, light, humidity, substrate, salt content of the water, and so on. In addition, it investigated the influences relative to life together with conspecifics, enemies, or other animals. Some species and genera are found very closely bound to a particular type of environment (biotope). Brown trout can only live in clear, cool, oxygen-rich streams; many blind, soft-skinned arthropods are adapted to living in caves; parasites have usually adapted to a particular host species. Such animals bound to a particular environment are called "steno-ecological forms." The "eury-

Environmental factors

ecological forms," on the other hand, are able to inhabit varying environments, and, because of this, are often widely distributed. This is the case, for instance, with many unicellular fresh-water animals, some insect species (such as the dungfly *Scatophaga stercoraria*), Norway rats, moorhens, and many other animals.

Some species are mainly restricted in their distribution by their adaptation to a narrow range of temperatures (stenothermic animals). Corals occur only in clear sea water which is warmer than 20°C; they belong to the stenothermic warmth-adapted animals. The brown trout, on the other hand, is only able to live continually in streams where the average temperature is below 15°C; it is a stenothermic, cool-adapted animal. Other animals, however, are able to stand a very wide temperature range, such as the puma, which ranges from the tropics to Canada and is also found in Patagonia, or such as the sea anemone *Actinia equina*, which occurs in the Atlantic Ocean from the Arctic to the tropics. Both are eurythermic species.

The temperature requirements of many species are determinable, since they generally select a definite temperature range when placed in a temperature gradient. The slowworm, which lives in woods, thus prefers a temperature of 28°C, the wall lizard, 38°C, and agama, which lives on plains, 45°C.

Similarly, the adaptation to other environmental factors can also be of varying ranges—in marine animals to salt concentration (stenohaline and euryhaline species); in land animals to air or substrate humidity (stenohydrous and euryhydrous species), or illumination level (diurnal and nocturnal animals).

Very often animal species from very different groups which inhabit the same environment develop similar structures. Many fishes, cephalopods, crustaceans, mollusks, and coelenterates that live in the lightless ocean depths have, during the course of evolution, each independently developed light-producing organs by which members of the same species are able to recognize each other or the reproductive partner. Numerous insect species which live in preference on green leaves have developed green coloration. Birds and mammals distributed over different climatic zones have developed larger subspecies in cooler climates (Bergmann's Law). This is advantageous, since a larger body has a relatively smaller surface-area-to-body-volume ratio, and thus loses less heat in cooler climates: the body volume increases in three dimensions, the surface area only in two.

All environmental influences determine the distributive ability (vagility) of a species. Its distribution area has its limits where one of the conditions necessary for life drops below the allowable minimum. The European fire salamander is not found in the Alps above 1000 to 1200 m, since the warm summer there is too short for its stream-inhabiting larval stage, which requires a long time to develop.

Distributive ability

Biotopes

Land environments with relatively constant environmental factors (biotopes), are more plentifully divided than in fresh water or the sea. Tropical rainforests, mountain bamboo thickets, coniferous forest zones, fields, marshes, moors, plains, deserts, particular heights, tundras, or caves represent special environments for many land animals, where they can find food, protection, and the possibilities for reproduction. In numerous cases such biotopes are, however, much more closely limited. Marmots only occur in mountains and foothills where they are able to construct their burrows. Bearded tits only live in European and Asian marshes and lake areas that have a thick reed growth. It is similar in the case of species restricted to a special diet (stenophagous). The oak egger (*Tortrix viridana*; see Vol. II) usually inhabits the crowns of oak species; many lice (*Mallophaga*; see Vol. II) live only in the feathers of particular bird genera.

Inland waters offer special biotopes such as springs, streams, rivers, puddles, ponds, and lakes. But even in these environments, the conditions for life are very different—depending on whether the lake is cold or warm, whether it is generally silted up (eutrophic) and containing abundant nourishment or is a clear lake containing little nourishment (oligotrophic), or is acidic bog water.

In the oceans, environments with generally uniform living conditions are usually more extensive. Two such great biotopes (biochores) can be distinguished: that of the sea bottom (benthic) and that of the open ocean (pelagic). The warm and cold illuminated upper zones (littoral and illuminated pelagic) and the lightless depths (abyssal benthic and pelagic) are present in both types, and offer special conditions for existence. Brackish waters also occur, differing mainly in their salt contents, which lie between that of sea water (3.2 to 3.5 percent) and fresh water (0.02 percent and less).

Man's effect on the environment

Very many environments have been basically changed by man. In all parts of the world, man has turned huge forest areas into cornfields, meadows, and building and industrial zones, drained marshes and bogs, controlled rivers and poisoned them with wastes. During this century these changes have reached such a large scale all over the world that undisturbed land is present only in a few areas, and, in part, as mere wildlife sanctuaries. These changes in the landscape have led to a basic change in the species surviving, and to a marked impoverishment of the animal world. Many species have already become completely extinct owing to ruthless hunting, and only a small number of species have been able to adapt to the new environmental conditions brought about by man by becoming "followers of civilization" such as, for example, the birds of parks and gardens, the rats, house mice, cockroaches, bedbugs, fleas, flies, clothes moths, etc., which inhabit houses. In addition, the animal world has been changed by deliberate or accidental introduction of alien species to established ecosystems. It is thus a very necessary measure for the nature conservationists to maintain the natural environments, at least in part.

The majority of environments are inhabited by distinctive associations of life forms (biocenoses). The species concerned are from very different groups of animals which have similar requirements with regard to particular factors present in the biotope. Cohabitation has, however, during the course of time, also led to a closer, more complex, many-sided relationship of the species to one another and to the plant world. Carnivores are dependent on the number of prey present, parasites on the presence of their host animals or plants. All species which live on the same plant foods or prey, or that utilize the same hiding places or nesting sites, such as holes in trees, are in closer association as competitors.

In general, a certain balance has developed in animal communities which is, however, subject to continual and sometimes great fluctuations. If, owing to favorable climatic conditions during one year, a particular insect species shows a terrific population increase, its parasites, especially Ichneumon flies and parasitic Hymenoptera, show a corresponding population increase which leads to a rapid decrease in the number of the insect species in question.

Since the bodies of animals and plants offer an especially advantageous food source, an extremely large number of species have developed into internal or external parasites during the course of evolution. This is especially true of unicellular animals, flukes, tapeworms, nematodes, hookworms, parasitic Hymenoptera, Ichneumon flies, feather lice, body lice, fleas, and ticks, but also for many other groups of animals. As a result of this particular mode of life, several of these different types of parasites have developed a migratory phase. Since the parasite lives continually in a single host, it usually dies when the host does. Its species can therefore continue only if an opportune transmittance of its offspring to a new host is ensured. In external parasites (ectoparasites), such as feather lice, body lice, and ticks, this usually occurs on body contact between the host animals and their young or through the communal use of nests or other living areas.

For internal parasites (endoparasites) it is, however, very difficult to reach the interior of another host. Because of this, a "host change" has been evolved by the introduction of an easily available "secondary host." The malaria parasites, (*Plasmodium* spp.), for example, which live in blood, are taken into the intestine of a mosquito after it has bitten the host; a differently formed generation develops there, partly by sexual and partly by asexual reproduction, of which the last phase (sporozoite) is injected with the mosquito's saliva into the normal host animal, including man, when the insect bites. In the liver flukes, the eggs reach the outside with the host's feces. The larvae, on hatching, bore into snails, developing in the sexual glands of these secondary hosts into a new, differently formed generation, which leave the snail again as the second larval form (cercaria) and are occasionally taken up by other host animals, where they then become sexually mature in the liver. Since only very few larvae

Organismic communities

Parasites

achieve such a change of host, the number of eggs released by flukes and tapeworms is exceedingly large. A beef tapeworm (*Taenia saginata*) contains about seven million eggs.

Symbiosis

In other cases, life in or on a host can be of advantage to both partners. In such symbioses, the instincts of both the host animal and the symbiont have adapted themselves to one another. Some hermit crabs (pagurids) which live in empty snailshells are, for example, associated with certain sea anemones (*Actinia*; see Vol. III) which sit on top of the shells and live predominantly on the scraps of the crabs' meal. On the other hand, the sea anemones camouflage the crabs and help to deter enemies with the aid of their stinging capsules. When the growing crab needs to occupy a new empty shell, it induces the anemone, by special tapping, to release its hold on the old shell and transfer to the new one. Among marine fishes there are a number of small species, distinguishable by their stripes, which are skin parasites on larger fish, or which remove food particles lodged in their teeth (cleaning symbiosis) and are tolerated by them. Numerous symbioses have developed between animals and flowering plants. Many insects, as well as certain species of birds and bats, take nectar from the blossoms and are of use to the plant in pollinating the stigmas. Other species of birds eat berries and distribute the indigestible seeds in their droppings.

The intestines of higher animals are inhabited by bacteria which exist on the intestinal contents but which, when they die, release their all-important vitamins to the host. In hoofed animals, rodents, lagomorphs, and certain groups of birds which eat foods rich in cellulose, bacteria are present in the appendix which exude an enzyme capable of dividing cellulose, which is otherwise indigestible, into sugars which can be taken up by the intestinal cells. In addition, bacteria and yeasts occur as symbionts in special organs (mycetomes) in the intestines of certain insects, permitting the digestion of blood or plant juices. Many of the deep-sea fishes, squids, and mollusks have special light organs which accommodate bacteria which produce substances necessary for light production.

Animal geography

The particular life requirements of every animal species, and the possibilities for distribution set by these requirements, determine, to a great extent, the inhabited distribution area; this is, on the other hand, also dependent on geographical factors and on the time available for distribution. As man's introduction of European animals to North America, Australia, New Zealand, and other countries has illustrated, many animals are also able to live in lands which, owing to the oceans separating them or other geographical obstacles, they were unable to reach previously. All distribution areas are stipulated ecologically as well as historically.

In many cases, the distributional history of an animal species or group of animals can be ascertained, especially when the geographical and climatic conditions in the geologically recent past are known. Conversely, it is often possible to draw conclusions about previous geographical and climatic conditions from the present distribution areas. This is especially convincing when the distribution areas are separated (disjunctions). When, for example, snow hares, ptarmigan, and many other ecologically restricted species of northern Europe are also found in the higher reaches of the Alps, one can assume that they also inhabited the middle and north European area lying between these two zones in earlier times, and that a colder climate must have existed there in recent geological history. Or, when many of the flightless, ground-living animal genera which inhabit Sardinia are also found on the Italian mainland, this shows that the island was joined to the mainland in earlier times.

Such land bridges as formerly existed should not, however, always be considered as direct connections between presently divided distribution areas. One cannot conclude from the fact that lungfish are now found in Africa, South America, and Australia that these areas were once connected by land bridges over the oceans. Fossil remains of lungfish are, in fact, also known from the northern continents. The lungfish have thus become extinct only in the north and have survived in the southern continents.

Since the distribution areas of many animals are often similarly limited, it is possible to subdivide the earth's land masses into animal-geographical regions, subregions, and provinces which are distinguishable through the presence of specific animal groups, especially those which occur there exclusively or predominantly. However, the boundaries in question do not apply in the same way for all classes of animals. In general, seven land regions, at the most, are distinguishable on the basis of mammalian and avian distribution.

1. The Holarctic region includes Europe, North Africa, subtropical Asia, and North America. This is subdivided into the Palaearctic subregion (Europe, North America, subtropical Asia) and the Nearctic subregion (subtropical North America). What the two subregions have in common is made clearly evident since the same species are present, to a certain extent, in the arctic north (arctic fox, polar bear, redpoll, Lapland bunting, snow bunting, magpie, and others) and that European-Asian species such as the red deer, elk, reindeer, European bison, lynx, beaver, and others are represented by very closely related American species. In the Nearctic subregion, however, quite a few species exist which have South American origins (e.g., opossum, armadillo and, in addition, two bird families, hummingbirds and tyrant flycatchers).

2. The Ethiopian region, that is, Africa south of the Sahara, is distinguished, above all, by the hippopotami, giraffes, and numerous species of antelope (which originated mainly from Lower Tertiary southern Asian ancestors), and, apart from these, the aardvark, ape genera, ostrich, turako, and mouse birds, as well as other species and genera which are only found there.

3. Characteristic for the Madagascar region (Madagascar and the island groups north of it) is the absence of monkeys, giraffes, antelopes, large

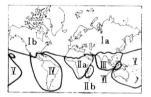

Fig. 1-9. The animal geographical regions. Ia and Ib Holarctic Region (Ia Palaearctic Subregion, Ib Nearctic Subregion), IIa Ethiopian Region, IIb Madagascar Region, III Oriental Region, IV Neotropical Region, V Australian Region, VI Oriental-Australian Region (Wallacea).

carnivores, and many other families of mammals. In addition, there are many genera of lemurs, tenrecs, and cuckoo rollers, among others, which are present there and are restricted to this region.

4. Only in the Oriental region, which includes lower and upper India, southern China, the larger Sunda Islands, and the Philippines, are the gibbons, tarsiers, tupaias, and lorises, as well as the dragon genera and other groups of animals. The Oriental region has the following groups of animals in common with the Ethiopian region: the pangolins, antelopes, Old World monkeys, sunbirds, and barbets; and the following in common with the Palaearctic region: deer, bears, and tits.

5. The Australian region, consisting of, apart from Australia itself, New Guinea, Melanesia and Polynesia, and New Zealand, with its markedly different fauna, is distinguished by the presence of the primitive egg-laying mammals (duck-billed platypus, echidna), the numerous genera of marsupials (which, like the platypus and echidna, are not present in New Zealand), and, in addition, by the kiwi, cassowaries, brush turkeys, cockatoos, lyre birds, birds of paradise, and honey eaters.

Celebes, the Moluccas, and the smaller Sunda Islands east of Bali comprise an Indo-Australian region. A few genera are only found there, the barbirusa, for example. The Hawaiian Island group also forms a relatively independent area in which animals of Polynesian origin as well as those of South American origin are present (Drepaniids).

6. The Neotropical region, which includes South and Central America, has a very large number of orders and families restricted (endemic) to it, such as armadillos, sloths, anteaters, New World monkeys, rheas, rock partridges, and toucans. The area has been enriched by certain deer, a bear, and several cat species, among others, from the Nearctic subregion.

7. Various penguin species breed in the antarctic region. The distinctive species are the king and emperor penguins, as well as a few invertebrates.

Seventeen regions of bottom-dwelling marine animals (littoral fauna) can be more or less clearly distinguished. The pelagic fauna, which lives continually in open water, is, in contrast, less clearly divisible across the oceans. Usually only a warm region stretching completely around the earth is separated from colder northern and southern regions. The two colder regions have, in addition to a few pelagic genera, several bottom-dwelling species and genera in common. This "biboreal" means of distribution is based on the fact that, during the ice ages, the present tropical seas were also colder, and, on the re-warming of the tropical belt after the ice ages, the area which once spread over all three regions was disrupted, which inhibited the marine fauna adapted to moderate temperatures.

The deep-water basins of the Atlantic, Indo-Pacific, Arctic, and Antarctic Oceans are, owing to their isolation from one another and their peculiar conditions for life, especially rich in families and genera which only occur within them (thus, endemic).

2 Animal Behavior

Movement is a very obvious animal characteristic. There *are* animals, which, like plants, are attached to one spot—the corals, for example; in general, however, animals are able to move around, mainly because of their means of feeding. While plants can build organic products out of such inorganic substances as water, carbon dioxide, and mineral salts, with the aid of energy from sunlight, animals must depend on eating organic substances which have already been synthesized. In short: animals must utilize plants or other animals as food, and this necessitates a high degree of mobility. Only those animals which strain their food from the water or which fish in currents can afford a sessile mode of life.

Movement in space requires a series of special arrangements. First, the animal must have the appropriate locomotory organs (pseudopods, flagella, cilia, muscles, fins, wings, legs) and the ability to move them in an orderly fashion; this requires special structures to coordinate these organs (coordination apparatus). The animal must be able to move. It must, however, also be able to stop. In addition, it must not run blindly into obstacles or its enemies; it is also a definite advantage if the animal is capable of recognizing and sensing its food from a distance. There are probably a few unicellular animals which swim about more or less randomly, colliding with obstacles instead of avoiding them, and only ensuring a prolonged stay in a food-rich area by swimming more slowly or settling there, instead of accelerating their movement as they would in unfavorable surroundings. Even the majority of unicellular animals, however, are capable of orienting themselves correctly in space. Paramecia, for instance, respond positively to touch and carbon dioxide, and these responses guide them to bacteria which serve as their food. They are also capable of avoiding obstacles before they collide with them.

Every oriented movement presupposes sense organs and, in addition, other organs which analyze the sensory data and transmit the corresponding orders to the locomotory apparatus. The behavioral repertoire with which an animal is equipped becomes more complex with an increase

Animal behavior, by I. Eibl-Eibesfeldt

in the level of organization. This encompasses the managing of plant or animal food in feeding as well as the very diverse social-behavior patterns of the more highly developed animals: courtship, brood care, attack, defense, and others.

The behavioral repertoire can be extremely fixed. Immediately on hatching, gypsy-moth caterpillars (see Vol. II) crawl up branches to the tips of twigs, where they find their food, which consists of leaves. If they are put in a test tube with leaves at the open end, pointed away from the light, they crawl away from the food and collect at the closed end of the tube which points toward the light. They finally starve to death, since they are simply incapable of turning around, because under natural conditions they are guided toward food by a positive light response. In nature, however, these caterpillars do not remain at the tip of a twig if they happen to have climbed up a leafless one. As soon as their forelegs lose contact with the substrate at the end of the twig, their behavior becomes temporarily reversed; they then climb down the twig, searching. Such a reversal does not occur in the test tube, since the animals there have continual contact with the substrate.

The fertilized female of the digger wasp (*Ammophila campestris*) cares for each individual egg in its own "nest." She digs a flasklike hole, closes the entrance with a carefully selected pebble which fits it exactly, captures a caterpillar and paralyzes it, puts it down in front of the hole, opens the entrance, crawls inside, explores the hole, appears afterwards head-first in the entrance and pulls the caterpillar in with her. If, during the meantime, the caterpillar has been moved about 20 cm away from the hole entrance, the digger wasp then goes searching for it. When she finds the caterpillar again, she brings it back to the hole, leaves it there, explores the hole once more, and then reappears head-first at the entrance in order to pull the caterpillar inside. If the caterpillar has again been removed in the meantime, she repeats the whole procedure once more. The digger wasp must explore the hole anew each time before she draws the caterpillar inside. One can repeat this experiment up to forty times, after which the wasp usually gives up. Similar examples of fixed behavior patterns have been collected by the French entomologist Jean-Henri Fabre (1823–1915). His observations on processionary caterpillars (see Vol. II) are well known. Incidentally, the behavior of the digger wasp described here is only fixed in so far as it proceeds when, on the "exploration" visit, the hole is found to be undisturbed. Further phases of brood care in this species are discussed in Volume II. The findings of the actual exploratory visit determine which of the three possible behavioral sequences are activated by the same nest.

The higher the animal is in evolutionary development, the more readily it is able to adapt its behavior to environmental factors by learning. Comparative behavioral investigations attempt to determine in detail why and for what purpose an animal exhibits a particular behavior pattern in a particular situation. The subconscious purpose behind each behavior pattern performed by the animal is a question of species survival. A woodpecker drums to deter competitors and attract a female. It is important to recognize the biological significance of a behavior pattern if one is to understand in which way it contributes to the continuation of the species. Apart from this, the natural scientist also asks how the behavior pattern in question has developed during the course of evolution and its development in the young. On doing this he discovers, among other things, that the woodpecker's drumming is a movement developed from chiseling (see Vol. IX) and, apart from this, that it is innate, since woodpeckers can drum without previous example. Finally, the behavioral investigator asks "why" to the causal factors. What releases this behavior pattern? What causes it to cease? Which physiological mechanisms are basic to it?

The variety of questions put requires various methods, observations, and experiments to be conducted on the intact animal and, apart from these, sensory, hormonal, and neurophysiological techniques to be used. Comparative methodology is of especial importance since it aids in the discovery of evolutionary relationships.

When one compares the behavior patterns of various animals, similarities come to light. They could have been brought about in completely different ways. In many cases, common environmental factors could have instigated similar adaptations. Throat pumping on drinking has developed in pigeons as well as in sand grouse, zebra finches, and other unrelated plains-living birds as an adaptation to life in dry areas; distant relatives have independently adapted their drinking behavior to a water deficit in their living conditions (so-called parallel development, or convergence). Other similarities can be shown to be associated with inheritance from a common ancestor: for example, the male courtship postures of various species of diving ducks (see Vol. VII). The comparison of as many closely related species as possible indicates gradations in the similarity of certain behavioral characteristics which can be used, in the same way as a morphological series, to reconstruct the course of evolutionary development; we refer to this as homology.

Comparative considerations

Nearly all ducks thus exhibit the behavior pattern called "inciting" in the female courtship movements (see Vol. VII). If a mallard duck has her eye on a particular drake, she tries to incite him to attack other drakes, and this serves to separate the two courting animals from the main group. She swims next to or behind her selected male, threatening other males with her beak over her shoulder, at the same time calling softly "queg-geggeggegg." We can determine how this behavior pattern came about from the courtship ceremony of the shellduck. Here, the female attacks neighboring pairs in threat display. If she approaches the enemy pair too closely, however, her fleeing drive becomes activated. The duck runs back to the protection of her mate; as she approaches him, she threatens over her shoulder toward the enemy pair. This threatening over the

Fig. 2-1. A mandrill threatens with its canines.

Fig. 2-2. A Kabuki player (Japan) mimes anger.

Fig. 2-3. An angry child.

shoulder has become an inherited event of firmly fixed usage (it has become ritualized to a fixed action pattern). A mallard duck also threatens backward over her shoulder when the threatened drake swims obliquely in front of her; the deflection of the movement in this case, however, is less pronounced than it would be if the strange drake was swimming behind her. In cases of weak excitation, the duck can also point directly toward the threatened drake.

It was possible, through comparison of various chickenlike and pheasantlike birds, to conclude that the origins of the peacock's courtship display comes from attracting the proposed mate to food. The domestic rooster (see Vol. VIII) attracts a hen by intimating that he has discovered a piece of food; he scratches, walks backward, pecks at the ground, picks up a pebble while giving the attraction call, and then drops it again. The hen hurries over and searches about in front of him, and copulation then takes place. The cock European pheasant (see Vol. VIII) uses similar means of attraction. A courting Monal pheasant bows deeply with his tail spread in front of the hen, and pecks at the ground with his beak. The hen runs up and searches around in front of him, while he bows with the primaries and tail feathers spread to their maximum extent and waves his tail feathers slowly backward and forward. The argus pheasant scratches like a rooster attracting a hen to food, and also bows in front of the female which is attracted under his raised wings and spread tail. If he is given food, he offers it to the female, which he normally does not do. In the peacock this courtship has developed so far as to be a fixed event (ritualized), the origins of which, without the knowledge of the intermediate phases described here, one would never connect with attraction to food. The peacock spreads the tailfeathers, shakes them, steps one pace backwards, the peahen, searching the ground in front of him, being the focus of the concavity formed by the tailfeathers. To a certain extent, the peacock's tail-spreading indicates "food not present." Young peacocks still scratch and peck during courtship, however, thus repeating in their juvenile development, in a manner of speaking, a part of their evolutionary one.

Movements behave very "conservatively" under the changes brought about during the course of evolution. Man, for example, when angry, opens the corners of the mouth and pulls them sharply downward. The reason why he does this is not immediately understandable. Here as well, comparisons with other mammals can offer suggestions. The mandrill, with the same gesture, exposes the entire length of its upper canines. In us, this gesture has survived the reduction of the canines. A similar case is illustrated by stags. Primitive forms, such as the muntjak, expose the knifelike canines of their upper jaw while threatening. Our red-deer stags, with their much reduced canines, also threaten by wrinkling their lips.

All behavioral complexes, such as feeding, but excluding play, terminate with fixed, species-typical or sometimes specific, innate behavior patterns termed fixed action patterns or inherited movement coordinations. [The term fixed action pattern (FAP) is most widely used today, although it is subject to misinterpretations. However, the word "fixed" should not be read to mean complete rigidity. In the original German term, Erbkoordination, (transl. inherited coordination), the word fixed does not occur, hence no misunderstanding is possible. Editor.] In an emergency, they are interspersed with orientation movements; we term both together "instincts." Fixed action patterns can be performed "in vacuo" when they are initiated by an external stimulus but lack a stimulus "follow up." In vacuo performance is much less common in orientation movements: they are much more dependent on the continuing action of external stimuli. This has been classically explained by Konrad Lorenz and Niko Tinbergen in the egg-rolling of the graylag goose. A graylag will return to the nest any egg lying in front of it by holding the egg at its far side with the underside of the beak and rolling it back into the nest. The beak moves sideways in such a way as to keep the egg on the right path. If the egg is removed once the goose has begun egg-rolling, it will continue to move without the egg—that is, "in vacuo"—toward the nest but in a straight line, without sideways deflections. Fixed action patterns and orientation movements which are intertwined to form an oriented behavior pattern thus behave with respect to one another as do a steamship's propeller and steering wheel.

Fixed action patterns do not have to be fully developed at birth. Some patterns mature gradually during the course of juvenile development. A squirrel enters the world blind and naked, and many of its behavior patterns first develop by degrees, such as food storing: adult squirrels bury nuts, acorns, and similar objects during autumn, as winter stores. First, with the nut in their jaws, they search for a hiding place; then they dig a hole with the forepaws, place the nut in it, tamp it down with the snout, scratch the excavated earth into the hole, and press it down.

If we wish to know whether this series of movements matures independently of experience or has to be learned, we must raise a squirrel so that it is unable either to observe a conspecific storing food or to try out the storage methods. For this purpose, squirrels are raised singly in wire cages and fed only mashed food. Thus they have no opportunity to carry something about or to bury it. If such a squirrel is later given nuts, it first eats its fill, but, instead of rejecting the remainder, it searches about in the room with a nut in its mouth, scratches mainly in corners or near table and chair legs, then drops the nut after several scratching movements, taps it down with the snout and finally performs the covering and pressing movements with the forelegs although it has not excavated anything and the movements thus occur "in vacuo." This shows clearly that we are dealing here with an innate movement program which matures during the course of development and "rushes out headlong" once the appropriate stimulus sets it in motion.

In humans as well, one can prove the existence of fixed action patterns by observing deaf-born and blind-born children or through culture comparison (see Vol. XI).

Animals are not only equipped with certain behavior patterns, but are also capable of giving meaningful species-specific responses to certain environmental stimuli, without previously having to learn them. Apart from innate ability, an innate recognition of a particular stimulus situation also exists. Immediately after metamorphosis, a frog is capable of capturing small prey without ever having to be taught. Experiments have shown that such a frog snaps at anything moving in front of it, even pebbles and leaves. Since in nature there is usually no experimental behaviorist to confuse the frog in this way, it succeeds with this simple response, since most things it finds moving in front of it are insects and small worms, that is, prey. In addition, a frog quickly learns to avoid unpalatable objects.

Innate releasing
mechanisms
Key stimuli

This ability of the frog presupposes special stimulus-filtering mechanisms to enable the animal to respond only to certain stimuli with certain behavior patterns. Such mechanisms have been compared to a lock which can only be opened by a particular "key stimulus." Since this mechanism is inborn, it has been termed the "innate releasing mechanism" (IRM).

Very many of the social responses made by animals are activated by such releasing mechanisms. In these cases, the partner has developed particular stimulus arrangements in adaptation to the stimulus receiver—termed "releasers." It is advantageous for the sender of a social signal when the animal for which it is designated receives it correctly. A frog's prey would thus never carry an additional signal flag to ensure that more of its kind were devoured. A male frog, on the other hand, *would* develop special signals for its female.

Releasers

Releasers are present in all sensory modalities. Moths are attracted to each other by sexual odor substances which they are capable of perceiving at astoundingly low concentrations (see Vol. II); minnows (see Vol. IV) and toad tadpoles (see Vol. V) respond with escape movements to an odorous substance exuded from the skin of damaged conspecifics. A turkey hen recognizes her chicks by their call alone; thus a deaf turkey will kill her chicks immediately after they have hatched. A normal turkey hen, on the other hand, will even brood a stuffed polecat containing a built-in apparatus which emits chick calls. Acoustical releasers include the frog's croak and insect and bird songs. The Viennese scientist J. Regen allowed male crickets to chirp into a telephone. In another room, females moved toward the transmitter as they would toward a male in order to mate under natural conditions. In the experiment described, they finally jumped into the telephone transmitter.

Visual releasing mechanisms have been especially well studied. During the breeding season, male three-spined sticklebacks maintain a small territory which they defend against male competitors. They court females, on the other hand, with a special dance (see Vol. V). The males "threaten" during the breeding season, using their now-red bellies; that of the female, in contrast, is silvery and swollen with eggs. Singly reared stickleback males respond to the most crude models with red undersides by fighting, and to other crude models which are silvery and slightly swollen, with courtship. Even a wax sausage-shape without fins but possessing these characters is sufficient. An accurate stickleback model without a red or a swollen silver belly is, on the other hand, ignored. Male fence lizards (*Sceleporus undulatus*) have blue stripes on their bellies, those of the females being gray. If blue stripes are painted onto a female, it is then attacked as a competitor by the male. Conversely, the males will court a member of their own sex if its blue stripes have been painted over with gray.

Several key stimuli often release exactly the same behavior pattern; under certain conditions, every "key" can open the "lock." The food begging of young chaffinches can be released by vibration stimuli, the parent's call, or simple visual stimuli. If these three stimuli occur simultaneously, they enhance each other's effect. Herring gulls have a red patch on the underside of the bill. The parent bird regurgitates food in front of the begging chicks and then offers them small scraps, which they take; later they peck at the food on the ground as well as at the parent's bill. Niko Tinbergen offered newly hatched chicks, which had never received food, two-dimensional models of a gull's head, and recorded how often they were pecked at. On the basis of 16,000 experiments, the following innate key stimuli were found to be effective: movement toward the chick, beaklike form consisting of the following characteristics: narrow, not too short, low, pointing downward, and with a patch, preferably red or contrasting with the beak color. The term "summatory" provides a very inadequate picture of the relationships which all these characteristics, with their highly developed structuring, have to one another.

As further experiments have shown, a model's effectiveness can be increased to such an extent that it releases a stronger response than the normal object. In this way, Magnus could present a female model to the male silver-washed fritillary (see Vol. II) to which they flew more often than to live females of their own species. **Superoptimal models**

Releasing stimuli are often—as termed by psychologists and behaviorists—"shaped." Males of the mid-European glowworm (see Vol. II) thus respond only to the light pattern of their own species. The light signal in *Lampyris noctiluca* consists of two parallel bars and two spots. If a corresponding cut-out is held in front of a torch, males of this species can be attracted. Many American glowworms communicate with each other by means of a "blink code"; the succession of light impulses is species-specific. A case of "signal forgery" is worth mentioning here: females of the glowworm genus *Photinus* are capable of copying the signal of the *Photurus* genus. If a male of the other species approaches, they show its species-specific signal and devour it as soon as it settles next to them. Just as the behavioral investigator "leads animals about by the **Signal forgery** nose" by playing on their innate releasing mechanisms through the use of models, so do certain glowworms.

Cases of "signal forgery" in the animal kingdom are not uncommon. Various carnivorous fish "angle" with prey models. In angler fish (Antennariidae), a skin flap or wormlike appendage is carried at the tip of the freely moving first dorsal-fin ray (e.g., *Phyrnelox scaber*). The cryptically colored angler lies on the bottom and moves only the fishing pole and its lure; by this means it attracts fish which are then engulfed. Deep-sea anglers even have lighted lures: in one species of the genus *Galatheathauma*, the lighted lure is present in the roof of the mouth so that the prey is actually attracted directly into the mouth cavity itself. A whole series of other animal species have also discovered the fishing pole. The catfish (*Chaca chaca*; see Vol. IV) fishes with its barbels. The alligator snapper turtle (see Vol. VI) lies on the floor of the stream and uses its tonguetip as a fishing lure, extending it to form a wormlike thread.

Mimicry

One terms such cases of signal forgery "mimicry." The most varied signals of living creatures can be copied. The cleaning wrasse (*Labroides dimidiatus*; see Vol. V) inhabits the Indian Ocean, living mainly on parasites it removes from reef fish. It even swims unharmed into the jaws of groupers and other carnivorous fish to do this. The hosts recognize their "cleaner" as much on appearance as on behavior. The cleaners have a distinctive dark stripe running the whole length of the body, and a blue back; in addition, while dancing, they wave their tails up and down obtrusively in front of their host. The knife-toothed blenny (*Aspidontus taeniatus*) copies the cleaner fish so perfectly in every detail that, during my dives in the Indian Ocean (Eibl-Eibesfeldt) it was a long time before I could tell them apart. At the beginning, I took this fish to be a cleaner; then I noticed that this presumed "cleaner" released flight in its host once it started to work on it. Only after I was able to capture one and it bit me so severely in the hand that I bled, did I realize that this was no cleaner fish that I was dealing with.

The knife-toothed blenny, like many other closely related species has specialized on tearing pieces out of the fins, soft-skinned body parts, and gills of other fish. The majority of these species must creep up on their prey and then attack it. To approach the prey more easily, the knife-toothed blenny disguises itself as a cleaner, dancing like one and bearing the same coloration. As soon as it has approached its prey, which opens its opercula or mouth invitingly, it then tears a piece out of its skin. It is worth mentioning that the mimic even copies the characteristics of its model's geographical subspecies. In the Tuamotu Islands of the South Pacific the cleaner is orange-red around the middle of the body, and its mimic there shows the same characteristic. In the Maldive Islands of the Indian Ocean a black fleck is present at the base of the cleaner's pelvic fins, and in the mimic's as well.

Even plants forge signals. The flower lips of the orchid *Ophrys* are similar in color and form to the females of certain wasps and solitary bees, and, at the same time, it copies these species' sexual attractant. When an attracted male attempts to "copulate" with these petals, it becomes covered with pollen which it then carries to the next blossom.

Expression movements aiding in interspecific or intraspecific communication have developed again and again. We have already described how they come into being, in the example of the peacock's courtship display. The information carried by such "symbolic behaviors" can be extremely complicated. Honeybees inform their hive mates about the position of a good food source through a special dance (see Vol. II). The fact that we humans are also well equipped with expression movements, physical releasers, and innate releasing mechanisms will be dealt with in more detail in Volume XI.

In general, animals do not wait passively for something to happen to them. They are often active themselves, searching for something more or less definite, depending on their mood. A nest-building bird thus searches solely for a nest site, a sexually active one only for a mate and—in some species—a spouse. Hungry animals search for food alone and aggressive ones for an antagonist. These varying readinesses to respond to one and the same releasing stimulus situation in different ways are based on special physiological mechanisms; these serve as an internal driving system which ensures that the animal responds correctly at the right time and, for example, does not starve if nothing palatable crosses its path. Certain of these mechanisms have been well investigated. Thirst, which releases searching for water, is instigated in mammals by a receiver in the brain stem which measures the tissue and blood fluid concentrations. If the tissue fluid is too highly concentrated, the receiver registers this and the animal becomes thirsty—it starts to search for water. If a saline solution with an osmotic value higher than that of blood is introduced into a vein, the animal also becomes thirsty. Intravenously injected water of low salt concentration, on the other hand, satisfies thirst, just as drinking does under normal conditions, requiring only slightly longer for the salt content of the blood to increase sufficiently. Until this occurs, various substances block thirst.

Dogs with an œsophageal fistula from which all the water drunk is drawn off before it enters the stomach, still drink only a certain quantity; their thirst appears to be satisfied for a short time. If a balloon in their stomachs is pumped up while they "sham drink," drinking ceases earlier. "Registration" of drinking movements and stomach inflation is obviously present; if the "stipulated requirement" is satisfied for both, then the animal has satisfied its thirst. Comparable mechanisms form the basis of hunger, but these, incidentally, are very different for different groups of animals. A blowfly takes up food until its upper intestine is full. This information is then transmitted by a nerve (nervus recurrens). If this is severed, the fly continues to suck until it is distortedly swollen and dies.

Apart from internal sensory messages, hormones also play a great

Moods

Drive

Hormonal control

part in the development of moods. The importance of sexual hormones for reproductive drive is well known. Many other behavior patterns, such as brood care, are similarly controlled by hormones. In female rats, the hormone progesterone increases nest-building activities during the second phase of pregnancy. This hormone, however, when artificially administered, also has an effect on non-pregnant rats. Progesterone secretion ends when the pregnant female gives birth; at this time the stimuli emanating from the tiny young have an increasing effect on nest building. In doves the same hormone releases brooding. If male and female doves are injected with progesterone and then put together in a cage containing an artificial nest and eggs seven days after the injection, they start brooding at once. In contrast, an untreated pair of doves is only ready to start brooding seven days after they have been put together. As experiments have shown, the partners excite each other through courtship, resulting in a level of hormone secretion comparable to that which was injected. A female only needs to see a courting male through a glass partition in order to be ready to brood within seven days. In contrast, a female does not respond to the sight of a castrated, non-courting male. Yet another hormone, prolactin, causes the secretion of "pigeon's milk" with which the doves feed their young (see Vol. VIII).

Movement impulsion

The reticular formation, a section of the brain stem, is, in part, answerable for what is termed "movement impulsion" or "exhaustion" in vertebrates. No animal existing must wait like a robot for an external stimulus to set it in motion or halt it again. Erich von Holst discovered the central nervous automatism underlying all innate movement patterns such as the locomotion types (jumping, climbing, swimming, and flying): these require no releasing mechanism and are self-organized. An eel in which all the distal spinal nerve roots have been severed (see Vol. IV), thus preventing impulses from the skin and muscles from reaching the spinal cord, is still capable of making coordinated movements; a toad treated in the same way is still able to swim and walk crossways. The constant impulses are not continually conveyed to the muscles in intact animals: many higher-order nerves hinder such a continuous discharge. They thus provide a clear pathway for motor impulses on the intervention of certain external stimuli.

Activity requirements differ from species to species. This is why normally fed wolves in zoo enclosures can be seen pacing up and down for hours on end, while lions usually lie about lazily during the day. This type of driving force seems to be basic to many instinctive behavior patterns. On the execution of this movement, the readiness to perform it is extinguished and the animal is no longer in this mood for the time being. Konrad Lorenz once owned a tame starling which was well fed but had no opportunity to hunt for itself. From time to time, it would suddenly fly upward from its perch, snap at "nothing," return to its perch, make movements as if it was battering something to death, and swallow. It then remained sitting quietly for a short time. Prey capture in starlings thus obviously includes certain mechanisms which are independent of feeding. This is also the case in many other animals. A satiated dog often comes into a hunting mood; when it has no opportunity to perform the hunting movements properly, it seeks out a substitute object to "shake to death" and carry about in its jaws—e.g., even its owner's slipper.

Investigations on the intraspecific fighting behavior of vertebrates have shown the presence of an aggressive drive underlying attack behavior which is capable of accumulating; the animal attempts to relieve this through fighting. If male cichlids of the species *Etroplus maculatus* are isolated for a long time, they even fight with females; pairing is only successful when another male is present to act as a "scapegoat" and deflect all the attack readiness (aggression) of the territorial male. Electric brain stimulation through fine electrodes sunk into the brain can cause clearly observable "moods" in an animal. On stimulation of a particular "center" in the brain, a chicken, for example, becomes pugnacious and wanders about searching restlessly until it finds an antagonist. Hunger, thirst, fear, courtship, or sleep can be activated from other brain centers.

The majority of animals can learn by experience and thus adapt to changing environmental conditions. In this way they gain knowledge as well as skill. A toad gradually ceases to snap at unpalatable things. If an animal senses pain, it notices the circumstances which have led up to this and avoids them in the future. Animals learn to find their way by "signposts," associating certain stimuli which they had previously ignored with particular events which are meaningful to them, if these events are regularly preceded by the stimuli. Dogs start salivating on seeing a piece of food, but not when they hear a bell. When, however, the bell rings every time a piece of food is offered, they will start to salivate in response to the bell alone. The experimental animal has thus associated the "conditioned stimulus," the bell, with the innate "unconditioned stimulus," something to eat. The Russian physiologist Ivan Pavlov (1849–1936) termed this the formation of a specific reflex; psychologists now call it "association."

Learning

Apart from such knowledge, animals—as mentioned—also develop skills. Squirrels crack hazelnuts with an efficient technique. They gnaw a short groove across the base up to the point of the nut, then place the lower incisors leverlike into the groove and crack the nut in half. As experiments with naïve squirrels showed, this technique must be learned: only the individual movements themselves are innate. Naïve squirrels first gnaw irregular grooves over the nut's surface. They repeatedly attempt to insert their incisors as a lever, which can lead to success only when the groove is correctly situated. They rapidly learn that it is easier to gnaw parallel to the fibers of the shell rather than transversely. These two skills enable them to open the nut, and, since they quickly retain how they have reached success, they "grasp" the cracking technique within a short

Fig. 2-4. A great tit lifting a milkbottle top.

Traditions

Fig. 2-5. The discoverer (a female macaque) washing sweet potatoes.

Prerequisites and precursors of our speech in animals, by O. Koehler

Imprinting

period of time. They learn, through "trial and error" or "self training," how the two innate movements, gnawing and cracking, work together as a functional entity.

Apart from learning on one's own behalf, one can also learn by copying others. In the Belfast area of Northern Ireland and in southern England, bluetits and great tits discovered how to open milkbottles left outside doors in the morning. During the course of twelve years, this "discovery" has become a true tradition, widespread on both islands, obviously through "observation."

We must thank Japanese investigators for especially accurate studies stretching over decades on the creation of true traditions in about thirty different troops of Japanese macaques (see Vol. X). These troops, which wander about in thick forest, were gradually accustomed to visiting observable feeding sites and thus adapted more or less quickly to human observers—depending on whether they had previously had bad experiences with humans. On Koshima Island an old female macaque discovered how to wash dirty sweet potatoes in a stream. The investigators watched carefully how the discoverer's baby, then more animals, and, after years, all of them—with the exception of the oldest males—followed this example. Soon the macaques were even washing the sweet potatoes in the sea, thereby salting them. This is especially interesting since monkeys, with very few exceptions, are considered to be extremely shy of water. A typical aquatic life developed, the monkeys swimming and bringing up "sea fruit" (edible marine animals) from the bottom and feeding on them. In Takasakiyama researchers gained an accurate insight into troop sociology as the troop became larger and larger, until it finally split up. The "pashas" here took care of a particular young female, which they allowed complete freedom with regard to themselves. Because of this, she was ensured a rise in social rank. The same type of behavior was also seen regularly in Takasaki, occasionally to rarely in six other places, and never in the remainder.

The best example of tradition formation is that which distinguishes man from all other animals—speech. Everything necessary to speak relevantly in words is already present in animals, such as the ability to copy accurately what has been heard. No mammal can compare with the African gray parrot (see Vol. VIII) in this respect. In averbal (wordless) thinking, the chimpanzee is the master among the animals, although it cannot "mimic" speech.

Konrad Lorenz has termed a very rapid type of learning occurring during a relatively short period of impressionability "imprinting." Sensorially perceptive thought assumes the presence of innate releasing mechanisms (IRM's) by which details are "impressed" during suitable moods. The IRM can thus be described psychologically as the inherited range of variation, the essence, so to speak, of all such releasing situations. IRM's release an instinctive behavior pattern associated with the correct mood after goal-oriented search (appetence) and discovery, and allow it to proceed to completion, at which moment the behavior pattern attains its full significance. The mood—that is, the particular readiness to behave—then ends, giving way to another mood. Newly hatched goslings (see Vol. VII) are thus imprinted in their "following behavior" on the first thing which they see which moves. In normal cases, this is the mother which has hatched them. Graylag goslings hatched in an incubator become imprinted on "man," and are capable of learning individual features if the same person leads them for a long time. If the newly hatched gosling does not see anybody, it can equally well be imprinted on a shoebox dragged across the grass, especially when a tape recorder inside the box plays a sound reminiscent in rhythm or tone to the species-specific following call "gang-gang."

Even humans cannot learn milking, mowing, skiing, flute playing, or typewriting just by watching: we have to practice them diligently. Only a few highly evolved animals are capable of watching movements which are not innate and then performing them afterwards. In contrast, many bird species are expert in copying sounds exactly, after having "listened" to them. How chaffinches learn the details of their innate species-specific "chaffinchese" during their second spring is described in Volume VII. True dialects can be brought into being through this skill. Once, in a copse where "correct" singers were absent, all the chaffinches, year after year, sang like tree pipits. The Californian white-crowned sparrow (*Zonotrichia leucophrys nuttalli*) becomes imprinted at the age of three weeks on its father's dialect; but it is only able to start singing after its voice has "broken," even when it is prevented from hearing its "model" during this time. It must thus have memorized its father's song. If, however, the bird is deafened before it starts to sing, its memory cannot help it, and it never becomes a "dialect" singer. When, in contrast, it has once sung correctly as a fledgling while it could still hear, it is still capable of singing correctly even if it is deafened immediately afterwards.

Listening and dialect singing

In general, one could say that the bird itself does not need to hear in order to copy its innate song, but it must be able to listen to itself to copy the sounds correctly, even when it has had to remember its model in detail, before the development of its singing powers and the presence of the correct mood made it possible for it to sing. When it has sung correctly only once, it can continue to sing correctly, even if it has lost its hearing. Deaf-born humans or those who have become deaf at different ages show exactly the same responses in their speech.

Although laboratory experiments are important and interesting in the determination of learning abilities and memory duration, the normal capabilities of an animal in its natural environment should not be ignored. Often enough, the wrong conclusions have been drawn from comparisons of "delayed choice," as it has finally been termed. To make these, absolutely comparable experimental methods were used in presenting food to a

Behavior in the wild, and delayed choice

monkey, pig, rat, and similar experimental animals. The food was presented in two or more inverted containers, and the hungry animal usually lifted the baited container first. If the animal tired of the experiment and started to search along the row of containers, it was punished. When the action of first lifting the baited container had reached a certain criterion level, the animal was prevented, for longer and longer time intervals, from performing, thus making it possible to determine for how long this delayed choice remained correct; this gave, in the best cases, only a difference of a few minutes. Under the circumstances, the rat performed better than the monkey. In nature, however, a stork must remember its birthplace for three years or more to enable it to return, without guidance, from South Africa to its East Prussian village (see Vol. VII). A pair of ringed plover (see Vol. VIII) bred for eight years in the same area of the same island in the North Sea; both had been separated at their winter quarters in northern Africa all year and had then flown home, still separated, and met each other again on the island—since the males arrive about a week earlier than the females.

The whydah birds (see Vol. IX), brood parasites which have been investigated by Jürgen Nicolai, have, in addition to their species-specific vocalizations, those of their host birds as well. Whether this is innate or learned during the time they are being raised in their host's nest must remain speculation until rearing of Kaspar-Hauser individuals (birds raised in isolation without being able to hear their conspecifics or host birds) is successfully accomplished. The brood parasite gives exactly the same call to attract its female to the nest as does the host bird, and only when this call is copied absolutely correctly does the female permit copulation.

Every type of territorial behavior implies "self training" to the area, the personal domain. Fidelity to the partner presupposes individual recognition. Young guillemots (see Vol. VIII) learn, while still in the egg, to distinguish the call notes of their own mothers and fathers from the hubub of noise from the cliff rookery; conversely, the parent birds recognize their single egg by sight, long before the chick within it starts cheeping. Herring gulls know the species-specific characteristics of their eggs but cannot distinguish their own eggs from those of conspecifics, and only recognize their own chicks several days after they have hatched.

In experiments, the investigator trains the animal by "reward" and "punishment," including, for example, food or water for the hungry or thirsty experimental animal when it has made the correct choice, chasing away or electrically shocking for the wrong choice. Punishment-free training, using reward only, is termed "positive" by the behavioral investigator; training in which only punishment is used, "negative"; and one in which both are used simultaneously, "double training." A highly evolved animal, however, learns tasks set him by man more quickly without either, since fear of punishment inhibits, and the most dependable rewards are pleasure in achievement, personal performance, and sometimes

Fig. 2-6. The female host bird (violet-eared waxbird) sitting on the nest and beneath her, the male, the song of which is copied by the long-tailed male of the shaft-tailed mhydah bird as he leads his female (uppermost) to the nest, on which copulation occurs.

Means of training
Inclination and
disinclination

praise. The "innate schoolmarm," as Konrad Lorenz terms it, is reward through the desire to find something searched for, to satisfy a drive and perform something which serves to preserve the individual and the species successfully. All types of "disinclination" serve as a "punishment," since no "reward" is forthcoming: disappointment caused by nasty tasting food or from eating food with nauseating after-effects, an unsatisfactory partner, the losing of a fight, the loss of living space, fear, and many others.

The more rigid the innate behavior linkages and IRM's, the less the species is capable of learning and the more it is endangered by environmental changes. In contrast to food specialists such as the koala (see Vol. X) and the clothes moth (see Vol. II), the "inquisitive" animals have an almost insatiable curiosity, or drive to investigate. In this way they develop into followers of civilization, such as crows, the black kites hovering above the sewer outlets of our cities, and the gulls which take their food from human hands during the winter, congregating on town bridges and following ships.

Inquisitive animals

The psychologist Wolfgang Köhler, who published his famous investigations on intelligence in the higher apes in 1917–1921, spoke about "insight" in animals when, instead of making continual tests and retaining the accidental successes, they immediately solved an unknown task by a sudden inspiration. Although such an inspiration comes suddenly, it is often preceded by a testing and weighing-up of thoughts which, in animals, including man to a certain extent, are always "averbal." Finally the penny drops, often after years of perseverance, and this is the greatest reward an investigator or inventor can experience. Following on from this comes the striving to share these thoughts with others, or to convert them into deeds.

Insight and the weighing-up of thoughts

The chimpanzees in Wolfgang Köhler's station on Tenerife have provided us with classic examples of insightful solutions, doubly valuable since Köhler restricted himself to posing the problems without showing the chimpanzees how to solve them beforehand and without guiding them, which is the usual case in circus training. Apart from the meaningful problems posed, Köhler's greatest merits were the inexhaustible patience he showed in observing and recording the slightest successes, the endurance of set-backs, and the sensitivity to and critical weighing-up of obvious "Aha" experiences in his experimental animals. For example, when a banana was hung high up in the open, it dawned on one chimpanzee to stand a pole about 4 m long on end and to climb up it as fast as possible to grab the fruit above. It then descended like a pole-vaulter in a wide arc. As far as I know, no human performer who can do the same thing has appeared to date; in any case, the chimpanzee had certainly not seen anybody do it. The "discoverer" was so pleased with his performance that this "sport" of climbing the pole, even without the banana present, became a generally popular game among the chimpanzees there.

Fig. 2-7. The chimpanzee has climbed up the unsupported pole and is about to swing down again with the banana in its right hand.

Fig. 2-8. The three-box and four-box towers built by the Tenerife chimpanzees.

Fig. 2-9. A primary solution without training made by Nellmann-Trendelenburg's rhesus monkey.

Tool improvement

The tower of boxes which the chimpanzees built under their goal in order to reach it is just as famous. The female "Grande" set the record with four stories after almost four months of perseverance, and this record has not yet been broken. The short-comings in their understanding of mechanics was made up for by the animals' outstanding balancing ability. Watching leads to copying, but none of the watching chimpanzees was able to grasp such a new behavioral principle immediately. Each had to go through the series of individual motions leading to success on their own account.

So-called "good" mistakes are sometimes surprising—such as when a chimpanzee pushes its box against the wall with both hands and then tries to climb up it. In these types of experiments, sad to say, there are also solutions which were at first meaningful but later degenerated into routine and finally were used in the wrong context.

On the other hand, such "dragging" of a formerly successful behavior pattern can also be of use to animals. We learn each new activity, such as sewing, piano playing, new paths, reading, and writing, completely alertly and aware, step by step and letter for letter. After a while, everything becomes "automatic," for instance, the puzzled search for a pair of spectacles which have already been put on. The advantage lies in that mechanical performance—such as combined instrument playing from a score—results in the thought processes being left free so that the others can be heard and the execution thus improved.

Köhler termed comprehension of an understandable connection between two things one of the main steps on the road to insightful solutions. When a revolving disc is put in front of a rhesus monkey's cage, and a cherry is placed on the side furthest away from the bars, well out of the monkey's reach, the animal will look at the unobtainable goal for a long time from the furthest corner of its cage. It will then approach, still undecided, squat down, reach through the bars and playfully put its hand on the disc's edge, thus making it move slightly. It will then immediately begin to turn the disc rapidly in the same direction (a) until it can take the cherry off the disc (b) and enjoy it.

Another monkey, which already knew how to use boxes, once found no box which it could put under the suspended banana. It then took its keeper's hand and led him under the fruit so that it could climb up onto his shoulders. When the keeper, following the investigator's orders, knelt down, the monkey also knelt down, stuck both arms under the keeper's rear, and tried to pull him upright again. This is an example of tool improvement comparable to forest chimpanzees fishing for termites and ants with plucked, leafless grass stalks which they have licked, or twigs broken off a tree (see Vol. XI). The chimpanzee "Sultan" gave an even more obvious example of such tool improvement: he stuck two sticks together to make a pole long enough to reach the banana. The chimpanzee was so entertained by this technique that he kept throwing the fruit up again and dismantling the sectioned pole so that he could put it together once more.

Sultan took the pole out of the hand of another chimpanzee unable to perform the task, showed him how to do it, and indolently tossed the banana to him afterwards. When no second stick was available, he took a slat and chewed the end of it until it fitted into the reed tube. Sultan must therefore have had some concept of how his stick should look and what was needed to improve it.

Fig. 2-10. Köhler's chimpanzee "Sultan," joining two sticks together.

The person who knows nothing of animals and considers himself too refined to have arisen from them is very ready to call the gulf between man and animals unbridgeable, since man, reputedly, is the only one which can talk, think, play, have culture, history, and tradition, use, make, and improve tools, and so on. But we have already given examples of all of these apparently purely human qualities—with the single exception, speech—in the previous sections. Speech alone, in the human usage of the word, seems to be lacking in animals. Speech makes something unique out of what man has in common with the animals; it forms the basis of a new life process which we call the "spirit," with its concommitant incomparable human accomplishments ranging from self-responsibility, morals, ethics, religion, art, and science to pondering over one's own thoughts—philosophy. However, human speech, which makes all these possible, did not drop from heaven, but developed, as all living things do.

The "unbridgeable gulf"

The most important prerequisite and condition for our speech is averbal thinking. This concept is a contradiction for those who consider thought to be bound to speech and the sole prerogative of humans. It is an unshakable fact that a human baby cannot speak. What sort of a mother would then admit that it could not think? Its smile thrills her on the first day that it appears. She smiles back, cuddles, carries, cradles, and rocks the baby until it crows for pure joy. When it cries, she guesses what it needs in order to make it smile again. The baby also smiles now and again when in deep sleep, maybe during a happy dream. Since records of the electric currents of the brain (encephalograms) show special wave formations characteristic of dreaming, it could be proved that the younger the child is, the greater the proportion of its sleep is spent dreaming.

Averbal thought in human infants

The child soon starts to babble, gurgle with saliva, and crow at tremendously high pitches. With time, new noise types appear continually, and the "babbling monologue," comparable to the juvenile song of many songbirds (see Vol. VIII), becomes richer and richer in content. Later, when single discrete syllables are repeated one after the other (the da–da stage), one gets the impression that the child is practicing. Deaf-born children babble like normal babies; probably all human babies, no matter what their skin color and nationality, go through this development. It appears that a number of vocalizations are innate in human children, and that, from these, all the conceivable world languages are constituted. They need, however, to learn the words of their mother tongue later. But how do they know what these words mean?

Babbling in small children

Copying

Some birds are capable of copying noises they hear—including human words—just as well, if not better, than we are (see Vol. VIII). No mammals can compare with us in this ability. Some birds and higher mammals are definitely superior to us in absolute hearing power and the retention of time measures and rhythms. These animals and man have averbal thinking, which presents us with the key to the understanding of words. Our own names and word groups of similar value, such as "Johann Sebastian Bach," "my house," "my firstborn grandson," describe

Concepts

something unique and inconfusable; they constitute so-called "ideas." One can duplicate them and record their development on film. It does not matter how dissimilar all possible houses and tables are to one another; they all have something in common by which we recognize them as a

Notions

house or table. We are, however, unable to duplicate these notions of "house" or "table," and can only describe them as single ideas. We can

Decisions

define these in general terms with certain limitations: table = to carry things, supported wooden plate. Our sentences carry information, for example, "This is my house," "This is the way home," SOS = "Help is needed, emergency."

Even children who are still unable to talk, certainly have definite concepts, such as familiar people and strangers, their toys, favorite and rejected foods. They also have ideas and opinions. A small child which has been away from a familiar room for weeks immediately shows familiarity with everything in it again, even down to the smallest details. The same occurs in the higher animals, in most cases to an even greater extent. Just consider what a migratory bird has to remember to be able to fly every year to the same winter quarters and then return to its previous breeding territory the next spring and, if it is true to its mate, to distinguish him or her from the thousands of migrants once it has arrived. Such examples of averbal rather than learned concepts can be cited ad infinitum.

Innate concepts

Innate releasing mechanisms (IRM's) are, subjectively expressed, innate and sometimes narrowly defined concepts. A tick, for example, will bite anything which is warm and has an odor of lactic acid; a moth caterpillar, on the other hand, selects keratin to chew. The releasing mechanisms for territorial behavior in the European stork consist of a place where a nest can be built from which a wide view of the surroundings can be obtained, together with an unobstructed flight path to the nest; those of the ringed plover consist of a sandy beach and salt water. The returning migrant searches for these in early spring, and once it has found them it makes itself familiar with the various characteristics of its surroundings, thus transferring the concept "territory" into "my territory" and declaring in effect, "Strange males must keep out, only my mate can breed here." IRM's are "imprinted" to a great extent, meaning that during a certain stage of development the animal learns rapidly and permanently which special characteristics are to be responded to with what behavior patterns. For instance, it is in this way that some hand-raised birds develop into "bird humans," which court the person who raised them instead of conspecifics, with which they want nothing to do.

In the unique language of the bees (see Vol. II), the angle at which the dance is performed with respect to the vertical on the dark, perpendicular honeycomb symbolizes the direction of the food source; the average speed of the dancer while dancing, the distance, and the odor of the dancing bee symbolize the type of food present. Dancing and compre-hending the meaning of the dance are innate in bees, but on occasion they must learn anew what the individual characters of the dance should communicate. Owing to this, they restrict the basic IRM's from dance to dance, resulting in variable "averbal concepts."

Apart from this, the higher animals are capable of developing new concepts from personal experience. A greater spotted woodpecker, for instance, became adept at chiseling holes in nest boxes inhabited by bluetits, at exactly the level where the nestlings were. This free-living bird thought all this out for itself. To analyze the events leading up to concept building, it is necessary to set tasks for an animal, and then test whether it is capable of using what it has learned in similar tasks presented in another form.

The elevated maze illustrated in Figure 2-11 is constructed out of aluminium strips 2.4 cm wide, and suspended at table-level by ten thin, criss-crossed piano wires stretched across a horizontal frame with a side length of 2.7 m. A mouse is released from a long-handled box at A; once the mouse has reached E, it is allowed to climb up a quickly extended wooden scoop, on which it is carried back to its home cage. The total length of the runway is 125 cm; the shortest distance to E, avoiding the nineteen "dead ends," is 402 cm. The observer changes his place behind a screen higher than a man after each trial. Mice release urine drops along unknown pathways, these aiding them in finding their way later. Within a closed room they can also orient themselves by the noises coming from conspecifics in their home cage and also echoes from the walls. But if the maze is dismantled after every trial, the sections deodorized and then put back together and, in addition, the maze rotated and situated in a different part of the room, normal as well as blind mice are capable of learning the shortest path within eight days, without the help of "false cues," provided that they are allowed to explore undisturbed. When, however, the mice are allowed to learn the maze without such precautions, they orient with respect to the false cues, and three months are required before they cease responding to them.

The mice are aided in their learning by a "muscle" or kinesthetic sense, meaning the impulses emanating from the nerve endings in muscles, sinews, and joints which communicate posture and movement sensations. Our brain also immediately registers every curve of a path we take so that, after sufficient repetitions, we "know it like the palms of our hands." Lorenz's example of the water shrews whose path from their nest box

Independently dis-covered concepts

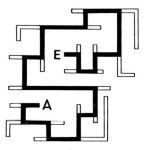

Fig. 2-11. View from above of a training maze: A-start and E-end of the "correct path" (black only in the diagram); white, the dead ends to be avoided.

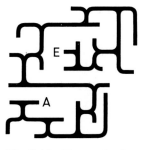

Fig. 2-12. Changes in the training maze, with curves instead of corners.

Labyrinth

Labyrinth

Labyrinth

ɥʇuᴉɹʎqoˈ

Labyrinth

Fig. 2-13. Types of print comparable t o the training maze and its changes (inclined forward, inclined backward, mirror image, and rounded form).

"Intelligent" dogs and horses

to their food passed over a box illustrates how such familiarity with a path can become deeply ingrained. When the box was removed on one occasion, they all jumped in the air at the point at which the familiar obstacle had stood and landed on the other side flat on their bellies.

To eliminate kinesthetic feedback, mice which had learned the path through the maze after training were presented with changes in the maze structure: 1) Inclined to the left; the first angle deceived them—instead of all angles being 90°, the first was 45°, the second 135°, and so on. 2) Inclined to the right; the first angle 135°, the second 45°, etc. 3) Unchanged angles but double the runway length throughout. 4) A mirror image of the original training maze. 5) All angles rounded off to arcs. At the start, all the mice balked at the first changed position they encountered, but "the penny dropped" very quickly; the mouse ran the new maze to the end almost without mistakes, without retraining. Soon the mouse took a shortcut at the third and all other obtuse angles following it, by jumping over the gulf between the runways instead of turning 135° like a vehicle. When all the new forms were presented to the mouse once, each in random order, it was familiar, without exception, with each one.

Should no other false cues have been overlooked, the conclusion reached is that the mouse had trained itself to the shortest path across the right-angled, original maze; in other words, it had learned to avoid all nineteen dead ends. It had thus acquired a memory image of the correct route, just as we, for example, are capable at any time of directing a visitor from some central point to our home. This memory image enables the mouse, without practice, to recognize immediately the changes described above, just as we can read the letters of our alphabet whether they are large or small, round or angular, vertical or inclined to the left or right, or even mirror images. Wolfgang Köhler termed such an ability "transposable path shaping." In our context it has been proven to be a developed averbal concept.

A further proof for averbal thinking is illustrated in the experiments in which animals are taught to "count." Unfortunately, this topic has suffered severely because of the continual revival of "intelligent" horses and dogs allegedly capable of "counting" and "spelling." In effect, all these "wonder animals" only learned to keep tapping their hoof or barking until a tiny—hopefully unconcious—movement from their trainer tells them to stop. Herd and group-living animals which register even the smallest changes in the moods of their conspecifics, never take their eyes off their human "group leader" at home, for obvious reasons. He, however, knows nothing of his unconcious expression movements, and wonders why his dog joyfully brings him the leash while he is still sitting at his desk, even before he himself has decided to stand up. "He knows me better than I know myself," is the usual response.

So-called "Kluge Hans errors" were first demonstrated on Mr. von Osten's "number-tapping" horse by the psychologist Pfungst; such mistakes are more difficult to avoid the closer the animal is, psychologically speaking, to its owner. In the circus, a continual interplay between trainer and animal is seen, rather than an example of continuous training. Only an experiment in which the trainer and all false cues are eliminated can show what an animal has learned and is capable of performing alone. This is why we have restricted ourselves in the following examples entirely to experiments with statistically significant results.

At the beginning, on a board at right angles to a pigeon's direction of approach, two wheat grains were placed on the left and one on the right, these positions being reversed randomly. The bird was allowed to eat the two grains but was continually chased away from the single one. Following on from this, the bird learned to distinguish three from two, then four from three, and the best performer could even distinguish five from four. Distinguishing six from five was not successful. To start with, the grains were laid out in a regular pattern; for example, three and four in a row or four in a square and five as on a die. Once the pigeon had grasped this, the "pattern aid" suddenly ceased; the grains comprising a group were arranged differently each time in every successive experiment, sometimes close together, sometimes far apart, and every time in a new pattern. These changes were too much for the birds, and all had to learn afresh each time.

The animal compared, as it approached, which of the two "simultaneously" visible groups contained the larger number of grains; it had to "see" numbers. In other experiments it had to "construct" numbers: for example, to pick out two grains, one after the other (successively) from a pile of grain, and to leave the remainder. If the pigeon stretched out its neck for the third grain, the experiment was brought to a halt immediately, until it had learned to master the task and stop by itself. This also succeeded up to number five. Simultaneous and successive abilities thus had the same upper limit. Once a pigeon could distinguish five without difficulty, it was presented with the board lengthways to its line of approach, with two grains in front and a pile of grain at the back. The bird which could distinguish five then took the two grains from the front and a further three from behind and walked off unperturbed. Individual training was unnecessary; even when the number of grains placed at the front was changed continually, the pigeon always picked exactly the right number of grains out of the pile at the back to complete the five. It could thus work out that $0+5=1+4=2+3=3+2=4+1=5+0$.

New methods, punishment-free and in the absence of the observer whenever possible, were continually developed to permit results to be recorded accurately without human influence. Experiments with seven other species of animals confirmed that both abilities had the same upper limit: pigeons could "see" and "construct" numbers up to five, budgerigars and jackdaws up to six, ravens, yellow-faced macaw, magpies, and squirrels up to seven. Zeier's newest experiments however, using a different technique, resulted in one pigeon reaching eight, with rhythm

Averbally counting animals

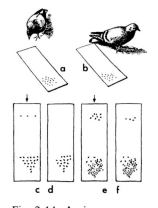

Fig. 2-14. A pigeon capable of counting to five discovers, without prior training, (a) one grain in front, eats four from behind, (b) leaves the rest and flies off. When (c) three grains are in front, it takes two (d) from behind. If nine (e) grains are in front, it will eat only five (f) of them and not go to the pile at the back.

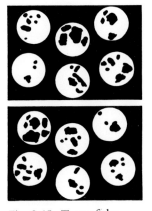

Fig. 2-15. Two of the hundred slides. The continually changing pattern is shown in the middle of the upper row, and, around this are five groups with three to seven spots whose shape, size, and arrangement vary each time. Those being the same in pattern and number are, in the upper picture, the group of four, upper right, and in the lower, the group of six, upper left.

aids. An African gray parrot reached the same sum, but with constantly changing rhythms in this case. It is certainly no coincidence that even humans cannot surpass these animals' feats in averbal counting.

A raven learned to distinguish groups of dots made up of varying numbers of thumb tacks. When, instead of thumb tacks, it suddenly discovered a large number of plasticine blobs of all different shapes and sizes on its pattern board as "information," it performed its task just as well as before, and even appeared to "stick to the point" to a greater extent. I presented exactly same task to a class of a hundred students. They were shown a hundred slides of such plasticine blobs one after the other, the time between the presentations being short enough to prevent them from counting the points. On each occasion, they were required to write down which of the five groups within the arc consisted of the same number of blobs as the pattern in the middle. Some were able to reach five, others six: not one of them beat the raven.

A clue as to how this successive "counting" works has been provided by a jackdaw. Several food dishes, all uniformly covered by a white lid, were placed in a row, and five baits were placed under these, arranged differently each time. When the series 1, 2, 1, 0, 1, was given, the jackdaw lifted up the first three lids and then walked away. The observer remarked "Wrong—only four baits taken." Then the jackdaw returned to the row of dishes through the sliding door—something which it had never previously done—tapped the first dish, which was now empty, once, the second one twice, the third once, opened the fourth lid, found nothing, then opened the fifth, took the last bait and left for good. Just as a child getting stuck in the middle of a recitation, it had started from the beginning again, and by doing so simultaneously avoided the pitfalls and made its "counting" visible by neck bobbing above the dishes (Fig. 2-16).

The two abilities, the simultaneous seeing of numbers and successive construction of numbers, obviously have nothing in common with each other from a sensory point of view. In comparison, our counting words can be used for anything which can be counted, and it is only because of this that six apples and six hammer blows are, of course, six, as far as we are concerned. Every normal child grasps this with little difficulty; when the mother raises five fingers, it understands that it can have five sweets, and when only two fingers are shown, then two. When, on the other hand, the mother asks "How many have you taken?" and the child shows her the right number of fingers, she then "sees" "constructed numbers." None of our experimental animals was able to accomplish this in such a general form, although they had learned in sequential training series to construct numbers from ones they saw, and to see constructed ones.

Different sensory modalities can be involved in such abilities. When an African gray parrot heard a single or bitonal sound, it learned to open a food dish with one or two spots on the lid, and, when, for example, it had learned to continue opening food dishes until it had found seven baits after having seen seven light flashes, it progressed directly to sound without having to be retrained, and responded to seven whistles or other sounds—on both occasions without rhythm cues—in the same way. Decades ago the Russian scientist Nadie Kohts gave her chimpanzee similar tasks. She took, for example, a circle out of a pile of variously shaped, flat children's blocks, and showed it to the chimpanzee, which then placed a similar block next to it, a triangle next to a triangle, and so on. Afterwards, she put them all into a bag into which the chimpanzee could not look, and showed it, on the palm of her hand, the shape which it was to pick out of the bag by touch alone. This succeeded immediately, without training. Such discoveries (technically termed "transmodal transpositions") leave no doubt that higher animals are capable of forming averbal concepts, in part dependent on sensory perception.

Ferster's chimpanzees have also shown astounding abilities in other fields. Two of them learned to count from one to seven on three switches comprising a counting mechanism of two units (binary), an experimental apparatus where 0 meant light and 1 meant dark. On command, they selected the following combinations: 001, 010, 011, 110, 101, and 111. In the seventh experimental series, after millions of individual experiments, they wrote the observed binary number pattern. Although they could write the observed number pattern from one to seven in binary form without a mistake, Ferster commented that "This is obviously not counting in the arithmetical sense of the word. Nevertheless, step by step, we have finally come this far. Out first task is to force the animals to complete the abstract process of counting upward in whole numbers, as a child would say the numbers from one to ten in an irrelevant situation. Once this chain of verbal behavior can be synthesized in chimpanzees, they should be capable of counting any number of objects correctly to the upper limits of their learned abilities." This, however, still remains to be seen.

Several attempts have been made to teach chimpanzees human words. The results, however, have been very disappointing, as they just do not have the ability to mimic sounds. Their gestures are much more expressive, as is the case in all monkeys and even humans—without us continually being aware of this—and play a very important role in communication. Because of this, an American couple, the Gardners, lit on the idea of teaching their chimpanzee, Washoe, the American deaf-and-dumb language. The results were astounding: in 1969 Washoe could "speak" and understand thirty words, delighted in showing them off, and could even form, spontaneously, in sign language, meaningful three-word sentences.

Our children learn to speak by listening to adults referring to things, which they already know averbally, with the same word time after time. They then try to form the word themselves. In the first attempts to speak, the averbal concept precedes every word; only when the child can speak is it also capable of asking the meanings of words which are unknown to it.

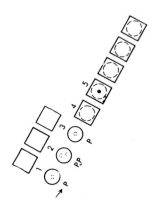

Fig. 2-16. Schiemann's jackdaw, which was able to count to five, approaches the row of dishes in the direction of the arrow, eats, after lifting the first three lids $1 + 2 + 1 = 4$ baits, and goes away, but returns immediately, bobs its head at the first three open dishes exactly as often as it had pecked there the first time (P), and finally progresses to five. The remaining lids were untouched.

Erich von Holst, F. A. Beach, and many other investigators, especially those from North America—see Gilbert and Sutherland's collected works—have recently provided us with valuable syntheses on behavior and brain physiology. The study of animal behavior via nervous processes (neuroethology) is, today, a rapidly developing discipline. Konermann's goose can serve as a good example of this. Owing to the complete crossing of its optic nerves, this goose learned to select a circle with its right eye and a square with its left when the same pair of patterns was offered it. Berhard Rensch rightly asked, "Would a mammal be capable of such 'schizophrenic' behavior?"

A very illuminating example of insightful planning in advance is offered by the five-year-old chimpanzee Julia, which J. Döhl, a student of Bernhard Rensch, taught, in more than 2200 trials of twelve grades of difficulty, to move an iron ring, which signified reward, along the single path leading to the exit point of a maze by means of a magnet, after the animal had been given sufficient time to examine the maze. The ring started at A, under the magnet in Julia's hand; the left arm of the maze ended in a dead end, and Julia had to aim toward her left thumb at the exit, along the right arm of the maze. The task was made more and more difficult in every experiment. Döhl only gave the chimpanzee each maze once. Soon the correct and incorrect arms of the maze crossed and became even more erratic, forming meanders, the number of dead ends increased, and, instead of only one exit, four or even five were given. In B, the ring had to be moved left across eleven points at which a choice had to be made and around thirty-three corners before the exit at the front right was reached. Six students who were given ten of the last hundred mazes presented to Julia, simplified the task by looking along the path backwards from the exit until they discovered the only clear way from the start—exactly as Julia did, as one could see from her eyes. Up to the first choice point from the start, the students often required more time than the practiced chimp and then went in the right direction with almost the same low frequency of entering dead ends as she did.

What is this, however, in comparison to the homing abilities of salmon, green turtles, terns, and seals, which travel over thousands of kilometers to their breeding grounds? Every creature is capable of ensuring its survival and that of its species without being taught. Body form, capabilities, and behavior work together to make a natural harmony, and this same never-ending miracle occurs at all levels of the zoological system and in all niches of the environment. It is true that many more species than are alive today have become extinct, but not one of them, through its own behavior, has worked so relentlessly toward its own destruction as does man. Only when we can be shaken into being sensible at the last minute and into learning from the animals something about how to ensure our individual survival and that of our species will Nature's rescue and, with it, our own survival, be made possible.

Fig. 2-17. Mazes from Döhl's fifth (upper) and twelfth (lower) series of experiments. M magnet, R iron ring.

3　Invertebrate Animals

The young Charles Darwin, during his cruise around the world with the British research ship *Beagle* from 1831 to 1836, had already collected the basic knowledge which was to make him the founder of modern evolutionary theory, which is still valid to a great extent. This knowledge was put forward in his book *The Origin of Species by means of Natural Selection*, published in 1859 and liberally illustrated with examples from studies he had conducted. Today, the concept of the evolutionary development of creatures has become something which is self-evident to biologists. However, opinions on the general trends in animal evolution remain varied.

In general, two opposing theories can be recognized: Ernst Haeckel's gastrula theory, named after the gastrula stage in development; and the ciliate or planula theory put forward by Steinböck, Hadzi, and Hyman (the planula is the typical larval stage of coelenterates).

Haeckel's theory, developed in 1866, is based on the following premise: "the series of developmental stages exhibited by an individual during the course of its development from a single cell to its completed form is a short, compact repetition of the extended series of forms which the ancestors of this organism itself or the prototype of its species followed from earliest times up to the present" (the basic biogenetic rule). Haeckel's gastrula theory, in this format, was finally put forward in 1874. According to this concept, multicellular animals have developed from colonial unicellulates and, in this case, the coelenterates (see Chapter 6) were the original multicellular animals.

Hadzi and Steinböck recently developed the ciliate theory in opposition to Haeckel's view. According to this, multicellular animals have been formed by simple cell division and cell formation processes (cellulation) from multinucleate unicellular animals (ciliates, class Ciliata; see Chapter 4), the individual nuclei forming cell membranes around themselves. This theory suggests that the original multicellular animals were not the coelenterates but the flatworms, which lack intestines (see Chapter 8).

Evolution, by
E. Thenius

Basic biogenetic rule

Even the coelenterates have developed from such flatworms, as indicated by their planula larvae. Numerous sound objections have been raised against the planula theory, especially those dealing with critical evaluation of basic coelenterate characteristics and those of flatworms (e.g., the subsequently degenerated body cavity). Supporters of the planula theory have not, however, acknowledged such criticisms: they hold that the coelenterate body form has become simplified later as a result of a sessile mode of life.

It is important to stress that, among present-day zoologists, opinions have hardly diverged from those of Haeckel. Of course, so-called "transition types" between unicellular and multicellular animals can be regarded as models leading to the creation of multicellular types such as, for instance, the charming *Volvox* colonies (see Chapter 4), but these cannot be looked upon as their ancestors.

By investigating the remains of creatures preserved in stone from previous geological ages (fossils), paleontology has offered valuable evidence for the evolutionary development of many groups of animals. In this case, however—the development of multicellulates from unicellular animals—it cannot help, since no fossil remains at this developmental level can be found. It is known that hard parts of the body fossilize, and furthermore, very few fossils exist from the Precambrian (more than 560 million years ago), during which period the multicellulates must have arisen. Actual fossil evidence starts with the Cambrian (560 million to 450 million years ago), from which several branches of invertebrate fossils are known. Which of the two theories, gastrula or planula, is therefore correct, can thus never be answered with the help of paleontology, but only by zoology.

General divisions of the invertebrates

Even the question of how the general divisions of the invertebrates came about, and the associated differences in opinion, cannot be resolved by paleontology, since the decisive larval forms are unknown as fossils. The "classical" binary division into the Protostomia and Deuterostomia is contested by the views of the Berlin zoologist Werner Ulrich, who suggests a triple division into the ancient animals with body cavities (ancient coelomates), those with a ventral nerve cord, and those with a dorsal nerve cord.

Fossil findings prove, however, not only the great geological age of the unicellulates and multicellulates but also the appearance of types of organization completely unknown in present-day animals and means of coordination not possible in present-day body forms (e.g., *Tribrachidium* and *Parvancorina* from the Australian Precambrian). The presence of unicellulates (Foramenifera and Radiolaria; see Chapter 4), sponges (stony sponges; see Chapter 5), coelenterates (Medusae; see Chapter 6), crinoids (see Vol. III), sipunculids (see Chapter 11), mollusks (see Vol. III), and segmented worms (polychaetes; see Chapter 12) has been proved by fossil findings, and, in addition to these, echinoderms and, among the arthropods, polychaets, trilobites, Chelicerata, and Crustacea. This indicates that the various branches of the invertebrates, with the exception of the unorganized animals, platyhelminthes, flatworms, and round-worms, which never or hardly ever fossilize, are known as fossil forms from the earth's most ancient times. Further information on the fossil forms and groups of other invertebrates can be found in the relevant chapters both in this volume and in Volumes II and III.

Natural scientists have tried, since earliest times, to create a system for the animal and plant kingdoms, to enable the various forms of life to be organized and placed in their proper order in the series. These arrange-ments, extending from those of Aristotle (384–322 B.C.) and Pliny the Elder (A.D. 23–79) to Konrad Gesner (1516–1565) and, finally, the founder of systematics, the Englishman John Ray (1627–1705), were all, without exception, "artificial" systems which paid little or no attention to the natural relationships between animals. Only when modern zoology began, in the second half of the 18th and first half of the 19th Centuries, did science endeavor to create a "natural" system, splitting up and arranging animal species and groups with regard to their relationships to one another. Such a "natural system" was made possible by George Cuvier (1769–1832), who founded paleontology, the science of extinct (fossil) creatures, and, above all, by evolutionary theory, whose protago-nists Jean Baptiste Lamarck (1744–1829), Charles Darwin (1809–1882), Thomas Henry Huxley (1825–1895), and Ernst Haeckel (1834–1919) recognized the evolutionary relationships and supplemented the natural system with family trees (or better, as we know today, "family bushes").

Of course, an arbitrary ordering system for the innumerable forms of the animal kingdom, even if clear and simple, is inadequate if not based on biological principles. Systematics—the branch of biological science which determines the evolutionarily important characters and behavior patterns of various animals and orders them according to these (classification)—does more than construct a register or inventory of the animal kingdom, as some non-scientists and even poorly instructed "scientists" still believe. It forms the basis of every zoological study.

How are the relationships determined which permit the construction of a natural system? These cannot be based on evolutionary theory, since evolutionary theory itself is supported by this system. Other evidence must be used to construct the system. Paleontology can offer the systema-tist only a fragmented series of suggestions and usually only the fossilized hard portions of extinct forms. Comparative science cannot depend on external similarities or on the form and internal structure of animals (comparative morphology) and conclude relationships from these, since even animals which are not related to one another can develop similarities associated with adaptation to the same environmental circumstances, as, for instance, the Kamptozoa (see Chapter 9) and Bryozoa (see Vol. III). In addition, parasitic life can change an animal's form to such an extent

Animal systematics, by H. Wendt

that the most remote similarities with free-living relatives disappear, as in the Rhizocephala, for example, among the barnacles.

Clarity is reached not only through a detailed comparison of mature animals but also of every condition in their development. Many parasites are able to give evidence of their host's relationships to other animals, since their evolutionary development mirrors that of the hosts. A good example of this is the flamingo feather lice (see Vol. II) and their hosts, the flamingos.

Evolutionary laws, however, are concerned not only with characteristics of form but also with characteristics in behavior patterns. The study of behavior patterns and their interrelationships is a recent development which will play a major role in the study and determination of relationships among animals. The chemists have also recently had their say: they have shown that the serial arrangement of amino acids in protein-molecule chains differs between species and can be used as an indicator of their relationships. In conclusion, the present-day animal world can only be understood as a whole when its distribution over the globe, not only the modern distribution but also that of bygone ages, is known. Land-masses and oceans in the past gave a far different picture than they do today. Continents which were once joined have separated, and the development of animals in each section progressed along its own lines. On the other hand, land bridges often came into being which, for varying periods of time, allowed an exchange between such separated animal populations.

The formation of a natural system, the goal of systematists, thus depends on numerous building blocks contributed by a diverse group of scientists: morphologists, paleontologists, behaviorists, biochemists, animal geographers, geologists, geneticists, and, last but not least, specialists in particular animal groups. Among the latter, apart from professional scientists, there are many enthusiastic amateurs who devote their spare time to this work.

The basis of modern animal systematics

The basis of modern animal and plant systematics was laid down by the Swedish scientist Carl von Linné (Linnaeus) (1707–1778) in the 10th edition of his most important work, *Systema Naturae* (1758), in which he introduced the present-day method of defining every animal species with two names, taken from Latin or Greek or Latinized languages, one of these being the genus name and the other the species name (binary nomenclature). Linné thus named the hare *Lepus timidus* (Latin: *Lepus*= hare, *timidus*= timid), and the rabbit *Lepus cuniculus*; thus both species are included in the same genus, *Lepus*. It emerged later that Linné meant the northern snow hare when he called the animal "hare," and even today it still bears the name *Lepus timidus*, while the European hare bears the species name *Lepus europaeus*, given it by the German biologist Peter Simon Pallas (1741–1811). The European rabbit has not been included in the genus *Lepus* for a long time, and now bears the scientific name *Oryctolagus cuniculus*.

Although, as this example illustrates, the scientific nomenclature has often changed drastically since Linné's time, and although differences in opinion still exist between the systematists with regard to relationships and thus to the ordering of certain species and groups of animals, Linné's binary nomenclature has proved itself so practical that it is still used worldwide. The systematic arrangement of animal forms is, of course, dependable only when exact information is available on the evolutionary development of the group in question. Since new and sometimes revolutionary evidence is continually coming to light, systematics thus adapts to comply with the views of science.

Various zoologists have, for years, called the same animal species by different scientific names. To end this confusion, the first scientific name given to the animal in question is given priority in every case (the priority rule). This provision was laid down in the "International Rules of Zoological Nomenclature." In scientific publications, the author's name is usually cited together with the scientific name, since he was the first to adopt it, and the year in which the name was given is also added. These have not been cited in the text in this work, but are given in the systematic survey at the end of each volume.

Many animals are given not just two, but three Latin or Latinized words (triple nomenclature). The European hare inhabiting the Mediterranean region is thus called *Lepus europaeus mediterraneus*. The third name, *mediterraneus* in this case, designates a subspecies. In various areas of their range, members of a species diverge from type in certain characteristics, thus making an additional subdivision necessary.

Sometimes subspecies are termed "geographical races" or simply "races." Although some zoologists still adhere to these terms, it is easy for confusion to arise between them and domestic races of animals bred by man or with various races of humans, which are certainly not, zoologically speaking, considered as subspecies.

In systematics, subspecies are assembled to form a species, and closely related species form a genus. Animals are then further ordered, on the basis of their characteristics, in progressively related groups according to a definite degree of development. Related genera are merged to form a family, related families to form an order. The rank relationships progress further through classes and phyla up to the animal kingdom itself. Since, on many occasions, these groups or categories are not sufficient, the systematists have included, where necessary, sub and super categories. Genera are thus merged to form generic groups (tribes), families divided into subfamilies or joined to form superfamilies; the same applies for arrangements such as sub and super orders, sub and super classes, subphyla, etc. Even though the best natural system is an imperfect human achievement, the systematists try their hardest in the arrangement of animals to do justice to the relationships under natural conditions, by delicately grading related groups with respect to one another.

Related groups

The animal kingdom, which is equal in rank to the plant kingdom, is divided into two subkingdoms: 1. Unicellulates (Protozoa; see Chapter 4); 2. Multicellulates (Metazoa; see Chapters 5 ff). Three divisions are distinguished among the multicellulates: A. Mesozoa (see Chapter 5), B. Sponges (Parazoa; see Chapter 5), C. True multicellulates (Eumetazoa; from Chapter 6 onward through the rest of the volumes). The division of the true multicellulates contains twenty phyla, listed below:

1. COELENTERATES (Cnidaria; see Chapter 6); 2. COMB JELLYFISH (Ctenophora; see Chapter 6), joined to form the phylum Coelenterata or Radiata; 3. FLATWORMS (Platy-helminthes; see Chapter 8); 4. KAMPTOZOANS (Kamptozoa or Endoprocta; see Chapter 9); 5. STRAPWORMS (Nemertini; see Chapter 9); 6. ROUNDWORMS (Aschelminthes; see Chapter 10); 7. POTAMIDS (Priapulida; see Chapter 11); 8. "PEANUT WORMS" (Sipunculida; see Chapter 11); 9. ECHIURIDS (Echiurida; see Chapter 11); 10. SEGMENTED WORMS (Annelida; see Chapter 12); 11. ONYCHOPHORES (Onychophora; see Chapter 13); 12. WATER BEARS (Tardigrada; see Chapter 13); 13. LINGUATULIDS (Linguatulida; see Chapter 13); 14. ARTHROPODS (Arthropoda; see Chapter 14, and Vol. II). The last five phyla are joined, forming the superphylum articulate animals (Articulata; see Chapter 12 ff, and Vol. II). 15. MOLLUSKS (Mollusca; see Vol. III); 16. COMB JELLIES (Tentaculata; see Vol. III); 17. ARROW WORMS (Chaetognatha or Homalopterygia; see Vol. III); 18. ECHINODERMS (Echinodermata; see Vol. III); 19. PROTOCHORDATES (Pentacoela; see Vol. III), animals with a body cavity divided into five sections; 20. CHORDATE ANIMALS (Chordata; see Vols. III through XIII). The vertebrates (Vertebrata) are a subphylum of the chordates, and are dealt with in Volumes IV through XIII of this work.

These numerous phyla of the animal kingdom are not completely unrelated to each other. Many connections exist between the ciliates (see Chapter 4) and the Rhizopoda (see Chapter 4). Craspedomonadids (see Chapter 4) and sponges (see Chapter 5) also appear to be related to one another. The flatworms, nemertines, sipunculids, mollusks, and arthropods all have a "spiral egg cleavage" in their early developmental stages. One can therefore correctly assume that these "spiral animals" are more closely related to one another than to other animal phyla. Nevertheless, each phylum has its own unique "structure," meaning the peculiar characteristics of form which remain when all common characteristics of the individual species making up the phylum have been subtracted, these being considered as special adaptations to their present mode of life. The phylum Annelida thus has a body form consisting of segments, each, except the clitellum, containing its section of coelom and a nerve ganglion; this is, however, characteristic of the body form of oligochaetes. The sucking mouthparts of leeches are not included in the typical body form: they are characteristic of the order Hirudinea. Body forms thus become richer and richer in characteristics the further we progress through the system.

Body forms within the phyla, by P. Rietschel

The number of species included in branches of the animal kingdom varies widely: the priapulids have only four species, in contrast to the arthropods with 800,000. Because of this, it is impossible to divide the twenty-six phyla equally between the thirteen volumes comprising this Animal Life Encyclopedia. Nineteen phyla are already included in the first volume. Therefore, Volume I contains many more sorts of body form than subsequent volumes. If more information is desired about animal forms, in addition to facts and figures about their "lives," this is illustrated in the marginal figure beside the text at the appropriate point.

It is not only due to the large number of phyla of the animal kingdom described in Volume I that this volume is of especial interest: we are prevented from observing many of these animals by their small size alone. Naturalists should try to see for themselves as much as they can of what is described. This is, in many cases, possible only with the help of a microscope. Most demands can be met with a "student's microscope" having the following objectives: $\times 3$, $\times 10$, and $\times 40$; a $\times 10$ is usually sufficient if it is standard. The small microscopes with special objectives, usually offered to children, are not to be recommended. Introductory literature can be found at the end of the volume. The number of species which can be found with the aid of the microscope, even in the city, can offer naturalists a substitute for the constantly decreasing numbers of more advanced animals.

It is not only the microscope, however, which can widen our view of the world of tiny animals: even a ten-power or twelve-power magnifying glass is able to do this and should always be carried, since animal life is generally encountered when it is least expected. On excursions into the country, the magnifying glass is a constant companion. Binoculars are also indispensable, larger animals being brought into range without encroaching within their flight distance. There are modern binoculars with a six-power or eight-power magnification and a medium light-collecting power that are also so small and light that they are no trouble to carry along on expeditions. The heavy binoculars with higher powers of magnification and greater light-collecting power are an expensive aid to special observations, but they are usually left at home when they are most badly needed. Unfortunately, the value of binoculars in zoos is much too little appreciated. Animal behavior is rarely seen in the areas nearest the visitors. Binoculars can fulfill the wish to observe distant animals as if they were right in front of us.

The desire to hold observed animal life in pictures is fulfilled by animal photography. For this, the ability to observe, patience, photographic experience, and, last but not least, rather expensive equipment are necessary.

▷
Body forms and means of locomotion among the unicellulates: A. Internal structure of a unicellulate (simplified): 1. Nucleus: 1a. Nuclear membrane (internal membrane only, the external to 2a), 1b. Nucleolus. 2. Cytoplasm: 2a. Reticulate protoplasm, 2b. Golgi apparatus, 2c. Mitochondria, 2d. Secretory vacuoles, 2e. Food vacuoles. 3. Flagellum: 3a. Flagellum base, 3b. Free flagellum, 3c. Transverse section of a flagellum showing microtubulae ("9 + 2" pattern). B. Movement and feeding in an amoeba: 1. Ectoplasm (transparent and transluscent). 2. Endoplasm (particulate). 3. Nucleus with nucleolus. 4. Above, engulfing; center, within the food vacuole; below, indigestible residue expelled. The blue arrows show the direction of protoplasmic streaming while creeping. C. Looping movements of an amoeba as seen from the side. D. Diagram of ciliary beat in a ciliate. The fast thrust in black, the slow return movement in blue.

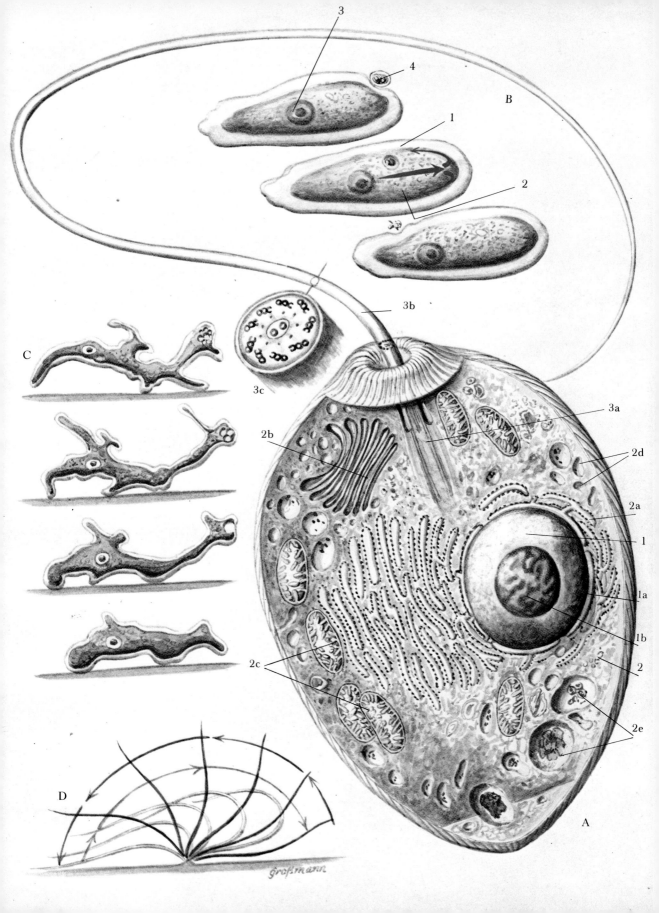

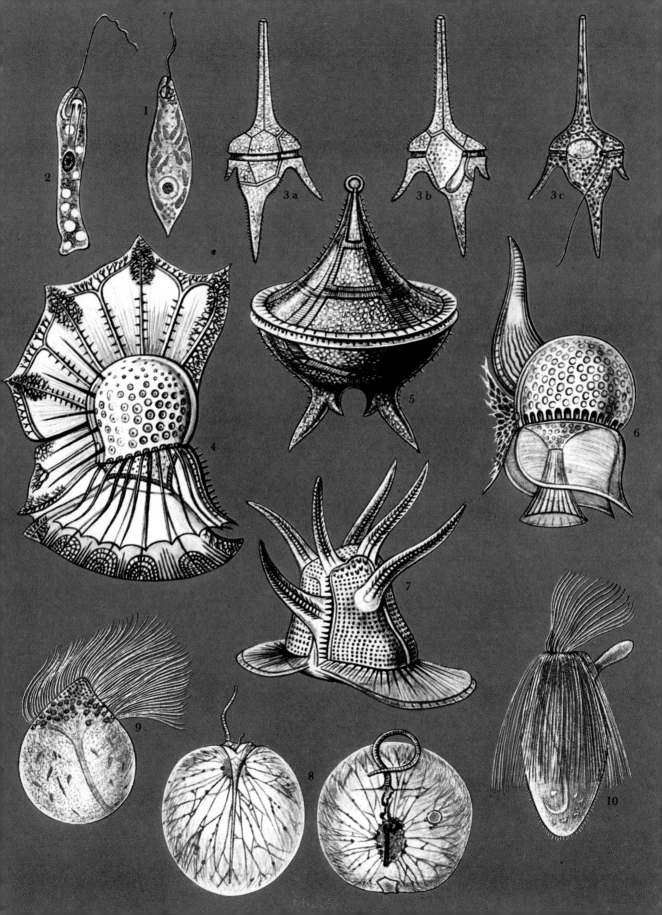

4 The Unicellular Animals

The subkingdom
protozoa, by
P. Rietschel and
K. Rohde

◁ Flagellates (class
Flagellata). Order Euglenoidina: 1. *Euglena viridis*
(plantlike mode of nutri-
tion); 2. *Peranema tri-
chophorum* (animallike
mode of nutrition);
3. Dinoflagellate (Dino-
flagellata) *Ceratium
hirundinela* (side view 3a
and 3b; living specimen 3c);
4. *Ornithocercus magnificus*;
5. *Peridinium divergens*;
6. *Histioneis remora*; 7.
Ceratocoris horrida; 8.
Marine phosphorescent
organism (*Noctiluca
miliaris*) Polyflagellates
(order Polymastigina,
termite symbionts);
9. *Calonympha grassii*;
10. *Joenia annectens*.

The whole of the animal kingdom, including man, and the whole of
the plant kingdom have developed by evolution from creatures which
consisted of only a single cell. These "unicellulates" (Protozoa) have
continued to exist, together with the multicellulate plants and animals
which arose from them, and it is not surprising that they have also
perfected themselves during their at least four million years of existence
on this planet. When the dividing line between plants and animals is
drawn on the basis of their modes of nutrition (see Chapter 1), unicellular
animals (protozoa) and plants (protophyta) can be distinguished from one
another. Of course, the division between the two becomes hazy among the
flagellates, since certain of them can obtain their nourishment by both
means. If one such species loses its plant-typical chromoplast, it switches
from being a member of the plant kingdom to the animal kingdom.
This transition to an animal has often occurred during the course of
evolution, and can also occur before our very eyes. Certain gold mona-
dines (*Rhizochrysis*, *Chrysarachnion*), which can feed either as a plant or
as an animal, sometimes divide so that one member of the succeeding
generation contains the entire chromoplast and lives as a "plant," while
the other contains nothing and it and its descendants are henceforth
"animals."

The name "protozoan" or "protozoon" (which has the same meaning) is misleading, since many of the present-day representatives are no
less highly organized than certain multicellulates; their paths have just
diverged since ancient times. The cells of multicellulates have become
part of various tissues and organs, each of which is specially developed for
certain tasks. They have thus developed into specialists which, because of
the onesidedness of their capabilities, are dependent on cooperation with
other similar and dissimilar specialists. The unicellulate, on the other hand,
has remained an "all rounder" and has perfected its internal structure to a
greater and greater extent during the course of time. With the aid of a
microscope, organlike structures can be recognized within the unicellulate,

these "organelles" (singular: organelle) fulfilling the same task as the cells comprising the tissues and "organs" of the multicellulates. Even the amoeboids, which are so changeable in form and whose protoplasmic body is talked about time and again as a "formless lump of slime," are much more than simply that, as the microscope and electron microscope divulge. How could they otherwise form such regular and wonderfully complicated skeletal structures? (See Color plates, pp. 93/94.)

The cell body of a unicellulate contains the same organelles as the cells of multicellular animals, and besides these they also have some which are lacking in multicellulates, for example the "cell mouth," which serves to take up nourishment and transport it to the cell's interior. It can be a permanent structure or can form only when necessary, disappearing again afterward. A cell covering with diverse structures gives the unicellulate a high degree of form, in the same way that it permits the close joining of cells within the tissues of multicellulates. While these can have only a single flagellum or a definite number of cilia restricted to a small area, the whole surface area of unicellulates can be used for this purpose; flagella and cilia in these show an inexhaustible profusion of arrangements and developments. In comparison to the cells of multicellulate tissues, the unicellulate living in fresh water finds itself in a special situation: while the multicellulate tissues are bathed in a fluid comparable in its content of salt and other compounds to the cell fluid, the unicellulates are in "hypotonic" water, poor in dissolved substances. The unicellulates therefore take up water continuously through their body surface, and would finally swell and burst if they had no means of excreting it again. By expenditure of energy, they collect the water in their bodies into bubbles which, after they have reached a certain size, are emptied from the body surface. These are the contractile vacuoles, found widespread among fresh-water unicellulates. They can also eliminate excretory products of metabolic processes, but this is definitely not their main function, since many unicellulates, particularly those parasitic in body fluids and those living in sea water which also have excretory products but do not need to excrete water, exist without them.

The "behavior" of many unicellulates is also much more highly developed than that of a cell which forms part of a multicellulate. The flagellate *Euglena*, for instance, swims toward or away from light, depending on external factors. The paramecia and many other ciliates respond to mechanical, chemical, thermal, and gravitational stimuli.

The role which unicellulates played and still play in nature shows how successfully the path driven by them through the earth's past and present history is and has been. Forty-five thousand species are known; of these, twenty thousand are fossils, species which have become extinct and are known only from their skeletons. Nevertheless, many new species are described every year so that a considerable increase in this total must be reckoned with. From the twenty-five thousand living species known,

Organelles

The richness in species

seven thousand (other authorities cite fewer than four thousand) are parasites, some of which cause dangerous diseases in man and animals. The large number of unicellulates which live in fresh water and sea water form an important source of nourishment for higher animals, and, despite their small size, many unicellulates play a decisive part in the formation of geological deposits. They are sometimes present in the ground in enormous quantities. One hundred thousand flagellates, fifty thousand amoeba, and a thousand ciliates have been counted in one gram of earth. They can flourish under the most extreme conditions; for example, snow and glacial ice are sometimes covered with a red film, consisting of red flagellates, and is called blood or red glacier snow. Many species of unicellulates have a worldwide distribution; in others, the distribution is limited, for instance as a result of dependence on a warm climate, such as the Radiolaria and Foramenifera, or owing to lack of a host, as is the case in numerous parasites.

Class: Flagellata

The class FLAGELLATA (from the Latin *flagellum* = whip), includes a great number of forms. Their characteristic is the presence of one or more flagella during the course of an extended development. The flagellates are considered the most primitive of the protozoa. The reason for this is that, in contrast to the other groups of protozoa, they contain both plantlike and animallike organisms. Plants, however, existed before animals; plants are more primitive than animals. This becomes clear when the difference between plantlike and animallike organisms is explained. Plants are organisms which form organic substances from inorganic ones with the aid of sunlight—that is, which are able to photosynthesize (from the Greek $\varphi\tilde{\omega}\varsigma$ = light and $\sigma\acute{v}\nu\vartheta\epsilon\sigma\iota\varsigma$ = put together). There are bacteria which do not utilize sunlight as an energy source for the formation of organic compounds, instead producing the energy required by chemical reactions (chemosynthesis), but they play a less important role in nature. Plants require a pigment compound for photosynthesis, for example the green plant-pigment compound chlorophyll (from the Greek $\chi\lambda\omega\rho\acute{o}\varsigma$ = green-yellow). The pigment compound required for photosynthesis is lacking in animals, and they obtain the organic substances necessary for their nourishment from other organisms, ultimately from plants. Plants must therefore have developed before animals during the course of evolution, or, as far as the unicellulates are concerned, the plantlike flagellates must have existed before the animallike flagellates.

It is probable that all other organisms have arisen from the flagellates, although naturally not from the species existing today. The fact that plants and animals, with few exceptions, and including the unicellulates, have flagella during a certain stage of development is proof of this. One has only to think of the human sperm cell. Even the amoebae, which, owing to their simple structure could be considered the most primitive of the unicellulates, sometimes have flagella. Only the majority of the seed-bearing plants have lost a flagellated phase during the course of evolution. In roundworms (nematodes), which for a long period of time were considered to be without flagella, flagella have recently been discovered. It is important to state that the flagella present in the whole of the animal and plant kingdoms, with very few exceptions, show the same structure under the electron microscope. A cross section of a flagellum always shows one pair of minute tubules in the center and nine pairs around the circumference. It can therefore probably be regarded as a structure which developed once during the course of evolution and which has been retained in its characteristic form by all higher organisms.

In our more detailed investigation of the flagellata (Mastigophora) we shall begin somewhat out of the systematic series with the order Euglenoidina, since the genus *Euglena* can be found everywhere in water puddles in such large numbers that they color the water green. The flagellates belonging to this order feed, depending on species and environmental conditions, either like plants or like animals, or like both. Species belonging to related genera only take up organic compounds. *Euglena viridis* (see Color plate, p. 88) is an example of a flagellate which uses both feeding methods. If kept in the dark, so photosynthesis cannot occur, it will live off organic substances dissolved in the water. *E. gracilis* loses its pigmentation and feeds in an animallike manner in a solution rich in organic substances even if light is present. If a small quantity of organic material is available, it turns colorless only in the dark and, if organic material is entirely lacking, the green coloration is retained even in darkness.

Euglena has a spindle-shaped body with the flagellum attached anteriorly, arising from the base of a flagellum sac. The nucleus is located near the pointed posterior end, and green bodies containing the photosynthesizing substance, the plastids, are found within the cytoplasm. The resulting products of photosynthesis are stored within the cell plasma in the form of starchlike grains. A red eye-spot is present in the wall of the flagellum sac, and plays an important role in responses to light. It is, itself, not sensitive to light but absorbs light rays from definite directions, thus allowing an ordered response to light stimuli. The actual light-sensitive portion of the cell is a thickening of the flagellum within the flagellum sac. A contractile vacuole, present near the eye-spot, fills with liquid and discharges this to the exterior into the flagellum sac. Around this are several smaller vacuoles which, on the disappearance of the large one, join together to form a new vacuole. The main task of these organelles is to remove water from the cell body. From this, it can be assumed that marine species, which live in an environment with a much higher salt content than fresh water has, and thus take up less water, generally have no contractile vacuole. In types introduced from fresh water to sea water, the rhythmic emptying of the vacuole slows down for the same reason. The surface of *Euglena* is covered by a rather firm but elastic envelope.

Since *Euglena* is very common, it is a practical object to investigate. If a microscope is at hand, the form and movements of this flagellate in a

Order: Euglenoidina

▷
Foraminifera (order Foraminifera, by E. Haeckel): 1. *Lagena interrupta* (view from above and from the side) 2. *Miliola striolata* 3. *Miliola reticulata* 4. *Lagena spiralis* 5. *Peneroplis planata* 6. *Nummulites orbiculatus* (diameter 25 mm). 7. *Bulimina inflata* 8. *Nodosaria spinocosta* (view from above and from the side). 9. *Bolivina alata* 10. *Frondicularia alata* 11. *Globigerina hulloises* 12. *Polystomella aculeata* (view from above and from the side) 1 with protoplasm; all others, skeleton alone.

MILLA

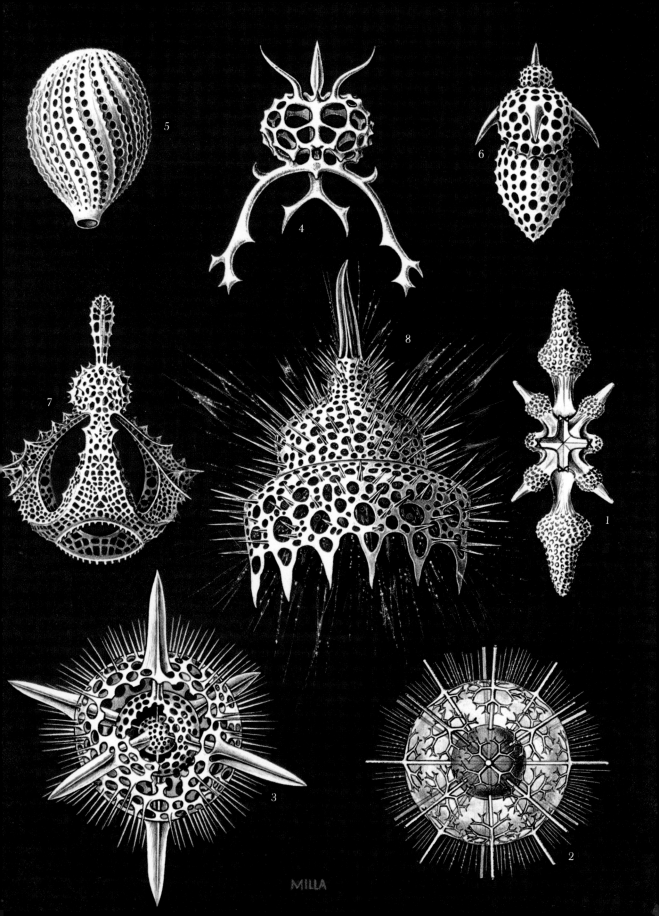

MILLA

◁
Radiolaria (order Radio-
laria, by E. Haeckel) suborder Acantharia: 1.
Acantholoncha flavosa, 2.
Stauracantha quadrifurca
(central capsule brown).
Suborder Spumellaria: 3.
Hexacontium asteracanthion
(the two external spherical
shells opened to expose the
third); suborder Nasselaria:
4. *Triceraspyris gazella*, 5.
Cyrtophormia spiralis, 6.
Pterocorys rhinoceros, 7.
Clathrocanium reginae, 8.
Calocyclas monumentum, 2
and 8 with protoplasm; all
others, skeleton alone.

drop of ditch-water can be followed under normal magnification. To observe actual movement, it is necessary for a little gelatin or paper gum (methyl cellulose) to be added or the slide cooled by ice. These decrease the speed of movement of this tiny protozoan. Under good conditions, it can be observed how the euglenoid is pulled forward by rapid, wavelike movements of the flagellum, the cell body turning along its longitudinal axis. Since the surface envelope is elastic, changes in form can also be observed occasionally. Division stages, in which the organism divides longitudinally into two daughter cells, can also be seen. Under unfavorable conditions, such as when the water drop starts to dry up, the euglenoid begins to encyst and this capsule (cyst) can withstand extreme conditions for a long time. Reproduction within the cyst also occurs in certain species.

A very simple experiment can be conducted without the use of a microscope. If a puddle with a large number of euglenoids is found—this being indicated by the green water color—and is illuminated from one side, it can be shown that, after a short while, all the flagellates have gathered at the lighted side. They thus show an ordered response to light. This is a positive light response, since they swim toward the source of light. Under certain conditions, such as intense illumination, they respond negatively, however, swimming away from the light source. The positive response is brought about by a different mechanism than the negative. When, during the positive light response, the light strikes the *Euglena* from the side, the light-sensitive portion of the flagellum is rhythmically shaded, since the *Euglena* turns on its longitudinal axis as it moves forward. The light-absorbing eye-spot, because of this, passes from time to time between the light source and the light-sensitive portion of the flagellum. Each time the shadow passes, the *Euglena* swims more toward the light until no further shading occurs, that is, until the light rays fall parallel to the direction of movement. During the negative response, the euglenoids swim backward, away from the strong light.

The genus *Phacus*, easily recognizable by its unique asymmetrical form, is often found together with *Euglena*. It is also green, and its extremely flattened body has one or more long processes at its posterior end. An eye-spot is present near the base of the flagellum, and the body surface is clearly striped. *Peranema trichophorum* (see Color plate, p. 88) can be mentioned here as a relative of *Euglena* which feeds in an animallike fashion; it is often found by looking through the microscopic fauna present in infusions and other bacteria-containing water. The anteriorly placed flagellum lashes only at its most anterior portion, the majority of the structure being held straight out in front. Owing to its width, it is visible under low magnification. The sac from which the flagellum arises serves as a gullet through which bacteria are taken into the body.

The order PHYTOMONADINA, which follows the euglenoids, contains, as the name (the plant monads) intimates, mostly green species which

Order:
Phytomonadina

feed as plants. However, among these we find unicellulates which indicate a trend toward a multicellular body form and thus attract the attention of not only the botanists but also that of a large number of zoologists. If a drop of ditch-water is examined under the microscope, spheres which turn on their axes are often noticed, these being constructed of numerous green organisms separated by intermediate spaces. This is a colony of green flagellates which have not separated after cell division, but have remained together in a gelatinous mass. This type of colony formation is very common among unicellulates. There are species which are joined by stalks, jelly, or pellicles, i.e., the non-living body portions, and in which every individual responds independently to external stimuli, and others in which individuals already have a division of labor and the colony can be considered as an entity. Some investigators distinguish between the first case, in which the individuals are less closely associated, terming this a "colony," and the second which is termed a "colonial individual," but all transitional stages between the two are present.

In this way, numerous unicellular species form similar simple colonies consisting of a few similar cells and finally similar spheres comprising a whole with an anterior and posterior end, responding as a unit and consisting of body cells (somatic cells; from the Greek $\sigma\tilde{\omega}\mu\alpha$ = body) and reproductive cells (generative cells; from the Latin *generatio* = begetting).

Now let us turn to the simplest flagellates, those included in the order GOLDMONADINA (CHRYSOMONADINA). They have yellow to brown plastids and only one or two flagella, which proves their relationship with the flagellates. Their body form can, however, vary considerably. Some species form pseudopods like the amoebae, and are able, with these, to flow around food particles and engulf them. They lie on the borderline between the plant and animal kingdoms, since they use both an animallike and plantlike mode of nutrition. When these forms lose their plastids, they belong entirely to the animal kingdom. Finally, they are able to resorb their flagella temporarily or completely as long as they still retain their plastids; without the flagella, they are considered as plantlike chrysomonadines, and, should the plastids disappear, we are left with an ameboid animal.

Order:
Chrysomonadina

Members of this venerable order can often be found: some quiet woodland pools can be covered with a brown, powderlike layer which makes the water surface look gilded under certain light conditions. This phenomenon is caused by innumerable chrysomonadines (*Chromulina rosanoffi*). In waters rich in nourishment, colonies of *Dinobryon sertularia* can often be seen under the microscope, forming cups on branched stalks. They also belong to this order, just as the smallest and, at the same time, the most numerous marine flagellates, the COCCOLITHOPHORIDS. Their envelope is covered with calcareous platelets furnished with motile processes, the "coccoliths." They remained undiscovered by marine biologists for a long time, since their minute size allowed them to pass

Calcareous platelets
with motile processes

through the meshes of the finest plankton nets. H. Lohmann first discovered them, at the turn of the century, by using the finest plankton nets known: the filter apparatus of tailed ascidians (Appendicularia; see Vol. III).

The minute skeletons of the coccolithophorids, the coccoliths, have recently attracted the attention of geologists. The tiny shells of ostracods and the equally tiny skeletons of the Foramenifera have been considered the main indicator fossils for the serial arrangement of prehistoric marine deposits, but the coccoliths lend themselves for this purpose to an even greater extent.

Other equally minute marine flagellates with a skeleton consisting of hollow silicious rods, the SILICOFLAGELLATA, are also included in this order.

Order: Cryptomonadina

The order CRYPTOMONADINA, with fewer species, includes biflagellate protozoa with a thicker pellicle and, because of this, a more permanent body form. The plastids can be yellow to brown, red-brown to red, or blue to green. *Chilomonas paramecium*, common in decomposing infusions, is colorless and takes up bacteria through its narrow gullet. Yellow marine species in symbiotic association with radiolarians and Foramenifera constitute a section of the "zooxanthellae"; dinoflagellates are, however, also involved in these associations.

Order: Dinoflagellata

A unique and widespread group of flagellates found suspended in fresh and salt water is the order DINOFLAGELLATA. They usually have two flagella which arise near to one another on one side of the body. A longitudinally extended flagellum points backward and delivers a pushing force; a horizontally extended one is wound around the body in a circle. Both flagella often lie in horizontal or longitudinal grooves. The movements of these flagella lead to a unique circling motion forward. The surface of the body usually consists of a rigid pellicle which is either uniform or formed from two or more plates, and often, as in the genus *Ceratium* pictures (see Color plate, p. 88), provided with processes of various types. The photosynthetic substance in dinoflagellates is usually not green but brownish. They can feed like plants, like animals, or like both.

Asexual and sexual reproduction are both present. When the original individual divides, each daughter cell receives half of the pellicle and builds the other half itself, or one daughter individual retains the complete pellicle and the others forms a new one, or both leave the old pellicle and each one builds anew. Multiple division also occurs, one flagellate giving rise to several daughter individuals. Some species are capable of existing for several years as cysts.

The species *Noctiluca miliaris* (see Color plate, p. 88) belongs to the dinoflagellates, this species being able to luminesce (thus *Noctiluca* = night light). It differs in form from typical dinoflagellates. There is no complicated pellicle, and only a short flagellum is present. An adhesive, contractile tentacle captures small organisms floating in the water; when

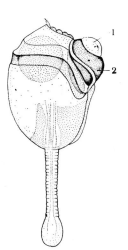

Fig. 4-1. *Erythropsis pavillardi*. 1. Lens, 2. Pigment. The taillike tentacle is below.

a large enough quantity has been accumulated, it is transferred by slow, waving motions of the tentacle to the middle of the cell body, where an opening specially formed for feeding purposes is present. *Noctiluca* has a worldwide distribution and appears in large numbers just below the surface of the water. Water movement—caused, for example, by people swimming or by the surf—results in *Noctiluca* producing an intense light, which a series of other marine organisms is also capable of doing.

Occasional massing of dinoflagellates in the ocean causes the water to appear colored, as in the "Red Tides." Among these are species, especially those of the genera *Gymnodinium* and *Goniaulax*, which release a nerve poison into the water, causing high mortality among fishes, crustaceans, and mollusks. In California there have been several cases of mass poisoning of people who had eaten mussels, the staple food of which had consisted of *Gymnodinium catenella*, another such notorious species. Rows of dead fishes along the beaches are a result of such red tides; collections of fossil fish remains in marine geological deposits could perhaps be attributed to red tides which had occurred ages ago. Dinoflagellates of the genus *Gyrodinium* live partly in fresh water and partly in sea water. *Gymnodinium paschari* obtains its red color from the color substance haemochrome, as does the phytomonadine *Chlamydomonas nivalis*. The color substance protects the dinoflagellate against the strong ultraviolet radiation on high mountains where this species is found. It forms brick-red patches on the snow and is also present in cold mountain water.

The order PROTOMONADINA includes all the colorless flagellates with one or two flagella which show no relationship to the previous orders which have a plantlike mode of nutrition. During the course of time, as investigators prove the existence of such a relationship, more and more animals are excluded from this group and included in others, especially the Chrysomonadina. Several groups of unknown evolutionary origin have remained in this "ragbag," although they have no relationship to one another. Hartmann believes that one member of this order, *Costia necatrix*, the instigator of *Costia* disease in fish, probably arose from a plantlike flagellate of the order Cryptomonadina. It covers the skin and gills of fish so thickly that it often causes severe damage, especially in the case of young individuals, where it can cause mass deaths. Aquarium fish can be freed of this dangerous parasite by repetitive baths in two to two and a half percent saline solution.

The family CRASPEDOMONADIDAE (choanoflagellates) has also found a temporary place in this order. On one end of the cell is a cylindrical or cone-shaped collar which encircles the flagellum; this produces a water current running from below to the margin of the collar. Food particles adhere to the collar and are carried along it to the cell body, where they are digested. Similar cells with a similar task are found in sponges. It is therefore possible that they have arisen from the choanoflagellates. Colony formation is also found in the collared flagellates. The species

Order:
Protomonadina

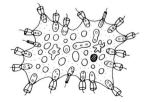

Fig. 4-2. *Protospongia haeckeli.*

Family:
Craspedomonadidae

Protospongia haeckeli (Haeckel's protosponge, from the Greek πρῶτος = first and σπογγιά = sponge) consists of a gelatinous mass with many collared flagellates around the rim and other types of cells in the middle. The name alone indicates that this species is considered a sponge prototype. It is, however, questionable whether this is a true ancestor of the sponges.

The members of this order that are by far the most important from the point of view of man are included in the family TRYPANOSOMATIDAE, the trypanosomes. These are usually elongate, spindle-shaped flagellates with only a single flagellum arising from a "basal granule." Near this is the so-called parabasal body or kinetoplast (from the Greek: movement shaped). Our present knowledge shows that this is not the instigator of flagellum movement but acts as a special mitochondrion (see Chapter 1) in cell metabolism. Strangely enough, it does not appear to be necessary for the survival of the trypanosome form in vertebrate blood, but only for the forms outside it.

Trypanosomes occur in four forms which can develop from one to the other to a certain extent:

1. The "Trypanosoma" form, in which the flagellum arises at the posterior end and runs the length of the body, ending free at the anterior. It is connected to the body by an extremely thin protoplasmic membrane which follows its wavelike motions and is thus termed the "undulating membrane." The trypanosomes live in this form in vertebrate blood.

2. The "Crithidia" form, distinguishable by the origin of the flagellum being near the center of the body on a level with the nucleus.

3. The "Leptomonas" form, with no undulating membrane, since the flagellum arises at the anterior end and is completely free. This and the crithidia form are found free in invertebrate digestive systems or attached to cells there.

4. The "Leishmania" form, round, with a basal granule and kinetoplast but no flagellum. This form lives in the tissue cells of vertebrates or attached to those of invertebrates. All these forms reproduce asexually by longitudinal division.

Dangerous instigators of human and domestic animal diseases are found among the trypanosomes. *Trypanosoma gambiense* (see Color plate, p. 107) and *Trypanosoma rhodesiense* cause the human disease "sleeping sickness," and a close relative, *Trypanosoma brucei*, causes nagana, an African cattle sickness. They all start their life history as blood parasites, and end by breaking through the brain membranes into the central nervous system, where they produce the symptoms from which these African diseases get their names. All three *Trypanosoma* species are transmitted by biting flies (tsetse flies) of the genus *Glossina* (further information on their developmental cycle is given in Vol. II). The South American Chagas disease (Vol. II) is also transmitted by insects, in this case by blood-sucking bugs in which the crithidia form multiples tremendously in the

Family:
Trypanosomatidae

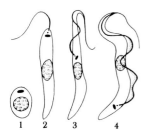

Fig. 4-3. The various body forms of the Trypanosomes: 1. Leishmania form (amastigote), 2. Leptomonas form (promastigote), 3. Crithidia form (epimastigote), 4. Trypanosoma form (trypomastigote).

Fig. 4-4. Distribution of *Trypanosoma gambiense* (1) and *Trypanosoma rhodesiense* (2).

mid-gut and especially the colon. They enter the human body through the bug's excreta on the skin, on wounds, and through the mucous membranes of the eyes. Here they multiply in various tissues into the leishmania form and then enter the bloodstream, where they take up the trypanosoma form again. In this way they can be taken up once more when a bug bites. Sandflies of the genus *Phlebotomus* are the carriers of Oriental *Leishmania*: they also multiply in the intestine of the carrier and do not penetrate the new host through rectal elimination but rather through the mouthparts. The "reserve host" from which the sandflies continually draw their leishmania is mainly the vagrant feral Oriental dog. The two species, *Leishmania tropica* and *Leishmania donovani*, are closely related to one another, and their distribution is almost the same, ranging through the warm Mediterranean belt of middle and western Europe and northern Africa. *Leishmania tropica* settles in the epidermal cells and causes the swelling known as "Oriental sore." The not un-common scars on the faces of Orientals are caused by them. *Leishmania donovani* is much more dangerous as the instigator of "Kala-azar" or intestinal leishmaniaisis (Blackwater Fever). They attack the tissue walls of the intestine, the liver, bone marrow, and spleen, which swell enor-mously. The disease is usually fatal without treatment. Other species cause dermal leishmaniaisis in South Africa.

"Oriental Sore"

"Kala-azar," "Blackwater Fever"

The presently worldwide *Trypanosoma evansi*, the instigator of "Surra" in horses and camels, is a close relative of *Trypanosoma brucei*, but is carried by horseflies. The parasite, however, does not multiply within them. They are only the "transport hosts." The same is true of the South American *Trypanosoma equinum*, the instigator of the feared "mal de caderas," pelvic paralysis in horses, which is also carried by horseflies. The capybaras (see Vol. XI) are also infected with this disease, and horse mortality is usually associated with a high mortality among the capybaras. *Trypanosoma equiperdum*, the cause of "dourine" in horses, requires no transmitting or transport host for its dissemination; it is transmitted directly through copulation. Finally, a native member of this family should be mentioned: *Trypanosoma melophagium*, found in the blood of European sheep and carried by the sheep louse *Melophagus ovinus*, in whose intestine it can be found in large numbers. The sheep show no symptoms of this disease, just as the wild antelopes show no symptoms with *Trypanosoma brucei*. They form a substance which prevents the multiplication of the trypanosomes in the blood. If the spleen is removed, the substance is no longer produced and the blood becomes full of trypano-somes within a short time. This is a good example of an equilibrium between host and parasite, the parasite not killing the host and thus not "cutting its own throat" at the same time.

Animal diseases

The eminent parasitologist, E. Martini, gives this description of the human suffering and need caused by trypanosomes: "The chiefs of tropical Africa know what swollen glands in the neck mean, just as doctors also

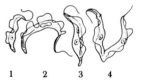

Fig. 4-5. Binary division in *Trypanosoma brucei* (the cause of nagana). 1. Trypanosoma form, 2. Division of the basal granule, flagellum, and nucleus, 3. Division of the blepharoblasts; the forma-tion of a second flagellum and undulating membrane, 4. Longitudinal division of the protoplasmic body, which divides from anterior to posterior.

know, after seeing and feeling them, that these are the first symptoms of a severe illness. They thus gladly give such still physically strong men, who would soon be a burden on the village, to recruiters of porters for caravans. A time of suffering then begins for these people. They soon have intermittent fever. Trypanosomes can be found in the glands or blood. They are also often present in the brain in large numbers. This can result in a highly agitated condition which is soon brought under control by the salve-dealer's whip, or a pathological sleepiness which is considered laziness and is treated with blows. There may be emaciation and transient swelling of the skin in different parts of the body. Strength fails. An increasing need to sleep follows on from this weakness. The horrible picture of sleeping sickness is fully developed. Finally the people are unable to stand; sometimes they die of this disease only after many months." In another section of his book, *Courses of Diseases*: "Sleeping sickness can wipe out whole populations to such an extent that the Belgian government recalled officials from sections of the Belgian Congo at the beginning of this year since there was no population to govern, and the trading companies had to close down since there was no more trade." The importance of the cure for this human-inhabiting flagellate cannot be over-emphasized. A cure was discovered by the German chemists Dressel, Khote, and Ossenbeck in 1916, and called "Germanin" (Bayer 205). It was tested, with success, on infected people by Professor Mühlens in 1920. The English biologist Julian Huxley says about it: "The discovery of Germanin is probably of more value to the Allies than all the reparation paid after the First World War."

Order:
Diplomonadina

The members of the following order have more than two, and sometimes a large number, of flagella or cilia. The order DIPLOMONADINA is a unique group: these are usually bilaterally symmetrical animals with eight flagella and two nuclei, thus being a double animal, to a certain extent. They could be considered as developing from flagellates with a single nucleus and four flagella, the organelles doubling during reproduction, but complete division having been suppressed. The species live partly free and partly in the gut of invertebrates and vertebrates, where they feed on bacteria and must thus be considered commensals rather than parasites. When the unicellulates which inhabit the frog's rectum are examined under the microscope, several other species can be found apart from the large, beautiful opalines. In the lower section of the small intestine and in the appendix of rats and mice, the parasites *Octomitus muris* and the more widely distributed *Giardia muris* can be found regularly; the human small intestine often contains *Giardia (Lamblia) intestinalis*, which is flat on the underside and rounded on the upper side. Its resistant form (cyst) is often found during fecal analysis; it is often quadrinucleate, since these animals divide within the cyst. They are found in large numbers in cases of diarrhea, but it is still not yet clear whether this mass multiplication is the cause or result of the illness.

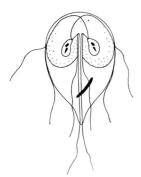

Fig. 4-6. *Giardia (Lamblia) intestinalis*.

Even the members of the order POLYMASTIGINA (from the Greek πολύς = many, and Latin *mastix* = flagellum) live mostly in the intestines of invertebrates and vertebrates. The genus *Trichomonas* contains many species, these having several flagella directed forward and a posteriorly directed flagellum which arises with them but is attached to the body by an "undulating membrane." A moveable "axostyle" runs the length of the animal. The electron microscope has shown that this is a bundle of many long threads (microtubulae), each one being about one forty-thousandth of a millimeter thick (240 Angstroms). *Trichomonas hominis* is found in the human intestine. *Trichomonas tenax* (= *elongata*) is present in the mouth, and *Trichomonas vaginalis* in the vagina of women and the urethra of both sexes. Although most species live on bacteria, the last mentioned one does not feed on live food. The high population of this unicellulate even in healthy women does not eliminate the fact that, under unfavorable conditions, it can instigate disease.

Among the members of the genus *Trichomonas*, as in other genera of the family Trichomonadidae, there are many species which inhabit the intestines of plant-eating insects. These are mainly the termites and leaf-eating lamellicorn beetles (ladybugs and rhinoceros beetles). Some families (Calonymphidae, Pyrsonymphidae, Hypermastigidae) have adapted completely to life in the termite's colon, which has made it possible for them to adopt a new mode of nutrition. They are by no means damaging parasites or harmless commensals, but essential symbionts. They inhabit the termite gut in enormous numbers, their total weight being estimated as a sixth to a third of the termite's total body weight. Wood-eating termites are dependent on the help of the flagellates in their gut, since they are unable to form cellulose-splitting enzymes for themselves. The flagellates, or maybe the bacteria which live within them, do this for the termites. It is still uncertain whether wood particles or bacteria are the true food of the termite-inhabiting flagellates. In any case, these multiply continually in the termite's rectum, and their excess population regularly infiltrates the midgut, where it is digested by the termite. Termites molt not only the cuticle covering their bodies but also that of the rectum, and, on molting, the contents of the rectum are thus sloughed off. Each termite must therefore, after molting, ensure a repopulation of its intestine with flagellates and bacteria. Licking the anus of another worker termite rapidly brings this about.

Within the family HYPERMASTIGIDAE is the sub-family HOPLONYM-PHINAE, which is worthy of special mention. Although *Hoplonympha* inhabits the intestine of termites, the presence of species from the closely related genera *Rhynchonympha*, *Urinympha*, and, especially, *Barbulanympha* in the primitive wood-eating cockroach *Crytocercus punctulatus* gives proof of the close relationship between the termites and the cockroaches. There is, however, one difference between the cockroach and termite symbionts: the formation of a non-pellicled cyst is associated with sexual processes

Order: Polymastigina

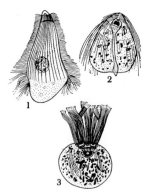

Fig. 4-7. Hypermastigina in the termite gut: 1. *Trichonympha turkestanica*. 2. *Microspironympha porferi*. 3. *Kofoidia loriculata*.

and both occur at the same time as the cockroach's molt, instigated by the molting hormone. The two participants in the symbiotic association are thus astoundingly well adapted to one another.

Order: Opalinina

As the most highly developed flagellates of the order POLYMASTIGINA, the opalines (Opalinina) also have an extremely large number of flagella, but these, in contrast, are short and similar to the cilia of ciliates. Owing to this, the opalines were, until a short time ago, included in the Ciliata. This mistake was easy to make since there is no basic difference between the flagella of the flagellates and the cilia of ciliates, and even the electron microscope can distinguish no difference in their microstructure. Basic to the inclusion of the opalines in the Flagellata is the homogenity of their many nuclei. The sexual processes of opalines progress in the same way as "copulation" in the flagellates and not as "conjugation" as is found in the ciliates. Opalines are either binucleate (genera *Protopalina* and *Zelleriella*) or multinucleate (genera *Cepedea* and *Opalina*); they are either circular in cross section (*Protopalina* and *Cepedea*) or flattened (*Zelleriella* and *Opalina*). The cilialike flagellae are arranged in definite oblique rows.

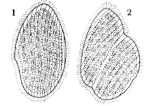

Fig. 4-8. *Opalina ranarum*: 1. Entire animal with numerous vacuolelike nuclei; 2. Division by falling into two equal-sized pieces.

All opalines inhabit the rectum of amphibia, although the most primitive genus, *Protopalina*, is not confined to this niche, certain species also being found in bony fish and reptiles. *Opalina ranarum* (L 0.6–0.7 mm) is easily seen with the naked eye, and can be found in every common frog, the smaller *Opalina dimidiata* (L 0.35–0.5 mm) in every edible frog. They reproduce in these animals throughout the year by binary division. In spring, however, they divide into four individuals with only a few nuclei, and form a capsule around themselves. This is excreted with the feces and enters the water, where it is then taken up by the young tadpoles. Within the tadpole, the animals emerge from the capsule and divide further into uninucleate ones, these consisting of larger female and smaller male animals (gametes) which join together in pairs on which a pellicle is formed. This "zygote" either leaves the intestine with the feces or remains within the same tadpole. The excreted opaline is taken up by another tadpole while feeding, and, on reaching the rectum, emerges from the cyst and continues to multiply there by binary division until the following spring. All opalines found in a frog's rectum have been there since its juvenile tadpole stage. Opalines do not have a cell mouth or anus, nor food vacuoles within them. They must therefore obtain their nourishment through the cell surface, feeding on substances in solution in the frog's rectum. The transparent opalines, which derive their name from their opallike shimmer, are especially popular objects of investigation by microscopists, since they can be easily obtained with a blunt pipette from any frog, without hurting it.

Class: Sarcodina

The class SARCODINA (Rhizopoda), which follows the flagellates, is distinguished by its ability to form pseudopodia. In the typical shell-less amoeba, the pseudopodia are changes in the outline of the protoplasmic body which from time to time are extruded and then resorbed into the cell body. Because of this, amoebae change their form constantly: at the pseudopod axis, the protoplasm pours in a rapid stream from the cell body; this stream then becomes laterally diverted, and finally halts. In this way, the whole animal gradually moves from the spot. In many species, the pseudopodia are formed singly; these animals are monopodial and flow in one direction on a substrate. Such amoebae which creep along like the garden snail (*Limax*) can be found with the aid of a microscope in practically any infusion of old hay. Other rhizopods have many pseudopodia and are able to change the direction of their creeping in that sometimes one and then another pseudopod increases its streaming at the expense of the others. In floating rhizopods, long pseudopodia spread in all directions in large numbers. These aid in floating, and only when one of them comes into contact with a solid object is the animal able to utilize them to move. Pseudopodia do not, however, simply aid in movement and floating, but also in food capture and engulfing. They flow around the food particle or lift it and, through the protoplasmic streaming, direct it into the cell body. Pseudopods can either be glassy and transparent or grained, they can either form in a fountainlike or baglike manner, and can build sheets, streamers, or nets. With their multiplicity and continual changes in form they make fascinating objects of study for any amateur microscopist.

Pseudopodia, however, are not the sole prerogative of the rhizopods: the white blood corpuscles found in the blood and tissues of multi-cellulates also move by means of pseudopodia and engulf intruding bacteria with them in the same way as the amoebae do. Certain flagellates are also able to form pseudopodia just as flagella sometimes appear for a short time in certain rhizopods. Unicellulates which have flagella and pseudopodia at the same time or which form them sequentially during the course of development are difficult to classify as either flagellates or rhizopods. They are included in the flagellates when they nourish themselves with the aid of plastids in a plantlike manner or when they show other signs of close relationship with the plantlike flagellates; if, on the other hand, they are colorless and show no signs of such a relationship, they are included in the rhizopods. This rule, however, should not be taken to imply that flagellates and rhizopods are not separated by a wide gulf from one another.

We shall start with the order AMOEBINA, the shell-less amoebae which are able to change their form continually. *Amoeba proteus*, found commonly on the floor of pools and puddles and measuring 0.2 to 0.5 mm in diameter, is recognizable by its numerous thick, stubby pseudopodia, which often constitute the majority of its body. *Verrucosa* amoebae (diameter 0.1 mm) are distinguished by their wrinkled body surface. The tiny monopodial "*Limax* amoebae" found in fresh water usually belong to the genus *Hartmannella*; they live on bacteria. The *Radiosa* amoeba is a

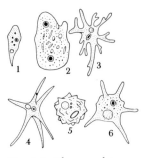

Fig. 4-9. Shape and type of pseudopodia in amoebae: 1. Limax amoeba 2. *Pelomyxa binucleata* 3. *Amoeba proteus* 4. *Radiosa* amoeba 5. *Verrucosa* amoeba 6. *Amoeba polypodia*.

Order: Amoebina

floating form found in fresh water. It has long, streaming pseudopodia. Our largest amoeba is fairly common in waterstands present on decaying earth: *Pelomyxa palustris* (up to 2 mm in diameter). *Pelomyxa* has a roughly egg-shaped outline since its whole body forms one single large pseudopodium. It usually includes a large number of nuclei and, in addition, many highly effective light-reflecting particles, termed "reflecting bodies," consisting of animal starch—glycogen—and, apart from these, numerous mud and sand particles. Two species of bacteria are also found within the body of *Pelomyxa* in such large numbers and so regularly that it can be concluded that this arrangement is of advantage to both organisms.

Colony formation in unicellulates

Species of shell-less amoeba found in decaying substances and especially in dung and excreta have a complicated reproductive cycle in which solitary and "collective" forms (those forming a colony) alternate. Individual amoebae feed and reproduce like normal shell-less amoebae. When no further food is available, however, they start responding negatively to moisture. Numerous amoebae begin crawling to a higher and therefore dryer place and form a perpendicular, towerlike structure; after a short time, this collapses and starts to creep about the substrate. Finally, the massed amoebae form a heap again, producing the so-called sporangia, with some of the amoebae forming cells in the stalk, and others building spores at the tip of the stalk. The spores are surrounded by an envelope which enables them to survive under unfavorable environmental conditions. When conditions improve, the amoebae crawl out of the envelope and take up their lives again as solitary animals. The cells forming the stalk, however, perish.

This is an example of colony formation in unicellulates, contrasting with the colonial flagellates mentioned above, which remain together after cell division, in that the cells creep together. As in many flagellate colonies, a division of labor is present here as well, in reproductive cells (the spores) which are capable of multiplying, and somatic cells (the cells forming the stalk), which are not capable of multiplying. The sporangia of "collective" shell-less amoebae occur in the most variable arrangements. The spores can either lie in a single spherical structure at the tip of the stalk, or several spore spheres can be present at the ends of lateral branches from the stalk, arising from a point at the tip of the "main stem" or from many centers lying at different levels, or in still more complex ways. Structures are formed which, superficially speaking, are amazingly like simple multicellular plants. It must be re-emphasized here, though, that this has nothing to do with transitional forms to multicellular organisms.

The genus *Naegleria* is an example of an amoeba which possesses flagella at a certain stage in development. The free-swimming flagellated form settles on a substrate, absorbs the two flagella, and transfers to an amoebalike form of motion by pseudopods. Under conditions of poor

Fig. 4-10. The distribution of amoebic dysentery in the Mediterranean area.

nutrition and other environmental factors, two new basal granules form from which two flagella develop. The amoeba thus develops into a flagellated organism once more.

Apart from free-living shell-less amoebae, there are also a number of species which live in man either as harmless commensals or dangerous parasites. The common species of amoeba, *Entamoeba coli*, is one of the commensals. This feeds mainly on bacteria in the human colon and also on other unicellulates present there. It reaches other hosts by forming cysts, the amoeba rounding off and forming an envelope about itself. Eight nuclei are formed within the cyst by triple division. The cyst is released with the feces and enters another host through the mouth, for example on unwashed fruit; the cyst opens and the multinucleate amoeba crawls out of the envelope and divides into eight daughter cells.

Certain other harmless species are also found in the human intestine apart from *Entamoeba coli*, differing only slightly in size, the form of the nucleus, and the number of nuclei present in the cyst. One species even inhabits the oral cavity of man (and probably also that of domestic animals and monkeys): *Entamoeba gingivalis*. To date, no cysts have been found, and it is probable that transmission occurs through the mouth-to-mouth contact of two people or through unwashed dishes or impure drinking water. This species also is probably harmless, although it is continually suspected of causing certain tooth diseases.

The dysentery amoeba, *Entamoeba histolytica*, is, in contrast, extremely dangerous, causing amoebic dysentery, usually associated with severe pain, blood-streaked diarrhea, and often further complications. This species occurs in the human intestine in various forms. In a person suffering from acute amoebic dysentery, amoeba which have ingested numerous red blood corpuscles can be found in the stools. These are relatively large (20–30 microns; 1 micrometer of μm = a thousandth of a millimeter) and are termed either the magna (from the Latin *magna* = large) or tissue form, since they are able to penetrate the intestinal tissue from the intestinal canal. Even the species name (*histolytica* = tissue dissolving) is derived from the ability to penetrate the intestinal wall and destroy the tissues. The appearance of blood in the intestinal canal is brought about by the destruction of the tissues and blood vessels. The amoebae can reach other organs through the blood, including the liver, lungs, skin, and brain, and can cause the formation of extensive, sealed-off abscesses there.

The smaller minuta or intestinal lumen form (from the Latin *minutus* = reduced) is more common than the disease-producing magna form, and is, in general, only 12–18 micrometers in length, and contains no red blood corpuscles. It lives within the intestinal canal and feeds on solutes and bacteria. The minuta form, in contrast to the magna form, is also capable of forming cysts. As in *Entamoeba coli*, the amoeba rounds itself off and secretes an envelope. In contrast to the former, however, with

▷
Unicellulates: Left, above to below: Blood smear: between the blood corpuscles are two flagellates of the species *Trypanosoma gambiense* (western and central Africa), the instigator of sleeping sickness. Amoeba (genus *Amoeba*) with food vacuoles. Two spores of the sporozoan genus *Henneguya* with extruded polar filaments. These animals are parasites in the gills of fish. *Myxobolus* spores in which the polar filaments can be clearly seen within the polar capsules.
Right, above to below: A flagellated amoeba (genus *Amoeba*) between algal filaments. The skeleton of the foraminiferon *Peneroplis pertusus*. Amoeba (genus *Amoeba*) with extended pseudopodia. A biflagellated flagellate (genus *Haematococcus*).

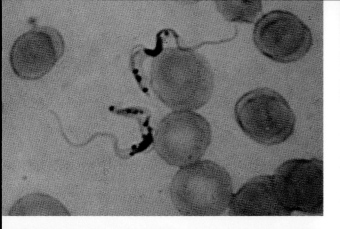
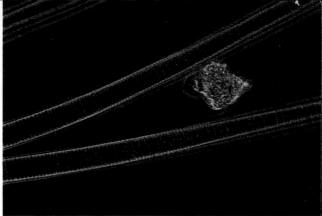
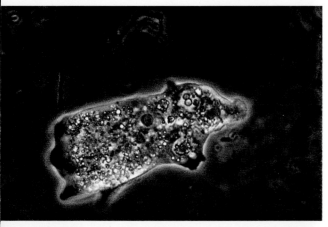
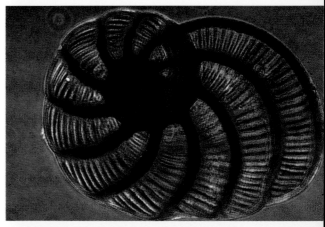
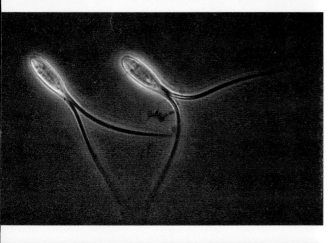
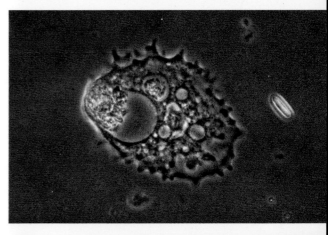
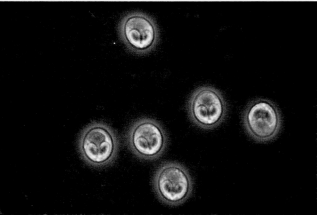
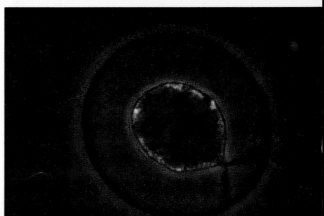

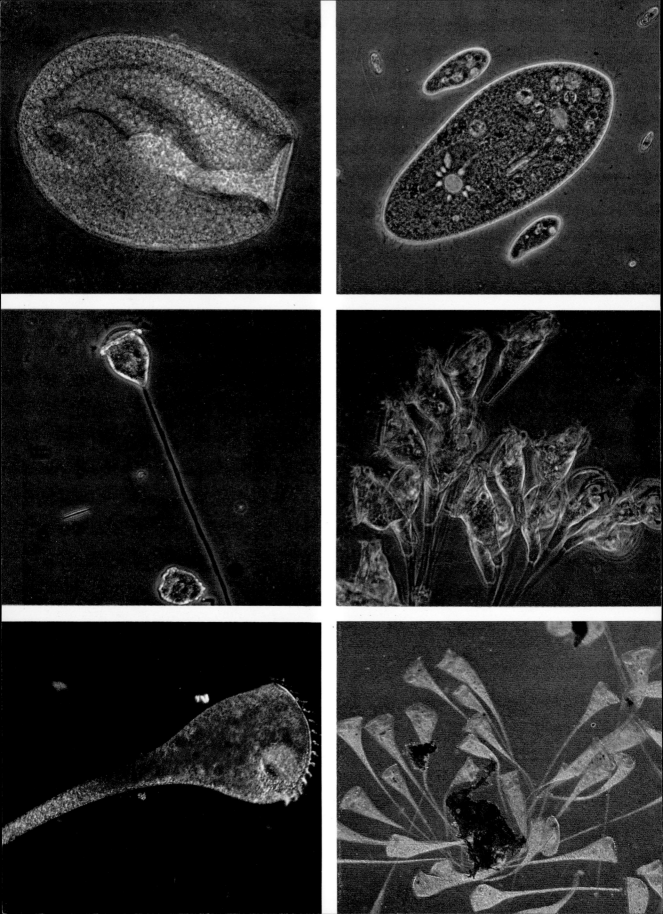

◁

Unicellulates. Left, from above to below: *Bursaria truncatella*, a ciliate with an oral cavity penetrating deep into the cell interior. At a length of 1 mm, it is a "giant" among its relatives. Solitary peritrich of the genus *Vorticella*. One individual with extended stalk, and, next to it, one with a retracted stalk.
Stentor (genus *Stentor*).
Right, from above to below:
Paramecium (genus *Paramecium*): The contractile vacuoles are clearly visible.
Peritriclo colony of the genus *Carchesium* with a branched stalk.
A collection of stentors (genus *Stentor*).

a few exceptions, only two nuclear divisions occur; thus a maximum of four nuclei are present in a single cyst. The cyst leaves the intestines with the stool and penetrates new hosts through dirt on unwashed fruit and vegetables or in dirty drinking water. Here the quadrinucleate amoeba emerges from the envelope and, following another nuclear division, separates into eight daughter amoebae. Active amoebae eliminated with the stool die within a short period of time and are thus not considered as transmitters of dysentery. Cysts can also be carried by insects, especially flies and cockroaches. They either remain attached to the outside of the insect's body or are devoured and excreted again in the feces. The cysts retain their infective qualities longer in water than in feces. Experiments have shown that *Entamoeba histolytica* cysts can survive for thirty-five days in feces-containing water and up to seven months in distilled water kept at 19–22°C. Epidemic outbreaks of amoebic dysentery are thus continually traced to drinking-water contamination. Kudo describes two such cases. During 1933, an epidemic occurred in Chicago which, during its course, claimed about 1400 people. It was traced to water contamination due to incorrect installation of pipes. In the same city a year later, about a hundred firemen caught amoebic dysentery from drinking impure water.

Although amoebic dysentery is found mainly in warmer climates, as, for example, along the whole length of the Mediterranean coastline, it can also occur in northern areas. Carriers of this parasite are not uncommon even in northern Europe and the U.S.A. The presence of the parasite in the stool does not always signify that the person suffers from amoebic dysentery. Even when the parasite is present together with a dysenterylike illness, this is still no proof that the parasite was responsible for the illness, since even bacteria and ciliates can cause dysentery. Only when the magna form is found together with engulfed red blood corpuscles, can it be considered with certainty a case of amoebic dysentery. A decrease in resistance brought about, for example, by a change to a warm climate, insufficient nourishment, and other infections plays a role in the transformation of the harmless minuta form into the disease-producing magna form.

Amoebae of the genus *Malpighiella* are found as parasites in the nephridia of insects. *Malpighiella mellificae* is seldom found alone in the honeybee; it is then relatively harmless. It only becomes dangerous to the bees when they suffer at the same time from *Nosema* disease. Beekeepers recognize the combined amoeba and *Nosema* attack by the presence of fluid, yellow excreta about pinhead size on the landing surface in front of the hive. The specks, individual or smeared together, retain their yellow color even when dry. Cases of such double infection are much more common than attacks by the amoebae alone.

Order: Testacea

The TESTACEA (shelled amoebae) differ from the shell-less amoebae in the possession of shells consisting of organic substances. The shell can be strengthened by silicic acid platelets, minute sand grains, or silicious algal platelets, and always consist of one chamber. The shelled amoebae multiply by binary division, the shell, however, usually not dividing. *Euglypha*, which is armored with tiny silicic acid platelets, already forms a reserve of such platelets within the cell body before division. Before the division starts, these are extruded from the shell opening and form a new armor around the emerging protoplasm. On microscopic examination of the minute fauna from a puddle of water, shelled amoebae are often found. These would include species of the genus *Arcella*, with a light brown, finely sculptured shell consisting of organic substances; *Euglypha alveolata*, which is armored with silicic acid platelets; the *Difflugia* shells consisting of foreign matter, especially that of the large *Difflugia pyriformis* (up to 0.5 mm long); *Centropyxis aculeata*, with its caplike shell, and *Lequeureusia spiralis*, with a retortlike shell. Peat pools are especially rich in shelled amoebae. Once the observer has seen the various forms and constructions of the shells, the differences in the protoplasmic cell bodies and, above all, the various pseudopodial forms, are no longer a secret.

The FORAMINIFERA (order Foraminifera) also live in shells they build themselves, just as the shelled amoebae do. Their shells are usually penetrated by pores through which the protoplasm is extruded. Another difference between them and the shelled amoebae lies in the chambering of the shells; they are also termed Polythalamia (many chambered) as well as Foraminifera. Nevertheless, there are also exceptions to the rule here (single chambered, Monothalamia). The foraminifera are further distinguished by the fact that they live entirely in the sea or salt-water lakes, and that, in many of them, sexual processes have been observed. The majority of the Foraminifera live on the floor of the sea, a few species inhabiting the surface waters in enormous numbers. These, especially the globigerines (genus *Globigerina*), play an important role in the formation of mud on the seabed: this "globigerina ooze" covers more than 100 million km^2, over a third of the present-day seabed, at a depth of 1000–4000 m. At more extreme depths the *Globigerina* shells are destroyed in large quantities by carbonic acid. During previous geological ages Foraminifera occurred in such enormous numbers that immense geological deposits consisted almost exclusively of their shells. The fusuline chalk of the more recent geological periods (Carboniferous and Permian) and the nummelite chalk of the Lower Tertiary (Eocene) are formed almost entirely from the chalk shells of large Foraminifera. The Egyptian pyramids are not only proof of man's work 4500 years ago, but also of the activities of innumerable unicellulates about 48 million years ago, since they were built of stone consisting of nummelite shells.

How could these nummelites have developed in such large numbers on the seabed? They are extinct today, but this secret has been revealed by surviving members of the genus *Heterostegina*, their closest living

Order: Foraminifera

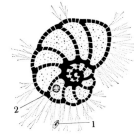

Fig. 4–11. Schematic longitudinal section through the multi-chambered *Elphidium* (*Polystomella*) *crispum*: 1. Food particle 2. Nucleus.

relatives: they grow and flourish in pure sea water without obtaining nourishment from the exterior. They live on algae which are present in their protoplasm. These grow and multiply using the mineral salts dissolved in pure sea water and carbonic acid. They naturally require sunlight to be able to do so, and therefore they, and the Foraminifera, are found in the well-lighted coastal areas of the sea. One puzzle remains to be answered, however: how the unicellulate managed to live on the algae without destroying the entire stock. There must therefore be a means of regulating the balance between the foraminiferan and its nourisher, the alga, in this apparently simple protoplasmic cell body.

Fossilized foraminiferan shells are also found in enormous numbers in the cores from borings made in the search for petroleum oil, and are more common than those of mollusks which are used as "dating fossils." Since the shell forms change in a relatively short period of time, geologically speaking, they make it possible for the serial deposits in the cores to be very accurately divided and their exact geological age to be determined. The numerous foraminiferan experts of the world working for oil companies have thus, during the seven-year period between 1949 and 1955, described over 4000 species, a veritable avalanche of "oil animalcules" which have even swamped the experts themselves!

"Oil animalcules"

Inland-living Foraminifera are found in the salty ground water of the central Asian Carachoum Desert, in the Deva salt-pools of Transylvania, in the Hungarian salt-water lakes, and the central German salt area stretching from Artern to Sangershausen.

Foraminifera shells either consist entirely of an organic substance secreted by the unicellulate or are strengthened by adhering chalk or foreign bodies (e.g., sand, sponge spicules). In the latter case, certain species show a definite ability to select. The shell form can be extremely variable: it is rarely a single chamber and is in most cases composed of many chambers which are either rectilinear—consisting of one, two, or more rows following on from one another in a horizontal plane— or spirally inclined. The size and arrangement of the pores and the surface sculpturing also varies considerably. The great variety of forms and the beauty of foraminiferan shells is illustrated to some extent in the Color plate on p. 93. The naturalist who has an opportunity to use a simple microscope or even a binocular microscope will find the study of these animals completely absorbing. They can be found in the clay of ocean beds after it has been mixed with water and then sieved. A practically inexhaustible source of present-day species is the layer of slightly larger sand grains on beaches. A field survey of this, with a good magnifying glass, can determine whether it is worth taking a sample. Foraminifera shells, after they have been sorted out at home, can be kept in "Franke Cells" or "DMW Cells" which can be bought especially for this purpose, and thus a collection of "Nature's Works of Art" can be built up.

Order: Heliozoa

With the HELIOZOA or SUN ANIMALCULES (Heliozoa), we return to rhizopoda which are mainly found in fresh water. Long, thin pseudopodia emerge in all directions from their almost spherical bodies, these pseudopodia being attached by "axial filaments." The axial filaments are anchored deep in the protoplasmic cell body. Tiny organisms become attached to these raylike pseudopodia, and protoplasmic streaming then conducts them to the cell body in which they are digested. The Heliozoa reproduce by division; sexual processes are present in the form of self-fertilization. The uninucleate heliozoan *Actinophrys sol* (diameter 50 μm; see Color plate, p. 305) reproduces in the following manner: the pseudopodia are withdrawn and the animal secretes a jellylike cyst. The nucleus and cell divide once, thus creating two daughter cells (gamonts). The nucleus of each cell divides once again, one of the daughter nuclei being resorbed; in this way, only one nucleus remains in each cell at the end, this containing only a single chromosome complement. These cells are the "gametes," which now join to form a "zygote," this containing, as in the original individual, a double chromosome complement. A strong inner cyst forms in which the new individual goes through a long period of quiescence. This process differs in the large heliozoan *Actinosphaerium eichhori* (up to 1 mm in diameter), which contains up to five hundred nuclei: this also builds a jellylike cyst at first and, within it, resorbs the majority of its nuclei. The remaining nuclei (about five percent of the originals), surround themselves with a portion of the protoplasm, and each builds a cyst. Then each of these cells divides into two daughter cells (gamonts), and mitotic division and copulation then proceed as in *Actino- phrys*.

The final rhizopod group is the strictly marine RADIOLARIA (order Radiolaria). In this order the protoplasmic body contains a thin-walled, porous "central capsule" separating the internal protoplasm and nucleus from the external vacuolated protoplasm. This external portion can grow as the animal grows and can disappear during reproduction. The radiolarian skeletons are included in nature's most beautiful structures. They consist of strontium sulphate in the Acantharia, and silicate in the remaining radiolarians. The numerous forms of these structures, visible only under the microscope, cannot be explained in the same way as the display structures of higher animals which are visible and aid in bringing the sexes together, indicating differences between species, warning, or camouflage. The reason for their existence probably depends on other behavioral aspects which we do not know. The Color plate on p. 94 can give a far better idea of the delicate filigree of these skeletons than words can; over 5000 living and 1000 extinct species are recognizable on the basis of these designs. The Jena zoologist Ernst Haeckel made these animals known all over the world in his treatise on radiolarians, *Art Forms of Nature*. The skeletons of dead radiolarians sink to the seabed and form "radiolarian ooze" there; this extends as a belt from east to west across the eastern Pacific north of the equator, an area of about 750 thousand km².

$\triangleright$
Ciliates (class Ciliata) 1. *Vorticella nebulifera* (a Peritricha) 2. *Paramecium aurelia* (a paramecium) 3. *Stentor polymorphus* (a multiformate stentor) 4. *Entodinium caudatum* 5. *Trichodina pediculus* (the polyp louse, from below) 6. *Diplodinium denticulatum.*

Order: Radiolaria

$\triangleright\triangleright$
The developmental cycle of a malaria parasite (*Plasmodium vivax*) Above: In the malarial mosquito *Anopheles.* Center: The malarial mosquito bites a human. Below: within a human. A. Infected mosquito bites a healthy person. 1. The form infectious to humans: the sprozoite in the mosquito's saliva. 2. "Schizogony" within the liver cells: penetration of the sporozoites (2a) or merozoites (2a′); their multiplication by division into merozoites (2b).
(Continued on page 115)

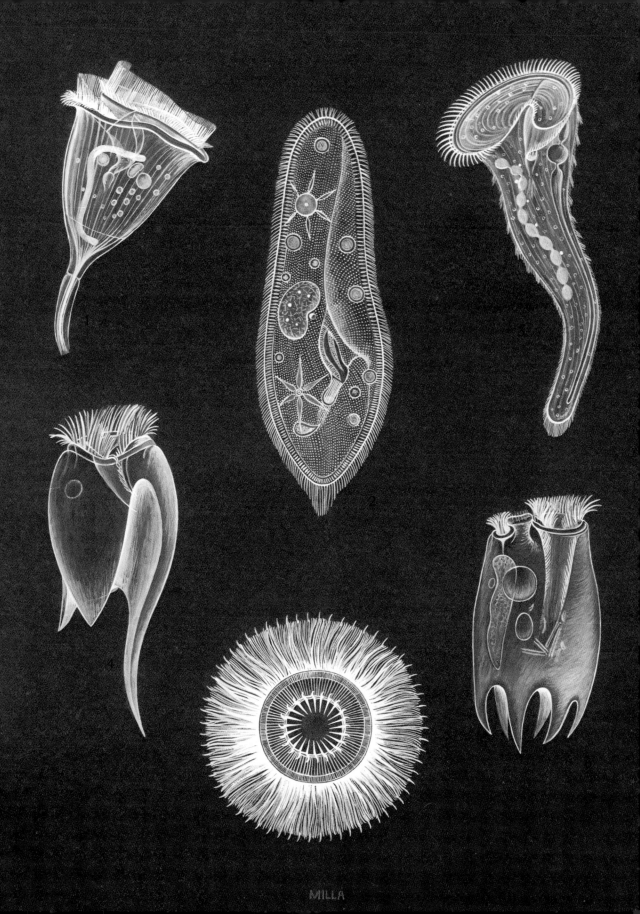

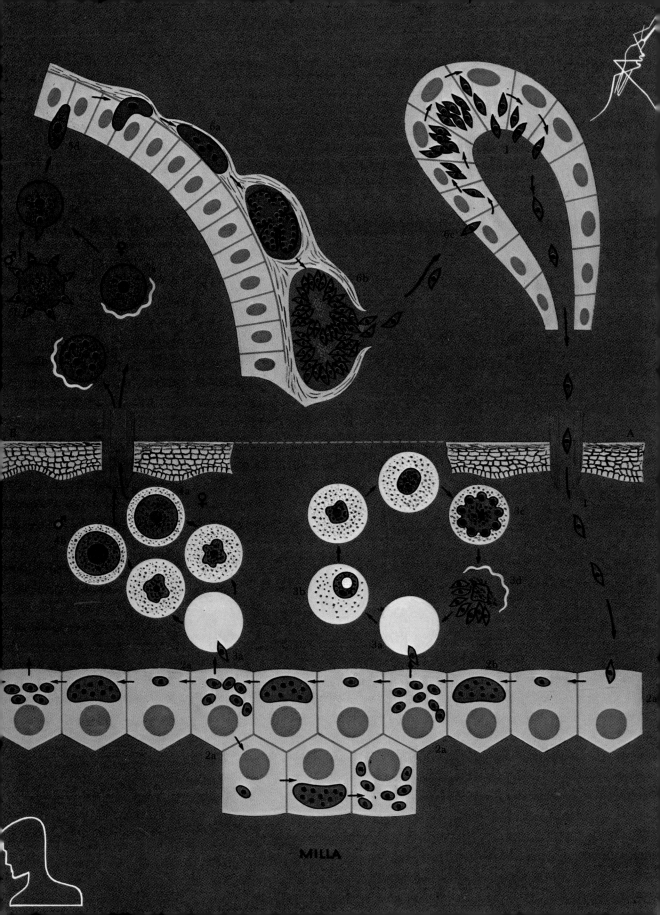

MILLA

3. "Schizogony" within
the red blood corpuscles:
penetration of the mero-
zoites (3a), the ring form
and growth (3b), multi-
plication by division (3c),
release by collapse of the
blood corpuscle (Fever
attack! 3d).
4. "Gametogony" within
human blood: formation
of the female macro-
gametocytes (4a) and the
male microgametocytes
(4b). B. Healthy mosquito
bites an infected human.
5. Continuation of "game-
togony" within the mos-
quito's stomach: female
sexual forms (macro-
gametes, 5a) by growth,
male microgametes (5b)
usually eight, by division.
Joining of a male with a
female (copulation) to form
a ookinet (5c) which
penetrates the wall of the
stomach. Doubling of the
chromosome complement
by fusion of the nuclei. 6.
"Sporogony" in the mos-
quito stomach: After the
first two division phases
(maturation division) a
single chromosome com-
plement again (6a), further
multiplication to numerous
sporozoites (6b) which
move to the mosquito's
salivary glands (6c) and
into the saliva (1). Cycle
continues from A.

Class: Sporozoa

Marine deposits from the Lower Tertiary (Eocene, about 40 million years ago) from Barbados and Haiti in the Antilles consist almost entirely of radiolarian skeletons. They are of economic value as raw material and as fertilizer, and at the same time offer the amateur microscopist an almost inexhaustible source of pleasure in observation and contemplation.

Radiolarians feed partly on tiny floating organisms which become trapped on their ray and netlike pseudopodia and partly through symbiosis with yellow-brown algae, the "zooxanthellae." The pseudopodia also aid in floating, as do the skeletal spicules. The skeleton itself is heavier than water and is compensated for by lighter-than-water oil droplets in the protoplasm. Radiolarians are capable of regulating their specific weight: when this is the same as the surrounding water, the animal floats; it if is higher, it sinks; if it is lower, the animal rises. Under calm weather conditions, they usually float near the surface where there is sufficient light for their essential zooxanthellae. Under conditions of great heat or waves, they sink deeper. Reproduction in radiolarians has been studied only sketchily. As a rule, they probably undergo multiple division into biflagellate motile stages which emerge from the skeleton and start to grow; very little is known, however, since this obviously occurs at great depth in the ocean. Multiplication by binary division has also been observed. Owing to the similarity between the motile phase and dinoflagellates, it has been concluded that the radiolarians have arisen evolutionarily from these flagellates. This observation can, however, be confusing; radiolarians are quite often infested by parasitic dinoflagellates (genus *Merodinium*), the motile phases of which could be taken for those of the radiolarian itself.

The SPOROZOANS (class Sporozoa) all are parasites on other organisms. Previously, forms which reproduce by alternate asexual and sexual processes were included in this class together with those which reproduce entirely asexually as a result of their adaptation to the host and which thus show no close relationships either between themselves or to other groups. As K. G. Grell has done, we shall restrict the class here to the two orders GREGARINES (Gregarinida) and COCCIDIANS (Coccidia), which show a close relationship to one another, the other groups being discussed subsequently.

Sexual processes in the sporozoans consist of the formation of male and female presexual cells (gamonts) from which male and female sexual cells (gametes) are produced. This is termed "gamogony" (from the Greek γάμοδ = marriage and γονή = beget). After the joining of the sexual cells to form a zygote (from the Greek ζυγόν = yoke, fusion), corresponding to the fertilized egg, asexual reproduction begins by division ("sporogeny"). This results in the formation of spore-enclosed sporozoites, which aid in transmission to a new host. Here a further asexual mutliplication occurs, this serving to populate the host: "Schizogony" (from the Greek σχίζω = I divide). These three types of reproduction are found in all sporozoans, differing extensively, however, in the various groups. Reduction division, however, always occurs at the beginning of sporogony, so that the double chromosome complement is found only during the short zygotic phase.

We turn now to the first order, the GREGARINES (Gregarinida), the presexual cells of which, both male and female, produce a larger number of sexual cells on division. In addition, two presexual cells (gamonts) are already joined. Sporogony results only in slight multiplication, since the zygote forms a spore case and produces only eight sporozoites. In the suborder Schizogregarinida, schizogony within a new host results in a great increase in the parasite's numbers. In the Eugregarinida, on the other hand, a period of rapid growth occurs instead of this multiplication and, finally, the large gregarines come together in pairs and form a "syzygy," in which the nuclei of the two divide several times. Each then surrounds itself with a portion of the protoplasm and becomes a gamete. The separating wall of the syzygy dissolves, and each gamete then joins with one from the partner. All gregarines are parasites of invertebrates.

By far the majority of gregarines belong to the suborder GREGARINA (Eugregarina), in which schizogony is replaced by trophozoites. In certain members of this group, the cell body is uniform, while in others the body is divided by a partition into a "protomerite" and a "deutomerite." In addition, the protomerite can carry a frontal "epimerite" which aids in attachment. The genus *Monocystis* belongs to those gregarines with a uniform body, this being found almost exclusively in all its developmental stages in the European earthworm. An earthworm killed by chloroform is opened dorsally so that the seminal vescicles present in the tenth to twelfth segments are exposed. Their contents, when examined under the microscope, contain all developmental phases and, especially during the spring, the adult gregarines themselves. Pairs of reproducing gregarines ("syzygies") which have surrounded themselves with an envelope and syzygies in which the nuclei of both partners have already divided and the separating wall has disappeared can both be seen. In a later phase, the gametes originating from both partners can be found at the edge of the cyst in the middle of which the residual protoplasm is visible. The fused pairs of sex cells surround themselves with a boat-shaped spore coat similar to the calcareous alga *Navicula*, and are thus termed "pseudonavicellae" (false Navicula). The cysts full of such "boats" are the most striking feature of the preparation. The three division stages of sporogony, from which eight "sporozoites" emerge, take place within them, a process in which, owing to the uniform light-reflecting properties of the spore contents, nothing can be seen in a live preparation. With luck, however, especially in spring, the *Monocystis* gregarines themselves, with their curious form of movement, can be observed. They contract their body envelope anteriorly and posteriorly in rapid alternation so that the body contents are alternately driven forward and backward, a highly interesting process! Some free-living gregarines are surrounded by thin threads on

Order: Gregarinida

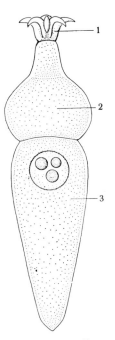

Fig. 4–12. *Corycella armata*. Entire cell (gamont): 1. Epimerite 2. Protomerite 3. Deutomerite.

Earthworm-inhabiting gregarines

their body surfaces, these being the earthworm's sperm threads. Some appear similar to these, but the threads in this case are products of their own cell surface. The various sizes of the boatlike spores indicate that usually several gregarine species inhabit the seminal vesicles of the worm at the same time. This is easily possible, for the common earthworm is parasitized by nine species of gregarines, and *Lumbricus rubellus*, which is just as common, by fourteen. The large number of species present and the frequency with which they are found raises the question of how these parasites penetrate the worm's seminal vesicles. Science, however, has not been able to answer this clearly. How do the spores get out of the worm? Naturally, in one case, after the worm's death and decay. Now and again, however, the hindmost segments are found filled with spores. These segments can be constricted and then die. Even here, however, the question of how the spores reach the hind segments from the seminal vescicles at the anterior end still remains unanswered. This unsolved problem presented by one of our most common animals is still a puzzle for our scientists.

Gregarines with a epimerite aiding in attachment are all intestine-inhabiting forms which, as least during their period of growth, are attached to the intestinal cells. In most animals the body is divided by an ectoplasmic partition into a protomerite and a deutomerite. They inhabit an amazing variety of hosts: annelids, echiurids, arachnids, crustaceans, centipedes, millipedes, insects, and mollusks. Gregarines are exceptionally common in insect intestines; *Gregarina blattarum*, for example, is practically always present in the intestine of the German and Oriental cockroach. Not everybody has access to these parasites, although the larvae of the meal beetle, mealworms, can be bought in pet shops as animal food; these contain the gregarine *Gregarina cuneata* (L 0.35 mm) in the anterior end of the midgut—*G. cuneata* is recognizable by the constriction between the protomerite and deutomerite—*Gregarina steini* (L 0.15 mm) in the posterior section of the midgut—recognizable by its pointed posterior—and also *Gregarina polymorpha* (L also 0.35 mm)—distinguished by its cylindrical body form. These gregarines have a unique means of movement which can be clearly observed under the microscope. It is quite different from that of *Monocystis*: they glide along without any visible change in body shape. When observed in a suspension of India ink particles it can be seen that the gregarine leaves a cylindrical slime sheath behind it. It has been suggested that the animal moves forward on this, but one fact which counts against this suggestion is that the sheath moves along with the gliding animal. Because of this, it is more probable that the gliding movement is brought about by minute wavelike movements of the surface protoplasm.

The sporocysts of these gregarines are also worthy of mention: they have a number of ducts (sporoducts) around the cyst which initially are inverted into the cyst interior. When the cyst envelope contracts, the ducts are extruded and release the spores. These spore ducts consist of the protoplasm remaining as a residue after the formation of the gametes.

The second sporozoan order is the COCCIDIA (Coccidia). They differ from the gregarines in that, during gamogony, only the male presexual cell, the gamont or gametocyte, multiplies and several—usually eight—small sexual cells (microgametes) are produced from it. The female presexual cell, on the other hand, usually grows to form a large macrogamete without undergoing division. While the gregarines are of no importance from an economic or medical point of view, being invertebrate parasites, the coccidians include a number of dangerous disease organisms infecting both man and domestic animals. These belong to the SCHIZOCOCCIDIA (suborder Schizococcidia), and undergo schizogyny. A few parasites, the sporozoites of which become presexual cells within a new host without further multiplication (schizogyny), infest marine annelids and are grouped together in the suborder EUCOCCIDIA.

The numerous instigators of the disease coccidiosis in domestic animals are found among the SCHIZOCOCCIDIA. They are members of the family EIMERIDAE, especially the genera *Eimeria* and *Isospora*. The two can be distinguished through their sporogyny: The "oocysts" found in the droppings of the sick animals contain, in *Eimeria*, four spores, each with two sporozoites, and in *Isospora*, two spores with four sporozoites, on maturation. Rabbit coccidiosis is caused by *Eimeria stiedae* in the liver; poultry coccidiosis by *Eimeria tenella* in the appendix, and other *Eimeria* species in the small intestine, this being a widespread and much feared disease; *Eimeria zurnii* and *E. bovis* are the chief causes of coccidiosis in cattle. *Isospora* species are mainly songbird parasites but are also found in carnivores, especially dogs and cats; two *Isospora* species are even human parasites.

A unicellulate parasite in man, *Toxoplasma gondii*, has until recently proved difficult to place within the system, but is now included among the coccidians. This causes toxoplasmosis in humans, numerous mammals, and certain birds. This animal was discovered as early as 1908 in a North African rodent, the gundi (see Vol. XI), and derives its name from this. When it proved to be a widespread and harmful human parasite, it was found to cause brain and lung abscesses, hydrocephalus, and severe retardation deformities—especially of the eyes—in children still "in utero," being transmitted through the mother. In the majority of cases, *Toxoplasma* causes no symptoms in man. To date, only asexual reproduction of this parasite has been discovered among most of the large number of hosts it infests. The sexual stages are known as they occur in a epithelium of the intestine of cats both wild and domestic. Experiments have shown that the disease is transmitted through eating the flesh of infested animals, but this does not explain its wide distribution in humans and herbivores (cattle, sheep, and certain birds). Another type of transmission is through infective cysts in the feces of cats. There may be other types of transmission.

Order: Coccidia

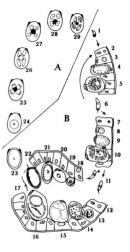

Fig. 4-13. Life cycle of *Eimeria stiedae* from the rabbit's intestine A. Development outside the host B. In the epithelial cells of the intestine. 1. Sporozoite 2-5. Schizogony 6. Merozoite 7-10. Further schizogony 11. Merozoites 12-16. Development of the microgametes 17. Microgamete 18-20. Development of the macrogametes 21. Macrogamete with protein particles 22. Fertilization 23. Oocyst 24-29. Sporogony (four spores, each with two sporozoites, form without the Oocyst).

At first dogs were considered the main carriers, but such persistent forms were looked for here without success; infectious "oocysts" were, however, found in the feces of infected cats. Each contained two spores which each contained four sporozoites, thus being similar to the persistent form of *Isospora*. The long-searched-for sexual reproductive phases, which produced the oocysts, were discovered later in the intestinal mucosa of the cat's small intestine. As far as we know today, the cat is the "final host," harboring the sexual forms and releasing the infective cysts. All the other hosts are "transition hosts" in which the parasite reproduces only asexually and from which it is now and again transmitted to carnivores through their prey, but which do not release the persistent form in their feces. Among these hosts, the mouse is the most important infection agent for cats. Cats, being clean creatures, could never become infected through the feces of other cats or by eating their flesh. Therefore a change of hosts must occur here between cat and mouse in most cases, *Toxoplasma* being closely associated with only a few hosts in the sexual phase and with a larger variety of hosts in the asexual one. This can be compared to the bean aphid and the peach aphid (see Vol. II), which also have a single "primary host" and numerous "secondary hosts."

Malarial parasites

While certain species of the family Eimerida are highly dangerous disease organisms, some members of the HAEMOSPORIDIA (family Haemosporidae) are, for humans, a much greater worldwide danger: the causal agent of malaria, genus *Plasmodium*, for instance. These are parasites in blood and tissue; it is thus not possible for them to be transmitted from host to host through the feces and in feeding. A second host, the blood-sucking mosquitoes of the genus *Anopheles* (see Vol. II), in which both gamogony and sporogony occur, transmits the parasites. Of course, they have sporocysts in which the sporozoites are produced and from which they migrate to the salivary glands for transmission when the mosquito feeds. During the transmission of the cells from mosquito to man, there is no danger of desiccation. For this reason, the plasmodia can be termed "sporozoa without spores." Their course of development is given in Vol. II as well as in Fig. 4-13. We shall restrict ourselves here to following their further course of development in man.

When a female mosquito bites (the male mosquitoes feed only on plant juices) the sporozoites enter the bloodstream with the mosquito's saliva, and are carried to the liver. They then embark on a phase of asexual division (schizogony) in the liver cells. They grow within the cells, and their original single nuclei divide into several nuclei. A portion of the protoplasm surrounds each one so that numerous daughter cells (merozoites) are formed. This occurs entirely in the liver cells at first, but later also within the red blood corpuscles. The merozoites released by the collapse of a red blood corpuscle then infect other red blood corpuscles, and schizogony continues. On the other hand, however, merozoites which have entered blood corpuscles without first having divided become presexual cells (gamonts) of both sexes. They can only develop further when they reach the intestine of another mosquito after it bites their host. Their further development in the second mosquito is described in Vol. II. These processes within the mosquito, up to the final phase of transmission of the mature sporozoite, require a certain period of time, dependent both on temperature and the type of malaria. In *Plasmodium vivax* (see Color plate, p. 114), about ten days are required at a temperature of 24°C for the mosquito to produce infective sporozoites. Sixty days after the mosquito has been infected, all the parasites are dead; the mosquito is therefore unable to transmit the parasite over an indefinite period of time. The majority of mosquitoes, however, do not live so long. When the temperature is too low, no multiplication occurs within the mosquito; this is probably the reason for malaria being found almost exclusively in warm countries.

The name "malaria" ("bad air") for the disease produced by the plasmodia is derived from the old superstition that the disease was caused by poisonous air from swamps. The German name, "alternating fever," describes the disease more accurately: in cases of infection with *Plasmodium vivax* and *Plasmodium ovale*, fever occurs every two days, and in the case of *Plasmodium malariae*, every three days. Including the day of fever, therefore, these are cycles of three and four days, termed "tertiary," with a 48-hour rhythm and "quaternary," with a 72-hour rhythm. Pernicious (i.e., dangerous) malaria, caused by *Plasmodium falciparum*, has a 40- to 48-hour development cycle in the blood, and continues over a long period of time. But it is not because of this that pernicious malaria is so dangerous. The multiplication phases in the blood occur mainly in the blood vessels of the internal organs, where the decaying blood corpuscles attach to each other and the walls of the blood vessels so that they finally block the blood vessel. This is a grave threat to the host's health, especially when it occurs in the brain and heart muscle. On primary infection on two subsequent days, the fever can continue without a break ("quotidian"). Even in tertiary and quaternary malaria, fever can occur daily when the parasite's multiplication cycle originates from several primary infections which overlap each other, and also in a case of a single primary infection.

Fever occurs every time a phase of asexual multiplication in the blood (erythrocitic schizogony) is completed and the merozoites formed are released. It is the response of the infected person to the destruction of the red blood corpuscles and to the metabolic products of the parasite released on their destruction. To enable the regularity of the fever attacks to be understood, it is necessary to know that the parasite's multiplication in the blood is temporally synchronized: the division of the plasmodia in the blood corpuscles occurs at exactly the same time. This is obviously not the case at the beginning; the first fever attacks are irregular, and it is the human diurnal rhythm in metabolism which generally influences the parasite to multiply at the same pace.

Alternating fever

Fig. 4-14. The present-day distribution of malaria.

Even when the fever attacks of tertiary or quaternary malaria no longer occur, this is no sign that the disease has come to an end: the forms present in the liver and sometimes in the blood vessels of the internal organs are able to survive for months or even years, and occasionally give rise to new attacks. This ability to survive lasts about two years in the case of tertiary malaria and decades in the case of quaternary. Only the causal agent of pernicious malaria gives no evidence of a relapse.

A doctor is able to prove the presence of a malarial infection by examining the blood under a microscope. The number of schizonts within a single blood cell, the shape of the gamonts, and the dots present in the blood cells give information on whether the type of malaria present is tertiary or quaternary. In pernicious malaria, on the other hand, usually only sexual forms and immature asexual forms—the so-called ring forms—are present.

Malaria is, without doubt, one of the most terrible of human diseases. Just after World War II, the number of people infected in a year was reckoned at 350 million, and those who had died of the disease, 3.5 million. Severe outbreaks have often occurred in past history. E. Martini, an investigator of tropical diseases states: "... malaria played a powerful role in the political balance of Italy in that it harassed the German Kaiser's army continually. The fever made it impossible for his army to occupy the flat coastal lands, and wrecked Konradin's last military campaign, thus becoming one of the main causes of the collapse of Staufer's power in Italy. It sounds almost like a saga of a powerful fight between malaria and earth's governors."

Anti-malarial measures

Fighting this human scourge is still an urgent task which is being carried on at various levels. Today a whole range of drugs is known which aid in curing the disease; quinine, obtained from the bark of the fever tree, is the oldest of them, the healing properties of this bark being known by the pre-Columbian Incas. Nowadays, quinine has been supplanted by more effective drugs with fewer side effects, such as Resochine and Paludrine. Travelers visiting malaria-infested areas take such drugs as a prophylactic measure. Mosquito bites are prevented at night by the use of fine-mesh mosquito nets around the sleeping area and across the windows. The greatest measures, however, are taken against the mosquito itself and its larvae. These larvae live in water; the removal of their breeding water is therefore of greatest effect. The use of water in rice paddies cannot be dispensed with, but the rice plant is not damaged by frequent drying out of the paddies, which prevents the full development of the mosquito larvae. A thin film of oil kills both mosquito larvae and pupae, which must come to the surface to breathe, but also kills a large number of other insects at the same time.

Certain live-bearing fish belonging to the toothed carp, especially some of the Gambusias (see Vol. IV), have proved to be valuable predators on mosquito larvae. The fight against mosquitoes in certain areas proved very successful when DDT and other substances were used, but these contact insecticides have proved to be a two-edged sword and are now banned in many countries. Also, the mosquitos have developed resistance or immunity to the chemical poisons!

The goal at the moment is a specialized fight against the fever mosquito alone, without damage to other animals and without indiscriminate poisoning. Pointers have been found in this direction, but the mode of life of the malaria-carrying mosquito species differs widely, as do the carrier species from area to area. A basic investigation of the present relationships by an expert is the basis for success in this case. The building of dams and irrigation ditches has increased the danger of infection, notably in Egypt.

The CILIATES (class Ciliata) possess numerous cilia (from the Latin *cilium* = eyelash). Only the Suctoria are aciliate in the mature phase; their juvenile stages, however, are ciliated. Cilia and flagella are similar in structure and consequently there would be no sharp dividing line between the multinucleate flagellates (Polymastigina and Opalina) and ciliates when no divergence was present in the ciliate nucleus: the twofold task of the nucleus—regulation of life processes and retention of genetic material—being, in this case, divided between two nuclei. The macronuclei, differing widely in form, are responsible for the first task; the spherical micronuclei are involved with the second. This division of labor within a single cell is comparable to multicellulate divisions of labor in the somatic "body cells" and germinal "germ cells."

Ciliates multiply, as do other unicellulates, by "division" into two similar cells or by "budding," one small cell forming out of a larger one. These processes are asexual and are thus termed "asexual reproduction and multiplication." Sexual processes, which will be discussed in more detail in the section on *Paramecium*, are termed "conjugation." In contrast to "copulation" (not to be confused with copulation in multicellulates), it ends with the partners separating, two animals thus being produced from two animals. In copulation, on the other hand, the two animals join permanently. Conjugation and copulation are "sexual" processes, since animals with different genetic material, new individuals, are formed from them.

Ciliates are also called "infusoria" since they were first discovered in infusions of hay, grass, or earth. Even today, this is the best way of obtaining them, but their small size requires the use of a microscope. To obtain them, the following instructions should be followed:

A handful of hay is boiled in 25 cl of non-chlorinated water together with about ten to twenty wheat grains or a little grass. After a few days, a little loam with leaf mold and some ditch-water containing water plants is added, once a fusty film has formed over the surface of the liquid. Within a short while, a rich fauna develops within the liquid, consisting of numerous ciliates of various species, shell-less and shelled amoebae, flagellates, and small multicellulates. The composition of this population

Class: Ciliata

Conjugation and copulation

Paramecia

of small animals changes with the age of the infusion. Small ciliates are the most numerous during the first few days, but they are gradually replaced by large numbers of paramecia (genus *Paramecium*; see Color plate, p. 108). If a drop of this water full of paramecia is placed on a glass slide and covered with a coverslip supported at the corners by tiny plasticine legs, little can be seen of these animals since they dash too quickly through the visual field of the microscope. If the cover slip is pressed closer and closer to the slide surface by compressing the plasticine legs with a needle, the lightly compressed ciliates are finally unable to move. They can then be observed in detail, without disturbing the beat of their cilia, the movement of their food vacuoles, or the filling and emptying of their contractile vacuoles. One can even see the micronucleus, which usually lies in a depression of the macronucleus, if this is in a favorable position. It is fascinating to observe the path of a single food vacuole from the cell mouth through the entire animal to the cell anus, as is the interplay between the two contractile vacuoles which are, in this case, constant features.

Simple experiments can be conducted on paramecia with very little apparatus. If a drop of vinegar is added to the water, the paramecia die, but before doing so, they shoot out numerous threadlike structures from the surface protoplasm: the "trichocysts." Their task has not been explained fully, but their microstructure suggests that they could be defense organelles. Their tips are arrow shaped and one could easily conceive of many of these "arrows" driving off tiny attackers. This could occasionally be necessary, since even among the unicellulates there are species which hunt other unicellulates. The water bear (*Didinium nasutum*; see Color plate, p. 305), for example, devours paramecia almost exclusively. It swims about in the water and, if it encounters a prey animal, it attaches its anterior end to the substrate and shoots out poisonous threads, similar to the trichocysts, which cripple the paramecium. Finally, the much smaller water bear engulfs its prey by enormous extension of its gullet. The paramecium defends itself by shooting out a dense layer of trichocysts, but to no avail (no weapon in the animal kingdom affords complete protection).

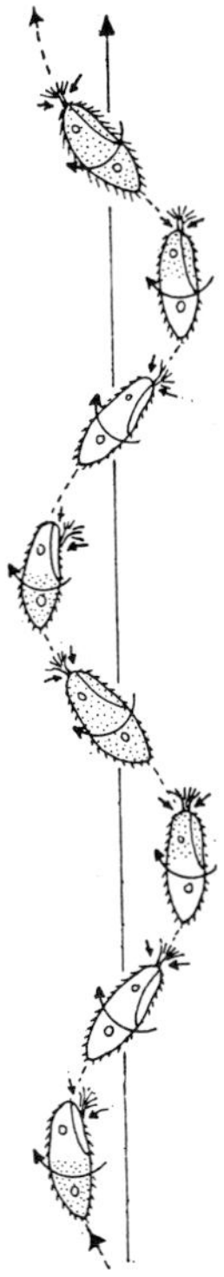

Another simple experiment allows the exact observation of feeding and defecation to be made. If a little India ink or carmine particles are introduced into the waterdrop containing the paramecia, the following can be observed clearly: the collection of the particles into a ball at the cell mouth, the formation of a food vacuole, how this food vacuole traverses the cell body along a definite pathway, and, finally, the expulsion of the indigestible remnants through the cell anus.

The beating movement of the cilia can also be clearly observed in these suspensions, since the particles indicate the current created by the cilia.

If paramecia are observed on a slide under the microscope without a cover glass, it can be seen that they swim in a spiral, and at the same time

Fig. 4-15. Path taken during swimming by a paramecium.

turn on their longitudinal axes. In this way, a final straightforward motion is achieved. If the paramecium collides with an obstruction, it first swims backward a short way, owing to the reversal of its ciliary beat, then turns the anterior end sideways in a cone-shaped path and swims forward once again: this process is repeated until the obstacle is passed.

Very similar "avoidance responses" occur to chemical stimuli, as a simple experiment will prove: Place a drop of water containing a large number of paramecia on a glass slide and cover it, as before, with a cover slip supported on plasticine legs so that the animals can still move around freely. Then place a drop of diluted acetic acid under the center of the cover slip with a hair-fine pipette. If the acid is dilute enough to be attractive to the paramecia, they will then collect in its center. If it is too strong, however, they form a ring about the drop in the zone where the dilution gradient approaches that of the attractive concentration. They ricochet backward and forward in their corkscrew path from the too concentrated and too dilute zones as if from walls. If the slide is warmed on one side, the paramecia will collect in the zone where the heat gradient is optimum. Under normal circumstances the common, colorless paramecia (*Paramecium aurelia* [see Color plate, p. 113] and *Paramecium caudatum*) do not respond to light.

An interesting exception are the GREEN PARAMECIA (*Paramecium bursaria*). Their green coloration is due to unicellulate green algae (zoochlorellae) which form organic substances with the aid of sunlight and, at the same time, produce oxygen. In symbiosis with these algae, the paramecium is able to exist without engulfing food, but only in the light. Owing to this, its behavior has become adapted to accomodate the light requirements of its algae: in a half-darkened container the paramecia collect in the lighted portion, at least when the oxygen content of the water is not very high.

Paramecia are not only able to respond to stimuli by the behavior patterns described—swimming backward, turning sideways, swimming forward—but also by turning on the spot. This is produced by the stimuli impinging on the body surface at different points, releasing variable activity in the cilia. The paramecium swims against the current in moving water when the current is not strong enough to sweep it away. Their response to gravity is worthy of note: in water with a high carbon dioxide content, the animal swims upward (negative geotaxis). It is probable that gravitational effects are registered by the pressure exerted by the contractile vacuole contents on the cell body protoplasm. This behavior pattern could be of use to the animals, since it guides them to the surface, which is richer in oxygen. Under natural conditions, the response of paramecia to an electric field is certainly of no adaptive significance: in this case they swim toward one of the electric poles, depending on the strength of the field, a result of a forced return stroke stimulated by the electric current on a section of the cilia.

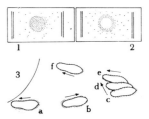

Fig. 4-16. Discriminatory or avoidance response in paramecium: 1. Congregation in a drop of 0.02% acetic acid. 2. The formation of a ring around a drop of more concentrated acid. 3. In response to the stimulus, the animal stops moving forward (a), swims backward for a short distance, the cilia reversing their beat (b), halts again and starts to turn (c, d, e); on this, the response ends and the animal swims forward again (f).

Paramecia respond clearly to an increase in the oxygen content of the water: they increase their swimming speed and the majority of the animals swim perpendicularly, some upward (negative geotaxis) and some downward (positive geotaxis). The increased rate of swimming is released at the same strength, if at all; this is an example of an "all or nothing" process, the same as many nervous responses.

A further behavior pattern present in paramecia can be observed in water drops containing stationary particles: the animals often press closely against these. This "contact response" is well known among multicellulates inhabiting cracks, and is termed "thigmotaxis."

It is an old question whether unicellulates are able to learn. This is in no way irrelevant, since the ability to learn has been proved among multicellulates such as the flatworms (see Chapter 8), the most primitive of the bilaterally symmetrical animals. It has been shown that memory is associated with ribonucleic acid, which is also present in unicellulates. Training experiments with unicellulates, especially with the paramecia owing to their ability to respond, have already been conducted. No clear-cut results have emerged yet, although more recent experiments on the ability of other ciliates to learn have been more successful. "Habituation" must not be confused with the ability to learn, habituation being a gradual decrease in responsiveness to a continually repeated stimulus and finally a cessation of response. Habituation also occurs in unicellulates.

Paramecia multiply by transverse division into two similar daughter cells. In certain ciliates the two daughter cells differ in size, this process being termed "budding." The cell body becomes constricted, and at the same time the micronuclei are in the process of "mitosis," which ensures that the hereditary characteristics are equally divided between both cells. This process is, however, somewhat modified in the macronucleus.

The processes within the nuclei during division are more complicated than those during conjugation, the fertilization process of ciliates. Conjugating pairs are quite often found in water having a population of paramecia. Conjugation begins with two apprently similar paramecia coming together side by side so that their cell bodies fuse with one another. Feeding ceases some time beforehand, but swimming continues. The two animals respond like a single individual in this respect. The pair, just as an individual, has a definite relationship between the number of turns on the longitudinal axis and the same swimming direction. Swimming behavior in response to mechanical and chemical stimuli does not alter, as, for example, faster swimming in oxygen-rich water, and neither does other behavior in general. Within the nuclei, however, involved basic processes occur: the two macronuclei degenerate and the two micronuclei divide into four, in two sequential divisions. This process is reminiscent of the maturation divisions in the eggs of multicellulates, but in these cases it is the whole cell which divides, not just the nucleus. In both processes, however, the number of chromosomes within the nuclei is reduced to the

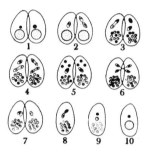

Fig. 4-17. Diagram of the nuclear processes occurring during conjugation in a ciliate: 1. Joining of the two conjugants by the cell mouth (cystostome). 2. The first pregametic division of the micronucleus in each conjugant. 3. The destruction of the macronucleus. 4. The second division of the micronucleus. 5. Three of the four micronuclei arising from the second division disappear (black dots); the fourth micronucleus divides again and produces a female or stationary nucleus and a male or migratory nucleus. 6. Exchange of migratory nuclei. 7. Joining of the male and female nuclei to form a zygotic nucleus in each conjugant. 8–10. Separated animals. 8. Division of the zygotic nucleus. 9 and 10. A new micronucleus arises from one daughter nucleus of the zygotic nucleus, and, from the other, a new macronucleus.

single complement (haploid) from the double (diploid). As in the maturation division of multicellulate eggs, three of the four nuclei formed in
paramecia also degenerate. Each partner thus posesses a single micronucleus once more, but this only contains a single chromosome complement. These nuclei, however, divide once again into a stationary nucleus
and a migratory nucleus, both with a single chromosome complement.
The two migratory nuclei then traverse the cytoplasm bridge between the
two individuals, and each fuses with the stationary nucleus of the other
individual. This is the true fertilization process from which a new nucleus,
with a double chromosome complement once more, is created. After
separating, each individual must form a new macronucleus again, since
without this it would be non-viable. This comes about by the nucleus
dividing, one half becoming the micronucleus and the other then doubling,
quadrupling, and finally multiplying its chromosome complement without external division until it becomes "polyploid" and forms the macronucleus.

There are several variations in the conjugation process described above,
which is typical of *Paramecium caudatum*, as, for example, in the case of
Paramecium aurelia (see Color plate, p. 113), which has two micronuclei,
and other ciliates which are multinucleate. In these, in contrast to the
above, similar processes have been observed without pair formation
occurring before: these are termed "self-fertilization" (autogamy). In the
case of non-paired individuals, everything progresses as in a normal
conjugation up to the point of formation of the stationary and migratory
nuclei, these then fusing with one another. The continuing processes are
the same as those following a complete conjugation. There is no exchange
of genetic material in this case; on the contrary, after self-fertilization the
animal is homozygotic in all its characters, since the double chromosome
complement of the micronucleus and the polyploid macronucleus all
arise from the single chromosome complement of the original nucleus
which had divided into the stationary and migratory nuclei.

Although the two conjugating partners appear to be absolutely alike,
it has been shown that in most cases only paramecia of different hereditary
origin pair. If a paramecium culture arises from a single individual,
pairing only occurs very rarely, even under the most favorable conditions.
In the case of two cultures originating from the two partners of a previous
conjugation, no pairing occurs within the cultures; when, however, these
two cultures are mixed, numerous pairings occur if the circumstances are
favorable. From this it can be concluded that two pairing strains are
present which pair with each other but not among themselves. A comparison with the sexes of multicellulates could be made, especially in the
case of those ciliates in which only two such pairing types occur. These
are termed the plus and minus types. This comparison breaks down, however, when more than two pairing types are present, as, for example, in the
green paramecia: certain cultures of this species show four pairing types,

Variations in
conjugation

identified by the symbols I, II, III, and IV. Pairing type I can conjugate with pairing types II, III, and IV, but not with I; in the same way, pairing type II conjugates with I, III, and IV, but not with II; and so on. Other cultures of this species have eight pairing types.

Reproductive communities

More detailed investigations on conjugation in ciliates, especially among the various species of paramecia, have given further interesting results: a single species can contain reproductive communities (syngens), each containing the pairing types mentioned. Not one of the pairing types of one reproductive community can, however, conjugate with any of the pairing types from another reproductive community. The various reproductive communities should be termed different species, since in nature a species is a group of animals which can reproduce only among themselves and produce fertile offspring. Since members of different reproductive communities cannot be distinguished from one another in morphology, there is no basis to describe them as species. In this case, *Paramecium aurelia* (see Color plate, p. 113), which is known to have fourteen reproductive communities, should be divided into the same number of species. Only small, statistically determinable differences have been found between these reproductive communities, and, similarly, differences in their geographical distribution. Two of the reproductive communities are worldwide; another is found only in Europe.

Symbionts of paramecia

The detailed investigations on the paramecia brought forth further highly interesting results. Some *Paramecium aurelia* with the same evolutionary descent contain bacterialike organisms within the cell body, with which they appear to live in symbiosis. One of these is termed the "Kappa-symbiont." The paramecia containing this organism secrete something which kills members of other paramecium cultures. Those which die are called "sensitives" and the others, "killers." The killer characteristic is hereditary (genetic), ensuring the killers' immunity to the deadly poison, and, owing to this, they have the ability to live with the Kappa-symbionts. A whole series of similar symbionts exist, which are also named with the letters of the Greek alphabet: the Lambda-, Sigma-, Pi-, and Mu-symbionts. They are termed symbionts because it has been shown that the Lambda-symbiont gives the paramecium folic acid, an essential substance which it is unable to produce itself. It has been shown that, in the absence of the Lambda-symbionts, the paramecia have to obtain this from their food. The Mu-symbionts have also proved to be "killers," but only in respect to their partner during conjugation. The strain of paramecia harboring this symbiont is thus termed "mate killers." Investigations with the electron microscope have recently afforded surprising information on the nature of the lethal "something": it is not the Kappa-symbionts themselves that are lethal; they are simply transmitters of deadly particles, probably viruses.

Numerous other ciliates can be found in the infusions where paramecia are present. If a microscopic transect is made through the soft upper mud layers of a puddle and its algal covering, a very impressive survey of the types of body form present among these delicate unicellulates can be made. The systematist requires a system based on relationships among this great variety; the most primitive forms, those in which the cell body is completely covered with cilia, are thus all included in the order HOLOTRICHA (from the Greek ὅλος = whole and νριχος = hair). These can be divided further by mouth form: the suborder GYMNOSTOMATA, which have no gullet, the mouth of which is only opened to hold and swallow the prey. The water bears (*Didinium nasutum*), one of the main enemies of the paramecia, belong to this group, having the mouth at the anterior end, at the tip of a non-ciliated cone, the "nose." The much smaller species *Coleps hirtus*, a tiny, barrel-shaped organism with the mouth on a broad, lopped-off anterior end, is also capable of swallowing prey much larger than itself. The flasklike, long-necked *Lacrymaria olor* has a mouth at the tip of the neck process, while that of the more thickset *Dileptus anser* is at its base.

Order: Holotricha

In the suborder TRICHOSTOMATA the mouth is at the base of a gullet fringed with rows of simple cilia. These guide the food to the mouth. A member of this order is *Colpoda cucullus*, a kidney-shaped organism found in rotting water stands and also on occasion in infusions. In the suborder HYMENOSTOMATA the cilia around the mouth are fused to form a fine membrane. The paramecium species (genus *Paramecium*; see Color plate, p. 305) are members of this suborder, as are their almost constant companions in infusions, the bean-shaped *Colpidium colpoda* (see Color plate, p. 305). One of the giants among the ciliates, ICHTHYOPHTHIRIUS (*Ichthyophthirius multifiliis*; L up to 0.8 mm), is feared by tropical-fish fanciers. This causes "Ich," or "white spot disease," in fresh-water fish, growing as a parasite within the skin to such a size that it is visible to the naked eye. It then leaves the host and multiplies within a gelatinous cyst in several—up to eight—steps, numerous—up to 256—juvenile stages resulting. These leave the cyst and attack the fish's skin once more. Continuous immersion in a quinine solution (1:50,000) kills the free-swimming juvenile stages; the treatment must be continued, however, until the fish is free of parasites, since the quinine has no effect when the parasites are within the skin.

White spot disease

The suborder ASTOMATA includes certain completely ciliated mouthless species which live as internal parasites particularly in the intestines of annelids. Loss of the mouth is an adaptation found in many unicellulates which are continually bathed in soluble nourishment and which take this up through the entire body surface. One need only think here of the trypanosomes, opalines, and the entire class Sporozoa. Among the multicellulates, degeneration of the feeding organ as an adaptation to life within soluble food is also found among the tapeworms (see Chapter 8) and the spiny-headed worms (see Chapter 10).

The order PERITRICHA (from the Greek περι = around) is distinguished from the Holotricha in that the cilia around the mouth are arranged in a left-hand spiral. Within the deep gullet, the vestibulum, they are fused to

Order: Peritricha

form a delicate membrane as in the mouth of the Hymenostomata, from which the Peritricha probably arose. Most species consist of a bell-shaped head attached to a stalk which is fixed to the substrate. Their name "Peritricha" derives from this. If a longitudinal muscle is present within the stalk, this can be contracted in a screwlike manner. The species of the genus *Vorticella* (see Color plates, pp. 113 and 305) are solitary and have a contractile stalk. Other genera form colonies with branched stalks. The stalks are not contractile in *Epistylis* and *Opercularia*, but can be contracted in *Zoothamnium* and *Carchesium* (see Color plate, p. 108). In *Zoothamnium* all the stalk muscles are bound together. In *Carchesium*, each "head" has its own stalk muscle which is not connected with those of its neighbors. *Ophrydium versatile* can finally reach clumps the size of a fist by the extrusion of gelatinous substances, the individual organisms here being green in color owing to the presence of algae (Zoochlorellae).

The sedentary mode of life of the Peritricha has had an effect on their multiplication and conjugation. In contrast to other ciliates, they do not divide horizontally, but longitudinally like the flagellates; often, especially in the solitary species, the daughter cells are unequal in size. The larger daughter cell then remains on the stalk while the smaller forms an extra ring of cilia at the posterior end, breaks away from the other cell, and then conjugates with a sedentary, fully grown individual as a free-swimming stage. Conjugation here is one-sided, since the free-swimming individual forms only a migratory nucleus which fuses with the single stationary nucleus formed by the large individual; its protoplasm is taken up by that of the larger partner. This copulationlike process is an adaptation to a sessile mode of life and will be met with again among the equally sessile Suctoria.

The "Peritricha," together with certain forms which build an exoskeleton, are included in the suborder SESSILE PERITRICHA (Sessilia). Only a few peritrichs are able to move freely about the surface of their hosts, moving from time to time by means of an adhering disc at the end of the stalk; these form the suborder MOTILE PERITRICHA (Mobilia). *Trichodina domerguei* can cause damage to the skin of fish in aquaria and aquicultures, while *Trichodina pediculus* (see Color plate, p. 113) very often attacks fresh-water polyps. Conjugation progresses in the free-living trichodines in the same way as in the sessile Peritricha, indicating their relationship to these.

Order: Spirotricha

The ciliated mouth area also forms a spiral in the order SPIROTRICHA. This, however, in contrast to the mouth spiral of the Peritricha, is clockwise, and in addition to this, the cilia are not fused to form a simple lamella but form numerous flickering platelets arranged in many short, oblique rows. The order HETEROTRICHA includes species whose body surface is more or less uniformly ciliated. Their name refers to the difference in ciliation of the oral region and body. The best-known of these are the STENTORS (genus *Stentor*), their generic name coming from that of the loud-voiced hero of the Trojan War, Stentor. Their trumpetlike form probably led L. Oken, a naturalist and philosopher of the Romantic Period, to make this comparison. The BLUE STENTOR (*Stentor coeruleus*; see Color plate, p. 305) is most often found in rotting waterstands, sometimes attached to the substrate and sometimes free-swimming. The thinner GRAY STENTOR (*Stentor roeseli*) has a gelatinous envelope, while the GREEN STENTOR (*Stentor polymorphus*; see Color plate, p. 113) is recognizable by its green color, which is due to symbiotic algae (Zoochlorellae). The large *Spirostomum ambiguum* (L more than 2 mm) is wormlike in form, and can be seen by the naked eye, being found in foul marsh water. The genus *Nyctotherus* is an intestinal parasite in invertebrates as well as vertebrates. The heart-shaped *Nyctotherus cordiformis* is found in the intestines of toads and frogs, its cysts having previously infected the tadpoles. The parasites undergo conjugation at the time of the host's metamorphosis. The descendants of the conjugated animals are much larger, and these are the only ones found in the fully grown amphibia. An occasional intestinal parasite of man also belongs to this group: *Balantidium coli* (L up to 0.15 mm). It is especially common in pigs, which show no disease effects owing to mutual adaptation between host and parasite. In man and chimpanzees, however, *Balantidium coli* causes ulcers in the intestinal tract: "balantidial dystentery." *Balantidium* forms a cyst which is excreted with the feces and transmitted in this way to a new host.

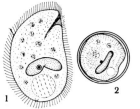

Fig. 4–18. *Balantidium coli*: 1. Trophozoite. 2. Cyst.

The suborder HYPOTRICHA is distinguished by a flattening of the body, which is divisible into an upper surface with a few stiff hairs and a flat undersurface. This is set with strong "cirri" consisting of fused cilia, in much the same way as the cilia platelets of the stentors. The animal uses these when crawling across the substrate. The most commonly encountered of these highly developed infusoria is *Stylonychia mytilus* (L over 0.3 mm; see Color plate, p. 305), which is often found together with the paramecia. On its underside it has eight frontal, five abdominal, and five anal cirri, and, at the tail, three strong posterior cirri. These very differently shaped organelles, formed of fused cilia, work together in mutual harmony, and their activity is so controlled that the animal is able to carry out delicate stepping movements. The degree of cooperation between them is amazing, especially since a comparable coordination in the organs of higher animals can only come about through the presence of a controlling nervous system. This is lacking *Stylonychia*, and all these processes are carried out by a single cell! Smaller, more flexible Hypotricha lacking the three posterior cirri belong to the genus *Oxytricha*. These are also common in infusions.

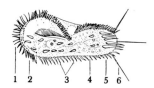

Fig. 4–19. *Stylonychia*: 1. Membranelles 2–6. Frontal, abdominal, anal, and posterior cirri.

The cilia have mostly degenerated in the members of the suborder OLIGOTRICHA. The tiny *Halteria grandinella* (L 0.03 mm) is often found in marsh water; it has only a few stiff bristles. The observer's attention is immediately drawn to this species, since it continually leaps jerkily through the field of microscopic vision. The slightly ciliated members of the family TINTINNIDAE form delicate envelopes about themselves. Only a few representatives of these, however, inhabit fresh water. Many of these species are marine, and have existed in the oceans since the middle geological ages; the presence of their envelopes in the Upper Jurassic chalk of the Mediterranean and Himalayan regions give the earliest records for prehistoric ciliates.

The suborder ENTODINIOMORPHA includes unusual forms with posteriorly directed body processes. They are mainly found in the stomach or intestines of hooved animals, but cannot be simply termed parasites. It has been shown that they split hermicellulose, one of the compounds comprising the cell walls of plants, and thus make it digestible for their host. In general, they feed on the starches present in the plant fragments. They are usually carried with the half-digested food through the three stomachs of cud-chewing animals (see Vol. XIII) and into the intestine itself, where they are finally digested. In this way they provide their host with an easily digestible protein. This, however, does not appear to be of great importance to the ruminants, since cows freed artificially from their protozoan symbionts fare no more poorly than others. The Entodiniomorpha are carried from cow to cow through the saliva during social cud-chewing. Other members of this suborder are present in the fermentation chambers of other animals, such as the colon of horses and pigs and the appendix of certain rodents such as the guinea pig and capybara.

Order: Chonotricha

The order CHONOTRICHA (from the Greek $\chi\omega\gamma\eta$ = cone) has only a few species, these being sessile forms almost devoid of cilia apart from a row running clockwise around the gullet. As in other sessile ciliates, reproduction is by budding, and fertilization is brought about by one-sided copulation. The transparent *Spirochona gemmipara* (L 0.08–0.12 mm) is very commonly found on the gillplates of the European crayfish.

Order: Suctoria

The order SUCTORIA consists of greatly modified ciliates. Their juvenile motile phases still move by means of cilia, and thus betray their geneological origin; the mature forms, however, are completely aciliate and therefore unable to swim actively. They are either attached to a substrate or float in the water, occasionally being moved passively by the water currents. Their mode of nutrition is unique: they have tentacles through which they suck out the contents of prey animals they have captured (from this, the name Suctoria). In some species, long capturing tentacles and shorter sucking tentacles can be distinguished. If a prey animal touches the swollen end of a capturing tentacle it remains stuck to it and is crippled; its cell body is then partly dissolved. Other capturing tentacles also bend toward the prey. All then contract, and in this way draw the prey toward the shorter sucking tentacles, which rapidly suck it dry. In species lacking the capturing tentacles, the sucking tentacles are used to capture the prey.

Reproduction in the Suctoria, as is often the case in sessile ciliates, is by budding. Either a single bud can be formed, or several are formed at the same time (multiple budding). The buds can either be on the surface or within the animal; these are distinguished as "external" and "internal" budding.

One species with simple internal budding is *Choanophrya infundibulifera*, the motile phases of which have several rings of cilia about the middle of the body. The reproductive processes of the marine species *Tachyblaston ephelotensis* are rather complicated. The juvenile developmental stages of this species have tentacles by which they bore into the cell bodies of *Ephelota gemmipara*, another member of the Suctoria. They continue to bore into the cell body of their host and suck its protoplasm. They then begin to grow and start to form, from their external surface, numerous motile phases, one after the other, these being diagonally ciliated at the anterior end. These detach and swim about, finally settling on the substrate. They then form stalks on their undersides, and an envelope forms around the cell body. By continual budding, new larvae, each with a tentacle, are formed; these leave the original animal to attack another *Ephelota* host. Two generations are thus present, both of which reproduce by budding but only one of which is parasitic.

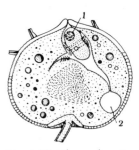

Fig. 4-20. *Choanophrya infundibulifera*: 1. Released bud; 2. Contractile vacuole.

Sexual reproduction in the Suctoria is by conjugation, as in all ciliates; the partners are sometimes, however, very different in size, and in some cases reciprocal fertilization has become one-sided, as has already been described for the Peritricha and Chonotricha.

Anyone who wishes to observe Suctoria under the microscope for himself can often find the delicately branched *Dendrocometes paradoxus* among *Spirochona gemmipara* on the carapace of the crayfish. The continual waterstream here provides both ciliate species with both nourishment and oxygen for respiration.

The majority of unicellulates can clearly be included in one of the four protozoan classes, even when, as is the case in parasites, their fundamental class characteristics have been lost. In many of the parasitic groups which were long included among the Sporozoa, relationships with the gregarines or coccidians have not been found; it is possible that they could have arisen from multicellulates which, owing to their mode of life, have undergone simplification. These are the myxosporidians (Myxosporidia), actinomyxidians (Actinomyxidia), and the microsporidians (Microsporidia). All have one characteristic in common: they form spores which contain one or more "pole capsules" in addition to one or more amoebalike germ cells. The pole capsules consist of tightly curled hollow threads which are capable of being shot out by internal pressure in the same way as the finger of a glove, the same process as is found in the nematocysts of coelenterates. It is easy to conceive of such a highly developed cell type coming about only once during the evolutionary development of animals. They are considered collectively as the Cnidosporidia, and this group—with reservations, of course—can be accorded the rank of a class which finds its place next to the Suctoria.

Class: Cnidosporidia

Order: Myxosporidia

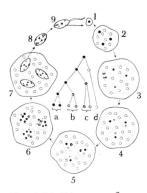

Fig. 4-21. Diagram of development in a Myxosporidian: 1. Uninucleate amoeboid germ. 2. Multinucleate plasmodium: Differentiation into generative (black) and somatic (white) nuclei. 3. Cutting off of the sporoblasts. 4–6. Various stages in nuclear multiplication within the sporoblasts, this being illustrated in the series of divisions shown in the middle of the diagram (envelope nucleus white, spore-case nuclei striped, polar-capsule nuclei dotted, gamete nuclei = genetic progression, black). 7. Plasmodium with spores. 8. Individual spores with binucleate amoeboid germ, a polar capsule with extruded polar filament. a = gametic nuclei, b = polar capsule, c = spore-case nuclei, d = envelope nucleus.

Among the MYXOSPORIDIANS (order Myxosporidia) are certain parasites of fish which are of importance to inland-water fisheries. The developmental processes of these species have been studied in detail. As an example of such types, *Myxobolus pfeifferi* (see Color plate, p. 107) has been selected. It causes boil disease in barbs. This species lives as a tissue parasite in almost all its host's organs. An amoebalike germ cell ("amoeboid germ") which has penetrated the host starts growing; its nucleus multiplies by division but its protoplasm does not divide. Such a multinucleate association which is not divided by cell walls is termed a "plasmodium" (not to be confused with the sporidian genus *Plasmodium*). Soon, however, some of these nuclei cut off a section of protoplasm around themselves, this leading to the formation of separated cells within the plasmodium. Since these later form spores, they are termed the spore-formation cells (pansporoblasts). The nuclei of the pansporoblasts then divide in a strictly ordered series of divisions until, finally, fourteen nuclei are formed which, with the surrounding protoplasm within the sporoblast, form two spores, each of which has two pole capsules.

These processes are more clearly understood in species in which the pansporoblast forms only one spore. Here only seven nuclei are formed from the original one: two reproductive nuclei, two pole-capsule nuclei, two spore-case nuclei, and one envelope nucleus. The envelope nucleus moves to the thin protoplasmic layer surrounding the mature spore, the two spore-case nuclei lie internally to the two halves of the spore case, the two pole-capsule nuclei with their associated protoplasm form the pole capsules, and the two reproductive nuclei fuse to form the nucleus of the amoeboid germ. Similarly stereotyped consequences of cell division are also found among certain classes of multicellulates. In this case "cell constancy" (see Chapter 10) is the result. Here a "cell geneology" can be made for each species, and, for the myxosporidians, a "nuclear geneology" along exactly the same lines. Since the myxosporidians show such conformity in their development with that of the multicellulates, they can hardly still be regarded as unicellulates. Whether, as in the case of the flagellate *Volvox*, they are independently on their way to becoming multicellulates or whether they branched off early from the roots of the multicellulate geneological bush is a question which has not been answered to date.

The ACTINOMYXIDIANS (order Actinomyxidia) differ from the myxosporidians in that their cell body forms only a single pansporoblast from which eight sporoblasts are formed. Each of these spore-forming cells turns into a spore which contains not two, but three spore capsules arranged in a row of three. The few known members of this order live either in the body cavity of sipunculids or in the body cavity and intestinal walls of tubificids (see Chapter 12).

The MICROSPORIDIANS (order Microsporidia) have much simpler spores than the members of the two other orders of the Cnidosporidia.

Order: Microsporidia

Each spore contains only a single pole capsule which lacks a firm cell wall; it thus appears as a vacuole attached to the amoeboid germ. The pole filament, when not extruded, lies not within the vacuole itself but is wound around it, and, owing to this, it is questionable whether the extrusion process is the same as in the case of the pole filaments of myxosporidians and actinomyxidians. Because of this, the relationship of the microsporidians to the two other orders and thus their inclusion among the Cnidosporidia is a matter of doubt.

The microsporidians' small size (their name means "those with the small spores") has made study of their development very difficult, and their sexual reproductive processes still have not been explained. They have, however, been much discussed, owing to the damage they cause in agriculture as parasites of domestic insects. *Nosema bombycis* causes pebrine (German: *Fleckenkrankheit*; French: *pebrine*; Italian: *Gattina*) in silkworm caterpillars. They attack all the caterpillar's organs, and it usually dies as a result. In the case of light infections, the caterpillar still metamorphoses into the moth, so that even the eggs from this insect can become infected. In silk-producing areas *Nosema bombycis* caused tremendous damage during the 19th Century, as in France, where a loss of more than a billion marks occurred during the period from 1845—when the parasite was first recorded—to 1867. *Nosema*-infected eggs are eliminated today, because the pairs separated for copulation and oviposition are carefully examined for the parasite, and the eggs of infected pairs are destroyed. This method of ensuring that only healthy eggs are used was discovered by the French scientist Louis Pasteur. Because of him, the disease is of lesser importance today.

At the present time, a disease of even greater agricultural significance than pebrine in the silkmoth is *Nosema* disease of the honeybee, caused by *Nosema apis*. It attacks almost exclusively the midgut of the imagoes (adults)—the workers as well as the queen and drones. The spores of the disease agent can be brought into the hive in water fouled by bee feces. The pole capsules are extruded within the gut of the insect, and the creeping amoeboid germ is released. This enters one of the cells of the bee's gut, where it grows and multiplies until new spores have been formed. Continually more intestinal cells are attacked, the same process being repeated. The attacked and dying cells are continually replaced by the intestinal endothelium, and the bee can either recuperate, hold a balance, or succumb during this race between the destruction and replacement of the intestinal cells. Healthy bees defecate outside the hive; even during the winter quiescent period they wait until a warm day allows them to make a "cleaning flight." Intestinally infected bees, on the other hand, defecate within the hive, and healthy bees, which clean the fouled combs with their mouths, become infected through the spore-containing feces.

Since a single bee can contain and excrete millions of infective spores, the disease can spread within a short time to all members of the hive:

the *Nosema* infection of one bee becomes a *Nosema* epidemic. This spread is enhanced by all factors which increase food consumption of wintering bees or which stress the hindgut region: any disturbance during the winter quiescent phase (the drumming of a woodpecker or tits at the hive entrance, the apiarist opening the hive, mice nesting within the hive), and also feeding on dark conifer-needle honey or leaf honey during the winter. Apiarists recognize the onset of the intestinal disease mainly from the formless, fluid, brown to yellow spots of feces on the hive's landing board. A continuous decrease in population instead of a spring increase is also an indication of the infection. Whether the infection is present or not can first, however, be proved by microscopic investigation. About thirty abdomens from collecting bees (not juveniles) are ground to a paste in a mortar, together with a little water; one drop of this examined under the microscope at moderate magnification can show the egglike, strongly refractory and thus light, shiny spores. Single spores are not important; in cases of infection, hundreds are present within the microscope's visual field. This investigation is carried out by departments of agriculture on samples sent to them. Today, effective treatments for pebrine exist, these being added to the food; they can also act as a prophylactic.

The HAPLOSPORIDIANS (Haplosporidia) are a group which included, at the beginning, the most variable forms which could not fit in anywhere else. Some of these forms have since been recognized as fungal types and thus excluded. Others are closely related to the microsporidians but lack a pole capsule. In some, however, a spore cap is present which opens to release the uninucleate germ cell. The species which are obviously members of this group are parasites in various fresh-water and marine invertebrates.

The SARCOSPORIDIANS (Sarcosporidia), whose position within the system was until recently a debatable point, must, on the basis of discoveries using electron-microscopic techniques, be included within the group although their position within it can not be clearly explained. They consist of a tube surrounded by an envelope which is divided into multi-surfaced chambers by numerous dividing walls. The spores develop within these chambers. These "miescherian tubes" are found in the muscle cells of numerous mammals, and also in certain birds and reptiles. The most common are *Sarcocystis miescheriana* in the pig, and *Sarcocystis tenella* in the sheep. It is questionable, however, whether the differences between the two "species" come about only through the differences in the hosts. The approximately 1-cm-long, egglike tube from the sheep can infect mice through ingestion, and, within this host, they form tubes several centimeters long, which are just the same as those of the mouse sarcosporidian *Sarcocystis muris*. When these, with their 13-15-micrometer-long spores, are fed to guinea pigs, tiny tubes only 0.1 to 0.04 mm are formed within the course of 50 to 100 days, and the spores here are only a third to a quarter the size of the original ones!

Miescherian tubes

The experiments show that sarcosporidians can be transmitted by eating raw, infected flesh. The rare appearance of *Sarcocystis lindemanni* in humans could be explained on this basis, but not the frequent infections of sheep with *Sarcocystis tenella* and the fairly common infection of cattle by *Sarcocystis fusiformis*, both of these hosts being pure vegetarians. Another method of transmission, as was the case in *Toxoplasma*, must also be present. The discovery of this could, perhaps, as in *Toxoplasma*, explain the complete developmental cycle and thus the sarcosporidians could be ordered within the system on the basis of relationships. Nothing is known of damage caused to man by sarcosporidians. Severe infections in pigs and sheep, on the other hand, can cause damage, especially in reducing the value of their meat. Severe infections in mice are lethal.

The position of the PIROPLASMIDS (Piroplasmida) within the system has not been fully explained. Their present inclusion among the Sporozoa is a matter of compromise, since no spores have been found. They are parasites within the blood cells of mammals and also within the salivary gland cells, intestinal wall, and other organs of ticks. The piroplasmids remain unchanged, no matter whether their host is a mammal or a tick; they therefore have no alternation of generations but simply a change of host. Sexual forms are unknown. Piroplasmids cause severe (in many cases, lethal) diseases in domestic animals, especially in tropical countries. Two families can be distinguished on the basis of their modes of life, the theilerians and the babesians.

Piroplasms

Parasites without an alternation of generations

The THEILERIANS (family Theileridae) reproduce only within the white blood corpuscles of mammals and the salivary glands of ticks. In contrast to the haemosporidians, their multiplication always occurs by binary division, never by multiple division. The tick's ovaries are not infected, and thus the disease does not get passed on to the next generation. The most dangerous species is *Theileria parva*, which causes East Coast Fever in African cattle, which appears as a devasting epidemic. The main transmitter is the tick *Rhipicephalus appendiculatus*. About half of the African cattle infected recover from the disease. A slight infection has a lifelong effect; however, it also protects the animal against further infection. European cattle which have been introduced usually do not survive an attack of East Coast Fever.

Theilerians

The BABESIANS (family Babesidae) have a different mode of life from that of the theilerians: they reproduce within the red blood corpuscles of mammals and, apart from this, within the salivary glands and other organs of ticks. They therefore infect the eggs and subsequent generations of the infected tick. Some of the transmitting tick species are "unihostal," remaining on the same individual host during their larval, nymphal, and adult stages. They are therefore unable to transmit the babesians to another mammalian host. Without the ability to infect subsequent tick generations, it would be impossible for the babesians to spread to new hosts.

Babesians

Babesian infections of
domestic animals

The babesians reproduce within the red blood corpuscles in the same way as the theilerians within the white ones: by binary division, although quadruple division also occurs. The blood corpuscles are destroyed by this; their red color substance, hemoglobin, is thus released and excreted in the urine. Blood in the urine is thus a characteristic symptom of severe babesian infection. It is found in all domestic mammals: cattle, goats, sheep, horses, pigs, dogs, and cats, and also in numerous wild ones, especially rodents and bats. The most dangerous species is *Babesia bigemina*, which causes Texas Fever in cattle, which is not restricted to Texas but is also found in Africa and all the warm countries of the world. Indigenous cattle are usually immune owing to previous infection; European cattle introduced to these areas, on the other hand, become severely infected with babesians, and many of them die. *Babesia canis* is also widespread in warm countries, whose indigenous dogs are usually immune owing to early infection; to introduced pedigreed dogs, however, it is often fatal. This disease is transmitted in the U.S.A. by the tick *Rhipicephalus sanguineus*, and elsewhere by others, including *Dermacentor marginatus*, which is also found in the warmer European countries. Despite the large number of mammalian hosts, neither piroplasm family has been found in man.

To complete this chapter, here are a few tips regarding getting to know the unicellulates from personal observation. One requirement is, of course, a microscope. Obtaining the unicellulates themselves is no problem. Every infusion contains innumerable individuals and a wide range of species which change continually during the course of time. The surface growth on plants and posts in ponds, in addition to the surface layers of mud on the bottom, which can be brought home in a Thermos flask and emptied into a Mason jar, provide other species of unicellulates over a long period of time. Samples collected during the winter require a few days before the warm room temperature causes them to emerge from their quiescent, resistant phases. Even parasitic unicellulates can be found by someone searching for them, since hardly any species of animal remains uninfected. One can almost always find the gregarines in mealworm intestines or in the seminal vescicles of earthworms, or the opalines in the rectum of European frogs. Anyone who has turned his attention to the unicellulates will be surprised time and again by their omnipresence, their large number of species, and the multiplicity of forms which can be seen in all their variations by means of the microscope.

5 Mesozoans and Sponges

The MESOZOANS (subkingdom and phylum Mesozoa) consist simply of a tube of cells. This surrounds the reproductive cells. These animals do not have an internal cavity which takes up nourishment. Asexual parasitic forms alternate with sexual, reproductive, free-swimming forms. The latter are always ciliated. The two orders can be distinguished from one another by their asexual parasitic phases: Members of the ORTHO-NECTIDS (order Orthonectida) are, during this phase, multinucleate protoplasmic masses in which the protoplasm thickens around the individual nuclei. These nuclei, together with the surrounding proto-plasm, divide and develop into the sexual animals. In the DICYEMIDS (order Dicyemida), in contrast, the parasitic phases are ciliated as well. The orthonectids are parasitic in various marine animals: flatworms, annelids, mollusks, and brittlestars; the dicyemids parasitize the bladder of cephalopods.

The position of the Mesozoa within the animal kingdom is not known. Their parasitism of extremely ancient marine animals suggests their great age, evolutionarily speaking. Their simplicity of form, however, is no proof of their primitive origins: parasites, in general, tend towards such simplification. Relationships to the Myxosporidia (see Chapter 4) are suggested in that during reproduction these cross the borders into multicellulism.

Another extremely simple animal should be mentioned here: *Trichoplax adhaerens*, consisting of two layers of cells with external cilia, enclosing a loose cell tissue. This animal creeps about in sea water as a plate of continually changing shape, and reproduces only by division, sexual reproduction not yet having been observed.

When one is not concerned with bath sponges either as a dealer or diver, one must—as the scientist A. Thienemann once said in a different connection—devote oneself to details without losing sight of the whole, if one is to concern oneself with animals which appear to be so unattractive. Whether sponges were even members of the animal kingdom was a

Subkingdom and phylum: Mesozoa, by P. Rietschel

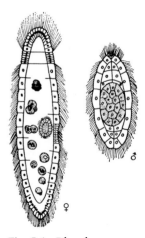

Fig. 5-1. *Rhopalura*: Sexual individuals. The swarmers develop within the female, the testes within the male.

Subkingdom and phylum: Spongia, by E. F. Kilian

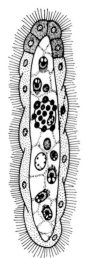

Fig. 5-2. *Dicyema* (Dicyemida): asexual form, ciliated cells externally, large axial cells internally, their cell nucleus a little below center, above this the formation point of the swarmers (according to other authorities, males, five of which are already mature).

Prehistoric sponges, by E. Thenius

Present-day sponges, by E. F. Kilian

debatable question as late as the 19th Century. The ancient Greeks, who were well acquainted with the sponges, called them σπογγοζ, a word from which the Latin *spongia* is derived and which has been adopted by most languages.

SPONGES (phylum Spongia) are divided off from the true multi-cellulates (Eumetazoa) as the subkingdom Parazoa. In these sessile multi-cellulates of usually only slightly differentiated form, the body wall consists of two layers (dermal layer and paragaster); only at the points where these meet is a tissue (epithelium) formed. It is penetrated by water canals originating from pores (*porifera* = hole bearers), which join together into a communal exhalent tube (oscular tube). True muscle and nerve cells are absent. Variations in cell form are present but no true organs are formed. The body size is from a few millimeters up to 2 m in diameter, but, within a single species, size can vary a hundredfold under certain circumstances. Their coloration sometimes is dull, but some species are bright yellow to deep black. Internal structure can be divided into three main types: the ascon grade, the sycon grade, and the leucon grade (see Color plate, p. 173). Diverse skeletal elements of calcium carbonate, silicic acid, or horn are present. Development can be either sexual or asexual, the first having the typical multicellulate larval form. About 5000 species are known.

Fossil silicious sponges—common sponges and glass sponges—exist from the Cambrian Era. The first calcareous sponges, however, are found from the Devonian. From this it can be concluded that the sponges are a geologically ancient group. They were especially common during the Triassic and Jurassic, where they formed geological deposits. The reefs of the south German Jurassic, called "sponge stumps," and those of Provence in France are well known. Evolutionarily speaking, it is interesting to note that the ancient relatives of the glass sponges, which are now deep-sea inhabitants (Hexactinellida), were shallow-water inhabitants during the Mesozoic.

The now extinct groups PLEOSPONGIANS or ARCHAEOCYTHIANS (class Archaeocyatha) are usually included among the sponges when they are not regarded as an independent phylum. They are known only from their calcareous skeletons, which usually took the form of a double-walled, sievelike, porous goblet. Even as early as the Lower Cambrian, they inhabited a broad ocean belt surrounding the whole planet; they became extinct during the Middle Cambrian.

The systematics of the present-day sponges is based on extremely variable skeletal elements and is not yet fully clarified. Three classes can be distinguished: the calcareous sponges (Calcarea), the glass sponges (Hexactinellida), and the common sponges (Demospongia). The most important orders of these classes are the Homocoela, Heterocoela, Tetraxonida, Cornacuspongida, and Dendroceratida.

There is no doubt that sponges are true multicellulates; there are no living or fossil forms, however, which can act as guides in determining from which other group of animals they arose. Similarly, there is no proof that other multicellulates arose directly from the sponges. Accordingly, it is probable that the sponges must have derived from an original, but as yet unidentified, common gastrula larva and developed as a side branch of the great multicellulate geneological tree. Since they were not capable of forming a central nervous system, they could not develop further than their basic form. As mentioned previously, the sponges are put in the phylum Parazoa, next to the true multicellulates (Eumetazoa), this being the opinion of most zoologists. There are grounds, however, for dispensing with this division.

The larval forms of sponges, especially those formed like gastrulae, do not differ in any way from those of other multicellulates. The sponge gastrula larvae, however, attach themselves by the mouth, while those of coelenterates do so by the opposite pole. Present-day sponges probably originated from free-swimming forms which initially had no skeleton but then became sessile, like the glass sponges. Sponges have sometimes been regarded as a community of choanoflagellates (see Chapter 4) and rhizopods (see Chapter 4); but certain findings contradict this theory. The cell-differentiation microstructure, visible only under the electron microscope (flagella, mitochondria, Golgi apparatus) is similar in choanoflagellates, rhizopods, and sponges, but this also holds true of the same structures in the more highly evolved multicellulates with which the sponges, as recent investigations have shown, have much in common in their protein composition.

Multicellulism associated with a division of labor appears for the first time in the systematics of the animal kingdom in the sponges—disregarding the Mesozoa. From this point onward, all the volumes of this work are concerned with animals whose cells are specialized, to a greater or lesser extent, for definite physiological tasks. In almost all cases, however, the cells retain the ability to metabolize by obtaining energy through the combustion of foodstuffs and, during their developmental stage, at least, to divide. Cells of mature sponges can be divided into two clearly differentiated layers: an internal layer (gastral layer) and a wall layer (dermal layer). In the dermal layer all cells are still capable of amoeboid movement. Their shape and tasks are extremely similar to those of amoebocytes. They creep over the whole sponge by means of pseudopodia of various lengths and are capable of moving approximately their diameter's distance in five minutes.

Cells of similar appearance which still contain food reserves (yolk granules) during their juvenile phase, can develop into any of the other cell types. In fresh-water sponges, this juvenile stage lasts about three to four days. These cells are thus termed primordial cells (archaeocytes, from the Greek ἀρχαῖορ = age-old, original). It is not certain whether the migratory and primordial cells are two different cell types or whether

Position within the animal kingdom

Fig. 5-3. Sponge cells. Young amoeboid cell with yolk granules.

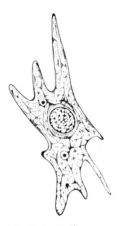

Fig. 5-4. Collenocyte.

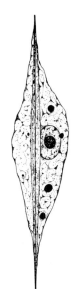

Fig. 5-5. Spicule-forming cell.

Size and form

this is a question of two different functional states. The primordial cells are similar to the collenocytes (from the Greek κολλα = glue) which usually have long, thin, protoplasmic processes and no nucleolus. They form a sort of connective tissue within the sponge (Fig. 5-4).

The spicule cells are concerned with the formation of the skeletal elements (scleroblasts, from the Greek σκληρός = hard, βλαστη = germ; see Fig. 5-5). The cells of the covering layer (pinacocytes) are flat, only 0.02 to 0.05 mm in size, and have no special protoplasmic inclusions; they form an epithelium. This is the only true tissue in sponges, and serves to limit the body externally. The sexual cells of sponges are indistinguishable from those of other multicellulates. The most highly differentiated cells are the collared flagellate cells (choanocytes, from the Greek χοατπ = cone; Fig. 5-6). They line the tubelike, sacklike, or spherical internal cavity (gastral layer) and cause the water currents within the sponge. They are the same in shape and size as the previously mentioned choanoflagellates, which are unicellulates, and are also similar to them in microstructure. The pear-shaped cell body is 0.005 to 0.01 mm in size, and from this arises a flagellum with a basal granule, which reaches a length of 0.04 to 0.06 mm and performs wavelike movements. The lower part of the flagellum is surrounded by a delicate transparent collar whose structure, under the electron microscope, shows it to consist of thirty to forty-five rods. Apart from these, tuft cells (lophocytes) and contractile cells (myocytes) have also been described from sponges; these could, however, simply be connective-tissue cells which have taken over special tasks. Between the cells is a gelatinous to fluid layer. Amoebocytes wander through this layer, which in many sponge species has been shown to contain fibers which are very similar in their chemical consistency and microstructure to collagen fibers of other multicellulates.

Body size within a single species is much less restricted in sponges than in any other animal group. Individual sponges growing next to one another can grow into each other to such an extent that a new organism is formed; in shape it is completely independent of either—or even of other—basic shapes. Consequently, zoologists have often pondered to what extent sponges are individuals or colonies. In their purest form, these two concepts are either inapplicable to a single sponge, or are only temporarily applicable. In addition, another phenomenon, unique in the animal kingdom, is also present: in completely mature sponges, all cells can be separated from one another by chemical processes or sieving through a fine gauze and then mixed with those of another animal of the same species; they then arrange themselves into a new organism with all the characteristics of the species in question. This ability makes sponges, among others, preferred objects for investigation by scientists since they promise diverse insights into the laws of developmental physiology.

Although the size and shape of sponges varies more or less, depending on the species, all animals of this branch can be traced to the three previously mentioned basic types on the basis of their microstructure. Starting from the gastrula larva and stretching it along the axis of the invagination, the external shape of an ascon form is obtained. Young calcareous sponges of the species *Leucosolenia coriacea* (see Color plate, p. 173) are extremely similar to this. The body wall is permeated by numerous pores, these being structures formed by special cells (porocytes). The water flows through these pores into the internal cavity which—with the exception of the pores themselves—is closed and lined with collared flagellate cells. The water stream produced by the flagella leaves the sponge through a single wide opening (osculum) at the uppermost pole. The whole organism is supported and attached to the substrate by triaxial calcareous spicules. Between the plate cells of the outer layer and the collared flagellate cells of the interior are found all the types of sponge cells previously mentioned. Only a few of the calcareous sponges are constructed along these primordial lines, and they are never broader than 2 mm and only a few millimeters high.

Within the same class CALCAREOUS SPONGES—and only among these—are found examples of the slightly more advanced sycon grade (from the Greek οὖπου = fig, referring to the shape of the body, see Color plate, p. 173). The wall of the still more or less cup-shaped body is thicker; the collared flagellate cells are restricted to "radial tubes" which direct the water into the internal cavity like pumps. The sycon grade also remains small and is rarely 10 cm or so high.

The formation of a rootlike, branched canal system containing the "pumping mechanism," the collared flagellate cells, made an increase in size possible, limited only by static relationships. This system is found in the leucon grade (from the Greek λευκός = white, after the old generic name *Leuconia*; see Color plate, p. 173), in which a typical "spongy" structure is developed between the external layer and the interior. Canals leading to the interior are branched into a capillary system and ensure that each individual flagellated chamber remains in contact with the exterior. From the flagellated chambers, a similar system leads in the opposite direction, coalescing more and more until it finally ends in the communal exhalent opening (osculum). The majority of sponges are of this type, even the calcareous sponges; the largest forms reach more than 1 m in diameter.

The form and function of a flagellated chamber can be especially well observed in fresh-water sponges (see Fig. 5-8). These sponges can be grown in the very flat space between a slide and cover glass. The flagellated chamber here is a basketlike structure of thickly compressed collared flagellate cells which form a hexagonal pattern like that of a honeycomb where they touch. All their collars and flagella are directed toward the "basket opening" (apopyle) in the wall of the canal tube. There is always only a single wide opening to the exhalent canal. The water enters the flagellated chamber through a few small cracks (prosopyles)

Fig. 5-6. Collared flagellate cell.

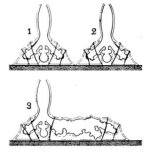

Fig. 5-7. The growing together of two originally separate sponges (1 and 2) to form one individual (3). Semi-diagrammatic figure.

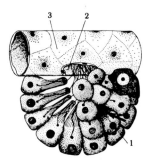

Fig. 5-8. Flagellated chamber of a fresh-water sponge (a number of collared flagellate cells have been removed in the left half): 1. Inhalent opening (arrow), 2. Canal opening, 3. Exhalent canal.

Fig. 5-9. Skeleton of a fresh-water sponge (section). The spicules are attached to one another by spongiolin (stippled).

between the collared flagellate cells. The speed of the flagellar movement is above our discernible limits; the waves move along the flagellum with a frequency of thirty to sixty a second.

A few statistics can illustrate the output of these "pumps": for a sponge of the species *Leucandra aspera* of the family Leuconidae, the number of flagellated chambers in a 7.1 centimeter-high individual has been estimated at 2.25 million, comparable to an area of 52.5 cm². The daily output was measured at 22.5 l. A sponge of the species *Suberites domuncula* has, for example, a daily output of 12 l from a volume of 60 cc. Fresh-water sponges of the same size have outputs of the same order of magnitude. From an exhalent opening (osculum) 0.08 mm in diameter of the fresh-water sponge *Ephydatia fluviatilis*, I measured a water flow of 4.5 cc in twenty-four hours.

This system of water flow, with low pressure in the inhalent and high pressure in the exhalent canals, would collapse immediately if the body were not supported by a special skeletal system. Certain unicellulates already have protective and supporting structures for the cell body; the substances used by these animals—calcium carbonate, silicate, and hornlike substances—are also found in the sponges, as well as in the rest of the animal kingdom, to a greater or lesser extent. The chemical composition of the skeleton was used earlier as the basis, without exception, for the systematic arrangement of sponges into the calcareous, stony, and horny sponges. The great variety of forms in the individual skeletal elements is exceptionally large, however, and there is no direct relationship to their chemical composition. For example, simple needles can consist of calcium carbonate, silicate, or hornlike spongiolin.

Although the majority of species have a very uniform and poorly characterized appearance, the skeletal elements are nearly always highly differentiated structures. They are termed "spicules" (from the Latin *spiculum* = spine) or sometimes sclerites (from the Greek σκληρός = hard). The spicules are formed within a single cell (scleroblast) or by several cells together. In calcareous sponges, their shape is derived from crystal axes and faces; calcareous spicules consist of about eighty-five percent calcium carbonate ($CaCO_3$). It must be imagined that the living cell "gnaws" off all parts of the crystal which are not part of the genetically determined spicule shape. The silicate spicules consist of about ninety percent silicon dioxide (SiO_2), deposited around an organic axial filament. The laws of spicule formation, in this case, are completely unknown.

For obvious reasons, this section is mostly concerned with sponges of impressive form. Calcium carbonate and silicate spicules can vary in size from 0.01 mm to 20–40 cm, and in the genus *Monoraphis* can even reach 3 m in length. In the BASKET SPONGES (order Cornacuspongida) the silicate spicules are cemented together by spongiolin, which consists mainly of a protein skeleton containing iodine (spongin with 1.5–14% iodine content). Spongiolin can also, with or without inclusions, vary in form from long filaments to a complete skeleton, as in bath sponges. Apart from its supportive function, the skeleton also provides a definite protection against enemies.

The color of sponges is involved either with algae which live in symbiosis with them—in the majority of cases—or with coloring substances. The following coloring substances have been found in sponges: spongioporphyrin, floridine (pinkish-red with violet or green florescence), uranidine (usually yellow, but can easily change into brown to blackish-brown substances), also lipochrome (usually red or yellow); and coloring substances of unknown groups.

All sponges are current or filter feeders. The water current caused by the flagellated cells carries particles suspended in the water and tiny organisms into the sponge's body, if they are capable of passing through the pores in the external skin. These pores are formed of ringlike pore cells and an aperture between the cells of the underlying covering layer. Their size varies among the smaller species and also, within certain limits, in every individual, usually being between 0.05 and 0.2 mm in diameter with extremes of 0.005 and 0.2 mm. They are capable of closing within a few seconds. In this way it is possible to prevent damaging substances from entering the body, at least for a few hours. The individual sponge cells are capable of distinguishing between digestible and indigestible substances. The details of feeding in sponges are not known exactly. Possible foods are bacteria, unicellulates, minute animal and plant plankton, organic particles, and also—as has been proved recently—soluble organic substances.

The path of the food particles can be followed clearly in fresh-water sponges. The particle-containing water flows swiftly through the external skin's numerous pores, which give it a netlike appearance, into the space below it (subdermal space). From here, the food-containing water stream divides and enters the canals on whose walls are the flagellated chambers. Amoebocytes, among others, are found within this inhalent system. They either take up the particles directly on contacting them as they flow through (phagocytosis)—these usually being the largest ones—or they obtain nourishment from the collared flagellate cells. In these, mainly the tiniest particles settle on the outer wall of the collar of these cells—never on the inner wall—and from here they are brought into the cell body by protoplasmic streaming, enclosed in vacuoles and usually extruded within a very short time (a few minutes to a few hours) or passed on to the amoebocytes which have the task of transporting and distributing them over the whole body of the sponge. Digestion within the cell follows the same path as in unicellulates. Indigestible remains are enclosed within an extremely thin protoplasmic envelope and extruded from the cell (especially by the amoebocytes which then move to the walls of the exhalent canals). True excretory organs are lacking in sponges.

▷
Axinella cannabina, a Mediterranean sponge growing in a treelike, branched form.

▷▷ and ▷▷▷
Spirastella cunctatrix, worldwide in distribution, attaches itself preferably to shadowed, rocky walls, especially caves (right). A close-up (left) shows the size of the canals radiating from the exhalent siphon.

▷▷▷▷
Left, from top to bottom: A live, microscopical preparation of the fresh-water sponge *Ephydatia muelleri* grown between two glass plates. Calcareous sponge (*Sycon*). Fresh-water sponge (*Spongilla lacustris*). *Hemimycale columella* is found as a jellylike cushion in caves and cracks, and under stones.
Right, from top to bottom:
The sea orange (*Tethya aurantium*); right, a bisected animal whose cut surface resembles an orange. *Myxilla rosacea*, a fibrous sponge.
Ephydatia muelleri.
Dalmatian sponge (*Spongia officinalis*).

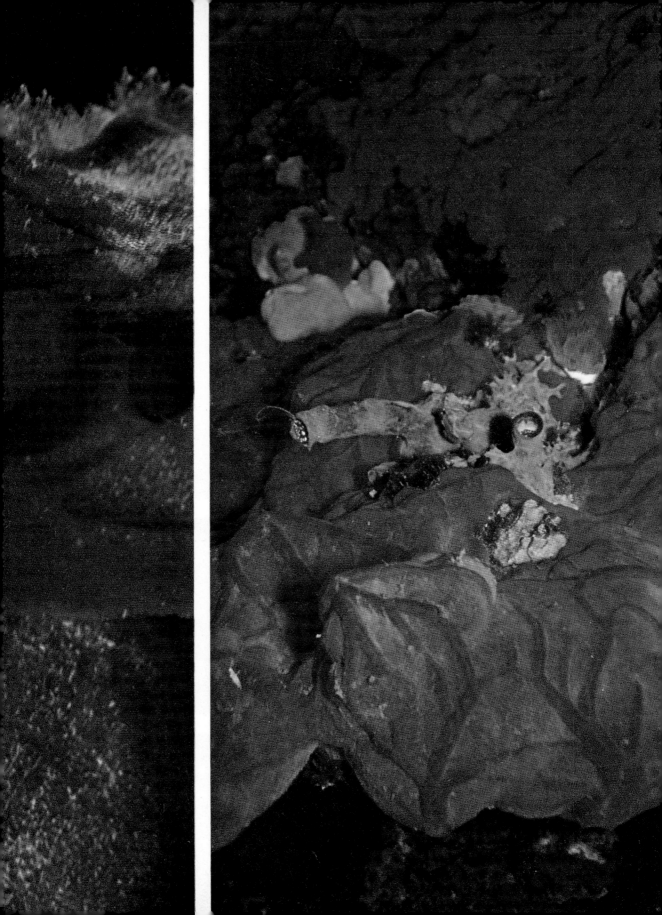

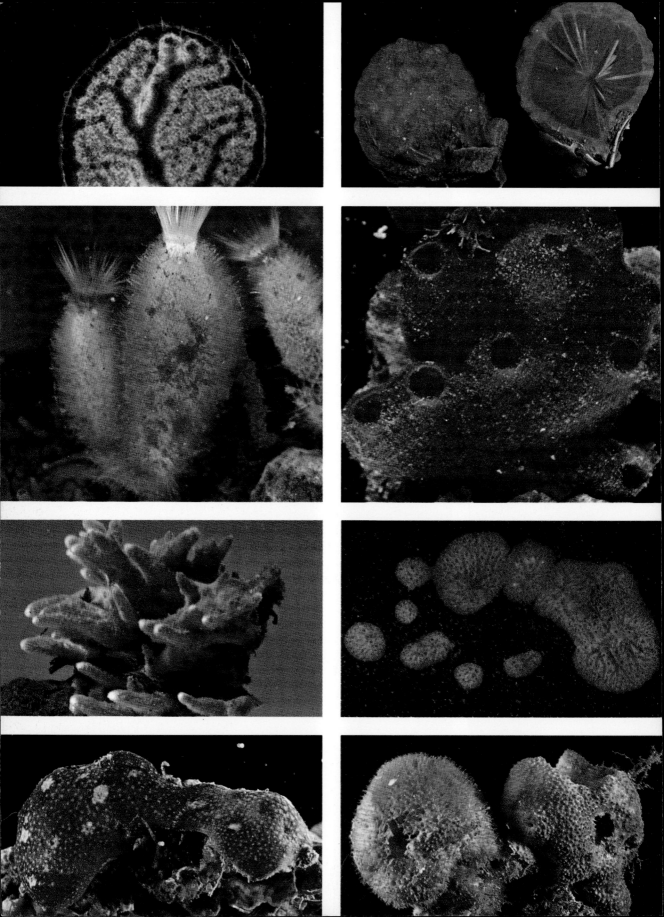

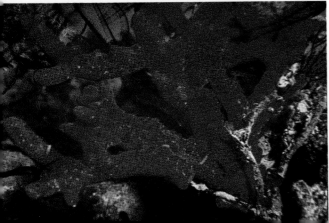
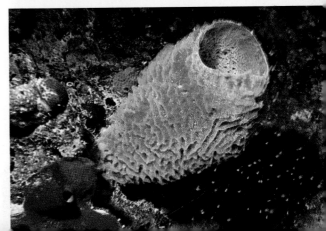

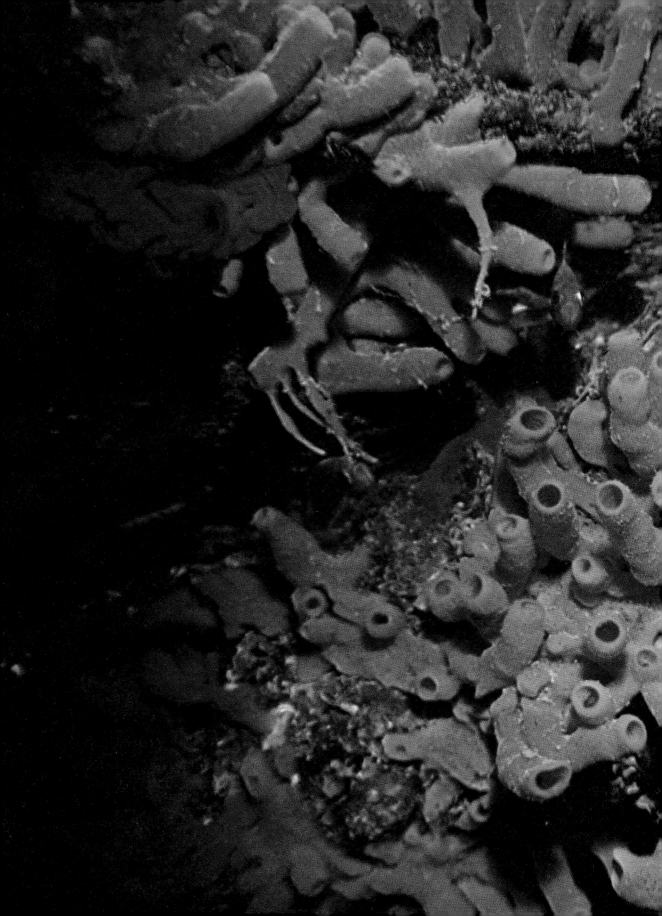

Practically nothing is known about the amount of substance turnover in sponges. It is only clear that fresh-water sponges prefer water with a high nourishment content. Suspended particles in high concentration are a danger to all sponges, since they can cause blockage of the delicate filter system for long periods of time. A continual stream of water is a prerequisite for the provision of oxygen and the carrying away of carbon dioxide. Every individual cell is self-sufficient in this respect, and no special respiratory organs are present. Digestive enzymes are produced by the individual cells; they have been shown to produce amylase, invertase, lipase, erepsin, labenzyme, and tyrosinase. Ammonia has been found as a metabolic residue in the few species investigated with respect to this product (genus *Suberites* among others), but the presence of urea and uric acid has not been proved to date. Some species are poisonous, such as the cork sponge *Suberites domuncula*, probably connected with the symbiotic relationship it holds with the hermit crabs. The juice of fresh-water sponges (*Ephydatia fluviatilis*, *Ephydatia muelleri*, *Spongilla lacustris*, and *Spongilla fragilis*) is deadly to mice (within three to twenty hours) and guinea pigs (in twenty-four to forty-eight hours) when injected intraperitoneally.

Although sponges have no true muscle cells and nerve cells, they respond to stimuli. On mechanical or electrical stimulation and with thermal or light stimuli either the whole body or at least a portion of it is contracted, the osculum and pores first and the canal system last. This movement occurs much more slowly, however, than in animals with a nervous system (a few seconds to minutes). Such responses to stimuli either are an ability of the protoplasm in general or the sponges have a hitherto undiscovered stimulus-conduction system. Young sponges respond to unilateral illumination, performing slow, creeping movements over short distances away from the light, as in the fresh-water sponge *Ephydatia fluviatilis*, among others; or, depending on the species, show a preference for the area of greater illumination by moving toward the light source, as in *Heteromeyenia*.

Investigations in submarine caves, especially in the Adriatic, have shown that many species live only in darkness or at least strongly shadowed environments. Sponges living in commensalism with algae grow in relatively well-illuminated places. Larvae of the calcareous sponges *Sycon raphanus* and *Sycon setosum* swim toward the light. The majority of larvae of the rest of the sponges, in contrast, respond negatively to light. There is proof that this response can reverse at a definite level of illumination. It has even been observed, on occasion, that both fresh-water and marine sponges in aquaria can change their position without any recognizable reason. Several weeks are required to traverse a distance of a few centimeters.

The majority of sponges can reproduce both sexually and asexually. In fresh-water sponges especially, asexual reproduction is of vast impor-

◁◁◁◁
Left, from top to bottom:
Calyx nicaeensis, a flasklike fibrous sponge.
Tropical sponge (Demospongiae) from the Australian coast.
A Red Sea sponge (Demospongiae) from the Jidda region at 5 m depth.
Sponge (Demospongiae) from the Gulf of 'Aqaba at 30 m depth.
Right, from top to bottom:
The calcareous sponge *Clathrina coriacea* covers the substrate like a yellow network. Above, *Spirastrella cunctatrix* spreads itself like a red crust.
Tropical sponge (Demospongiae) from the Australian coast.
Tropical sponge (Demospongiae) from the Australian coast.
Tropical sponge (Demospongiae) from the Caribbean Sea.

◁◁ and ◁◁◁
Tubular sponge colony (*Leucosolenia*) from the Corfu coast.

◁
A sponge (Demospongiae) from the Caribbean Sea.

tance. Our knowledge of sexual reproduction in the sponges is very sketchy. Among fresh-water sponges, for example, both separate sexes and hermaphrodites are known among the various species. Fertilization of the egg occurs within the female's body; the sperm cells are carried to it by the water current.

Within the individual groups of sponges, development occurs in different ways. In the beginning, the first cleavage stage develops from the eggs, and, from this, the blastula. In the calcareous sponges and primitive common sponges, for example *Oscarella*, a ciliated coeloblastula is formed (0.05 to 0.08 mm in size in the calcareous sponges, 2 mm in *Oscarella*), which leaves the female's body; the swimming larva is aflagellate on the posterior body portion. This larva attaches itself by the anterior flagellated pole to the substrate, and a pocket pointing inward is formed at this point, connecting with the blastula lips, which are next to form. In the homocoels, members of the calcareous sponges, the gastrula larva grows upward like a chimney, the invaginated cells becoming collared flagellate cells which form the uniform gastral layer.

The heterocoels also belong to the calcareous sponges. They build a ring of invaginations which become the radial tubes. In some species of the genus *Leucosolenia*, cells already wander from one or several points into the blastula cavity (unipolar or multipolar migration) during the swimming phase. The larva thus has, on settling, a large supply of cells for the formation of the dermal layer.

Preparations for the formation of the dermal layer are even more advanced in many species of Cornacuspongida. Here a sereoblastula (complete larva with a diminished cavity) is formed within the female sponge, from which an evenly flagellated larva (parenchymula larva; up to 0.8 mm in size) is produced, already containing amoebocytes and skeletal elements. Flagellated chambers are even present in the parenchymula larvae of fresh-water sponges (about 0.75 mm in size), by the time they leave the maternal sponge. The sponge larvae usually settle after a swimming period of twenty-four hours; this swimming phase can, however, be extended to from two to five days. The complete development from larva to juvenile sponge lasts only about three days.

In asexual reproduction, either buds of varying shape or rootlike runners are formed, these then separating from the original animal; resistant buds (gemmulae) are also present in the majority of fresh-water sponges and certain marine sponges, for example the cork sponge *Suberites ficus*. Such resistant buds are lumps of body cells which cannot be differentiated from one another, surrounded by a hard spongiolin sheath. The resistant buds of fresh-water sponges are highly variable in form; they are used in the identification of individual species and are a great aid in systematic determinations.

Very few observations are available on the rate of growth of sponges; they are not often comparable since size relationships come about through

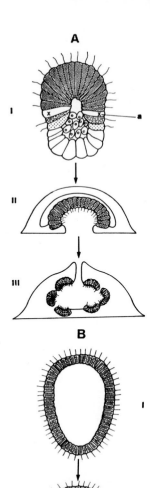

◁
Fig. 5-10. The meta-
morphosis of a calcareous
sponge larva:
A. In *Leucosokenia*, *Sycon*,
and related genera.
B. In *Clathrina*: I. Blastula
with aflagellate
blastoderm cells (a–x);
II. Paremchymula larva;
III. Sedentary larva (early
gastrula stage); IV. Young
ascon form.

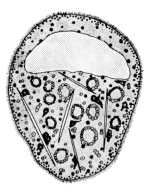

Fig. 5-11. Swimming
larva of a fresh-water
sponge.

Fig. 5-12. *Lophocalyx
philippensis* with buds.

Sponge parasites

budding or division of individuals or their combining, which is difficult
to survey. Even water temperature and variations in the amount of
available food play a great part. The less massive representatives of the
ascon and sycon grades in the calcareous sponges grow relatively rapidly.
Sycon ciliatum reaches a height of 35 mm in fourteen days on the Dutch
coast. In the Berlin Aquarium, a *Dysidea spinifera* grew, within a year, from
13 by 10 cm to 32 by 20 cm. Salable bath sponges 30 cm in circumference
require about seven years to grow. Fresh-water sponges arising from a
larva or resistant bud require approximately two months to cover an
area of about 1 cm².

It is equally difficult to determine age and natural death in sponges.
There is evidence that bath sponges 1 m in diameter are fifty or more years
old. The nerve cells, which in higher animals usually stipulate senility, are
lacking in sponges but, despite this, they also appear to die "of old age."
Such a "death of old age" usually has the following characteristics:
the thinning of the central part of the animal increases until holes appear.
The edges can continue to live for weeks or months but eventually break
off in increasingly smaller pieces. There is hardly ever a change in color.
External effects cannot be the cause of this, since other sponges of the
same species can continue to grow in the aquarium at the same time. In
the majority of fresh-water sponges the formation of resistant buds starts
after three to five months, independent of environmental factors, and the
remainder of the sponge dies as a result.

Marine and fresh-water sponges have relatively few enemies. Pieces of
sponges and spicules are sometimes found in fishes' stomachs, but sponges
appear, in general, to be avoided. They are, however, often destroyed by
strong water currents or by snails or other animals crawling over them;
their ability to regenerate, however, is so great that such damage, no
matter how great it is, heals rapidly in the majority of cases. Sponges are
especially resistant to radioactive irradiation. *Leucosolenia complicata* and
Leucosolenia variabilis are able to stand doses a thousand times greater than
man can for fifteen to seventeen days: 100 kr of β-radiation (90 Sr + 90 Y).

Members of various animal groups live in company with sponges,
and it is often hard to determine whether or not this is accidental. Regular
relationships between sponges and other animals are, even today, known in
only a few cases. Nineteenth Century investigators of sponges knew that
larvae of the lacewing *Sisyra*, a green animal about 1 cm long, were present
in and on fresh-water sponges, as were the eggs and larvae of the fresh-
water mite *Unionicula crassipes*, which can be considered a true parasite.
Parasitic mites of the family Halacaridae have also been described for
marine sponges. In the field, but especially in aquarium cultures of fresh-
water sponges of the genera *Ephydatia* and *Spongilla*, annelids of the
genera *Aelosoma* and *Naïs* (order Oligochaeta), only one to a few millimeters long, are often found; they creep around and about the body of the
sponge, and sometimes even enter the larger canals. It is obvious, however,

that the sponges in question are not very viable. This also seems to be the case in certain predatory ciliates which are found on sponges. Fresh-water gammarids of the family Leucothoidae are known as parasites on fresh-water sponges. Colonies of bryozoans (see Vol. III) and fresh-water sponges sometimes grow closely intertwined; the remarkable Polish investigator of fresh-water sponges, Wierzejski, observed that in this case the sponges thus saved on skeletal material and grew more luxuriously.

Sponges have twice been observed to serve as a breeding locality for fish. In the Chilean lakes a small indigenous perch, *Percilia gillissi*, lays its eggs only in the canals of *Spongilla igloviformia*, and the male guards the clutch until it hatches; on the coasts of the South Florida Keys, the blenny *Paraclinus marmoratus* lays its eggs on or in the sponge *Verongia fistularis*. In both cases, the eggs are protected and continually bathed in a water current.

A specific life association occurs between the hermit crabs of the genus *Eupagurus* and the sponge *Suberites ficus*, which gradually grows over the whole snail shell and serves as an increased living space for the crab. A similar relationship is present between *Suberites domuncula* and the hermit crab (*Paguristes oculatus*) in the Mediterranean. The sponge often continues to grow in the same spiral shape as the snail shell. The association between sponges and certain unicellular algae, which were previously grouped together as Zoochlorellae and Zooxanthellae, is especially striking. A series of observations and investigations is available on green fresh-water sponges but no conclusive explanation has been found. The original theory, that the algae benefit from the sponge's excretory products and provide the sponge with oxygen, could not be proved. Resistant buds (gemmulae) containing algae open faster and at a higher percentage rate than those of alga-free sponges. *Spongilla lacustris* very often is green in color. There are microscopically small algae of the genus *Chlorella* (0.003–0.008 mm in size) within the cells of the sponges, but these are also found free-living in the water. Alga-free sponges can be experimentally infected with *Chlorella* which even reproduce within the cells of their host if it is kept under sufficiently illuminated conditions; green sponges lose or digest their algae after being kept in darkness for a few weeks.

The cosmetic effect of the "badiaga" substance from fresh-water sponges, used excessively, can lead to skin diseases affecting fishermen and swimmers in the Balkan lakes and ponds and in the Amazon basin. In the fresh-water-sponge spicular disease, the microscopically tiny sponge spicules pierce the skin on contacting it through being stirred up by the water currents or blown with the dust from dried-out stream beds; this leads to an extremely itchy skin infection which sometimes forms lumps in the skin which can become worse should infectious agents enter them. The spicules and skeletal particles of the sponge *Drulia (Parmula) brownii*, distributed over the whole Amazon basin, are especially unpleasant,

Fish eggs in sponges

Fig. 5-13. Resistant bud (gemmula) of a fresh-water sponge.

Associations with algae

Fig. 5-14. Uniaxial calcareous spicules.

Class: Calcareous sponges

since they are sometimes spread by the wind from dried-out sponges and irritate the mucous membranes of the eyes and nose.

In Brazil the natives rub fresh-water sponges with fat and flour and give them to dogs in order to rid them of tapeworms; but the dog as well as the tapeworm often dies. It should not be surprising, therefore, that this home cure is sometimes added to people's food for criminal purposes. Fresh-water sponges have drawn attention to themselves in a number of cases in which they cause problems in water supply systems and in water purification establishments by blocking tubes and filters.

Among the marine sponges there are a few West Indian species, known as "fire sponges," which cause a very painful type of nettle rash simply upon being touched. *Fibula nolitangere* is an especially unpleasant member of this group. This gray-brown structure about the size of a human head breaks easily; the irritating effects are due to an unidentified chemical substance. Skin divers in the Mediterranean sponge fisheries fear "Zervos disease"—named after its hardworking investigator, Skevos Zervos. This condition was known about the middle of the 19th Century under the name "Vromo disease" in Crete, and was falsely blamed on the sponges. This irritating skin disease, which can lead to infected sores and burning, is caused by the sea anemone *Sargartia elegans*, which is often found attached to bath sponges and horse sponges.

Many sponges are found worldwide, and it is only a matter of sufficient test probes and observations to determine their distributions. The preferred habitats will be mentioned in the descriptions of the individual orders and families. The sponges, as well as all other marine and inland water life, are being threatened by man through the increasing pollution and contamination of water with chemicals and poisonous substances. Synthetic sponges are a preliminary means of saving the bath sponge.

As among the majority of classes and orders of the invertebrates, varying arrangements for sponge systematics are found simultaneously in printed books. A definitive classification has not been found, since systematists either follow the evolutionary arrangement in as much detail as possible, or begin with more practical differentiating characteristics. An example of this is the class CALCAREOUS SPONGES (Calcarea).

At present, about 500 species of calcareous sponges have been described. In a study appearing in 1963, these are divided into twenty-two genera and the number of species is reduced to forty-six. The skeleton consists of calcareous spicules; in the tri-radiate ones, the angle between the axes being 120 degrees. These skeletal elements lie separately in the tissue. Only in the order PHARETRONIDA—which according to the above work should not be included among the calcareous sponges—are they connected to one another by a calcareous cement. All calcareous sponges are restricted to life in shallow water, and are rarely found deeper than 150 m.

One of the most common species in clear, moving water belongs to the HOMOCOELS (order Homocoela, family Homocoelidae; see Color plate, p. 173), *Leucosolenia botryoides*, an ascon-grade sponge which can vary greatly in form, depending on the environment; the ascons are always between from a few millimeters to 2 cm in height. These are soft, thin tubes, whitish or yellowish in color, which can form a spreading, netlike structure. This species is found at a depth of about 20 m, under stones and especially in small caves and niches from the low-water mark downward, in the Mediterranean and Atlantic, Antarctica, and New Zealand.

Order: Homocoela

Members of the HETEROCOELS (order Heterocoela, family Sycettidae) portray the sycon grade; the collared flagellate cells here are restricted to the so-called radial tubes. *Sycon ciliatum* (see Color plate, p. 148) usually grows as a single individual or as two to four individuals which arise by budding at the base of the original one. The body is sacklike, the exhalent tube lying at the end, surrounded by a circle of long spicules. These sponges, generally 0.5–2 cm, have a bristly surface, and algae and stones are often found attached to them. They avoid sunny places. *Sycon ciliatum* is distributed worldwide, and is found from the tidal zone to a depth of 250 m.

Order: Heterocoela

The LEUCONIDS (family Leuconidae) are members of the same order, and have a leucon grade of structure; they are capable of massive growth, and form either thick-walled encrustations or bulbs, or bushy individuals. *Leuconia nivea* covers the undersides of stones as a thinly spread crust like a lichen, or is attached to shadow-inhabiting algae. The surface of this sponge is smooth and white and has a spread of up to 30 cm, a thickness of 5 mm at the most, and up to fifty oscula. This species is found down to depths of 50 m in the North Atlantic, North Sea, and Mediterranean.

Family: Leuconidae

The bulblike growth form of the leuconids is represented by *Leuconia aspera*. Individuals or pieces are either irregular, egg-shaped, or bulblike in form. They reach a height of about 4 cm. Their coloration varies: pure white, yellowish, golden-yellow, or dark brown. The surface is bristly and pricklelike. This species also appears to be restricted to the Mediterranean and North Atlantic; it is found to depths of 150 to 180 m.

By far the majority of sponges of prehistoric seas belong to the class GLASS SPONGES (Hexactinellida) and the COMMON SPONGES (Demospongia). Despite this, a few species of calcareous sponges, among them the order PHARETRONIDA, existed from Devonian times to the end of the Cretaceous. They have recently been found as "living fossils," with seven species in the Indo-Pacific and one, *Petrobiona massiliana*, in the Mediterranean. This species, varying in shape from a plump finger form to a half-sphere, reaches a size of a few centimeters and is only found in submarine caves. As in all pharetronids, the greatest portion of the body consists of the massive dead, calcareous skeleton which is almost always attacked by boring sponges. The sponge itself is found on the upper portion of the skeleton, like an almost pure-white cap, which penetrates slightly into the interior by means of rootlike canals. The cells of the intestinal and

Order: Pharetronida

dermal layers are not basically different from those of other calcareous sponges. The triaxial and quadriaxial spicules lie free in the much reduced mesenchyma.

Class: Glass sponges

The GLASS SPONGES (class Hexactinellida) are representatives of a world which is now hidden both in time and space; only a few of the sponge investigators have ever seen living specimens. They live in the lightless deep-sea depths, protected from strong water movement and changes in temperature. When brought to the surface in deep-sea trawl nets, the body is completely collapsed, but the skeletal elements suggest an unknown regularity and delicacy of form. The skeletal elements consist of tri-radial silicate spicules with their axes at an angle of ninety degrees to one another, the axes then extending beyond the point of crossing to form a hexiradial spicule. They terminate in simple points, plates, or spicular bundles and are one of nature's works of art.

The majority of the glass sponge species do not branch but retain a tubelike, beakerlike, or flasklike structure for their entire lives, with a more or less extended central cavity and a large osculum. The typical body form is clearly seen in the Venus's-flower-basket. The connective collencytes lie far apart and their processes are longer and finer than those of shallow-water sponges. Plate cells are relatively rare. These white or chrome-yellow, quiet deep-sea inhabitants require no hard outer covering. Their means of reproduction is practically unknown; only the larva of *Farrea occa*, which grows down to depths of 150 m, has been discovered.

The Venus's-flower-basket

The most beautiful of all sponges is the VENUS'S-FLOWER-BASKET (*Euplectella aspergillum*), order HEXASTEROPHPORA, which generally grows to a height of 30 cm and sometimes double this. Like all glass sponges, it is anchored to the substrate by a crown of fine, flexible spicules. It rises like a cornucopia, the long, right-angled spicules, one on top of the other like a chain, look like fine lacework, thus forming a white, cagelike skeleton. Within this open, water-washed structure the sponge tissue hangs like a cloud, soft and coarse-meshed, rarely thicker than 2–3 mm; it forms the wall around the large internal cavity which opens above into a broad osculum. This is vaulted above by a flat, ceilinglike mesh. The wall of the internal cavity bulges outward into numerous raylike, short canals in which the thimblelike flagellated chambers are situated. The Venus's-flower-basket is found in the waters around Japan and the Philippines. It not only looks like a fish trap or cage; it is one, for the isopod *Aega spongiophila* and for the shrimp *Spongicola verusta*, which often are found, protected but caught, within the internal cavity. In the case of the shrimps, usually a pair is present, which the natives take to be a symbol of fidelity in marriage; it is for this reason that the "Venus's basket" is given as a wedding present to young couples.

The Venus's-flower-basket has been known in Europe since the beginning of the 19th Century, and was described by Richard Owen

(1804–1892). The first specimens were an expensive treasure in zoological collections until it was discovered that they formed a carpet on the sea bottom in the vicinity of Cebu, Philippines, and along the Japanese coasts at depths of 200–300 m.

The present-day glass sponges are descendants of the VENTRICULITES (Ventriculita) which were especially common during the Cretaceous, especially in England. This fossil group mainly contained beakerlike representatives, with netted or regularly perforated walls. The genus *Hyalonema* is closely related to them (see Color plate, p. 173); its members are found mostly on the Shetland Islands coasts, the Portugese coast, and Japan's, at depths of 500–800 m. They belong to the order AMPHIDISCO-PHORES (Amphidiscophora). In *Hyalonema thomsoni* (Fig. 5-16), a beaker-like body about 12 cm long and 8 cm in diameter sits on a long "stalk" of spirally arranged spicules which is about 1 mm thick, L up to 40 cm. The shrimp *Spongicola* also inhabits this sponge, and the zooantharid *Epizoanthus* lives on the curving rootlike structure, the latter causing much confusion for a long period of time regarding the zoological relationships of its host. *Hyalonema*, known in Japan since ancient times, is often sold there for high prices as trinkets. One notable fact is that the *Hyalonema* grounds have a large population of sharks about them.

The anchoring methods used by the glass sponges reach their culmination in *Monoraphis chuni* (Fig. 5-17), a species belonging to the same order, dredged up from 1644 m in the Indian Ocean by the German deep-sea *Valdivia* expedition. Apart from small spicules of various forms, thousands of the sponge's cells form a transparent "stake" needle, about the thickness of a pencil, sometimes reaching a L of 3 m, approximately the lower third of which is imbedded in the sea bottom. A more or less irregular spongy body is attached to the upper half, looking like flax on a spindle. How such a relatively unorganized association of cells, as is found in sponges, can be capable of forming a regular structure such as this remains just as much of a puzzle as the laws of formation for all the other variously shaped skeletal elements of silicious sponges.

The class COMMON SPONGES (Demospongiae) contains far more species than any other group of sponges. They are all leucon types in structure. Insofar as the spicules are quadri-radiate, these rays all issue from a common point but do not lie on the same horizontal plane. Usually one ray is longer than the others; it is possible that the commonly encountered uniradiate types may have come about through the complete degeneration of the shorter rays. The spicules are attached to one another by spongiolin fibers and in some genera even the uniradiate spicules have disappeared and the skeleton consists entirely of spongiolin (ceratinous sponges).

The common sponges, like the glass sponges, are a very old group; the majority of prehistoric species, however, are known only from isolated spicules, since the body of the sponge disintegrated on the dissolving of the organic spongiolin. The "stone sponges" (Lithistida) are an exception,

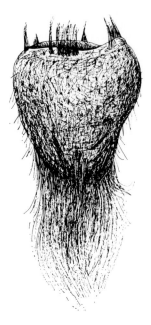

Fig. 5-15. *Pheronema raphanus.*

since their quadri-radiate spicules form irregular, knotty, twigged branches. They are firmly hooked to one another and thus give the body a constant form even without the previously present spongiolin. According to their structure they are tetraxonid sponges (order Tetraxonida), but many paleontologists give them ordinal status. Next to the glass sponges, which also retained their shape owing to their skeletal formation, these sponges were the reef-forming species of the Jurassic and Cretaceous. Even today they are distributed worldwide, especially at depths from 100 to 350 m, although they include far fewer species than do other genera.

Order: Tetraxonida

Fig. 5-16. *Hyalonema thomsoni.*

In the order TETRAXONIDS (Tetraxonida), the large spicules (megasclerites) are quadri-radiate. In certain species the large spicules are lacking, or they are completely aspiculate. The latter is the case in *Oscarella lobularis*. It is soft and meaty in consistency, and forms flabby, 2-6-mm crusts on stones or runners of the Neptune grass (*Posidonia oceanica*). This red, green, blue, or violet-colored sponge ranges in size from about that of a pinhead to that of the palm of a hand; on its upper surface there are manifold twisting furrows similar to those of a brain, and the flagellated chambers lie within their walls. The oscula lie at the top of especially large folds. This species is found on the Atlantic coasts, in Antarctica and the Arctic, and also in the Mediterranean and Black Seas, from tide level down to depths of about 150 m.

As do all the species of GEODIIDS (family Geodiidae), *Geodia cydonium* also forms spherical bulblike growths, the flagellated chambers lying beneath a barklike layer about 0.5 mm thick, containing small spicules. This bulky sponge, found mainly on muddy sand substrates in all seas at depths of 20 to 50 m, is white to grayish in color, and can reach a diameter of 80 cm. *Geodia* is colonized by a whole series of other animals, especially by the red crab *Pilumnus hirtellus*. Owing to its disgusting smell, this sponge has gained a German name meaning "stinking anchor sponge." It is advisable to break up large pieces of this species with a spade, not with the hands.

One remarkable form is the spherical SEA ORANGE (*Tethya aurantium*; see Color plate, p. 148), a member of the DONATIIDS (family Donatiidae). This has a brown to orange color and has short stumpy papillae on its external surface, giving it an amazing similarity—even with respect to its size—to an orange. The sea orange forms resistant buds and is common in all seas on sandy and gravelly substrates, as well as on deep rocky bottoms down to 400 m below sea level.

The representatives of the CHONDROSIIDS (family Chondrosiidae) have neither spicules nor true spongiolin fibers, and only possess fine threads (fibrillae) of an undetermined chemical nature in the barklike external layer. Tiny light-refracting granules (granulae) are present beneath the surface layer. The whole aspect of this sponge is thus viscous and fatty. The KIDNEY SPONGE (*Chondrosia reniformis*), so called because of its shape, is the size of a fist, and is dark brown in color on the side exposed to the light and white on that away from the light. It is found in the Atlantic, Pacific, and Indian Oceans as well as in the Mediterranean at moderate depths down to 40 m. On the Adriatic coast it is eaten by the inhabitants.

Polymastia mamillaris (family Polymastiidae) forms coarse crusts with 1–4 mm-long conelike papillae which are white or pink against a yellowish-orange background. This sponge is also found in almost all the oceans of the world, usually in shallow water; individual animals have, however, been found at depths of more than 1000 m.

The members of the CORK SPONGE group (family Suberitidae) are very widely distributed and vary considerably in shape: they can be crustlike, lobed, fingerlike, or bushlike. They are nearly always conspicuously colored yellow or orange-red. They have a hard but very thin surface layer. Small spicules are absent and the large spicules are uniradial. The cork sponges form resistant buds. The FIG SPONGE (*Suberites ficus*, previously termed *Ficulina ficus*), chiefly grows to the shape which its name connotes, and can reach up to 30 cm in size. It is sometimes found on snail shells inhabited by hermit crabs, this being the rule in the case of *Suberites domuncula*. The two species can be definitely separated only on the basis of their spicule arrangement: in the fig sponge these lie at random in the tissues of the sponge; in *S. domuncula* their arrangement is netlike. Both species have a strong phosphoric odor. Cork sponges are mainly found in shallow water.

The most important role in nature has been played, and continues to be played, by the BORING SPONGES (family Clionidae). The bath sponge is, doubtless, better known, but—as already stated in *Brehm's Animal Life Encyclopaedia* and still valid today—"if the bath sponge had not existed, the world and man would have the same form and possess the same culture as that of the present day. There would be no sponge fishers and the wholesale dealers would not be able to grow rich at the cost of these poor, troubled people"—who then, most probably, would have been even poorer. "If the boring sponges had not existed and worked since ancient times, however, the chalk and limestone layers of the earth's crust, and the coasts of present-day seas consisting of these deposits, would have a completely different extent and form than they do."

On all sea coasts, especially those of the Mediterranean, in all places where calcareous rock occurs, their peculiar raggedness is much admired by present-day tourists. Large and small stones litter the tidal regions and the bottom of the cliffs. Hardly one of the bivalve or snail shells found there is without tiny holes and tunnels, just like the pieces of limestone: this is the work of the boring sponges. There are several species, all belonging to the same family. The larvae, which are free-swimming at first, settle, and wherever they encounter limestone, chalky shells, and skeletons, they set to work to a far greater degree than other marine borers. They attack not only the calcareous structures mentioned above

Fig. 5-17. *Monoraphis chuni.*

Fig. 5-18. Quadriaxial silicious bodies.

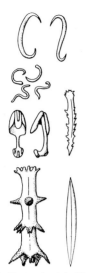

Fig. 5-19. Uniaxial silicious bodies.

Order:
Cornacuspongida,
Family:
fresh-water sponges

Fig. 5-20. Boring-sponge tunnels in an oyster shell.

but also the shells of living animals; one can hardly ever find an old oyster whose outer shell is not somewhat riddled with holes. Mature boring sponges and those which have already bored out the interior of their substrate almost completely continue to grow on its surface as lobular or encrusting masses; these are a few millimeters or centimeters thick. In the boring sponge *Cliona celata* the surface is lemon-yellow to orange-red, and the interior, orange.

Boring sponges are the forerunners for the destruction of extensive coastal areas and coral reefs by waves, one link in the cycling chain of the formation and destruction of marine calcareous deposits. How do these sponges bore? There is a whole series of hypotheses about this, but few observations and experiments. The sponges do not appear to form any appreciable quantity of acid. On the other hand, the water stream issuing from inside the hole containing the sponge carries with it large numbers of calcareous fragments about 0.02 mm in diameter. Their surfaces are more or less rounded, which never occurs in the case of mechanical pulverization. The first microscopic observations during the last few years have shown that the amoeboid cells of boring sponges have long, thin or flattened processes which flow around the calcareous particles and take them into the cell body. Ultimately, however, a chemical dissolving process must occur at very close quarters; this could be compared to the boring of a series of holes in a rockface, the whole of which must be broken away. The "bit" in this case is most probably a pseudopodium containing acid (carbonic acid?).

With the family FRESH-WATER SPONGES (Spongillidae) we shall start with the order CORNACUSPONGIDA. Among these, eighteen genera form typical resistant spores (Fig. 5-13) of fresh-water sponges, while in nine other genera the systematic arrangement is still not certain, since no resistant spores are present or, at least, have not been discovered. In total, about 150 species have been described, of which a whole series should certainly be considered simply as sub-species. The fresh-water sponges, apart from the cases previously mentioned, are of very little practical importance.

Previously, the young ladies in rural Russia used these sponges as a substitute for rouge; they were sold in chemist shops as "Badiaga substance." Sponge powder, when rubbed into the skin, causes a reddening brought about through irritation by the microscopically tiny spicules. *Lubomirkia baicalensis*, which forms large, cakelike coverings in shallow water, is extremely hard when dried. The silversmiths of Irkutsk use it under the name "Morskaja Guba" (lake sponge) to polish silver, brass, and copper objects. A polishing slate has been obtained from the Lower Miocene near Bilin in northern Bohemia, consisting of a 27-cm-thick layer composed almost entirely of spicules from fresh-water sponges. Certain Indian tribes of the Amazon basin add pulverized sponges, mainly those of the large species *Drulia browni*, to their clay during pot-making, to increase the silicon content and therefore the strength of the pots.

Spongilla lacustris (see Color plate, p. 148), distributed worldwide, can serve as an example of the characteristics of this family. This sponge resembles fresh-water plants very closely, because of its usual green color, resulting from an association with algae, and its frequently twiggy mode of growth; it is commonly confused with them. In this species, as in the majority of fresh-water sponges, a penetrating odor similar to that of technical-grade calcium carbide is present. In comparison with other native fresh-water sponges, the body is relatively hard. In spring, new sponge "tissue" grows from the resistant buds which have lain dormant during the winter, filling the old spicule structure, which is rapidly enlarged. Animals reaching 50 cm or more can thus be formed. The larvae, which appear in early summer, first become tiny sponges and serve in the further distribution of the species.

Fig. 5-21. *Drulia browni.*

Ephydatia fluviatilis differs from the aforementioned species of this family because of the presence of sclerites with star-shaped plates at each end (amphidiscs) of the resistant bud envelope. This sponge is also distributed worldwide, and grows more or less as clumps or crusts.

The relatively uncommon species *Trochospongilla horrida* has, as do all the other members of the same genus, reellike amphidiscs. The species *Ochridaspongia rotunda*, from Lake Ochrid, Yugoslavia, is unusual in comparison to other members of the usually shapeless fresh-water sponges. It forms no resistant buds. This apple-shaped sponge usually has a single osculum above the clearly delineated atrium within the skeleton; it is probably a survivor of the animal life which inhabited this ancient lake at the end of the Tertiary.

In commerce, about 400 types of BATH SPONGES (family Spongidae) are recognized. All of them belong to the genera *Spongia* and *Hippospongia*, which contain a maximum total of six species and a few sub-species: 1. DALMATIAN SPONGE (*Spongia officinalis*; see Color plate, p. 148), 2. FINE LEVANTINE (*Spongia officinalis mollissima*), 3. ELEPHANT EAR SPONGE (*Spongia officinalis lamella*), 4. ZIMOCCA SPONGE (*Spongia zimocca*), 5. YELLOW SPONGE (*Spongia irregularis*), 6. HORSE SPONGE (*Hippospongia communis*), 7. VELVET SPONGE (*Hippospongia communis meandriformis*), 8. GRASS SPONGE (*Hippospongia communis cerebriformis*), 9. WOOL SPONGE (*Hippospongia canaliculata*).

Family: bath sponges

The bath sponges are pure keratinous sponges, without spicules and with only spongiolin fibers. The dead fibrous skeleton is used for human and technical purposes. Their general internal structure is that of other members of their order, the Cornacuspongida. The joined, netlike fibers have an average diameter of 0.02 mm, and the much less numerous so-called main fibers which usually run raylike through the sponge, about double this. Tiny foreign bodies (sand grains, particles from other sponges, etc.) are usually deposited within these fibers; the horse sponge, especially, has numerous such coatings spread over the entire fibrous net.

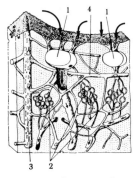

Fig. 5-22. The internal structure of a bath sponge (semi-diagrammatic); arrows indicate the direction of the water currents: 1. Main canal, 2. Group of flagellated chambers, 3. Radially deposited keratinous filaments, 4. Lateral filament net.

Fig. 5-23. Sponge-fishing areas.

Fig. 5-24. *Dendrilla rosea.*

The value of bath sponges is concerned with their large internal surface area (25 to 34 m² in a sponge skeleton 3 to 4 g in weight); in addition, although being relatively firm, they are also soft. They are able to absorb water equal to twenty to thirty-five times their weight (fifteen times more than linen), and these characteristics have attracted man's interest since early times. The use of sponges has been verified from at least the Bronze Age, and Cretan frescoes dating from the Minoan Period (1900–1750 B.C.) have shown that they were used during this time just as paint rollers are used today. In ancient times they were used not only to clean the body, as paintings on ancient vases show, but also to stop bleeding, as compresses for the cervix and boils, as pessaries, and also a type of gas mask as a protection against diseases. The high iodine content (up to fourteen percent of the dry weight) of dried or charred sponges was used as a medicine in fumigation, which was commonly practiced during these times.

During the Middle Ages sponges were used, apart from medicinal purposes, almost entirely by the Church as "Liturgical Sponges"; the High Altar could only be washed down with a sponge. The "sleeping sponge" was impregnated with vegetable anesthetic compounds and used in the easing of pain and calming of patients. Today about seventy percent of the sponges brought up are used in industry: for car care, filtering, sanding and polishing, in graphical businesses, and for painting. Despite the great competition from synthetic sponges, during 1967, 14,400 kg of marine sponges with a value of 1,365,000 DM were imported into Germany, mainly from Greece and Cuba. During 1959, 105 collecting boats with 1186 crew members, half of them divers, were registered in Greece, the land of origin of the sponge divers. Their catch consisted of about 100 thousand kg, and was worth about 2 million dollars.

The history of sponge diving is bound up with the thrill of chance and a great deal of human suffering. Earlier, sponges were simply taken from shallow water by hand or with a trident (Kamaki) from a boat. They were brought up from the deeper fishing grounds by a diver, at that time naked—a method hardly ever used today among the sponge fishers. Weighted down by a marble stone and attached to the boat by a rope, these divers worked at a depth of down to 30 m for two to three minutes. The accident rate was high, and incidents of the "bends" (caisson disease) were common. Victims who survived can often be seen today on the Greek islands, sitting with paralyzed legs by the harbors. Working with trawls is only possible in certain areas, and usually quickly results in exhaustion of the fauna. The Greek divers are prominent not only in the Mediterranean but also in the Caribbean area.

The natural color of a living bath sponge is dark brown to black. Only after they have been dried and bleached, either by the sun or chemically, do they attain the usual yellow-brown tone of commercial sponges.

Experiments in growing sponges commercially were begun as early as the late 19th Century. Such culturing is basically possible if large sponges are cut into pieces and then planted out in the sea again. They are attached to stones or slates. Such cultures are commercially profitable in only a very few cases. Natural bath sponges are still superior to synthetic ones in quality, but it is questionable whether they will be able to keep pace with them over a long period of time. Then another business, full of tradition and romance but also dangerous, will die out.

The BREAD CRUST SPONGE (*Halichondria panicea*), a member of the CIOCALYPTID family (Ciocalyptidae), is the most common sponge along the German coasts, where it grows mainly within the tidal zone. It is variable in form, and usually builds thick crusts with a large surface area. It, and related forms, are mainly responsible for the brightly colored yellow, red, and green surfaces of submarine caves. Its name derives from the crumblike texture of dried sponge pieces.

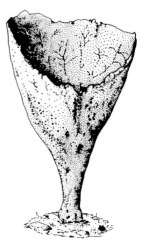

Fig. 5-25. *Phakellia ventilabrum.*

The species *Verongia aerophoba*, from the family APLYSINIDS (Aplysinidae), is especially remarkable owing to its chimneylike processes and rapid change from yellow to blackish-green on exposure to air. In the Mediterranean this sponge often grows between sea grass and on stony substrates; it lives at depths of 2 to 10 m, and reaches a size of up to 40 cm.

The few species of DENDROCERATIDS (order Dendroceratida) have either no supporting skeleton, such as the JELLY SPONGE (*Halisarca dujardini*), or simply a skeleton of treelike branched spongin fibers and the previously mentioned keratinous spicules. These sponges are small types growing in crustlike masses, the treelike branching of the skeleton only being recognizable in dried specimens. *Aplysilla rosea*, about 5 cm high and growing perpendicularly, has a skeleton which particularly justifies the name of the order, this species being rather uncommon but present in most seas.

Order:Dendroceratida

We shall close our description of the sponges with a survey of the reef community. Not only sponges, of course, belong to the reef formers and reef inhabitants; the principal inhabitants are members of the phylum Coelenterata, dealt with in the next chapter, and members of many other groups of animals. Reefs are not only the most richly inhabited marine environments; they are, at the same time, an extremely constant calcareous formation. In the interplay between reef formation and destruction, the living reef is constantly in a state of flux; the dead reef, as a geological entity, has endured through the geological ages.

Prehistoric reef communities, by S. Rietschel

The reef is built by living creatures which—growing on a firm substrate—excrete a calcareous skeleton; we term them generally reef builders. The most important reef builders are the STONY CORALS (Madreporari; see Chapter 6) and CALCAREOUS ALGAE. These form the reef scaffolding, and other animals, such as the OCTOCORALLIA (see Chapter 6) and BRYOZOA (see Vol. III), play a subordinate role. Large numbers of plants and animals of all the marine phyla live within the reef skeleton erected by the reef builders. They are not principally concerned with reef building, and are termed reef inhabitants. Either by boring into the reef, growing upon it, or living free either within the reef cavities or above it, these

Reef builders

animals make use not only of the highly varied and rich food supply, but also of its varied geography.

Reef inhabitants

A reef can be divided into various zones which differ in depth, water movement, illumination, food supply, the animal communities present, and their resulting deposits. According to the position, a reef can be roughly divided into: central reef area (true reef), pre-reef area, and post-reef area. In this way, admittedly arbitrary boundaries are drawn between a series of environments with gradual transition areas. In the post-reef area, the water is usually shallow with very little water movement, restricted to surface or channel currents. Polluted water or water of high salt concentration can occur. According to the general reef form, the post-reef area can contain beach, lagoon, or an extensive reef platform. The central reef area is usually well delineated, and lies near the surface. The seaman's concept of "reef" is based on this area. It is characterized by surf, which cause turbulent currents with great destructive powers. The central reef area usually contains large quantities of rubbish.

In the pre-reef area the reef face drops down into the depths. Rising or passing oceanic currents provide this area with a high oxygen content and planktonic nourishment, used both by reef builders and reef inhabitants, especially in the upper illuminated portion of the pre-reef. In each of these zones, microhabitats of every description occur in large numbers, and are the basis of life for many animals adapted to just these conditions. This adaptation of animals and plants to the micro- and macro-habitats of the reef is shown clearly in their growth forms and skeletons. Typical examples of this are the variety of growth forms found within a single coral species under different environmental conditions. It is even possible to determine previous environments by examining the skeletal growth forms of reef builders from now-dead reefs.

Fringing reefs, barrier reefs, and atolls

Three general reef shapes can be distinguished, depending on their situation relative to the mainland: fringing reef, barrier reef, and atoll. The fringing reef surrounds coastal regions; the barrier reef is along coasts or islands, but at a distance; the atolls are oceanic circles of reef usually formed upon sunken islands, and have a more or less ring-shaped form surrounding a "lagoon." The conspicuous South Pacific atolls are—as Charles Darwin realized—the end result of a reef development starting with an island, at first surrounded by a fringing reef, then by a barrier reef, the island then submerging. They are thus an indicator of a submerging sea bed. Prehistoric reefs can rarely be definitely classified into one of these main types, since they are usually not broadly enough excavated.

If a modern reef is bored through, thus "traversing time" as the bore penetrates deeper, the history of the reef can be determined. Such bores usually show that the reef community has not changed much in the last 30 to 50 million years. During this period, several hundred meters of reef have been built up by animals and algae which only flourish at depths of down to about 40 m. Enormous reefs thus show the movement of the sea bed. It is also clear from such a bore that even during the Tertiary period, corals and algae formed the majority of the reef skeleton, and, as today, snails, echinoderms, and fish inhabited the coral forests.

An exact investigation shows, however, that during the "fossilization" of the reef, only a certain number of the reef inhabitants were involved. Only the reef builders themselves left more or less true fossils. The great majority of the reef inhabitants are not present in dead reefs, since only hard skeletal parts remain. Even these, when not consisting of a single piece which had grown in that form, fragment and are ground into detritus. In addition, at certain depths the skeletal minerals change their crystalline form, and even the calcareous skeletons are changed almost entirely into dolomite (calcium-magnesium carbonate). Only a very imperfect picture of a once rich living community can thus be gained from fossil reefs. Despite this, it is usually possible to recognize individual depth zones and environments of the fossil reef by the growth forms of the reef builders and the type and distribution of the deposits. A short progression through the earth's history can show how and by which organisms prehistoric reefs were formed and inhabited.

The oldest reefs known are from Precambrian times, and are 600 million to a billion years old. These were formed by algae, and it is likely that they already possessed a rich animal fauna. The Precambrian animals usually lacked hard skeletal parts, however, and owing to this there is no real proof of their presence in the algal reefs. Animals first started to be active in reef building during the Cambrian (500 to 600 million years ago). The spongelike ARCHAEOCYATHIDS (class Archaeocyatha) lived, together with the algae, in reef meadows which were distributed worldwide in an obviously climatically favorable belt. The archaeocyathids had a calcareous skeleton consisting of two sacklike walls, one inserted in the other. The porous walls were connected to one another by struts, longitudinal walls, and, in part, also by horizontal floors. It is still uncertain whether these reef builders were true sponges or coelenterates, or whether they should be considered as a separate phylum. They became extinct during the Middle Cambrian.

During the Ordovician (500 to 440 million years ago), coelenterates, as well as true sponges, took part in reef formation. The most important reef builders during this time were the calcareous algae and bryozoa. Within the reef skeleton formed by these, not only SOLITARY CORALS but also COLONIAL CORALS grew and spread, thus aiding in the reef's growth. CALCAREOUS SPONGES and STONE SPONGES (Lithistida) were also present, but formed a large proportion of the reef community in only a few instances. The first STROMATOPORES (Stromatoporoidea) also appeared, becoming the commonest reef builders during the Silurian and Devonian. Prehistoric snails, echinoderms, and CYSTOIDS (Cystoidea) were widely distributed reef inhabitants during the Ordovician.

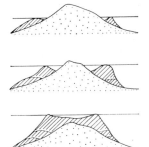

Fig. 5-26. Atoll formation according to the Darwinian theory: an island gradually becomes flooded and the previous fringing reef (above) becomes an atoll (below).

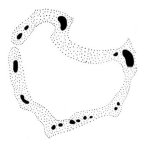

Fig. 5-27. Stewart Atoll: Reef (stippled); islands lying above the reef (black).

Fig. 5-28. *Archaeocyathus* (diagrammatic), Middle Cambrian.

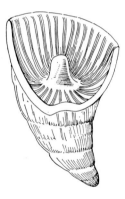

Fig. 5-29. Rugosa coral (*Dalmanophyllum*), Silurian, Gotland Island.

Fig. 5-30. Stromatopore (*Trupetostroma*), Upper Devonian, Belgium.

Expansive and massive coral reefs appeared for the first time in the earth's history during the Silurian (400 to 440 million years ago). Coelenterates and the disputed stromatopores, included here among the fossil coelenterates, were the main reef-skeleton builders, and among them the reef inhabitants led an extremely rich life. By this time colonial as well as solitary corals played a great part and grew to form large coral heads. They also grew on muddy bottoms, and thus afforded other reef builders a firm substrate to attach to. Algae and bryozoa were superseded by the coelenterates in their role as reef builders during the Ordovician but, nevertheless, played a substantial part.

The Silurian corals were members of the now extinct orders TABULATA and RUGOSA. The MADREPORARIA arose from the Rugosa during the beginning of the Mesozoic. The Rugosa had extremely large polyps sitting in single cups or next to one another in extensive heads. The Tabulata, on the other hand, had small polyps which, by budding, usually formed colonies, often of enormous size. Horizontal plates within the cups allowed rapid vertical growth by the colony.

The previously mentioned STROMATOPORES probably also formed colonies. Their skeleton consisted of aragonite and was built of thin layers supported by struts; canals permeated this structure. Since living space was only afforded to small polyps, it was supposed that the stromatopores were related to the Hydrozoa. They existed from the Cambrian to the Cretaceous but their period of maximum growth was during the Silurian and Devonian. Recently, unique living sponges with a skeleton similar to that of the stromatopores have been found in reef caves in Jamaica. New investigations of both groups should show whether stromatopores are really sponges, something which was supposed earlier for a certain period of time. The stromatopores, together with the tabulata and Rugosa, were the most important reef builders during the Silurian. They probably lived symbiotically with algae and, like the Tabulata, colonized various reef depths with different growth forms.

The Silurian reefs were inhabited by a large number of reef animals. Among the mollusks, the snails were the most widely distributed; among the echinoderms, the crinoids (see Vol. III). The cystoids surpassed them all and grew in extensive fields on the reef flanks. Brachiopods (see Vol. III) and Bryozoa (see Vol. III) were found almost everywhere. Among the Crustacea, the ostracods (see Chapter 16), and, among the other arthropods, the trilobites were the most common reef inhabitants.

The Devonian reefs (350 to 400 million years ago) were very similar to those of the Silurian. During Devonian times as well, Rugosa and Tabulata corals, together with the stromatopores, were the main reef builders. The Rugosa were even more varied in form and also more important in their role as reef builders. The stromatopores spread into all reef environments; variation in environments can be determined from the changes in their growth forms. In areas with strong currents, such as the reef edge, they were platelike crusts flattened against the substrate; they were spherical in form in the surf area between the pre-reef and post-reef regions; in areas of good water exchange on the pre-reef and post-reef regions, they grew as "meadows" of delicately twigged bushes, and in slack water areas they grew vertically in skittlelike or funguslike forms. Some of the Silurian genera of the tabulata had already become extinct; the Devonian species also had growth forms similar to those of the stromatopores. The variation in growth form increased still further among the reef inhabitants of the Devonian. The echinoderms had developed to a greater extent, mollusks became more widely distributed in certain reef environments, and the prehistoric sharks appeared as the first vertebrates to inhabit this area (see Vol. IV).

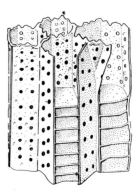

Fig. 5-31. Tabulata coral (*Favosites*), Devonian, Europe.

During the Carboniferous (270 to 345 million years ago) the calcareous sponges started to play an important role in reef building next to the corals, bryozoans, and calcareous algae. Echinoderms, brachiopods, and mollusks left plentiful fossil remains. During the Permian (225 to 270 million years ago) the calcareous sponges reached a peak as the most important reef builders, which lasted until the Triassic. Apart from these, bryozoans and the unique, corallike brachiopods (*Richthofenia*) in particular played a part in reef building. Local reefs, consisting almost entirely of Bryozoa, also occurred. Great variation in reef-area distribution also occurred during the Permian in connection with climatic changes. The crinoids began, in general, to abandon the reefs and colonize other environments.

During the Triassic (180 to 225 million years ago) the ancient corals began to die out to a large extent, and the stony corals to take their place, rapidly developing into prominent reef builders. At the beginning, however, they were secondary to the calcareous sponges. The community of reef inhabitants incorporated all the important present-day groups, differing, naturally, from them in details. In central Europe the Triassic reefs are a well-known part of the landscape, since the Calcareous Alps and Dolomites consist, to a great extent, of these reefs (see Chapter 6).

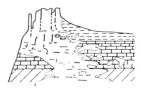

Fig. 5-32. Jurassic sponge reef, southern Germany. The lump of unlayered calcareous reef stone lies in the layered deposits of the reef surroundings.

Even the Jurassic reefs (135 to 180 million years ago) play an important part in certain European landscapes. They form part of the French and Schwäbish Alps and the Jurassic Alpine chain. Coral reefs which appear to be "modern" appeared as a characteristic of shallow waters. Next to the stony corals, calcareous sponges, bryozoa, hydrozoa, and calcareous algae built the reef skeleton together with hornlike mollusks (*Diceras*). Within the reef was a rich fauna of sea urchins, brachiopods, crustaceans, mollusks, snails, and fish. Ammonites and belemnites—reflecting their frequent occurrence during the Jurassic—also appeared, with a few species as reef inhabitants. Apart from the animal groups mentioned, silicious sponges also took part in reef building. Their proportions increased with increasing depth, and the other reef builders more or less disappeared.

In deeper water the reefs consisted almost entirely of sponges, TRI-

Fig. 5-33. Stony coral (*Meandrina*), Tertiary to present, tropics.

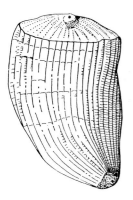

Fig. 5-34. Reef-building mollusk (*Hippurites*), Upper Cretaceous, Austria.

AXONID and TETRAXONID silicious sponges, inhabited by brachiopods, foramenifera, and algae. Fine calcareous particles collected between the sponge skeletons so that the reef grew as a layerless lump between the inter-reef deposit layers. The silicious sponges had few enemies, and when they died their organic parts were imperfectly removed by scavengers. Owing to this, gases released by decomposition precipitated large quantities of calcium carbonate from the water within the reef area, which coated the dead silicious sponges and "mummified" them; one important factor was that, due to the depth, there was very little water movement. A great variety of transition stages was present between the richly inhabited shallow-water coral reefs and the silicious sponge reefs with their monotonous life communities at about 50 to 150 m depth.

During the Cretaceous (70 to 135 million years ago) extensive sponge colonies were also present in deep water. They did not, however, develop into such conspicuous reef formations as did those during the Jurassic. Every sponge colony or sponge-formed structure termed a reef during the Mesozoic would hardly be termed so today, since these are deep-water formations. They are true reefs, however, according to geological terminology, since they are constructed from deposits laid down by living organisms. Extensive coral reefs were present during the Cretaceous, similar in many respects to those of today. One exception was the Rudista reefs, consisting of uniquely formed, corallike mollusks. Reefs consisting mainly of bryozoans were also widely distributed during the Cretaceous. It is possible to show how, since the Jurassic, the coral-reef belt gradually shifted more and more toward the present-day equator. The Lower Jurasic seas in central Europe had moderate temperatures ranging from 17° to 23°C at the surface. The general shifting of the coral-reef belt was certainly due to climatic factors.

During the Tertiary (2.5 to 70 million years ago) the coral reefs, with respect to the composition of their communities, did not differ appreciably from those of today. They can be generally considered as gradual transitions to these modern formations.

The investigation of prehistoric reefs is not only of interest to the paleontologist and zoologist. A knowledge of them is also important in obtaining answers to questions regarding the arrangement of seas and the climatic relationships of prehistoric times (paleogeography and paleoclimatology). Definite conclusions can thus be drawn when it is known that certain reef communities also preferred shallow, warm water during prehistoric times. Under these conditions, prehistoric reefs can draw, within favorable climatic zones, the outlines of coasts, islands, and shallow waters, and also give indications of the equatorial position. This, of course, is not true of all reef communities.

The growth of reef organisms has led, time and again during the course of the earth's history, to calcareous deposits which now play important roles as enormous and extensive structures in the upper geological layers. Petroleum oil and natural gas move into porous sediments and are quite frequently trapped in the extremely porous reef deposits. Reef investigation is thus also of importance in the search for oil. Prehistoric reefs, apart from being carriers of petroleum oil and natural gas, are also of great economic importance as limestone beds.

Present-day reefs are restricted to the tropical and subtropical zones; fossil reefs, on the other hand, can, colloquially speaking, be found on the doorstep of people living in the temperate or even arctic areas. The Silurian reefs are raised once again in the Gotland cliffs, and the steep slopes of the French and Schwäbish Alps consist of those from the Lower Jurassic. Valley walls and quarries in the Ardennes, Eifel, Sauerland, and Harz areas consist of Devonian and Carboniferous reefs, and, in the Alps, those from the Triassic to the Cretaceous. Many of the calcareous reef stones are polishable and are thus used as technical marble. Evidence of prehistorical reefs from many areas can thus be recognized in house fronts, steps, tables, and window ledges.

▷
Sponge body structure (Spongiae):
A. The three basic sponge forms, Ascon (A_1), Sycon (A_2), and Leucon (A_3). The Gastrallager (collared flagellate cells) are ochre-yellow, the Dermallager brown. Thin blue arrows indicate the direction of the water stream through the pores and flagellated chambers into the central cavity, the thick blue arrows, the passage of the water leaving the sponge through the osculum (or several oscula).
B. Skeletal structure: B_1. The silicious spicule arrangement (triaxial, hexaradiate) of a glass sponge (*Hyalonema*), B_2. Triaxial hexaradiate spicule of a glass sponge, B_3. Triradiate calcareous spicule with the cells building it, from a calcareous sponge.
C. Collared flagellate cells: C_1. Three cells from the flagellated chamber of a sponge, C_2. The collared flagellate animalcule *Salpingoeca amphoroideum*, a protozoan very similar in structure.

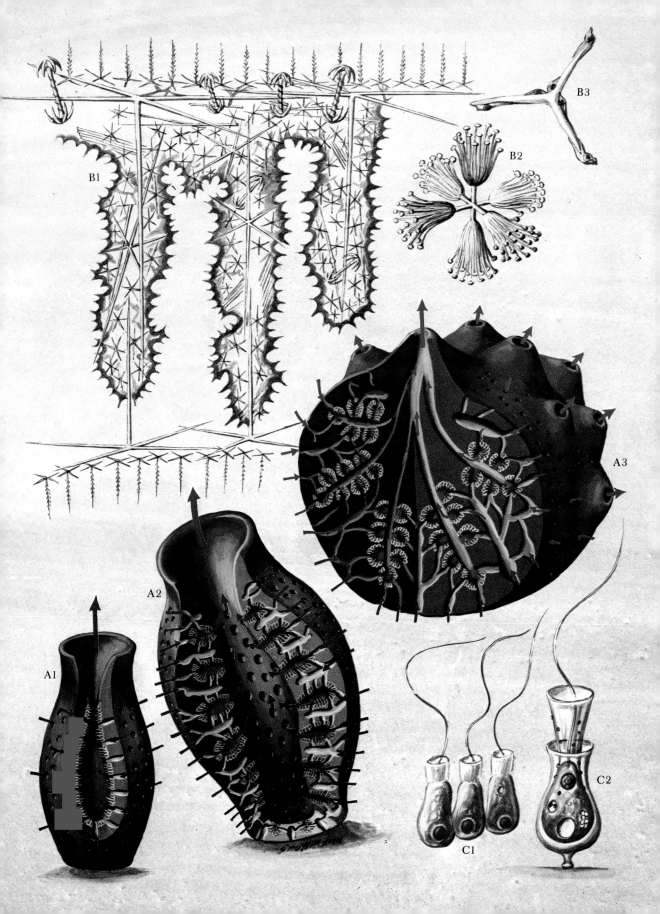

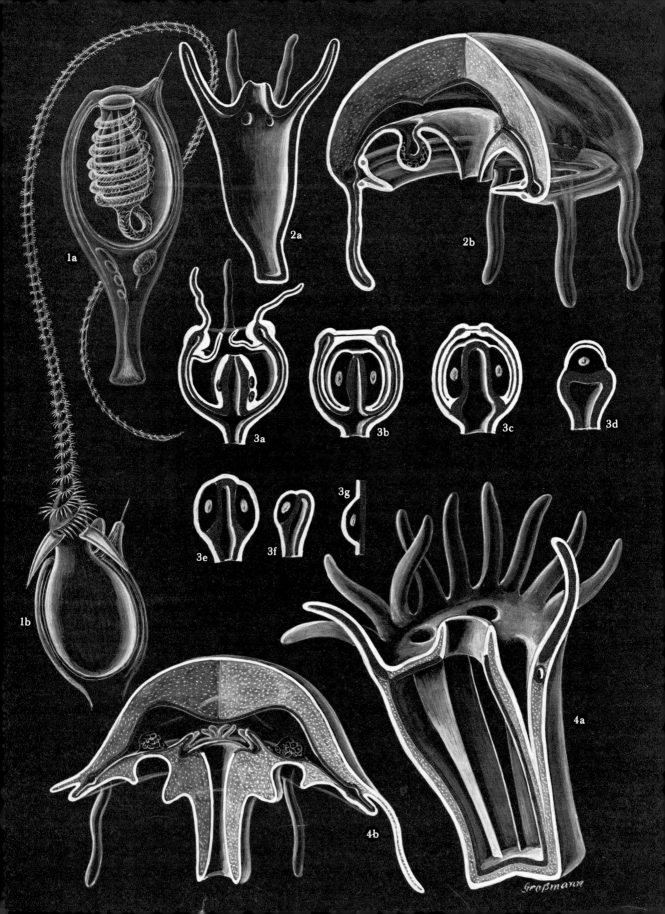

1a 1b 2a 2b 3a 3b 3c 3d 3e 3f 3g 4a 4b

Großmann

COELENTERATE BODY FORMS

1. Nematocyst

1a. Within the nematocyst cell with the trigger (cnidocil). Ready to be extruded with the fiber coiled and inverted. The nematocyst capsule is closed.

1b. Nematocyst discharged, fiber uncoiled and everted. The cap of the nematocyst has been sprung open.

2. Hydrozoan

2a. Hydropolyp with basal disc, trunk, arms (tentacles), oral disc, oral tube, and mouth. 2b. Hydromedusa (Leptomedusa with sexual cells in the radial canals) with umbrella tentacles, velum, oral tube, mouth (directed downward in contrast to the hydropolyp), enteron, radial canals, and circular canal.

3. Sexual buds (gonophores or sporosacs).

3a. Complete medusa not yet released from the polyp colony (anthomedusa with sexual cells in the oral tube). 3b-3f: "Incomplete" medusae with sexual cells. 3b. Eumedusoid. 3c. Cryptomedusoid. 3d. Heterome-dusoid. 3e. Styloid gonophore. 3f. The same with unilateral proto-ovum. 3g. Sexual cells within the wall of the polyp on complete medusal regression (fresh-water polyp).

4. True jellyfish or scyphozoans.

4a. Scyphopolyp: enteron divided into four digestive pouches by four gastral septae, joined to four depressions in the oral disc (septal depressions). 4b. Scyphomedusa: umbrella with a broad enteron, the sexual cells being present in its internal cell layer; umbrella edge without velum; four bunches of gastral filaments within the enteron.

The polyps and medusa 2a, 2b and 4a, 4b are sectioned to show the enteron.

Schematic coloration of structures (not corresponding to the natural coloration): white = ectoderm, yellow = endoderm, red = female reproductive organs, light blue = nervous system.

6 The Coelenterates

Two completely different phyla are included under the name COELENTERATES (subkingdom Coelenterata or Radiata): the Cnidaria and the Acnidaria. The representatives of these two phyla have the following in common: their bodies consist of two cell layers (epithelia), the external covering of the body (ectoderm), and the layer lining the interior of the body (endoderm). Muscle cells are mostly formed from ectodermal cells, although they can also be formed from many endodermal cells; in general, the former develop into longitudinal muscles and the latter into circular muscles. The sudden contraction which can be observed in many coelenterates is concerned with coordination between these muscle fibers. Other epithelial cells have developed into nerve cells which are joined to form a nerve net stretched over the whole body. A supportive substance is found between the two layers of cells, capable of developing into an additional, third, cell-containing layer (mesogloea) by infiltration of ecto-dermal cells, or, more rarely, endodermal cells. The body has only a single cavity which is commonly divided into subcavities, and a single body opening which serves as both the mouth and the anus.

The coelenterate body form can best be understood in connection with individual development (ontogeny). After fertilization of an ovum by a sperm cell, a compact, spherical cell mass develops by complete and fre-quent division of the yolk. It resembles a mulberry and thus is termed a morula (from the Latin *morula* = mulberry). A cavity then develops within this sphere, so that a hollow ball is finally formed, the walls of which consist of a single layer of cells. This stage is termed the blastula. In some species the blastula is already free-swimming. The blastula then extends to form a long oval, and, at the same time, a second cell layer is deposited within it by various means, lying tightly against the external cell layer (ectoderm) and forming the internal cell layer (endoderm). Finally an opening appears at the so-called vegetative pole, the pros-tomium. The cavity enclosed by the two cell layers (enteron) takes over the function of both the stomach and the intestinal system. The structure

Subkingdom: Coelenterata, by H. R. Haefelfinger

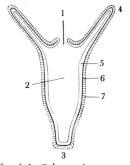

Fig. 6-1. Schematic coelenterate: 1. Mouth opening, 2. Eneteron, 3. Basal disc, 4. Tentacle, 5. Endoderm, 6. Supportive layer, 7. Ectoderm.

consisting simply of an external and internal cell layer is generally called a gastrula but, in special cases among the coelenterates, may also be termed a planula larva. This free-swimming individual usually settles after a certain period of time; small processes, the capturing arms or tentacles, are formed around the prostomium. Among certain coelenterates, the planula larva remains free-swimming, however, and while swimming metamorphoses into the actinula larva, also by tentacle formation. The basic coelenterate type is represented by these two similar developmental stages; from these the numerous other forms present among the about 9000 species of the phylum Cnidaria develop.

Evolutionary development, by E. Thenius

Fossil remains have been found from the Hydrozoa, jellyfish (Scyphozoa), and sea anemones (Anthozoa) among the Cnidaria. The STROMATOPORES (Stromatoporoidea, genus *Stromatopora*; from the Ordovician to the Devonian) are usually included among the Hydrozoa. They were the most prominent reef builders of prehistoric times, and had a massive calcareous skeleton. The most variable growth forms (bulbous, banklike, and twiglike) can be distinguished among them. During the Mesozoic—in the Jurassic and Cretaceous—the SPHAERACTINOIDS (Sphaeractinoidea, genus *Sphaeractinia*) outnumbered the hydropolyps while the hydrinids (Hydrina) became more common during the Tertiary without, however, reaching the same level of importance as the extinct groups. The jellyfish, whose presence during the early Precambrian was proved by fossil imprints, gives no evidence of a definite evolutionary arrangement; it appears that, in the prehistoric CONULARIA (Conulata), we may be dealing with the sessile phase of a free-swimming jellyfish—an interpretation which, at present, seems rather improbable.

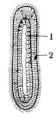

Fig. 6-2. Planula larva (section): 1. Endodermal cells, 2. Ectodermal cells with cilia.

The CORALS are well represented by fossils. The best-known fossil history is that of the skeletized group, the Rugosa or Pterocorallia, and stony corals (Madreporaria or Scleractinia) among the hexacanthid corals (Hexacorallia), and the Tabulata. The hexacanthid corals first appeared during the Ordovician as rugosa corals. Originally, the rugosa were solitary animals, later becoming colonial. They represent the prehistoric stony corals and reached their peak in species number and form variation during the Silurian and Devonian—for example *Streptelasma* from the Ordovician and Silurian, *Calceola* from the Devonian, "*Cyathophyllum*" = *Hexagonaria* from the Devonian and Carboniferous, and *Lithostrotion* from the Carboniferous and Permian. The massive forms played a part in reef-building. A more or less definite bilateral symmetry is shown in the sometimes pinnate arrangement of the intra-cavity walls (septa).

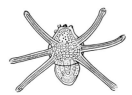

Fig. 6-3. Actinula larva.

The rugosa became extinct toward the end of the earth's prehistory and were replaced during the Tertiary by the stony corals, which, even then, were present as reef-builders and which today are numbered among the most important of the reef corals. As a result of the regular arrangement of the septa, the skeleton of individual animals (polyps) has a radial symmetry. Opinions are divided with regard to the evolutionary origins of the stony corals; certain investigators consider the rugosa to be their precursors; others believe they developed from an unknown, skeleton-less ancestral form. An evolutionary developmental series which started during the Triassic led, in the Upper Cretaceous, to an enormous number of species and forms which was never quite reached again during the Tertiary.

Whereas the rugosa were distinguished in the skeletal formation of individual animals by the laying down of complicated and additional skeletal elements in the septal region (trabeclulae), the stony corals are distinguished by an increasing porosity of wall and septa as well as a degeneration of the septa to ridges or spikes together with an increase and strengthening of the hard tissue between the individuals (coenesteum).

Among the OCTOCANTHID POLYPS (Octocorallia, Alcyonaria) the presence of SEA PENS (Pennatulacea) has been proved as early as the Precambrian; their systematic relationships are, however, still questionable.

The most important characteristic of the CNIDARIA (phylum Cnidaria) is definitely their capture and defense organ, the nematocyst, which is formed in enormous numbers by the ectoderm. Cnidaria can be solitary or colonial. They appear in two forms, the more or less sessile polyps and the usually free-swimming medusae or jellyfish which are umbrella- or mushroomlike in appearance. The polyp is tubelike in shape and closed at the end, by the basal disc, by which it attaches itself to the substrate. The body column is formed by the "tubular wall" and terminates at the opposite pole in the oral disc (peristome); the tentacles are formed between these two sections of the body. The mouth lies in the center of the oral disc, which in certain species is elongated to form a trunklike oral tube. While the polyp appears to be "pulled lengthwise," the medusa is compressed along this axis but can still be developed from the polyp form. The body column and basal disc form the umbrellalike, convex outer wall (exumbrella). The oral disc retains its basic form and changes into the concave underside of the umbrella (subumbrella). The stomach is re-modelled in a unique way: the usually supportive layers (supportive lamellae) of the polyps are, in the medusa, tremendously thickened; between the external and internal cell layers of the umbrella walls an enormous jellylike umbrella is formed which compresses the stomach to a great extent. In most cases, only a series of raylike (radiate) canals remains, which runs from the base of the tubular mouth where the "residual stomach" is found to the edge of the umbrella; here they are usually connected together by a circular canal. A unicellular endoderm lamella remains, however, between the radial canals.

Although the two forms—polyp and jellyfish—show certain structures in common with respect to their body form, they are basically different with respect to their modes of life. The polyp—apart from a few exceptions—is always attached to a substrate or can only change its position to a very slight extent. The medusa, in comparison, swims freely (plank-

Phylum: Cnidaria, by H. L. Haefelfinger

tonic) in the water; in part through its own efforts and in part by water currents it is moved from place to place. This difference is emphasized by the massive calcareous skeletons of certain polyp forms, for example, the reef corals; the jellylike medusal structure, which often consists of ninety-nine percent water, stands at the other extreme. Both forms can, however, adapt to certain conditions. Polyps, for example, can become planktonic, while medusae, on the other hand, can become creeping or even sessile.

Both forms are distinguished by their nematocysts. They are extruded from special cells (cnidoblasts) and are among the most highly developed structures extruded by animal cells. Although their main task is to overwhelm prey, they also serve as an effective weapon in defense against enemies. The nematocysts are double-walled, egg-shaped capsules; at one pole, the internal layer protrudes into the cavity and is extended to form a long, hollow tube arranged like a coil of rope. At the point of the protrusion an opening is present which is closed by the outer layer as if by a lid. The capsule remains within the cell forming it and, from this, a fine process (cnidocil) protrudes into the water. The capsule contains a poisonous and corrosive substance, and the tube is often covered with a sticky secretion. The previously mentioned process releases the entire procedure of discharge. The size and form of the nematocysts are so different that they are used in systematics as characteristics. According to their tasks, the following capsule types can be differentiated: penetrating or stenotele capsules, wrapping, volvent, or semoneme capsules, and adhesive, glutinant, or isorhiza capsules.

Spinelike structures, called barbs, are found at the base of the tube in penetrating capsules. When the processes protruding above the ectodermal layer (cnidociles) are stimulated by touch, the following responses occur within 0.003 to 0.005 of a second, without influence by the nervous system: The capsule contracts, the lid above it springs open, and the tube or stinging thread is extruded toward the exterior like the finger of a glove. The tip of the filament first enters the victim's body and, during extrusion, rips a large wound in it—even in the hard armor of tiny crustaceans—assisted, of course, by the corrosive action of the released secretion. The path is thus made for the filament itself, which is also capable of penetrating the victim's body. The poison released through the pores in the tube cripples the prey. The wrapping capsules are barbless; on extrusion, the filaments quickly wrap themselves in a spiral around any body process, such as the bristles of the prey, and hold it fast, thus helping further penetrating capsules to discharge. The adhesive capsules, whose filaments have a sticky surface, have a similar task. The extrusion process has not been explained to date. Each extruded nematocyst dies, together with the cell which formed it; new ones are formed by undifferentiated (interstitial) cells which are often found beneath the ectodermal layer. The various types of capsules are often joined to form a battery of nematocysts.

Nematocysts

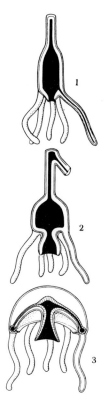

Fig. 6-4. Coelenterate body forms, the development of a medusa from the polyp form: 1. Hydra type (hollow tentacles), 2. Hydropolyp type (solid tentacles), 3. Hydromedusa type.

The ectodermal cells in the tentacle of a fresh-water polyp can contain up to two penetrating, three adhesive, and twenty-eight wrapping nematocysts. The adhesive nematocysts are also used by polyps (*Hydra*, for example) in a looping form of locomotion, in addition to their use in prey capture. Nematocysts are usually distributed over the whole body but are often concentrated in one place, such as the tentacles.

The Cnidaria reproduce sexually but also frequently asexually, usually by budding. These often complicated relationships will be dealt with in more detail in the description of the three classes: 1. Hydrozoa, 2. True jellyfish or Discomedusae, 3. Actinozoa.

The HYDROZOA or HYDROPOLYPS (class Hydrozoa) are usually relatively small; their bodies are divided into a stalklike tube (hydrocaulus) and a head (hydranth) with the mouth opening at its tip. The head can be club-shaped with tentacles along its whole length; in other types it is more disclike, with one or more rings of tentacles around its edge. The enteron is usually baglike in form and not subdivided. The hydrozoan sexual individual, the hydromedusa, is generally very small, with a more or less concave umbrella and containing a cell-less jelly. At the edge of the umbrella, the ectoderm fringes the bell cavity like a diaphragm (velum) and thus constricts it. The diameter of the opening can be varied by muscle fibers. During swimming, the water expelled by contraction of the bell produces a thrust the strength of which depends on the size of the opening; the speed of swimming is thus regulated in this way (the propulsion principle). The quadriradial symmetry of the hydromedusa is very conspicuous and is especially clear in young individuals (four tentacles, four radial canals). The ectoderm forms the gonads.

Reproduction in the Hydrozoa has certain unique features. Asexual reproduction, carried out in various ways, is very common. Longitudinal or horizontal division occurs rarely. A "lump" or bud formation, on the other hand, is frequently found on polyps. Elliptical or round buds appear on the wall of the body; later they are cut off from the trunk by a deep dividing cleavage, and begin a separate life as an aciliate, compact body. They are similar in structure to the planula larva but are capable of motion only by muscle contraction. These buds then turn into polyps by forming tentacles and an enteron. Budding is widely distributed among the polyps and medusae, and is the typical means of asexual reproduction. In this, all layers of the body protrude outward into a swelling and develop rapidly into a new individual. The daughter individual formed in this way then either detaches or remains on the parent individual for the rest of its life, so that large colonies come into being, consisting of hundreds or even thousands of individuals.

There are several types of colonies. In various hydropolyps, the original polyp (primary polyp) forms a system of tubes at its base, the stolon net (hydrorhiza), covering the substrate; new polyps then bud from these rootlike processes. Under certain conditions, the stolon net can also

Nematocyst batteries

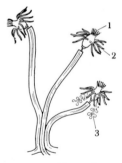

Fig. 6-5. Disclike polyp heads: 1. Oral tentacle whorl, 2. Aboral tentacle whorl, 3. Sporosacs.

Class: Hydrozoa

Fig. 6-6. Club-shaped polyp head: 1. Tentacle, 2. Medusal bud, 3. Rootlike stolon.

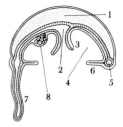

Fig. 6-7. Hydromedusa (schematic section): 1. Mesogloea, 2. Enteron, 3. Mouth opening, 4. Bell cavity, 5. Circular canal, 6. Velum, 7. Tentacle, 8. Gonads.

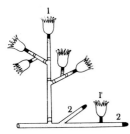

Fig. 6-8. Monopodial branching (schematic): Black: growth zone at the base of the polyp head. Stippled: budding zone: 1. First main axis, 1′. Second main axis, 2. Stolon.

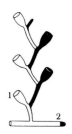

Fig. 6-9. Sympodial branching of a polyp colony: 1. Primary polyp, 2. Stolon.

Fig. 6-10. The direct development of a medusa (*Aglaura hemistoma*): 1. Planula larva, 2. Actinula larva in metamorphosis, 3. Medusoid larva shortly before the final medusal state.

grow perpendicularly, forming a type of branching colony which branches irregularly and carries the polyps. Usually buds are formed at the base of the primary polyp, which then anchors itself to the substrate with a hydrorhiza. Depending on the type of branching, colonies with monopodial and sympodial branching can be distinguished. In monopodial branching, every branch formed develops from a growth zone at its tip; in this case, two types of formation are possible. If the planula first forms a polyp, the growth zone is presented at the base of the head, and this zone becomes the budding one, producing new heads. The oldest polyp is thus present at the tip of the branch. Colonies based on this growth plan are relatively simple in structure. Complicated colonies come about when the planula larva does not immediately grow a head (hydranth), but forms a growth zone; the underlying budding zone then forms the first head. The tip of a branch in these colonies consists of a growth zone, and the polyps lying nearest to it are the youngest. The delicate sea moss colonies, for example, are constructed according to this system.

In sympodial branching, the primary polyp grows for only a very short time; the budding zone produces a new polyp from which, after a short period of growth, buds are produced again. The colonies built on this principle also vary greatly in form. The ectoderm of the stalk secretes a chitinlike, elastic covering on the exterior (cuticula), which affords the colony some measure of protection against enemies; this is termed the periderm. New layers are continually added from the interior so that, in the case of old branches, a relatively thick, barklike layer can form which gives the colony a definite degree of strength, especially in flowing water. The periderm layers of the stolon system can often join together; instead of a chininous secretion, calcium carbonate is sometimes deposited. It is for this reason that really massive colonies can come into being, even among the hydropolyps which are similar in appearance to those of the reef corals.

Polyps can also produce medusae by asexual means; the medusae then produce ova and sperm cells, from the joining of which a new polyp is formed. The developmental cycle from sessile polyp to free-swimming medusa and then to sessile polyp again, i.e., between asexually and sexually reproducing animals, is termed alternation of generations (metagenesis). This is almost always considered the normal state of affairs in Hydrozoa. It is therefore amazing that only a third of all the species reproduce in this way; the other two-thirds have changed this process appreciably. For example, the development of the medusal buds can be terminated at an early stage so that animals without a velum, tentacles, or a mouth are present. The sexual cells are formed within the polyp colony from interstitial cells and migrate into the medusa; such simplified forms (medusoids) often detach and swim freely in the water for a few hours. If the termination of development occurs even earlier, before the radial canals and subumbrella have been formed, these "jellyfish" do not detach from the polyp but remain attached to it as undeveloped medusae (sporosacs). Depending on the level of medusal simplification, the sexual cells are released into the water, fuse, and develop into planula larvae which, after a short swimming period (abour twelve to twenty-four hours), sink to the bottom and, with the aid of the adhesive capsules at the vegetative pole, attach themselves and metamorphose into polyps.

Some hydropolyps release planula or actinula larvae directly; a few species have suppressed all planktonic forms. In rare cases, in contrast, the asexual form (polyp generation) is supressed; here actinula larvae are produced from the fertilized eggs, developing directly into medusae. The reproductive relationships even within the systematic groups are very irregular and are often dependent on environmental factors. Usually, a hydropolyp colony consists of a single polyp type which not only captures food but produces the sexual generation by budding. In certain forms, however, a division of labor is found within the colony. Certain polyps (trophozooid, feeding polyps) are concerned with the taking in of food. Other polyps, with a reduced mouth and tentacles, receive their required nourishment through the canal system of the endoderm which joins every individual within the colony. These sexual polyps (blastozooid or gonozooid) only form medusal buds; threadlike defense polyps (dactylozooid) carry effective batteries of nematocysts by which they drive off enemies and can also free the colony of algae and refuse anchoring on it. Eight orders can be distinguished among the about 2700 hydropolyp species, of which 700 are "free-swimming" medusae:

1. Athecates-Anthomedusae (Athecata-Anthomedusae) 2. Limnohydrinids-Limnomedusae (Limnohydrina-Limnomedusae); 3. Hydrinids (Hydrina) 4. Halammohydrinids (Halammohydrina); 5. Siphonophorids (Siphonophora); 6. Thecaphorids-Leptomedusae (Thecaphora-Leptomedusae); 7. Trachymedusae (Trachymedusae); 8. Narcomedusae (Narcomedusae).

The order HYDRINA is ranked here as the first of the hydrozoans since its most important representatives, the fresh-water polyps, are well known to fresh-water naturalists and aquarists, and are the animals by which coelenterate characteristics can be most easily learned. This part was also played in the historical discovery of the coelenterates by the fresh-water polyps. Their simple structure, however, is not primitive but the result of degeneration, this being especially pronounced in the case of the medusa generation. Today, they are mainly fresh-water inhabitants and thus exceptions to the rule on two counts. Nevertheless, they gave the name Hydrozoa, meaning "water animal," to the entire class.

Weed-filled ditches with clear water are the preferred habitat of fresh-water polyps. Although they are widely distributed, they are not so easy to find. Owing to their color and form, they disappear easily among the weeds. This fact caused certain academic arguments during the last centuries about the true nature of these animals. More than 200 years ago,

▷
1. Cyanea lamarckii, 2. Chironex fleckeri, 3. Rhizostoma pulmo, 4. Nausithoe rubra, 5. Chrysaora hyoscella, 6. Aurelia aurita, 7. Nausithoe punctata.

Order: Hydrina

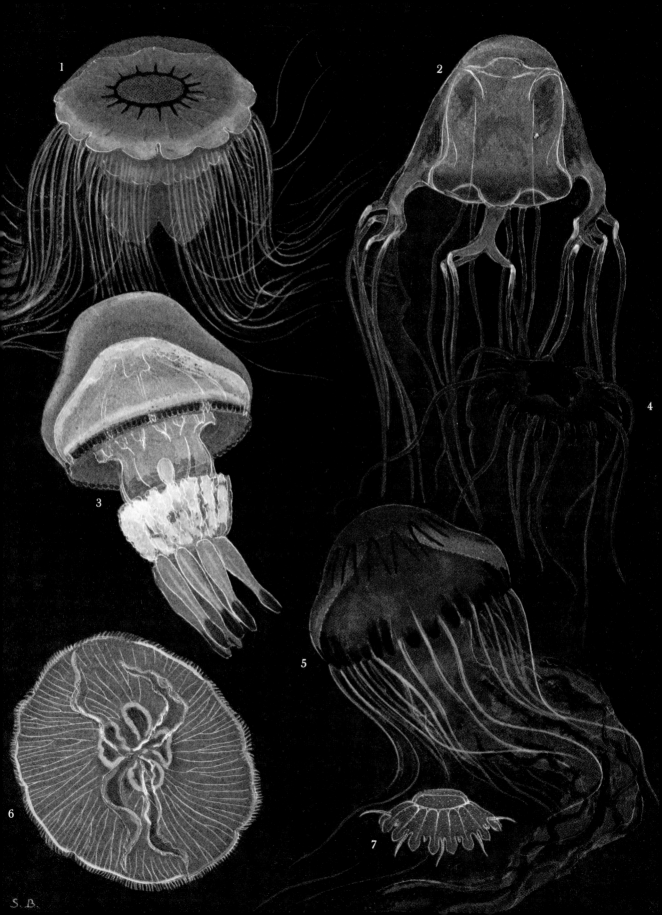

◁

1. *Corymorpha nutans*, 2.
Steenstrupia nutans, 3.
Leuckartiaria nobilis, 4.
Perigonimus, 5. *Eleutheria
dichotoma*, 6. *Eleutheria*,
7. *Geryonia proboscidalis*,
8. *Solmundella bitentaculata*,
9. *Solmissus albescens*,
10. *Oceania armata*, 11.
Köllikeria fasciculata, 12.
Euphysa aurata, 13. *Algura
hemistoma*.

the Swiss naturalist Abraham Trembley (1700–1784) published his carefully conducted investigations on fresh-water polyps. His zoological studies were well recognized by his contemporaries, but then became forgotten; modern zoologists recognized their great value again and honored Trembley's pioneering work.

One day, Trembley brought his new captures home from a ditch near the Hague. He later discovered flowerlike structures attached to the stems of water plants among the host of tiny crustaceans and insect larvae. His first thought was that the plants were blooming, but after some time he saw that the "flowers" contracted violently when the glass container was shaken and that only tiny green lumps remained. He then left the glass container undisturbed for a time and the creatures slowly extended themselves once more; first a stalk appeared, then the "flower head" unfolded, and finally the threadlike "petals" hung in the water again. With this observation, however, Trembley had discovered nothing new, since the first important student of the microscopic world, Anthony van Leeuwenhoek (1632–1723) had already described these creatures some time before, without, however, concerning himself with their life forms and systematic relationships. Trembley experimented with fresh-water polyps for more than three years to determine whether they were plants or animals. The color of the GREEN HYDRA (*Chlorohydra viridissima*) and its mode of attachment first led the investigator to the conclusion that the creature was a plant, although its own movements and the elasticity of the body were more like that of an animal. Trembley's conclusions were shattered when, one day, he proved that his hydras made actual migrations. He found all these animals on the illuminated side of the aquarium; if he turned the aquarium 180 degrees, within a few hours all the hydras had moved to the illuminated side once more. In doing so, they made looping movements: the oral disc with the tentacles curved down to the substrate and attached itself; the basal disc was then released from the substrate, the body contracted and the basal disc put down again near the crown of tentacles. Once this had attached, the tentacles released themselves from the substrate, the polyp extended and was then ready to take a new "step."

Trembley then conducted a new experiment: he wanted to use the power of regeneration as a means of proving the creature's membership in the animal or plant kingdoms. He thus bisected numerous hydras at various parts of the body and observed what happened to the cut sections. If they developed into new individuals, which is the case with plant cuttings, the creature must be a plant, according to his reasoning. If they died, it must be an animal. We must not lose sight of the fact, however, in considering Trembley's train of thought, that during his time there had been no investigations on the reformation of destroyed parts of the body (regeneration) in the animal kingdom. It was owing to this that Trembley was completely confused with regard to his results: the bisected polyps remained alive and, even more than this, the upper section of a horizontally bisected hydra formed a new foot, the lower half, a new ring of tentacles. Within about two weeks, Trembley had two complete individuals within his container, and even smaller sections of hydras completely regenerated. Trembley had, with this, stumbled upon one of the rare examples of complete regenerative ability within the animal kingdom.

At about the same time, he discovered that undamaged polyps were capable of building buds on their bodies which finally developed into tiny hydras and separated from the parent animal. This was, as far as he was concerned, a definite proof that the creature, like a plant, could reproduce asexually. This fact confused Trembley anew; the divergence in his views was not made any smaller by this experiment. He finally decided to send a few hydras to the great French naturalist Réaumur (1683–1757) so that he could test them. Réaumur proved that the creature must be an animal, and gave it the name "polyp," since the tentacles of the hydra were reminiscent of the "sea polyp," the eight-armed octopus *Octopus vulgaris*. The name "hydra" was created by Linné since this animal, like the many-headed water snake of the Greek myths, could grow two new heads for every one which had been cut off, an amazing regenerative ability.

Trembley himself finally became convinced that his polyps must be animals. He discovered, for example, that the creatures used their long tentacles to capture small water fleas swimming by, and carried these to their mouths; after a time, the indigestable remains of the swallowed prey were expelled into the water through the mouth opening. Trembley concerned himself in further experiments with the animal's ability to move from place to place and with its self-initiated division; he propped two cut halves of animals together and observed their fusion. In addition, he turned them inside out with a pig's bristle and painstakingly discovered that ectodermal cells which were now on the inside migrated again to the exterior and that the endoderm, now on the outside, regenerated again within. Trembley further discovered that the green color of the polyps he was investigating was due to the presence of tiny algal cells imbedded in the ectoderm as symbionts. He even investigated, to a certain extent, the structure and tasks of the nematocysts.

During the course of the past two hundred years, certain observations of hydras have been made which complete Trembley's research, so that today we know a great deal about these small fresh-water polyps. Well-fed animals usually reproduce only asexually; in this, both layers of the body bulge outward, forming a wartlike growth. The tip of this structure becomes the mouth area with its ring of tentacles as the whole structure elongates; the two enterons remain connected to one another at first so that the daughter polyp obtains nourishment from the soluble food in the enteron of the parent. Several buds can be formed at the same time by the parent. Sexual reproduction also occurs, usually under poor environ-

Fig. 6–11. A strongly contracted and bloated polyp (ab), c and e, engulfed gnat larva, bd pig's bristles used to invert the polyp, touching the basal disc (b). From this position, the posterior of the polyp is pushed inward to turn it inside out. 2. An inverted polyp (ab), attached to the tip of the pig's bristle. 3. Another polyp (cab) at the tip of a pig's bristle (bd), but not yet completely inverted. Since the basal disc (ab) is pushed inward, it emerges through the mouth, the most anterior portion; the oral disc (ac) has not yet been inverted.

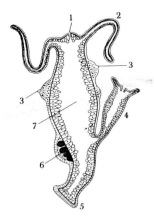

Fig. 6–12. Fresh-water hydra (schematic): 1. Mouth, 2. Tentacle (hollow), 3. "Testes," 4. Bud, 5. Basal disc, 6. Ova, 7. Enteron.

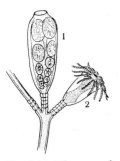

Fig. 6-13. Thecate polyp (medusal budding): 1. Blastozooid with medusal buds, 2. Feeding polyp.

mental conditions, such as a lack of food. In the upper third of the body, swellings occur in the ectoderm, within which sperm cells are formed from undifferentiated cells. Another swelling occurs in the lower half of the body. This contains a single ovum, also formed from undifferentiated cells. The sexual cells are released into the water, where the sperm and ova meet. A polyp then develops from the fertilized ovum (zygote). Usually both male and female sexual cells are formed on the same polyp; the animals are thus hermaphrodites.

As Trembley observed, fresh-water polyps have an amazing regenerative ability. These tiny creatures have been cut into 200 parts, each barely 0.2 mm in size, and even under these conditions all grew to form new polyps. The only requirement for regeneration is that each piece contain ectodermal, endodermal, and undifferentiated cells; undifferentiated cells can then be produced according to requirements.

Various species of fresh-water polyps can be found in European waters, although many are difficult to distinguish from one another. The various forms of ovum development, the structure and coiling of the adhesive capsules, and the sexual relationships serve as distinguishing characteristics. The most important representatives are: 1. GREEN HYDRA (*Chlorohydra viridissima*; L 1–2 cm); color dependent on symbiotic algae; usually eight tentacles that are always shorter than the body. 2. BROWN HYDRA (*Hydra vulgaris*; L 1–2 cm); stalk clearly distinguishable from the head; color changes dependent on food taken in; usually six tentacles about three times as long as the body. 3. GRAY HYDRA (*Hydra oligactis*; see Color plate, p. 198); size and body part relationships dependent on food availability and the degree of contraction. This hydra can extend its tentacles a good 25 cm and rapidly contract them to less than 5 mm. The astounding flexibility of its body allows it to capture and engulf large prey, causing a balloonlike swelling of the trunk.

Protohydra leuckarti, found in brackish water along the North and Baltic Sea coasts, is a near relative. The 1–2-mm-high, tentacle-less polyps live mainly in sandy regions, capturing their prey by means of the numerous penetrating nematocysts scattered all over the body. *Protohydra* can reproduce asexually by horizontal division; under unfavorable environmental conditions, it can also reproduce sexually, the sexes being separate.

Order: Athecata-Anthomedusae

The multiformed ATHECATES-ANTHOMEDUSAE (order Athecata-Anthomedusae) are rarely solitary individuals; they are usually found in groups or form twigged branches. In this order, an asexual budding polyp generation alternates with a sexually reproducing medusa generation. The polyp stalk either lacks or at the most has a short external envelope (peritheca), which does not enclose the head (thus "athecate"). Since the majority of the forms are barely visible to the naked eye, the wanderer along the shore has little opportunity to know these animals. He might encounter, on harbor piles or sunken rope, groups of *Tubularia*, various species of which are found in the seas surrounding Europe. Here, a tiny head sits at the end of a 2–3-cm-long stalk, with two rings (whorls) of tentacles, one of which is found immediately around the mouth opening and the other at the base of the head. Between the two rings are the sexual buds which form sporosacs rather than free medusae, producing only sexual cells. From the fertilized egg, a free-swimming actinula larva develops directly into a polyp.

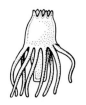

Fig. 6-14. Actinula larva of a *Tubularia polyp.*

The long branches—up to 10 cm—of the EUDENDRIDS (family Eudendridae) are especially well known to divers. These animals are most common in the upper part of the beach (littoral zone), where they often form diminutive forests and are a favorite source of food for certain invertebrates such as sea slugs and ghost crabs. The medusa generation is also simplified in the eudendrids. The sporosacs release ova and sperm into the water; the planula larva develops into a primary polyp which, by lateral budding, forms another branched colony.

The CORYNIDS (family Corynidae) belong to the typical representatives with a medusa generation, for example, *Coryne sarsi.* The polyp, only a few millimeters in length, is spherical in shape; the tentacles, with knoblike thickenings at the end, are scattered over the whole body with the exception of the stalk. Between the tentacles, the medusae are budded off. The free-swimming medusae, which were given the name *Sarsia tubulosa,* are relatively high-crowned bells, 12 mm in diameter; the opening of the bell is partly closed by a ringlike layer (velum). Four relatively short tentacles hang down from the edge of the umbrella; the gonads are found on the pipelike oral tube (manubrium). At the base of the tentacle is a small, light-sensitive organ (ocellus). The anthomedusa *Sarsia gemmifera,* found in the Mediterranean, is of some interest because its polyp generation is unknown. It is similar to *Sarsia tubulosa* externally, and medusal buds are often found on its highly extended oral tube. The polyp of *Cladonema radiatum* (L almost 5 mm) has four threadlike tentacles near the base of the oral disc and four with a swelling at the tip near the mouth. The medusae budding on the oral disc are of especial interest. These creatures, hardly 4 mm high, do not swim about, but creep along the substrate by means of their many-branched tentacles. *Corymorpha nutans* (L to 12 cm; see Color plate, p. 184) lives as a solitary polyp especially in areas of shallow water. The hydrocaulus "roots" in the muddy sand; medusae, named *Steenstrupia nutans* (see Color plate, p. 184), with only one tentacle, bud from the head of the polyp. The medusal bell is extended to form a spike. The best-known representative of the athecates, *Branchiocerianthus imperator* (L to 232 cm), is found at a depth of about 5000 m, its long body anchored to the mud by rootlike hydrorhizomes. The tentacles grow closer together at the tip of the animal, around its mouth, forming a ring; next to this are long, branched stalks with sexual-cell carriers (gonophores) and finally another ring of hundreds of long tentacles.

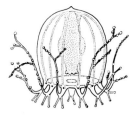

Fig. 6-15. *Cladonema radiatum,* an anthomedusa.

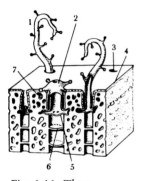

Fig. 6-16. Three-dimensional diagram of a fire coral: 1. Defense polyp, 2. Feeding polyp, 3. Retracted polyp, 4. Ectoderm, 5. Endoderm, 6. Basal layer, 7. Sectioned stolon tube.

Very similar to the stony corals in external appearance are the branches of the FIRE CORAL (*Millepora*; see Color plates, pp. 198, 237, and 238), which has an exceptionally painful sting. A tubular network (stolon network) first develops from the planula larva, growing in three directions and covering the substrate (rock, dead coral, and similar materials) with a layer hardly 1 mm in height. The walls of the stolon tubes secrete calcium carbonate so that the network fuses to form a calcareous crust; only the surface uppermost to the water does not calcify, but is covered with an ectoderm in which a multitude of nematocysts are embedded. The polyps arise from the lowermost stolons of the network and are thus sunken in tiny pits into which they can withdraw themselves extremely rapidly. The feeding polyps are very simple in structure; they have four knob-tipped tentacles around the mouth. Around a single feeding polyp are often five to eight defense polyps, which are tubelike with simple tentacles. The stoloniferous layer thickens by growing new networks across its surface. The underlying layers then die and the polyps extend; to prevent them from elongating beyond a certain point, they form a new basal layer from time to time so that the inhabited tube becomes chambered. After a certain period, a thick calcareous crust develops on which true branched colonies several tens of centimeters in height form; fire corals are therefore of relatively great importance in the building of tropical coral reefs. In contrast to other hydropolyps, the sexual cells are formed in the walls of the stolons and migrate from there into the polyp; this then metamorphoses into a simplified medusa without tentacles or velum. Once it has been released, it remains free-swimming for only a few hours.

The STYLASTERIDS (Stylasteridae; see Color plate, p. 237) are another skeleton-building form which previously was lumped together with the fire corals into the HYDROCORALS (Hydrocorallidae). They also develop a perpendicular, branching stolon network which calcifies and appears extremely similar to that of the fire corals which occur in the same environment. The two have another thing in common: the feeding polyp is surrounded by a ring of defense polyps; their tubes, however, are attached directly to that of the feeding polyp so that, at a rough glance, they resemble the tubes of stony corals which are chambered by thecae. In contrast to the fire corals, the stylasterids have a very thick, living, outer envelope. They are distributed from the Arctic Sea to the tropics.

Another type of athecate has penetrated brackish water and can even occur in fresh water: *Cordylophora caspia* (L up to 10 cm), found in the Baltic and in the estuaries of the Elbe and Weser. It has even been found near Berlin. *Cordylophora* consists of a delicately structured stalk arising from rootlike stolons. The sexual polyps bear only simplified medusae (sporosacs); these release planula larvae which develop immediately into a new colony.

Certain athecates are attached to snail shells, for example the species *Podocoryne carnea*, found in the Atlantic and Mediterranean. The periderm of the stolon network fuses to form a uniform chitinous plate from which the feeding and defense polyps arise. These coelenterates colonize the shells of the snail genus *Nassa* extremely frequently or snail shells inhabited by hermit crabs. The medusae of this species form on minute feeding polyps; their umbrella is barely 2 mm high and has four to eight tentacles. The related species *Hydractinia echinata*, found in the North Sea and especially in the Baltic, attaches to snail shells inhabited by the hermit crab *Eupagurus bernhardus*. The planula larvae produced by the sexual polyps settle only on substrates which move rapidly enough, that is, on shells that are inhabited. They also form a firm stoloniferous plate from which up to thousands of roughly 15-mm-long feeding polyps arise. The sexual polyps lie between these, their extremely simplified medusae releasing fully developed planula larvae. Defense polyps, with a great number of nematocysts in the ectoderm, form along the opening of the snail shell. Since the calcareous shell layers beneath the stoloniferous plate dissolve, the polyp colony slowly takes over the whole task of protecting the crab, as will be found later in the case of the zoantherian anemones.

The LIMNOHYDRIDS-LIMNOMEDUSAE (order Limnohydrina-Limnomedusae) are unobtrusive bottom-living polyps which form no colonies and often lack tentacles. Their medusae have hollow tentacles, and the gonads are present on the radial canals and sometimes on the oral cone. This order is best known for the fresh-water medusae it contains, especially *Craspedacusta sowerby*. Until 1880 it was believed that medusae were only found in the ocean; then tiny jellyfish 0.6 to 15 mm in diameter were found in the Kew Botanical Gardens in London. They were living in a tank planted with the Victoria water lily (*Victoria regia*) from Brazil. At that time, it was assumed that the medusae were brought in with the plant; this was proved shortly afterward, when *Craspedacusta* was found under the same conditions in Lyon. Within a few years, *Craspedacusta* had been found in various other European inland waters, for example in 1929 in the Tropt, a small tributary of the Garonne. The medusae appeared here in such numbers that the water looked milky and opaque. Within eight days, however, all the medusae had disappeared and only appeared again between 1932 and 1934. *Craspedacusta* prefers small artificial ponds, ditches, quarry ponds, blocked rivers, and occasionally also small lakes. In previous years they appeared quite unexpectedly in various parts of central Europe. Large jellyfish of this species were, for example, found several times in the water-lily tanks of the Frankfurt Palm Gardens, which were filled only during the summer; small jellyfish were regularly found in the Main before it became polluted.

In the late 1960s an aquarist wanted to catch live food for his animals from a fish pond in Alsace, France; he dragged his net through the water and was very surprised to find it filled with a jellylike mass when he pulled it out. On closer examination, he discovered that thousands of fresh-water medusae were swimming around in the pond. These had

Order: Limnohydrina-Limnomedusae

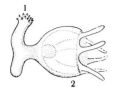

Fig. 6-17. Fresh-water medusa: 1. Polyp head, 2. Medusal bud.

a diameter of 20 mm across the middle, and appeared to be sexually mature. A week later, only a few dying medusae could be found.

The search for the associated polyp was a wearying task; finally a tentacleless, unobtrusive creature barely 2 mm in length was discovered. With the exception of the oral disc, the polyp is surrounded with a sticky layer of slime and attaches itself to stones of mollusk shells in running water; in this way it circumvents the danger of becoming covered with silt. The batteries of nematocysts around the oral disc capture the prey wafted toward the polyp by the water currents, especially tiny worms such as oligochaetes (see Chapter 12) and threadworms (see Chapter 10). Asexual reproduction can occur in two ways. Swellings can be formed in the walls of the trunk or at the oral pole which, after detachment, form new polyps. Lateral budding of polyps has also been observed; the daughter polyps regularly remain attached to the parent polyp for a long time. In the Rhine Valley, double polyps and even colonies of up to seven individuals have been found. If the water temperature increases, medusae can be formed directly from the buds which, on detaching, are barely 1 mm in diameter and have eight tentacles; this number increases with increasing size, and 614 tentacles have been recorded on a medusa 22 mm in diameter. It has been proved in the aquarium that medusae are sexually mature at a diameter of 9 mm.

Gonionemus vertens is a small medusa of about 20 mm in diameter found in the Atlantic and Mediterranean. I have often caught dozens of them while fishing for bottom-living plankton in the sea-grass beds around Nice, since, during the day, they settle on the blades of the sea-grass, usually with the mouth pointing upward, almost certainly feeding in this manner. Usually, however, feeding occurs in the following way: the jellyfish swim at the surface, turn upside down, and, with relaxed bell and widely spread tentacles, sink slowly to the bottom; in this way they can, of course, strain a relatively large area of water. A similar method of food capture has been observed in *Craspedacusta*. The umbrella of *Gonionemus* is hemispherical; there are over 100 tentacles around its edge, and the gonads lie on the radial canals in the form of folded ribbons, yellow to orange in color; the oral cone is also colored. The solitary-living polyps are minute (barely 0.5 mm high) and flasklike in form. They are found in a mucous envelope, and extend their three to five tentacles, which are only 2 mm in length, across the substrate like lime twigs. Asexual reproduction occurs through occasional swellings in the body wall from which polyps or medusae can develop. The much larger *Olindias phosphorica*, with an umbrella diameter of a good 5 cm, is also found in the sea-grass beds. Apart from the four radial canals, it has numerous blind canals running from the edge of the umbrella toward its middle. The umbrella is yellow to bluish-pink, the muffled gonads, red to blue.

Order: Halammohydrina

The unique order HALAMMOHYDRINIDS (Halammohydrina) was first discovered in the 20th Century. It contains only bottom-living forms similar in appearance to fresh-water polyps. A larva develops from the fertilized egg, metamorphosing into the adult animal by forming two rows of tentacles at the basal pole, at the side opposite to the oral pole. An umbrella is no longer formed. Owing to this, the tentacles and oral cone are thickly ciliated. According to Remane, these creatures "glide" with fluttering movements close to the surface of the sand and even between the sand grains. *Halammohydra octopodides* (L barely 2 mm), an inhabitant of the sandy-bottomed Kiel Bay, serves as a typical representative of this group. Further species have recently been described, because of more thorough studies of sand fauna.

Under the rather prosaic concept "free-swimming polyp colonies of the high seas consisting of medusae and polyplike individuals," then are numerous, unique and, in some cases, enchantingly beautiful creatures. The SIPHONOPHORES (order Siphonophora) is the name given to representatives of this order, the individuals of which remain attached to each other for the whole of their lives and show a division of labor for the good of the whole. Depending on their present task, the individuals are adapted in form; it is often unclear whether the individual in question has developed from a medusa or a polyp. The individuals of the colony will be termed here "persons," as the most famous investigator of this group, Ernst Haeckel, also termed them. Division of labor between several types of polyp on a branch has already been mentioned in the athecates; *Hydractinia echinata*, for example, includes feeding, defense, and sexual polyps arising from a common rooting surface. The colonial jellyfish have several characteristics which indicate their development from the athecate Hydrozoa. It was not the formation of an animal colony which resulted in the colonial jellyfish achieving more than their ancestors, but rather detachment from the substrate which enabled them to inhabit the high-sea environment.

The colonial jellyfish developed in two separate ways in their adaptation to a life on the high seas: the majority, with about 145 species, have, since Ernst Haeckel's time, been put together in the SIPHONANTHS (suborder Siphonanthae), while the DISCONANTHS (suborder Disconanthae), with which they have little in common, contains only about ten species.

The structure of a siphonanth colonial jellyfish can be best understood when the development of the colony is followed, although it is very difficult, owing to the many types of body form present in this group, to give a general description. A free-swimming planula larva develops from the fertilized egg. An infertile medusa without tentacles and oral cone first develops from a lateral thickening of the ectoderm, this being a swimming bell. The first tentacles develop from a wartlike structure beneath this swimming bell. The interior of the original planula larva becomes the enteron and, finally, the mouth opening breaks through and the siphonula larval stage is reached. A growth zone then forms at the base of the first tentacles and, lateral to this, the budding zone. The growth zone elongates

Order: Siphonophora

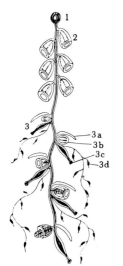

Fig. 6-18. Colonial jelly-fish (schematic):
1. Pneumatophore,
2. Swimming bell,
3. Cormidium, 3a. Hydro-phyllium, 3b. Sexual medusa, 3c. Feeding polyp, 3d. Tentacles with nematocyst batteries.

Fig. 6-19. Eudoxy from *Abylopsis*.

Portugese man-of-war

visibly and finally becomes a long, tubelike axis which is none other than the common elongated branch (hydrocaulus) of the primary polyp, since the polyp's mouth and tentacles are at its end. The budding zone then forms a sterile, mouthless, and tentacleless hydromedusa which takes over the task of the larval bell. Shortly after this, the larval bell degenerates and the newly formed bell takes over the task of moving the colony.

The budding zone forms, from now on, almost exclusively other types of buds which, owing to the continually active growth zone, are pushed away from the swimming bell and form a long stalk. Each of these buds becomes a short lateral branch of the main axis, a so-called group of persons (cormidium). This group of persons consists of a feeding polyp (trophozooid) and a long tentacle. At the base of the feeding polyp is a wide, jellylike plate which is curved inward. Within this thick mesoglea, the polyp extends a process to the main enteron. At the polyp's base are also one or more tubes on which the sexual medusae bud. On disturbance, the colonial jellyfish is capable of withdrawing the entire axis together with the groups of persons within the cavity of the swimming bell.

In some species the budding zone produces several swimming bells, one of these being cast off at regular intervals so that only two active bells are present at any time. Other species have several swimming bells which are usually arranged in two rows. These colonial jellyfish are put together in the group CALYCOPHORES (Calycophorae). The second group has a characteristic gas-filled bladder (pneumatophore) above the swimming bells, and is termed PHYSOPHORES (Physophorae). A "gas bag" forms by a complicated means, and is filled by a special gas gland.

Sexual reproduction resembles in most points that which has been described already for the athecates-anthomedusae. The sexual medusae produced are often free-swimming; sometimes, however, they are simpli-fied and remain attached. In other species the oldest group of persons detaches and swims away as an eudoxie, the thickened basal part and the oldest and largest sexual medusa taking over the function of swimming. If the sexual cells are already ripe, this medusa detaches and the next medusa, still immature, takes over its task.

Colonial jellyfish are found only in the ocean. They are continual swimmers which usually occur from sea level to a depth of 200 m. In experiments with "plucked" colonial jellyfish it has been proved that certain parts of the body, such as the gas bag, the swimming bells, and the basal plates, have a lower specific gravity than sea water and conse-quently bouy up the other parts. Larger colonial jellyfish are also capable of hanging horizontally in the water, the efforts of the swimming bells thus going entirely into forward motion.

The suborder SIPHONANTHS is divided into various groups depending on their swimming direction. One of the best-known colonial jellyfish with a gas bag is the Portugese man-of-war (*Physalia physalis*; see Color plates, pp. 198 and 271), distributed worldwide. The founding polyp grows to form an enormous structure, L almost 30 cm. This horizontally lying branch has two rows of person groups on its underside which, for their part, bear second- and third-order person groups. Each person group consists of a feeding polyp, several sexual polyps, and tentacles, so that a thick, almost beardlike tangle develops. The whole "colony" has a huge gas bag on the water surface. The bag is an irregular oval in shape and is about 20–30 cm long and 10 cm in width; the two poles extend to a point and on the upper surface a comblike structure runs in a longitudinal direction so that the whole structure is similar to a sail. The gas produced by a gland consists mainly of nitrogen, ten to fifteen percent oxygen, and, surprisingly, up to 1.18 percent argon. This inert gas, which occurs in nature, is extremely concentrated here, by an unknown method. Since the Portugese man-of-war lacks both basal plates and a swimming bell, it can only move forward by "sailing."

Often thousands of Portugese men-of-war are carried before the wind, appearing like an armada; like their namesakes, they also carry weapons. Their tentacles can extend to a L of 50 m, and their many nematocysts are extremely poisonous. It is dangerous to pick up a floating or stranded Portugese man-of-war, as hours of burning pain and heart disturbances can be the result. The prey captured while floating is hauled upward by the contracting tentacles and brought to the mouths of the feeding polyps. Very often fish up to 8 cm in length (*Nomeus gronovei*; see Vol. V) are found swimming between the tentacles, tearing—as stomach-content analyses have shown—whole bundles of person groups from the animal. The fish, however, is not immune to the poison in the tentacles, since it is sometimes paralyzed by it and eaten as prey.

Although the Portugese man-of-war is the only one of the physophores which show true shoaling, shoaling can also occur in other species, although very rarely. One of the most impressive experiences was the view of hundreds of colonial jellyfish of the species *Forskalia contorta* moving just below the surface in Villefrancha-sur-Mer Bay. These large colonies can reach a good 1 m in length. Below the small gas bag are many swimming bells, arranged in several rings around the main axis. They contract simultaneously or, on changing direction, also independent of one another, and are capable of moving the colony rapidly in any direction. The individual person groups (cormidia) are protected by a large basal plate. It is rarely possible to capture one of these extremely delicate colonies, to study it in the aquarium, without damaging it. The majority of representatives of the FORSKALIDS (family Forskalidae), to which the species described above belongs, live in warm seas. Atlantic species are occasionally carried northward by the Gulf Stream and thus can be found in the North Sea. This happens, for example, to *Physophora hydrostatica* (see Color plate, p.271) and *Agalma elegans*, which are similar in structure to *Forskalia*. These colonial jellyfish, which are often tens of centimeters in length, look like delicately tinted flower garlands.

Extremely beautiful forms are also found among the calycophores. Chun, a natural scientist, gave one species occuring in the Canary Island region the name *Stephanophyes superba*; according to him, it belonged to "one of the most brilliant creations among the delicate pelagic organisms." At the tip of this 25-cm-long creature, three or four bells, 4 cm in length, are arranged as a horizontal circle; above this circle is a minute supplementary bell. The basal plates touch each other and cover the whole axis; each person group contains a sterile swimming bell. The tentacles bear nematocyst batteries of variable appearance, dotted with nematocysts of all sizes and types (each battery contains over 1500 nematocysts). *Hippopodius hippopus* is commonly found as macroplankton in the Mediterranean. This colonial jellyfish, only a few centimeters long, has six or seven swimming bells one inside another, arranged in two rows. In a resting position the colony floats perpendicularly in the water; it is able to swim in both its horizontal and vertical plane. The most surprising thing is that only the two lower, oldest swimming bells are employed in swimming actively.

The DIPHYIDS (family Diphyidae) are more unpretentious in appearance. They have only two angular bells, arranged parallel to one another; the water current expelled by the two on swimming thus flows in the same direction. This, together with the torpedolike form of the bells, results in rapid swimming; in some species it could almost be termed "shooting about." *Chelophyes appendiculata* (L 15 mm; see Color plate, p. 271), from the Mediterranean, is a good example. The colony axis arises at the point of junction of the two bells. A frequently encountered species with only one bell, also from the Mediterranean, is *Muggiaea kochii* (L 5 mm), which often occurs in extremely large numbers in the plankton and is visible to the naked eye. In *Muggiaea* the oldest person group often detaches as an "eudoxy" and becomes independent.

Sailors-before-the-wind

Once, on an expedition into the Mediterranean, I suddenly discovered rows of little triangular "sails" protruding from the flat, calm surface of the sea: thousands of sailors-before-the-wind (*Velella velella*; see Color plates, pp. 198 and 271) which, carried by the current, wafted past. The structure of these most highly developed colonial jellyfish, belonging to the suborder DISCONANTHS (Disconanthae), is unique. The primary polyp on the axis becomes a huge, centrally placed "feeding person" (trophozoid) with a single stomach tube extended longitudinally over the oval jellyfish. Several rows of sexual polyps (blastostyles) surround this and are surrounded, for their part, by several rings of tentacles. These blue-colored persons hang from a thick central disc permeated by numerous ectodermal canals and air tubes. The uppermost surface finally forms a transparent swimming disc of chitinlike material. A section through this shows that it consists of concentric air-filled tubes. The air cavities are connected to the atmosphere by fine pores and also to each other. Beneath this, many-branched canals penetrate deep into the body, sometimes extending as far as the feeding polyp and the sexual polyps. Since they provide these tissues with atmospheric oxygen, they are termed tracheae, as are the breathing organs of various articulated animals; this is rather unusual in coelenterates. The upper, air-filled section of *Velella* shines silvery in full sunlight; the typical triangular sail is formed by extension of the upper body layer.

Velella thus consists of a large central feeding person and sexual persons which bud from this in several rows. Compared with the polyp of *Tubularia*, it swims axis-upward in the water but bears, however, a sail instead of a stalk and, instead a chitinous basal disc, an inverted chitinous breathing-canal system. It therefore consists not of a single person but rather the inverted apical organ of the feeding person. The sexual persons surrounding this produce male and female medusae in large numbers. These are similar to the athecate (Anthomedusae) medusae—an important indicator of the evolutionary origin of *Velella* and its relatives. The male and female medusae sink to depths to 1000 m; here they release their ova and sperm which, as in the Anthomedusae, are found in the oral tube. The embryo formed after fertilization slowly floats to the surface again, rising on fat droplets in the internal germ-cell layer.

The adult Mediterranean *Velella* is formed from the larva through complicated metamorphosis within six weeks. Owing to the enormous number of hydromedusae produced by a single *Velella*, huge flocks of new animals are produced although each female medusa carries only a single ovum. Reproduction here consists simply of asexual multiplication, while the sexual means, depending on the number of offspring per female, results in a reduction of numbers. In the Mediterranean, after long periods of onshore winds, wind-drifts of these jellyfish 50 cm deep and wide and over 1 km long have been found. In the Atlantic, an expedition ship sighted a swarm of *Velella* over 260 km in length.

Only scanty information is available on feeding in this species; feeding has never been observed. On occasion, half-digested copepods have been found in the enteron; even the sexual polyps around the mouth, which also have nematocysts, can contain tiny shrimps, which suggests that these reproductive individuals can also take part in feeding. It is noteworthy that the sailor-before-the-wind makes no self-induced swimming movements, but allows itself to be carried by the wind and water currents. In this, the jellyfish positions its sail at an angle of about forty degrees to the wind direction. The animals can sail very elegantly and are completely at home in their element, the water. As soon as the air chambers have developed during metamorphosis, it is impossible for them to sink, and *Velella* which have turned over are unable to right themselves again.

Apart from *Velella*, huge shoals of round *Porpita porpita* are also found. Their swimming disc, which has a diameter of about 5 cm and consists of up to 100 concentric tubes, is flat and bears no sail; the tentacles are club-

▷
A rhizostomid jellyfish (*Cotylorhiza tuberculata*) which often appears in the Mediterranean in shoals, usually accompanied by juvenile fish.

▷▷
Upper row from left to right:
Hydra with three daughter individuals. The sessile stauromedusoid *Craterolophus tethys*. *Cassiopeia xamachana*, a rhizostomid jellyfish. It lies on its umbrella and raises the oral arms with which it captures plankton.
Middle row from left to right:
Fire corals (*Millepora*) produce a hard calcareous skeleton similar to that of the stony corals. *Olindias phosphorica*, a Limnomedusa with numerous tentacles and an elongated oral tube.
Lower row:
Part of a *Sertularella* colony (left). Large, bloated sexual buds can be seen between the tiny polyps. Portugese man-of-war (*Physalia physalis*; above right). This floats on the water by means of its gas-filled float and "sails" along by raising the longitudinal comb. A shoal of sailors-before-the-wind (*Velella velella*; lower right) is not a rare occurrence in the Atlantic and Mediterranean. Violet-blue coloration: characteristic for planktonic animals of of the high seas.

▷▷▷
Metridium senile. The almost 1000 tentacles give this animal a flowerlike appearance.

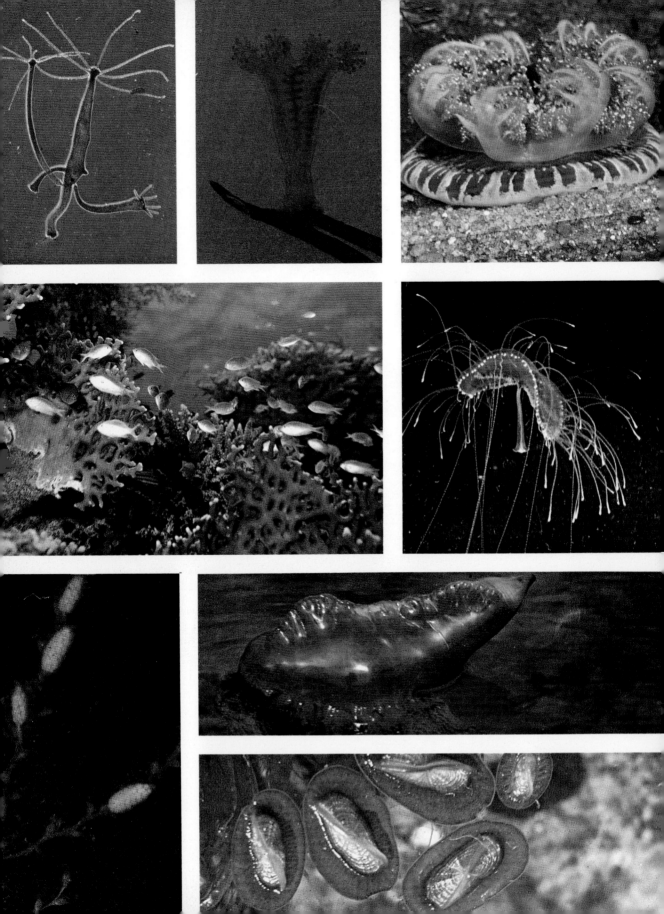

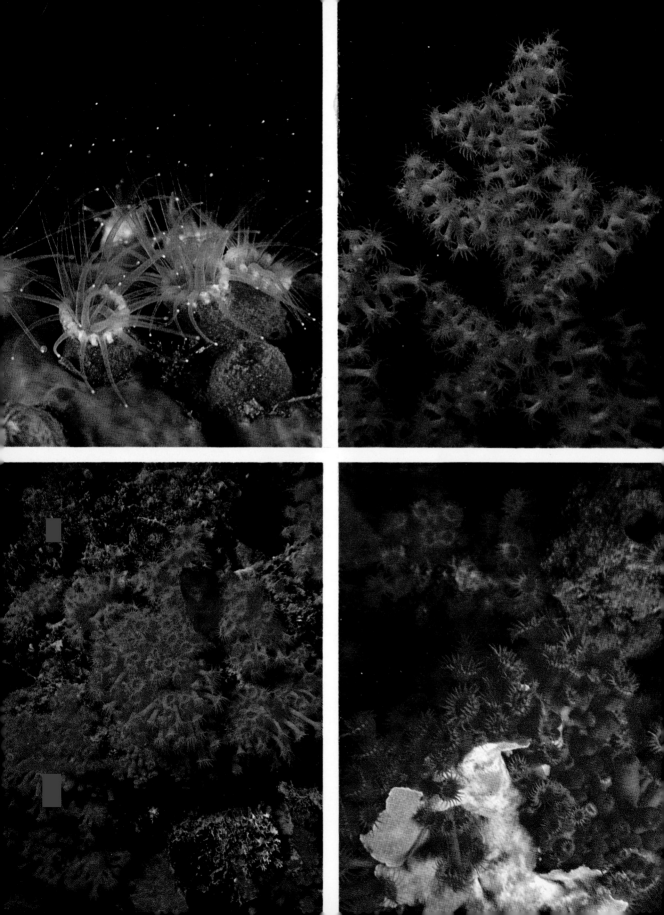

◁

Above (From left to right):
Colony of the zoantherid
anemone *Epizoanthus
arenaceus*. Zoantherids
overwhelm the skeleton of
a horny coral. Both lower
pictures: The yellow
zoantherid (*Parazoanthus
axinellae*) often forms
clumplike colonies to-
gether with sponges on
shady rock walls and in
caves. Their color varies
from light yellow (left)
to orange-yellow (right).

◁ ◁

Sea anemone (*Cerianthus
membranaceus*) in a field of
Caulerpa (alga).

◁ ◁ ◁

Left (from above to
below): Hermit crab
anemone (*Calliactis parasitica*) on the snail shell of a
hermit crab (*Eupagurus
bernhardus*) with which it
lives in symbiosis. Close-up
of a tentacle from a sea
anemone showing the
nematocyst batteries.
Right (from above to
below):
Cereus pedunculatus. Adhesive protuberances—re-
cognizable as white spots—
attach themselves to foreign
bodies, such as grains of
sand.
Tropical giant anemone
with anemone fish.

◁ ◁ ◁ ◁

Bunodactis verrucosa.

◁ ◁ ◁ ◁ ◁

The sea dahlia (*Tealia
felina*). The four photo-
graphs give an impression
of the variation in coloration and patterning.

shaped. This species is also found in the Mediterranean, Atlantic, and
Indian Oceans. Useful animals of various types follow the shoals of jellyfish. Two species of snail, one belonging to the prosobranchs, the violet
snail (*Janthina nitens*; see Vol. III), and one belonging to the Opistho-
branchia, *Fiona pinnata* (see Vol. III), feed exclusively on *Velella* and can
be regularly found in such shoals. The unusual sunfish (*Mola mola*; see
Vol. V), often encountered dozing at the water surface, apparently feeds
exclusively on coelenterates, especially *Velella*. Since it is sometimes
encountered together with *Velella* shoals far north of its normal distribution range, this hypothesis could be correct. About ten species of disconanths are known, the largest roughly 8 cm in diameter.

The THECAPHORES and LEPTOMEDUSAE (order Thecaphora-Leptomedusae) always have a flasklike envelope (theca) into which the polyp
heads can contract. Their tentacles are always in a whorl around the
mouth opening. The branches grow multiaxially in the "sympodial
pattern." The medusae never grow free on the feeding polyps but always
on a sexual individual which can be considered a metamorphosed polyp.
These are also surrounded by a periderm envelope (gonotheca). As in the
athecates, all varieties of medusal simplification are present from a completely developed free-swimming medusa to a medusoid or a sporosac.
Usually, the medusae which are released have an almost flat umbrella,
at the most, hemispherical. The sexual cells are formed within the radial
canals; the gonads are often easily identifiable by their bright colors. In
the majority of cases, balance organs (statocysts) are present in the velum.

The best-known representatives of this suborder belong to the genus
Laomedea (previously *Obelia*). The rootlike stolons cover a large variety of
objects; for example, seaweed, stones, bouys, and flotsam. From this
stolon net, tiny branches grow perpendicularly, reaching, depending
on the species, L of 4 cm. They are conspicuous owing to their zig-zag
mode of growth; a theca enclosing a polyp is at each angle. The sexual
individual (gonotheca) is urn-shaped, and buds in the branch-axes; it
buds tiny medusae which were once termed *Obelia*, since the polyp
generation was unknown. These flattened discs, about 4 mm in diameter,
were often found in the plankton, and their jerking movement is easily
recognizable. The velum has completely degenerated, and 16–100 strong
tentacles are found around the umbrella edge. Four radial canals run from
the central enteron to the edge of the umbrella where they are joined to
one another by a circular canal. The gonads are present as obvious thickenings along the radial canals. As a result of their easy availability and the
clarity of their structure, these medusae are very popular as objects of
investigation, although it is still not certain to which polyp colony they
belong.

The dried polyp branches, commercially termed "sea moss," dyed
and made into artificial flowers, garlands, and wreaths, are equally well
known. These are mainly colonies of CYPRESS MOSS (*Sertularia cupressina*;

compare Color plate, p. 198) and CORAL MOSS (*Hydrallmania falcata*) which grow to a height of 20–30 cm, but can even reach 70 cm. The colonies usually grow on stones or shells. In the German seas, especially, real banks of sea moss have formed at depths from 1 to 14 m. Further large collections have been found on the Dutch, English, and eastern American coasts. Sea moss often occurs in such quantities that, at the end of the last century, a proposal was made to harvest it for fertilizer. These hydropolyps have been gathered for decades by specially equipped boats. German concerns, especially, have been involved with this catch and bring several tons of sea moss to the surface every year; in 1958, for example, 48,818 kg were gathered, valued at over 300,000 DM. To prevent the sea moss from being wiped out, closed seasons have been declared since 1911. The branches grow up to 25 cm in height per year, and reach an age of about three years. The envelopes (hydrothecae) enclosing the polyps can be closed, after the polyp has been retracted, by a bivalved system. Free-swimming medusae are no longer produced. The sexual polyps form so-called gonangia which release fully developed planula larvae.

The colonies of PLUMULARIDS (family Plumularidae) look like delicately constructed feathers, as they only branch in the horizontal plane. The envelope (theca) surrounding the polyps is very short and lies close to the periderm on one side. The gonangia are found on short stalks and are enveloped, broochlike, by its lateral branches; this structure is termed a corbula. *Litocarpia myriophyllum*, which looks like a brown ostrich feather, is sometimes brought up by fishermen from depths between 30 and 100 m. The colony is coarse in growth form and can reach a length of up to 150 cm. Mussels (*Avicula hirundo*) and barnacles (*Scalpellum scalpellum*) are often found attached to the stalk. The smaller species *Aglaophenia pluma* is more common and can often form beds on solitary rocks at depths of 1–2 m.

The CAMPANULINID family (Campanulidae) also has covered thecae. *Campanulina* forms tiny colonies with polyps hardly 1 cm in height; one of the most difficult species to delineate produces the largest leptomedusa, *Aequorea forskalea*, which reaches a diameter of 20 cm, and, in exceptional cases, even 40 cm. Fifty to 300 radial canals run unbranched from the enteron to the umbrella edge where almost the same number of fringing tentacles are present. The tentacle length can vary considerably and reach triple that of the medusa's diameter. The gonads, distributed along almost the whole length of the radial canals, are very brightly colored: milky white to pink, red, violet, or even brown. The oral cone is blunt but opened wide. There is hardly any other medusa which shows such variation in gonadal color as *Aequorea*, found in the Mediterranean and Atlantic. The polyps of *Cuspidella* are even smaller than those of *Campanulina* and are barely 1 mm in height. They form creeping colonies and often live on other hydropolyps; owing to their small size, they are often overlooked. The associated medusa, *Laodicea undulata*, is often found in the

Sea moss

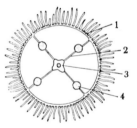

Fig. 6-20. Medusa of *Obelia* (Anthomedusae, schematic): 1. Circular canal, 2. Radial canal, 3. Enteron, 4. Gonads.

Fig. 6-21. *Litocarpia myriophyllum*: 1. Gonangium.

Fig. 6-22. Thecate polyp with blastozooid (*Campanulina*, schematic): 1. Extended polyp, 2. Blastozooid with medusa, 3. Polyp retracted and theca closed.

plankton. Its hemispherical umbrella has a diameter of up to 25 mm. The yellowish to pink gonads are found on the radial canals. Apart from numerous fringing tentacles, which are usually spirally contracted like a corkscrew, club-shaped tentacles are also present. The polyps of *Campanopsis*, about 3 mm in height, no longer have a theca. The associated medusa, *Octorchis gegenbauri*, is very conspicuous; its oral cone is extended to form a tube and at its free end—the mouth opening—are liplike structures. The likeness to a bell and clapper is thus even more emphasized. The gonads of this medusa are found not only on the radial canals but also on the oral tube.

Order: Trachymedusae

The order TRACHYMEDUSAE (Trachymedusae) has, at present, been united with the order Narcomedusae, since both no longer have a sessile polyp stage. The medusae of both orders are often found in the high seas, where they usually develop directly from swimming actinula larvae. Certain species have developed, secondarily, into bottom-dwellers again. The members of both orders can be distinguished by the length of the gonads, the structure of the radial canals, and the form of the umbrella edge.

The umbrella diameter in Trachymedusae lies between 1 mm and about 10 cm; in contrast to the Narcomedusae, the umbrella edge is smooth and a well-developed velum is present; this hangs hemlike beneath the umbrella. The tentacles have no cavity and are therefore rather stiff. Several radial canals run from the enteron to the umbrella edge where they are joined by a circular canal. A few of the fifty species are also found in European seas. *Rhopalonema velatum* is found in the Mediterranean and tropical oceans, where it floats mainly near the surface. The umbrella is roughly 15 mm in diameter, with eight long tentacles and, between these, eight to twenty-four short ones. The balancing organs (statocysts) hang free on the umbrella edge. Since this medusa is transparent and colorless it is often overlooked. *Aglantha digitalis*, found in the North Atlantic, can be considered a typical cold-water form. The umbrella is thimble-shaped; the oral stem is extended and hangs like a clapper beneath the umbrella. Long, trunklike oral stems are also found in *Liriope tetraphylla* and the TRUNKED JELLYFISH (*Geryonia proboscidalis*; see Color plate, p. 184), both of which are found from time to time in the Mediterranean; they are most commonly found in warm oceans, however. The Trachymedusae have very reduced oral stems, and inhabit great depths, such as *Pantachogon rubrum* (down to 1800 m) and *Haliscera papillosum* (down to 4000 m).

Order: Narcomedusae

The most important characteristic of the NARCOMEDUSAE (order Narcomedusae) is the absence of radial canals; the enteron pouches are, however, greatly extended, often reaching the umbrella edge. Owing to this, the gonadal position has been changed; the organs are found directly above the enteron, no longer on the radial canals. Apart from this, the tentacles do not form directly at the edge of the bell, but a little distance away on the upper side of the bell (exumbrella). Since the point of attachment is connected with the umbrella edge by a groove, the rim is clearly lobate. The Narcomedusae usually develop from ova through an actinula larval stage to a medusa again.

Of the sixty-four species known, some are parasites during the larval stage; a medusa does not develop directly from the actinula larva but, instead, several medusal buds form at the vegetative pole. The animals usually obtain their nourishment during this period through parasitism. The larvae of *Pegantha* remain within the enteron of the parent, which could be considered extended parental care. The aciliate planula larva of *Polypodium* is especially interesting, since it lives in the eggs of the sturgeon species *Acipenser ruthenus* (see Vol. IV); it feeds on the yolk mass from June to September, growing to a stolon with over thirty buds. During the following spring the tentacles grow, and, on the sturgeon's spawning, the whole colony is released. The buds detach from one another and become sexually mature medusae. How the planula reaches the ovaries of the sturgeon has not been explained. The larvae of *Cunina octonaria* attach themselves by their four tentacles to the oral tube of *Turritopsis*, another hydromedusa species, and thrust their single, extremely long oral tube into that of their host. On this, several medusal buds form and, after their release, develop into sexually mature jellyfish. The solid umbrella of *Solmissus albescens* (see Color plate, p. 184) is found very commonly in the Mediterranean plankton. The medusae, 3 cm in diameter, reproduce, in contrast to the other Narcomedusae, exclusively sexually. *Solmundella bitentaculata* (see Color plate, p. 184), characterized by its two tentacles which stand perpendicularly upward, is occasionally also found in the Mediterranean.

One can often find hundreds of large jellyfish stranded on the coast, or, far out in the oceans, "jelly umbrellas" of impressive size are seen moving past a ship. Almost without exception, these are representatives of the TRUE JELLYFISH (class Scyphozoa). In contrast to the Hydromedusae, they show the following characteristics: apart from the fact that they are practically always larger, they have no velum. The jelly mass of the umbrella contains cells. The umbrella edge is divided into eight obvious lobes by eight indentations, these, for their part, having second order indentations. On the lobes there are sense organs with odor pits, balance organs (statocysts), and simple light receptors (eye spots, ocelli). Numerous elastic tentacles hang from the edge of the umbrella. The umbrella can either be a flat disc or have a high-crowned bell shape. The jelly mass formed between the ectoderm and endoderm contains cells and can reach an enormous thickness, for example, in the genus *Rhizostoma*. The cells originate in the ectoderm, simultaneously forming a third body layer (mesogloea) which, however, is not comparable to the third germ-cell layer of bilaterally symmetrical animals (mesoderm). Owing to the cells it contains, the jelly has an almost cartilagelike consistency and the body of the jellyfish can thus better resist the mechanical stresses put upon it.

Class: Scyphozoa

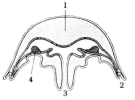

Fig. 6-23. Scyphomedusa (schematic): 1. Cell-containing jelly layer, 2. Lobes, 3. Oral tube, 4. Gonads.

Despite this, these animals have a water content of about ninety-four percent.

The mouth in scyphozoans is cross-shaped, with its edges drawn out and often subdivided, fringed, or curled; owing to this, these processes have also been termed oral tentacles or oral branches. The jellyfish capture their prey with their tentacles and sometimes with the oral tentacles; the prey consists of small planktonic creatures and, in certain circumstances, also small to middling fish. The rather large enteron is divided by four internal walls (gastral septae) into four pouches; bushy gastral filaments arise from the septa, the glands of which produce digestive juices. The gastral pouches and canals run from the enteron to the umbrella edge and, in contrast to those of the Hydromedusae, can be multi-branched. The majority of true jellyfish have separate sexes and are thus either male or female. The gonads are present in baglike projections of the stomach wall and thus originate from endodermal cells. They can be seen through the jelly layer as usually brightly colored, ribbonlike, and greatly convoluted structures. The sexual cells (ova and sperm) are released through the mouth and contact each other outside the body of the jellyfish. Free-swimming planula larvae develop from the fertilized eggs, and live in the plankton for up to ten days, then settling on the substrate. The rather small polyp of the jellyfish (scyphopolyp) then develops; externally it resembles the fresh-water polyps.

Since the scyphopolyps are rather inconspicuous, they are often overlooked and are rarely found. In the center of the oral disc is a four-cornered oral cone and, around this, are four—in older polyps up to sixteen—tentacles. These are not hollow, but are filled with endodermal cells. As in the medusae, the enteron is divided into four pockets by four protruding longitudinal folds (gastral septa); the four septal cones merge into these septa from the exterior. The scyphopolyps, L usually less than 1 cm, are solitary, and are found attached to stones, seaweed, shells, or rock, usually at a depth of 1 to 2 m and, on occasion, down to 70 m. *Nausithoe* is an exception to this, the branching polyps being sunk into sponges so that only the oral disc is visible; this species is found to depths of 8300 m. As in the case of the hydropolyps, the scyphopolyps capture their prey with the tentacles, paralyze it with the poison from the nematocysts, and bring it to the mouth where unicellulates, copepods, snail larvae, and arrow worms are engulfed. Supplementary food is carried by ciliary currents which flow from the tentacles to the mouth; this extra nourishment consists mainly of the tiniest organisms such as diatoms and related forms.

Scyphopolyps reproduce only asexually, by horizontal division (strobilation). A deep, ringlike depression develops from time to time in the body wall near the oral disc and, in this way, a disc is formed which becomes the ephyra larva on detachment. Depending on the polyp's species and nourishment, a polyp can form up to thirty of these ringlike

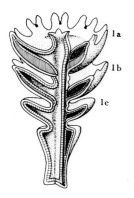

Fig. 6-24. Scyphostoma polyp showing constrictions (start of strobilation).

Fig. 6-25. Strobilation (schematized). A 120° section removed from the body. Left, the gastral pockets, right, the septum interrupted by septal slits (septal ostia, black). 1a, 1b, and 1c: Protomedusae.

depressions. This type of polyp is termed a strobila and can be compared with a pile of plates. The oral disc is the underside of the first ephyra larva formed; the tentacles previously present are resorbed and eight fringing lobes take their place, a sense organ forming at the split tip of each. The enteron septa are also more or less resorbed, and finally the uppermost disc detaches and swims away as an ephyra. Despite the loss of the tentacles, the polyp can continue to feed, exclusively, however, by ciliary currents, since the enteron is not affected by this horizontal division (strobilation). Once the last disc has detached, new tentacles can form on the polyp within a week.

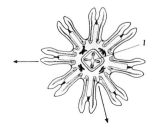

Fig. 6-26. Ephrya larva: 1. Mouth opening. The arrows show the section removed in Fig. 6-25.

Even today, the duration and frequency of strobilation is inadequately known. In the COMMON JELLYFISH (genus *Aurelia*) a definite dependence on temperature could be determined in Kiel Bay. In water temperatures of 2° to 4°C, the polyp forms ephyra from mid-January until the end of February, the whole process lasting fifty to sixty days. Even when the life expectancy of a scyphopolyp is only a year, the rate of reproduction is enormous, because of the tremendous number produced by the medusae. Apart from strobilation, bud formation has also been discovered in various scyphopolyps. The buds usually detach rapidly and become strobila. Only in *Nausithoe*, as previously mentioned, is a true polyp colony formed.

Apart from the "classical" means of reproduction—scyphopolyp-strobila-ephyra-jellyfish-planula-scyphopolyp—a few cases of maternal care are known. In certain forms, the jellyfish retains the eggs until the planula larva develops, holding them in the oral tentacles, for example, in *Cyanea*; other species retain the eggs in specially formed pits in the endoderm until this stage is reached (e.g., *Aurelia*). In *Chrysaora* the development of the planula larva occurs within the ovary, and in *Stygiomedusa* fully developed jellyfish form within the mother's body. In the metamorphosis from ephyra larva to scyphomedusa, new lobes form from the oral disc and grow between the original fringing lobes so that a closed umbrella forms from the previous eight-rayed star structure. The tentacles bud on these velar lobes. Growth is relatively rapid. In the North Sea, the common jellyfish reaches 1 cm in diameter within a month and, within three months, 20 cm.

Brood care

The fully grown jellyfish are continual swimmers some of which can reach speeds of several kilometers per hour. The water is pressed out of the hollow of the umbrella by contraction of the umbrella edge, the jellyfish swimming umbrella-surface forward. They usually cover great distances with the aid of marine currents. Shoal formation is known in certain jellyfish species, in which thousands of jellyfish can be involved. The LION'S MANE JELLYFISH (*Cyanea capillata*), in kilometer-long assemblies, have been sighted off the Norwegian coast, and even in the North Sea, areas can be packed with jellyfish. *Pelagia noctiluca* sometimes appears in shoals 45 km long; once, near Trieste, it was estimated that over 40,000

Rhizostoma pulmo were collected within a square kilometer. Sometimes fishermen must give up fishing because the nets keep filling with jellyfish.

In certain species, such as *Rhizostoma* and the common jellyfish, nematocysts are rarely found; in other species, such as *Cyanea* and *Chiropsalmus*, extremely effective nematocysts are present. Not only the tentacles attached to the jellyfish, but also its upper surface and even torn-off, floating tentacles can be dangerous to swimmers. Shortly after they have touched the skin, it starts to itch and burn. Later, the area becomes red and swellings may form. Under certain conditions, muscles may cramp, which is especially dangerous to swimmers. High fever can also be a result of a collision with such stinging jellyfish. On several occasions, these stings can lead to hypersensitivity in the person affected, even reaching allergic-shock proportions. Cases of death caused by jellyfish occur repeatedly, especially in tropical seas. Storms can drive huge numbers of jellyfish onto the shore. Even these supposedly dead animals are capable of stinging. It is best not to touch these "jelly lumps" with the hand, since it is not easy to distinguish harmless from dangerous species at a glance. If the animals are to be observed more closely, however, it is best to turn them over with some instrument.

The true jellyfish are divided systematically into five orders: 1. Stauromedusae, 2. Cubomedusae, 3. Deep-sea jellyfish (Coronata), 4. Semaestomae, 5. Rhizostomae. About 250 species are known, the largest being the giant arctic jellyfish, with an umbrella diameter of over 2 m.

The sessile Strauromedusae, which contains about thirty species, are probably the most primitive jellyfish. Scientists still argue whether these are polyps or secondarily sessile medusae. In the middle of the umbrella surface is a stalk, the tip of which develops into a basal disc. The jellyfish attaches itself to seaweed, sea grass, and stones by this means. If the development of the individual (ontogeny) is followed, the uniqueness of form can be better understood. A planula larva develops from the fertilized ovum, turning directly into a tiny, stalked polyp from which, in turn, an adult stauromedusa develops without the strobilation characteristic for true jellyfish. The medusal stalk is therefore the same as the polypal stalk.

The oral disc of the polyp forms a funnel or beaker; owing to this, the common name "beaker jellyfish" is also in use. The entire organism can be circular, but usually is pointed in a typical way, the oral disc forming eight arms arranged in pairs. At their tips are wartlike protuberances with 20 to 100 short tentacles. In the depression between each pair of arms is a short, strong tentacle, the marginal anchor, which produces a sticky secretion and serves as an organ of attachment. The Stauromedusae, like the fresh-water polyps, are able to move with a looping motion, using the pairs of arms and the basal disc. Either the arms and disc are attached to the substrate alternately or the polyp moves along on the tips of the

The effects of stinging

Order: Stauromedusae

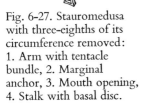

Fig. 6-27. Stauromedusa with three-eighths of its circumference removed: 1. Arm with tentacle bundle, 2. Marginal anchor, 3. Mouth opening, 4. Stalk with basal disc.

arms; one pair of arms detaches, extends, and attaches anew to the substrate, after which the other arms are pulled after it and the process begins anew.

Color substances can be taken up by the basal disc from algal cells present in the substrate and retained within the jellyfish's body. In this way, it slowly aquires the color of the background. The Stauromedusa feeds on small bivalves, snails, brittle stars, and crustaceans, using its tentacles; the prey are quickly "touched," adhered to, brought to the mouth opening, and swallowed. Feeding residues, such as mollusk shells, are later eliminated through the mouth.

Stauromedusae live almost entirely in cold seas and especially near the shores; only a single form, *Lucernaria bathophila*, is found at depths of about 1000 m. The individual jellyfish are about 3 cm in height and have a disc diameter of up to 8 cm. They are usually found mouth upward on the substrate. In the North Sea and its channels two species occur, *Craterolophus tethys* (L 3 cm, diameter 2.5 cm; see Color plate, p. 198) and *Haliclystus octoradiatus* (L 5 cm, diameter up to 3 cm); apart from these, *Lucernaria quadricornis* (diameter to 6 cm) is distributed from the North Sea to the Baltic.

The CUBOMEDUSAE (order Cubomedusae) of the tropical seas belong to the most feared of all the jellyfish. Sixteen species are known. They are not called "fire jellyfish" for nothing: a touch of their tentacles or even the bell can lead to severe burns. In these jellyfish, the umbrella is shaped like a die with more or less rounded corners. The underside is very concave; therefore the simple oral tube is hardly visible. On each of the four corners of the umbrella is either a single tentacle or a bundle of tentacles. The base of the tentacle changes during the course of growth into a strong lamellum (pedalium) which serves as a steering organ during swimming by being extended from the bell cavity. The bell cavity is partly closed by a circular membrane (velarium), which takes over the same task as the velum of the Hydromedusae. The umbrella is capable of contracting two or more times a second owing to the strong musculature present; thus the animal moves forward by the jet principle.

Order: Cubomedusae

All these factors lead to the Cubomedusae being the fastest and nimblest swimmers among the coelenterates. The ova are fertilized within the enteron and develop there into planula larvae. These are free-swimming for a few days and then settle and metamorphose into scyphopolyps. This process has only been observed in the aquarium. Little is known yet regarding strobilation and medusal development. It has been observed, however, that juveniles of the genus *Charybdea* show no resemblance to an ephyra larva at a diameter of 12 mm; it is therefore probable that the medusae develop directly from the scyphopolyp. The ways and means by which Cubomedusae reproduce are therefore still largely unknown.

Cubomedusae are relatively small animals with a bell height of barely 5 cm; they reach a size of 25 cm only rarely. They are found almost exclusively in shallow tropical seas and prefer dirtied water, such as harbor areas and river mouths. They are found only occasionally in the high seas, where they have probably been carried by currents. One single species, *Charybdea marsupialis*, is found in the Mediterranean at depths of 500–1000 m. The other members are definitely tropical animals, various species of which inhabit all the topical seas. The SEA WASP (*Chiropsalmus quadrigatus*), found in the Pacific, is the most feared of these, having a bell height of about 10 cm. Its nematocysts can cause blisters and swellings on human skin. Cramps and weakness also often occur, sometimes leading to the death of the victim. It is therefore surprising that this species is pickled and sold as food in eastern Asia, in the Philippines, for example.

Recently (1956) another Cubomedusa was discovered and described is Australia: *Chironex fleckeri* (see Color plate, p. 183), which until then had always been confused with *Chiropsalmus quadrigatus*. Microscopic investigations showed, however, the differences in the two forms. *Chironex* is distributed in the Australian region from the equator south to the Tropic of Capricorn; it is not fully clear whether it is present in other parts of the Pacific. Measurements have shown that these jellyfish can reach a speed of 5 kph and, if danger threatens, dash away at 9 kph. Because of the astounding steering mechanism present (pedalium), they are able to change direction extremely quickly.

Chiropsalmus and *Chironex* are attracted by weak light sources; they avoid harsh and strong light, however. *Chironex*, as J. H. Barnes has stated, is a great threat to the Australian swimming beaches. In their search for bottom-dwelling shrimps (*Acetes australis*) as food, they move in relatively near to the shore in bays which are also frequently visited by bathers. According to the cases registered, fifty people have been killed by these jellyfish in the previous few years. Only immediate help (treatment of the area affected with methyl alcohol, mouth-to mouth resuscitation, injection of medications to raise the blood pressure) gets the patient over the worst. It takes weeks, however, for the burnt areas to heal, and the person is usually scarred for life.

Order: Coronata

The DEEP-SEA JELLYFISH (order Coronata) are also termed bag or crown jellyfish. As their name suggests, they mainly live at great depths, usually in the tropical oceans. They have a simple mouth without long processes, and a deep horizontal cleavage in the umbrella, dividing it into a central cap and an outer ring. The umbrella can be a pointed, bell-shape, thimblelike, or conical, and, in certain species, even flat. It is usually brightly colored; the deeper the species lives, the brighter the coloration, which can vary from red and a rust color to deep violet and blackish-brown. The fringing lobes are usually deeply indented. The larger species have an umbrella diameter of up to 20 cm but, on an average, umbrella diameter only reaches 5 cm.

The genus *Nausithoe* is probably the best known, the branching scyphopolyps of which are found down to depths of 8300 m. The widely distributed species *Nausithoe punctata* (umbrella diameter, 1.4 cm; see Color plate, p. 183) is found in all the subtropical and tropical seas and sometimes in the Mediterranean. Its color varies from green to light brown, the gonads being yellow, red, or brown; there are reddish spots on the upper side of the umbrella. *Periphylla regina* is one of the most beautiful types, with a rust-red umbrella up to 20 cm in diameter. It is found from sea level to depths of 4500 m. In a similarly brightly colored species, *Atolla*, it has been observed that the majority of the animals living near the surface (250 to 500 m) are glassy and transparent except for their black enteron and weakly colored gonads. Between 500 and 1000 m deep the jellyfish have colored circular muscles. If the upper side of the umbrella is colored but the gonads still visible through it, the animal in question inhabits depths of 750 to 1500 m. If the *Atolla* jellyfish are completely opaque, they live at depths greater than 1500 m. There are naturally a great number of gradations in color but, in this genus especially, the dependence of color on water depth can be clearly followed.

The best-known jellyfish of the European seas belong to the order Semaestomae. The umbrella is relatively flat, and earlier they were also termed disc jellyfish. Only during swimming does the umbrella contract regularly into a half-sphere. The edge of the umbrella is fringed with small lobes. The four corners of the oral tube are extended to form four long, elastic, oral arms. The radial canals extend outward from the enteron, but no circular canal is present. The body is usually transparent and glassy but often has an extremely brightly colored patterning in various colors. The largest known jellyfish belong to this order, which contains about fifty species.

Chrysaora hyoscella (umbrella diameter up to 30 cm; see Color plate, p. 183) is often found in large shoals in the Atlantic, the North Sea, and the Mediterranean. It has sixteen yellow to red-brown radial stripes, giving it its German name, meaning compass jellyfish, since it resembles the face of a compass. Twenty-four long and rather thick tentacles hang from the edge of the umbrella. In this interesting hermaphroditic species the jellyfish are usually male at first, then male and female simultaneously, and finally only female; self-fertilization can therefore occur during a certain period. The ova are not released, but develop within the enteron and become free-swimming as planula larvae. Further development occurs through a scyphopolyp and the formation of ephyra larvae to the sexually mature jellyfish. In 1905 M. L. Delap stated that within a year a *Chrysaora* almost 23 cm in diameter developed from a scyphopolyp within his aquarium. The jellyfish was fed with plankton, showing a preference for small medusae.

Pelagia noctiluca (umbrella diameter up to 8 cm), a phosphorescent jellyfish, is very common in warmer parts of the Atlantic and in the

Order: Semaestomae

Phosphorescent jellyfish

Mediterranean. The umbrella ranges from bell-shaped to hemispherical; the color varies from weak purple to brownish-red. The eight delicately tinted tentacles are extremely elastic. The jellyfish phosphoresce brightly on mechanical stimulation, hence their German name meaning lamp jellyfish. A stream of water or the wake behind a ship is sufficient to release this phenomenon. Nothing precise is known yet regarding the adaptive significance of such phosphorescence or the chemical processes preceding it.

Chrysaora is exceptional in its means of reproduction. A planula larva develops from the fertilized ovum but this metamorphoses directly—bypassing the polyp stage and strobilation—into an ephyra larva. Thus, *Chrysaora* can also be distributed in the high seas. Enormous shoals of such jellyfish can sometimes be observed, having been carried by the current as far as the shore, where they form huge windrows. Bathers often receive burning weals by coming into contact with the extended tentacles, up to 10 m long, which, owing to their thinness and transparency, are not noticed; such contact can cause blisters and lead to high fever. I observed a mass invasion of *Chrysaora* in Villafranche-sur-Mer Bay, near Nice, in 1959. The plankton nets and gill nets put out by fishermen were completely blocked with jelly, and a windrow of stranded jellyfish tens of centimeters high formed on the shore. From the bay point, dozens of animals of all sizes, from ephyra larva to sexually mature jellyfish, could be fished out of the water in a bucket. Twenty-four hours later there was hardly a jellyfish to be seen.

Giant forms

The LION'S MANE JELLYFISH (*Cyanea capillata*) and the ARCTIC LION'S MANE (*Cyanea arctica*) are members of the giant jellyfish. The systematic independence of the latter species is still questioned, and some authors consider it a subspecies of the lion's mane. These giant jellyfish can have an umbrella diameter of more than 2 m, and an average lion's mane up to 1 m. Fringing tentacles are lacking in the family Cyaneidae, but they have eight groups, each of 150 tentacles, on the underside of the umbrella. These can extend to 40 m or more, and can contract to a tenth of this size within a second. The tentacles hang like a huge poison-spreading and death-bringing net beneath the animal. In prey capture, the jellyfish sinks slowly with its net of tentacles spread in a circle around it. It has been determined that an arctic lion's mane is capable of covering an area of a good 500 m². Luckily, these giants only occur in the wastes of the Arctic Seas. It is hardly to be thought of, what would happen to a swimmer caught in such a "net."

The lion's mane inhabits the cooler regions of the Atlantic, Pacific, North Sea, and Baltic. The oral arms are purple, the tentacles reddish but mainly yellow, hence its name, lion's mane. The umbrella can be delicate pink to reddish-gold or brownish-violet. As this only reaches 30 cm in diameter, the danger is not so great as that of the arctic lion's mane but, despite this, some swimmers have had unwholesome contact with the lion's mane. It has even been immortalized in literature because of its poisonous effects; Conan Doyle, in one of his Sherlock Holmes series called *Adventure with the Lion's Mane*, tells of a criminal offense brought off with the help of this poison. The related BLUE LION'S MANE (*Cyanea lamarckii*; diameter about 35 cm; see Color plate, p. 183) is also termed the cornflower jellyfish because of the beautiful, shining blue color of its umbrella. This species occurs in the Atlantic and North Sea. Its tentacles can also sting painfully and cause unpleasant burning.

Adventure with the Lion's Mane

The best-known semaestomid is probably the COMMON JELLYFISH (*Aurelia aurita*; see Color plate, p. 183), distributed worldwide and found from the equator to near the poles, also occurring in enclosed seas such as the Mediterranean and Baltic. Despite its tremendous distributional area, the forms occurring seem to be constant. Its body has a water content of practically ninety-eight percent and is colorless and transparent. Thus, the more horseshoelike than ear-shaped gonads, which are violet or light red in color, shine through the jelly. The body resembles an inverted plate about 25 cm in diameter, exceptionally, up to 40 cm, and can hardly be termed a "medusal bell." The fringing tentacles are very numerous but relatively short; the oral arms, in contrast, being well developed and brightly colored. A system of radial canals which branch and coalesce again leads from the enteron to the umbrella edge where all are joined by a circular canal. In contrast to certain other semaestomids, *Aurelia* is harmless; its nematocysts appear incapable of penetrating human skin.

At certain times, the common jellyfish appears to feed exclusively on plankton up to 6 mm in size, such as flagellates, snail larvae, crustracean larvae, and copepods. The small fringing tentacles capture only a small amount of the prey, the main prey-catching apparatus being the cilia of the entire body surface, which are also present in other species but are not employed by them in feeding. Ciliary action directs the plankton landing on the upper surface, which is enveloped in mucus as it impinges, to the edge of the umbrella and, from there, to eight special "food pits." The oral arms collect it from these and carry it to the mouth. The oral arms and tentacles can reject food particles or unusable components carried to the pits by the cilia and return them to the water. Apart from this microscopical feeding, the common jellyfish can also capture larger food; annelids up to 5 cm long and amphipods up to 1 cm long have been found within the enteron. *Aurelia* 2.5 cm in diameter can capture fish 12 cm in length.

The development of the common jellyfish was observed in the wild at the beginning of this century by English scientists. The sexually mature medusae appear on the western coast of Scotland in early summer. By the end of July free planula larvae are found between their oral arms. At the beginning of August masses of scyphopolyps appear on the leaves of kelp (*Laminaria*), and they release ephyra larvae from November to the end of January. During February all the polyps have usually dis-

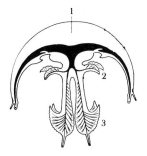

Fig. 6-28. Rhizostome jellyfish (schematic):
1. Jellylike umbrella,
2. Shoulder ruffle,
3. Oral arms.

Order: Rhizostomae

appeared. At this time, the plankton contains ephyra and juvenile jellyfish (up to 1 cm in diameter) in large numbers. By April the medusae have reached a diameter of a good 7 cm and, after three months, about 20 cm. By the beginning of summer the reproductive period begins again, ending with the death of the jellyfish; the average life span lies at about four and a half months. At the end of summer the last damaged jellyfish can be found. In nature, therefore, a yearly cycle appears to be present; each year both a polyp and a medusa generation comes into being, each dying off after its period of asexual or sexual reproduction. Common jellyfish have lived up to four years in aquaria and formed strobila several times; the conditions leading to this are not clear to date.

The reproductive relationships of the genus *Stygiomedusa* are especially worthy of mention. The atentaculate jellyfish has an umbrellar diameter of a good 30 cm. The umbrella is brown to purple-red. This jellyfish lives at depths of about 3000 m. Cysts are formed from endodermal cells and gonadial cells, hanging from tubelike projections into the enteron. Bedded within these cysts is a double-walled sac, probably a degenerated scyphopolyp, which is nourished by the cyst. Finally, a medusa about 10 cm in diameter develops from this sac; the way in which it leaves the parent animal, however, is unknown. This is certainly an exceptional case of maternal care for the coelenterates. *Dactylometra quinquecirrah*, from the Atlantic, is an especially feared species. It has an umbrellar diameter of up to 30 cm. The beautifully colored umbrella has nematocysts on its upper surface in wartlike prominences. Four long, pink oral arms hang under the umbrella, and about forty yellow-gold tentacles, which can extend up to several meters and can sting rather violently, are present at the umbrella edge.

Occasionally, jellyfish about the size of a human head can be seen in marine aquariums, swimming about by strong beating of the umbrella. The absence of fringing tentacles is conspicuous, as is the high-arched umbrella and the presence of delicate ruffles on the oral arms and underneath the umbrella. Whether one likes it or not, one is reminded of the ruffles and lace of old-fashioned petticoats. These are RHIZOSTOME JELLYFISH (order Rhizostomae), members of the genus *Rhizostoma* (umbrellar diameter up to 80 cm) which inhabit all the European seas. The umbrella is shaped like a high-crowned toadstool and, depending on the species, ranges from white to a delicate cream color. About eighty tiny bright cobalt-blue to violet lobes are present around the edge.

The metamorphosis from ephyra to medusa is associated with extremely complicated changes in the oral area. At first, the free edges of the oral arms grow together, at the same time forking. Numerous horizontal attachments form as the horizontal folds of the oral arms coalesce creating the previously mentioned ruffles. The original oral opening is usually closed completely and, replacing this, the many branched enteron system opens to the exterior through numerous pores.

At the base of the oral arms another pore system can be formed as a "shoulder ruffle." Since there is no longer a large mouth opening, food consists of the most minute of the planktonic organisms, those which are able to be taken in through the pores. Whether larger food particles can be digested between the oral arms and taken up through the pores as a "soup" is not completely certain yet. Sixteen radial canals lead from the enteron to the umbrella edge where they are connected by a circular canal.

Two species are present in the European area: *Rhizostoma pulmo* (see Color plates, pp. 183 and 272) in the Mediterranean and the closely related species *Rhizostoma octopus* in the Atlantic and North Sea. Both can be kept in aquaria for long periods of time. Large numbers of juvenile fish can be found underneath the umbrella, usually mackerel species (*Trachurus trachurus*) and goldline (*Boops salpa*). The small, ineffectual nematocyst batteries of the jellyfish can hardly protect the fish; it is more probable that they use the large umbrella and oral arms as a retreat. Some investigators doubt this, however, believing that the fish swim about between the arms for other reasons. Pieces of the jellyfish's body are very frequently found within these fishs' stomachs. In the case of jellyfish which had been introduced into the aquarium together with their "companions," I could observe that the fish attacked them.

Cotylorhiza tuberculata (see Color plate, p. 197) also occurs in the Mediterranean, often forming large shoals. The umbrella is very flat, the oral lobes are extremely ruffled, and violet-blue tentacles can be seen between them; shoulder ruffles, however, are completely lacking. The umbrella is often brown to greenish because of the presence of symbiotic algae (zooxanthellae); these are also present to a greater or lesser extent in other rhizostomes, but their presence there is still not explained. *Cotylorhiza* is also commonly accompanied by young fish of the two species already mentioned.

An interesting transference to a sessile mode of life is shown by the shallow-water medusa *Cassiopeia xamachana* (compare Color plate, p. 198), which lives in the mangrove swamps of tropical seas. It swims about until it reaches an umbrellar diameter to 2 cm at the most, and then settles in muddy lagoons "on its back," raising the oral arms upward. It can suck itself fast onto the substrate by means of a concave area on the umbrella surface, which prevents it from being swept away by the currents. These jellyfish can be found by the thousands in quiet water, moving their oral arms in the plankton- and oxygen-bearing water currents created by contraction of the umbrella edges. They are capable of pulsing twenty times a minute. Innumerable symbiotic algae (zooxanthellae) are present in the jelly of the umbrella, which, through the oxygen they produce, enable the jellyfish to inhabit this biotope where no other jellyfish is capable of living. This has made the return to a "polyp state" possible in such infertile mangrove swamps. *Cassiopeia*

Transition to a sessile mode of life

only leaves the substrate when severely disturbed; it swims a short distance and settles again immediately. Nothing is yet known about the development from egg to scyphopolyp, although it has been proved that the polyp forms only a single ephyra during strobilation.

When one has touched and felt the jellylike umbrella of the rhizostomes, it will come as no surprise that these forms are eaten in eastern Asia. Usually, these are members of the genus *Rhopilema*; in China, for example, the whole of *Rhopilema esculenta* is eaten, while in Japan, only the umbrella area is eaten. The jellyfish are either preserved in rock salt and alum or stored in leaves of an oak species (Kashiwa oak). Before eating, the medusa is softened in water, washed, cut into small pieces, seasoned, and then served. It is said to taste like salted gherkins. Rhizostomes are also often used as bait in fishing, making the theory that the jellyfish "protect" juvenile fish appear questionable.

Class: Anthozoa

The ANEMONES (Anthozoa), with a good 6500 species, is the largest class of coelenterates. These types are also the first ones to be known by the layman; large sea anemones are on show in practically all display aquaria. Many people also know the precious corals and reef-building corals. It is easy to see how these animals were given the name anemones once underwater color photographs of them, or especially the live animals themselves, have been seen. They are entirely marine and live either as solitary polyps or colonies. Medusae are no longer formed. The polyp form reproduces both sexually and asexually. The body of the polyp is more or less cylindrical and has one or more whorls of tentacles at the oral pole. The enteron is subdivided by at least four longitudinal walls (mesenteries) into lateral pockets. Processes also lead to the tentacles. The oral tube, which is laterally compressed, is continued deep into the enteron. The mesenteries, the number of which is important in the systematics of the Anthozoa, are a fold of the endodermal tissue; the cavity between the folded tissue layers is filled with cell-containing mesogloea. Longitudinal muscles form a type of roll on one side of the mesentry. The free edge is thickened to form a pad, and usually is strongly convoluted, as it is longer than the mesentery. This edge, termed the mesenterial filament, contains collections of digestive and absorptive cells; it is only in this part of the body that digestion and absorption of foodstuffs takes place.

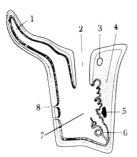

Fig. 6-29. Section through an anthozoid polyp: 1. Tentacle, 2. Gullet, 3. Radial canal, 4. Mesentery, 5. Gonads, 6. Acontia, 7. Enteron, 8. Pores.

After the prey has been swallowed, it is completely enveloped in the extremely motile filaments, and dissolved. The sexual cells form within the endoderm, between the filaments and longitudinal muscles, and, when mature, are released into the enteron. The mesenteries are attached to the trunk, basal and oral discs, and, in part, also to the oral tube. As already described, the enteron pouches formed by the mesenteries not only serve in digestion but also in respiration, transport of foodstuffs, and, under certain conditions, care of the offspring. The stream of water which circulates from one pole of the mouth, through these pockets, and out of the other mouth pole is created by beating cilia. The trunk musculature consists of ectodermal longitudinal muscles and endodermal circular muscles. Depending on the species, these are developed to a greater or lesser extent. Sense cells are scattered over the whole body surface; sense organs, however, are completely absent. Stimulation is transmitted through both an ectodermal nerve net and an endodermal one.

Two subclasses are distinguished: 1. The hexaradiate corals (Hexacorallia); 2. Octoradiate corals (Octocorallia). Two extinct subclasses are also included: the Tabulata corals, relatives of the Octocorallia, and the Rugosa, precursors of the hexacorallia.

The HEXACORALLIA, which includes over 4000 species, contains the various sea anemones which are often admired in aquaria, the tiny coral polyps which can build whole islands, the treelike corals reminiscent of the Octocorallia, and the flat, crustlike coral colonies. Both solitary and colony-forming types are included in this group, as are those with a calcareous skeleton and those without it. As the group's name states, they all have a six-rayed symmetry; there are thus six mesenteries and six enteron pouches, or multiples of these. Quinquiradial symmetry is rare. The tentacles are, almost without exception, feathered. There are five orders: 1. Actinians, or anemones, 2. Stony corals, 3. Anthipatharians, 4. Ceriantherians, 5. Zoantherians.

Subclass: Hexacorallia

The body of the ACTINIANS or SEA ANEMONES (order Actinaria) is a squat or elongate cylinder often bright yellow, green, red, blue, and even multitoned. The tentacles are arranged in one or more circles around the mouth opening. These are hollow, the tentacle cavity connecting with the enteron. Usually the polyp can attach itself to the substrate extremely quickly. The wall of the trunk is usually relatively thick and contains a well-developed cellulate mesogloea. The body shape can be changed drastically by interplay between the circular muscles of the endoderm and the longitudinal muscles of the mesenteries. The well-developed nervous system allows rapid contraction on perception of mechanical stimuli. No sense organs are present apart from sense cells on the tentacles and in the mouth area. The mesenteries or septa are always arranged in pairs; from the first twelve of these, the primary mesenteries, each pair attaches itself to the lower end of the oral tube and forms a stomach bag; these are termed directive mesenteries. The remaining mesenteries also attach to the oral tube and form other stomach bags termed endocoelic spaces when the longitudinal muscles of the mesenteries lie within them. If the opposite is the case, they are called exocoelic spaces. During the course of growth, the animal forms further mesentery-pairs, termed secondary or tertiary mesenteries, but always within the exocoelic space. They also enclose an endocoelic space, above which a new tentacle is formed within a short period of time. These and subsequent mesenteries are always smaller than the primary ones and thus are unable to reach the oral tube.

Order: Actinaria

The majority of sea anemones have separate sexes. Male animals

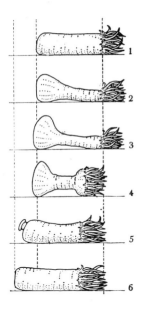

Fig. 6-30. Creeping movements in *Aiptasia*:
1. Quiescent position,
2. Fluid is compressed into the foot area, 3. Attachment of the edge of the basal disc, 4. Drawing in of the tentacular area, 5. Fluid is pressed back again by the musculature of the foot, 6. Quiescent phase.

release the sperm threads from the mouth, tips of the tentacles, and trunk pores; occasionally the dispersing "sperm cloud" can be observed in the aquarium. Depending on the species, fertilization can occur within the female's enteron or in the water. In certain species, care of the offspring occurs. The genus *Actinia* (see Color plate, p. 277), for example, releases from its interior tiny polyps with twelve tentacles. The majority of species, however, have a free-swimming Actinula larva which is planktonic for up to two weeks and then settles.

Sea anemones occur in all seas from the tidal zone to depths of 10,000 m. Various species can endure hours of exposure and greatly varying salt content in the water, caused, for example, by heavy rain at low tide; some "specialists" even inhabit brackish water. Although the majority of sea anemones are sessile, some are able to move around by wavelike movements of the basal disc; this activity, however, is poorly developed and the speed attained never exceeds more than a few centimeters an hour. Looping movements as described for the fresh-water polyps have also been observed. *Aiptasia* (see Color plate, p. 271) crawls by lying on its trunk in the following manner: the circular muscle layer near the oral cone contracts, this contraction slowly proceeding toward the basal disc. The body fluids in the oral area are thus put under pressure, so that the basal disc enlarges balloonlike and the body becomes slightly elongated. The animal then attaches itself to the substrate with the edge of the basal disc, and the longitudinal muscles of the mesenteries contract; this causes the tentacle crown to be drawn toward the basal disc, and in this way a "step" is taken.

Certain species can even be observed using true swimming, by rhythmical beating of the tentacles. The flight behavior shown by *Stomphia carneola* on the approach of enemies, such as a starfish and the sea slug (*Aeolidiap apillosa*; see Vol. III), is an example of this. As soon as this sea anemone is touched by the enemy, it detaches from the substrate and moves away quickly. The body is bent into a U-shape alternately to the right and left, up to forty times a minute. This behavior can also be elicited by skin extracts of the enemy.

The nematocysts of sea anemones are usually collected on the tentacles. In some species, these are capable of penetrating human skin. In the group ACONTIARIANS (Acontiaria), batteries of nematocysts are concentrated on special threads (acontia), which are attached to the base of the mesenteries and, by contraction of the body, are shot out through the mouth or through special pores in the trunk. Even the blue, nematocyst-containing lateral pockets which lie between the most exterior ring of tentacles and the trunk wall aid in defense. Should a conspecific approach too closely during its wanderings, these pockets are pressed against the body of the unwelcome neighbor. The skin of the pockets remains attached to the competitor and causes tissue collapse; should the intruder not move away, it can be killed by further pockets being brought into use.

Despite these various means of defense, sea anemones are eaten by a variety of animals such as opistobranch mollusks (opistobranchia; see Vol. III), starfish (Asteroidea; see Vol. III), pantopods (Pantopoda; see Chapter 15), and fish such as flounders, haddock, and eels. The sea anemones also live only on animal food, feeding in two different ways: as "fishers" (*Actinia*, *Tealia* [see Color plate, p. 200], *Calliactis*, *Anemonia*, etc.), and as "particle feeders" (*Sagartia*, *Gonactinia*, *Halcampa*). The first group captures prey with their strong tentacles which have numerous virulent nematocysts. They attach to the prey, paralyze it, and bring it to the oral tube; within the enteron of such species the remains of fish, mollusks, and crustaceans have been found. The members of the second group feed on tiny organisms; when such particles impinge on the sea anemone, they are coated in mucus and brought to the mouth by ciliary movement. Not only live organisms but also dead substances serve as food. The sea anemone can, however, distinguish well between useful organic food and inorganic particles. Digestion proceeds rapidly owing to extremely strong enzymes (proteases), and within a short period of time the indigestible remains are extruded through the mouth.

Many anemones harbor symbiotic algae (zooxanthellae), whose function has not been fully explained yet. Sea anemone coloration, however, usually does not depend on the algae, with a few exceptions, but on carotinoid substances. Since these carotinoids can become green and blue, depending on their joining with different proteins, the color is not especially uniform and can vary considerably within a single species.

The associations which various actinoids have with crustaceans and fish is especially worthy of note. Certain hermit crabs which can be seen in marine aquaria carry sea anemones on the snail shell protecting their armorless abdomen. A variety of levels of dependency in the interplay between anemone and crab are present, as exact observations have shown. That between *Calliactis parasitica* (see color plate, p. 202) and *Pagurus arrosor* or *Pagurus bernhardus* is rather loose. The anemone can live as a solitary individual on stones or empty snail or bivalve shells without any trouble; only rarely does it settle on a shell inhabited by a snail. If an anemone is detached and placed on any substrate, it begins to attach by one edge of the basal disc. By the interplay of various muscles and the help of adhesive secretions, it is finally able to attach the whole basal disc and stand upright. The attachment to a shell inhabited by a crab occurs quite differently and is much more complicated. First the anemone touches the shell with its tentacles; some of these attach—probably by adhesive nematocysts—to the shell surface so that the anemone is carried along by the crab in case it should move away. A few minutes later the entire oral disc is pressed against the shell; once this has occurred, the anemone, by bending its body, starts to lower the basal disc onto the shell. When it has attached firmly, the tentacles and oral disc release their

Associations with fish and crustaceans

Sea anemones and
hermit crabs

hold and the anemone takes up its normal stance. This whole process takes roughly from fifteen to thirty minutes.

It was previously assumed that the hermit crab excited the anemone to detach and transfer to the snail shell by "pinching" it. Recent investigations have shown, however, that the crab has nothing to do with this and the whole process is carried out entirely by the anemone, since it lives on empty snail shells as well as occupied ones. Up to eight anemones can live on a single crab. The crab thus has a certain amount of protection against enemies, but I have observed, in an aquarium containing a large number of hermit crabs, that under certain conditions the anemones attempt to attack the crabs as prey. It also proceeded the other way around, the crabs trying to eat the anemones. On transference to a larger shell, the crab abandons the anemone; it is a question of chance whether the anemone changes its habitation at the same time.

Adamsia anemones

The relationship of *Adamsia palliata* to *Eupagurus prideauxi* is quite different. This crab regularly carries a single anemone on its shell. By a means still unknown, the young anemone attaches itself to the underside of a shell inhabited by a juvenile crab. The basal disc increases enormously in size and finally envelopes the snail shell so that little can be seen of it. The mouth opening with the ring of tentacles is ventrally and anteriorly placed; it is directly behind the mouth of the crab. As the crab grows, so does the anemone, and its horny basal disc forms a tubelike covering around the crab's body. Therefore, the crab has no need to change its shell. The calcium carbonate of the original snail shell is often dissolved and nothing remains of primary habitation. Although the hermit crab cannot retract entirely into its "house," it appears to be well protected by the nematocysts of the anemone's tentacles and especially the acontia; it is rarely attacked by fish and even octopi retreat after making contact with the protective anemone, abandoning the tasty morsel which it envelopes. This is amazing since *Eupagurus*, removed from its "house," is used in great quantities as fishing bait. If the crab leaves its house, either on purpose or if forced to do so, the anemone dies within a short period of time.

It appears that other crabs have recognized the protective value of anemones; crabs of the genus *Melia*, from the Indian Ocean, remove anemones (especially *Bunodopsis*) from the substrate and carry one in each claw. If they are attacked by an enemy, they hold the virulently stinging anemones toward it. These crabs never use their claws as weapons or to attack prey but solely as carriers for the anemones.

Giant anemones and
anemone fish

The symbiosis between certain fish and anemones of the genus *Stoichactis*, which inhabit tropical seas, is of another kind. The fish belong to the genera *Premnas* and *Amphiprion* (see Vol. V), which live as pairs or sometimes with their young in a defended area around "their anemone"; depending on the species, they never move further away from the anemone than a few decimeters, to 4 m at the most. When danger threatens, they hide immediately in the long, fleshy tentacles, which would mean certain death for other animals. These "anemone fish" stick close to their tentacular retreat and, should the anemone be removed, they fall prey to carnivorous fish within a short while. Certain problems regarding this symbiosis between anemone and fish are still not explained and are under investigation by many biologists.

A few actinians should now be mentioned briefly. One of the most primitive forms is *Gonactinia prolifera*, which occurs in the Atlantic and Mediterranean down to depths of 75 m. This is one of the few species which reproduce by horizontal division. Apart from this, the animal moves around frequently by swimming, beating its tentacles or shooting water out from the mouth, or by creeping on the basal or oral discs. Sea anemones of the genera *Edwardsia* and *Peachia* live buried in the sand up to the level of the tentacles. They have no basal disc, and also occur in European seas. *Actinia equina* is distributed worldwide and often is found in large numbers in the tidal zone. *Bunodactis verrucosa* (see Color plates, pp. 201 and 271), from the Atlantic and Mediterranean, has many suctorial protrusions on the exterior of the trunk, which attaching to sand grains and broken mollusk shells; this probably helps to camouflage the animal. This species is a live bearer and can easily be kept in the aquarium; it is always surprising when dozens of tiny anemones appear there overnight. *Anemonia sulcata* (see Color plate, p. 271) can be found in large numbers in shaded areas of rocky cliffs and boulders in the Mediterranean. The long tentacles, only capable of contracting slightly, wave like grass stems in the water. It is advisable not to approach the animals too closely, since the tentacles can sting virulently and easily break off. The yellowish-brown to green color is caused by symbiotic algae. When these are absent, the animal is almost white. The tentacles, especially at their tips, have a delicate violet tint.

Cereus pedunculatus (see Color plate, p. 202) is also widely distributed; these anemones attach themselves very firmly to the substrate and can rarely be removed undamaged. Over 7000 tentacles, arranged in several rows, surround the oral disc. *Cereus* do not capture large fish, but take in tiny organisms which impinge on their tentacles. One of the most elegant forms is *Metridium senile* (see Color plates, pp. 199 and 271) which occurs near the coasts down to depths of about 100 m in the Atlantic, North Sea, and the western part of the Baltic. The slender body has a L of up to 30 cm, and above this, over 1000 delicate, short tentacles are set close together, forming a crown a good 20 cm in diameter. The polyp can be white, yellow, light to dark red, cobalt-blue, or brown in color; the tentacles are usually lighter. If several of these anemones lie close together, one is reminded of a huge bunch of carnations.

The order STONY CORALS (Madreporaria) includes over 2500 species, most of which are reef builders, similar in structure to the actinids. They differ from the relatives of the anemones mainly in that the stony coral

Live bearers

Order: stony corals

polyps secrete an exoskeleton from the basal disc. The polyps are usually very small (1–30 mm); members of the genus *Fungia* (about 25 cm in diameter) are one exception to this rule. Stony corals have a more delicately structured body than the actinians. The trunk musculature is weakly developed, as are the longitudinal muscles of the mesenteries and the mesogloea. The skeleton, whose construction is a very complicated process, is secreted in the form of fibrous aragonite (a mineral related to calcite); ninety-eight to 99.7 percent of this consists of calcium carbonate. In young polyps, the basal disc first secretes a calcareous plate. Six radially arranged ridges (sclerosepta) are then formed, pressing the basal disc inward into the enteron. Each is positioned within one of the stomach bags formed by the primary mesentery filaments; later, a second series of sclerosepta develops, lying between the mesenteries. In the next phase of development, twelve sclerosepta form simultaneously between the ones already present. From this point onward, a regularity in scleroseptal formation cannot be definitely proved.

At the edge of the sclerosepta a circular wall (theca) forms at the same time, joining the external edges together. This calcareous structure then grows continually upward until the pit inhabited by the polyp becomes relatively deep. Owing to this, a horizontal plate (dissepiment or tabula) forms from time to time, closing like an iris diaphragm and completely separating the lower part of the polyp from the upper. The tissue of the lower part then dies. The enterons of individual polyps are not directly in connection with one another, as they are in octocorallia, but this connection is maintained by the living tissue completely covering the coral colony. It is owing to this that growth also occurs in polyp-free sections of the colony.

The sexual relationships in stony corals are not uniform; hermaphroditic species and species with separate sexes are both present. The ova are fertilized in the enteron of the female or the hermaphoriditic enteron, and the planula larvae develop there. They are often released through the mouth opening in enormous numbers, appearing almost like a cloud in in the water. They can move relatively rapidly, 14 cm per minute. The larvae remain in the water from one to eight weeks, finally settling and growing into a new colony. The huge coral heads develop, however, through budding and also by asexual reproduction. Sometimes certain corals can be formed by millions of individual animals, but these have all developed from a single tiny planula larva.

The appearance of a coral colony is influenced deeply by two types of budding. In one case the buds are formed within the tentacles (intratentacular budding), and in the other case, outside them (extratentacular budding). In extratentacular budding, the pear-shaped planula larva, L about 2 mm, is free-swimming. It already has a mouth, six mesenteries, and nematocysts. Within twenty-four hours, it settles on the substrate and, within the next twenty-four hours, forms a basal plate and the first

Fig. 6-31. "Extratentacular budding": 1. Primary polyp, 2. Secondary polyp.

sclerosepta. Twelve tentacles then form and the polyp starts to grow. Within three weeks, four swellings appear at the base of the trunk and develop into daughter polyps which have already formed a complete skeleton. Second-grade daughter polyps develop at the base of these daughter polyps, and so on. By the end of eleven months, a colony of twenty to thirty individuals has already formed. Depending on the way in which the basal disc of the daughter polyp is arranged (on the same plane as that of the parent polyp or at a certain angle to it), the colony can be encrusting, vertical, or angled upward, forming coral branches such as are found in the STAG'S HORN CORAL (*Acropora*).

The processes in intratentacular budding are more complicated: the daughter polyps form within the ring of tentacles on the oral disc of the parent polyp. The soft tissue parts of this grow laterally so that it has an oval horizontal section. The skeleton, round at first, follows this growth form higher up and also becomes oval. Irregularities occur in the mesenterial arrangement; those at the edge of the extended portion bend toward the second focus of the oral disc. At this point, another oral tube develops and a constriction can occur between the two mouths, leading first to the formation of two separated oral discs and finally to two separate polyps. The skeleton of these polyps, however, indicates their history, since they lack a normal scleroseptal system; in this way, such polyps can be differentiated from those having extratentacular budding.

In some species, the separation of mother and daughter polyp does not occur. As division follows division, a meandering, ribbonlike oral disc is formed in which oval mouths occur, parallel to the longitudinal axis. One or more rows of tentacles surround the oral disc. The calcareous pits of neighboring polyps coalesce, thus forming typical ridges between which parallel continuous rows of paired septae can be seen—a proof that the polyps formed by budding have not been divided by a theca wall. Because of these meandering calcareous bands, the skeleton is unique in appearance and reminds one of the sulci in the mammalian brain. They have thus been termed "brain corals" or "niggerheads." If the polyps are completely separated from one another, a lump of calcareous matter is formed, covered with the thecae of numerous individual polyps; its appearance has led to the name "star coral."

Apart from the special growth forms of the species in question, external influences also have an effect on the growth of the colony as a whole. The following types can be distinguished: 1. Crustlike growth, 2. Growth in horizontal or vertical plates, 3. Egglike or hemispherical chunks and mushroomlike growth, 4. Tree or branching colonies, which, depending on their habitat, show great variation in form. They are delicate with fine branches in quiet water, thick and trunklike in moving water, crustlike in the surf area. This can be clearly proved by experiments with the stag's horn coral. This coral forms crusts or thick cups on the wave-whipped edges of the reef. If a piece is broken off from there and

Fig. 6-32. "Intratentacular budding" (schematic): 1. Parent polyp, 2. Growth of the parent polyp in one direction; the mouth opening of the daughter polyp is indicated, 3. Mouth opening of the daughter polyp complete, 4. Start of constriction and separation of the daughter polyp.

placed in calm water, the rounded crusts suddenly grow vertically and begin to branch. Both types of growth form can develop on the same coral colony depending on depth, a proof that these are one and the same species. Longevity and rate of growth also depend on various conditions.

Fig. 6-33. Skeleton of a star coral (*Favia*). The polyps are separated from one another.

Experiments on growth rate were made by C. M. Yonge in the late 1920's in the Great Barrier Reef area of the Australian coast. He broke off small pieces of coral from the substrate and cemented them on previously prepared concrete blocks. The colony was then carefully photographed and measured. The blocks were sunk in various submarine areas for six months, then raised to the surface, photographed again, and this picture compared with that at the beginning of the experiment. Yonge restricted himself, however, to branched colonies which showed the greatest rate of growth. They reached twice their original size easily within the six-month time span. A few examples from the literature should show thoroughly the growth rate of various corals.

A ship sunk off the American coast in 1792 had a tree coral 5 m in height growing on it sixty-five years later. Saville-Kent, one of the pioneers of science on the Barrier Reef, measured and photographed a series of corals on the south coast of Thursday Island in 1890. His descriptions were so accurate that Dr. Mayor, twenty-three years later, could definitely identify every coral; he found that a brain coral (*Symphyllia*) had increased its diameter from 75 to 185 cm. A huge *Porites* colony was 5.8 m wide in 1890 and twenty-three years later it measured almost 7 m; this is a growth rate of almost 5 cm a year in this rather massive form. Branched coral showed a yearly growth rate of 10–20 cm. On the pumice stone formed during 1883 after the eruption of Krakatau Volcano (Indonesia), coral crusts of 10 cm in surface area were found two years later. Charles Darwin reported that, in the Persian Gulf, a 60-cm-thick stony coral layer was deposited on a copper bottom of a ship within twenty months, a growth rate of 36 cm a year. Weinland made the following observation on the Haitian coast: branches of *Madreporia alcidornis* stuck a distance of 12.5 cm out of the water even at high tide. Since the sea level only reaches such a height during December to February, Weinland had to accept the fact that the corals there had a yearly growth rate of 50 cm.

Fig. 6-34. Skeleton of a brain coral. The theca of individual polyps are combined with one another and form "tentacular paths." The ridges formed are stippled.

Similar examples can be presented in large numbers. As a result of this extremely rapid rate of growth, voyages by ship in coral reef areas are always associated with great danger, since within a few years a huge coral reef can grow in a location which the chart shows to be free for passage. A reef channel in the reefs of the Andamen Islands was measured in 1887 as being 12 m in depth; in 1924 it was not quite 30 cm. Many wrecks have thus occurred in coral reef areas, the remains often being covered thereafter by a thick coral layer. A reef grows, on an average, 0.5 to 2.8 cm a year, an enormous rate when one realizes that the chalk is formed by cell layers only a few millimeters thick. Corals do not grow regularly, however, but in "spurts." The reasons for such spurts are not fully explained; it is hypothesized, however, that this is associated with the formation of the tabulae in the polyp tube.

In order to thrive, the majority of corals require a firm substrate such as rock or wrecks; only a few, for example, members of the genus *Fungia*, are able to prosper on soft substrates. Corals are found in a vertical direction from the surface down to depths of 6000 m or more. Reef corals which are dependent on the presence of symbiotic algae within their tissues can only flourish, however, down to depths of less than 40 m; below this, growth is reduced, and at depths of 90 m death occurs, because the zooxanthellae can only perform their task when light is present. Also in this connection, it has been mentioned that coral colonies are phototropic; they grow toward the light. If a broken coral branch is placed horizontally in the water, the tips of the branch start to grow vertically again within a short period of time, forming an obvious bend. Wave direction and water currents also influence the direction of growth. The colonies grow toward these two influences, since they carry food in the form of tiny organisms.

In contrast to the fire corals, which are members of the Hydrozoa, stony corals are incapable of stinging humans, fish, or large vertebrates; their nematocysts thus provide them with little protection and serve only as an aid in feeding. When danger threatens, their only protection is to retreat completely into their calcareous pits. They have, however, few enemies. Parrot fish, which have extremely strong jaws, can break off whole branches of a coral colony and eat them. Recently, a new enemy has grown in number to such an enormous extent that whole reef areas are threatened with destruction. This is the "Crown of Thorns" starfish (*Acanthaster planci*; see Vol. III), which suddenly appeared in huge swarms, destroying square meter after square meter of living coral. The origins of this starfish invasion have not yet been explained. Originally, it was thought that the mass collection of triton sea snails (see Vol. III)—one of the natural enemies of this starfish—was at fault; recent investigations have shown, however, that there are also other reasons.

Under certain circumstances, the damage caused by the starfish could lead to a real catastophe. Coral reefs not only protect islands and the mainland from the surf and sudden flooding, but they also provide an environment for fish and their food. If the reef is destroyed, one of the most important sources of nourishment for the island inhabitants is lost. The first damage that the Crown of Thorns starfish caused was in the Red Sea in 1963. Enormous damage was brought about on Guam, in the Marianas, one of the island groups of the Pacific; here ninety percent of a reef 3 km in length was destroyed. A Swiss diver informed me at the end of 1969 that in the Cairns area of North Queensland, Australia, large areas of reef were dead. A Crown of Thorns destroys 300 to 400 cm² of

Reef corals

Crown of Thorns starfish threatens stony corals

coral a day. All that is left is the dead calcareous skeleton, which is then covered up by algae and other growths. Further colonization by coral is thus made very difficult. The areas destroyed per day appears small; when one takes into account, however, that up to 100 starfish can be present within 1 m² of reef, one can imagine the extent of the damage. A practical means of combating this menace has not yet been found.

Anyone who has had the opportunity to dive in a coral reef or to observe one from above through a diving mask is astounded to find that, during the day, the minute polyps are rarely visible; they start to emerge as darkness falls, and retract again into their calcareous calyxes when morning breaks. This activity is associated with food availability; during the night, very many planktonic organisms, the entire source of food for corals, are present, moving from the depths up toward the water surface. The tentacles and cilia bring the tiny organisms toward the mouth. The "tentacle fishers" stretch their bodies enormously, thus increasing the size of their enterons which normally, owing to the numerous calcareous septa and mesenteries, are rather constricted. If a prey animal contacts a tentacle, it is captured by the nematocysts in the usual way. If the prey is so large that there is no room for it within the enteron, the mesenterial filaments are extruded, enveloping the exposed portion of the prey and digesting it outside the body. Some polyps only ingest the most minute of the planktonic organisms. These "microphages" conduct the attached plantonic organisms to the mouth or tentacles by ciliary currents. Vegetable foods are rejected by the tentacles or mouth. Corals are thus purely carnivorous; their food is digested extremely rapidly. Within about twelve hours the polyp has digested the nourishing portions of it, and the indigestible remains are then extruded through the mouth.

Significance of algae

Reef corals are often very brightly colored; the polyps can, for example, be bright red, green, or blue in color. The majority, however, are brownish, yellowish, or olive-green. This color is due to the symbiotic algae (zooxanthellae) which live within the endoderm. Even in a planula larva 1 mm long, over 7000 of these tiny algal cells have been found. There has been extensive discussion about the meaning and usefulness of these symbionts, but even to date the questions in connection with them have not been completely answered. Certain things have, however, been proved: during skeletal formation, the algae play an extensive part, since during calcium carbonate formation they bind the carbon dioxide (CO_2) by their assimilatory activity. As experiments have shown, corals with symbiotic algae can form much more calcareous deposits than those lacking them. They also provide the coral with oxygen, which it requires for respiration. In addition, the algae remove metabolic wastes (phosphorus and nitrogen compounds) from the polyp and also aid in its growth by providing hormones and vitamins.

Means of nourishment and symbiotic algae have led zoologists, however, along a completely wrong path. The first investigators on coral reefs never found food within the coral's stomachs; they thus asked themselves how the polyps survived at all. They finally thought that they had found the answer; according to them, the algae, which found extremely good conditions for growth within the polyp, were destroyed in part by the polyp and used as food. This was an example of a food chain closed upon itself—the coral colony being a self-sustaining unit. The experimenters then decided to observe the coral colonies at night; they then found the enterons of the polyps full of plankton, bottom-living crustaceans, and worms. Thus, the puzzle was solved.

The systematic arrangement of the stony corals is extremely difficult and is almost exclusively concerned with the skeletal structure. In all seas, and even in cooler waters, solitary coral polyps are endemic. In the Mediterranean and Atlantic, members of the genus *Caryophylla* are found. These are 1-3 cm high, delicate calcareous cups which colonize rocky walls, stones, and sea shells. The polyps of *Caryophylla clavus* (Mediterranean) and *Caryophylla smithii* (Atlantic) have thin, colorless, transparent bodies. If the polyp is rectracted, the animal looks as if it is simply a skeleton. The organism extends through water intake, and the twelve tentacles expand. These have knoblike thickenings at their tips and can stretch more than 1 cm over the edge of the calyx. *Leptopsammia pruvoti* inhabits grottos and overhangs of rocky walls in the Mediterranean. The base of the calyx is much broader than that of the two species mentioned previously. The bright orange to yellow soft parts of the body are very striking. All these corals are easy to keep in the aquarium.

The FUNGUS CORAL (*Fungia fungites*) is especially worthy of mention as a solitary polyp, being found in warm seas. A beakerlike feeding polyp first develops from the planula larva. Once it has reached a certain size, the oral region expands like an unfolding flower; the side walls no longer grow vertically, but horizontally. Finally, the skeletal section between the cuplike base of the polyp and the disclike mouth area disintegrates and the oral disc is carried away by the water currents and begins a life of its own, lying free on the substrate. The polyp is capable of building new discs at intervals; this process is reminiscent of strobilation in the jellyfish polyps. The released disc represents the sexual generation of the polyp; it never attaches to the substrate. Thus the coral is sometimes turned over by wave action. The polyp can usually turn itself over again through movements of the soft body parts, and can clean itself of attached sand and detritus within twenty-four hours. The fungus corals, with a diameter of up to 25 cm, are the largest of the solitary corals. Their brightly colored body is thickly strewn with tentacles up to the mouth area. If the polyp is disturbed, it is capable of contracting almost completely within the large number of calcareous mesenteries. The discs are of separate sex but are also capable of reproducing asexually by intra- and extra-tentacular budding.

The genus *Balanophyllia* is similar in structure to the fungus coral,

Fungus coral

Fig. 6-35. Skeleton of a balanophyllid coral as an example of a solitary coral.

and inhabits muddy bottoms in the Mediterranean, Atlantic, and Californian coasts. These are flat, usually attached discs, oval in form and with a thickened, porous edge to the calyx. When the polyp is fully extended, it is much larger than the calyx. Wartlike batteries of nematocysts are present on the transparent tentacles, giving the group its German name meaning "warty corals." The smooth, yellow-brown *Balanophyllia italica* is found in the Mediterranean, the reddish-yellow *Balanophyllia regia*, in the Atlantic. *Balanophyllia elegans*, an orange-red polyp found along the Californian coast, is especially beautiful. The genus *Flabellum* includes some large fanlike solitary corals. These are usually found on muddy bottoms in the Atlantic (*Flabellum angulare* and *Flabellum goodei*) and Mediterranean (*Flabellum anthophyllum*); since they are found only at great depths, they are occasionally brought up by trawls.

The cool-water representatives of the colonial corals will be mentioned here first. *Cladocora cespitosa* (see Color plate, p. 238) is often found by divers in the Mediterranean. Sharp eyes are needed to spot the tiny, delicately brown-tinged polyps, since they are easily overlooked. The colony forms small bushes, flat sheets, or even high, arching cups up to 50 cm in diameter. The calyxes are about 1 cm across and have obvious longitudinal ribbing on the exterior. Side buds, which grow out of the base of the existing polyp, increase the colony size. The buds extend until they have reached the height of the parent polyp. Owing to this, the structure not only increases in size but also becomes more dense, and the impression of a lawn is obtained when the colony-polyps are extended. *Cladocora* often grows in shallow water, at depths of 3 m, but is usually more commonly found at depths from 10 to 70 m.

Lophelia pertusa is widely distributed in both the Atlantic and the Mediterranean, especially in shallow water. It forms huge banks along the Norwegian coast, together with *Madrepora* and hydrocorals. *Lophelia* is stocky in build with irregularly branched twigs which are relatively thin and very breakable. The colonies, which can reach a height of 60 cm, are found along the Scandinavian coast at depths from 60 to 600 m, and in other areas to depths of 2,000 m. *Madrepora oculata* has a similar distribution-range. As previously mentioned, it occurs together with *Lophelia*. It grows in the form of regularly branched colonies up to 40 m in height. The skeleton is snow-white and the whitish to orange polyps are about 4 mm in diameter. The coral formations built by these two species differ from tropical coral reefs because of their position in deep water; since their polyps do not contain zooxanthellae, they are not restricted to lighted zones. *Dendrophyllia ramea* is coarser in structure than the two previous species. It lives in the Atlantic and Mediterranean, usually at depths of 200 m or more; occasionally it is found in shallow water. The colonies can reach a height of 1 m and are stumpy and compact. The polyps, which are relatively widely spaced from one another, emerge only slightly above the surface of the skeleton; they are yellow-gold. All these branched

Fig. 6-36. A colony-forming coral, *Lophelia pertusa*, from the Atlantic and Mediterranean.

colonies are usually covered with thick growths of other invertebrates, the larvae of which are brought by the water currents and attach themselves to the skeleton.

Astroides calycularis, (see Color plates, pp. 238, 271, and 277), a warm-water form, is now and again found in the southern Mediterranean, forming rounded cushions on shadowed rock walls and in caves. *Astroides* resembles *Cladocora* in some respects, but the individual calyxes are joined together at their upper edges; the polyp is bright orange. Similar forms are also found on the American Atlantic coast, such as the encrusting growths of *Astrangia danae,* which are about the size of a hand. The calyxes of this species are close together; the delicately formed polyps are transparent and whitish to pale pink, stretching only barely 15 mm above the calyx. Relatively few knotted tentacles are arranged around the oral disc; however, its wartlike prominences carry very efficient batteries of nematocysts with which the organism is capable of catching small crustaceans and fish.

In conclusion, the most important reef-building corals of the tropical seas should be mentioned. The TREELIKE CORALS (family Acroporidae) play a large role among the reef builders. Their extratentacular budding results in a very rapid growth rate. It is reckoned that within a decade the colony generally reaches its full growth, and after this it slowly dies. The largest branches can cover an area of a good 2 m square. The polyps are extremely small, having a diameter of only 1 to 2 mm. One hundred and twenty-five species of present-day corals are known and fossils of species from the genus *Acropora* can be followed down to the Lower Tertiary (Eocene). Among the massive growths, those of the family Poritidae, genus *Porites,* which form huge blocks and thick-branched colonies, should be mentioned. The polyps are 2 to 3 mm in diameter; these tiny creatures can, however, build blocks of up to 6 m in breadth and height. It is estimated that the creation of such a coral "hill" takes more than 200 years. As in the case of *Acropora,* the genus *Porites* is also known since the Eocene. Related forms have existed since the Jurassic.

The colonies of *Favia* and *Gonastrea* grow by means of intratentacular budding and separation of the oral disc. The first usually forms crusts and leaves, although sometimes it also forms blocks; the second can grow in the form of hemispherical clumps up to 2.5 m in diameter. *Lobophyllia* shows only slight constriction between the individual polyps. This species occurs in the Red Sea and forms domelike colonies more than 6 m in diameter and 3 m high. The description of this group will be closed with the BRAIN CORALS (*Diploria,* among others) in which the oral discs are not separated. The long, parallel, bandlike oral disc forms more or less convoluted paths with parallel ridges, giving the colony its meandering appearance. The colonies are usually hemispherical or semi-egg-shaped structures about the size of a human head.

The order Anthipatheria are colony-forming, usually attached,

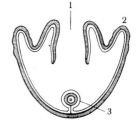

Fig. 6-37. Vertical section through an anthipatherian coral: 1. Mouth opening, 2. Tentacle, 3. Axial skeleton.

Order: Anthipatheria

hexaradiate corals similar to plants in their mode of growth. Their branches have a dark, acalcate central skeleton which is horny in structure and carries fine thorns; in extreme cases, this can reach a diameter of 15 cm. The skeleton is covered by a thin white "bark." The individual polyps are minute, scarcely 1 mm in size; they have a crown of six, and, in exceptional cases, eight, short, thick tentacles. The body axis is at right angles to the skeletal axis; the polyps are joined to one another by a common, living basal layer (coenosarc). They live on the most minute vegetable plankton, from algal cells. The growth form differs appreciably; the colonies can be branched on one plane, and they then have a fanlike shape. Neighboring twigs often join together so that a mesh is formed. Other species are branched in whorls, the side branches occurring all the way around the main axis. Such colonies are similar to bottle-brushes. Depending on the species, the colonies can be a few centimeters to up to more than 1 m in height. *Cirripathes rumphii*, the largest form, grows up to six meters in height. Anthipatherians are most widely distributed in tropical seas where they are found at depths of 100 to 1000 m. A few of the over 100 known species are also found in Europe, such as the BLACK CORALS (family Anthipathidae, species *Anthipathes subpinnata* and *Paranthipathes larix*) which, together with other species, were eagerly fished for in earlier times. Similar to the precious corals, the black corals are turned into all kinds of jewelry, and entire colonies are also sold as souvenirs. Even today, the black corals are used in eastern Asia as medicines or amulets against all sorts of illnesses, just as they once were used in Europe.

Order: Ceriantharia

The delicate CERIANTHARIDS (order Ceriantharia) belong to the most beautiful of the hexaradiate corals, inhabiting soft substrates such as sand and mud. They live in a tube which they build themselves, sometimes reaching more than 1 m in depth. The ceriantharid lines this tube with a leatherlike envelope which consists of mucus, the finest substrate particles, and nematocysts. When danger threatens, it can retreat into the tube instantly. In contrast to the majority of the coelenterates, the ceriantharids have no basal disc; the pole opposite the mouth is pointed and has a hole through which water is shot out as the sausagelike body contracts. This is by no means an anus, since the indigestible food remains are expelled through the mouth, as in all coelenterates. Ceriantharids can thus move freely within their tubes. Around the mouth opening are two whorls of tentacles: in the middle the short oral tentacles and, around this, over 100 long, fringing tentacles arranged in up to four rings. The tentacles are unable to contract; they are withdrawn in their entire length by the animal retreating into its tube. When the ceriantharid extends its tentacles maximally, they are reminiscent of water spouts from a fountain, falling in gracious arcs to the ground. An animal whose body is 2 cm in breadth can cover a circle almost 60 cm in diameter in search for prey.

Hydromedusae and small crustaceans are the main sources of food for the ceriantharids. In captivity, they also take small pieces of meat and

worms. Their beauty and resistance have made them favorite aquarium
animals. When well taken care of, they can live for decades in aquaria and
grow appreciably. A Mediterranean species, *Cerianthus membranaceus* (see
Color plates, pp. 203 and 271), was put in the display aquarium in Naples
in 1882; it lived and grew there for more than fifty years. The roughly
fifty known species are found in all seas and at all depths; they are able to
prosper even in polluted harbor areas. *Cerianthus lloydii*, a small form, is
found in the Atlantic and North Sea; the tube of this species can extend
up to 40 cm into the substrate. The tentacles are a delicate brown to gray.
The previously mentioned species, *Cerianthus membranaceus*, which is much
larger, is found in the Mediterranean. Its fully extended body can reach
a L of 20 cm. There are two color forms of this species: *Cerianthus membranaceus fuscus*, which has a white to brownish body and ringed, greenish,
shimmering tentacles; and *Cerianthus membranaceus violaceus*, which has a
violet to black body and gray tentacles. Hardly any individual resembles
another, however, and they differ not only in color but also in pattern.
Cerianthopsis americanus, found along the American Atlantic coast, is also
a favorite aqaurium animal. Ceriantharids are rarely caught by fishing
boats, however, since they retreat into their tubes as the trawl net
approaches. Divers are usually able to dig out younger and thus smaller
animals from their tubes without too much laborious and time-consuming
work.

Anyone who has dived in the Mediterranean has certainly come across
caves or vertical and overhanging rocky walls on which colonies of yellow-
gold to orange-colored polyps, only a few centimeters in height, and
densely packed, can be seen. These are members of the order Zoantharia,
Parazoanthus axinellae (see Color plate, p. 204), which often colonized
sponges of the genus *Axinella* as well. This species is also found on bivalves
(*Microcosmus*), and only rarely grows directly on the rock. The polyps are
unable to form their own calcareous skeleton. As a substitute, they in-
corporate sponge spicules, sand grains, forameniferan skeletons, and
similar things into the body tissues. About 300 species of zoantherians are
known in all; externally, these resemble small sea anemones. In contrast,
however, they have no basal disc but attach themselves by inserting their
pointed aboral pole into the substrate. They usually do not live as solitary
individuals but in small or large colonies. Their main distribution-range
lies in warmer seas and, depending on the species, shallow or deep water.
The polyps of colonial species bud from a common cushion-, strip-, or
lamellalike body mass.

Epizoanthus arenaceus (see Color plate p. 204) lives on snail shells (e.g.,
Murex and *Aporrhais*), bivalves, and horny corals in the Mediterranean.
The tiny grayish-brown to sandy brown polyps are about 1 cm in height
and have up to twenty-four whitish tentacles; they form small colonies,
which, however, are never cushion-shaped. The dull coloration results
in their often being overlooked. Related species such as *Epizoanthus*

vatovai and *Epizoanthus incrustatus* cover shells inhabited by hermit crabs, and gradually dissolve the snail shell so that, finally, the crab lives within a mass of these zoantherians.

Subclass: Octocorallia

The most important characteristic of the usually branching OCTO-RADIATE CORALS (subclass Octocorallia) is the various parts of the body or organs which appear in eights or multiples thereof. Eight septa (mesenteries) are arranged as longitudinal walls within the enteron, dividing it into the same number of pockets. In the upper portion of the polyp, the mesenteries grow together onto the central oral tube. Near the base, they bulge outward owing to the presence of sexual cells formed by the endoderm; these protrude into the enteron cavity as grapelike structures. Eggs and sperm are extruded through the mouth. The oral disc is surrounded by eight delicate, featherlike tentacles. Their nematocysts are too weak to penetrate human skin. Apart from feeding-polyps (autozooids), polyps without tentacles or mesenterial filaments are also present (siphonozooids); these are unable to capture or digest their own food. These, owing to their well-developed ciliary fields, have the task of washing out the internal tube system with fresh water and providing the tissues with oxygen. They also play a part in expansion and contraction, a process which can be observed in certain colonies. Ectodermal cells which have wandered into the mesogloea are the only ones which secrete the skeleton. The skeletal elements formed are partly ceratinous and partly calcareous. They are in the form of tiny needles (spicules or sclerites) of about 0.01 to 10 mm in length. They vary in form and surface area; systematists use them as species characteristics. In certain species, the sclerites fuse to a formless mass which is relatively stable; such colonies can play an important part in the building of coral reefs. Even when the calcareous spicules are individually embedded within the mesogloea, they give the fleshy octocorallian colonies a definite stability; despite this, the colonies remain elastic enough to expand and contract, which is important for certain forms. The endodermal canals are also embedded in the mesogloea, joining the individual polyps to one another. About 2500 species are known, divided into four orders 1. Alcyonarian corals, 2. Gorgonids, 3. Blue corals, 4. Sea pens. The largest forms are found among the gorgonids, which are able to form colonies up to 3 m in length.

One of the commonest inhabitants of the North Sea fishing grounds is the DEAD MAN'S HAND or FINGER (*Alcyonium digitatum*; see Color plates, pp. 240 and 271), one of the alcyonarian corals; it is usually attached to empty mollusk shells, stones, or similar things. The fleshy colony forms fingerlike or lobed, branched structures which are 20 cm or more in height; the color is quite variable—white, yellowish, orange-red, to purple; the state of contraction of the colony has a marked influence on the intensity of the color. The polyps are pure white. It is a wonderful sight for a diver to see dozens of these colonies close together in the twilight gloom, all with their polyps extended to the maximum and looking

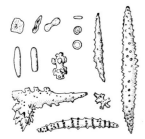

Fig. 6-38. Skeletal spicules from octocorallians.

like submarine flowers. When the polyps are retracted, the colonies can often hardly be distinguished from the background.

The calcareous spicules are loosely distributed in the central jellylike layer; the body is coarse, leathery, and elastic. By taking up water, a contracted dead man's hand can increase its size to a multiple of the original; body form and size are thus extremely variable. One must be patient to observe this process, and sometimes wait for hours in front of the aquarium. At the beginning, the warty-surfaced colony, which appears crumpled and contracted, looks smoother, the body wall becomes more conspicuous, and, in especially favorable conditions, the calcareous spicules can be seen against the light much like human bones on an X-ray plate. When the colony is fully swollen and the warts visible as obvious prominences, the polyps start to appear as tiny white spots; the polyp body then extends, elongates, and finally the feathered tentacles unfold. The colony now looks like a bouquet of flowers. During this phase the polyps take in food in the form of tiny planktonic organisms. The colony must not be disturbed at this point, since groups of polyps or even the entire animal can contract completely on the slightest stimulation, and will not extend again for a long time. Contraction is a complicated process: the tentacles disappear first, with the aid of special muscles, then the rest of the polyp inverts like the finger of a glove; expansion, in contrast, appears to be brought about entirely by internal pressure (turgor).

A large number of siphonozooids are present, apart from the feeding polyps; these control the taking in and expulsion of water. The enterons of the oldest polyps stretch to the base of the colony as numerous internal cavities. They are joined to one another by numerous endodermal canals embedded in a common mesogloea. Younger polyps usually bud from these connecting canals. It is cited in the literature time and again that the colonies contract and expand twice a day. This process is necessary not only for respiration but also for feeding, since the polyps can only take up food when expanded. Observations and experiments both in the aquarium and in the field have not been able to substantiate this statement, however. Contraction and expansion occur at extremely variable time intervals, often not occurring for one to two days, or happening several times in one.day. The polyps do not expand each time the colony does; in such a case it must be assumed that the swelling only provides the enterons with fresh water.

As an indicator of the numbers of these dead man's hands present on the sea bottom, the amount brought up by trawlers on each fishing trip can be quoted: an average of 500 kg. Owing to this, attempts were made to make some commercial use of this animal, since to throw them back into the sea would only mean that the next trawler would bring them up again. By salting and drying them, the water, which made up eighty-two percent of the colony's weight, was removed; the remains were ground to make fertilizer (Alcyonium fertilizer). Approximately 100 kg of

▷
Two families of hydrozoans—milleporids and stylasterids—form calcareous skeletons very similar to those of the true corals. Illustrated is a stylaster species from the Caribbean Sea.

▷▷ and ▷▷▷
Left from above to below: Star coral (*Astroides calycularis*; see Color plate, p. 271). *Cladocora cespitosa*. Fire coral (*Millepora platyphyllus*). Right: Coral reef: Flattish areas of soft coral spread in the foreground between high-branching colonies of stony corals and fire corals.

▷▷▷▷ and ▷▷▷▷▷
Left from above to below: Soft corals: the alcyonarian *Xenia* showing clearly the plumed tentacles of an octocorallian. Red Sea.
The dead man's hand (*Alcyonium digitatum*) from the Mediterranean. Polyps in full "bloom."
Sarcophyton trocheliophorum (see Color plate, p. 242). Polyps retracted. Great Barrier Reef.
Sarcophyton ehrenbergi with extended polyps. Red Sea. Right: The exceptionally brilliantly colored, tree-like, and branched soft coral *Dendronephthya* from the Red Sea.

◁
Part of a fanlike, branched horny coral from the Indian Ocean. The polyps are snow white.

◁◁ and ◁◁◁
Above left: A giant anemone with anemone fish in the right foreground. Left, *Sarcophyton*, center, *Dendronephthya*. Red Sea. Center left: Precious coral (*Corallium rubrum*). Its red calcareous skeleton is used in jewelry. Mediterranean. Below left: *Dendronephthya* in close-up. The calcareous skeleton spicules can be clearly seen within the jellylike layer covering the stem and branches of the colony.
Right: Horny coral fans: left and in the background, *Paramuricea chamaeleon*; right, the yellow horny coral *Eunicalla cavolini* (compare Color plate, p. 262). Mediterranean.

this coral is required to make 15 kg of fertilizer. Alcyonarian corals have also been used in medicine since the 13th Century. Roasted dead man's hands were even prescribed until recent times as a cure for goiter, owing to their high iodine content. The name "dean man's hand" is also very ancient, and was introduced into the literature in 1623 from the posthumus works of the Italian naturalist Ulisses Aldrovandi (1522–1605).

A closely related species, *Alcyonium palmatum*, found in the Mediterranean, is very similar to the North Sea form. It lives at depths of 20 to 200 m; in appropriately shady areas I have found them even at 2 m depth. *Alcyonium brioniense* is much less common, however. It has yellow polyps with only eight featherlike processes on each side of the tentacle. The genus *Gersemia*, which replaces *Alcyonium* in the North Atlantic and American coasts, is very similar in shape and color to the European genus. *Anthomastus grandiflorus* is also found in the same area, and has a mushroom-shaped form. On the upper side of the "cap" are a few very large polyps almost 2 cm in diameter. Although certain alcyonarians live in cold water and even penetrate into the far north, the large majority of them is found in sub-tropical waters. The greatest variety of forms is found in shallow water on coral reefs, usually as flat cushions or short branches in the most brilliant colors. The upright, large growth forms are usually found in deep water.

As end points of a developmental series, the tiny solitary polyps of the genus *Haimea*, which can reach a size of 10 mm, and the colony-building forms of the CORNULARIDS (family Cornulariidae) can be taken as examples. In the cornularids, a primary polyp develops from the planula larva; this forms a horizontal network over the substrate from which secondary polyps are budded off, the end result being a lawnlike colony. There is no central skeleton and calcareous spicules are entirely lacking. *Cornularia cornucopeia* (see Color plate, p. 271) is found in the Mediterranean from the surface down to depths of a few meters. The polyps, which are about 1 cm in size, live in delicate brownish tubes. This species is very commonly seen in aquaria, but the polyps are only rarely extended.

A further development is the genus *Clavularia*, since it shows tendencies to form a common central jelly-mass (coenochyme). The jelly envelope in several species joins together in the rootlike areas and forms a plate covered with ectoderm, from which the polyps arise. Massive growths are found among the ORGAN PIPE CORALS (Tubiporidae; see Color plate, p. 271) of the Indo-Pacific reefs. These are lumpy colonies about the size of a human head, constructed of a number of red tubes about 20 cm long and 2 mm in diameter. The polyps are greenish, and, when the colony "blooms," give it a most beautiful color contrast. The polyps are not only joined together by the original rooting area at the base but also higher up at several points by a tube system formed by the endoderm and enveloped in mesogloea and ectoderm. Red calcareous skeletal elements are found in the mesogloea, formed by invading ecto-dermal cells. In the oral area, these remain single, but below this they join with one another to form a hollow tube which includes the endodermal canals. A new horizontal floor forms from time to time within the tube, also becoming calcified and bringing about the death of the lower part of the polyp. This gives the colony its typical appearance: rows of parallel tubes joined to each other every few centimeters by a horizontal plate. The largest of the over 800 alcyonarid corals known to date is *Sarcophyton*, (see Color plates, pp. 240 and 242), a member of the family Alcyoniidea, living on tropical coral reefs and reaching a height of up to 1 m.

Every sport diver knows the plantlike structure of the GORGONIDS (order Gorgonaria). The branches are brought up in the hundreds from the sea bottom and dried, later finding their way into the homes of marine enthusiasts as souvenirs and decorations. These coral colonies, almost without exception, have a central skeletal axis constructed of a hornlike material, gorgonin. This envelops the tiny solitary calcareous spicules (sclerites) in a threadlike mass. This is the case in the suborder Holaxonia. Under certain circumstances, gorgonin is entirely lacking or present only in a very small quantity. In such forms, the calcareous particles can be cemented together, as in the suborder Scleraxonia. The soft, elastic axis and calcareous skeleton are enveloped by a soft outer sheath in which the polyps are embedded. The colony growth forms vary tremendously and can be branched to a greater or lesser extent. There are colonies with a strong main axis and finer lateral branches which resemble a bird's feather. Others have several main axes and look like many-branched bushes or little trees. In some species, the branches join each other over large areas so that a netlike structure is formed. In areas where strong currents are present, the colonies usually grow so that their branching plane—if one is present—is at right angles to the direction of water flow. Owing to this, the food, which usually consists of small planktonic organisms, can be caught as in a vertical net.

This type of structure has certain disadvantages. There are a number of planktonic larval forms which require a solid substrate for their further development and thus settle on such gorgonid corals. Divers are always amazed at the variety of strange growths that can come about in this way. Algae, sponges, coelenterates, bryozoans, sabellid worms, and even barnacles can be found attached to the corals. This conglomerate can reach such proportions that the coral itself finally dies. The larvae of certain crustaceans and worms are also able to bore into the living covering of the gorgonid coral and stimulate diseased growth within the colony. As in the case of plants, tumorous swellings develop, forming "galls." Gorgonids also have direct enemies which feed on the polyps and the barklike cover-ing. An opisthobranch snail, *Duvaucelia odhneri*, apparently lives entirely on the coral *Eunicella verrucosa*. Its body form mimics that of the polyps, and its color is exactly the same as that of the branches; this is an example

Order: Gorgonaria

Fig. 6-39. A precious coral with part of the "bark" removed: 1. Polyps, 2. "Bark," 3. Axial skeleton.

of food-directed color adaptation. The about 1200 species of gorgonids are divided into two suborders: the Scleraxonia and Holaxonia.

In the SCLERAXONIANS (suborder Scleraxonia), the calcareous spicules are either loosely bound to one another by gorgonin fibers or completely cemented together by calcareous substances. The best-known genus is definitely the PRECIOUS CORALS (*Corallium*), among which is the RED PRECIOUS CORAL (*Corallium rubrum*; see Color plate, p. 242). Their skeleton contains no gorgonin and is brittle and thus as hard as a rock. Usually, the colonies are 10 to 40 cm in height; the branches are up to 4 cm thick at the base. Related species are present in Japanese waters. They often reach a height of 1 m or more and weigh up to 40 kg. The colonies, which are branched in all directions, resemble tiny trees. A comparison to a leafless, blooming bush is very easy to make when the polyps expand, since these are white and have eight feathered tentacles, thus appearing like tiny flowers.

The living tissue covers the skeleton as a thin layer, similar to the bark on a branch. This layer contains the many-branched net of rootlike stolons which join the individual polyps to one another; the longitudinal canals are visible on the skeleton as thin, longitudinal ridges. The skeleton can vary in color from white to deep red; brick red, rust red, dark brown, and rarely, even black colonies have been found. Flecked forms are also found occasionally. The edge and center of the branch often differ in color, as is the case in tree trunks. A Japanese coral, for example, had a whitish center and a reddish periphery; another had a reddish center and a whitish periphery. The skeleton is constructed of red or other colored calcareous spicules fused together; tentacle-less siphonozooids are also found together with the tentacle-carrying feeding polyps which are deeply embedded in the outer layer and appear almost like pores. They are concerned with the water-circulatory system within the colony. New polyps bud from the thick stolon net which joins the enterons of individual polyps with one another.

The development of the precious corals was studied by the well-known French zoologist Lacaze-Duthiers in 1864 on the African coast of the Mediterranean. He proved that the colonies are, as a rule, of separate sexes and thus carry either male or female polyps. Occasionally, however, polyps of both sexes can occur in the same colony; individual polyps can even be hermaphroditic. The ova are fertilized within the parent polyps and continue their further development there until the elongated, delicate planula larvae are released. Within a few days, these attach to a rocky substrate and develop into primary polyps which, through budding, grow to form a new colony. Precious corals require certain conditions for the best possible growth: apart from a solid substrate, only slight fluctuations in water temperature and salt concentration, and a low rate of water flow and indirect sunlight. Because of these requirements, the main distribution area lies between 20 and 300 m in depth. Above this, they occur only rarely in protected, shadowy areas, beneath overhangs, in cracks in the rocks, in caves, or in breaks in the so-called "Coralligène." At greater depths they can be found growing at right angles or even horizontally on rocky walls. Apart from solitary forms, which can often reach enormous sizes, precious corals also occur in small groups and can, if conditions are right, form actual banks.

I have found precious corals in shallow water down to about 10 m in depth, not only in Villefranche-sur-Mer but also on the Costa Brava and the Côte Vermeille. These areas are, however, plundered completely by ruthless snorklers and scuba divers. The corals have begun to grow again in these areas because strong protective measures have been taken, and now whole beds of knitting-needle-sized corals can be seen there, although it will take decades before reasonable colonies have grown again.

Precious corals have been used for centuries or even thousands of years as decoration and worked into jewelry. For example, artifacts such as brooches, helmets, arm rings, and parts of horse's trappings decorated with coral have been found in La Tene graves (400–58 B.C.). Precious corals came into general use within the European area, however, during the 15th Century; they appeared in place of other precious stones and were worked into necklaces or earrings or used in inlaid work. Long before this, however, they were used in China as jewelry, and from there their use spread to Japan. Since the precious corals present in Japanese waters were only fished for during the middle of the 19th Century, it must be assumed that a large proportion of the raw coral worked here must have originated from the Mediterranean area and that the rest was obtained from the Malayan Islands. We know definitely that the Romans had a lively trade in precious corals with the states of Asia Minor. According to the reports of the famous traveler Marco Polo (1254–1324) and to Chinese sources, the Tibetans used precious corals as currency during the 13th Century.

Precious coral as jewelry

Despite the growth of synthetic substances and fashion jewelry, precious corals have been able to hold their own in the jewelry industry. Large numbers of necklaces, bracelets, brooches, and other articles of daily use are made from them. It is no wonder, therefore, that illegal attempts are continually made to rob the coral banks. In earlier times, harvesting the precious coral fisheries was a rather dangerous practice which could be carried out only in especially favorable areas—vertical or horizontal rocky walls—with any hope of success. A special capturing apparatus was used, consisting of two metal or wooden rods, each about 3 m in length, which were bound to one another to form a cross. These were often weighted with stones or iron. Several bundles of wide-meshed net were attached to the rods and the apparatus let down to the bottom on a long rope and then dragged across the coral colonies or up the rock wall. The heavy metal or wooden rods broke off the coral from the substrate, and it was then caught in the meshes of the net. Naturally, many corals were lost by this means, and the coral banks were severely damaged. Since the corals usually inhabit rocky areas, the nets often became caught and could only be released by a series of difficult maneuvers or were abandoned. This type of harvesting apparatus is still in use in some areas.

Nowadays, the majority of the coral is brought up by divers; this reduces loss on the one hand, and on the other, allows areas which could not be fished before to be invaded. Divers can find caves and cracks and bring up from these protected places enormous colonies which have taken decades to grow. During the last few years I have had the opportunity to observe how divers search the bottom square meter by square meter and remove even the tiniest coral colonies. Extensive areas of the French, Italian, and Yugoslavian coasts have been so completely stripped that the precious corals have disappeared completely from certain spots. Planula larvae colonize these areas again, of course, but, if no colonies capable of reproducing are present, no larvae can be formed, and even protective measures are no longer of any use. These corals were also, during earlier

Healing powers and magical properties

times, attributed with a great healing value. Up until the 19th Century, pulverized colonies were regarded as a universal medicine against all kinds of illness. Neckbands and earrings were worn as amulets, since they "drew" the illness out of the person or "protected" them from the "Evil Eye." Regarding magical properties, which the corals were considered to have, they were used as protection and built into many implements of war such as helmets, scabbards, and sword hilts.

Apart from the about twenty species of precious corals that live in shallow water, another species, *Corallium abyssorum*, found in Hawaiian waters, lives at depths of 1800 to 2400 m. Among the forms in which the calcareous spicules are joined together by small amounts of gorgonin, is the species *Paragorgia arborea*, found along the Norwegian coast and in the Atlantic. This species forms treelike colonies up to 2 m in height, and the base of the stem reaches a diameter of 4 cm. The basal living tissue is reddish-white to brick red, and the polyps are either blood red or yellowish-white. These are so closely packed together that the branches look as if they are thickly covered with flowers. It is worthy of note that the feeding polyps in *Paragorgia* are sterile and the sexual cells are formed within the siphonozooids. *Paragorgia* is found together with *Lophelia* (a stony coral) and *Stylaster* (a hydropolyp with a skeleton) at depths of 60 to 2000 m and can form high, long "hedges," comparable to coral banks. These corals can live at temperatures of down to 4°C. The FALSE CORALS (*Parerythropodium coralloides*) also belong to the scleraxonid corals. These form irregular, crustlike colonies, especially on the branches of horny corals, but also on tunicates (*Microcosmus sulcatus*) and, on occasion, algae. Small pieces of such colonies are very easily confused with the precious corals. When the living tissue is scratched away, however, the skeleton, which is of another color, can be seen. These species survive well in aquaria.

In the HOLAXONIDS (suborder Holaxonia), the axis consists of flexible, elastic gorgonin fibers between which calcareous matter is laid down, but not in the form of spicules. Enormous numbers of SEA FANS (*Eunicella*) can often be found in shallow waters around the coasts of Europe, appearing to the diver almost like a low shrubbery. The WARTY CORAL (*Eunicella verrucosa*), found in the North Sea and Atlantic, is only slightly branched. The dark brown to black axial skeleton is covered with a whitish or pale orange to reddish "bark" from which innumerable whitish polyps emerge. Some divers have been delighted with this delicate coral and have broken off branches and brought them to the surface to be dried and used as living-room decorations. These people have been bitterly disappointed, however, since the branches retain their form but lose their color completely. In contrast to other horny corals and especially precious corals, the color is not impregnated in the calcareous skeletal parts but is found in the form of carotinoids—compounds similar to carotine found in carrots—within the living cells of the "bark." When the animal is dried, the cells die and the color is lost.

A related species, the WHITE HORNY CORAL (*Eunicella stricta*; see Color plate, p. 262), is often found in large clumps in the Mediterranean. The branches are all on one plane, and, since this plane is usually at the same level as that of neighboring animals, this gives the impression of a deliberate planting. The "bark" of this coral is pure white and retains this color on drying. The olive to light brown polyps are extremely small, even in an expanded state, and are hardly noticeable. *Eunicella verrucosa* and *E. stricta* are usually found at depths of 10 to 30 m. I have found small colonies of *E. stricta* even at 2 m depth in certain shadowed areas. The preferred areas of attachment are vertical rocky walls or large boulders. These corals sometimes occur on soft bottoms, but here they are always attached to stones and bivalve shells.

The beautifully colored colonies of the VIOLET HORNY CORAL (*Paramuricea chamaeleon*; see Color plate, p. 243), up to 1 m in height, are first met with by the diver at about 20 m depth. These colonies have a stronger growth form than the previously mentioned species and are much more richly branched but, like them, they grow on one plane. The polyps are extremely crowded, especially on the tips of the twigs, where knoblike protuberances are visible. The colony's color varies from carmine-red to violet; individual twigs or their tips are often bright yellow. Since the "bark" contains only very little calcareous material, only the dark axial skeleton remains after the animal has been dried.

The YELLOW HORNY CORAL (*Eunicella cavolini*; see Color plate, p. 243) is found only rarely, in comparison with the white horny coral. Its colonies are very strong and irregularly branched. This impression is strengthened since the lateral branches are not all on one plane. In comparison with the violet horny coral, this horny coral is very delicately structured; despite this, it often forms colonies which are much more extensive than the others. I have found colonies 70 cm or more in height. Reidl has found colonies of yellow horny coral in caves at depths of less then 1 m; the normal distribution range, however, lies between 15 and 100 m. Since this coral remains relatively attractive when dried, and especially since it, retains its orange-yellow color, it is plundered completely by divers in certain localities.

VENUS'S FAN (*Rhipidogorgia flabellum*) occurs in subtropical waters. The individual branches of the colony are partially joined together, forming a flattened network. The Venus's fan can be compared with a huge tennis racquet, having a height of 2 m and a breadth of 1.5 m. The colonies are quite variable in color: yellow, purple-violet, and even blue. Since the color substance is embedded in the calcareous parts of the colony, the coloration is retained on drying. Relatives of this species from the Pacific Ocean are the violet-red *Eugorgia rubens* and the coral-red *Lophogorgia chilensis*, which forms small bouquests with white polyps.

Order: blue corals

Previously, only the fossil remains of the BLUE CORALS (order Helioporidae) were known. Nowadays they are also represented by the single living species *Heliopora coerulea*, from the Indo-Pacific areas. The rootlike stolon net formed by the polyps joins to form a plate from which vertical tubes are developed; the outer envelope thus looks like a palisade. The skeleton does not consist of calcite in the blue corals, but of united fibers of aragonite, another crystalline form of calcium carbonate. Massive blocks of up to 50 cm in height can thus be formed, with lobelike or fingerlike branches, thus resembling the stony corals. The living tissue is only 2 to 3 mm thick and envelopes the skeleton in a blue-gray "bark." The minute polyps are chocolate-brown, and measure barely 1 mm in diameter. Only the uppermost part of the polyp extends above the surface of the colony. The "bark" layer, which covers the calcareous skeleton, is formed from ectodermal cells, and secretes the aragonite lamellae. Within the calcareous mass are narrow tubules containing the blind-ending, rootlike processes and the wider and broader tubes for the polyps. From time to time the polyps are cut off from the deeper parts of the tube by a horizontal plate. The tissue separated by this process continues to live for a short time, and secretes bright blue calcareous material colored with iron salts. This reduces the tube diameter, and the calcareous mass in general becomes more compact.

Fig. 6-40. The living "bark" of a blue coral removed from the skeleton.

Order: sea pens

The SEA PENS (order Pennatularia) are most definitely one of the most usual forms among the octocorallians. Their fleshy colonies, in contrast to the remainder of the octocorallians, are not attached to a firm substrate but are only loosely embedded. The axial skeleton is reduced to a horny axial pole which is only rarely calcareous in the lower third of the colony. The colony is clearly divided into two sections: a polyp-free stalk (pedunculus) and a polyp-carrying quill (rachis). The sea pens are anchored to the sandy or muddy substrate by the strong stem. The polyps are either individually arranged on the quill or in "featherlets."

The strange colonial structure becomes more understandable when the development of a sea pen is followed. The free-swimming planula larva metamorphoses into an elongated primary polyp, the base of which becomes the stem; large numbers of secondary polyps grow laterally to this; these are normal feeding-polyps (autozooids) and siphonozooids, able to take up and exhale water. In most cases, the primary polyp loses its crown of tentacles and mouth. Individual calcareous bodies are found in the jellylike layer between the ectoderm and endoderm, often being brightly colored, giving the entire colony its yellow, orange-red, brown, or even purple tone. Sea pens are also known for their luminescence in darkness. No true light-producing organs are present, however, and the process comes about through tiny grains embedded in the mucous covering of the sea pen which luminesce when extruded. Even if it has been dried out, the mucus glows when it comes into contact with water. It has been proved that sea pens respond to touch and also probably to chemical stimuli by producing mucus, and this luminescence is thus mainly the result of such stimulation. The light-producing process spreads from the stimulated zone over the entire colony at a speed of 25 cm a second. It is controlled by the nervous system and inhibited by sunlight.

Phosphorescing abilities

Scientists of the 19th Century all thought that sea pens could move from place to place by swimming; members of the family which is about to be described, the PTEROIDIDS (family Pteroididae), were even termed the "swimming polyps." In actual fact, swimming has never been observed in these animals; sea pens can, however, move slowly about the substrate by creeping, or revolve on their vertical axes while still embedded in the substrate. If the animals are observed over a long period of time, it can be proved that the colonies expand and contract and that waves of expansion move along their axes. Certain of these observations came about through slow-motion filming of the animals. Some zoologists still believe that rhythmical expansion and contraction occur in connection with the tides or have a diurnal rhythm. I have observed sea pens in the aquarium for periods of from a few days up to two weeks and recorded their movements by slow-motion film; I was unable to prove any connection between these movements and external conditions.

The shallow-water sea pen *Scytalopsis djiboutiensis* appears to be an exception to the rule. D. Magnus was able to prove a definite nocturnal activity period in a population of this species at about 1 m depth. As soon as night fell, the colonies emerged from the substrate, and as soon as the light intensity at the water surface reached approximately 100 luxes, the colonies expanded completely. The feeding polyps began taking up food. This condition lasted until morning; then the sea pens retracted completely into the substrate. Temperature and water movement and also the tides had no effect on them.

Two or more sea pens kept in the same aquarium can behave completely differently. A certain level of expansion can last from a few minutes to several hours. Sometimes only partial elongation or contraction occurs. The sea pens move under their own power over short or long periods of time, and can move distances of up to 40 cm. Digging into the soft substrate occurs by coordination between water intake and contraction of the circular and longitudinal musculature. A sea pen lying on the sand and about to dig in will bend the tip of the stalk at right angles to the substrate and, by contraction and expansion of this tip, make a tiny hollow in the sand. Then, by various other movements, the stalk is sunk deeper and deeper into the substrate. From time to time the rachis lifts itself from the bottom, each time sinking back down again. The deeper the stalk is embedded in the substrate, the more stiffly and for longer periods of time the rachis holds itself above the substrate until finally, with a jerk, it takes up a perpendicular position.

If the sea pen is held horizontally, within a short time the lower portion of the stalk forms a right angle to the earth's surface; burying is therefore not brought about by tactile stimulation but by gravity. The rhythmical contraction waves can be clearly seen in a suspended sea pen, progressing over the whole length of the stalk, the rachis remaining motionless at first. Burying processes last between half an hour and several hours, depending on the consistency of the substrate and the sea pen's level of activity. When the burying activity is completed, the colony is by no means fully expanded; the animal usually reaches its full expansion much later. This proceeds so slowly that it can hardly be perceived with the naked eye. The contraction of an expanded colony, however, can take place within about thirty seconds. If a colony is taken out of the water so that this contraction can be observed, one can prove conclusively, while the animal is still held in the hand, that water is expelled through the siphonozooids and that longitudinal muscles also play a part in this process. Not only embedded but also freely lying sea pens can expand completely; digging in, however, has never been observed while the animals have been in this condition, and is probably impossible then.

Certain of the sea pen species have separate sexes. This refers not only to the individual polyps but to the colony as a whole. It appears that the number of female animals is much greater than the number of males. As in the majority of the octocorallians, the sexual cells are formed within the mesenteries of the feeding polyps. In the majority of species, fertilization occurs freely in the water; in only a few species does development of the planula take place within the parent polyp. After a few days of free swimming, the planula metamorphoses into a primary polyp.

The GRAY SEA PEN (*Pteroides griseum*; see Color plate, p. 271) is found in the Mediterranean and along the European Atlantic coast. A fully grown individual, depending on the degree of expansion, reaches a size of 10 to 30 cm. A relatively short, thick rachis is attached to the compact orange-red stalk. Two rows of "featherlets" are attached to the main axis, one on each side; there can be up to forty of these, supported by ten or eleven

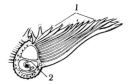

Fig. 6-41. Sea pen, horizontal section through the stem and "featherlet":
1. The polyps of the "featherlet," 2. Enteron cavity.

large, calcareous spines. Each of these lateral branches is formed of 300 to 500 individual polyps (feeding polyps and siphonozooids); a single gray sea pen therefore has up to 40,000 individual polyps. The feeding polyps bud on the lower edge of the "featherlet," which is nothing more than the fused, elongated enterons of these polyps. All are connected with the enteron of the primary polyp. The siphonozooids are found mainly on the lower part of the colony and are especially concentrated at its base near the main axis.

The main distribution area of the red-tinged PHOSPHORESCENT SEA PEN (*Pennatula phosphorea*; see Color plate, p. 263) is the Atlantic; solitary individuals are also found in the Mediterranean. It is thinner and more delicate in structure than the gray sea pen, and the lateral branches have no calcareous spines. The number of whitish individual polyps per featherlet is much smaller, being approximately twenty per row. Between the lamellae are the siphonozooids, which also cover the upper edge of the rachis except on a narrow central strip. The ability to phosphoresce has been extensively studied in this species, whose strong, greenish-blue light has resulted in its being given the specific name *phosphorea*. Related forms of a similar growth type are found in the Indo-Pacific area—for example, *Acanthoptilum* in the Indian Ocean and even in Antarctica. They are usually found in coastal waters, and have never been discovered at depths of more that 200 to 300 m.

Apart from these truly featherlike types of growth, another family, the VIRGULARIIDS (family Virgulariidae), is also known. These have the polyps on lateral protuberances or on very short "featherlets." The rod-like *Virgularia mirabilis* is occasionally brought up by fishermen in the Mediterranean and Atlantic from depths of at least 40 m. This species is about 50 cm long and has small lateral featherlets, each with up to sixteen polyps. They are similar in shape to a fern. These creatures are yellowish, orange, or flesh-colored in tone. The similarly structured *Stylatula* is found in the Pacific. According to a report by Ricket and Calvin, these colonies can form lawns within the tidal zone in certain areas, and the investigators compared such collections to a field of wheat. *Veretillum cynomorium* (family Veretillidae; see Color plate, p. 262) can be found near the coasts of the Mediterranean and Atlantic. These sea pens live at depths of from 30 to 100 m on substrates which allow them to dig in but that are, at the same time, solid enough to provide a good anchorage. The colony is divided into a stalk, swollen beneath the substrate to form an anchor, and a club-like, polyp-bearing upper portion. The secondary polyps are scattered irregularly over the whole surface of the club; their axes are perpendicular to the main axis. The siphonozooids lie between the feeding polyps. Such a colony can reach a L of 50 cm when fully extended and, when contracted, only 5 cm. An axial skeleton is completely lacking. The polyps are capable of withdrawing completely, and are then only visible as pores on the surface, the colony resembling a pinkish sausage. In extended form, the individual polyp's L measures up to 2 cm; because of their size and transparency, they serve as good demonstrations of octocorallian structure. The internal organs can be easily seen with the naked eye.

The SEA PANSIES (family Renillidae), found in West Indian waters, are unique in structure, although related forms also occur in the Indo-Pacific. The upper portion of the primary polyp is dislike in shape and heart-shaped or kidney-shaped in outline. Polyps bud only on the underside of this disc in regular arrangements; the oral side has no polyps and the siphonozooids are found between the feeding polyps. There is no axial stalk. Sea pansies are either violet or red and form an exceptionally beautiful picture when thousands of them are found together on the muddy sea bottom. Their food consists entirely of minute organisms which become caught in the mucus covering the upper surface; from there they are taken up by the feeding polyps.

The SEA WHIP (*Funicula quadrangularis*; see Color plate, p. 271), occurring in the Mediterranean, Atlantic, and Indo-Pacific Oceans, is another animal which is unique in structure. It prefers muddy bottoms at depths of from 40 to 400 m. The lower portion of the very slim, elongate colony (up to 150 cm long) is a skeletal axis. This is stiff in comparison to the flexible upper portion. The feeding-polyps are found on both sides of the stalk. In the sea whip, there is no great difference in structure between the two types of polyp; the less common siphonozooids differ from the feeding-polyps only in that they lack featherlets on the tentacles. Their low numbers and relatively small differences from the feeding-polyps is probably associated with the fact that the sea whip has a relatively general type of body structure.

The UMBELLULIDS (family Umbellulidae) are definite deep-sea inhabitants, for example *Umbellula antarctica* (see Color plate, p. 271), found in the Antarctic Ocean at depths of about 500 m. Its stalk is embedded in the substrate, and the rachis is elongated (up to 60 cm) and thin. Above this is a group of very large polyps which are bright orange-red to purple. The creatures thus resemble a carnation. An umbellulid of this description was brought up by the Danish deep-sea expedition *Galathea* from a depth of 5000 m. The few examples known indicate the great variety of forms present in the order, which includes about 300 species.

Phylum: Acnidaria, by H. R. Haefelfinger

The ACNIDARIA form the second branch of the coelenterates. Two basic characteristics differentiate these from the Cnidaria. As their name indicates, they are completely lacking in nematocysts. In their place, numerous adhesive cells are often formed which are also capable of attaching themselves to prey but which are unable to paralyze it. The second difference is that the acnidarians have no radial symmetry; the body is divided into two different symmetrical halves by two planes at right angles to one another (the tentacle plane and the gullet plane). The system is best compared to a walnut which can be divided by two divisions into two bilaterally symmetrical halves. The body of acnidarians, as is the case in the cnidarians, is composed of three layers: ecotderm, endoderm, and a layer composed of migrated cells, the mesogloea. The acnidaria are exclusively marine animals. Their single class, the COMB JELLYFISH (Cteno-phora), is divided into the subclass Tentaculifera—those bearing tentacles—and the tentacleless Nuda. About eighty species are known; the largest is the Venus's girdle, with a L of 1.5 m.

Despite the small number of species, an exceptional variation in form is found within this group, with all gradations between free-swimming species in the high seas (pelagic) to sedentary ones. The basic structure of comb jellyfishes can be well exemplified by the SEA GOOSEBERRY (*Pleuro-brachia pileus*; see Color plates, pp. 271 and 277). This glassy creature, which looks like a berry, has the mouth opening at one pole of the body and at the other, a sense-organ center. When quiescent, the mouth is directed downward. The surface of the body is clearly divided by eight longitu-dinal rows of ciliary plates arranged lengthwise along the body. These are the "combs" from which the class derives its name. In juvenile animals, these plates are not fully developed; they consist simply of a row of thin, elongate cilia thickly packed together. During the course of growth the individual cilia fuse with one another, forming a "comb plate." The number of comb plates increases with age. The animal is capable of swim-ming actively by the up-and-down beating of these structures. The stroke always starts with the plates nearest the sense organ and proceeds along the row in the direction of the mouth. The individual plates move independently and their coordination is controlled by the nervous system. It is always suprising that the beat is short and strong in the direction from sense organ to mouth but slow and weak in the opposite direction; because of this characteristic, the comb jellyfish always swims mouth-forward.

The play of light along the delicate, beating combs causes a beautiful range of rainbow colors to appear (interference effects), moving in waves along the length of the comb plates. Since certain of the comb jellyfish are completely transparent and colorless, this brilliant color play is the only thing which betrays their presence. Comb jellyfish can change their swimming direction instantly by reducing or increasing the rate of beat of the individual comb plates; they can therefore move in any direction and also in circles. Their water content sometimes rises above ninety–nine percent, and since the specific weight of the body fluids is also less than that of sea water, the rate of sinking is curtailed appreciably so that the animal is able to float at a certain water level without a great deal of effort. Pulsing swimming movements, known in other jellyfish, are not present in comb jellyfish. Some of the lobed comb jellyfish have developed addi-tional means of locomotion by the addition of swimming lobes.

The presence of sense cells is very difficult to prove in comb jellyfish; it is probable, however, that they are distributed over the whole of the body surface. The oral zone, for example, responds strongly to chemical

Class: comb jellyfish

Fig. 6-42. Section through a comb jellyfish.

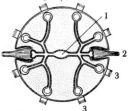

1. Central enteron,
2. Tentacle, 3. Comb plates.

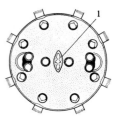

1. Gullet.

Color on color

stimuli. At the aboral pole is a well-developed organ (statocyst) which controls the animal's position in relation to the environment and has a direct connection to the nervous system controlling the motile organs. If the granule in the statocyst (statolith)—a mulberry-shaped sphere of calcium phosphate—is removed, the comb jellyfish can no longer maintain its normal position in space, but the movement of the comb plates continues as normal. Two pockets are present in the upper part of the body on a level with the tentacles, from each of which a long, branched tentacle emerges. They are able to retract completely into these hollows (tentacular pouches), which are constructed of ectoderm.

Fig. 6-43. Adhesive cell from *Coeloplana*: 1. Main section with adhesive granules, 2. Spiral filament.

Within the ectoderm of the tentacles are adhesive cells (colloblasts), consisting of a hemispherical main portion covered with adhesive granules. They are so firmly attached to the base of the ectoderm by a strong spiral filament and an elongate, thin central filament, that the prey can hardly detach itself from the adhesive cell or tear the latter from the tissue of the comb jellyfish. These structures have, however, nothing in common with the nematocysts of the Cnidaria. For example, a nematocyst extruded by a cnidarian is worthless afterward and must be replaced; an adhesive cell, on the contrary, can perform its task several times, since, after use, the spiral filament can be returned to its original state. The prey captured by the adhesive cells on the tentacles is brought by them to the horizontal, oval mouth (elongated in the direction of the oral plane), where it is sucked in. The prey is then carried to the oval gullet and from there to the laterally compressed stomach, which lies at an angle of ninety degrees above the mouth. The longitudinal axis of the stomach thus lies on the plane of the tentacles. Endodermal canals then conduct the nourishment direct from the stomach to the main areas of usage (sensory pole, oral zone, and combs). One unusual fact is that both canals leading to the sensory pole have an opening through which certain metabolic wastes can be expelled—the first hint of an anus, which is present in almost all bilaterally symmetrical animals. The majority of the waste material is, however, extruded through the comb jellyfish's mouth. The entire endodermal system is embedded in a huge cell layer (mesogloea), lying between the ectoderm and endoderm. This layer includes numerous ectodermal cells which have migrated into it, and contains both connective tissue and longitudinal and circular muscles.

Comb jellyfish are hermaphroditic; the endodermal cells produce the sexual cells which are released through the mouth via the stomach. In the majority of species, the gonads are visible through the jelly. A minute comb jellyfish develops directly from the fertilized egg, without an intervening planula stage. Strangely enough, these 0.5- to 1.5-mm-long animals can reproduce for the first time immediately after hatching from the egg or shortly thereafter. The gonads then regress and the animal becomes reproductively mature again only when fully grown. The ova formed the second time are, however, twice as large as the first ones. Such a phenomenon, found exclusively in the comb jellyfish with tentacles and a few other coelenterates, is termed "dissogony" (from the Greek δισσός= double, γονή= production).

Another peculiarity is also met with for the first time in the comb jellyfish: the future determination of individual body zones is set earlier than in the majority of other animal groups. Immediately after fertilization an entire unalterable mosaic appears; this is the "germ mosaic." If the two cells are separated after the first division, each of these develops into a half-larva with four instead of eight combs; on being separated at the four-cell stage, a quarter-larva forms with only two combs. A similar type of early differentiation is found in the rotifers (see Chapter 10), nematodes (see Chapter 10), mollusks (see Vol. III), and certain insects (see Vol. II). In the majority of animals, however, the germal zones differentiate later, so that cells separated in the two-cell stage do not form "half-larvae" but complete "identical twins" (for example, in humans occasionally) and those separated in the four-cell stage form "identical quadruplets" (as in the armadillos). Such germ layers are termed regulation germ layers; it must be emphasized, however, that even in them, the differentiation of individual organ systems does not occur simultaneously.

Comb jellyfish are found mainly in the high seas; only a few inhabit coastal zones or have gone over from a pelagic mode of life to a sessile one. Of the roughly eighty species known, seventy are found in warm seas; only three inhabit arctic waters and three others live in the ocean's depths. Two species are cosmopolitan, found in all oceans—the SEA GOOSEBERRY and the MELON JELLYFISH (genus *Beroe*).Since the comb plates can produce only a relatively small amount of swimming energy, the comb jellyfish are dependent on currents to carry them over long distances. Huge swarms of comb jellyfish can thus be brought together; they feed avidly on fish ova and, apart from this, deplete the food sources of the young fish by devouring large quantities of plankton. Stormy seas can be a great danger to these delicately constructed animals if they are unable to retreat to calmer depths quickly enough. After a storm, undamaged comb jellyfish are rarely found for a long period of time. As a result of the delicate body structure, observation and display of living comb jellyfish is exceptionally difficult because they are usually badly damaged during the capturing processes. Owing to the process of early determination, they are unable to heal wounds. It is rarely possible to keep comb jellyfish alive for a long period of time; melon jellyfish are kept most successfully.

The main characteristics of the TENTACULATE COMB JELLYFISH (subclass Tentaculifera) are the pair of tentacles present in the oral half of the body, and, apart from this, a tubelike gullet. The name Tentaculata is used for this subclass in various zoological works, although this term has been given for a long time to a phylum of bilaterally symmetrical animals, the kamptozoans. The newly formulated name Micropharyngea (small gullets; from the Greek μικρός= small, φαρυγξ= gullet) is also not very appropriate, since the majority of forms have a very large gullet. Therefore, Kaestner proposed the term Tentaculifera (tentacle bearers; from the Latin *tentare*= touch, and the Greek φελ= bear), and introduced it into the literature. The subclass is divided into five orders: 1. Cydippids, 2. Lobates, 3. Venus's girdle, 4. Compressed comb jellyfish, 5. Sessile comb jellyfish.

The CYDIPPIDS (order Cydippidea) are spherical, pear-shaped, or berry-shaped. They have two long tentacles which can be withdrawn completely into the tentacular pouches. Their food consists of a variety of planktonic organisms such as copepods, crustacean larvae, arrow worms, planktonic snails, fish ova, and fish larvae. It is believed today that the herring catch, which varies considerably from year to year, depends on the number of herring ova and larvae which fall prey to exceptional swarms of comb jellyfish. The SEA GOOSEBERRY (*Pleurobrachia pileus*; L about 15 mm, sometimes reaching 30 mm) can be taken as a typical example, as it is the best-known form and is distributed worldwide. This species can appear in enormous numbers in the North Sea, especially in spring, and can even penetrate the Baltic—despite its lower salt content—and reach as far as the Finnish coast. The sea gooseberry is also found in the Atlantic, extending northward in range as far as the Arctic Ocean. The cydippids are among the smallest of the comb jellyfish.

The SEA NUT (*Mertensia ovum*; L about 50 mm) is a definite cold-water form. Its main area of distribution is near the Labrador coasts, but during winter it moves southward to the areas near New Jersey. In contrast to this species, *Hormiphora plumosa* inhabits the warmer waters of the Mediterranean and tropical Atlantic; it can also be met with, however, on the English and North American coasts, having been brought there by the Gulf Stream. *Callianira bialata* (L about 20 mm) is a true Mediterranean form. The body is rectangular in cross section, and has two characteristic long "wings" at the aboral pole. Various comb jellyfish of the genera *Hormiphora* and *Mertensia* are found in the Pacific. The larvae of *Eulampetia pancerina* is worthy of note, since it lives as a parasite in salps (planktonic tunicates; see Vol. III). This species was first described under the name *Castrodes parasiticum*. The minute larvae bore into the jellylike envelope of *Salpa fusiformis*, live on the blood corpuscles and body fluids of the host, grow within it, and finally leave by a means which is still unknown. All the cydippids described are transparent, exceptionally delicate creatures which are, at the most, delicate pink or yellowish.

Order: Lobata

In the LOBATES (order Lobata), the gullet plane ultimately is larger than the tentacular plane; as a result, the body is laterally compressed. A thin swimming lobe hangs downward from each corner of the elongated oral slit. The broad side of these swimming lobes is parallel to the tentacular plane. A minute jellyfish first develops from the egg, resembling a tiny sea gooseberry and carrying two long tentacles. This stage is thus termed the cydippid larva. During the course of post-embryonic development the tentacles become very reduced and the tentacular pouches disappear completely. Parallel to this, the formation of the swimming lobes occurs together with the formation of two fingerlike, ciliated attachments (auricles) to the corners of the mouth. In contrast to certain other comb jellyfish, the lobates are unable to open the mouth to any large extent to swallow prey, and even searching the surroundings for prey with the tentacles is not possible. They thus live almost entirely on small planktonic organisms such as molluscan larvae and larvae of various other invertebrates which they capture partly with the short tentacles and partly with the aid of the ciliated attachments near the mouth, which waft these organisms straight into it. The lobata are able to progress not only in a regular manner by the beating of the comb plates but also in jerks by rhythmical beating of the swimming lobes. In this, they change the direction of swimming and move with the aboral pole forward.

Leucothea multicornis (L up to 60 cm) is an inhabitant of the Mediterranean Sea. The body is very delicate in structure, and it is an uncommon occurrence to be able to observe an undamaged specimen in the aquarium. After storms, only damaged animals float about in the water for days afterward. This species, along with the Venus's girdle, is one of the largest of the comb jellyfish. *Leucothea* is a delicate pinkish-brown and completely transparent. Four extensible protuberances are found on the body surface, covered with adhesive cells (colloblasts). These protuberances are also used in prey capture. The prey captured by the adhesive cells is moved from protuberance to protuberance until it can be reached by the mouth. Species related to *Leucothea* are also found in the Atlantic and Pacific Oceans. *Bolinopsis infundibulum* (L up to 15 cm; compare Color plate, p. 263) is widespread in the Atlantic (as far as the Baltic) and in the Mediterranean. In feeding, it spreads the swimming lobes and captures plankton with them, similar to the way in which fish are caught with a trawl net. This food is wafted to the tentacles by cilia, attached there by adhesive cells, covered with mucus, and then brought to the mouth.

The white, opaque *Mnemiopsis leidyi* (L up to 10 cm; see Color plate, p. 271) is very common along the North Atlantic coast of America down to about the level of South Carolina. It is pearlike in shape and the swimming lobes are about two-fifths longer than the body. They are able to produce a very bright, greenish light during the nighttime, especially in summer. The adaptive abilities of this delicate creature to temperature and the salt content of the water are astounding. Animals have been observed whose swimming lobes have continued to beat until they have been frozen completely. *Mnemiopsis* feeds mainly on copepods and molluscan larvae. If these comb jellyfish appear in large swarms in the vicinity of oyster beds, this usually has the result that thousands of oyster larvae are devoured and the future oyster population is considerably endangered.

A few years ago, I filled a container with fresh sea water far out from the Cote D'Azur, near Nice. This stood for days in the cold room at a temperature of less then 10°C; the clear water was to be used to fill an

▷
A horny coral from the Great Barrier Reef with very closely-packed polyps. A crinoid (see Vol. III) has settled upon it.

▷▷ and ▷▷▷
Above from left to right: The widely spread fans of a horny coral from the Great Barrier Reef. The fans are at right angles to the water currents so that the polyps are able to fish the largest section of the current stream possible. The white horny coral (*Eunicella stricta*), one of the Mediterranean gorgonids. A typical characteristic of this species is the white branches running almost parallel to one another. The phosphorescent sea pen (*Pennatula phosphorea*). Below from left to right: A branch of the horny coral *Eunicella cavolinii* with thickly packed polyps. Close-up of a polyp from *Veretillum cynomorium*. *Paralcyonium elegans*, a Mediterranean soft coral with a definite diurnal activity rhythm in which it pumps itself full of water and then contracts again. Anterior view of the sea pen *Pennatula phosphorea* showing the bilateral symmetry in polyp arangement. A radially symmetrical sea pen (*Veretillum cynomorium*) with widely spaced polyps in whorls around the axis. *Pennatula phosphorea* with the polyps fully extended. A comb jellyfish (*Bolinopsis vitrea*).

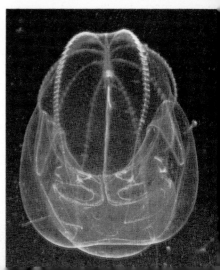

aquarium for photographic purposes. One day I filled a flat laboratory tray with the water and, as I switched on the dark field light, I discovered that a large creature was swimming about in it. Despite transportation, pouring out, and cooling, an undamaged specimen of *Bolina hydatina* had managed to survive. The swimming lobes in this species are relatively small, so easily one could take it for a cydippid at first glance. The tentacular pouches are lacking, however, and the tentacles are short, and closer observation shows clearly that the animal is a lobate. The comb jellyfish swam langorously around in the laboratory tray. I fed it with freshly caught plankton and it lived for some days under my care. Certain visitors were especially impressed with the comb plates which iridesced in all conceivable colors. These seemed to hang free in the water; as soon as the light was switched on, however, the jellyfish appeared in the dark field.

Order: Venus's girdle

The VENUS's GIRDLE (Cestidea) comprise the third order of comb jellyfish. During a daily trip to collect plankton, in which not only tiny planktonic organisms (microplankton) are caught with the usual plankton net but also larger creatures (meso- and macroplankton) with small, flat, dish nets, I stumbled across a mass collection of VENUS's GIRDLES (*Cestus veneris*; width a maximum of 8 cm, L up to 150 cm; see Color plate, p. 271) at Villefranche-sur-Mer, near Nice. The transparent, jellylike ribbons snaked in hundreds through the water. It is hard to believe that these elongated creatures also belong to the comb jellyfish. The body form of a Venus's girdle can, however, be very easily drawn from that of a cydippid larva since early Venus's girdle larvae are also very similar to this stage. After the embryological period, the body grows enormously in the gullet plane, as if the larvae had been taken by the corners of the mouth and drawn outward. A shallow oral groove runs the length of the oral edge of the Venus's girdle; it is depressed in the middle of the body to form a deep gullet. A groove is also present on each side of the mouth slit, from which the lateral branches of the tentacles hang like curved cilia. The lateral branches of the tentacles capture the prey, which consists mainly of copepods and other small planktonic organisms, and carry them to the oral groove from whence they are brought to the gullet in the center of the body by wavelike movements of the edges of the mouth.

Owing to the enormous elongation of the oral plane, the four rows of comb plates are extended over the entire upper edge of the aboral zone. The extremely contracted tentacular plane and its four rows of comb plates are also shortened and consist of only four to six plates. The eight radial canals are recognizable as two fine lines on the upper edge and two lines in the center of the ribbon. They extend from the tentacular plane to the lateral edges. An oral vessel is present on each side of the oral rim. All these vessels combine at the short-based edges.

Although the Venus's girdle is usually illustrated in the process of snakelike movement, this state is an exception. When fully extended with

◁
Above:
Reef builders, primarily stony corals and calcareous algae, have, during the course of geological time, formed huge masses of chalk such as the Langkofel Group in the Dolomites (shown here), which was deposited during the Triassic (about 200 million years ago).
Below:
The stony corals still are active reef builders, under favorable conditions in the tropical seas. They form fringing reefs, barrier reefs, and coral islands (picture).

the gullet plane upward, they hang motionless in the water, being carried by the currents; fishermen have thus given the Venus's girdle the name "sea sword." The body is so transparent that only the upper row of comb plates, with their beautiful greenish-gold shimmer, betrays the animal's position. As soon as the Venus's girdle is disturbed, however, it swims away with snakelike movements, and the glassy body takes on a bluish to deep ultramarine tone. The Venus's girdle is able to phosphoresce brightly at night. Watching these glowing ribbons in the aquarium is one of the most impressive sights that can be offered by the coelenterates.

Once one has become familiar with the planktonic, transparent comb jellyfish, it is hard to believe that the creeping organisms included in the COMPRESSED COMB JELLYFISH (Platyctenidea) can be related to them. A compressed creature develops from the cydippid larvae, a "platelet," which resembles a polyclad worm in appearance and mode of life. This external similarity is so close that early zoologists believed that they had discovered, in the compressed comb jellyfish, a transitional stage between the coelenterarates and the ancient flatworms (see Chapter 8). Exact anatomical investigations showed, however, that this could not be the case. When a compressed comb jellyfish is compared with a sea gooseberry, it is easy to see that the main body axis between the oral and sensory poles has been compressed to a few millimeters. A horizontal section through the body (or better, the body circumference) is oval, since the tentacular plane is more strongly developed than the gullet plane. The mouth appears to be, to a certain extent, wide open and the gullet wall, the interior of the gullet, has become an organ for creeping. These compressed comb jellyfish are able to creep about over the substrate by means of their ciliated epithelium.

Order: Platyctenidae

Ten species of these curious comb jellyfish are known. They range in distribution from the Red Sea through Indochina to Japan. They live exclusively in warm waters. The creatures, which usually measure 10 mm across the tentacular plane but which can reach 70 mm in certain species, live on encrusting algae or octocorallian coral colonies, where they ingest the living "bark" (coenosarc). They vary in color from milky-white to gray, yellow, olive-green, or red. The genus *Ctenoplana* resembles the normal comb jellyfish shape to the greatest degree, the adult form retaining its short row of comb plates from the larval stage. *Coeloplana* (see Color plates, pp. 271 and 277), however, loses its comb plates on going over to a creeping mode of life. This species is able to swim for short distances, however, by beating together the two halves of the body along the tentacular plane. Only the eggs are released through the mouth in the compressed comb jellyfish; special outlets are present for the sperm cells.

In 1908 a sessile comb jellyfish, *Tjallfiella tristoma* (L 6.5 mm), was discovered in Umanak fjord in western Greenland. It was attached to a sea pen (*Umbellula lindahlii*) which had been dredged up from a depth of 475 to 575 m. This species, the only representative of its order the SESSILE

Order: sessile ctenophores

COMB JELLYFISH (Tjallfiellidea), has a cydippid larva. It attaches itself with its wide open mouth and gullet—similar to the compressed comb jelly-fish—to the upper side of its host. A chimneylike tube is present on the aboral pole of the oval *Tjallfiella*, and the tentacles protrude from this. Since this tube is also connected with the enteron, it can be termed a secondary mouth. The comb plates present in the larval stage disappear as the animal matures. The eggs develop in a special brood pouch which is an extension of the digestive system; tiny larvae with comb plates leave the parent animal. After a short period of time they settle on a host animal and metamorphose, as previously described, into the adult form.

Subclass: Nuda

Order: Beroidea

The subclass Nuda contains a single order, the MELON COMB JELLYFISH (Beroidea; BL up to 20 cm). The body form is thimblelike, caplike, or barrel-shaped; the main characteristic is the complete absence of tentacles and adhesive cells in both the adult and larval forms. The body is laterally compressed and the gullet plane extended with respect to the tentacular plane. The mouth opening is extremely large and the point of fusion with the gullet encompasses almost the whole width of the body and six-sevenths of its height.

If the interior of a melon comb jellyfish is looked at through the mouth, one has the impression that the entire animal is hollow, and has no internal organs. One asks how the melon comb jellyfish are able to feed, since they have no tentacles. The wide-open mouth and the adjoining gullet are used to capture prey like a plankton net. To prevent the prey which has been captured from escaping through the large mouth opening, the jelly-fish has a large number of hooked cirri in the lower fifth of the gullet, directed toward the aboral pole. Poison-secreting glands are found in this area, and their poison paralyzes the prey. The true stomach is very small and is situated directly beneath the aboral pole. From here, longitudinal canals run beneath the comb plates to the edge of the mouth. These structures give the animal a unique marbled appearance. The branches of these canals even penetrate the rather firm jelly layer and provide the extremely strong muscles with nourishment. Even the two gullet vessels, which end at the edge of the mouth, send lateral branches to the muscula-ture.

The melon comb jellyfish are one of the worst predators among the comb jellyfish. Such a "feeding lust" is not expected in these delicate creatures but has been proved by observation. As early as 1843, F. Will in Trieste showed that the favorite food of melon comb jellyfish was related species from the order Lobata. Chun, one of the best investigators of Mediterranean comb jellyfish, made the following observation in Naples: he was about to draw a *Leucothea*, and had put it in an aquarium which also contained a melon comb jellyfish which had had nothing to eat for some time. Within a short while, the melon comb jellyfish—probably by means of its chemical receptors—had perceived the presence of a prey animal. It began swimming around vigorously with its mouth wide open. Searching about continually in circles, it finally approached the *Leucothea*. It captured it in its mouth after a rapid turn and held fast to this prey, which was at least twice the size of the melon comb jellyfish itself and which defended itself vigorously. Within a quarter of an hour, it had swallowed the *Leucothea* and lay balloonlike and pulsating on the floor. Witnesses who were called in could hardly believe their eyes when they saw that such a small animal could overwhelm such a large one.

Experiments have shown that melon comb jellyfish take no food except other comb jellyfishes. In this respect they could be of great advantage to the fisheries because they would reduce the numbers of their destructive relatives. Members of the genus *Beroe* are among the only comb jellyfish which eat readily and otherwise do well in aquaria; certain individuals have lived as long as 200 days. A *Beroe* eats, for example, three other comb jellyfish in two days. *Beroe ovata* (see Color plate, p. 271), found in the Mediterranean, is about 10 cm in height. These whitish or pinkish animals sometimes appear in large swarms. They can phosphoresce bright blue to blue-green at night. The larger *Beroe cucumis* (height up to 16 cm) is probably present in all seas. About fifty species of melon comb jellyfish are known.

7   The Bilaterally Symmetrical Animals

Subdivision:
Bilateralia,
by P. Rietschel

In contrast to the coelenterates, whose bodies are more or less radially arranged around a central axis, the remainder of the multicellulate animals are bilaterally symmetrical: the body is divided into two equal halves, which are mirror images of one another, about an imaginary middle plane (the symmetry plane). Even in animals such as the secondarily radially arranged echinoderms, e.g., the starfish and sea urchins, the larva remains bilaterally symmetrical. All these animals are included in the "BILATERALIA."

The bilaterally symmetrical animals repeat during their ontogeny the simple coelenterate body structure and consist of two cell layers, the ectoderm and endoderm. The cavity between these two layers is termed the "primary body cavity"; the cavity enclosed by the endoderm, which is equivalent to the enteron of the coelenterates, is termed the "archenteron." It opens at the point of junction of both cell layers to the exterior in the "blastopore." The mouth and anus come into being within the blastoporic region either by remaining open after the blastopore has closed or by breaking through after it has closed. In the majority of animals, the mouth is a remnant of the blastopore. These are termed the "PROTOSTOMIA." In certain phyla of animals (see Vol. III and all subsequent volumes), the mouth breaks through the blastopore region later in development. In this case, the anus can be a remnant of the blastopore which has remained open, or a newly formed structure. All such phyla are put together in the "DEUTEROSTOMIA." In contrast to the coelenterates, the bilaterally symmetrical animals are not composed of only two cell layers. Between the ectoderm and endoderm an interstitial tissue layer (parenchyma) can occur and, in many cases, a "secondary body cavity" (coelom) is present, surrounded by mesoderm. Its formation differs in the protostomia and deuterostomia: in the protostomia, it is formed from mesoblasts; in the deuterostomia, it comes about by folding of the endoderm.

The division into the main branches, protostomia and deuterostomia, occurred early in the evolution of animal life. The animal phyla existing today which stand near the point of this division are the kamptozoans (see Vol. III) and the arrow worms (see Vol. III). The protostomia reach their peak of development in the arthropods and cephalopods, the deuterostomia in the vertebrates. Up until Volume III, we shall be dealing with the protostomia.

▷
Above left: 1. Portugese man-of-war (*Physalia physalis*), 2. Sailor-before-the-wind (*Velella spirans*), 3. *Praya diphyes*, 4. *Halistemma rubra*, 5. *Chelophyes appendiculata*, 6. *Physophora hydrostatica*, 7. *Abylopsis tetragona*. Above right: 1. Star coral (*Astroides calycularis*), 2. *Corynactis viridis*, 3. *Anemonia sulcata*, 4. *Aiptasia mutabulis*, 5. *Caryophylla clavus*, 6. *Ragactis pulchra*, 7. *Bunodactis verrucosa* (see Color plate, p. 201), 8. *Cerianthus membranaceus*, (see Color plate, p. 203), 9. *Tealia crassicornis*, 10. *Metridium senile*, (see Color plate, p. 199). Below left: 1. Sea whip (*Funicula quadrangularis*), 2. *Renilla amethystina*, 3. *Umbellula antarctica*, 4. *Eunephthya rosea*, 5. *Tubipora purpurea*, 6. Dead man's finger (*Alcyonium digitatum*; compare Color plate, p. 240), 7. *Cornularia cornucopeia*, 8. Gray sea pen (*Pteroides griseum*). Below right: 1. Sea gooseberry (*Pleurobrachia pileus*), 2. Melon comb jellyfish (*Beroe ovata*), 3. *Eucharis multicornis*, 4. Venus's girdle (*Cestus veneris*), 5. *Mnemiopsis leidyi*, 6. *Callianira bialata*, 7. *Coeloplana gonoctena*.

8   Flatworms and Gnathostomulids

Phylum:
Platyhelminthes,
by P. Rietschel

The FLATWORMS (phylum Platyhelminthes) have an external ectodermal layer and an endodermal layer lining the intestine, as in the coelenterates, but between these is a loose interstitial cell layer, the parenchyma.
Within this layer are nerve cords, muscle cells, and excretory and reproductive organs. This tissue is not simply a packing tissue, since its cells also
store foodstuffs (fat and the carbohydrate glycogen.) A blood-circulatory
system and respiratory organs are, however, lacking in the flatworms,
but the fluid between the parenchymal cells serves as a carrier for respiratory gases and food and excretory products. A coelom is absent; there is,
however, evidence that this was once present during phylogeny but has
since disappeared. A supportive and packing tissue has already been met
with in the coelenterates (mesogloea), such as the jellylike mass of a
medusal umbrella; this, however, is derived from ectoderm, in contrast
to the parenchyma of the flatworms. During the course of embryology,
this loose median tissue derives from the same cells (mesoblasts) as the
mesoderm of echiurids, sipunculids, mollusks, and annelids. The parenchyma of flatworms can thus be regarded as the equal of the coelomate
protostomian's mesoderm, even when, in this case, it does not enclose a
body cavity.

The nerve cells of flatworms are concentrated at one pole of the body
as a "brain"; light-receptive organs and gravity receptors are also present
in this region. This pole is thus the "head" and therefore in flatworms,
as in almost all other animals, an anterior (cranial) end, a posterior (caudal)
end, a back (dorsal side), and an underside (ventral side) can be distinguished. One or more pairs of nerve cords run the length of the body
from the brain. The nerve cells of these cords are not concentrated to form
ganglia, but are distributed over the whole of the system. Such nerve
cords are termed "nerve strands." Longitudinal and circular muscles are
present beneath the simple ectodermal cell layer, which is either ciliated
or covered with a cuticle; these form the external muscle sheath. Other
muscle cells run from the dorsal side to the ventral. The intestine, packed

◁
The bluish rhizostome
jellyfish (*Rhizostoma pulmo*;
see Color plate, p. 183)
often occurs in large shoals
in the Atlantic and Medi-
terranean. The oral arms
form eight lobes which
close together to form a
tube internally. For half of
their length and at their
point of attachment to the
bell, they are so ruffled
and folded that a bundle
of matted roots appears to
hang beneath the umbrella.

in the loose parenchyma, opens to the exterior on the ventral side through the mouth; in the tapeworms this has degenerated as an adaptation to a parasitic mode of life. There is no anus; indigestible food remains are expelled through the mouth.

In contrast to the coelenterates, flatworms have a tissue fluid within the parenchymal network which often has a salt content higher than that of the surrounding water. Water thus continually enters the animal from the exterior and must be removed, together with the waste products of metabolism dissolved in the tissue fluid of the parenchyma. This task is fulfilled by excretory organs (protonephridia) which are found in many other higher invertebrates up to *Amphioxus*. Their task is similar to that of the contractile vacuole of unicellulates. They consist of a "flame cell" with a long bundle of cilia which protrude into an excretory canal. The water and the soluble excretory products given out by the cell are driven, with the aid of other ciliated cells, to the exterior.

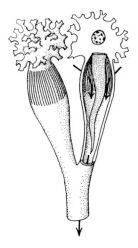

Fig. 8-1. A platyhelminth protonephridium. External view left, cut open, to show the "flame" cilia on the right. Arrows indicate water direction.

With the previously mentioned organ alone, the flatworms are far in advance of the coelenterates. Their usually hermaphroditic reproductive organs are among the most highly developed of the entire animal kingdom. Certain of the flatworms have a unique method of providing the eggs with nourishment, which has been maintained in the flukes and tapeworms which have arisen from them: usually yolk is present within the cells as nourishment during the first stages of development; in these animals, however, yolk is provided in many thousands of yolk cells which surround the ova and which are broken down during the course of embryology. The ovum and yolk cells are enclosed in the same envelope, and, owing to this, the ova must be fertilized internally. Coelenterates depend on the surrounding water for the fertilization of the ovum by the sperm; in the case of the flatworms, however, fertilization must occur within the mother, and thus copulation must be present. Despite their generally simple body structure, flatworms have extremely well-developed sexual organs. The original single ovary is divided into an ovary proper and a yolk sac; one produces the ova and the other their nourishment, the yolk cells. The male seminal canal can end in a copulatory organ (penis) or the tip can evert during copulation to form an appendix (cirrus).

Evolutionary history of the flatworms, by E. Thenius

The evolutionary development of the flatworms is hidden in the darkness of prehistoric times, since only rare fossil remains, of little evolutionary worth, can be found of their delicate bodies. Without doubt, the turbellarians are of low position in the evolutionary tree of the Bilateralia, but more and more scientists now believe that many of the simplest forms could have regressed into this simplicity at a later date. All other flatworms have arisen from the turbellarians with their separate ovary and yolk sac. The time at which this occurred can be estimated from the evolutionary age of these parasites' hosts. The subclass Monogenea, which mainly parasitize fish, are thus, according to Cameron, traceable back to the Paleozoic. On the other hand, although the subclass Digenea are parasites of fishes, amphibians, and reptiles, as well as birds and mammals, they arose well before the Mesozoic. Of the six orders of tapeworms, the members of three of them are parasitic in sharks and rays and those of a fourth order in similar ancient fish; the tapeworms, therefore, although many of them parasitize birds and mammals, are probably an evolutionarily very old class of animals; this is also true for Digenea.

The flatworms have a noteworthy relationship to other, higher phyla of animals: the simplest turbellarians, which have no yolk sac, have a first cleavage of the egg similar to that of the strapworms, echiurids and priapulids, the annelids, and mollusks, and this is so obvious that it cannot be a matter of chance. This similarity, which associates all these groups, is "spiral cleavage." The large group of related animals which contains those with spiral cleavage or those which have arisen from them is termed the "Spiralia." They comprise about nine-tenths of the entire animal kingdom, and, of them, the flatworms lie near the base of this evolutionary tree.

Class: Turbellaria, by P. Rietschel and P. Röben

The TURBELLARIANS (class Turbellaria) are the most primitive of the flatworms; the two parasitic classes, the flukes and tapeworms, have definitely arisen from them. They are the most primitive of the Bilateralia, or at least of the Protostomia. Today, they are the center of much argument among evolutionists; it is a lucky chance that this ancient and thus informative group of animals has survived to the present with about 3000 known species. The majority of them are marine; many inhabit fresh water (about 150 species in Germany), and a few have invaded wet land areas. The majority of turbellarians are hardly 0.5 cm long: the smallest are smaller than some of the larger unicellulates; the largest is the greenhouse planarian with a L of 60 cm. The turbellarians are usually flattened and wormlike in shape, the longer species ribbonlike, the very small ones, when viewed from above, egg-shaped or almost spherical.

The Turbellaria receive their name from their ciliation, which is present at least on the lower surface; smaller species move entirely by these cilia. The turbulence thus produced in the water resulted in their scientific name (Latin *turbo*). The ectoderm contains mucus-secreting cells as well as ciliated ones; the mucus produced also aids in movement. It also protects the delicate body from damage, from invasion by bacteria, and also from fungal attack. Experiments have also shown that fish avoid this mucus—another means of protection for the worm. Rodlike structures (rhabdites) are formed in cells beneath the ectoderm; these enter the ectodermal cells and, on extrusion, expand to several times their original length. These structures also serve as a protective measure. Finally, turbellarians which feed on coelenterates transport the coelenterates' nematocysts into their own ectoderm, where they are used in defense. These are termed "kleptocnidians," meaning "stolen nematocysts."

The turbellarians have a subdermal muscle layer consisting of both longitudinal and circular muscles and also muscles running from the dorsal to ventral surfaces. With these, the animals can extend, contract, and roll

Body Forms of Coelenterates.
1. Sea anemone (order Actinaria), purple sea anemone (*Actinia*) dissected: Oral tube lined with ectoderm and inverted into the enteron, enteric mesenteries with longitudinal muscles (horizontal sections anteriorly), reproductive cells within the mesenteries, and, below these, threadlike "acontia." The enteron pouches are joined to one another above through openings in the mesenteries (stomata, singular stoma).
2. Stony coral (order Madreporaria), star coral (*Astroides calycularis*): 2a. Young polyp; 2b. Colony with one polyp dissected, a

calcareous skeleton to the right. The hard mesenteries (sclerosepta) are between the soft mesenteries and joined by a circular wall (in 2a). A hard basal plate (in 2a) and, above this, a second horizontal floor laid down later (in 2b). A column (columella) is often present at the central axis.
3. Comb jellyfish (class Ctenophora): 3a. Sea gooseberry (*Pleurobrachia pileus*). Entire animal in horizontal section, upper half with the sensory pole, lower with the oral pole (directed upward during life). Two symmetrical planes: one the plane of the drawing (the tentacular pouches present within this, thus termed the "tentacular plane"), the other through the sensory and oral poles, at right angles to this ("pharyngeal plane" since the pharynx is laterally compressed along this plane, not visible in the drawing). The sexual cells are organized with the same "dissymmetry" (hermaphroditic), the comb plates, on the other hand, in octoradiate symmetry. The intestine opens in the tentacular plane at the sensory pole in two pores. 3b. Sensory pole with gravity receptor: ciliated bell with nodule (statolith) in contact with four ciliary processes. 3c. Adhesive cell from a comb jellyfish (*Coeloplana*). Drawings 1, 2a, and 3a with the same schematic color scheme as in the figures on p. 175: white—ectoderm, yellow—endoderm, cinnibar red—female reproductive organs, light blue—nervous system, carmine—muscles, violet—male reproductive organs.

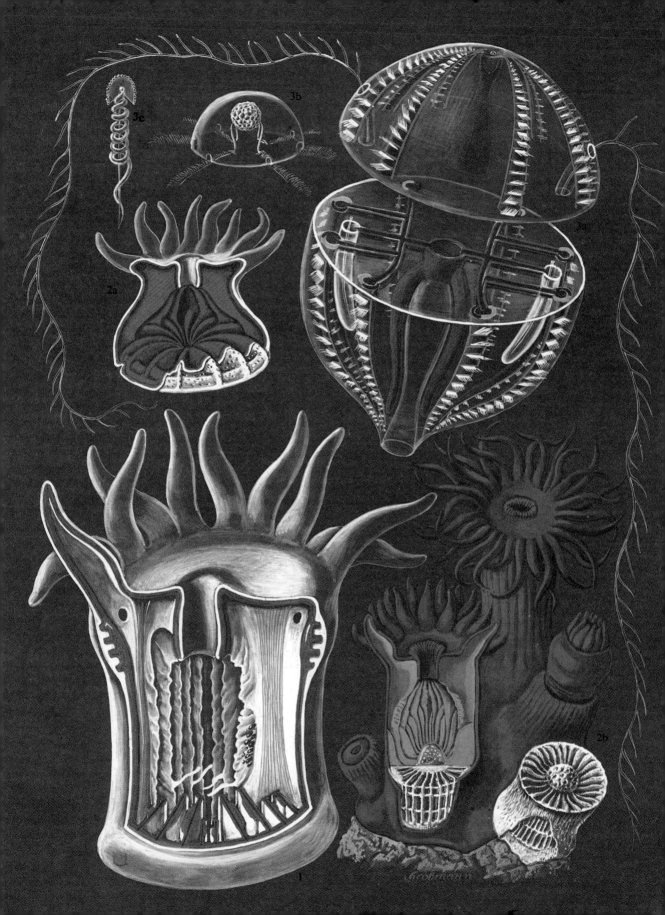

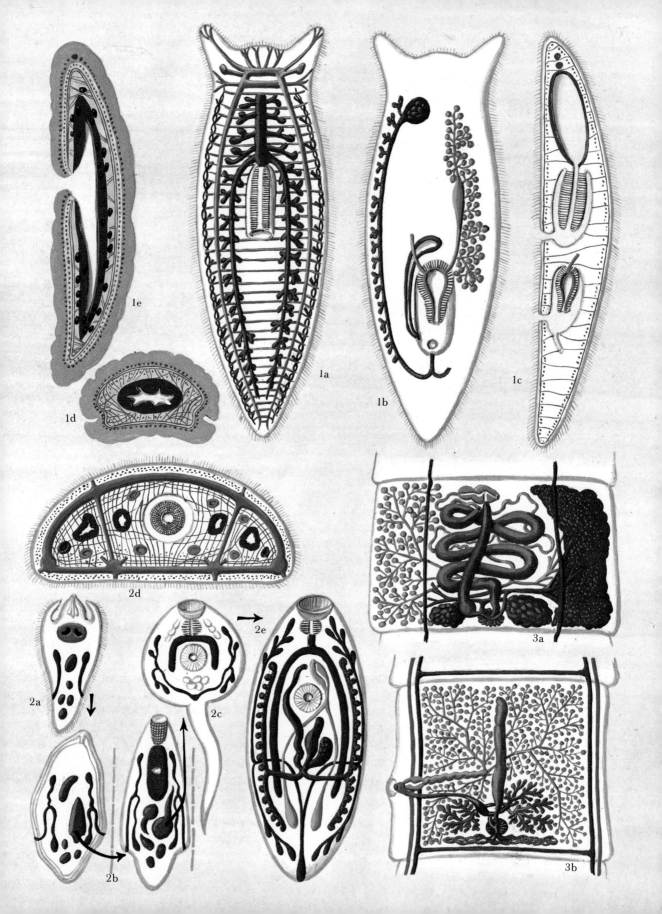

1a

1b

1c

1d

1e

2a

2b

2c

2d

2e

3a

3b

Body forms of Flatworms

1. Class Turbellaria: 1a and 1b in surface view, 1c in longitudinal section, 1d in horizontal section, 1e the genus *Xenoturbella* in longitudinal and horizontal section. 1a shows the pharynx, intestinal canal, nervous system, and excretory organs; 1b, the female reproductive organs to the left, male to the right; the vitellaria are shown lateral to the ovary on the left. Both open into the genital atrium from behind, as does the uterus from the side; the testes to the right with vas deferens and muscular copulatory organ, which often lies within the genital atrium. In 1c and 1d, the longitudinal, circular, and vertical muscles are shown. 1a and 1d show the right-angled (orthogonal) nervous system with longitudinally running nerve strands (two dorsal, two lateral, and two ventral) together with their horizontal connections. 1d, *Xenoturbella* with an epidermal nervous net without nerve strands, ova with-

in the mesenchymatous packing tissue (parenchyma: pink) and intestinal layer (yellow), the ova being scattered singly, not concentrated to form an ovary. 2. Class Trematodes; the three generations of a fluke (subclass Digenea): 2a. Ciliated miracidial larva with cerebral glands, brain, and protonephridia, the redia developing within it; 2b. Sporocyst developed from the miracidium always lacks an intestine; 2c. Redial larva with pharynx and simple intestine, the cercariae developing within it, and the birth opening in the body wall (in front of the intestine); 2d. Cercarial larva with oral and ventral suckers, pharynx, forked intestine, and lashing tail, glandular organ but lacking reproductive organs; 2e. The trematode which has developed from the cercaria by loss of the tail and maturity of the sexual organs.

3. Tapeworms (class Cestoda). Mature proglottid: 3a

of the fish tapeworm (*Dibothriocephalus latus*), the beef tapeworm (*Taeniarhynchus saginatus*); the vitellarium. (left) and testes (right) removed to clarify the drawing.

Male organs: Numerous testes, nodules on branched canals, vas deferens opens in the center (3a) or to the side (3b) with an eversible "cirrus." Female organs: paired ovaries (consolidated in 3a, branched in 3b) open by a canal into the ootype, the vitellarium (in 3a paired and lateral, in 3b single and posterior) also opening into the ootype. Uterus in 3a coiled and opening to the exterior (eggs released from the proglottid), in 3b a central, blind-ending tube (eggs released by decay of the proglottid) acting as an egg-formation chamber. The vagina runs from the genital atrium to the oviduct and joins it shortly before it expands into the receptaculum seminis. Copulation (including self-copula-

tion): cirrus of the younger (male mature) proglottid in the vagina of the older (female mature) proglottid. Ovaries divided into a small germ-cell-producing organ and larger vitellarium. Yolk cells, in this case, move down the vitelline canal to the oviduct and, within the ootype, 1000 to 2000 of these surround each fertilized ovum and are enclosed with it within an envelope, which is mainly secreted by the yolk cells themselves. Fertilization occurs at the junction of the fertilization canal and oviduct. Color key as in illustrations on p. 277: gray (white beforehand) —epidermis (ectoderm), yellow—endodermis (endoderm), carmine—muscle, cinnibar red—female reproductive organs, violet—male reproductive organs, blue—nervous system, green—excretory organs, white—parenchyma (except in Fig. 1e where it is pink).

up; some are also capable of leechlike movement. Some species, owing to this musculature, can form contractile waves, which run from anterior to posterior, and move forward by this means.

In turbellarians the mouth is always present on the ventral surface, either at the center of the body or near the anterior or posterior end. It is usually joined to the gut by a pharynx which varies in structure and is capable of being extruded. Some species have several mouths and pharynxes; a true anus, however, is lacking. As in the case of the coelenterates, the Turbellaria expel indigestible remains of their prey through the mouth. The gut can be rodlike, forked, or branched; in the Acoela, no gut cavity is present and the gut is filled with a loose tissue without cell boundaries (syncytium). This, however, is very likely a secondary development since the much simpler coelenterates already have a digestive cavity lined with cells. The food is only roughly broken down by digestive fluid within the turbellarian intestine; the particles are then taken up by the cells of the intestinal wall in the same way as unicellulates, and digestion is completed within the cells. Digestion is thus intracellular, as in the sponges and coelenterates. The main task of the excretory organs (protonephridia) is the elimination of water permeating the body in fresh-water forms. These organs are thus only slightly developed or completely absent in marine forms.

The turbellarians, as free-living and free-moving animals, are richly provided with sense organs. The sense of touch, especially at the edges of the body, is perceived by sense cells; chemical receptors are present in skin areas at the anterior and in the pharynx. Water-current direction is perceptible by all turbellarians; in *Mesostoma*, four pairs of bristle-bearing cells are present at the edge of the body for this purpose. Gravity receptors (statocysts) have also been discovered in the acoels, some catenulids, and many more highly evolved turbellarians. A large number of turbellarians have ocelli which, owing to their distribution, are capable of determining the direction of light but are not capable of perceiving images. These are the most conspicuous of the sense organs; their number and arrangement are thus used as a species characteristic of turbellarians. Even when eyes are absent, turbellarians can still respond to light through light-sensitive ectodermal cells.

A nerve net immediately below the ectoderm is reminiscent of the state present in coelenterates. In contrast to these, however, the turbellarians, as an adaptation to a free-living existence, have a "head" end richly furnished with sense organs. This concentration of sense organs results in a concentration in the nerve cells supplying them; this is the first encounter in the animal kingdom with a type of brain. From this, up to four pairs of longitudinal nerve strands run the length of the body, joined to one another by laterally running annular nerves. This right-angled (orthogonal) nervous system is the basis for the nervous systems of other animal groups and gives the key to their construction.

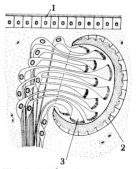

Fig. 8-2. The pigment-cup eye from a planarian. 1. Ectoderm (epidermis), 2. Pigment cup, 3. Light-sensitive nerve cells.

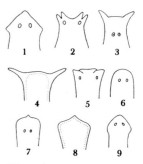

Fig. 8-3. Eye arrangement as a species characteristic in the most common European planaria. 1. *Dugesia gonocephala*, 2. *Bdellocephala punctata*, 3. *Crenobia alpina*, 4. *Polycelis cornuta*, 5. *Dendrocoelum lacteum*, 6. *Planaria torva*, 7. *Fonticola vitta*, 8. *Polycelis nigra*, 9. *Dugesia lugubris*.

All these turbellarian organs are at a relatively low stage of development. It is thus even more astounding that their reproductive organs have reached such a high developmental stage. Whereas in sponges and coelenterates, fertilization of the ovum by the sperm is left to the water, here we encounter for the first time a copulatory process. The genus *Xenoturbella*, which is the simplest in structure, does not have special organs for this; according to Beklemischev, however, copulatory behavior is already present; the sperm cells, which are released through the mouth, adhere to the ectoderm of a conspecific, bore through it, and wander into the parenchyma to the individual ova, which they then fertilize. The evolutionary step forward in behavior is thus, in this case, ahead of that in structure—something which can thus have a wide range of application.

In other Turbellaria, the ova and sperm cells are not scattered throughout the parenchyma but form closed groups or organs—testes and ovaries. Both organs are always present in all turbellarians—they are hermaphroditic. In many turbellarians, a division of labor has occurred in the ovarian complex: one section of the cells forms ova which are deficient in nourishment; another section forms yolk cells which contain large quantities of nourishment. These surround each ovum before it is enclosed in a shell and give their nourishment to it during the course of embryological development. In the genus *Prorhynchus*, the yolk cells surround the ovum when it is still in the ovary; in the majority of the higher turbellarians, however, the yolk cells form in a "vitellarium" separate from the ovary itself. These turbellarians are termed "Neoophora" in contrast to the more primitive "Archoophora," the eggs of which are dependent on their own yolk content for nourishment. Flukes and tapeworms show the same separation of ovary and vitellarium as is found in the turbellarians; one can therefore presume that they arose from neoophorans.

A higher level of development can also be found in the male reproductive organs. In the acoels and polyclads, the most simple copulatory organs are present; they were derived from the worm's defense organs. The so-called "pear-shaped organ" and "stalked gland organ," originally associated with defense and prey capture, take over the task of sperm transmission; the sperm cells collect beneath them and migrate into the other animal when its skin is penetrated by these organs. In higher turbellarians, a true sexual organ is formed by various means.

The early embryological development of the Archoophora, for example, is of basic importance in the understanding of relationships between the turbellarians and other animal groups. Four cells (macromeres), almost equal in size, are formed after the first two cell divisions of the fertilized ovum. In the following four cell divisions, each of these gives up a smaller cell (micromere); these are not positioned directly above the macromere of origin but alternately left and right in the cavities between the macromeres created by the inclined cleavage axis. These divisions proceed extremely regularly, and the fate of the micromeres follows a definite

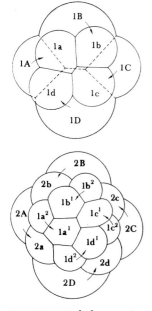

Fig. 8-4. Spiral cleavage in *Discocoelis tigrina*. The arrows show the direction of the cleavage spiral. a. The first quartet of micromeres is formed. b. The second quartet of micromeres. The first has divided into daughter cells. $1a^1$ and $1a^2$ has formed from 1a, etc. The macromeres are now numbered 2A. etc.

plan. The same process of "spiral cleavage" is found in the strapworms, mollusks, and annelids, together with some of the smaller groups, and the fate of the micromeres is the same in each—a definite proof of the evolutionary relationship of all these animals. Even the free-swimming larva (protrochula), which develops from the germ cells in turbellarians, has much in common with the trochophore larva of the groups mentioned above.

Apart from the production of new generations through eggs, a sexual reproduction by divisions is also widespread among the turbellarians. Members of the genera *Microstomum*, *Stenostomum*, and *Catenula* found in European inland waters and even aquaria, can form chains of animals from two to sixteen individuals long by horizontal division. When the animals are artificially divided, entire individuals can form from the pieces. By slitting the bodies of fresh-water planarians, extra anterior or posterior portions can be formed. This almost limitless ability to regenerate is brought about by cells capable of all types of development (neoblasts) which lie, waiting to be called upon, within the parenchyma.

No complete conformity is present in the systematic division of the turbellarians. In this review, the more primitive archoophorans are ranked before the more highly developed neoophorans.

The first animal to be described is *Xenoturbella bocki* (L 2–3 cm), found on the muddy sea bottoms of Scandanavia's western coast. Whether this extremely simplified creature is a member of the turbellaria is a matter of doubt, since the structures which it has in common with them are the simplest body structures, which could be expected to be found in a representative of the oldest, most basic forms of the Protostomia: the sexual cells do not develop within an ovary or testis but occur as single cells scattered throughout the parenchyma. There is no brain or nerve strands and the nerve net lies within the ectoderm and not below it, as does the statocyst. Whether spiral cleavage occurs during embryology is still not known.

The most simply constructed turbellaria belong to the ACOELS, in which intestinal walls, intestinal cavity, excretory organs, and oviducts are all lacking. The ova are released either through the skin or the mouth, the latter leading to a loose parenchymatous tissue. This simplicity, however, is probably not original but a later simplification of these extremely small animals. The genus *Convoluta* is the best investigated of this group. The green members of this genus live in symbiosis with the flagellated algae (chlamydomonadines). These flagellated algae degenerate within the parenchyma of the worm to such an extent that they are no longer carried over into the worm's eggs. The juvenile, hatching from the egg, must therefore obtain its own algae. Soon afterward, they stop feeding for themselves and are completely dependent on these green "zoochlorellae," living on their metabolic products. When they are older, they finally digest their symbionts and then die them-

▷
The alternation of hosts in the intestinal blood fluke (*Schistosoma mansoni*). The blood fluke (6) lives, the slender female within the ventral (gynaecophoral) groove of the male, inside the intestinal veins of humans, the final host (1). The ova, characterized by lateral spines (2), leave the human through the stool. They hatch within the water into miracidial larvae (3). These bore into an aquatic snail, the intermediary host (4). Here they transform into a sporscyst which produces daughter sporocysts. Within these, numerous cercarial larvae are formed (5), leaving the the snail in swarms. In the water, they bore through human skin (1) into the body and grow there to maturity (6).

2a

2b

1

2c

2

3b

3a

4c

4b

4a

MILLA

selves. On the northern coast of France, *Convoluta roscoffensis* collects in masses on the surface during low tide, but digs in instantly as soon as the tide turns. In this way, the worm offers its symbionts the greatest amount of necessary light but protects itself from drifting away. These animals retain this tidal rhythm for about one to two weeks even when held in the aquarium.

The members of the CATENULIDS (order Catenulida) already have an unpaired male sexual opening on the upper surface, but no female one, and the ova are thus released through the skin. These animals reproduce mainly by horizontal division, and chains of two to four, rarely eight, animals can be formed. *Catenula lemnae* (L up to 5 mm), which floats perpendicularly in quiet European inland waters, is fairly common, as is another chain-building planarian, *Stenostomum leucops*, which is often encountered in putrifying infusions for *Paramecium* culture. All the "chain-building" worms are fresh-water inhabitants.

The MACROSTOMIDS (order Macrostomida) include fresh-water, brackish-water, and marine worms. Some species can be found in all types of water, for example *Macrostomum appendiculatum* (L up to 2.5 mm) and the chain-forming *Microstomum lineare* (L of a sixteen-individual chain up to 15 mm). When these species feed on fresh-water polyps, they use the nematocysts of their prey as kleptocnids.

The POLYCLADS (order Polycladida) are recognizable by their extremely branched midgut. In contrast to the rest of the Archoophora, they have a fully extensible pharynx and extremely complicated sexual organs, although a vitellarium is absent. They often develop from a free-swimming larva which bears a whorl of ciliated lobes around the mouth. This Müller's larva (protrochula) is very similar to the trochophore larva (see Color plate, p. 352), but a separate anus is lacking. The polyclads are, almost without exception, marine; large, brightly colored species are present, especially in tropical waters. The few planktonic species, in contrast, are transparent and colorless. The majority of polyclads inhabit coastal sea bottoms and swim for only short distances. Numerous polyclad species are commensal in other marine animals; *Stylochus zebra*, for example, lives with a hermit crab within its snail shell. Other *Stylochus* species attack oysters within their shells and cause considerable damage to oyster beds, e.g., *Stylochus pilidium* in the Mediterranean and *Stylochus frontalis* in Florida. *Graffizoon lobatum* (L less than 1 mm), an inhabitant of the Californian coast, is noteworthy in that it becomes sexually mature at the Müller's larva stage. This neoteny, as it is termed, is widespread throughout the animal kingdom.

In the orders which follow, which comprise the NEOOPHORA, the germ cells do not obtain nourishment from their own yolk but from the yolk cells surrounding the ovum. The order LECITHOEPITHELIATA is intermediate, in that the yolk cells are not formed within a vitellarium but within the ovary itself. Its species are partly marine and partly

◁

The developmental cycle of the liver fluke *Dicrocoelium dendriticum*. 1. The parasite: the mature fluke (BL 8-10 mm). 2. The final host: sheep (2a), roedeer (2b), and wild rabbits (2c) in dry chalky areas. 2 to 4c. The developmental cycle: 3. The first intermediary host: a chalk-area snail; the zebra snail (3a) or the heather snail (3b), shown here during aestivation, becomes infected by eating the eggs within the feces of the final host (3a above). The parasite reproduces within the snail until numerous cercaria are formed; these are released from the pulmonary opening inside a ball of mucus (3a below). The second intermediary host: ants eat the mucous ball (4a) or carry it to the nest where other ants also eat it (4b). The ants infected in this way attach themselves by biting the tips of plants once the parasite has matured (4c), where they are eaten by the final host as it grazes (2). Within this, the fluke larvae attain sexual maturity in the gall-bladder ducts of the liver.

fresh-water and terrestrial inhabitants. *Prorhynchus stagnalis*, (L 6 mm) distributed worldwide, is found in European inland waters with muddy bottoms.

Within the order SERIATA, the places of origin of the ova and yolk cells are separate. The vitellarium forms small swellings which lie on each side of the oviduct and open into it. The pharynx is strong and can be everted; the intestine has numerous lateral branches. The best-known suborder and the best-known turbellarians are the PLANARIANS, also termed TRICLADS owing to the shape of their intestinal tract. From the mouth and pharynx, which lie at approximately the midpoint of the lower surface, two intestinal branches run posteriorly and one anteriorly. Their distensibility allows the planarians to swallow relatively large prey; they also serve as storage areas during long periods of starvation. Apart from this, starving planarians are also able to break down their own organs and tissues, thus reducing them, thereby, while still retaining their form and internal structure, decreasing to a fraction of their original size.

Scientists distinguish between MARINE PLANARIANS (Maricola), FRESH-WATER PLANARIANS (Paludicola), and TERRESTRIAL PLANARIANS (Terricola) on the basis of each animal's environment. Among the marine planarians, species of the genus *Bdelloura* are commensal on horseshoe crabs along the northeastern coast of the U.S.A. The transparent animals attach their stalked egg capsules to the horseshoe crab's gills. In *Procerodes lobata* (L 6 mm), which inhabits the Black Sea and Mediterranean, the intestinal pockets are arranged at regular intervals, the testes and vitellaria lying in the spaces thus formed. Such organ arrangement is reminiscent of the tapeworms but has developed completely independently.

The FRESH-WATER PLANARIANS are well known to aquarists; relatively large species of this group can be found in European streams. The northern subspecies of *Crenobia alpina* (*Crenobia alpina septentrionalis*; L up to 16 mm) lives in extremely cold springs and flourishes at temperatures of 6–8°C and is unable to stand temperatures of even 15°C. The southern subspecies, *Crenobia alpina meridionalis*, which occurs in the alpine region, is far less sensitive to warmth. Following on in the downward course of the brook from these cold-water planarians, come *Polycelis cornuta* (L up to 18 mm), whichis less warmth-sensitive and has numerous eyes, and *Dugesia gonocephala* (L up to 25 mm; see Color plate, p. 295), which is brownish. This species is not particular about the temperature or purity of the water.

A closely related species, the AMERICAN STREAM PLANARIAN (*Dugesia dorotocephala*; L 10–24 mm), has caused a great deal of interest because of its feats of memory. The American scientist McConnell was able to train this species by punishment with electric shocks to select a white or black tube as the correct escape route in a T-maze. Even the posterior end of a worm which has been thus trained could remember the task; this shows that it is not only the brain in these animals—as is the case in humans—

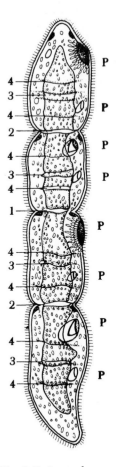

Fig. 8-5. Asexual reproduction in *Microstonum*. 1-4 Points of division; 1.-4. Arrangement, P Pharynx anlager of the individuals formed.

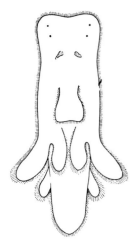

Fig. 8-6. "Müller's" larva from the ventral surface.

which is the seat of memory. Although this result was surprising enough, it went one step further: when a trained worm was eaten by an untrained one, the latter was then able to perform the task correctly. Even tissues removed from trained worms and injected into untrained worms were able to produce the same effect. The nature of these processes enabled a ribonucleic acid to be pinpointed as the "memory substance." When other scientists attempted to repeat McConnell's results with other planarian species, they could not even demonstrate the ability of the worms to be trained, and doubt was cast on McConnell's results. When the same species (*Dugesia dorotocephala*) was used, however, the results were found to be true. It is thus necessary to study the differences in behavior and take these into account when working with related species which are separated by systematists only on the basis of their external characteristics.

The planarians, being hermaphroditic, inseminate each other simultaneously, usually raising and pressing their posterior ends together while doing so. The ovum, surrounded by numerous yolk cells, is enclosed with these cells in an egg capsule (an eikokon), which encloses up to forty eggs in all. The tiny, transparent juveniles, L about 1 mm, hatch from this without an intermediary larval stage. The planarians which are occasionally introduced into aquaria with live food for the fish, are very unpopular with aquarists, who term them "plate worms," since they devour fish eggs and fry in addition to the food for the fish. They can be baited and caught by hanging a piece of meat in the water.

The TERRESTRIAL PLANARIANS are usually tropical and live mainly in damp forest floors; a layer of mucus covering the body protects them from desiccation. They hunt during the night when the humidity is at its highest. As a result, the European species *Rhynchodemus terrestris* (L 10 mm) is discovered only rarely and then is usually taken to be a slug. Its distribution range is thus not known completely. The GREENHOUSE PLANARIAN (*Bipalium kewense*), on the other hand, is much more conspicuous with its seven gray or green longitudinal stripes on a yellow background. This species has probably been introduced with plants from a tropical climate. It has prospered in the gardens of California, Louisiana, Florida, and the West Indies. Its true place of origin is unknown, however. In Europe, the greenhouse planarian reproduces only by division.

Turbellarians which have a saclike or branchlike intestine, such as the macrostomids and catenulids, but which, in contrast to these, have a separate vitellarium, are included in the order NEORHABDOCOELA. The majority are fresh-water inhabitants but some occur both on land and in the sea. In the suborder TYPHLOPLANOIDA is one species, *Mesostoma ehrenbergi* (L up to 15 mm), which is of great interest to naturalists because of its size, transparency, and means of prey capture. The animals lie quietly, attached to the glass of the observation chamber by mucus. If a water flea approaches, however, they dash after it with amazing speed, throw themselves upon it, glue it down, and, within a short while, the struggles of the prey cease altogether. The worm builds a web of mucous threads through the water, on which it often hangs. Even before a growing worm has formed a copulatory organ, it already has about 400 ripe ova ("summer" eggs). These are self-fertilized within the animal, and the young are released two to four weeks later. Adult worms copulate with one another simultaneously and produce only a few large, thick-shelled eggs ("winter" eggs). These are released when the female dies and decays; they require a cold period for their further development. The large number of rapidly developing "summer" eggs therefore aids in the reproduction and distribution of the species during the favorable time of the year, while the "winter" eggs, which are fewer in number, carry the species over the winter and summer drought periods and aid in distributing the species to previously uncolonized waters.

The suborder DALYELLOIDA is characterized by a mouth transposed far toward the animal's anterior. *Dalyella viridis* (L 5 mm), which lives in wet hollows, is light green because of the presence of zoochlorellae. Many members of the suborder KALYPTORHYNCHA, which live in the system of holes in the sand along sea coasts, have been discovered only recently. *Gyratrix hermophrodita* (L up to 2 mm), a member of this suborder, inhabits not only the sea but also brackish water and fresh water and even occurs on wet land. This worm paralyzes its prey by stabbing it with its copulatory organ, which is furnished with a chitinous tip and poison glands; this is reminiscent of the two-fold usage of the stalked gland-organ in certain polyclads.

The TEMNOCEPHALIDS (suborder Temnocephalida) are either aciliate or only slightly ciliated and much transformed from the basic turbellarian type. They were first thought to be associated with the flukes. They are either commensal or parasitic on other animals. The minute *Scutariella didactyla* is found in Yugoslavia in the gills of the shrimp *Trogocaris*, whose body fluids it sucks out.

The FLUKES (class Trematodes; BL usually only a few millimeters) are entirely parasitic. They are characterized by organs of attachment—either sucking discs, hooks, or pincherlike structures—which provide them with a secure hold on the host's body. Nearly all the species are hermaphroditic. The sexual organs are extremely complicated in structure. The general body structure is relatively simple; the blind-ending intestine usually has two branches. The class contains two subclasses which differ greatly in mode of life and course of development and have probably developed independently from free-living ancestors. These are the 1. MONOGENETIC FLUKES (Monogenea) and 2. DIGENOUS FLUKES (Digenea).

The MONOGENETIC FLUKES are usually external parasites. Only a few species live within the body of their hosts. They are usually found attached to the skin and gills of aquatic vertebrates, e.g., fish, newts, or terrapins. They retain a secure hold on the host by means of a huge attachment area at the posterior of the body, furnished with numerous hooks,

▷
The alternation of hosts in a fish tapeworm (*Dibothriocephalus latus*).
1. Portions of the fish tapeworm strobilus within the human intestine, or that of a dog or cat (C) after ingesting a cysticercus (9 from A or 10 from B). Freshly laid egg with a cap from feces of the final host. Three-to-four-week-old egg containing onchosphere. Free-swimming ciliated coracidial larvae after hatching in water. Water flea with the coracidium (now lacking cilia) within the body cavity (first intermediatry host). 6 to 8: Metamorphosis of the onchosphere (6) to a procercoid (8) within the water flea. Plercocercus within a fresh-water fish (A) after eating the water flea (second intermediary host). 10. Plercoercus within a carnivorous fish (B, transport host) after eating the fresh-water fish. Humans, dogs, and cats (C) can be infected with the tapeworm by eating either the fresh-water fish (A) or the carnivorous fish (B) raw.

Class: Trematoda, by W. Hohorst

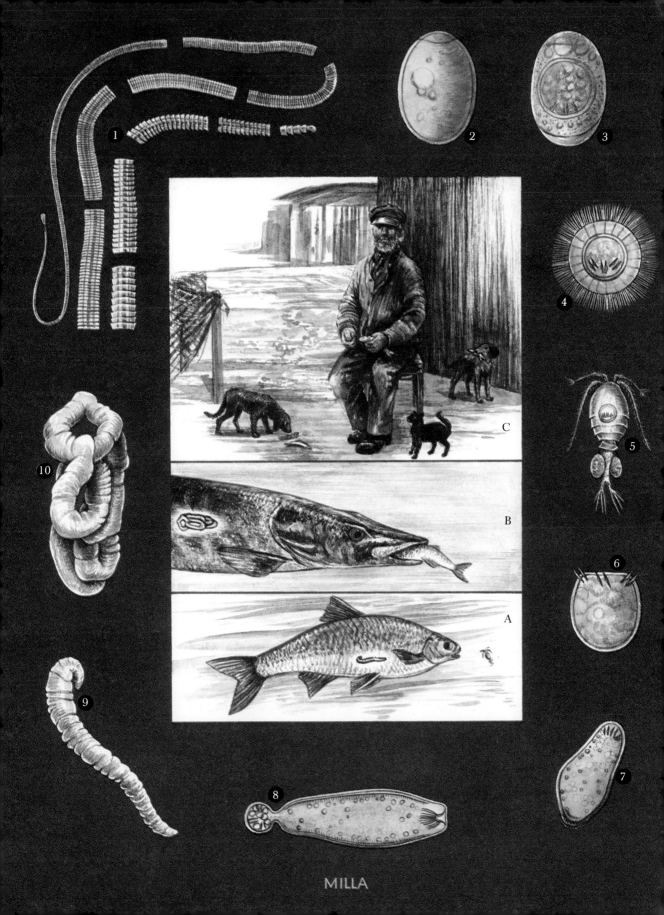

MILLA

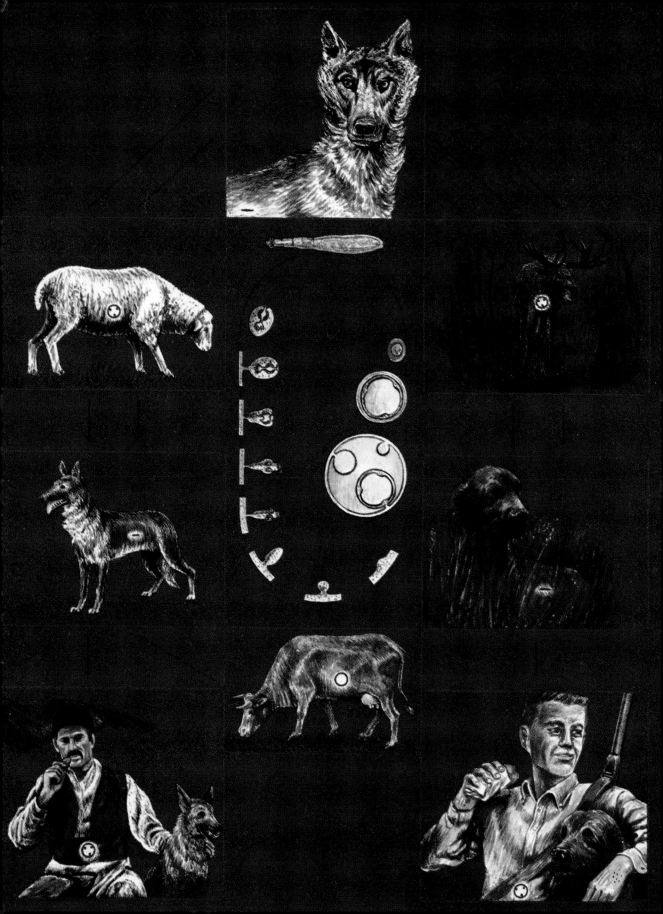

◁

The alternation of generations in *Echinococcus granulosus*. Left, from above to below in grazing land (i.e., the Balkans); right, from above to below, in woodland (e.g., northern North America). The wolf, the original final host (E_1), becomes a tapeworm carrier by eating ruminants infected with the tapeworm, and releases, with its feces, eggs contining onchosphere larvae. Ruminants (sheep in pasture land, deer, e.g., elk, in woodland) take these up in their food and become cysticercus carriers (intermediary host Z_1). The most common final hosts at the present time are dogs (E_2, here sheep dogs and hunting dogs). Further infection between Z_1 and E_2; the cow as an occasional intermediary host (Z_2) usually harbors sterile larvae. By contact (unhygenic) with infected dogs, humans (here herders and hunters) also become cysticercus carriers (Z_3). In the center: the developmental process of *Echinococcus*: 7 and 1 within the dog, 2 first in the open and then within a ruminant, 3 to 7 within the ruminant. 1. The mature tapeworm (L only 3 to 6 mm). The onchosphere within the egg envelope, ready for development within the intermediary host. 3. The bladder worm developing from this. 4. Bladder worm with daughter cysts, within the largest, tapeworm heads beginning to develop. 5. Immature tapeworm head. 6. The former inverted. 7. Detached, free head from the bladder worm, ready for development within the final host.

suckers, or pincers. Development is direct; the juvenile forms hatch from the egg, attack their host after a short period during which they are free-swimming, and develop on the host to sexual maturity.

One of the best-known species, which has become famous because of its unusual means of reproduction, is *Gyrodactylus elegans* (L barely 1 mm), a skin-and-gill parasite of European carp. It is a live-bearer and always brings a single young into the world; the offspring then attaches itself immediately to the same host and, after a short period of time, produces another single young itself; owing to this process, an infestation can come about relatively rapidly. This rapid sequence of generations is possible only because the young are enclosed one within another inside the mother—similar to the well-known Russian doll-toy. While the daughter is within the mother, it already contains the grandchild, and this a great-grandchild in which the great-great-grandchild is enclosed. All four larvae develop from a single egg and are true sisters. Since they develop one after the other in time and take over the parental care process for one another, they appear as "generations." The development of several larvae from a single egg is a widespread means of reproduction (polyembryony) in parasitic animal species. This method causes a definite increase in the number of young produced, which is especially important for parasites.

Diplozoon paradoxum (L 1 cm) is also a parasite on the gills of carp. It consists of two single individuals which have grown together in a cross shape. The young worms first live as solitary individuals, pairing only when sexually mature. After copulation has been completed, they grow together at the middle of their bodies and remain joined in this way for the rest of their lives. Young worms which have been unable to find a partner never become sexually mature.

The BLADDER FLUKE (*Polystoma integerrimum*; L 13 mm) is a parasite on frogs. During the juvenile phase, it attaches itself to the gills of tadpoles; as a mature fluke it lives only within the frog's bladder. This is an adaptation to the unusual metamorphosis of the frog. It is known that frogs change their habitat several times during the course of their lives. They live in water as tadpoles and breathe through gills; as frogs, they are terrestrial and breathe through lungs, nearly always returning to the water, however, during the reproductive period. As soon as the tadpole begins to metamorphose into a frog and the gills start to regress, the larvae of the bladder fluke changes its behavior drastically. It moves down through the intestine, into the cloaca, and from there to the bladder, where it is able to live through its host's period of terrestriality as if in a cistern. The flukes start to grow within the bladder, and reach sexual maturity at the same time as their host, at about three years of age—an extremely long period of time for a fluke. In this way, the flukes, whose development can only take place in water, attain the ability to reproduce at exactly the time that the frog returns to the water to lay its eggs.

The larvae which hatch from the eggs attack the tadpoles which have, in the meantime, hatched from the spawn; they prefer older tadpoles with internal gills. When the larvae settle on the external gills of very young tadpoles, they become tiny worms, which reach sexual maturity within about three weeks; they lay eggs immediately but die at the time of the tadpole's metamorphosis. The worms within the frog's bladder, however, live until their host dies, reproducing each year during spring, at the same time as their host. The secret of the synchrony between host and parasite is probably associated with the fact that the worms are influenced by the hormones they take in with their food, the host's blood.

The DIGENETIC FLUKES are internal parasites; their larval development is always associated with a change of host and an alternation of generations. As adults they inhabit various organs of terrestrial and aquatic animals; many species parasitize humans and thus are medically important. They develop as larvae within an alternate host, especially aquatic snails and bivalves but also in insects, crustaceans, or fish. The majority of species have two secondary hosts; others have only one. Each species has its particular secondary host which lives in close association with the final host; this facilitates the transmission of the larvae from one to the other. The adult worms reproduce sexually; asexual reproduction occurs during the larval generation within the secondary host, leading to several larval generations (polyembryony) being produced.

A miracidium-larva first hatches from the egg, and then transforms within the secondary host to a sporocyst which, in turn, produces new larvae; these larvae can either form new sporocysts or develop into redia-larvae. The redia have a short intestine, and produce new redia-larvae. Finally, cercaria-larvae develop in the sporocysts and redia, leaving the secondary host and frequently entering a paratenic second secondary host. Here they metamorphose into metacercarial larvae, this being a "waiting stage" until they are taken up by the final host. In many species, these metacercaria are present on the surface of plants; in others, the schistosomes, the cercaria bypasses the waiting stage and enters the final host under its own power. By asexual reproduction a large number of young liver flukes, sometimes in the thousands, can be produced from a single egg.

The GIANT LIVER FLUKE (*Fasciola hepatica*; L up to 4 cm; see Color plate, p. 295), distributed worldwide, is a common parasite in the gall bladder ducts of the liver in many domesticated animals. It attacks sheep and cattle especially frequently, but also goats, pigs, horses, hares, deer, and occasionally even man. The economic damage which it can cause is extremely great. The larval development occurs in water snails (Lymnaeidae; see Vol. III). This was discovered in 1882, by the famous German parasitologist Rudolf Leuckart and the English scientist Thomas, both working independently of one another. In Europe the small water snail (*Galba*

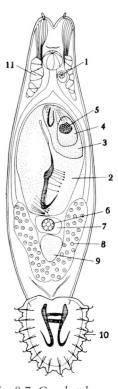

Fig. 8-7. *Gyrodactylus elegans*, an external parasite on European carp. Within the uterus, the following individuals are encapsuled within one another: child (2), grand-child (3), great-grandchild (4), and great-great-grandchild (5). 1. Penis, 6. Ovum, 7. Oviduct, 8. Ovary, 9. Testes, 10. Attachment disc with suckers, 11. Cephalic glands.

Fig. 8-8. *Diplozoon paradoxum*, a gill parasite on carp.

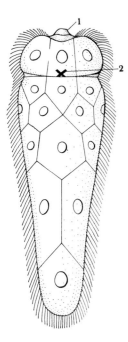

Fig. 8-9. The free-swimming miracidial larva of the giant liver fluke.
1. Boring apparatus,
2. Eyespot.

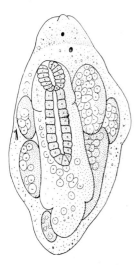

Fig. 8-10. Sporocyst of a giant liver fluke containing redia-larvae in various stages of development.

truncatula) acts as a secondary host; it is therefore also termed the "liver fluke snail." Other indigenous snails can be infected experimentally but play a negligible part as natural carriers. In non-European lands, numerous related aquatic snails act as the natural secondary hosts for this fluke.

The eggs of the giant liver fluke develop only in damp environments. The miracidial larvae hatch in the water and bore into the secondary host snail. After metamorphosis into a sporocyst they produce redia-larvae which, after growing for a short period of time, produce a further generation of redia-larvae. Cercaria finally develop within the redia, leaving the snail through its respiratory opening. After swimming about for a little while, they attach themselves to plants. Here they form a spherical metacercarian cyst which may, someday, be taken in by the host as it feeds. The young liver flukes bore through the abdominal cavity to the liver and become sexually mature there within seven to eight weeks.

The environment of the giant liver fluke is damp meadows; it is usually in these areas that infection of our domestic animals takes place, man being infected usually by eating water cress. This species is termed the "common liver fluke" outside Europe, since even larger species occur in these regions. *Fasciola gigantica* (L up to 7 cm) is found in camels and buffaloes throughout Africa and Asia. The AMERICAN GIANT LIVER FLUKE (*Fasciolopsis magna*; L up to 10 cm) usually infects North American deer and cattle. It has been recorded in Europe—Italy and Czeschoslovakia—as a parasite in red deer, probably introduced there through importation of the North American wapiti deer.

The SMALL LIVER FLUKE (*Dicrocoelium dendriticum*; L 8–10 mm; see Color plate, p. 284) is a common parasite of ruminants. Its main area of distribution is the temperate zone of Europe, but it also occurs in Asia, Africa, and the Americas. It attacks sheep very often, and also cattle, goats, and wild rabbits. Humans are rarely infected by this species. Its development occurs through two secondary hosts, and its environment is dry chalky regions used by sheep and goats as grazing land. The first secondary hosts are land snails such as *Helicella*, *Zebrina detrita*, or *Cionella lubrica* (see Vol. III). These snails take up the eggs of the fluke with their food; a miracidium develops from the eggs, and, after developing into a sporocyst, produces numerous daughter sporocysts; the cercarial larvae develop within these, and are furnished with a small boring bristle. On rainy days, the cercarial larvae—packed within a ball of mucus—are extruded through the snail's respiratory opening; the balls swell to about the size of a grape, each containing up to 6000 cercaria.

The second secondary hosts are ants of the genus *Formica* (see Vol. II) which eat the mucous balls. In this way, the cercaria enter the ant's stomach. They bore through the stomach walls and close the holes thus made with a special wound-closing substance. Within the ant's body cavity, the cercaria surround themselves with several cyst walls and complete their development in forty days or longer. In each ant attacked, one cercarian always migrates to the host's brain and bores into the sub-esophageal ganglion. Here it develops into a peculiar-looking cyst termed a "brain worm," which is not capable of infecting the final host.

As soon as the cysts have matured, the infected ants show an unusual change in behavior. They bite into the tips of plants near their nests and hold on there. This behavior is temperature-dependent and appears during cool weather. It is for this reason that the attached ants are most commonly found during the early morning and evening; they also spend the night away from their nest, still biting fast to the plant tips. When the temperature rises, for example during the sunny midday, the mandibles release their hold and the infected ants join their nestmates again. This behavior, which is extremely unusual for ants, is very useful for the liver fluke's further development, since the final host can thus easily obtain large numbers of ants together with the plants it eats. This behavior is probably controlled by the "brain worm," most likely neurophysiologic-ally. Its position immediately next to the nerves supplying the ant's mouthparts suggests this.

The CAT LIVER FLUKE (*Opisthorchis felineus*; L 5–12 mm) is mainly present in eastern Europe and Asian Russia. Apart from cats, it also infests dogs, foxes, and frequently humans as well. It requires two second-ary hosts for its development, a fresh-water snail (*Bithynia leachii*; see Vol. III) and carp (see Vol. IV). The spherical cysts are found in the muscles and fins of the fish and are eaten together with the raw meat. Close relatives with a similar development are the UPPER INDIAN LIVER FLUKE (*Opisthorchis viverrini*) from Thailand and the CHINESE LIVER FLUKE (*Clonorchis sinensis*) from China, Korea, and Japan. These also infest humans in addition to various domestic and wild animals, especially cats.

Many flukes, which are mainly Asian in distribution, live in human intestines and are termed INTESTINAL FLUKES. One of these is the GIANT INTESTINAL FLUKE (*Fasciolopsis buski*; L 5–7 cm), one of the largest flukes infecting humans, which also occurs frequently in pigs. It has only a single secondary host. The cercarian larvae develop in snails and form cysts on water plants. Human infections usually come about through eating the fruit of the water nut (*Trapa natans*), a popular food in eastern Asia, which is shelled with the teeth. The cysts enter the stomach by this means. In the case of other human intestinal flukes, two secondary hosts are always necessary, e.g., *Echinostoma ilocanum* (L about 1 mm) in snails and mollusks and *Metagonimus yokogawei* and *Heterophyes heterophyes*, the smallest intestinal fluke with a BL only slightly greater than 1 mm, in snails and fish. Human infection by all these species occurs through the eating of uncooked mollusks or fish.

Some flukes are lung parasites, such as the EAST ASIAN LUNG FLUKE (*Paragorimus westermani*; L about 8–16 mm). It attacks not only humans but also dogs, cats, pigs, and numerous wild animals such as mink, fox,

▷
Above left: The giant liver fluke (*Fasciola hepatica*), which parasitize especially the liver of numerous domestic animals, can also be dangerous to man. The dark areas are intestinal branches visible through the body.
Center: The brightly ringed, thickly swollen antennae of both fresh-water snails are caused by the sporocyst bladder of flukes. A sporo-cyst from *Leucochloridium paradoxum* has infected the worm on the left, and on the right, two sporocysts from *Leucochloridium fuscum* have bored into the snail's antennae. Below left: Fresh-water planarian (*Dugesia gonocephala*) with clearly visible pigmented eyecups.
Above right: *Yungia aurantiaca*, a marine tur-bellarian. Below right: A turbellarian from tropical seas, order Polycladida.

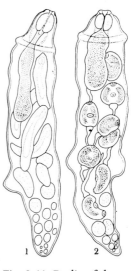

Fig. 8-11. Redia of the giant liver fluke. 1. Parental redia containing second generation redia-larvae, 2. Daughter redia contain-ing cercarial larvae.

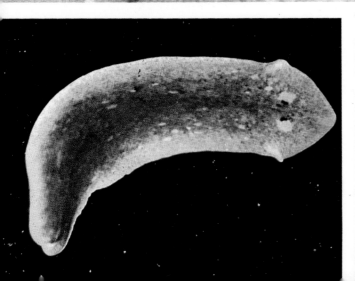

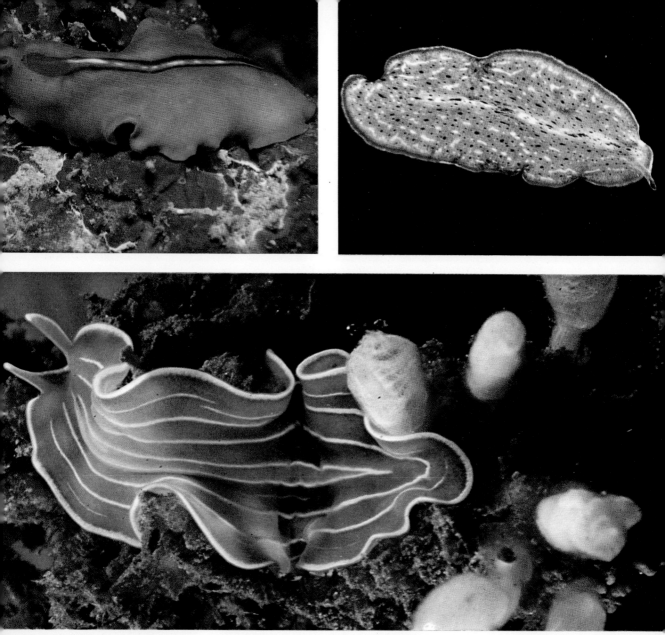
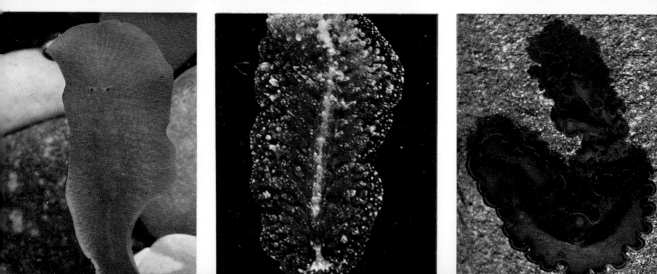

◁
Above left: A turbellarian from tropical waters, order Polycladiad. Center: *Prostheceraeus roseus*, Mediterranean. Below left: *Notoplana longestyletta.* Below center: *Thysanozoon brochii*, Mediterranean and Atlantic. Above right: *Prostheceraeus moseley*, Mediterranean. Below right: Flatworms of the order Polycladida, from tropical oceans.

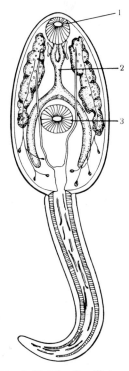

Fig. 8-12. The free-living cercarial larva of the giant liver fluke. 1. Oral sucker, 2. Forked intestine, 3. Ventral sucker.

wolf, and tiger. Two secondary hosts are necessary for its development —water snails and fresh-water crustaceans. Its main range of distribution is China, Korea, and Japan. In the New World a related species occurs, the NORTH AMERICAN LUNG FLUKE (*Paragonimus kellicotti*).

The BLOOD FLUKES (family Schistosomidae; L 10–16 mm) are of great importance medically, since they cause the tropical disease bilharziasis (the name being attributable to the German doctor Theodor Bilharz [1825–1862]) .The "classical" bilharziasis area is Egypt; the discovery of eggs in mummies shows that even the builders of the pyramids were sufferers of this disease. The blood flukes hold a special position among the flukes in general; they are of separate sexes and always live as a pair, the threadlike female lying within an abdominal groove in the male. Blood flukes live in the vessels of the venous system; the eggs reach the exterior through either the gut or the bladder, causing a great deal of tissue destruction. The miracidial larvae hatch in water and bore into a snail, the secondary host. After their metamorphosis into a sporocyst, they produce numerous daughter sporocysts in which fork-tailed cercaria develop. Thousands of these larvae swarm out of the snail and bore into the body of their final host under their own power. On entering the blood vessels, they are carried to the liver where they remain until sexually mature. An infestation with blood flukes can therefore take place only in water, during bathing or during farming procedures such as digging irrigation ditches or working in rice fields. Even today, the number of people infected by this parasite is more than 100 million; the mass accumulation of eggs from the parasite within the tissues causes infections, bleeding, swellings, and carcinogenic lumps. Until a few decades ago, there was no means of curing infected patients, but today several new medicines, in addition to the traditional antimony compounds, are available.

Three species are of special importance to humans. The BLADDER BLOOD FLUKE (*Schistosoma haematobium*), the cause of bladder biharziasis, occurs in Africa and the Near East. The eggs are provided with a large spine at the end, and are released with the urine. The secondary hosts are snails of the genus *Bulinus*. The INTESTINAL BLOOD FLUKE (*Schistosoma mansoni*; see Color plate, p. 283), the cause of intestinal bilharziasis, is distributed throughout Africa and is also present in South America and the West Indies, where it is was introduced through the slave trade. The eggs are provided with a lateral spine, and are released with the feces. The secondary hosts are various species of water snails. The JAPANESE BLOOD FLUKE (*Schistosoma japonicum*) occurs in eastern Asia, in China, Japan, Formosa, and the Philippine Islands; this also causes intestinal bilharziasis. Its eggs, which are also released with the stool, have only pronounced lateral swellings. The secondary hosts are amphibious snails of the genera *Oncomelania* and *Katayama*.

Related species (such as *Trichobilharzia szidati*) are parasitic in water fowl; their fork-tailed larvae also commonly bore into the skin of bathers and produce an extremely itchy rash termed "bathing dermatitis." *Leucochloridium* has an especially interesting developmental cycle. It is present as a mature worm in the cloaca of various species of birds, and its secondary host is the water snail *Succinea*. Within the snail it develops into a many-branched sporocyst, L about 1 cm; these have conspicuous green or brown rings and pulse rhythmically. The colorful sporocysts, filled with worm larvae, bore into the snail's antennae. In this way, they attract a hunting bird, which pecks the sporocyst out of the snail's antenna and swallows it (see Color plate, p. 295). The worm larvae thus enter the bird by this means, completing their development there.

The flatworm parasites reach their peak of development in the TAPE-WORMS (class Cestoda). In contrast to the flukes, an alternation of generations is present only as an exception to the rule, but an alternation of hosts, through one or more secondary hosts, always occurs. The larval host ("secondary host") is either an invertebrate or vertebrate; the larvae infect various organs of these animals. The "final host" of the mature tapeworm is always a vertebrate, however, and it is always the intestines which are parasitized. Here the worm is surrounded by the digestive fluids of its host and also by digested foodstuffs; the tapeworm thus needs only to take up the nourishment required through its entire surface area. It therefore requires no digestive system, and has none.

A young tapeworm is faced with the fateful question of how to find the way to the right host at least once, and sometimes several times, during the course of its development. It is unable to do anything about this personally, and the chance of "hitting the jackpot" is extremely small. Only the production of an enormous number of young can even the odds, thus ensuring the continuation of the species. The organization of the tapeworms, which is unique among the animal kingdom, aids in achieving this goal: on the head (scolex, plural scolices) are organs of attachment (rings of hooks, suckers, or sucking pits); immediately behind this is a region of continuous growth (collum). This is unjointed and grows longitudinally, forming rows of new reproductive organs, one after another. In the ribbon so formed, sections separate off, these usually having both male and female sexual organs within them. A jointed band (strobila) thus forms consisting of segments (proglottids) which ripen sexually as they are pushed away from the collum by the formation of new segments. They are sexually mature first as males and then as females. The central, male proglottids, can therefore frequently fertilize the end, female proglottids. This allows tapeworms which live as solitary individuals to reproduce sexually. The large number of sexual organs following on one after another first led scientists to believe that the tapeworm was not a single animal but a colony. The fact that the nervous system, the musculature, and the excretory organs form a single entity throughout the length of the worm's body, however, contradicts this point of view.

Class: Cestoda, by P. Rietschel

Fig. 8-13. The ring of hooks of a tapeworm viewed from above.

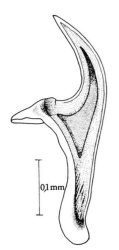

Fig. 8-14. A hook from the ring of hooks of the cat tapeworm.

Fig. 8-15. A ripe proglottid and the head of a beef tapeworm.

The most common human tapeworm will be taken here as an example: the BEEF TAPEWORM (*Taeniarhynchus saginatus*, earlier termed *Taenia saginata*; L up to 10 m; see Color plate, p. 278). Since its head carries no ring of hooks, it is also termed a "weaponless tapeworm." Its pear-shaped head is furnished with four egg-shaped suckers and has a diameter of 1–2 mm. This is followed by a neck zone of about half the breadth of the head. The first segments are wider than they are long; only the last segments, which are filled with eggs, are longer than they are broad. The openings of the sex organs on the edge of the segment are either to the left or the right in irregular order. The uterus of the last segments, which is filled with eggs, consists of a central longitudinal cavity and fifteen to thirty lateral branches on each side.

The beef tapeworm lives as an adult in the anterior portion of the human small intestine; man is its only "final host." The larvae, in contrast, can grow in various secondary hosts. The main secondary hosts are cattle and, in southern regions, zebus and buffalo. Occasionally, the larvae enter other ruminants which are termed "additional secondary hosts." The larvae, about 1 cm in size, are found in the fat bodies around the heart or in the muscle; they are not easy to discover, since they usually occur as single individuals and are difficult to distinguish from the surrounding fat tissue. They are able to withstand temperatures of 45°C and 1–2°C (for up to six weeks). They are thus not killed if they are in the middle of a large roast; eating such "rare" beef can thus lead to infection, as can eating "steak à la Tartar." The larva everts within the human intestine and attaches itself to the wall of the gut. The tapeworm is fully grown within two to three months and releases ten to twelve ripe proglottids per day, each containing over 100 thousand eggs. The parasitologist G. Osche calls animals which have such large numbers of offspring "egg millionaires." The tapeworm earns this title for a single day's production. Since it can reach an age of twenty years, it must be considered an "egg multimillionaire" by the time it dies.

Usually only a single tapeworm is found within the intestine of an infected person. The secretion of defense substances prevents other worms from attaching. This is not only a protection for the host, but also for the parasite, which would not be able to survive if the host were severely damaged. As soon as one tapeworm has left the intestine, however, the path is made free for another to colonize it. In rare cases, however, several larvae can enter the intestine simultaneously, and all then grow within the host. An infection with beef tapeworm is not easy to overlook: the ripe proglottids leave the host through the anus and creep around actively in the warmth of the bed or under the clothing, taking up a flattened egg-shaped form. They act like self-sufficient animals and have caused a great deal of worry to many doctors who are not well versed in parasitology. When pressed between two glass slides, however, the branched uterus is clearly recognizable.

Once they have reached the exterior, the proglottids die within a short time and release their large number of eggs. Cattle take up the eggs individually by accident with their food; their infections are thus slight. The "egg" contains a larva within the egg envelope, furnished with three pairs of hooks. The pancreatic juice within the cow's intestine dissolves the egg envelope and the released "onchosphere" larva burrows through the intestinal epithelium with the aid of its moveable hooks. The onchosphere larvae reach their final destination through the host's blood or lymphatic system, and develop there into a "bladder worm" (cysticercus). This carries the future head of the new tapeworm everted within it.

The human PORK TAPEWORM (*Taenia solium*; L 2–3 or even 8 m) also develops through a "bladder worm" (cysticercus). In other tapeworms the larva develops into a cysticercoid with a minute bladder cavity, or into a large bladder worm with numerous heads (coenurus and echinococcus). Other larval stages (procercoid and plerocercoid) are found in more primitive tapeworms.

As was the case with the beef tapeworm, the pork tapeworm also has man as its final host. Both species are extremely host specific and could only have developed because man is a meat eater. Man's habit of eating meat must therefore be at least as old as the two species in question. Since species require thousands of years for their development, the mere existence of these two tapeworms refutes the hypothesis of vegetarians that man was originally not a meat eater.

The pork tapeworm can be distinguished from the beef tapeworm externally by its smaller head, furnished with circular suckers and bearing a double ring of hooks; it is thus termed an "armed tapeworm." Within the ring of hooks is an extrusible, fingerlike process, the "rostellum." In this tapeworm, the segments are longer than they are broad from about the midpoint of the proglottid chain. The uterus of the ripe distal proglottids has only seven to ten lateral branches on each side. The proglottids are not released singly but as groups, and do not leave the anus under their own power, but passively, with the stool. The tapeworm thus remains undiscovered by its host for long periods of time.

The main secondary host of the pork tapeworm is the pig. There are, however, a large number of additional secondary hosts, man being one of them. He infects himself, through poor sanitation, with the eggs within the stool or through eating vegetables fertilized with latrine wastes. The larvae in man usually settle in the eyes, or just as often in the nervous system, as well as in many other organs. They can thus be much more dangerous to humans than the adult worm in the intestines. They appear in pigs—if ever—in large numbers, since the animals, on eating human feces, swallow whole groups of proglottids filled with eggs. The large larvae (L 15 mm) are more easily discovered in the meat than are those of the beef tapeworm. This parasite has thus become a rarity in central Europe,

▷
Left, from above to below: Encapsulated trachina between the muscle fibers, as a meat inspector would see them under a microscope. Chicken intestine with a severe infestation of *Ascaridia galli*. The ringed nemertine (*Tubulanus annulatus*). This bootlace worm lives on muddy bottoms and under stones in the Mediterranean and Atlantic.
Right, from above to below: A mature (brown) cyst, four young females, and several males of the potato nematode (*Heterodera rostochienis*) on a potato root. A female is in the process of copulation. The Mediterranean bootlace worm *Lineus geniculatus* can reach a L of 60 cm. Ribbon nemertine (*Derpanophorus crassus*).

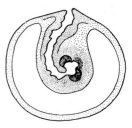

Fig. 8-16. Cysticercus larva with an inverted head.

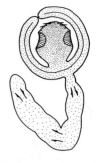

Fig. 8-17. A cysticercoid larva without a bladder.

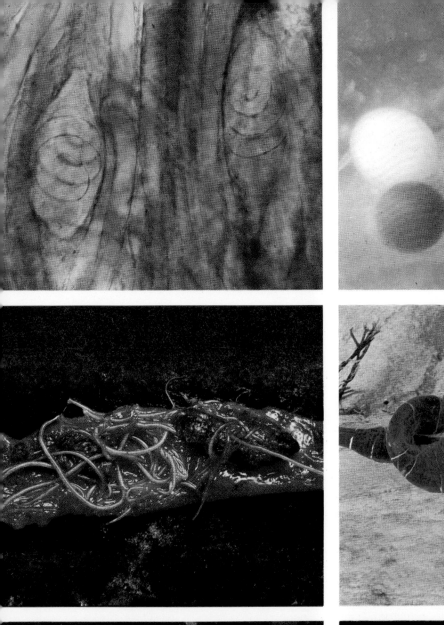

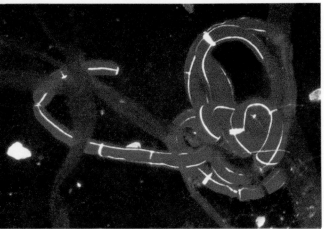
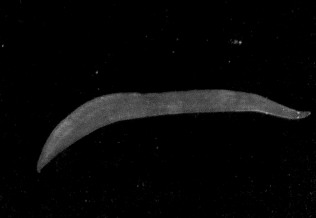

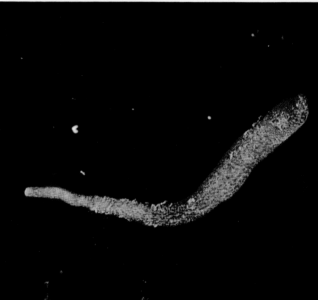

Left from above to below:
The nemertine *Cerebratulus marginatus* is able to swim extremely well because of its flattened body, and can equally easily dig into sandy and muddy substrates.

A female echiurid, *Bonellia viridis*. The plumlike body is usually protected within a crack, while the long, forked, and very motile proboscis searches the substrate for food. Two body sections can be clearly differentiated in the sipunculid *Sipunculus nudus*: A long, thick posterior attached to a thinner, extremely extensible anterior, which is able to be withdrawn into the posterior. Short tentacles protrude from the mouth opening.

Right, from above to below. The fresh-water nemertine *Prostoma graecense*. *Physcosoma granulatum*, a sipunculid, distributed throughout the Atlantic, North Sea, and Mediterranean. *Echiurus echiurus*, common in the North Sea, is also a member of the echiurids.

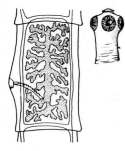

Fig. 8-18. Ripe proglottid and head from a pork tapeworm.

partly because of the routine examinations of meat for trichinosis infections.

The beef and pork tapeworms belong to the order CYCLOPHYLLIDEA, the most highly advanced of the tapeworm classes. Other members of this order live, as sexually mature individuals, in the intestines of other mammals, many birds, lizards and, on rare occasions, newts, but never fish. The one most dangerous to humans is the smallest one when fully grown: the DOG TAPEWORM or BLADDER TAPEWORM (*Echinococcus granulosus*; L 3–6 mm; see Color plate, p. 290). It lives as a mature individual in the dog's intestine, and has only two or three proglottids. Dogs infected with this species usually contain large numbers of individuals. The larva (cysticercus) is not choosy about its secondary host. It flourishes in the intestines of ruminants, pigs, horses, rabbits, rodents, kangaroos, carnivores, monkeys, and man. In contrast to the small size of the adult worm, the larva is enormous and can reach the size of a human head. Daughter cysts bud off from this, each containing ten to thirty tapeworm heads. A new tapeworm can develop from each of these within the final host, but each is also able—when released in the secondary host—to form a new cysticercus. The cysts released through the wall of the bladder reach the exterior as "hyatid cysts." A single mature cysticercus can thus contain 1.2 to 2.4 million tapeworm heads.

This enormous reproduction by budding compensates for the small amount of sexual reproduction in this tapeworm. In addition, the cysticercus can form daughter cysts externally as well as internally. The cysts most commonly infect the liver, but there is hardly an organ of the body which cannot be infected by them. The symptoms produced by the parasite are therefore quite varied. Only complete removal of the cyst can produce a cure; operating on it, however, can result in worm heads being released into the body fluids and being carried to another region where they form a new cyst. Since this tapeworm is so dangerous to humans, it is advisable to maintain strict sanitary measures, show care in handling dogs, and also to keep up a continual survey of their feces to determine whether a tapeworm infection is present. Dogs usually become infected by eating infected butcher's wastes—but only in areas where the slaughtered animals are not under continual control against infection. This is the case in most regions on earth. In Iceland, for example, keeping a dog was forbidden for a long time owing to the plaguelike tapeworm infections which occurred.

Another tapeworm infecting dogs and their relatives has included a period of asexual reproduction in its larval development: this is *Multiceps multiceps* (L 40–100 cm). It produces far more eggs than does its smaller relative; in contrast to this, however, the asexual reproduction of its cysticercus larva, which is about the size of a hen's egg, is kept within limits. Only a few hundred tapeworm heads bud from the interior of the cyst; this type is termed a "coenurus" rather than an "echinococcus,"

Microscopic Pond Life
Left: Fouled water with blue-green algae (a) and polluted water fungi (*Sphaerotilus* = *Cladothrix*) (b); right: fresh water with the green algae *Spirogyra* (c) and *Zygnema* (d). 1-18: Unicellulates (subkingdom Protozoa): 1. *Dinobryon sertularia*, 2. *Synura uvella*, 3. *Pandorina morum*, 4. *Amoeba proteus*, 5. *Arcella vulgaris*, 6. *Dif-flugia urceolata*, 7. *Lecque-ureusia spiralis*, 8. *Euglypha alveolata*, 9. *Actinosphaerum eichhorni*, 10. *Actinophrys sol*, 11. *Coleps hirtus*, 12. *Didinium nasutum*. Water bears attacking a paramecium. 13. *Lacrymaria olor*, 14. Paramecium (*Paramecium*; compare Color plate, p. 108), 15. *Colpidium colpoda*, 16. Vorticella (*Vorticella microstoma*), 17. Blue stentor (*Stentor coerulus*), 18. *Stylonychia mytilus*, 19-37: Multicellulates (subkingdom Metazoa): 19. Gastrotrichan (*Chaetonotus maximus*), 20. *Philodina citrina*, 21. *Rotaria neptunia*, 22. *Rotaria rotatoria*, 23. *Rotaria macroceros*, 24. *Limnias ceratophylli*, 25. *Collotheca gracilipes*, 26. *Epiphanes senta*, 27. *Brachionus calyciflorus*, 28. *Platzias quadricornis*, 29. *Keratella quadrata*, 30.*Euchlanis deflexa*, 31. *Mytilina mucronata*, 32. *Proales fallaciosa* on a dead *Brachionus* (27), 33. *Cephalodella forficula* within a sponge tube, 34. *Itura myersi*, 35. *Synchaeta pectinata*, 36a. *Macrobiotus macronyx*, 36b. Empty exuvia with eggs, 37. Nauplius larva from *Cyclops*.

in which the heads are budded on a hyatid cyst. The main secondary hosts are sheep; the larvae prefer to settle in the brain. If the larvae infect the brain on only one side, the sheep show the symptoms of "staggers." This has also occasionally been observed in humans.

Other tapeworms infecting dogs and their wild relatives are found in the same order. The longest of these is *Taenia hydatigena* (L up to 5 m). Its narrow-necked larva grows within the intestinal mesenteries of ruminants and pigs within seven to eight weeks. This tapeworm infects dogs fed with these meat wastes, especially butchers' dogs. The smaller *Taenia pisiformis* (L 1–2 m) develops as a larva in the intestinal mesenteries of hares and rabbits. Hunting dogs and farmers' dogs in areas where rabbits are present are therefore the most common final hosts.

Dipylidium caninum (L 15–40 cm) is generally widely distributed in the small intestine of dogs and cats; occasionally it is also found in children. Its main characteristic is its cucumber-seed-shaped proglottids; the openings of the paired sexual organs, which are doubled in each proglottid, lie in the middle of each of the longitudinal proglottid edges. This tapeworm is not adapted to meat-eating in its host; the larva (L 0.6 mm) and the cysticercoid, which is barely visible to the naked eye, live in dog fleas and dog lice. The fleas, being blood suckers, thus can not take up the tapeworm eggs with their food, but their larvae do so, living, as do fly larvae, in dirty areas near the dog's sleeping place and thus easily coming into contact with the extruded tapeworm proglottids. The dog takes up the adult flea when it bites and swallows the insects plaguing it. Children can contract these tapeworms by eating dog fleas which have accidentally gotten into their food.

Domestic cats are infected more often with the CAT TAPEWORM (*Hydatigena taeniaeformis*; L 15–60 cm) than with *Dipylidium caninum*. The secondary hosts are the house mouse and the rat, in the liver of which the pea-sized cysticercus larva is frequently found. If this is opened, a tapeworm up to 20 cm long is found coiled inside. The proglottids, however, do not contain sexual organs. A larva containing a chain of segments (strobia) such as this, is termed a "strobilocercus." The mouse or rat takes up the tapeworm eggs when it eats grain fouled with cat's feces or is fed with fouled grain as a laboratory animal.

As has been seen already in the case of *Dipylidium caninum*, tapeworm infections do not always have to come about by eating meat. Horses, sheep, and rabbits also have their tapeworms although they are vegetarians. How they ever managed to become infected remained a mystery for a long time. We know today that the larvae of these tapeworms from the family ANOPLOCEPHALIDAE develop in free-living, ground-inhabiting mites which are accidentally swallowed with the plant food. Invertebrates are very frequently secondary hosts; the larvae of the CHICKEN TAPEWORMS of the family Davaineidae live in slugs, earthworms, houseflies, beetles, butterflies, and ants; those infecting ducks and geese, members of the genus *Hymenolepis*, live in waterfleas, aquatic crustaceans, and copepods. This genus, which contains many species, includes tapeworms of small to middling size, the mature distal proglottids of which are much wider than they are long, with the sexual organ openings always on the same proglottid edge. The majority of these inhabit the intestines of birds; a few, however, parasitize mammals. *Hymenolepis diminuta*, for example, is common in rats and mice. Its larva, a cysticercoid, lives in mealworms. *Hymenolepis nana* (L 10–25 mm) can also develop as a larva in mealworm beetles. The cysticercoid also flourishes, however, in the intestinal walls of its final host, so the usual alternation of generations occurring in tapeworms can be dispensed with. Mouse infections with *H. nana* is much more commonly the case. Owing to this, the human species (*Hymenolepis nana nana*) and the mouse species (*Hymenolepis nana fraterna*), despite their outward similarities, must be considered subspecies. In humans, rare infections are usually overlooked, since the tapeworm segments are usually digested within the intestines; the uniquely structured eggs, however, are easily discovered by microscopic examination of the stool.

Fig. 8-19. Egg from *Hymenolepis nana*, internal egg capsule with polar filaments, and, within it, the hooked larva (onchosphere) with six hooks.

The members of the other tapeworm orders are mainly fish parasites as mature animals, especially in cartilagenous fish and ancient bony-fish families. Their heads are furnished with various (often fantastic) series of attachment organs, but not with the four suckers and rings of hooks present in the higher tapeworms. The larval form of these worms is the "plerocercoid," forming no cysticercus and not being inverted at the anterior end. The developmental cycle of the majority of these tapeworms is still unknown. The order PSEUDOPHYLLIDEA is, however, well investigated. It contains the FISH TAPEWORM (*Dibothriocephalus latus*; L up to 12 m; see Color plate, p. 289), which also parasitizes man. It has also been found from time to time in dogs, cats, foxes, bears, pigs, and other mammals. Its head is furnished with two sucking pits which lie opposite one another, which gives it its Latin species name which means "two-pitted head." The segments are broad and much wider than they are long, even at the distal end of the animal. The sexual organs open in the middle of the undersurface. The eggs are extruded from these; they are not released by detachment of the entire segment, as is the case in the majority of the higher tapeworms. The feces contain large numbers of eggs, but no empty segments. The intermediary hosts are firstly copepods and secondly, freshwater fish.

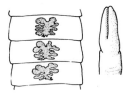

Fig. 8-20. Ripe proglottids and head from a fish tapeworm.

The start of development in the fish tapeworm larvae indicates their primitive level; the onchosphere is covered with a ciliated envelope (coracidium) which swims freely in the water. During the course of a few days, it is eaten by a copepod, within which it escapes from the ciliated envelope and bores through the wall of the intestine into the body cavity with the aid of its six hooks. Here it grows to form a procercoid larva. If this intermediary host is then swallowed by a fish, the larva bores through the fish's intestinal wall and migrates into its organs, where it metamorphoses

Fig. 8-21. *Gnathostomaria lutheri*: body form and organization. 1. Anterior pharyngeal region, 2. Intestine, 3. Ovary, 4. Bursa, 5. Testes, 6. Sperm sac, 7. Copulatory organ.

Class: Gnathostomulida, by P. Ax

into a pleocercoid larva (L 1–2 cm). A third intermediary host is sometimes included: a larger carnivorous fish swallows the second intermediary host, which is then termed a "transport host." The larva—now a plerocercoid—also bores through this host's intestinal wall and lodges in its organs, but does not metamorphose further. This third intermediary host is thus termed a "resting host" or "deposition host." The plerocercoid enters dogs or cats when they eat fish wastes, or humans when they eat raw fish. The pleocercoid finally develops into a mature tapeworm within these hosts. Raw fresh-water fish are usually not eaten by man. In areas where pike roe, raw-fish salad, and similar fish dishes are traditional, however, humans can become infected with the fish tapeworm. In Europe, such cases are most commonly met with around the Baltic Sea, the Danube delta, the Swiss and northern Italian lakes, and once even in the Starnberg Lake in Germany. In 1929, H. Vogel found all the fish he investigated from inland health resorts to be infected with fish tapeworms.

The fish tapeworm often causes a high degree of blood loss in its host (pernicious anemia). This probably comes about because the tapeworm removes large quantities of vitamin B12 required for blood building from its host. Vitamin B12 is present in high concentration in the fish tapeworm. When compared with a beef tapeworm of the same size, the fish tapeworm contains more than five times the quantity of vitamin B12.

Among the pseudophyllids, of which the fish tapeworm is a member, there is a tendency for sexual maturity to occur during the larval phase (neoteny). *Ligula intestinalis* (L up to 75 cm), which is dreaded by carp breeders, begins its larval development, as does the fish tapeworm, as a coracidium in a copepod. It also develops in a fish into a plerocercoid, which is, however, already sexually mature. If the fish is eaten by a water bird—usually grebes in Europe—the tapeworm remains within this final host for only a few days; it leaves as an entire, imperfectly segmented worm filled with ripe eggs. Sexual maturity is attained even earlier in *Caryophyllaeus laticeps* (L 4 cm). Its procercoid larva inhabits redworms (genus *Tubifex*) which live in muddy bottoms. Here the sexual organs, which are not arranged serially, develop. If these worms are eaten by bream, the tapeworm eggs then mature within it. The closely related tapeworm species *Biacetabulum sieboldii* (L up to 3 mm) needs no fish host to reach sexual maturity, since the eggs ripen within the *Tubifex* in the procercoid larva. This tapeworm has thus reached the highest stage of neoteny.

Even today, zoology can discover new types among the great variety of animal structures. The GNATHOSTOMULIDA (see Color plate, p. 315) have existed within the zoological system only since 1956. I first described them as a new flatworm order and, a little later, gave them their own class among the trematodes. In this work, too, they will be included in the trematodes. Some zoologists consider these animals to be a phylum in their own right, separate from both the trematodes and the nematodes.

I published the first study dealing with the two species *Gnathostomula paradoxa* from the Baltic and *Gnathostomaria lutheri* from the Mediterranean fifteen years ago [1956]; now more than twenty papers on the Gnathostomulida have appeared in German, English, and Russian. Numerous genera containing more than forty species are now known from this newly discovered class of animals, whose distribution stretches from the Arctic Ocean into the tropics, along the coasts of various oceans.

The gnathostomulids are microscopically small creatures (BL usually less than 1 mm; in *Nanognathia exigua*, according to R. Riedl, only 0.4 mm; largest representative, from the Bahamas, is *Pterognathia grandis*, according to Kirsteuer, at 3.5 mm). The body is slender and wormlike. The anterior end is distinguishable as a tiny head set with tactile bristles; in *Pterognathia* it is extended to form a long snout (rostrum). The body is cylindrical and rounded posteriorly (in *Gnathostomaria* and *Pterognathia*) or ends in a thread-like tail (in *Gnathostomula*).

Fig. 8-22. Epithelial cells from *Gnathostomaria* showing a single cilium per cell.

The first important characteristic of this group is its body covering. At medium magnifications, long and unusually loosely arranged threads are visible along the outline of the body. By regular serial beating, they produce the slow gliding and swimming movements performed by these animals. If a section of ectoderm is examined under the high power of the microscope, a great difference between this type of body covering and that of normal ciliated ectoderm (ciliated epithelium) is obvious. In the latter, the individual cells always have large numbers of short cilia; in the gnathostomulids, however, each cell has only a single flagellum. On the basis of this structure, I have termed the epithelium of the gnathostomulids "flagellated epithelium." Sterrer, regarding the functional difference between cilia and flagellae, has termed it "uniciliate (monociliate) epithelium."

The foregut of gnathostomulids shows two further unusual characteristics present in these animals. The floor of the oral cavity always has a basal plate originating from the external body layer (cuticular), varying greatly in form. The jaws lie at the anterior section of the pharynx, giving the animals their Latin name. This jaw apparatus is only 20 to 30 micrometers in length (1 micrometer $= 1\mu = 1/1000$ mm) and is of two basic types, clasper jaws and lamellae jaws. In the clasper jaws, the jaw pair is joined, similar to a pair of forceps (as in *Gnathostomaria* and *Pterognathia*); in the case of the lamellae jaws, both jaws are completely separate (as in *Gnathostomula* and *Austrognathia*). In the latter, up to three rows of minute teeth are present on the inner edges.

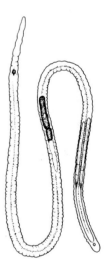

Fig. 8-23. Body form in *Pterognathia simplex*.

Gnathostomulids are hermaphroditic. Their male sexual apparatus consists of a pair of laterally placed testes, a copulatory organ of various shapes present in the center, and a posterior opening. The female organs are, in the simplest cases, only a row of ova along the dorsal side, as in *Pterognathia*. In the majority of genera, however, a storage organ (bursa) of complicated structure follows the last ovum within the ovary; it is used

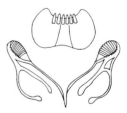

Fig. 8-24. Clasper jaws and basal plate from *Gnathostomaria lutheri*.

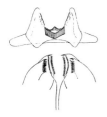

Fig. 8-25. Lamellae jaws and basal plate from *Gnathostomula paradoxa.*

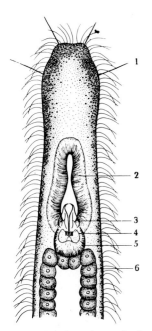

Fig. 8-26. Anterior end of *Gnathostomaria lutheri.* 1. Tactile bristles, 2. Pharyngeal wall, 3. Jaws, 4. Basal plate, 5. Pharyngeal sac, 6. Intestine.

to retain the sperm from the sexual partner. Only the genus *Austrognathia* has a female sexual opening on the dorsal surface.

After fertilization, the eggs are simply pressed out of the dorsal surface. Development, according to R. Riedl, occurs through spiral cleavage without a larval stage. Riedl observed gnathostomulids devouring bacteria, fungal mycelia, and algae. The animals shovel them up from the bottom with the basal plate in the oral cavity, and grasp them with snapping movements of the jaws.

Gnathostomulids are found from the tideline down to depths of 20 to 30 m. They mainly live in the spaces between sand grains (mesopsammal). The most richly inhabited areas are fine-sand beaches containing rubbish (detritus). The vertical distribution of these animals within the sand is very unusual: along the North American Atlantic coast, the number of species present and their concentration increases with the depth from the sand surface to the black-colored layer stained with iron sulphide and smelling of sulfur dioxide. The gnathostomulids, as Riedl has shown, reach their peak of development in this area which is exceptionally poor in oxygen or even oxygen-free. On the North Sea coast, *Gnathostomula paradoxa* also preferentially colonizes this gray sand layer, which is about 10 cm below the sand surface.

How the gnathostomulids are related to other worms is still an unanswered question. Owing to their spiral cleavage, they are included in the large group of phyla, the Spiralia. In common with the flatworms, they lack a blood system and a secondary body cavity (coelom) and have an intestine without an anus as well as a hermaphroditic sexual system. The pharyngeal structures, however, can not be extrapolated from those of flatworms. On the other hand, there is no close similarity to the chewing structures present in the rotifers, members of the phylum Nematoda. Apart from this, the gnathostomulid epithelium, which has only a single cilium per cell, is unique; among the whole of the bilaterally symmetrical animals, there is no phylum of animals in which the adult organisms have a comparable type of epithelium. These animals thus offer great scope for future investigation—gnathostomulids remain, as before, a puzzling group of worms from sandy shores.

9 Kamptozoans and Bootlace Worms

The KAMPTOZOANS (phylum Kamptozoa or Endoprocta; L up to 0.5 cm; see Color plate, p. 315) are small marine animals; only one family lives in fresh water. The stalk, which is attached to the substrate, bears a head which makes violent nodding movements when the animal is disturbed. The phylum was given its Latin name, Kamptozoa, which means "nodding animals," because of this behavior pattern. The head bears a ring of ciliated tentacles which waft food toward the animal; these structures are similar to those of the bryozoans, which feed in the same manner. The larvae of both groups are also similar. Therefore, the Kamptozoa were once included among the Bryozoa. Zoologists had not, however, noticed a peculiarity about the kamptozoans: in contrast to the Bryozoa, their anus opened within the ring of tentacles. They were therefore termed the "Endoprocta" ("animals with an internal anus"), in contrast to the bryozoans, which were termed the "Ectoprocta" ("animals with an external anus"). This name has been retained, together with Kamptozoa. They are called "goblet worms" in German, since they resemble a stemmed goblet in shape; but their means of feeding is certainly not that of a "worm."

As is the case in the flatworms, the kamptozoans do not have a true body cavity (coelom). The stalk and head are filled with parenchymatous tissue, similar to the body of a flatworm. In contrast to the latter, however, a great improvement is present: the mouth no longer serves as both mouth and anus; the prostomium closes from posterior to anterior, except for the remaining mouth opening, and the anus breaks through afterward in the posterior region. The intestinal canal thus becomes a "one-way street" for the ingested food. The kamptozoan larva is a trochophore (see Color plate, p. 352), similar to that of the segmented worms; the earlier processes of spiral cleavage are also very similar to those of the segmented worms and other members of the Spiralia. Certain stages in their development are therefore very similar to those occurring in the coelomates, although the kamptozoans themselves have no coelom. Their excretory organs are thus protonephridia, similar to those of the flatworms.

Phylum: Kamptozoa, by P. Rietschel

Fig. 9-1. *Micrura alaskensis.*

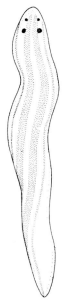

Fig. 9-2. *Tetrastemma quadrilineatum.*

Class: Nemertini, by H. Friedrich

The oral and aboral poles of a kamptozoan larvae are directly comparable to the same poles of the trochophore larva; during metamorphosis, the larva attaches itself by the oral pole (ventral surface). The organs within the larva then change position completely so that, finally, the ventral side with the mouth and anus forms the head and the concave side of the U-shaped intestine turns upward.

The eggs of the partly hermaphroditic and partly dioecious animals are laid on the upper surface of the head, within the ring of tentacles, and develop there into swimming larvae. Early in their lives, they also feed on the particles wafted along by the mother animal. The maternal epidermis near the sexual opening also produces cells rich in nourishment, which are also used by the larva as food. Finally, the larva detaches from the egg stalk and from the maternal oral-anal region. In addition to sexual reproduction, all kamptozoans can reproduce asexually through budding. In the LOXOSOMATIDS (family Loxosomatidae), in which the stalk widens gradually toward the head, the buds form on the wall of the head itself; in the PEDICELLINIDS (family Pedicellinidae), in which the head is well separated from the stalk, the stalk's basal plate either buds or produces budding cones. The FRESH-WATER KAMPTOZOANS (family Urnatellidae), including *Urnatella gracilis*, which is found in lakes in the Berlin area, reproduce entirely by budding. The head area dies in autumn, and new heads form on the naked stalk in spring.

There is no fossil kamptozoan record, for although this is a very old phylum, these animals have no hard parts. Today, approximately sixty kamptozoan species predominately inhabit marine coastal waters. There are a few fresh-water species.

BOOTLACE WORMS (class Nemertini; see Color plate, p. 315) are not often seen, since they usually live hidden in various marine environments. Only a few inconspicuous species have been able to adapt themselves to life in fresh water, and a small number of terrestrial species are restricted to tropical areas. Some of these have been introduced into greenhouses. The name "bootlace worm" comes about through the body form of the most conspicuous types. They are ribbonlike or stringlike, sometimes as thin as sewing thread. There are, however, robust, rather broad, and short forms; others, owing to their small size, do not seem threadlike. The newly introduced name RHYNCHOCOELIA comes from the organ system unique to members of this class: a rhynchocoelom (longitudinal proboscis cavity) is present on the dorsal side above the intestine, and within this is a proboscis extrusible through an opening at the anterior end. Four orders are distinguished:

1. PALAEONEMERTINES (Palaeonemertini); proboscis not armed, body-wall muscles with a circular muscle layer (external) and longitudinal muscle layer (internal). Nervous system within the epithelium (Protonemertini; see Color plate, p. 315) or the longitudinal muscle layer (Mesonemertini; see Color plate, p. 315). Lacking dorsal vessel,

*The Body Forms of
Gnathostomulids, Kampto-
zoans, and Nemertines*

1. Gnathostomulids (class Gnathostomulida).
The position of the organs as seen from above: anterior end with tactile flagellae, brain, and nervous system still within the epidermal layer; below this the longitudinal muscle sheath. Intestine lacking an anus, starting with a jawed pharynx; ovary with seminal receptacles above the gut; behind this and to the side, testes in two longitudinal rows, the copulatory organ lying between them.

2. Kamptozoans (phylum Kamptozoa).
2a. A larva which has just settled (previous to this a swimming larva);
2b. Adult animal in longitudinal section: Larva with two sensory organs: aboral organ (above) and oral organ (left). Mouth (left) and anus (right) situated internally owing to the previously developed tentacular ring. The tentacular groove, which closes off below, becomes the attachment point. After the internal organs have rotated 180°, the adult animal (2b) shows the mouth (now right) and anus (now left and raised on a protrusion) directed upward; both open within the ring of tentacles, as do the reproductive and extretory organs ("Endoprocta," Greek for "those with an internal anus"). The stalk attachment area is fully developed.

3. Bootlace worms (class Nemertini).
3a. Female from above; 3b–3e: Horizontal sections. In the white region of 3a: the last third of the proboscis cavity with proboscis contains, from posterior to anterior, the poison glands (U-shaped), poison receptacle (spherical), and stylet. Next to these, the pockets with accessory stylets; nervous system with brain and two or three longitudinal nerve cords lying at different depths within the body (3b–3e); the glandular "cerebral organs" opening lateral to the brain. Gut (here more or less covered by the proboscis sheath) usually with a mouth opening ventrally and a posterior anus, midgut with numerous lateral blind-ending sacs. Closed blood-vessel system consisting of two lateral vessels and often a dorsal vessel, also vessels to the proboscis organs. Ovaries arranged in lateral series with short tubes leading to the exterior; excretory organs (protonephridia) in the anterior third of the body. 3b–3e: The various positions of the nerve cords: 3b, in the epidermis (original situation, protonemertines); 3c, in the two-layered body-wall musculature, within the longitudinal muscle sheath (mesonemertines); 3d, in the three-layered body-wall musculature between the external longitudinal muscle layer and the circular muscle layer (heteronemertines); 3e, beneath the body-wall musculature (hoplonemertines).

Color key as in the Figures on pp. 175, 276, and 279: gray—ectoderm, yellow—endoderm, carmine red—muscle, cinnibar red—female reproductive organs, violet—male reproductive organs, blue—nervous system, green—excretory organs, white—parenchyma, red-brown—blood vessels.

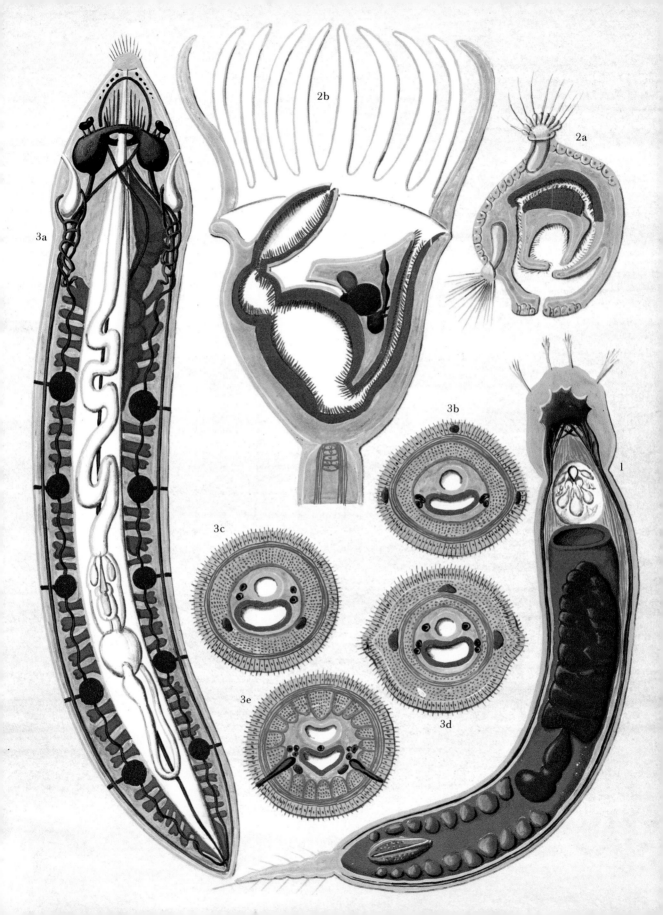

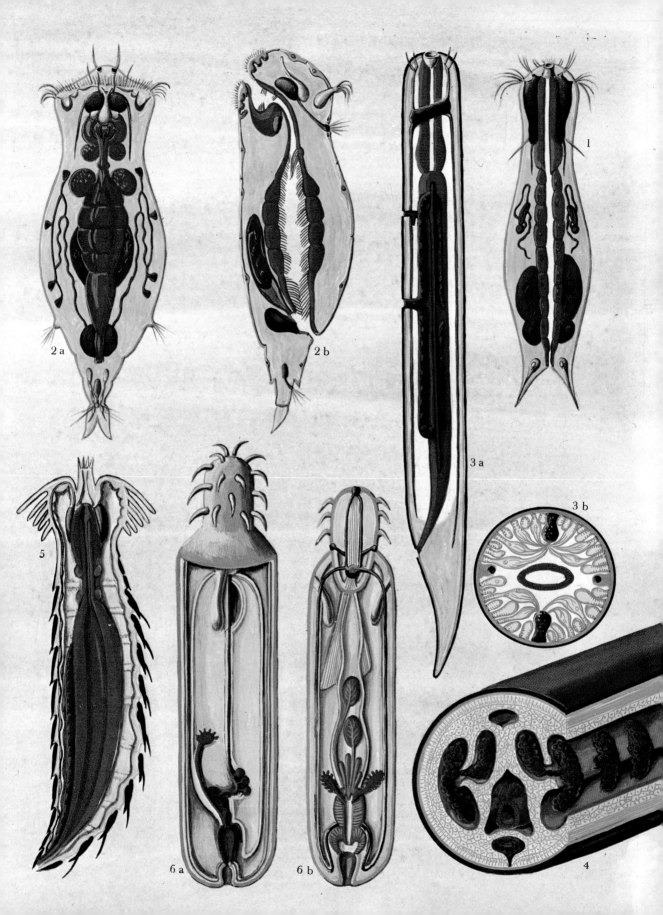

*Body Forms in the Aschel-
minths*

1. Gastrotrichan (class Gastrotricha): Head with large brain and tactile hairs, intestine with oral tube and anus, posterior end with two toes and adhesive glands. Ciliation of the flat ventral surface not shown.

2. Rotifers (phylum Rotatoria):

2a. Viewed from above; 2b, In longitudinal section from the left. Head with aboral sensory organ; below this, fringing the mouth, the cingulum. Behind the brain, a tri-lobed anterior adhesive organ (retrocerebral organ); antenna in the neck region. Stomach with a seven-sectioned chewing apparatus (mastax), muscles, digestive glands, and ganglia; gut with an internally ciliated layer and anterior, paired gastric glands. Vitteline gland and urinary bladder open ventrally into the cloaca, which opens to the exterior ventrally above the foot. Extensible foot with two toes, pedal glands, and tactile organs, which have their own ganglion. 3. Roundworms (class Nematoda):

3a. Longitudinal section from the left; 3b. Horizontal section: Anterior end with anterior sensory organ and mouth opening at the tip, esophagus with radiating muscle fibers, intestine single-layered and lacking glands, hanging free in the false coelum and ending in a ventral anus. The excretory organ (not a protonephridium), consisting of a single cell, comprises two lateral branches joined by a horizontal branch, and opens in the center of the ventral surface. Body-wall musculature consisting only of longitudinal muscles which send processes to the nerve cords (one dorsally and one ventrally, see 3b: usually always the reverse, the nerves sending processes to the muscles)

4. Nematomorphs (class Nematomorpha): Section through the body, dissected laterally: In this case also, body-wall musculature consisting only of longitudinal muscles and body cavities without their own lining. In contrast to the round-worms, however, the longitudinal nerve cord is only present ventrally and excretory organs are lacking.

5. Kinorhynchs (class Kinorhyncha): Longitudinal section from the left: esophagus, gut, and brain as in the round-worms, nerve cord only ventral with ganglia corresponding to the external segmentation.

6. Acanthocephalids (class Acanthocephala):

6a. Female from the side; 6b. Male from above, here opened: Hooks at the anterior region and lacking a gut, as in tapeworms (the same adaptation to the same mode of life); proboscis retraction muscles (above) and proboscis sheath (below) only shown in 6b, as is the nervous system (also present in the female); the pockets present at the anterior are the "lemnisci," which take up nourishment collected by the epidermis and pass it on to the body-fluid which washes around them, thus carrying it to the internal organs. Schematic color key as in the Figures on pp. 276, 279, and 314: gray—ectoderm, yellow—endoderm, carmine red—muscles, cinnibar red—female reproductive organs, violet—male reproductive organs, blue—nervous system, green—excretory organs, white—paren-chyma, red-brown—blood vessels.

cerebral organ, and eyes or possessing very simple ones: primitive group.

2. HETERONEMERTINES (Heteronemertini; see Color plate, p. 315); proboscis not armed, body-wall muscles with an additional external longitudinal muscle layer, nervous system lying between this and the circular muscle layer; possesses dorsal vessel, cerebral organ, and eyes.

3. HOPLONEMERTINES (Hoplonemertini; see Color plate, p. 315): proboscis armed, body-wall muscle layer consisting of a circular muscle layer and a longitudinal muscle layer with the nervous system beneath them.

4. BDELLOMORPHS (Bdellomorpha); proboscis not armed, posterior end with a sucking pit; body-wall muscles and nervous system as in the hoplonemertines. Parasites in mollusks.

The smallest repesentatives are animals living in the spaces between the sand grains of sandy sea bottoms, only reaching a L of 2–3 mm; in contrast to these, the giants of the class can reach a L of 30 m, with a width, however, of only a few centimeters. Of the about 750 species known, the average size is about 10 cm.

The bootlace worms are not segmented externally. In the majority of species, the anterior end is separated from the posterior by a slight constriction. The anterior is characterized by the presence of dark eyespots or longitudinal grooves. Appendages are present in only a few cases; one genus, *Malacobdella* (L 3–4 cm), which lives in the mantle cavity of marine mollusks, has a sucker at both ends of the body. Certain of the species which swim continually or are planktonic bear a pair of short, tentacle-like appendages.

Bootlace worms vary greatly in their coloration; among the planktonic species are colorless, gelatinous, transparent types; in bottom-living forms a whole range of colors is present from pure white to black or patterned with stripes, rings, spots, etc. Little is known regarding the extent to which the coloration can change.

If a bootlace worm is kept in a dish of water, during the day it lies within a slime envelope and is more or less contracted. The animals start moving only at night or when they are disturbed. Then they extend lengthwise, sometimes to twice their length in the quiescent stage, and become, at the same time, thinner, while gliding about the substrate. On stronger stimulation, the posterior contracts rapidly while the anterior simultaneously extends and narrows. This extension progresses from anterior to posterior along the length of the body, so that the anterior is pushed rapidly away from the source of stimulation. Frequently, this contraction and extension of the body areas alternates to such an extent that wavelike movements occur along the body. If the animals contact a stimulus during the process of moving forward, they either turn sideways, jerk backward by contracting, or creep away backward.

This behavior comes about through the structure of the body-wall musculature. The epidermis itself consists of only a single layer of cells; in the adjacent narrow, supporting-cell layer there are numerous gland

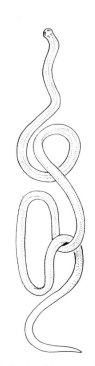

Fig. 9-3. *Amphiporus exilis.*

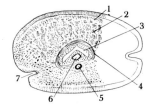

Fig. 9-4. Horizontal section through the head anterior to the brain. 1. Distribution of the epidermal glands, 2. The highly developed head glands, 3. Blood lacuna, 4. Circular musculature, 5. Glandular opening to the exterior, 6. Rhynchodaeum, 7. Anterior grooves.

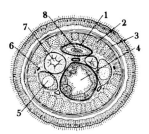

Fig. 9-5. Horizontal section directly behind the brain. 1. Proboscis cavity, 2. Longitudinal musculature, 3. Circular musculature, 4. Esophagus, 5. Lateral nerve cord, 6. Ocellus, 7. Gut, 8. Proboscis.

cells which produce the slime. The supportive cells bear tiny cilia on their external surfaces. Resistance to the beating of these cilia is provided by the body's slime covering, and in this way the animals are able to move forward without changing their shape. Extension and contraction of the body is brought about by the circular and longitudinal muscules which lie beneath the epidermis; the circular muscles are external, and the stronger longitudinal muscle layer lies beneath. This simple muscular wall has a very complicated structure in many species, in that an additional longitudinal muscle layer is present between the epidermis and the circular muscle layer, and the majority of the glandular epidermal cells are embedded in this longitudinal layer.

In certain species, the longitudinal muscles of the dorsal and ventral surfaces are more strongly developed than those of the sides. The animals are capable of making up-and-down lashing movements of the body by alternate contraction of these muscle layers; this movement aids in swimming. The ability to swim is present in only a few bottom-living forms, and is much commoner among the planktonic species. Apart from the longitudinal muscles, other muscle bundles which compress the body dorsoventrally also play a part; these are often very well developed in free-swimming species. The cavity enclosed by the muscle layers contains the other organ systems and connective tissue, which fills the remaining space. As in the case of the flatworms, a coelom is lacking.

The unique organ system possessed by the bootlace worms is their tubelike proboscis, which lies within the proboscis cavity and is attached to its anterior wall. If the fluid pressure within the proboscis cavity is increased through contraction of the muscles, the proboscis is extruded from the posterior of the canal to the anterior like the finger of a glove. The external surface of the proboscis is covered with slime glands and therefore is adhesive. The proboscis is constructed of longitudinal and circular muscles. If the bootlace worm encounters prey such as a small annelid or crustacean, it shoots out the proboscis, which then sticks to the prey and envelopes it. The proboscis shortens by contraction and pulls the prey toward the mouth opening.

One order of bootlace worms, the hoplonemertines, is characterized by a special arming of the proboscis. This consists of either a defensive stylet on a cone-shaped socket, or a large number of smaller stylets on sickle-shaped sockets, or of glands which lie within the proboscis wall and open in the immediate vicinity of the daggers. The structure of these proboscis organs varies, and thus subclasses, orders, and suborders of these worms can be distinguished by these characteristics. The proboscis can bear a poisonous stylet or one which is not poisonous; the proboscis cavity can be of different lengths and its walls of varying structure, the proboscis itself paired, single, or without projections. Recently, bootlace worms have been discovered in which the original single proboscis has, through continual division, become bushy, with about sixty-four free ends.

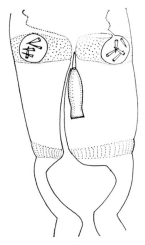

Fig. 9-6. Outline of the stylet apparatus and proboscis.

The intestinal canal starts either with a mouth opening on the ventral surface of the anterior end, leading into a pharyngeal tube, or with the tube connecting with a canal at the tip of the head through which the proboscis is extruded. The walls of the pharyngeal tube are usually furnished with numerous glands; the tube's posterior often widens to form a stomach. The foregut consists of the pharyngeal tube and stomach: it is connected to the midgut, a tube furnished with numerous paired lateral sacs of various forms. The midgut is often visible externally through the body, especially when full. The lateral sacs become smaller posteriorly; the short hindgut lacks these and opens to the exterior by an anus at the end of the body. Apart from these paired lateral sacs, single sacs can be present in both the midgut and foregut. The mouth and pharynx are extremely extensible, enabling bootlace worms to swallow relatively large prey.

The excretory organs lie along the animal's sides; they consist of a canal which, during its course, opens to the exterior by one or several pores—in exceptional cases, however, it can open into the intestine. The anterior of this canal connects with the true excretory cells. This anterior organ can be a system of closed canals, glandular, or even open, funnel-like structures. In the majority of cases, exact details are lacking and the ontogenetic development of the excretory organs is not known.

Although among the marine forms the excretory organs are usually restricted to the anterior of the body, in brackish, fresh, and terrestrial water species, they extend the entire length of the body. The organs are thus probably not only functional as means of removing metabolic wastes but also in osmoregulation by the removal of water.

The true excretory section of this organ has a close association with the blood-vessel system. The latter consists, basically, of three longitudinal vessels; two lie on either side of the intestine, the third lies above the intestine, between it and the proboscis cavity. The three vessels unite anteriorly, at the level of the brain, and also posteriorly, slightly in front of the anus. Horizontal connecting vessels are present along the rest of the body in different areas. In front of the brain and in the region surrounding it, the blood vessels show extremely variable arrangements. In many bootlace worms, the blood-fluid is red, because of the presence of hemoglobin, and the blood-vessel system can be seen externally in worms with slight or no body pigment.

Bootlace worms usually respond negatively to light. Since eyes are not present in many forms, this behavior appears to be associated with a general epidermal sensitivity to light. Other forms have simple eyecups, often lying deep within the tissues and serving simply to detect light direction. The sense of touch is concentrated at the sensitive anterior end, which sometimes has slightly longer, stiffer cilia than are found on the rest of the body. Statocysts are found exceptionally in a few sand-living species.

Epidermal sensitivity to light

The cerebral organs are characteristic of the bootlace worms. In their simplest form, these are pitlike depressions in the epidermis, provided with a nerve from the brain. Extremely complicated cerebral organs exist, however, consisting of a large collection of epidermal, glandular, and nerve cells connected closely to the brain and opening to the exterior by a ciliated canal. Whether these organs—as is supposed—are chemoreceptors, has not been proved. They are probably also of importance with regard to internal glandular secretion.

Central nervous system

The central nervous system in bootlace worms consists of a brain at the anterior end of the body, but not at the actual tip of the body, and two lateral nerves which proceed from the brain down the entire body length. The two brain halves are connected by dorsal and ventral commissures; these form a brain ring through which the anterior of the proboscic cavity, the proboscis, and blood vessels pass. Each brain half is divisible into a dorsal and a ventral ganglion, more or less fused with one another. In section, the structure shows the main nerve connections lying centrally in both the ganglia and commissures and also the lateral nerves, which look like cables. Around these are nerve cells, often mixed with glandular cells, forming a thick external layer.

The lateral nerves consists of connections arising from both the dorsal and the ventral ganglia. These connections remain separate in exceptional cases, and a section through the cord shows two nerve fiber nuclei. Nerves from the brain lead to the organs at the tip of the body, the proboscis system, and pharynx; the other organs are ennervated by the lateral nerve cords.

The position of the nervous system and body-wall musculature relative to one another has a special importance. In the simplest cases, the nervous system lies entirely within the epidermis. This is probably a primitive condition. Arising from this, a change in position toward the body center has come about in other forms. The nervous system lies between the epidermis and body-wall muscles, within the muscle layer itself or in the body cavity below the muscle layer in the various orders. This varying position of the nervous system is used as an important characteristic in distinguishing orders (see Color plate, p. 315).

Reproductive organs

The reproductive organs are very simple in structure. They form, at the onset of sexual maturity, as small pockets in the connective tissue between the intestine and the body-wall musculature. In many cases, they are arranged regularly between the midgut sacs; they can, however, be so extensive that they fill the whole internal cavity. The bootlace worms are usually of separate sexes, and the sexual pockets thus contain either ova or sperm. The number of ova present per ovary varies greatly. External sexual characteristics are lacking; the reproductive cells reach the exterior through pores or slits appearing in the body wall. Observations, when available, have shown that fertilization occurs within a slime envelope secreted by both sexual partners, which then forms a cocoon. Hermaphrodism has been observed frequently, and some species are viviparous. Very little is known about copulation, brood care, and other types of reproductive behavior.

Development is extremely variable. It starts with spiral cleavage, and, owing to this, the bootlace worms indicate their relation to the flatworms, arthropods, and mollusks: they are members of the Spiralia. A juvenile bootlace worm can develop directly, following on from this, or a larval stage can intervene. The basic bootlace worm larva is the "pilidium." This is similar to the Müller's larva of the flatworms but has two long flaps (lappets) hanging down to the right and left of the prostomium, giving the larva an appearance similar to a Roman helmet, the pilidium. This larva is planktonic and feeds through the prostomium. Further development is unique: three paired and two unpaired depressions, whose walls thicken, form around the mouth.

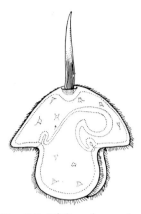

Fig. 9-7. Pilidium larva of a bootlace worm. The wide mouth can be seen through the lateral lappets, together with the intestinal cavity. The pockets anterior and posterior to this are imaginal discs.

Various parts of the adult worm form from these structures and develop separately, finally uniting to form the entire animal. Some species avoid the danger of a planktonic pilidium larva; their eggs are laid within a gelatinous envelope, each egg possessing an egg stalk. The larvae develop within the envelope and are either nourished by the yolk provided (Desor's larvae), or only a few larvae complete their development, the individuals which hatch earliest feeding on their more retarded brothers and sisters (Schmidt's larvae).

Certain species appear to reproduce asexually at more or less regular intervals. The body divides into four sections. Each section grows into a new individual by forming an anterior and a posterior end. Such a marked ability to regenerate seems to be widely distributed among bootlace worms, and is also necessary, when the length and fragility of the body are taken into account.

Regenerative ability

Bootlace worms are found in all oceanic areas; they live not only in coastal regions but also at great depths. Planktonic species appear to be restricted to deep-sea areas of the large oceans. A few species have penetrated fresh water, as has the hoplonemertine *Prostoma graecense* (L up to 12 mm; see Color plate, p. 302) in Europe. Others have invaded the land; one of these, *Geonemertes chalicophora* (L up to 12 mm), is occasionally brought into European greenhouses with imported plants, and has already been found living free in southern England.

10   The Roundworms

Phylum:
Aschelminthes,
by P. Rietschel

In contrast to the flatworms, the ROUNDWORM phylum (Aschelminthes or Nemat-helminthes; see Color plate, p. 316) has no characteristics which are common to all of its six classes, separating them from the rest of the animal kingdom. A definite relationship to other animal phyla is difficult to determine. During the early development of the rotifers and acanthocephalids, spiral cleavage appears to be present, but later this becomes severely modified. These two groups indicate a connection with the Spiralia.

All roundworms have a body cavity, to a greater or lesser extent, but, in contrast to the true body cavity (coelom), it is not lined with epithelium, and consists simply of a space between the ectodermis and endodermis. All roundworms also possess an upper skin layer (cuticula) formed by the epidermal cells; this, however, is also found in other groups of animals. In addition to this, the gastrotrichans and rotifers also retain the original epidermal ciliation in certain body areas. With the exception of the parasitic acanthocephalids, which lack an intestine, the majority of male rotifers, and a few female rotifer types, all roundworms possess an anus. Three of the six classes—the rotifers, the nematodes, and the acanthocephalids—show further unusual factors in common; the number of cells constituting their bodies is relatively small and predetermined for each species (cell constancy). Cell pedigrees can be made for the species belonging to these classes which, starting with the fertilized egg, lead through a fixed series of cell divisions to the mature animal. None of the cells of the body is capable of an extra division; these animals therefore have no means of repairing body damage. They are neither capable of healing wounds in the body nor of regeneration.

One could assume that this high degree of specialization during the course of evolutionary development would prevent any type of new adaptation, that, owing to this cell constancy, they have entered a developmental dead end. Acanthocephalids and nematodes definitely are conspicuously uniform in structure; the parasitic acanthocephalids are so

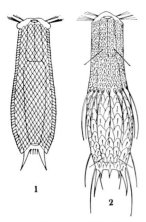

Fig. 10-1. Specialization in the cuticle (cuticula) in *Aspidiophorus* (1) and *Chaetonotus* (2).

because of their parasitic mode of life. The nematodes, however, are astoundingly variable and adaptable. They inhabit not only land and inland waters, but also marine habitats, where they live in enormous numbers in spaces within the substrate. Apart from this, they parasitize both plants and animals, the latter from the lower invertebrates to the birds and mammals. The rotifers also show great variation in their form and ways of life. They are also found in waters of all types as well as interparticular spaces in the ground. Some of them even change form during the course of the year. A marked predetermined development and cell constancy do not have to mean that the animal possessing them has reached a developmental dead end, evolutionarily speaking.

Another notable fact is that these three classes of roundworms show the tendency to lose the cell boundaries during the course of development so that the original number of cells, from this point onward, can only be determined by the number of nuclei present. This type of cell fusion is termed a "syncytium."

Since the roundworms form a phylum whose boundaries are rather difficult to define, it is a problem to determine which classes should be included within it. In general, five classes are present: 1. Gastrotrichans (Gastrotricha), 2. Rotifers (Rotatoria), 3. Nematodes (Nematoda), 4. Nematomorphs (Nematomorpha), 5. Kinorhynchids (Kinorhyncha). The acanthocephalids (Acanthocephala) are included here as a sixth class; certain investigators, however, considering them as a separate phylum. The priapulids (Priapulida), which are also sometimes included in the roundworms, are considered here as a separate phylum.

The GASTROTRICHANS (class Gastrotricha; see Color plate, p. 316) first developed in the sea, and from there invaded the fresh-water habitat. Even today, barely 200 extremely small species (L 0.07–1.5 mm) are known; species measuring more than 1 mm in length can be considered exceptional.

The first definite description of a gastrotrichan can be found in a study by O. F. Müller, *Animalicula infusoria fluviatilia et marina*, published in 1786. Here they were termed *infusoria*, that is, considered to be unicellulates. This is perfectly understandable for that time, since the gastrotrichans are not only the size of unicellulates but are also ciliated. In 1838, Christian Gottfried Ehrenberg termed them the Ichthydina ("ciliated fishlets") and included them among the rotifers. It was H. Ludwig who in 1875 put them in a class of their own, next to the rotifers and nematodes, where they have remained to the present. Our knowledge of the gastrotrichans was greatly improved by Adolf Remane during the 1920s and '30s, when he started his investigation of the coastal sand fauna; this was found to contain gastrotrichans belonging to both orders within the class, uniquely adapted to this mode of life.

The gastrotrichans are elongated, flattened, and practically colorless animals, although occasionally they appear colored because their gut

▷
Rotifers (class Rotatoria):
1. *Trochosphaera solstitialis*, 2. *Trichocera* (three animals from the side), 3. *Microcodon clavus* (from below), 4. *Ascomorphella volvocicola* (within a Volvox sphere), 5. *Notommata allantois* (from above), 6. *Monommata* (from above), 7. *Asplanchnopus multiceps* (from the side), 8. *Rotaria neptunia*, 9. *Eudactylota eudactylota* (from the side), 10. *Ascomorpha ecaudis* (from above), 11. *Limnias melicerta* (from below), 12. *Collotheca campanulata* (from above), 13. *Squatinella* (from above), 14. *Octotrocha speciosa* (two animals), 15. *Collotheca hoodii* (from the side), 16. *Trichocera* (*Diurella*; from the side), 17. *Macrochaetus subquadratus* (from above), 18. *Resticula nyssa* (from above), 19. *Colurella* (from the side).

Class: Gastrotricha, by P. Röben

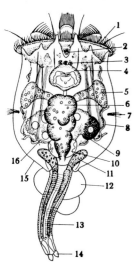

Fig. 10-2. Organization of *Brachionus rubens*, from above: 1. Cingulum, 2. Antenna, 3. Eye, 4. Mastax, 5. Digestive glands, 6. Stomach, 7. Lateral antenna, 8. Protonephridium, 9. Ovum in formation phase, 10. Bladder, 11. Pedal glands, 12. Male ova, 13. Foot, 14. Toes, 15. Cuticle, 16. Vitellarium.

◁

Rotifers (class Rotatoria): 1. *Trichotria pocillum* (from the side), 2. *Conochilus unicornis* (colony), 3. *Collotheca ornata* (two animals), 4. *Limnias* (two animals), 5. *Dicranophorus* (from the side, about to capture a *Trichocera*), 6. *Notommata copeus* (from the side), 7. *Synchaeta* (from above), 8. *Stephanoceros fimbriatus*, 9. *Rotaria*, 10. *Floscularia ringens* (male), 11. *Floscularia ringens* (females from above and from the side), 12. *Lepadella* (from above), 13. *Brachionus quadridentatus* (female from above), 14. *Brachionus quadridentatus* (male).

contents show through their transparent bodies. The body surface is covered with an external layer (cuticula) which can form plates, bristles, or spines of extremely variable shapes. The ventral surface bears ciliated bands with which the animals glide about over the substrate. In certain forms, the groups of cilia are joined together to form strong bristles; such forms "walk" on these "legs" in a manner similar to that of the hypotrichan ciliates. In many species, the anterior cilia are extremely long and enable the animal to swim for short distances.

Nearly all gastrotrichans possess adhesive glands which secrete an adhesive fluid, enabling the animal to attach itself to the substrate extremely rapidly. Such adhesive tubes can be developed in large numbers (up to 250); they are present on the ventral surface, at the anterior, and especially on the two pronglike projections at the posterior, which are often called "toes." The mouth of gastrotrichans is present at the front of the "head" and leads to an elongated, simple tube, the internal cavity of which is triangular in section, similar to that of the nematodes. The gut forms a straight line between mouth and anus and lacks lateral pouches.

Unicellulates form the main source of food for gastrotrichans, especially calcareous algae, flagellates, forameniferans, and ciliates, but bacteria and decaying organic matter are also devoured. The majority of species creep about while searching for food, and take it in by strong pharyngeal sucking movements. Others attach themselves by their adhesive glands to the substrate and waft their food toward themselves through the action of the cilia present on the ventral surface and head.

The nervous system consists of a paired ventral nerve cord with thickenings at the pharyngeal region, connected by commissures with supraesophageal ganglia. Tactile stimuli are perceived by tactile hairs, bunches of cilia, and sensory pits at the side of the head. Certain species possess extremely simple light receptors in the form of pigmented bodies within individual brain cells, capable only of distinguishing between light and dark. In many forms, connections between the brain cells appear to act as gravity receptors. No uniform layer of body wall muscles is present and muscles are arranged in individual longitudinal or circular muscle chains.

The species belonging to the order MACRODASYOIDEA are hermaphroditic. Dissolved metabolic wastes are excreted through a "ventral gland" which appears to be homologous to the excretory organ in nematodes. All macrodasyoids are marine coastal animals, where they mainly occur in sand and more rarely on plants. They are capable of moving in a leechlike manner by alternately attaching themselves to the substrate by their anterior and posterior adhesive tubes as well as gliding by means of their cilia. Their distribution is almost completely unknown owing to insufficient investigation.

In the order CHAETONOTOIDEA, the male reproductive organs have degenerated (except in the genera *Neodasys* and *Xenotrichula*). The animals reproduce by parthenogenesis. A pair of protonephridia serve as excretory organs. The chaetonotoids occupy both marine and fresh-water habitats; many species are distributed worldwide and are found regularly in aquaria and unicellulate cultures. Adhesive glands are restricted to the posterior processes except in a few cases. Even these are lacking in the families NEOGOSSEIDAE and DASYDYTIDAE which, in contrast to all other types, are fresh-water planktonic organisms; their long, movable spines aid in floating. The most common environments for gastrotrichans are in the neighborhood of plant colonies and, next to this, the substrate. Wet, mossy areas are also commonly colonized by these animals. Whether gastrotrichans can inhabit dry areas for short periods of time has not yet been proved conclusively.

The ROTIFERS (class Rotatoria; see Color plate, p. 316) are the most interesting of all the roundworms from the point of view of the naturalist. Anyone examining aquatic microorganisms under the microscope cannot fail to encounter these tiny animals and be continually amazed at their great variety of form, their tremendous activity, and their transparency, which allows a glimpse into some of their life processes. The person wishing to study them more thoroughly enters a field of investigation which is practically inexhaustible.

Despite the great variety of body forms, the rotifers are all sectioned into three parts, the head, trunk, and foot, the foot, however, having degenerated in some forms. Upon the head is a ciliated organ which surrounds the mouth except for a naked central field (apical field). Ciliary beating progresses in waves and gives the impression, in some species, of turning wheels; this characteristic has given the entire class its name. The cuticular layer covering the body can be soft and flexible; it often forms a stiff armor of various shapes and surface types, however. If a rotifer is examined under the microscope, the mastax is probably the most conspicuous body organ, often containing, hard, pointed "jaws" (trophi). These chop the food and, in some species, can even be projected forward to grasp the prey. A thick-walled stomach follows the mastax, leading to a gut and finally opening by an anus into the posterior cloaca. The protonephridia, which remove metabolic wastes and water, thus being similar in function to the contractile vacuoles of unicellulates, open by paired excretory channels into a urinary bladder on the animal's lower ventral surface, which empties rhythmically. Owing to their size and transparency, members of the genus *Asplanchna* offer the rare opportunity of observing the action of flame cells *in situ* under the microscope.

The rotifer body also contains a curious "ovary-vittelarium" which is paired in the digonontids and single in the monogonontids. It develops from a small cell layer which originates from the earliest cleavage cells and a large, usually octonucleate "vitellarium." In both of these, the cell walls have disappeared; they are thus "syncytia." The individual ova arise from the germ-cell layer and are provided with yolk from the vitellarium,

Class: Rotatoria, by J. Donner

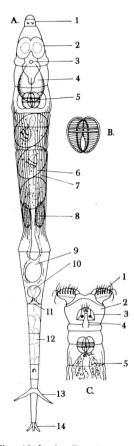

Fig. 10-3. A: *Rotaria rotatoria* with usual ciliated head (proboscid eyes, proboscis ciliary fields) extended in creeping, the mastax (B) and head with neck (C) during ciliary activity. A: 1. Proboscis with eyes, 2. Ciliary (wheel) organ, 3. Dorsal antenna, 4. Brain, 5. Mastax, 6. Three developing ova, 7. Cuticular folds,

causing them to grow appreciably. Their further development is a strongly predetermined mosaic one, animals with an exact, predetermined number of cells being produced (cell constancy). The trunk organs are located in a body cavity which is filled with fluid, not lined by a layer of cells, and thus not a true coelom. The rotifer foot usually consists of two "toes" with adhesive glands, both being used to anchor the animal.

This description is applicable only to female rotifers. The males, in contrast, apart from those of the seisonids, are much more simply constructed (Monogononta) or are completely absent (Digononta). Even the small monogonontid males are hardly known. Their degeneracy is especially pronounced in the intestinal tract and excretory organs; their apical cingulum no longer has the task of wafting food toward the mouth. It has become adapted completely to form a locomotive organ. The males are able to swim very rapidly in a straight line, which aids them in their search for the females. They are not capable of feeding during their lives, which span only a few hours or days. The sense organs, which assist in discovering the females, are less degenerate. The only organ which is well developed in the males is the testes. During copulation, the male inserts a copulatory organ or its foot into the female's cloaca or breaks through the female's body wall with these organs.

Rotifers have several means of locomotion. Those inhabiting damp earth and waterlogged moss move by "looping" in the same way as a leech; they are capable of performing this motion mainly by the presence of an elastic body and a snout at the anterior end together with the adhesive glands at the posterior, both serving as attachment organs. Other rotifers glide along the substrate by means of their cilia; others possess body processes which increase their ability to glide. The majority of species, however, swim in spirals. A few rotifers settle after the free-swimming larval stage and remain attached to the substrate for the rest of their lives, either naked or in a delicately constructed shell.

In the mode of feeding, the following types can be distinguished: filterers, graspers, and capturers. The filterers produce a water current with their cingular cilia, this current usually containing the decaying remains of plants and animals. The graspers capture unicellulates and tiny multicellulates by extending the mastax. The capturers are sessile and thus no longer require their cingulum for locomotion. It has become, together with the mouth and pharynx, a type of funnel, the edge of which is fringed with long, stiff cilia. These cilia serve as a "fish-trap" from which captured prey cannot free itself. On each occasion that prey is captured, the edge of the funnel contracts, the fish-trap cilia relax, and the prey is carried past a ring of cilia into the vestibulum and then to the mastax, where it is chopped up.

Reproduction in the rotifers is unique. The digonontids reproduce only parthenogenetically, since no males are present. The monogonontids, on the other hand, alternate between asexual and sexual reproduction

Reduced or degenerate males

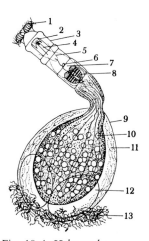

Fig. 10-4. *Habrotrocha pusilla textris brevilabris*, a flukelike rotifer within its slime envelope during ciliary feeding. 1. Ciliary (wheel) organ, 2. Proboscis, 3. Head, 4. Dorsal antenna, 5. Esophagus, 6. Esophageal tube, 7. Brain, 8. Mastax, 9. Envelope, 10. Food pellets within the stomach wall, 11. Unformed food pellets in the envelope, 12. Ova, 13. Soil particles.

8. Vitellarium, 9. Rectum, 10. Bladder, 11. Anus, 12. Pedal glands, 13. Spines, 14. Toes, C: 1. Ciliary (wheel) organ. 2. Head, 3. Proboscis, 4. Dorsal antenna, 5. Stomach.

(alternation of generations, heterogony). Females hatch from hard-shelled resting eggs after a quiescent period, these in turn producing females which reproduce parthenogenetically (amictid females). In this way, the species can reach its highest population density within the shortest period of time. After a while, females which resemble the other females externally are produced; these, however, produce only tiny eggs from which males develop (mictid females). These males are capable of distinguishing between the two types of females; they copulate with the mictid females and ignore the amictid ones. If the mictid females are not copulated with, they continue to produce males; should copulation occur, however, they produce thick-shelled resting eggs which, in their turn, produce amictid females.

Males and mictid females can be produced several times yearly; the production of the latter is brought about by several environmental factors. In some species, the environment even has an effect upon body form. In *Brachionus calycifloris*, for example, the anterior and posterior processes on the armor develop extensively only under conditions of poor nourishment and cold, or when rotifers of the genus *Asplanchna* are present. Even the filtered water from an *Asplanchna* culture can stimulate this development. A similar case of body modification through environmental influences (cyclomorphosis) is known from water fleas.

The majority of the approximately 2000 rotifer species now known are worldwide in distribution; each species, however, prefers its own environmental conditions. Only a few seisonid species are marine, the remainder living in fresh water, either being planktonic or living in algal growths or among the organic detritus on the floor of the water. The digonontids are the most adapted to unusual environments. Those which move by looping prosper in the thin film of water covering moss or in the interparticular spaces of the earth; others survive in the layer of thawed water produced by the sun's rays on antarctic ice. Others inhabit hot springs. The ability to desiccate "in their living bodies" has made the survival of certain digonontids possible in temporarily dry areas. They and their eggs are scattered far and wide by the wind. For example, the volcanic island Surtsey, which emerged from the sea near the southeastern coast of Iceland in 1963, had hardly cooled before the rotifer *Habrotrocha constricta* was found there—a pioneer of life! Rotifers are capable of withstanding five years of desiccation when within a cyst. They have been able to survive a temperature of minus 271°C for three weeks and a temperature of plus 78°C for even longer. Cysts are found everywhere that there is a chance of life; even a single cyst can, owing to parthenogenetic reproduction, become the founding female of a whole population: genetic homogeneity of a clone!

From the large numbers of rotifers existing, only a few can be more closely considered here. The order SEISONIDS (Seisonida), which contains only a few species, lives in marine habitats; little is known about its mode

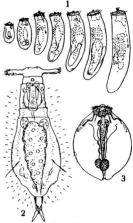

Fig. 10-5. Rotifers with a soft outer covering. 1. *Asplanchna priodonta*, between 30 May and 9 July; during this period, each subsequent generation lives longer; 2. *Notommata copeus* with extended cephalic lobes as it swims within its jelly envelope; 3. *Horaëlla brehmi*, a very delicate Indian species.

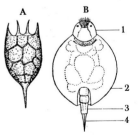

Fig. 10-6. A. The exoskeleton of *Keratella cochlearis* in dorsal aspect; B. *Lepadella patella* while swimming, ventral aspect: 1. Head, 2. Exoskeleton, 3. Foot, 4. Toes.

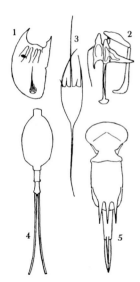

Fig. 10-7. 1. *Trichocerca taurocephala* from below; 2. Its asymmetrical mastax; 3. Exoskeleton of *Kellikottia longispina*; 4. The strengthened cuticle of *Eudactylota eudactylota*; 5. Exoskeleton of *Squatinella lamellaris*, a diadem animalcule, from above.

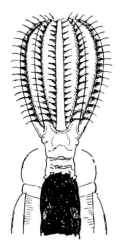

Fig. 10-8. Crown and anterior portion of the envelope of *Stephanoceros fimbriatus*.

of life. The males are similar to the females, the reproductive organs are paired, and the females possess no vitellarium. The size of these animals (*Seison nebaliae* can reach 3 mm) and their attachment to the most primitive of the higher crustaceans, the leptocostacans, give evidence of their own primitive origins. Since the *Seison* species are attached to the gills of their hosts and obtain their nourishment from the water current present there, their ciliated organ is only poorly developed.

The DIGONONTIDS (order Digononta), which reproduce entirely parthenogenetically and contain a single suborder, the LEECHLIKE ROTIFERS (Bdelloidea) with their telescopic bodies, are more similar to the impression of a tubelike "worm" than any other groups of rotifers. The previously mentioned inhabitants of extremely unusual environments and the ground all belong to this order. The slender *Rotaria neptunia* (L over 1 mm; see Color plates, pp. 305/306 and 325), found only in highly polluted waters, is also a member of this order. The over 100 species which live in the interparticular spaces of the ground also belong to this order. H. Franz calculated that, in the upper 5 cm of 1 m² of grass or cultivated land, up to 300,000 rotifers could be present! They live here partly on other microorganisms and partly on decaying substances, thus certainly aiding in decomposition. Many digonontids form a protective envelope. *Mniobia incrassata* secretes a jelly which solidifies between the folds of its epidermis, preventing it from moving. *Macrotrachela insulana* contracts to form a sphere and, within a few minutes, secretes a glassy, transparent envelope, constituting a cyst within which it can remain quiescent for months. *Habrotrocha flaviformis* fastens an adhesive thread produced from its snout to the substrate, pulls it across its dorsal side to the other side of the body, attaches it there, and continues this "weaving" until it is enclosed within a beautiful, symmetrical net. Some species even use their own fecal pellets in the walls of their mucous envelopes. Others inhabit completed structures such as the skeletons of testacid amoebae or the water cavities of moss plants.

The MONOGONONTIDS (order Monogononta) possess a single reproductive organ and never move by "looping," as do the digonontids. Their more or less completely degenerate dwarf males (the smallest of which measures only 0.04 mm) are by no means known for all species. The variety of forms present in the females is enormous; the extreme beauty of some of these species is shown in the color plates on pages 305/306, 325, and 326. The suborder PLOIMIDS (Ploima) contains the largest number of species. Many of these live in stream-bank areas where plant stands are present and are either free-swimming, creeping on the submerged portions of the plants, or temporarily attached to them. Continuously sessile forms are not present. A few species, which, however, contain large numbers, are found in the plankton. The large, gutless genus *Asplanchna*, which also possesses no anus, and the skittle-shaped *Synchaeta* (see Color plate, p. 326) are found in this environment because of their low density. *Brachionus* (see Color plate, p. 326) and *Keratella* possess short or elongated spines at the anterior and posterior of their armor, which aid in floating (see Color plates, pp. 305/306). In *Kellikottia longispina* a central anterior spine and posterior spine serve as floating organs. The small, armorless, and footless *Polyarthra*, in contrast, possesses six swordlike structures on each side of the body. The majority of planktonic forms carry their eggs around with them at the base of the foot for long periods of time; the footless asplanchnids, however, retain the eggs within the mother's body until they hatch, and are live-bearers.

The beautiful, glassy *Cephalodella gibba* lives in moss along the banks of slightly polluted streams. Its body is humped on the dorsal side and its foot possesses two long toes which are pointed upward. This species hunts by grasping, and is capable of swallowing nematodes about twice its own body length. It, in turn, is hunted by another rotifer, *Dicranophorus forcipatus*. This extrudes its mastax, which is equipped with sharp, pointed teeth, out of the mouth when grasping its prey. By this means, a food chain consisting of several links is formed: decaying matter, nematodes, *Cephalodella*, *Dicranophorus*, insect larvae, fish fry, pike, and finally man. Some of the ploimids have special requirements with regard to the water they live in: *Epiphanes senta* (see Color plates, pp. 305/306) requires that a certain amount of urine be present; *Encentrum oxyodon* and *Encentrum incisum* require wood. *Proales werneckii* causes club-shaped or saclike galls in the filaments of the green alga *Vaucheria*, and inhabits these together with its eggs. *Hertwigella volvocicola* is parasitic in *Volvox* colonies, and *Albertia* species are internal parasites in the intestines of annelids.

The suborder FLOSCULARIACEANS (Flosculariacea) possess on their foot—when present—a ciliated cap rather than toes. *Filinia*, which possesses a single posterior and two lateral "jumping bristles," is fairly commonly found in plankton, as is the colonial species *Conochilus* (see Color plate, p. 326). In the latter, the cone-shaped individuals are attached to one another by their elongated foot in the middle of a jellylike communal sphere. The strongest animals are below, and they hold the whole colony up by their ciliary beating. The eggs are found in the center of the sphere; young animals either increase the size of the colony or abandon it to form new colonies.

The casing of the sessile forms is probably the most amazing structure among the rotifers. The larva of *Floscularia ringens* (see Color plate, p. 326) forms a slime bowl with its foot glands and remains attached within it permanently. Its four ciliated lappets continually waft mud particles toward the animal; these are turned into tiny balls by cilia in a pit beneath the lip, and are then added, in a definite order and in exactly the right place, to the tube being constructed. *Ptygura pilula* uses its own fecal pellets in exactly the same way. *Limnias melicerta* (see Color plate, p. 325), in contrast, lifts its trunk over the edge of the tube under construction

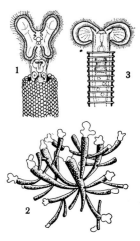

Fig. 10-9. 1. *Floscularia ringens* within its tube, from above; 2. A colony of the same, consisting of animals of different ages; 3. *Limnias melicerta* within its firm tube.

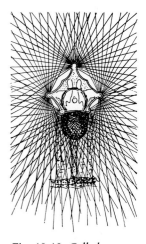

Fig. 10-10. *Collotheca ornata cornuta* within its mucous envelope. The prey-capturing funnel possesses numerous cilia which act as "fish traps," here foreshortened. View from in front and below.

and secretes a new layer of construction material along the edge, which is then compressed by the body.

The suborder COLLOTHECACEANS (Collothecacea) contains only a few sessile species. Their heads consist of huge prey-capturing funnels whose use has been previously described. The animal is enclosed—attached by its long foot—in a jellylike casing, as is *Stephanoceros fimbriatus* (see Color plate, p. 326), whose funnel resembles delicate filigree.

The observation of rotifers can be undertaken by anyone who owns a microscope and knows how to use it, offering the observer an almost unending source of pleasure in discovery. There is never a lack of animals to be found in the environment throughout the year; in winter, the quiescent creatures can be brought to activity again within a few days by placing them in a jar of water in a warm room. In our own surroundings there are enough forms of rotifers to keep a naturalist continually occupied with observing and investigating them. In the Rotifera, especially, the sense of beauty has not been lost.

The THREADWORMS (class Nematoda; see Color plate, p. 316), the commonest of the multicellulate animals, are also termed "eelworms" because of their snakelike movements. A handful of loam may contain over 1000 animals. They are also found, however, on the bottom of water stands, in deserts, in antarctic ice, in hot springs, and as parasites of plants and animals, including man. If the entire material which constitutes the world disappeared, leaving only the nematodes—as the American parasitologist N. A. Cobb stated fifty years ago—the world would still be recognizable as a ghostly hollow ball consisting of nematodes. Mountains and valleys, rivers and oceans, deserts and forests could be distinguished by the nematode layer which once lived there. Even towns would still be recognizable, and the trees lining the streets would still be visible as ghosts, represented by the threadworms which inhabited them.

Despite their abundance, which has been so well illustrated by Cobb, most people know nothing about nematodes. Only a few species have common names; the majority are identified only by their scientific names. A few belong to the worst scourges of mankind and were mentioned in the Old Testament and in ancient Egyptian documents over 3500 years ago, such as the maw-worm (*Ascaris*) and the medina worm (*Dracunculus*).

The nematodes are usually threadlike and elongate (L 0.1 mm to 8.4 m, usually 0.5 to 5.0 mm); body round in transverse section; transparent; almost complete cell constancy; only longitudinal muscles present in the body-wall musculature, each muscle cell with a process to the nerve cord; reproduction either unisexual or bisexual; strictly determined development.

Despite the fact that they have colonized almost every type of habitat, the nematodes are very uniform in structure. Their body is usually spindle-shaped, the posterior being more strongly pointed than the

Class: Nematoda, by B. Weischer

Fig. 10-11. The majority of nematodes are elongate and threadlike in form; other body forms are, however, present.

anterior, sometimes even becoming extended to form a thread. The anterior end is less often threadlike as, for example, in the WHIPWORM (*Trichuris trichiura*). In certain plant and animal parasites, the females swell and become immotile. The smallest nematode is the plant parasite *Sphaeronema minutissimus* (L 0.1 mm); the largest are *Placentonema gigantissimus*, a parasite in the placenta of the sperm whale, and the medina worm, which can reach a length of 1 m. The body diameter is a maximum of one-thirtieth to one-three-hundredth of the body length. Males are usually smaller and more slender than females.

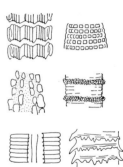

Fig. 10-12. The cuticle with characteristic patterning.

The body covering is a so-called "epidermal muscle layer," which consists of a covering layer (cuticle), a skin-cell layer (epidermis or hypodermis), and a muscle-cell layer. The lamellated cuticle consists of protein (collagens and keratins). Its surface is often patterned peculiarly, especially in marine forms. The epidermis consists of a single cell layer. In some cases, the cell walls have degenerated and the entire layer forms a multinucleate giant cell (syncytium). This is an important storage organ for fat and animal starch (glycogen). The majority of nematodes are whitish and transparent. Marine forms often have a pink or bluish shimmer. Resting forms are usually brown. In certain species, the intestinal contents visible through the body make the animal appear colored (green to brown in the nematodes living on algae, red to black in blood parasites).

Nematodes possess no internal or external skeleton. Body solidity is brought about by an interplay between the elastic body-wall musculature and the internal pressure of the body fluid (hydrostatic skeleton). The internal pressure is extremely high; in the ascarids, for example, it reaches 225 mm of mercury, that is, 0.3 atmospheres. When the animal is damaged, therefore, the body fluids literally squirt out. This high-pressure system is almost unique among the animal kingdom. Only longitudinal muscle cells are present in the nematode body-wall musculature. These are spindle-shaped and arranged parallel to the longitudinal axis of the body in four main fields. A single muscle cell can reach amazing proportions; in the ascarids, for example, it can reach a length of 1 cm. As in the rest of the animal kingdom, the cells are activated through the nervous system; however, in the remainder of the animal kingdom, the nerve cells send processes to the muscles along which the impulse is carried. In the nematodes, on the contrary, each muscle cell sends a process to the central nerve cord and receives the impulse from it—a completely unique arrangement.

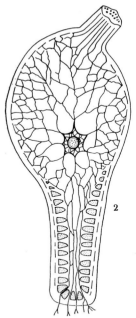

Fig. 10-13. Muscle cell from the body wall in longitudinal (1) and horizontal (2) section. The process to the nerve strand above, fibers for anchoring in the epidermal layer below.

The most common form of locomotion is a snakelike movement in the dorsal-ventral plane, alternate contractions of the dorsal and ventral longitudinal muscles: "draconid" movement. Nematodes perform "looping" much more rarely, and likewise only rarely move with the aid of bristles. Each form of active motion, however, is of service to the species only in moving over short distances; passive transport is more important in general distribution. Wind, water, and animals, including

Fig. 10-14. The oral cavities vary in type depending on the mode of life:

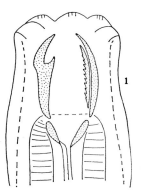

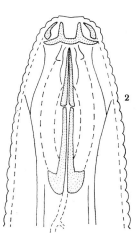

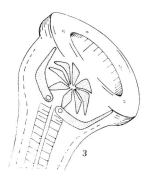

1. Hunter with teeth; 2. Plant parasite with oral spine, 3. Animal parasite with teeth (*Cont. p. 336,*)

humans, with their various means of transportation, aid nematodes in covering both short and long distances and colonizing new environments. For example, the nematodes inhabiting cow dung crawl out of a dropping when it dries, and attach themselves to the legs of dung beetles, which then carry them to a new dropping. Plant-parasitic nematodes are easily carried in the soil on people's shoes or on the tires of cars from infected to healthy fields. For some species, such as the potato eelworm, special laws are brought into action to prevent it from being transported and thus more widely distributed.

The LUNGWORM (*Dictyocaulus viviparus*), which parasitizes the ruminant respiratory system, is spread in a rather unusual way. The infectious larvae crawl out of the dung, in which they live at first, onto the fruiting bodies of a feces-inhabiting fungus (genus *Pilobolus*), which is capable of shooting its spore containers for a distance of up to 2 m by means of a special mechanisms; the nematodes attached to them are spread with the spores. In this way, they avoid the grass zone which is also avoided by the grazing animals—that immediately around the feces—where they would have little chance of being taken up by their host when it grazes. Changes in organ infestation within the host, which the parasite makes during the course of its development, also occur passively, the larva being transported by the blood, lymph, or food streams.

Free-living nematodes mainly feed on solids, such as small organisms; parasites, on the other hand, live mainly from the fluids of their animal or plant hosts. The digestive system's structure is the same in all nematodes; it consists of a mouth cavity, pharynx, midgut tube, and rectum with an anus. The mouth cavity varies greatly in structure, depending on the animal's mode of life, and is an important characteristic in distinguishing between forms. The pharynx is a simple tube with muscular walls, that is used as a pump to press in the food against the internal body pressure. The midgut, a thin-walled, straight tube, removes the digestable portions of the food. In a few animal parasites, which live in surroundings containing large amounts of nourishment (blood, gut contents) and which take up soluble food through the epidermis, the midgut is adapted to form a storage organ. The short, muscular rectum joins the midgut and anus together and serves as a mechanism preventing the food from being expelled from the body too rapidly, owing to the high body pressure. As far as is known, food cycling occurs extremely rapidly, especially in the case of parasites which live continually within their food source. In *Ascaris*, for example, defecation occurs every three minutes. A whole series of digestive substances and enzymes have been found in nematodes. Their structure and function varies with the means of nourishment.

How nematodes excrete the residues of protein and carbohydrate metabolism has not yet been completely explained. The two longitudinal tubes on each side of the body, which possess a large glandular cell and an opening to the exterior, are usually considered to be the excretory system. It is more probable, however, that the large majority of these wastes are passed out through the intestine. More can be laid in the covering layer of the body (cuticle) and the egg-capsule wall; the lateral tube system, which is lacking in several groups, probably serves simply as an additional means of excretion. According to its structure and occurrence in the nematodes from various environments, its main function appears to be the control of the body-fluid concentration under conditions of salt or water loss or gain (osmoregulation).

Nematodes have no special respiratory organs. The oxygen required for energy release is taken up through the epidermis. Species living in environments of low oxygen content, such as decaying matter and gut contents, have a means of energy production through chemical processes in which free oxygen does not play a part (anaerobic respiration). Since this type of respiration has a much lower productive effect, these animals possess large stores of animal starch. A blood-circulatory system is also absent in nematodes. The substances taken up by the intestinal cells from the gut contents and the oxygen taken in through the epidermis are released into the body fluids and distributed throughout the body by this means. Hormones and glands which release internal secretions have not yet been found in nematodes; it is suspected, however, that hormonal processes are involved in ecdysis.

All nematodes possess processes—such as warts (papillae), bristles, lateral organs (amphidia), and similar structures—which, according to their structure, could be receptors for tactile and chemical stimuli. Such structures must be present, since nematodes respond to both tactile stimuli and various chemical substances. The males are also frequently attracted by special sexual substances released by the females, especially in those species where the females are motile. A definite allocation of these organs to certain sensory functions is, however, not possible. Many nematodes respond to light stimuli although light receptors have been proved present in only a few species. These are frequently simple pigment spots at the anterior end, as in the insect parasite *Mermis subnigrescens*. Sometimes, however, complicated structures are present, as in *Dilaimus denticulatus*, a marine species; these consist of a lens above a pigment spot. Very little is known about the purpose of such light receptors. In the previously mentioned insect parasite *Mermis*, the perception of light is associated with egg laying. Only the mature females possess a pigment spot. They creep up the stalks of plants during a definite period of time, and lay their eggs there. It is hypothesized that they orient to light during this migration. Females kept in darkness do not lay eggs.

A few nematodes have a very delicate ability to differentiate between temperatures. Larvae of the potato eelworm respond to a temperature-gradient drop of $0.02°C$ per centimeter. Organs for temperature reception have not yet been found.

The most important part of the nervous system in all nematodes is a

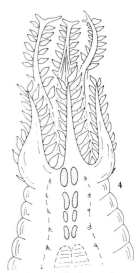

4. Bacteria feeder with a "fish-trap" apparatus.

Fig. 10-15. A schematic section through the body. The two main nerve strands above and below, with connections to the muscle cells, the intestine within the body cavity above; below this, the reproductive organ.

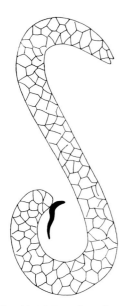

Fig. 10-16. In *Sphaerularia bombi*, the female (black) is, at the end of its development, only an appendage to the sexual organ.

Fig. 10-17. The outer covering of the nematode egg is often provided with typical patternings and processes.

thickened nerve ring in the pharyngeal region, the ganglia attached to this, and six nerve strands which run parallel to the longitudinal axis of the body from the ganglia to the tip of the tail; in the region of the anus, these are connected to one another by annular horizontal nerve strands and ganglia. Apart from this thinner individual nerve strands pass to the various sense organs. Generally speaking, the head region does not play such an important role with respect to the nervous system in general and the control of behavior as it does in higher animals.

Reproduction is always unisexual or bisexual, never asexual. The majority of species have separate sexes; a few are hermaphroditic. In a few cases, unisexual and bisexual reproduction alternate (alternation of generations, heterogony). The female reproductive organs are paired, with a common opening (vagina). Sometimes, the posterior, and more rarely the anterior, ovary has degenerated and, depending on which has degenerated, the vagina becomes elongated either anteriorly or posteriorly. In a few forms, the sexual organs increase greatly in size during the course of development, until they fill the whole animal. Development is especially unusual in *Sphaerularia bombi*, a parasite in European wasps and bumblebees. In this species, the sexual organs of the female are everted from the genital opening as they grow, increasing in size to such an extent that the female itself is finally only a small appendage attached to them.

The eggs of nematodes have an outer layer consisting of protein; they are often covered with patterns and attachments which can be used in species identification. If feces are examined for the presence of certain nematode eggs, a parasitic infection can thus be determined; this is difficult to do on the living animal or person. This process is of great importance in parasite control for domestic and zoo animals, as well as in human medicine. The number of eggs laid by a female varies. Parasites whose larvae have only a small chance of survival produce enormous quantities of eggs; a female *Ascaris*, for example, lays 70 million a year. Nematodes living in a relatively constant environment, whose eggs hatch in the immediate vicinity of food, produce, on the contrary, only twenty to thirty eggs. A few species are live bearers. In old age, or as a normal occurrence, many nematodes are able to delay egg laying so that the larvae hatch within the female, thus producing a live birth. In the final stage, the larvae no longer leave the mother but remain within it as parasites feeding on the organs. Only when the mother has been completely consumed do they leave. In a few nematodes, the larval development within the mother progresses to the point at which the larvae become sexually mature. Even copulation can occur within the mother's body, the fertilized females being the first to leave the mother, which has died in the meantime.

The male sexual organs (testes) are usually single. The sperm are carried to the genital opening by the vas deferens, the genital opening and anus having a common exit (cloaca). The sperm cells are relatively large and, in contrast to those of the majority of animal phyla, possess no flagellum by which they are able to move. During copulation, the male usually wraps around the female, presses its genital opening against that of the female, and spreads the female's genital opening by means of special spikelike structures (spicula). The sperm cells are then carried from the vas deferens into the female's genital opening by means of muscle pressure. The males often possess a winglike or umbrellalike structure (bursa) in the region of the cloaca, aiding in attachment to the female. In a few species, the males remain continually attached to the females, as in the species of the genus *Syngamus*, which live as parasites in the respiratory tubes of mammals and birds. In *Trichosomoides crassicauda*, which parasitizes the urinary system of rodents, the males remain in the larval stage and live within the oviduct and uterus of the females.

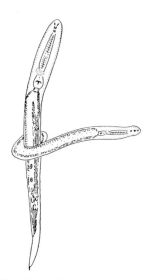

Fig. 10-18. During copulation, the male usually twists the posterior end of its body around the female.

The ratio of males to females varies, depending on the species; it can even vary within a species depending on the population density. In populations with only a few males, the females lay unfertilized eggs; such males are thus of little importance with regard to reproduction. If large numbers of males are present, there is copulation. The sexual ratio also often depends on the environment. Unfavorable conditions usually lead to an increase in the number of males. For example, more males develop from the larvae of the grasshopper parasite (genus *Mermis*) if the host is heavily infected, that is, when less nourishment is available for the parasite. A similar case occurs in the root-gall eelworms, where even a change of sex can take place. If, during the course of development, a decrease in the available food occurs, larvae which are obviously female develop into males. In the plant parasites of the genus *Heterodera*, which form cysts, the number of males also increases with a lack of nourishment or available plants; this, however, is not caused by a change in sex but by a higher death rate in the females under unfavorable conditions.

During the development of the fertilized egg of a nematode, the function and future development of the individual cells are already determined at the first cleavage. When embryological development has ceased and the larva is still within the egg, all the cells of the adult body are already present. The total number of cells or the cells comprising individual organs is the same for every member of the same species; this phenomenon is termed cell constancy or eutery. During the further development to the adult animal, cell division occurs only within the sexual organs. The increase in size which occurs comes about through an increase in the size of the cells themselves. The muscle cells of the HORSE WORM (*Oxyuris equi*) for example, increase from 0.03 mm in length in the larva to 6 mm in length in the adult animal, that is, by a factor of 200. The fully developed larvae remain within the egg for a relatively long period of time, especially in the case of parasites. They are finally stimulated to hatch by the effect of environmental influences, and then assume their active life. The parasites have adapted their life cycles as closely as possible to those of their

Fig. 10-19. In the *Syngamus* species, the males are attached to the females in continuous copulation.

hosts. If large numbers of infectious larvae hatch when the correct host is present, they have the best chance of survival. It is for this reason that the larvae of the potato eelworm and other cyst-forming nematodes remain within their hard capsule until they are activated by the root secretions of a host plant and are attracted to it.

In the sheep parasite *Nematodirus battus*, the larvae go through two ecdysial periods while still within the egg, afterward remaining at this stage of development. They are still not capable of infecting, even when they are taken up by the sheep with its food. Only when the larvae within the egg are chilled to about 0°C and then warmed again, do they hatch and have the ability to infect the sheep. This dependence on a definite temperature progression results in large numbers of larvae hatching during the spring, when the number of sheep available is also increased by the presence of numerous newborn lambs, and thus the chance of entering a host is particularly good. In other nematodes, hatching is brought about by humidity or by increased carbon dioxide in the surroundings or other chemical substances which indicate the presence of favorable conditions.

Despite their quite varied modes of life, the development of the nematodes from the embryo to the adult animal is very similar. Four larval stages can be distinguished, together with the adult stage, each being separated from one another by ecdysis. This is reminiscent of the development of insects, but is not comparable to it. In nematodes, the increase in size does not come about in short spurts, but gradually. Certain organs can, however, increase terrifically in length within a short period of time, for example, the oral spine. At each ecdysis, the entire covering layer (cuticle) is lost, together with the linings of the buccal cavity, the gland openings, and the anus. Certain organ changes also occur at the same time. The exact processes involved in ecdysis are only sparsely known. In some cases, definite environmental chemical stimuli result in the larva secreting a fluid which causes ecdysis to occur. Whether this is a hormonal substance, as is the case in insects, is not known. Sometimes, the stimulus releasing the ecdysial fluid comes from the host. The growth of nematodes is—in contrast to that of the insects—not completed after the last ecdysis. The females of some plant parasites swell after this to multiples of their original diameter, and female ascarids grow within the few weeks following their last ecdysis from 8 to 30 or 40 cm long.

There are very many variations to this basic developmental plan which have come about as adaptations to definite environments or special modes of life. These are especially well illustrated in the parasites. It must be assumed that parasitism has arisen during the course of evolution from many groups of nematodes at varying times, totally independent of one another. When the nematode class is surveyed, parasitic forms are found in many orders, in various stages of specialization. A few of these will be discussed subsequently, more from the point of view

Ecdysis within the ovum

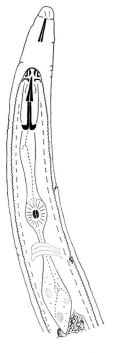

Fig. 10-20. On each ecdysis, the whole cuticle is shed, including the lining of the oral cavity and other body openings.

of their modes of life rather than the systematic group to which they belong.

Nearly all plant-infesting nematodes live for short or longer periods in the ground and attack their host plants from there. In attacking, they bore through the cell walls with their hollow oral spine and release enzymes or poisons which chemically change the cell contents and loosen the plant tissues. Damage, the loss of food substances, and the effects of the enzymes affect the plant tissue's normal processes and lead to poor growth. The damage caused by nematodes to agricultural plants is estimated at 372 million dollars a year in the U.S.A. Severe infestations can lead to catastrophes. For example, about a century ago, the whole sugar industry in Germany's Magdeburg area collapsed completely after years of good crops, because the sugar-beet eelworm had increased in such enormous numbers over the years of intensive sugar-beet cultivation that finally only scrub plants could be harvested. The unknown reason why the Mayas left their fertile fields in Guatamala in A.D. 700 was probably the fact that their agricultural areas had been the victims of a mass infestation of plant nematodes resulting from continual plantings of the same crops. The Mayas had to leave their old environment and live in the Yucatan highlands simply because of the lack of food. The most effective means of preventing such infestations are regular crop rotation, the planting and development of resistant strains, and—if neither is possible—treating the plants or ground with a substance which kills the nematodes (nematicide). Biological control of the nematodes through special introduction of their enemies has thus far been effective only in the laboratory, not in the field.

Migratory and sessile plant nematodes can be distinguished on the basis of their modes of life, several hundred species now being known. In the simplest case, the worms remain in the ground for their whole lives, sucking root cells from there and being generally restricted to the outer cell layers of the plant root. Species of the genera *Tylenchorhynchus*, *Helicotylenchus*, *Rotylenchus*, *Paratylenchus*, and others are members of this group. The genera *Xiphinema*, *Longidorus*, and *Trichodorus* are worthy of special note, since they are carriers of virus diseases infecting agricultural crops. Species of the genera *Pratylenchus* and *Radopholus* are one step nearer to being specialized parasites. They retain their slender bodies and motility, but reproduction occurs within the plant tissues. Both larvae and adult animals are free to leave the plant at any time, however, in order to move to new roots. They are capable of surviving for long periods in the ground without food, as are almost all plant-infesting nematodes.

Those nematodes infecting the arial portions of plants are more strongly bound to their hosts; these include, for example, *Ditylenchus dipsaci*, the LEAF NEMATODES (genus *Aphelenchoides*), and the COCONUT NEMATODE (*Rhadinaphelenchus cocophilus*). Their active lives occur entirely within the plant, within which they migrate from one part to another. The ground serves only as a place in which they can survive unfavorable conditions. At present, over 400 host plants of *Ditylenchus* are known, including rye, maize, red clover, onions, daffodils, and phlox. *Ditylenchus* also has a series of races which are differentiated on the basis of their host plants. Some are able to reproduce in many plant types; others appear to have only a single "proper" host, such as the white-clover race. Leaf nematodes, of which several species are known, live within the leaves and flowers of strawberries, rice, chrysanthemums, gloxinias, ferns, and other cultivated and wild plants. When the plants are wet with dew or rain, or from being watered, they move out of the ground, climb the stem, and bore through the spiracles into the leaf tissue; here copulation and egg-laying take place. Under favorable conditions, they continue to migrate or become quiescent until re-activated by moisture. The coconut nematode enters the interior of the palm stem from the ground but also has another means of entering its host plant. When the palm borer beetle— a weevil—bore into palms infected with the nematode, the worms attach themselves to its legs and mouth-parts and are thus carried to new palms. Recent investigations have shown that the coconut nematode can even bore into the body of the beetle and live there as a parasite.

Sessile plant nematodes

The sessile plant nematodes have obtained their name from the fact that the adult females, at least, and often earlier developmental stages, lose their motility. In *Tylenchulus* and *Rotylenchulus*, all four larval stages and also the young females are still capable of movement. They regularly change their point of attachment to the plant, either in or on the roots. It is during the course of further development that the females attach their anterior ends to the plant tissue, their posterior ends hanging free and gradually swelling irregularly. The eggs are released to the exterior of the plant. Species of the genus *Nacobbus* are a step further along the path to becoming a sessile internal parasite. All the larval stages are capable of movement, but an increasing tendency to remain in the same position can be shown at each ecdysis. The final ecdysis always occurs within the plant's tissues, in which the females remain for the rest of their lives. In these species, the center of the body starts to swell and only the tip of the tail protrudes from the root.

A further adaptation to a parasitic mode of life is found in the cyst-forming species of the genus *Heterodera*, which includes important parasites of cultivated plants such as the POTATO EELWORM (*Heterodera rostochiensis*), the BEET EELWORM (*Heterodera schachtii*), and the WHEAT EELWORM (*Heterodera avenae*). Only the second-stage larvae are motile. They leave the egg under the influence of certain weather conditions, and migrate through the ground into the appropriate roots. All further larval stages, and the females, are immotile. They swell first into a flask shape and later become pear-shaped, breaking through the root cuticle to the exterior. The mature males retain the slender nematode form and search—

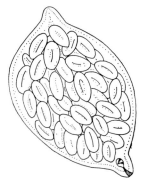

Fig. 10-21. The females of the cyst-forming nematodes are, on completion of their development, nothing more than egg-filled capsules.

attracted by definite sexual-attractant substances—for the females, in order to copulate. Several hundred eggs develop within the female. The females then die, their body wall hardening until, finally, they form cysts filled with eggs which fall from the root and remain in the ground. The larvae within the cyst remain viable for years.

The ROOT GALL NEMATODES (genus *Meloidogyne*), which are especially common in warm climates, are similar externally and in their mode of development to the cyst-forming nematodes; these, however, never leave the plant tissues at all. After hatching, the larvae usually settle immediately on the root. They leave the plant only to search for a new host when the original plant dies and its infected roots rot away.

A larger number of species and a greater variation in mode of life are found among the animal parasites than the plant parasites. They either enter their host with the food or bore through its skin. When the final host, that is, the host in which reproduction occurs, is infected immediately, this is termed direct development. If parts of the larval development occur in a secondary host before the final host is attacked, this is termed indirect development. An animal in which the nematodes remain temporarily, without further development occurring there, is called a "transport host." Both invertebrates and vertebrates can be infected, usually, however, by parasites of different groups.

Animal parasites

Some of the parasites of invertebrates will be considered first. In many forms, only a part of the parasite's life takes place within the host; for example, members of the genus *Parasitylenchus* lay their eggs within the body cavity of the insect host, and the larvae hatching from them parasitize the host until their second ecdysis. They then leave the insect and develop into sexually mature animals. Copulation also occurs outside the host; only the fertilized females bore into the insect again in order to lay their eggs. In the MERMITHOIDIDS (superfamily Mermithoidea), which parasitize insects and snails, the eggs enter the host with its food. The larvae then hatch and complete three ecdyses there. Shortly before the last ecdysis, the larvae leave the host and migrate into the ground, where they winter. In spring, another ecdysis occurs and the mature animal develops. The fertilized females remain in the ground and crawl up wet plants during periods of warm summer rain. Here they lay their eggs. Such a rainfall after a long period of drought can result in a mass migration of these worms, which are up to 30 cm in length and very conspicuous —this was termed a "worm rain" in earlier times when nothing was known about the life cycle of these animals.

Invertebrate parasites

"Worm rain"

An alternation of generations is found in species of the genus *Heterotylenchus*, which parasitize beetles and flies. The females of the bisexual generation lay their eggs, after fertilization, within the body cavity of their host. The females of the unisexual generation develop from these eggs. These lay unfertilized eggs from which, in turn, males and females of the bisexual generation develop. Another form of parasitism is found in the species of the genus *Neoaplectana*. The larvae bore into caterpillars or are taken up by them with their food. Within the host, they release certain bacteria (genus *Achromobacter*), with which they live in association and from which they feed. These bacteria kill the caterpillar and use its body as a source of food. Owing to their mass development there, they offer the nematode the food basis for its further development and reproduction. The use of *Neoaplectana* species in the biological control of insect pests has had good results in a few cases.

Vertebrate parasites

Several thousand nematode species which parasitize animals, including humans, are known. They injure their host by robbing it of nourishment or by damaging it or destroying organs, through poisons and similar substances. A heavy nematode infestation can lead to the death of the host. Twenty years ago, it was estimated that about 644 million people were infested with ascarids alone, 457 million with hookworms, and 356 million with whipworms. Increased sanitation measures and an improvement in hygiene have resulted in a decrease in these worm species; despite this, however, threadworms are still worldwide in distribution and, in part, remain dangerous parasites, since only a few can be fought with medicines. The majority of vertebrate nematode parasites do not settle in their final situation immediately, but make rather complicated migrations through the body. In this, they move through the blood system, the heart, the lungs, and often other organs. During this time, they continue their development and store food reserves. These migrations are considered by certain investigators to be a partial repeat of their evolution;

Ontogenetical repetition of phylogeny

they show, for example, that the ancestors of the ascarids, which now inhabit the intestinal tract, were originally blood parasites or parasites in the lymphatic system, then in the respiratory organs, and only later came to parasitize the intestines. Some nematodes which parasitize vertebrates are capable of entering a developing animal or child from the blood system of the mother through the placenta, the newborn animal thus entering the world already infected with the parasite.

The DWARF THREADWORMS (genus *Strongyloides*) are parasites of vertebrates and are worldwide in distribution. A unisexual parasitic life and a bisexual free-living generation can be distinguished. The parasitic females live in the intestinal mucosa. Their eggs reach the exterior in the host's feces. The larvae hatching from these live free in the ground and develop into males and females. From the fertilized eggs, larvae develop which enter the host and, after migrating through the lungs, attach themselves to the small intestine and develop into females. In the hookworms, of which *Ancylostoma duodenale* and *Necator americanus* (both: L 8–18 mm) are still dangerous human parasites in certain parts of the world, the sexually mature animals live as blood-sucking parasites in the intestinal mucosa. The larvae hatch from the eggs, which reach the exterior with the feces, pass through two ecdysial phases without feeding, and then bore through their host's skin. After migration, which again takes place through the lungs, they settle in the intestine and become sexually mature. The hookworm *Ancylostoma duodenale* requires a temperature of 25–30°C in order to develop, in addition to sufficient moisture. These conditions are present in the plantations of tropical lands and also in temperate climates, but only in mines and tunnel construction works. They have thus received a German name meaning "trench worms." In America, the "death worm," *Necator americanus*, is present in addition to *A. duodenale*, and, despite its scientific species name, is of African origin, having been brought to America with the slave trade.

"Trench worms"

Species of the genus *Syngamus* live as parasites in the respiratory organs of birds and mammals. *Syngamus trachea* is a common parasite of wild and domesticated chickenlike birds, pigeons, crows, and thrushes. The birds become infected by picking up the worm eggs which reach the exterior in the droppings of infected animals. If an egg is taken up by a host which is not appropriate, such as a snail or earthworm, the larvae which hatch are encapsulated within the host's body cavity and can remain viable there for years without developing any further. If the transport host is then eaten by a bird, development resumes.

The THREADWORM (*Enterobius vermicularis*) is especially common in children. Unlike many other species, it undergoes a direct development without migration. The fertilized females, filled with eggs, migrate nocturnally to the anus and leave their egg capsules in the anal region. The itching stimulus these produce causes the child to touch the anal zone, infecting its hands; when a lack of sanitary measures is present, the eggs are carried by the hands to the mouth, and from there they are passed directly into the intestinal tract. They can, however, fall off and dry. They are so light that they can be carried in the air as dust, settle, and be taken up again by a playing child.

The threadworm

The MAW-WORM (*Ascaris lumbricoides*; L 20–40 cm, thickness 5 mm) is one of the most easily visible human parasites, and has been known for a long time. The sexually mature animals live in the small intestine, where the female produces masses of eggs which reach the exterior with the feces. In areas lacking a sewage system and where the contents of cesspools are used as fertilizer for lettuce and other vegetables, the path is laid open for the maw-worm to spread. If the maw-worm eggs are ingested with impure drinking water or unwashed raw salads or vegetables, they hatch within the small intestine. The larvae, however, do not remain there, although it is their final environment, but first migrate through the bloodstream, the right side of the heart, and the pulmonary arteries; here they break through the walls of the lungs and enter the alveoli. They are carried into the bronchial and pulmonary tubes by ciliary action. In this way, they enter the throat; when swallowed, they go down the esophagus into the stomach, and finally into the small intestine again, where they grow to sexual maturity. This migration, which lasts approximately seven weeks, is essential for the development of the larvae.

The maw-worm

The medina worm

An alternation of hosts occurs in numerous parasites of vertebrates. The intermediary host is almost always an invertebrate, as in the case of *Dracunculus* and *Filaria* worms; only rarely is it a vertebrate. The MEDINA WORM (*Dracunculus medinensis*; L, ♀♀ almost 1 m, thickness only 1.7 mm) is a human parasite of tropical and subtropical areas. The females cause painful swellings, especially on the legs. When stimulated by cold, such as would occur during bathing or wading through a river, the female extrudes its head through the swollen ulcer and releases large numbers of larvae into the water. These bore into their intermediary host—small copepods—developing within it into infectious forms after two ecdyses. When they enter their final host through impure drinking water, they begin their development into sexually mature animals. The whole course of development lasts approximately a year. The female's response to cold, that of extruding its head, was used even in ancient time as a means of combating the parasite. "Doctors" would clamp the extruded tip in a forked stick and slowly wind the animal around it, at the same time carefully removing it from the patient. Such a stick with the threadworm wound around it may have been the original portrayal of the Aesculapian baton, which has now become the symbol of medicine.

The FILAROIDEA, which belong to various groups of threadworms, such as the order SPIRURIDA, were until recently considered the scourge of the tropics. The intermediary hosts are blood-sucking insects. The mature filaria live in the human lymphatic system. The larvae (microfilaria) are produced and released there, and migrate from the lymphatic system into the bloodstream. They are taken up by mosquitoes in the blood they suck out when biting. Since they are found in large concentrations in the blood within the mosquito's gut, it is assumed that they are attracted by the mosquito's saliva which is injected into the wound when it bites. They continue their development within the mosquito, and when their host bites another person, they enter their final host's bloodstream. In areas where the intermediary host is a nocturnally active mosquito species, the microfilaria collect during the night in the tiny blood vessels of the host's skin. During the day, they move toward the interior of their host. In the Polynesian Tokelau Islands, where a diurnally active mosquito, *Stegomyia variegata*, is the main intermediary host, the microfilaria migrate to the epidermal capillaries during the day. This behavior in the microfilaria is not dependent on an internal rhythm of their own, but is directed by chemical changes in the blood which initiate the day-night rhythms of humans. An infection by the HAIR WORM (*Wuchereria bancrofti*) is indicated in two to five percent of the cases in certain areas by misshapen swelling of the limbs or scrotum (elephantiasis); in other areas, despite heavy infestations, these horrible symptoms are entirely lacking, strangely enough.

Trichina

The TRICHINAS (*Trichinella spiralis*; L, ♂♂ 1.4–1.6 mm, ♀♀ 3–4 mm; see Color plate, p. 301) are well known but rarely a cause of illness in Europe.

This has come about by a law passed in Prussia in 1877 and the remainder of Germany in 1937, that every pig slaughtered must be inspected for "trichinosis." In the U.S.A., where only inspected pork may be sold, trichinosis of humans is quite rare, although it was a serious problem until the early years of this century. For the rest of the world, an estimated 7 million people are infected with trichinosis. Despite these numbers, not every infection with trichina causes illness. Trichinosis is caused by eating raw, smoked, or incompletely cooked meat; thus only carnivores can be affected. Among the latter, however, the trichina is not selective about its final host; it is present in domesticated and wild pigs, in foxes, badgers, dogs, mink, cats, and especially rats, in addition to man. In the point of origin of the trichina, the far north, polar bears, arctic foxes, seals, and walrus are its main hosts. Pure vegetarians never become infected by ordinary means; therefore there is no meat inspection of ruminants, hares, or horses.

The encapsulated larva in the flesh of one of the previously mentioned hosts is released by the digestive juices of the carnivore's stomach after being swallowed, and matures within a few days in the small intestine. The sexes copulate here, after which the males die. The females, in contrast, bore through the intestinal wall, and within a few weeks produce up to 1500 larvae. From the lymph canals of the intestine, they progress over the chest region and finally into the bloodstream. Here they are carried with the blood through the left side of the heart, the lungs, the right side of the heart, and finally into the body's main circulation system. They bore through the walls of the blood vessels supplying the striped muscles and make their way between the striped-muscle fibers, where they continue to grow for seven weeks, without, however, becoming sexually mature. From the third week onward, the host tissue surrounding them begins to encapsulate the larvae, the capsule measuring 0.4 by 0.25 mm in diameter. This later becomes calcified. The larvae within the capsule can remain alive for years or decades. Domesticated pigs usually become infected through rats which enter the sties and are eaten by them. The rats become infected by eating the remains of their dead companions.

The danger of humans contracting trichinosis does not come only from pigs: during 1898, the three members of the André Polar Expedition died of it because they lived on polar bears and seals during their march over the ice; even in 1930, the smoked ham obtained from a circus polar bear in Stuttgart, which had had to be killed, was fatal to several people. During times of food shortage in which large numbers of pigs are slaughtered, trichinosis cases can also increase in numbers. The symptoms of trichinosis follow the course of the parasite's development: when the female bores into the intestinal wall, severe intestinal disturbances occur; the migration of the larvae to the muscles results in stiffness and pain as well as circulatory and metabolic upsets. In the third stage, anemia, rheumatic pains, and abnormal thirst occur. Finally, a general reduction in

The development of Trichina

Trichinosis

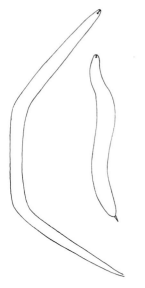

Fig. 10-22. Mutants of the stem eelworm (right) often differ markedly in body form from the normal type (left).

strength results; this can lead to death, as can the symptoms of the second phase.

The life expectancy of a single worm is not yet fully known, since the animals are so difficult to observe in their normal environment. During the active phase, the majority of species survive for only a few weeks, the males having a shorter life expectancy than the females. A few ground-living threadworms live for three or more years, and certain parasites of vertebrates can survive for fifteen years or more. During the resistant or quiescent stage, which is especially common in parasites as a bridging period between hosts, threadworms are able to survive for decades. Encapsulated trichinas have been dissected, still alive, out of human muscle after twenty-five to thirty years, and it is known that the larvae of cyst-building plant parasites retain their ability to infect after at least ten years. A tiny threadworm species, *Tylenchus polyhypnus*, holds the longest record yet; this was isolated from a rye plant which had been preserved dry in a herbarium for thirty-nine years.

Threadworms have not been studied in detail by geneticists or used frequently in experiments. Only the maw-worm has been studied in depth during the last fifty years. The chromosomes of threadworms are very small and difficult to study. Subspecies are known from many thread worms, differing in definite, genetically inherited characteristics from the typical animals of the species in question. In some cases, these are corporal characteristics, in others, host preferences which differentiate these subspecies. It can be proved that subspecies often come about during a single mutation in one or more hereditary characteristics. Even new species are created in some instances by a single mutational step, especially through multiplication of the nuclear number (polyploidy). It was recently discovered that substances were present in the embryos of certain cucurbid plants which caused an increased number of mutated animals (mutants) in plant eelworms. The changes in outward appearance were sometimes so great that no expert could identify the animals. It is noteworthy, however, that the eelworms were capable of recognizing each other, although mutated, and reproduced normally with the type species.

As mentioned previously, practically no part of the earth exists where threadworms are not present. They require only a little moisture, since without a thin water layer present over the body surface, they are not capable of active life. The upper layers of the ground, to about a depth of 20 cm, are the most thickly inhabited. They live on the rotting plant material there, but also on living plant cells which they penetrate with their oral spines. Some ground-living nematodes hunt other threadworms and other tiny ground-living creatures. The population density of nematodes decreases rapidly with increasing depth. They are not capable of living in air or the free water of large lakes and seas. They are, of course, carried by wind and flying birds and insects, and also transported by water currents, but they are not able to survive for long in these media.

Despite their tremendous numbers, the distribution of individual groups or species is very poorly known. Even in modern textbooks on animal geography one can find—if anything—only extremely sparse information. Only in the case of human parasites is the distribution more or less well known.

Certain Filaroidea, such as *Loa loa* and *Dipetalonema streptocerca* are found only in Africa. The medina worm also occurs mainly in Africa and Arabia, but sporadically in Asia Minor and India as well. During the period between the 16th and 19th Centuries, several species of filaria were introduced into the Americas through the slave trade and have survived in Central and South America for some time. In contrast to these, the HAIR WORM (*Wuchereria bancrofti*), also a filarian, is found in almost all tropical and subtropical lands between 30°N and 30°S. The point of origin of plant parasites is more difficult to determine because travel and the transportation of plants and plant products over the whole world resulted in their distribution in large numbers. Forms which are free-living in the ground or in the mud at the bottom of lakes, for example, have been the least affected by this, but it is these types which have been the least studied. The majority of species do not appear to be distributed worldwide, but inhabit only restricted areas. The factors which have led to the present distribution patterns are largely unknown. The most important are, naturally, climate and food; geological developments may also have played a part. For example, one threadworm genus (*Ohridia*) is present only in the region of Lake Ohrid, in Yugoslavia. This is probably a remnant of a Pleistocene form (about 6 million years ago) which was once worldwide in distribution and which, owing to geological changes, was able to survive in the upper layers of the earth around Lake Ohrid, a very deep-lying waterstand, having been wiped out over the rest of the world.

The number of threadworm species presently existing is not known. Almost 15,000 have been described, but it is estimated that almost 500,000 species exist. The discrepancy between these two numbers indicates how poor our knowledge of the threadworms is, even today. About 250 new species are described yearly. With regard to the evolution of the threadworms, their position within the animal kingdom, and their division into large groups, new opinions are continually expressed. Since, apart from a few animals embedded in amber, fossils are lacking from which the appearance of primitive forms or even a proto-nematode could be extrapolated, development can be determined only from the numerous species alive now. By the consideration of various characteristics, different types of groupings can be made. Those given in systematic arrangements are developed from the systematics of the nematodes put forward by the Belgian nematologist L. de Coninck in 1965, and of these groups, a few of the commonest genera have been described.

The NEMATOMORPHS (class Nematomorpha; see Color plate, p. 316)

Class: Nematomorphs,
by P. Rietschel

owe their name to their body form: they are found as hard, threadlike, coiled and rolled worms (L up to 80 cm, diameter up to 1 mm) in shallow water, even in springs and puddles, sometimes in large numbers. Their English name, "horsehair worms," is attributable to their body form. Peasants call these creatures, which appear to have arisen from the water itself, "water calves." It is obvious but puzzling to even the naturalist that only adult animals are found in the water. This puzzle is soon solved, however, when it is discovered that the entire development from larva to mature worm occurs in insects, in the case of fresh-water species, and crustaceans in the case of marine ones. The mature worm does not feed. A mouth and intestine are present but more or less occluded. The false coelom is full of reproductive organs, with a large quantity of loose connective tissue in the cavities between them. Excretory organs are lacking, and the nervous system consists of an esophageal ring and a ventral cord.

The development of the horsehair worms starts in open water, where the female lays its eggs, numbering into the millions, in heaps or strings. Unique larvae, less than 1 mm in length, hatch from them. They possess an extrusible proboscis at the anterior end, furnished with three spines at the tip and three rings of hooks beneath them. The larvae are capable of developing into a mature worm only within the body of a large insect; the means by which they reach the host vary considerably. The most direct method is when the larva manages to bore through the thin skin between the segments of a *Dytiscus* waterbeetle larva. The odds are, however, that the larva bores into a mosquito larva or a tadpole. Even here, there is hope of further development if these are eaten by either the waterbeetle itself or its larva; in these cases, however, the larva must bore through the gut wall into the body cavity. If the water into which the larva hatches happens to dry out, it is capable of surrounding itself with a capsule; in this phase it can also occasionally be taken up by a terrestrial insect or millipede. The means of entering these other hosts are not known in detail, but it is not uncommon for large, terrestrial, vegetarian insects (grasshoppers), as well as carnivorous ones (long-horned grasshoppers, carabid beetles, praying mantes), to be the final host.

As far as the growth of the worm within a water insect is concerned, the mature worm manages to enter its aquatic environment again without damaging the host. Those parasitizing terrestrial insects, however, differ: they leave their hosts after they have induced them, by unknown means, to seek water. In this, they utilize their host's drive to their own advantage —a process reminiscent of the drive modification instigated by the small liver fluke larvae which are parasitic in ants.

Over 200 species of horsehair worms are known, mainly members of the superfamily GORDIOIDS (Gordioidea). Their larvae are parasites in insects, those of *Gordius aquaticus* infesting members of the *Dytiscus* waterbeetle family, the larvae of *Gordius dectici* parasitizing long-horned

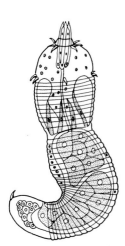

Fig. 10-23. The larva of a horsehair worm (*Gordius*): anteriorly the three stylets, behind the opening of the poison gland; next to this the retractor muscles of the proboscis (stippled). In the posterior part of the body, the poison gland anteriorly; behind this, the midgut, which is closed anteriorly.

Body form and
development of *Bonellia*
(echiurid worm).
1. Presexual larva:
Completely ciliated and, in
addition, with an anterior
and posterior ciliary ring
and two eyes; intestine
still lacking mouth and
anus.
2. Female larva:
Ciliated only at the anterior
(future proboscis), intestine
with mouth and anus, a
pair of bristles ventrally;
anteriorly, a pair of
protonephridia (closed by
the flame cells) posteriorly,
two pairs of metanephridia
(open with a funnel
opening into the body
cavity).
3. Mature female:
3a. From the exterior,
about natural size, bilobed
proboscis with a ventral
groove to the mouth,
three males attached to
this. Body covered with
epidermal warts (papillae),
a pair of bristles below the
mouth, behind them, the
sexual opening, anus at
posterior pole. 3b. Body
opened dorsally: body
wall, the body cavity with
its own lining (coelom)
containing the extremely
long, coiled intestine
which is suspended from
fibers, the midgut with a
diverticulum which opens
into it anteriorly and
posteriorly. The excretory
organs are two structures
opening near the anus and
consisting of a tube with
numerous funnels opening
into the coelom (metane-
phridium). The uterus is
present in the center of the
intestinal coils. 3. The body
after removal of the mid-
gut and right excretory
organ. In the center,
ventrally and connected
to the body wall, is the
ovary. It releases its ova
into the coelom. The
uterus opens to the
exterior below the mouth
and by a funnel into the
coelom; the ova are taken
up by this. The mature
males are present in the
anterior portion of the
uterus, and fertilize the
eggs; the presexual larvae
develop in the posterior
part of the uterus (shown
here with three bulges).
4. Mature male:
Body (L 1–3 mm) still
completely ciliated (as in 1),
without mouth or anus;
only midgut present;
foregut changed to form a
seminal tube with a funnel
opening into the coelom
within which maturation
of the sperm takes place.
These "dwarf males" form
from presexual larva after
they have become
attached to the female's
proboscis, wandered into
the foregut, and, on
maturity, entered the
sexual opening.
Color Key to 1, 2, 3b, 3c,
and 4 as in the Figures on
p. 316, i.e.: gray = ectoderm, yellow = endoderm,
carmine red = muscle,
cinnibar red = female
reproductive organs,
violet = male reproductive
organs, blue = nervous
system, green = excretory
organs, white = paren-
chyma, red-brown = blood
vessels.

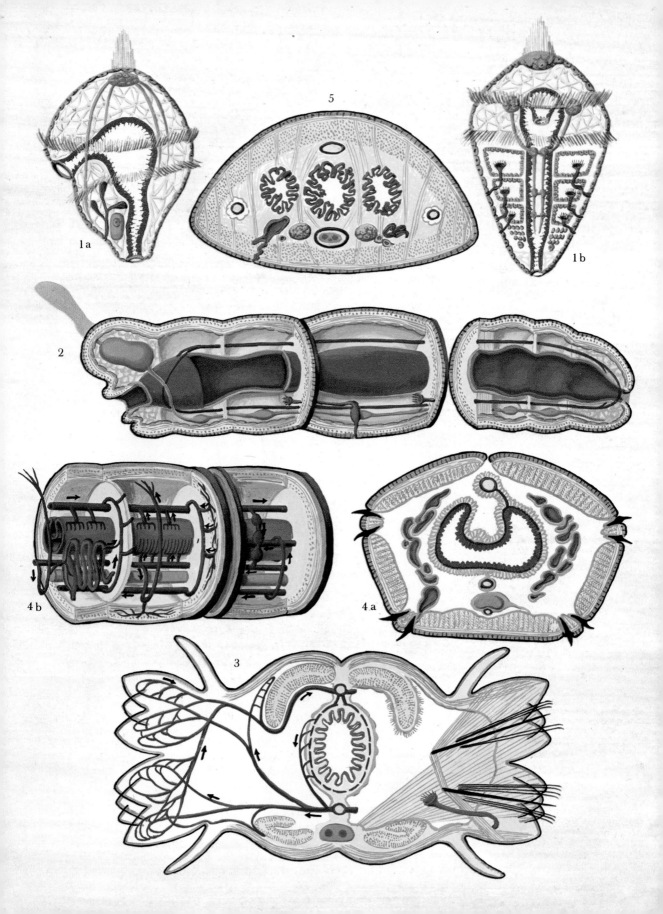

1a

5

1b

2

4b

4a

3

Development and body form in the segmented worms.

1. The larva:

1a. Trochophore larva from the side, with two ciliated bands: the upper, the "prototroch" and the lower, the "metatroch." The "episphere" above the prototroch, with an apical organ and ciliary tuft (sense organ) forming the prostomium. The "hyposphere" below the prototroch, with mouth, intestine, anus, excretory organs (protonephridia), and the cells which will form the future mesoderm, all forming the rest of the mature worm's body.

1b. Metatrochophore larva from the front: the endoderm forms the walls of three pairs of body cavities, behind which is the formation zone for more body cavities. The first segments (larval segments, deutometameres) are formed simultaneously; other segments bud posteriorly (post-larval segments, tritometameres), one after the other. The segments with a body cavity possess "metanephridia" as excretory organs, which have a funnel opening into the body cavity.

2. Structure of a segmented worm (phylum Annelida): From the left, removed in part: left, the prostomium without a body cavity, formed from the episphere of the trochophore larva, with brain, tactile organs, and, basally, the mouth. The following rings (segments) with a body cavity which is lined completely with mesoderm (also covering the intestine, yellow here, visible in section as a thin red line). The excretory organs (metanephridia) extend over two segments, with the funnel in the anterior one and the opening to the exterior of the posterior one; only the dorsal and ventral blood vessels are shown, the ventral nerve cord with a pair of ganglia in each segment; the body ends in a pygidium without ganglia or body cavity and thus, like the prostomium, is not a true segment.

3. Class Polychaeta: Horizontal section: Both sides of the body with tactile hairs (cirri), bifid "bristle feet" (parapodia), the precursors of the segmented and jointed appendages of the arthropods; left, the blood vessel with the direction of blood flow (the blood-vessel net present in the parapodia aids in respiration); right, the bristles and muscles causing their movement; the body wall (circular muscles externally, longitudinal ones internally) separated into muscle bundles.

4. Order Oligochaeta:

4a. Section through an earthworm. No parapodia but the bristles are still arranged in four groups. Body wall complete, body cavity open above through dorsal pores, this connection to the exterior via the excretory organs (metanephridia); intestinal wall with a longitudinal fold (typhlosole, the larger surface area allowing greater absorption).

4b. A block model of the blood vessel system in the earthworm (the anterior shown to the right, the posterior to the left); blood flow in the dorsal vessel from posterior to anterior, in the ventral vessel, from anterior to posterior, as in all the arthropods. Lateral connections between these two vessels: right, by pulsating sections (hearts), left, by branches to the body wall, in the center, by branches back from this.

5. Leech (order Hirudinea): Section through a leech: strong body wall and almost complete disappearance of the body cavity; the remains replace the completely vanished blood-vessel system in the Gnathobdellae and Pharyngobdellae. The space between the body wall and internal organs (about the intestine in the center and its lateral sacs) filled with packing tissue (parenchyma similar to that of the flatworms) and traversed by muscles. The excretory organs (metanephridia) strongly modified at the funnel end owing to the disappearance of the body cavity, but still with a urinary bladder and excretory tube. Nervous system within a "blood vessel" (the original body cavity). Color Key as in the previous diagrams i.e.: gray = ectoderm, yellow = endoderm, carmine red = muscles, cinnibar red = female reproductive organs, violet = male reproductive organs, blue = nervous system, green = excretory organs, white = parenchyms, red-brown = blood vessels.

grasshoppers, especially the wart-biter, and those of *Parachordodes tolosanus*, carabid beetles. The few marine horsehair worms of the genus *Nectonema*, which are the sole representatives of the superfamily NECTONEMATOIDS (Nectonematoidea), develop in shrimps and hermit crabs.

The KINORHYNCHIDS (class Kinorhyncha; L up to 1 mm; see Color plate, p. 316) are tiny marine creatures inhabiting the sea bottom; their epidermis is segmented externally into thirteen rings (zonites), the first forming the "head," and the second, the "neck." The body cavity penetrating all segments is itself unsegmented. The brain forms an esophageal ring as in the nematodes. A ventral nerve strand runs posteriorly from this, forming a ganglion at every second segment. The nerve strand between two ganglia consists only of nerve fibers. A pair of protonephridia serves as excretory organs. The esophagus and midgut are similar to those of the nematodes. The early development of the kinorhynchids could possibly give indications of their relationship to other animals but, unfortunately, this is still completely unknown. The young larvae are unsegmented; with each ecdysis, they increase their segmentation and the number of zonites present. These zonites should not be termed true "segments," since they are not comparable to the segments of the segmented animals. Of the roughly 100 species known, a few are present in European waters.

Class: Kinorhyncha, by P. Rietschel

The class SPINY-HEADED WORMS (Acanthocephala; see Color plate, p. 316) contains about 400 parasitic species. Their larvae live in invertebrates; the adults, however, live in the intestine of vertebrate animals. They have only very general characteristics in common with the rest of the roundworms, these being the wormlike body, separate sexes, and the presence of a cuticle and a body cavity. They are thus frequently considered as a phylum in their own right. They have, in common with the rotifers and nematodes, cell constancy and the union of cells to form a syncytium. The proboscis, which is set with regularly arranged backward-pointing hooks, is reminiscent of the head of a tapeworm with its hook rings; it is also used as a means of attachment to the host's intestine.

Class: Acanthocephala, by P. Rietschel

The spiny-headed worms also feed in the same way as do tapeworms; and, as in the latter, no gut is present. The food, which is already broken down by the host's digestive juices, is taken in through the skin. This, however, is completely different in structure from that of the tapeworms: its cells are joined to form a syncytium, which is penetrated by a body cavity. The nourishment taken up through the skin is distributed through the body with the body fluids through these "lacunae." At the end of the neck section of the proboscis there are two or more structures (lemniscae) formed by the epidermis. These protrude far into the body cavity and are also penetrated by lacunae. Nourishment enters the body cavity through them, and reaches the organs, especially the reproductive organs.

As is the case in all parasites, the spiny-headed worms must even out the chances of a single animal reaching its final host by producing large num-

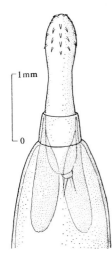

Fig. 10-24. Anterior end of the acanthocephalid *Acanthocephalus lucii* (L 12 mm). Anterior, the proboscis armed with hooks; behind this, the two lemniscae.

bers of offspring. At least the larger members can be considered "egg multimillionaires." The reproductive organs in both sexes open at the posterior. The male possesses an extrusible pocket at the base of which is a tube with a short sexual process. The tube contains the vas deferens, the tubes from the excretory organs, and the openings of the adhesive glands, the secretions of which the male uses to attach itself to the female during copulation.

The female's reproductive organs are unique. At an early stage of development, the ovaries detach so that the primal ova form in masses within two large sacs which run the whole length of the body. The uterus possesses a sorting mechanism constructed of four pairs of cells, through which only the ripe, slim, spindleshaped, and already-shelled eggs reach the exterior. The immature, round, and shell-less eggs, on the contrary, move back into the sac.

The spiny-headed worms are no rarity among European animal life. A member of the most primitive order, the ARCHIACANTHOCEPHALA, and, at the same time, one of the giants of its type, the GIANT SPINY-HEADED WORM (*Macracanthorhynchus hirudinaceus*; L up to 65 cm) can be found in the intestines of pigs. Its larvae develop within the larvae of chafer beetles (the cockchafer, June bug, and rosechafer), remaining there during the insect's pupal stage and metamorphosis into the flying beetle; this aids in the parasite's distribution. Pigs, rooting through the earth, devour the infected larvae; even carnivores and monkeys, which also occasionally eat insect food, sometimes serve as hosts of the giant spiny-headed worm. Occasionally, it even parasitizes humans, but the means by which this comes about is still unknown. Another smaller species, *Moniliformis moniliformis* (L up to 11 cm), a parasite in rodents, is also occasionally found in humans. Its larvae develop in various beetles and cockroaches.

Species of the order PALAECANTHOCEPHALA are more common. Two species are found in the intestines of ducks: *Polymorphus boschadia* (L up to 2.5 mm) and *Filicollis anatis* (L up to 3 cm). The larvae of the first species develop in copepods; the second, in aquatic woodlice. The infected crustaceans are easily recognizable by their orange-red color, that of the larval parasite visible through the walls of the body. When the larval development has been completed, the "juveniles" encapsulate themselves within their last larval skin, with the proboscis inverted. The excretory products collect between the parasite and its capsule; this lessens the danger to the host but does not prevent it completely. The majority of copepods infested with *Polymorphus* larvae have degenerate reproductive organs and are thus infertile.

Spiny-headed worms are found in the intestines of fish as well as those of ducks, one being *Echinorhynchus truttae* (L up to 2 cm), present in trout and vendace. Their larvae also develop in copepods. Other species of the same genus, as well as the genus *Pomphorhynchus*, are common parasites in the intestines of barbs.

11   Priapulids, Sipunculids, and Echiurids

In this chapter, three animal phyla will be considered which, until fairly recently, were put together in the class Starworms or Gephyrea. The scientific name Gephyrea means "bridge animals." These animals were considered as a bridging group between the phyla termed "worms" and the sea cucumbers, which are members of the Echinodermata (see Vol. III). Nowadays we know that the priapulids, sipunculids, and echiurids are not related to one another and also do not constitute a "bridge" between the true worms and the echinoderms. External similarities had, in this case, led to false conclusions regarding systematic relationships. Each of these groups, which contain only a few species, must be considered as having the same status as the arthropods (see Vol II), mollusks (see Vol. III), and chordates (see Vols. III through XIII), although the latter contain very many more species.

Only four species of PRIAPULID WORMS (phylum Priapulida) are known, but these have been enough to cause problems for zoologists. They are usually only a few centimeters long, cylindrical, and, like the Kinorhynchida and Acanthocephalida, have an extrusible proboscis at the anterior end of the body. They live in mud in colder-sea regions and move by alternate extension and contraction of the body or by protrusion and withdrawal of the proboscis. The priapulids hunt, preying upon lower marine organisms (annelids, crustaceans, brittle stars, and mollusks) which live in the same environment. They capture and swallow these, as well as members of their own species.

The priapulids possess a uniform body cavity. Nothing is known about the way in which this is formed, since the early development of the larva has not yet been studied. The systematists who consider this body cavity to be a true coelom (secondary body cavity) place the priapulids near the sipunculids (Sipunculida). The body cavity, however, is not lined by a cell layer, but by a thin dermis; this indicates that it must be considered a primary body cavity (pseudocoelom). The priapulids should thus be placed nearer the roundworm phylum. The presence of a hooked

Phylum: Priapulida, by P. Rietschel

Fig. 11-1. The unicaudate priapulid (*Priapulus caudatus*), from the ventral surface.

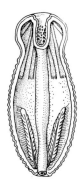

Fig. 11-2. Body structure of the priapulid *Halicryptus spinulosus*. Anterior of the body (inverted) with retractor muscles, intestine straight; next to this, the organs suspended by a ribbon (mesentery), the excretory organs (protonephridia) and sexual organs being fused.

Phylum: Sipunculida, by P. Rietschel

Fig. 11-3. Distribution of *Priapulus caudatus*.

Fig. 11-4. "Urn cell" from the body-cavity fluid of the sipunculid *Phascolosoma granulatum*, filled with metabolic wastes.

extensible-and-retractable proboscis, however, is no proof of their close relationship to the roundworm classes Kinorhynchida and Acanthocephala. For this reason, the priapulids are best considered as a phylum in their own right, even though the number of species belonging to it is far fewer than those of the other animal phyla.

The distribution of *Priapulus caudatus* (L 8–20 cm) is unique. On the one hand, it inhabits the North Atlantic, the North Sea, and the western Baltic and on the other, the cooler waters around Antarctica; it is absent from the warmer waters between these two areas. There are two explanations for this type of distribution, which also occurs with a few other species: the species could have inhabited the areas between south and north previously and then disappeared when the water temperature became too high. According to this view, the present-day distribution areas must be considered relics of a more extensive one. In contrast to this, even Charles Darwin suggested that the warm belt which separates the two distribution areas today, has been much colder, at least for a short period of time, and thus could be traversed by certain species. In this way, they were able to colonize the same climatic zones on opposite sides of the world. The second ice age offered the best opportunity for this to occur (Mindel Ice Age, about 450,000 years ago). This "migration hypothesis" is thought to be the more likely one.

In the phylum SIPUNCULIDA, we first meet animals whose body cavity in lined with a special cell layer. These cells are ciliated. This body cavity is similar in its development to the coelom of the annelids and mollusks. However, as in the other organs, little evidence of segmentation is present; this is a characteristic of the annelids. The adult sipunculids live buried in the sea bottom. Their larvae are trichophores, as in the annelids. It is therefore assumed that the sipunculids and annelids had common ancestors, and that segmentation arose after the two phyla had separated in their evolutionary development, the sipunculids remaining unsegmented.

The cylindrical body bears an evertible proboscis with a terminal mouth surrounded by cirri. The intestine is straight at first, but then coils posteriorly, this coiling continuing as the intestine reverses upon itself to open at the anus, at the base of the proboscis. The nervous system consists of a brain, an esophageal ring, and a ventral nerve cord which runs posteriorly, sending nerves to the strong, muscular body wall. Metabolic waste products are removed from the body cavity through metanephridia, opening both into the body cavity and to the exterior. The reproductive cells also reach the exterior through these organs. Cells which line the body cavity also store waste products, thus reducing their toxicity. Other cells of this lining layer—the ciliated "urn cells"—detach and swim freely in the fluid in the body cavity. They carry foreign particles to the opening of the metanephridia and ensure the flow of fluid within the cavity by means of their ciliary action.

Of the 250 species living today, a few occur in European waters, such as *Sipunculus nudus* (L 25 cm; see Color plate, p. 302) and species of the genera *Phascolosoma* and *Golfingia* (L only a few centimeters). The largest sipunculid, *Siphonomecus multicinctus* can reach a length of 50 cm. An Indonesian sipunculid, *Phascolosoma lurco* (L 6 cm) has invaded the land, its adults living in the muddy ground of mangrove swamps and breathing air. Its larva, however, is still marine.

The phylum ECHIURIDA (see Color plate, p. 351) shows some similarities with the sipunculids: both are marine, with a wormlike body form which bears a proboscis at the anterior end. In contrast to the sipunculids, however, this proboscis in the echiurids is not retractable, but can be extended and contracted, and when fully extended can reach relatively great lengths in some species. In *Ikeda taenioides* (L 40 cm), which inhabits Japanese coastal waters, the proboscis can reach a length of almost 1.5 m. The ventral side of the proboscis possesses a broad, longitudinal band of cilia which ends at its base, where the mouth is. The intestine starts at this point; it opens in the anus at the posterior pole of the body. The uniform, unsegmented body cavity is a true coelom, connected to the exterior only through the metanephridial canals, through which the mature reproductive cells are also released. The larva, as a characteristic trochophore (see Color plate, p. 352), indicates connections with the annelids; in addition, the bristles (chaetae) present in the epidermis are similar in structure and development to those of the annelids. The nervous system consists of an esophageal ring and a ventral nerve cord. A brain separated from the esophageal ring is not present, and higher sense organs are lacking. In contrast to the sipunculids, the echiurids possess a simple blood-vessel system.

Of the about 150 known species of echiurids, two will be mentioned here: *Echiurus echiurus* (L 15–30 cm; see Color plate, p. 302), an inhabitant of shallow water around the North Sea coasts, and the GREEN BONELLIA (*Bonellia viridis*; L ♀♀ 8–15 cm without proboscis, proboscis up to or greater than 1 m; ♂♂ L about 1 mm; see Color plates, pp. 302 and 351), which colonizes the gravelly bottoms of shallow bays and submarine clefts in the rocks of the Mediterranean Sea. *Echiurus* digs U-shaped tunnels from which the proboscis, which is only a few centimeters in length, protrudes. With this, it searches the immediate surroundings for food, which the ciliary tract then carries to the mouth. In contrast to this, *Bonellia* keeps its body hidden in clefts or under stones, and extends its proboscis, which bears two long terminal processes, over a much larger area.

The green bonellia has become well known among zoologists, because of the means by which its future sex is determined. In most animals, the decision as to whether the future organism is to be male or female is made at the time of fertilization; in *Bonellia*, however, this decision remains completely open even at the time that the creature is a free-swimming larva. If such a larva happens to contact the proboscis of a female, it

Phylum: Echiurida, by P. Rietschel

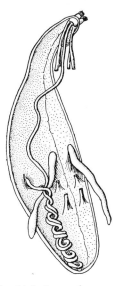

Fig. 11-5. Internal structure of the sipunculid *Phascolosoma granulatum*: the four retractor muscles of the anterior portion have been cut in the center; behind these, the spiral intestine around a central muscle spindle; next to this, the excretory organs (metanephridia). The brain is dorsal to the intestine at the anterior end; the longitudinal nerve cord is ventral between the four points of muscle attachment.

Fig. 11-6. *Echiurus echiurus.* The four openings of the excretory organs (metanephridia) below the anterior ventral chaetae.

develops within a few days into a minute, sexually mature dwarf male which bears no proboscis and is more similar to a turbellarian than an echiurid. Numerous males can thus be present on the proboscis of a single female at the same time; they migrate along the ciliary band into the anterior intestine, in which as many as eighty-five of these dwarf males have been discovered. Some of them continue to migrate from there, and bore into the oviduct of the female, where they fertilize the eggs emerging into the body cavity.

The larvae which do not contact the proboscis of a female grow and metamorphose in the next year into mature female animals. If larvae are carried by the water currents to areas which are not yet colonized, they are thus forced to become females. Larvae which are carried by the same currents later thus have the chance to develop into males on contact with the probosces of these females, and thus to fertilize the eggs. In this unusual way it is possible for males to be produced only in areas where females are already present and where they can aid in the maintenance of the species. The nature of the substance emanating from the female's proboscis which causes this metamorphosis in the males is not yet known.

In this chapter, we have met for the first time with the sipunculids and echiurids, animals having a "secondary body cavity" (a coelom). In contrast to the "primary body cavity," the space between the ectoderm and endoderm, it is lined by a thin layer of cells (epithelium) which originates from the mesoderm. This "mesothiel" also covers the organs present within the body cavity. The coelom forms an elastic supportive skeleton as it is filled with fluid, and is termed a "hydrostatic skeleton." It has, however, been retained by many animals which possess a hard skeleton, and in these cases it has taken over other functions, e.g. in the echinoderms (see Vol. III) and the vertebrates (see Vols. IV, ff). On the other hand, the primary body cavity took over the role of a hydrostatic skeleton in the nematodes in a most perfect way. With the appearance of the secondary body cavity, the type of excretory organ changed completely within the animal kingdom: instead of a closed protonephridium which began with a flame cell, the metanephridium appeared with a ciliated funnel opening into the body cavity itself. The reproductive cells frequently release their products into the secondary body cavity; these are then carried to the exterior either through the metanephridia or by special body-cavity tubes (coelomoducts). The coelom, with its cellular lining and various passages to the exterior (metanephridia and coelomoducts), is thus not simply a cavity, in contrast to the primary body cavity, but an essential "organ" for those animals possessing it.

Secondary body cavity

12   The Annelida

When the great systematist of the animal and plant kingdoms, the Swede, Carl von Linné (1707–1778), developed his system for the animal kingdom, he divided it into only two classes: the insects and the worms. During the course of time, more and more groups of animals have been removed from the latter "rag-bag" and given their own class; the group "worms," however, has stubbornly persisted in certain textbooks even to the present. This includes the flatworms, the roundworms, the kamptozoans, the sipunculids, the echiurids, and the annelids, with the arrowworms and acorn worms, that is, protostomids with deuterostomids. All, more or less, have in common only a worm-shaped body; the concept "worm" thus means as little to a zoologist as the concepts "tree," "bush," or "plant" to a botanist. It was for this reason that the zoologist Karl Vogt in the 19th Century poked fun at the unscientific "worm rag-bag" with a quotation from Wilhelm Busch: "The length of the worm varies."

In 1812 an important French zoologist, Georges Cuvier (1769–1832), arranged the systematics so that the jointed worms (Annelida) were considered to be more closely associated with the arthropods; he thus joined these two groups together to form the ARTICULATES (Articulata). This concept means that an annelid worm, such as the earthworm, is much more closely related to the spiders and winged insects than to the roundworm, which is much more like it in shape. What is it which thus associates the annelids with the arthropods, which are so unlike them in shape? In both groups, the body is divided into serial segments: it is "segmented." In the lower animals, these segments are either the same or very similar to each other; their segmentation is "homonomous." Among the arthropods this also holds true for the posterior segments of millipedes and the lower crustaceans as well as for the abdominal segments of many insect larvae. All groups of articulates, however, show evidence of the fact that, with increasing evolutionary development, the segments differ from one another: these animals thus show

The segmented animals, by P. Rietschel

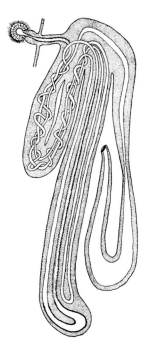

Fig. 12-1. Excretory organ (metanephridium) of an earthworm. The ciliated funnel above left; at the center right, the opening to the exterior.

"heteronomous" segmentation. In addition, groups of segments combine to form body sections ("tagmen" or "tagmata," singular "tagma"). Depending on their number and arrangement, these are distinguishable as head and abdomen; head, thorax, and abdomen; or anterior and posterior body sections.

Each of these segments is primitively identifiable through the possession of a pair of coelomic cavities, a pair of appendages, its own muscle system, and a pair of nerve ganglia. The latter are joined to one another within the segment by a horizontal connection (commissure), and between segments by a nerve cord (connective), forming a ladderlike ventral nerve cord. This type of nervous system is thus termed a "nerve ladder" among the segmented animals. At the anterior of the nerve cord the two longitudinal nerve cords from the anterior ganglia extend around the esophagus (esophageal connectives) to the dorsal side, thus joining the ventral nerve cord to the brain.

Apart from the ladderlike nervous system, the appendages of segmented animals also serve as a relatively important characteristic, even though they exhibit enormous developmental steps from the unjointed appendages (parapodia) of the annelids to the highly developed jointed appendages of the arthropods. Even in the latter case, however, the appendages of individual segments may be reduced or absent.

More basically, however, the articulated animals have altered with respect to their body cavities: the closed blood system of the annelids has developed from the primary body cavity. Within this, the blood flows from posterior to anterior above the intestine, and below it, from anterior to posterior. In the higher segmented animals, the blood-vessel walls tend to disappear more and more until, finally, only the dorsal vessel remains. This becomes a pulsating heart for various distances along its length. Even the walls of the secondary body cavity disappear in the higher arthropods, so that both body cavities join to form a "myxocoel." Within this, the blood circulates freely.

Hand in hand with the changes in the body cavities goes a degeneration of the excretory organs (metanephridia), since these are closely dependent on the presence of a secondary body cavity. They are still found in the annelids in a full state of development, but in the arthropods, only degenerated remains can be found. The malphigian tubules, attachments of the intestine, take over their task not only in the Arachnida (spiders and their relatives), but also in the millipedes and insects.

The most primitive segmented animals have a spiral cleavage, and thus connect the segmented animals, with their thousands of species, to the Spiralia. The segmented animals, with over 825,000 species described, constitute over three-quarters of all the animals on earth. They are usually divided into five phyla: 1. SEGMENTED WORMS (Annelida), 2. ONYCHOPHORA, 3. TARDIGRADA, 4. PENTASTOMIDA, 5. ARTHROPODA.

Within the superphylum Articulata, the phylum Annelida is placed

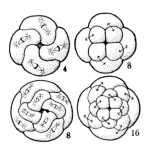

Fig. 12-2. Spiral cleavage in a segmented worm embryo showing the four to sixteen cell stages.

at the lowest level. The remaining articulate phyla have not arisen from present-day Annelida but from forms which were at a lower evolutionary level. Even though fossil remains of these ancestral types have not been found, many of their characteristics have been preserved by present-day annelids. Among them are the more or less "homonomous" segmentation of the body, the paired coelomic cavities in each segment, the body-muscle layer consisting of an external circular layer and an internal longitudinal layer, paired excretory organs (metanephridia) in each segment, with a funnel opening into the coelomic cavity, and a ladderlike nervous system with a pair of ganglia in each body segment. Early development (cleavage) in the lower annelids is the complete spiral type, and thus connects all other Articulata to the Spiralia. The trochophore larva (see Color plate, p. 352) is also a primitive characteristic of the annelids, which they possess in common with the mollusks, and which is closely related to the larval forms of other members of the Spiralia. Even the epidermis, which consists of a single layer of cells, is ciliated in the lower annelids, but usually a cuticula is secreted, consisting of protein compounds. In contrast to the cuticle of higher Articulata, this cuticle does not contain chitin, a special substance which plays an important part in the external skeleton of arthropods, and doubtless aided in their rapid evolution. Chitin, however, although not present in the cuticle, is already present in the bristles of the Oligochaeta. Even though primitive characteristics are more common among the annelids than in the higher Articulata, this group also contains highly evolved forms which show various modifications as adaptations to a multitude of environmental conditions. The almost 9000 species are divided into three classes: 1. Polychaetes (Polychaeta), 2. Myzostomids (Myzostomida), 3. Earthworms and leeches (Clitellata).

The POLYCHAETA (see Color plate, p. 352), because of the variable development of their appendages, show the most diverse body forms among the Annelida. The prostomium bears slender tentacles (antennae) and more squat palps. The prostomium of sessile polychaetes can also bear two processes which often end in an enormous antennal wreath, the "tentacle crown." Two or three true segments also form part of the head in addition to the prostomium, these together forming the "peristomium." The first is the mouth segment without antennae (metastomium); one or two body segments are attached to this, the parapodia of which are transformed into antennalike "cirri."

The remaining body segments of polychaetes also bear a pair of highly variable parapodia. These are divided into an upper and a lower branch, each of which bears a tuft of bristles arising from ectodermal pockets. Their form varies considerably: hooked, shovellike, comblike, knifelike, and simple bristlelike; their function is equally variable: they serve as grappling hooks, swimming bristles, digging attachments, and weapons of defense. The bristle bundles, together with the entire parapodium,

Phylum: Annelida, by P. Rietschel

Class: Polychaeta

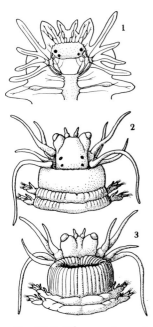

Fig. 12-3. The anterior end of a polychaete.
1. Nervous supply to the cephalic appendages;
2. Head from above, showing the four eyes;
3. Head from below, with oral opening.

Family: Aphroditidae

can be moved by muscles. The parapodia bear processes termed cirri, on both the upper and lower surfaces; in the aphroditids, they are transformed into scales (elytra) which protect the gills. Both the upper and lower branches can carry gills as extra appendages; these may be thread-like, leaflike, comblike, or branching. The variability can also be increased through the subdivision of the posterior section into tagmen where the parapodia differ in their development and can even be absent.

The systematic subdivision of the polychaetes into orders is still a matter of dispute. The usual division will be given here even though it provides no clear picture of the relationships between the groups. According to this, they are divided into the free-living polychaetes (Errantia), the sessile polychaetes (Sedentaria), and the Archiannelida. Since the division of the first two orders is not a natural one, the following must be kept in mind: a sessile mode of life has, without doubt, developed from a free-living one on several occasions within this group. It has also been shown that the primitive structure of the "archiannelid" is not really primitive but a secondary simplification which has come about independently several times in various branches of the annelids.

The order FREE-LIVING POLYCHAETES (Errantia) includes forms with more or less uniform segmentation of the entire posterior. All the segments are furnished with parapodia of the same structure and with excretory organs. The suborder AMPHINOMORPHS (Amphinomorpha) includes small, squat species. *Hermiodice carunculata* (see Color plate, p. 373), found in the Mediterranean Sea, is feared by fishermen because it possesses bristles with recurrent hooks which break off easily and cause painful burning of the skin by the poison they contain.

The suborder NEREIMORPHS (Nereimorpha) includes many unusual forms. The type genus is *Nereis*, for which the suborder is named, the name coming from that of a Greek nymph; other nereimorphs also bear the names of nymphs and goddesses, alluding to their beauty. The SEA MOUSE (*Aphrodite aculeata*; see Color plate, p. 373), which is a member of the SEA CATERPILLARS (Aphroditidae), is widespread in European seas. The gills are covered by a double row of fifteen dorsal scales (elytra), which are transformed cirri, and these are covered in turn by a thick, gray bristle layer. The lateral bristles, in contrast, shimmer all colors of the rainbow. It is owing to them that this animal received its name of Aphrodite, after the goddess of love, who was born from foam. Despite its concealed mode of life and the protection offered by the dorsal bristles, this worm is devoured by many species of fish. Less conspicuous, and having only twelve pairs of dorsal scales, is *Lepidonotus squamatus*. Its short bristles are of little aid in protection, but the scales produce light flashes when touched, and may aid in defense.

The family EUNICIDS (Eunicidae) includes both tiny and giant polychaetes: *Eunice gigantea* (L up to 3 m) is the largest of all the polychaetes and, together with the giant earthworm, the largest annelid and the longest of all the Articulata. The so-called Palolo worms are well known because of their mode of reproduction. The SAMOAN PALOLO WORM (*Eunice viridis*; L 40 cm) is found on the coral reefs of Samoa and the Fiji Islands; the ATLANTIC PALOLO WORM (*Eunice fucata*; L 70 cm), on the reefs of the Bermudas and the West Indian islands. Strangely enough, their period of reproduction is closely associated with the phases of the moon: in the case of the Samoan Palolo worm, breeding occurs in the second or third day following the third quarter of the moon in October (small Palolo time or *Mblalolo lailai*) and November (big Palolo time or *Mblalolo levu*); the Atlantic Palolo, on the other hand, breeds during the three days prior to the third moon quarter between 29 June and 28 July. A dependence on the moon phases with respect to reproduction has been shown to be present in other polychaetes; in these two species, however, they can be considered a true "wonder of nature," and their reproduction is accompanied by a festival among the natives. The Palolo worms are negatively phototropic and leave their cases only during dim light or darkness; this determines the length of time they spend feeding. When sexually mature, the worms produce eggs and sperm only in the last segments of the body, which, in contrast to the rest of the animal, becomes active only under light conditions, when it separates off and swims toward the light. The swarms of posterior body portions release their eggs and sperm at the surface of the water until sunrise, thus forming a writhing layer several meters thick. They are fished for by the natives using baskets and are considered a delicacy.

The posterior end of the worm which is constricted off and contains the sexual organs cannot, however, be considered an independent animal. Certain members of the family SYLLIDAE, however, constrict off posterior portions of the body which grow their own heads before detachment and thus become self-sufficient, free-swimming animals. They reproduce sexually, producing further bottom-living forms which then reproduce asexually. This, then, is an example of a generation which reproduces sexually alternating with one which reproduces by budding. In *Odonto-syllis* the two sexes are brought in contact by means of a pattern of light organs. The female leaves its tube in the mud during darkness, and rises from a depth of from 4 to 6 m up to the water surface, where it swims in circles and produces light flashes from its light organs. A male, having perceived these, also rises from the depths to the surface and emits flashes like a firefly. As soon as it enters the cruising circle of the female, both release their reproductive cells. The whole process lasts only a few moments.

The NEREIDS (family Nereidae) live, for the most part, on the sea bottom, feeding there mainly on algae and small benthic animals. These are slender worms with numerous parapodia which, at the beginning, are similar in structure along the whole length of the body. As the ovaries or testes mature, the parapodia of the posterior end of the body change,

Palolo worms

in the genus *Nereis* into scooplike oars. This transformed animal was once given its own genus, termed *Heteronereis*, and even today the sexually mature animal is termed a "heteronereid." The body-muscle layer of the transformed segments degenerates so that the sperm and ova are able to be released into the water when the thin, fragile body wall breaks. Only the swimming muscles of the parapodia remain; these enable the heteronereid to swim to the surface of the water. *Nereis virens*, which is usually a benthic species which feeds on algae, thus forms shoals during the reproductive stage. *Nereis diversicolor* (see Color plate, p. 373), a benthic inhabitant of the North Sea, is much more of a bottom-living species, and even its trochophore larva and later stages of development remain in the benthic environment. This species also invades brackish water to a great extent, and indicates, with this developmental direction, one in which the Indonesian nereids have followed to a still greater extent: *Lycastopsis raunensis* lives on the muddy banks of bodies of fresh water in this region, and progresses just like a millipede with its dorsally swiveled parapodia. *Lycastis vitabunda* and *Lycastis terrestris* are found in a terrestrial habitat in clay-containing sand in Sumatra. *Lycastopsis amboinensis* is even found in damp humus layers within the leaf axils of coconut palms. The tidal mangrove swamps and especially the high humidity of the tropical rain-forest areas there offer the best opportunities for the development of a terrestrial way of life; hand in hand with this goes the high level of adaptability found among the nereids. That characteristic which must show the greatest degree of change is the embryonic development through the trochophore larvae, but, unfortunately, nothing is known about this aspect of the animal's life history.

The members of some polychaete families have a free-swimming mode of life. The many species belonging to the family PHYLLODOCIDAE are elongated and narrow, with many segments whose parapodia has become reduced to such an extent that the dorsal branch consists of only a cirrus. The *Phyllodoce* species remain hidden on the seabed during the day, but during the night they swim about with horizontal wavelike motions of the body which, together with the variegated play of colors they produce, are an extremely beautiful sight.

Members of the two families ALCIOPIDAE and TOMOPTERIDAE have become completely free-swimming. The alciopids, which are perfectly circular in section, are not particularly well adapted to this mode of life from the point of view of body shape. Only the presence of leaflike dorsal and ventral cirri permits them to maintain a planktonic mode of life, and these cirri must keep up a continual swimming motion. The alciopidian eyes, however, are perfectly adapted to a hunting mode of nutrition. In the free-living benthic forms, these are mostly ocellar pits but there are also lensed eyes which, using the same principle as the pinhole camera, allow almost complete picture vision. The two large eyes on the head of alciopids are true lensed eyes, however, whose perform-

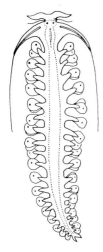

Fig. 12-4. The planktonic polychaete *Tomopteris*.

ance almost equals that of the cephalopods. In the relaxed condition, they are adapted for far vision; through the function of a gland, the lens can be moved away from the retinal portion, bringing about close-up vision! The strongly compressed tomopterids are more efficient planktonic animals: their bristleless parapodia are elongated, and each ends in two plates which probably originate from the cirri. Alciopids and tomopterids, as are many planktonic animals, are completely transparent.

Many free-living polychaetes live as commensals with other marine animals; the species *Polynoe scolopendrina*, a member of the aphroditids, is even commensal with sessile polychaetes. Certain of these polychaetes gain an advantage from this commensalism in that they feed on the scraps of food dropped when the other partner is feeding. In most cases, however, no such advantage can be shown, as in the case of those worms commensal with sea cucumbers which devour sand. A few polychaetes, however, have become true parasites and have changed in form to such an extent that the systematists had to produce two new families to include them, the ichthyotomids (Ichthyotomidae) and the histriobdellids (Histriobdellidae). The single species of the ICHTHYOTOMIDS is *Ichthyotomus sanguineus* (L up to 1 cm), a parasite on the fins of marine eels. Its mouth is adapted to form a sucker, and the pharynx, which includes two dagger-like teeth, forms a pump. The entire equipment required for cutting the fish's fin and sucking out the blood is thus complete. In the other family, the HISTRIOBDELLIDS, which contains several species from 0.25 to 1.5 mm in length, somatic transformation has proceeded to a far greater extent. Parapodia are absent, and the anterior and posterior tips of the body bear two processes ending in suckers. Two daggerlike jaws are also found within the mouth, which also contains four pairs of hard, toothed structures. *Histriobdella homari* lives within the gill cavities and on the eggs of the lobster, and progresses, in the absence of parapodia, by means of a sparse ciliary covering. A related species, *Stratiodrilus tasmanicus*, lives within the gill cavities of crayfish in Tasmanian waters.

The second order of polychaetes, which contains a large number of species and is certainly not evolutionarily uniform, is the SEDENTARY POLYCHAETES (Sedentaria). They are "sessile" only in that they built tubes of epidermal secretions, or use those secretions to cement burrows, and do not leave them except for short periods of time. They are unable to hunt their prey, but either waft it toward themselves, search the bottom in the region of the tube, or swallow mud or sand. Depending on the mode of nutrition, the head varies in structure: in the sand- and mud-eaters it has no appendages; in the fishers and searchers of the bottom it is furnished with a crown of tentacles, which can reach large proportions. Those of the terebellomorphs are developed from cerebral flaps, those of the hermellimorphs and serpulimorphs, from the oral segment. The body behind the head usually is divided into two sections (tagmen), in a few cases, into three. The excretory organs are present only in the anterior

Order: Sessile polychaetes

rings, that is, near the entrance of the tube. There are five suborders: spiomorphs, drilomorphs, terebellomorphs, hermellimorphs, and serpulimorphs.

One of the most unique creatures among the polychaetes is *Chaetopterus variopedatus* (L up to 25 cm), a member of the suborder SPIOMORPHA. The oral segment surrounds a small, funnellike cerebral flap without extremities. The body is divided into three sections. The first segment of the middle section bears a pair of tentacles, these being the dorsal branches of the parapodia, which secrete a mucous bag which is swallowed by the worm when the bag has become covered with planktonic particles, and is then formed again anew. The third, fourth, and fifth (last) segments of the central section bear large plates on their dorsal surface; their movement causes the current within the tube. These plates have developed from the dorsal branches of the parapodia on the appropriate segments. The posterior part of the body is equipped with numerous pairs of parapodia. *Chaetopterus* lives in the Mediterranean Sea and Atlantic Ocean, in U-shaped tubes within the substrate; these are parchmentlike, with constricted openings. They are continually washed by a current carrying food particles and oxygen-rich water, created by the dorsal plates. The food particles are sieved out of the water by the mucous net; oxygen is taken up through the whole surface of the body since gills are not present.

The LUGWORM (*Arenicola marina*; L up to 40 cm) is a member of the suborder DRILOMORPHA. The funnellike holes and piles of feces exposed between the tidelines at low tide are well known by every vacationer and beachcomber in the North Sea area, and the worm itself is prized as bait by anglers. Even though its spoor is well known, its means of nourishment was, for a long time, wrongly interpreted. According to the findings of F. Krüger, the lugworm is by no means an eater of sand, but rather a filterer. Its burrow consists of a U-shaped tube with hardened walls (Fig. 12-6). One leg of the tube is open, but usually is covered by the spiral strand of feces which the worm excretes every forty minutes or so by creeping backward toward the opening. Usually the animal is found in the lower section of the tube, where it ingests the sand particles which are continually descending through the other leg of the tube. At the same time, however, its continual forward and backward movement within the tube produces a water current, which runs from the tube opening to the bottom. This provides the animal with oxygen for respiration, and also draws in planktonic particles which become trapped in the descending sand, since sand itself is an excellent filter. The lugworm larva first lives on the surface of the substrate, taking up the adult mode of life at about 8 mm in length.

The TEREBELLOMORPHS (suborder Terebellomorpha) also inhabit the substrate. They, however, have a crown of tentacles developed from the cerebral flap (prostomium). *Pectinaria koreni* (L 5 cm) lives within—and never leaves—a quiverlike tube, constructed of cemented sand grains,

Fig. 12-5. Anterior end of *Chaetopterus*.

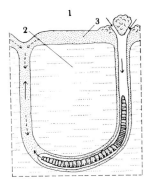

Fig. 12-6. The lugworm *Arenicola* in its burrow. Sand funnel to the left, right, the pile of feces above the tunnel. Straight arrows: water current. Broken arrows: sand current. 1. Water. 2. Mud, 3. Sand.

which widens anteriorly. It is set obliquely in the sand, with the widest opening downward. At the posterior end, which protrudes from the sand, there is a "chimney" lined with balls of feces. The worm's anterior bears a comb of fifteen to twenty strong, golden digging bristles (paleens), with which it loosens the sand around its anterior end. It then probes with the tentacles for small organisms. From time to time, it produces a water current through the tube by means of contractions of its body. These disperse the loosened, examined sand in front of the tube. Thus, a cavity is gradually formed about the anterior end of the tube, which instantly collapses and opens when an object touches it. This causes new organisms and settling particles to come within reach of the tentacles, thus providing fresh nourishment. In addition, the worm, in the same way as the lugworm, produces a respiratory current through the chimney and the quiver, and into the collapsing cavity, but this is not filtered. *Lanice conchilega* (L up to 30 cm, diameter 5–6 mm), the "shell collector," on the other hand, builds a delicate branchwork of sand particles and bits of shell across the anterior part of its tube: this serves, on the one hand, as a net to capture small organisms which the worm then takes up by the tentacles or with the mouth, and, on the other, as a support for the long, delicate tentacles.

The "shell collector"

The HERMELLIMORPHS (Suborder Hermellimorpha) bear a crown of tentacles arising from the oral segment. Of these, *Sabellaria spinulosa* (L up to 3 cm) will be mentioned. Its tubes, consisting of cemented sand particles, are joined together into organlike structures, and even form small reefs (sand-coral reefs). For instance, during 1944/45, the breakwaters west of Norderney Island, in northern Germany, were surrounded on both sides with "sand corals" for a distance of 60 m. The fringing reef around each breakwater consisted of about 75 million worm tubes. They remained until 1946, and were probably the result of a unique, extensive larval settlement. The tubes of the genus *Phragmatopoma* form similar reefs in the tropics, sometimes up to 1 m in height. It is suspected that settled worms attract new larvae toward them by means of chemical stimuli.

The SERPULIMORPHS (suborder Serpulimorpha) also possess a crown of tentacles springing from the oral segment which more or less covers the reduced prostomium. This group includes the most common polychaetes of all, the POTAMIDS, such as the genus *Spirorbis*. Their calcareous, snaillike tubes, which are sometimes wound to the left and sometimes to the right, are found in enormous numbers on seaweeds, stones, and similar surfaces. These calcareous tubes are often discovered as fossils in marine deposits. The serpulimorphs are mainly current filterers whose tentacles are covered with ciliated feathers; they use these to waft the plankton, which forms the main part of their diet, toward themselves. For a vacationer in the Mediterranean, it is always a fascinating sight when *Spirographis spallanzanii* (L 30 cm; see Color plates, pp. 374 and 375) spreads its huge, brightly

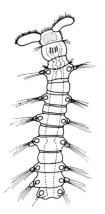

Fig. 12-7. The groundwater polychaete *Troglochaetus*.

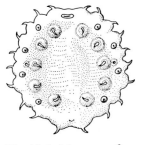

Fig. 12-8. *Myzostoma* from the ventral surface. Mouth anterior, anus posterior; on each side, four glandular "lateral organs" and five "parapodia"; the sexual openings at the level of the third pair.

colored tentacle-crown at his feet. At the slightest disturbance, the tentacles vanish immediately but then slowly re-emerge from tubes that look like old gas pipes. But another much plainer worm can also capture an observer's attention: if he searches through seaweed put into a shallow dish of seawater, he can almost always find a small worm which, unlike its relatives, often leaves its tube and swims around freely in the water. This is *Fabricia sabella* (L 1–8 mm). *Fabricia* usually swims backward, dragging its crown of tentacles behind it like the head of a broom. As it possesses a pair of eyes posteriorly as well as anteriorly (eyes are a rarity among the tube-dwelling worms), it could be a problem deciding which end of the worm is anterior and which posterior. This was a difficulty encountered by Oskar Schmidt, writing in the section on the lower animals in *Brehms Tierleben*, about his first encounter, in 1848, with the tiny *Fabricia* (termed *Amphicora* at that time). "We . . . strongly recommend to those interested in such animals, or in visual beauty, the study of the lively *Amphicora*, during a stay at the seaside."

As in the case of the other two, the third order, ARCHIANNELIDA, is also not an evolutionarily homogeneous group. The only thing which the members of this group have in common is that they are simplified in structure or have remained in a larval stage of development. Their nervous system is often subcutaneous, the parapodia are poorly developed, the bristles are single or lacking altogether, the number of body segments is often small, and the epidermis is still ciliated in many cases. The majority of these small worms live within the interstitial spaces of the sand of the sea bottom. The tiny *Troglochaetus beranecki* (L 0.5 mm), which lives in groundwater, is a relict from the fresh-water fauna of the Tertiary epoch. It was first discovered in Switzerland and then more commonly in the vicinity of Strasbourg, Since then, *Troglochaetus* has frequently been recorded from numerous springs fed by groundwater in the Main Valley between Karlstadt and Hanau, from Schlesien, Darmstadt, and Bonn. It is, however, absent from central and upper Franconia. Further discoveries regarding its distribution may provide more information about the geological past.

The MYZOSTOMES (class Myzostomida) are metamorphosed annelids according to their body form, but little more can be told regarding their evolutionary past. Their branching off, probably from the polychaetes, lies far in the geological past, since they are parasites on the most primitive of the present-day echinoderms, the sea lilies and brittle stars, and have been for more than 300 million years. This has been proved by the finding of galls caused by these worms on fossil sea lilies from the Devonian. They are totally unlike annelids in shape: a flat disc whose thin edge is furnished with long processes (cirri). If such a creature is observed from below, its bilateral symmetry is recognizable by the presence of a posterior anus and an anterior transverse opening. From this, a proboscislike structure can be everted, containing the mouth and esophagus. In the large order PROBOSCIFERS (Proboscifera), which contains over 100 species, a brain is also present, thus designating this region as a "head"; in the smaller order PHARYNGIDS (Pharyngidea), which contains about twenty-five species, only the mouth and esophagus are everted; the structure is thus a "proboscis." On the ventral surface there are five pairs of unsegmented protrusions, each of which bears a strong, curved bristle; these are homologous to the parapodia of polychaetes, and correspond to five body segments which are no longer recognizable externally. Internally, the unsegmented body cavity shows no sign of segmentation, but this is recognizable in the nervous system. The midgut has lateral pouches, a characteristic also present in other parasites—leeches, ticks, and fish lice; its structure, however, is an indication of the animal's mode of life, not of its evolutionary relationships. Individual (ontogenetic) development does not indicate a relationship with the polychaetes, since it progresses through spiral cleavage to a true trochophore larva.

The PROBOSCIFERS live mainly on their host's skin, to which they attach by means of their hooked claws. In this way, they include themselves in the stream of food leading to their host's mouth, and are more commensal than parasitic. *Myzostoma cysticolum* causes the host to form galls, each of which contains two parasites. Although myzostomes are hermaphroditic, the larger individual of each pair within the galls takes over the role of the female, the smaller, that of the male. The PHARYNGIDS have become parasitic to a greater degree, and are mainly internal parasites. They can often be found in large numbers in the gonads of brittle stars.

While the polychaetes, to a great extent, and all the myzostomes are marine, the oligochaetes and leeches are mainly fresh-water inhabitants. Many have also invaded moist land, but only a few have returned to the original habitat of the annelids, the sea. All have in common an extremely glandular epidermal zone, extending over several body segments and being termed a "saddle" or "clitellum." The class, which includes the oligochaetes and leeches, is thus termed the Clitellata on the basis of this structure. All clitellates are hermaphroditic.

The best-known of all annelids are the OLIGOCHAETES (class Oligochaeta; see Color plate, p. 352). The earthworms belong to this group and are typical examples of what people understand under the concept "worm." The clearly segmented body bears no parapodia, but the bristles have been retained on each body segment, arranged in four groups. Reproductive organs are present only in a few segments relatively far forward on the body, the testes always lying anterior to the ovaries. The sperm cells ripen in sperm sacs, and from there reach the partner through a sperm funnel leading into the vas deferens during copulation. They are stored by the partner in a spermatheca. During oviposition, the clitellum secretes a mucous tube around the worm. The worm retreats backward out of this tube and releases within it the ova from the oviduct and the partner's sperm from the spermatheca. The slime tube then

Class:Myzostomida

Class: Clitellata

Order: Oligochaeta

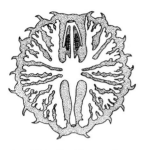

Fig. 12-9. Position and branching of the intestine in a myzostomid.

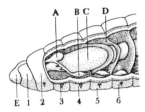

Fig. 12-10. Anterior of an earthworm from the side. A. Brain, B. Ventral nerve cord, C. Pharynx, D. Septum, between two body segments, E. Prostomium; 1 to 6, the first to sixth body segments.

hardens to form a capsule (cocoon). Within it, the sperm fertilize the ova; the young worms develop inside without passing through a larval stage. This is a unique case, since, following a true copulation, fertilization occurs outside the animal.

The over 3000 species of oligochaetes have been divided into three suborders distinguished mainly by the position of the male sexual openings relative to the testes. In the suborder PLESIOPORES (Plesiopora, from the Greek πλησιος = "near") these pores are immediately behind the segment containing the testes. One family with extremely primitive characteristics is the AELOSOMATIDS (Aelosomatidae), worms of only a few millimeters in length, found in puddles. They travel mainly by the use of the cilia on the underside of the prostomium. The body cavity is a uniform space, since the internal walls between these segments have broken down into loose muscle fibers. They reproduce mainly asexually through budding, and chains of animals can be formed, consisting of up to ten individuals. Strangely enough, the budding zone does not lie in the body segment anterior to the pygidium, but in the pygidium itself. In *Aelosoma hemprichi*, a purely asexual reproduction has been observed over a period of fifteen yesrs.

For the microscopist, the study of the tiny (L up to 10 mm) NAIDIDS (family Naididae) has always been a source of interest, since these animals are transparent and all the organs are visible in a live individual. These creatures are often found among water plants in pure water stands, where they live on algae, especially diatoms. The ponderous *Chaetogaster* species are hunters, and *Chaetogaster limnaei* even lives within the pulmonary openings of water snails, especially the giant water snail (see Vol. III) and the ram's horn snail (see Vol. III). These snails are often attacked by fluke larvae, parasites which form the main source of nourishment for the *Chaetogaster* worms.

The best swimmer among the naidids is *Stylaria lacustris*, which snakes through the water in horizontal waves. The prostomium is extended to form a long, pointed, antennalike process. This naidid reproduces asexually for most of the year, and can produce chains of individuals up to 2 cm in length. In autumn, however, it reproduces sexually, forming egg capsules covered with a layer of slime, which it attaches to water plants. *Stylaria* also occurs in brackish water and, in the waters around Helsinki, Finland, even in estuaries with a fifty-seven percent salt content. It is also present in Lake Constance and Lake Lugano at depths of 200 m.

Ripestes parasita (L up to 7.5 mm) lives in slender, transparent tubes about 1.5 cm long, in unpolluted inland waters. The peristomium is extended to form a long, antennalike process. Very long, mucus-covered bristles are found in bundles on the sixth to eighth body segments. The anterior end, which is extended from the tube, moves violently from side to side, thus causing water currents which waft tiny planktonic particles toward the animal. These become attached to the slime on the bristles.

From time to time, the worm passes the bristles of the left and right sides of the body alternately through its mouth and sucks off the adhering plankton. *Ripestes* is not capable of taking food in through the mouth in the usual manner. The current produced by the swinging anterior part of the body also aids in respiration. *Ripestes* leaves its tube from time to time and swims about freely, the long bristles serving as organs aiding in floating. If it contracts into its tube, the long bristles fold forward.

Certain naidids live in the upper layers of mud at the bottom of standing water, which consists mainly of decomposing matter. Although this layer presents an ample food supply, it contains very little free oxygen, as the oxygen is all utilized in the process of decomposition. To enable the worms to breathe, they protrude the hind ends of their bodies into the free water above. In naidids of the genera *Dero* and *Aulophorus*, the posterior part of the body has developed into a respiratory cavity which contains four (*Dero*) or six (*Aulophorus*) gill fibers covered with cilia and provided with a good blood supply. The posterior does not move, but the cilia ensure replenishment of the water supply.

The family of TUBIFEX WORMS (Tubificidae) includes small to medium-sized (L up to 20 cm), slender worms, which usually appear red, because of the hemoglobin in solution in their blood. They live in the upper mud layers of water stands, in tubes consisting of mucus secreted from the body and mud particles. They feed on decomposing matter in the mud. The most common is the COMMON TUBIFEX (*Tubifex tubifex*; L up to 8.5 cm), known to aquarists because it is easily available on the market as live food for fish. Its life in decomposing mud poses the same problem as that facing the mud-living naidids, lack of oxygen, but *Tubifex* has found another answer to the problem: the posterior, which protrudes from a craterlike depression, keeps up a continuous lashing movement which brings down water from layers above the mud. The frequency of the lashing and the extent of the body's protrusion from the mud are clear signs of the oxygen content of the water: the lower the oxygen content, the farther the body protrudes and the higher the frequency of the lashing. *Tubifex* possesses no gills and no blood vessels near the epidermis which are able to take up oxygen from the water. Such vessels, however, surround the gut, and thus make ample intestinal respiration possible. The gut continually takes up fresh water with ten "pulses" per minute. In cases of great oxygen deficiency, the gut no longer functions as a digestive organ, food is no longer taken in, and the empty intestine becomes a purely respiratory organ. The feces of *Tubifex* worms contain mineral particles which become stuck together through bacterial action and produce a loose, finely grained sediment. In the presence of ample organic material, subaquatic floors can be formed, the organic layers of which have a spongelike structure. Ancient deposits of this type, which now lie above the water level of the Rhine, ensure good drainage because of their fine, porous structure.

▷ Above: The bristles of the fire worm (*Hermodice carunculata*) are arranged in thick tufts. They remain embedded in the skin when the worm is touched, and produce severe irritation.
Above left in the center: The sea mouse (*Aphrodite aculeata*) from the ventral side. The parapodia extend laterally, the dorsal surface is covered with leaflike parapodial processes which are covered with a thick layer of bristles.
Above right in the center: *Nereis diversicolor*, an inhabitant of mud along the European coasts. The dorsal blood vessel is visible through the skin.
Below left in the center: *Syllis cornuta* with long parapodial processes which look like a string of pearls.
Below right in the center: *Lagisca extenuata* is a relative of the sea mouse, but the leaflike parapodial processes covering the back have no bristles.
Below: *Eulalia viridis* is able to swim gracefully with snakelike movements, the leaflike parapodia serving as oars.

▷▷ In the sabellid *Spirographis spallanzanii* (see Color plate, p. 375), one of the two tentacular processes is far larger than the other, and is wound spirally.
Below: The burrowing worm *Megalomma vesiculosum* possesses large black eyes at the tips of the tentacles.

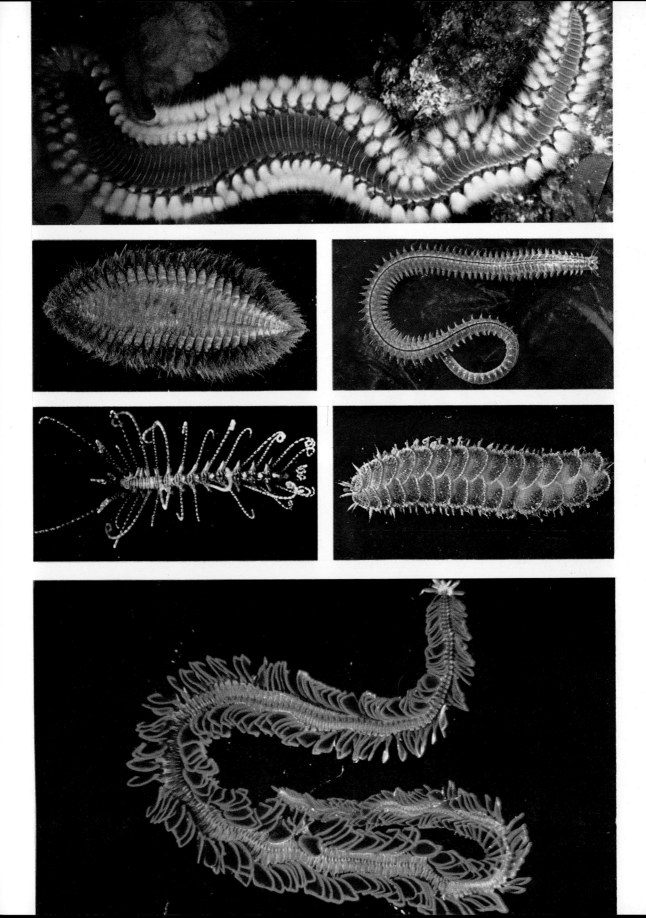

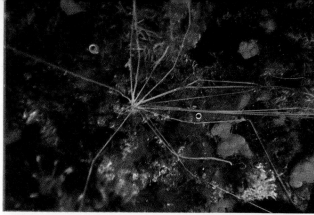

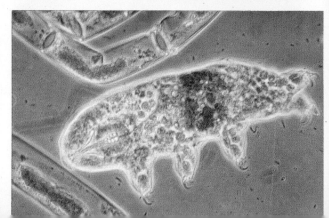

Family:
Branchiobdellidae

The mud-living worm *Branchiura sowerbyi*, faced with the same problem as *Tubifex*, solved it in another manner: it developed gills at its posterior end, like the naidids, but not inside the anus. They are long, delicate, gill fibers on the dorsal and ventral surfaces of the posterior segments, giving the impression of extremely delicate feathers. This species, which has probably been introduced from southeastern Asia, is mainly found in warm water stands in greenhouses, but has even been found in the wild in France and, more rarely, in Germany.

The family ENCHYTRAEIDAE contains animals which vary considerably in length (1 mm to 5 cm) and have very different modes of life. The majority of them live in damp earth and feed in the same way as earthworms, changing the consistency of the earth in the same way as they do. Some species live in the banks of lakes or on the seashore. *Enchytraeus albidus* (L up to 36 mm), prized by aquarists as live food, is found among the rows of seaweed washed up on the beaches. It is also, however, found in compost and in flowerpots and, in the latter case, can cause severe damage to the roots of the plants.

The suborder PROSOPORES (Prosopora) includes all oligochaetes in which the male sexual openings are present in the segment containing the testes or, when several segments contain the testes, in the last segment. Two families which differ in both structure and mode of life comprise this suborder: the lumbriculids and the branchiobdellids. The LUMBRICULIDS (Lumbriculidae) are externally similar to the earthworms, and also have a similar chaetal arrangement, but are smaller (from the Latin *lumbricus* = earthworm, and *lumbriculus* = tiny earthworm). They can be easily distinguished from earthworms, however, by their tendency to break into pieces on the slightest stimulation. Since these pieces are able to regenerate into complete worms within a short period of time, this autotomy represents a case of asexual reproduction. The best-known species is *Lumbriculus variegatus* (L 8 cm), an inhabitant of rotting leaves in woodland pools and the banks of inland waters. If the pool dries up or the water level drops, the worms retreat into the moist ground beneath.

Members of the family BRANCHIOBDELLIDAE look more like leeches than earthworms. Four species of the genus *Branchiobdella* (L 3–12 mm) are found in Europe, all being parasites on crayfish. They are thick-bodied, and the last three body segments form a sucker which lies behind the anus. The oral opening is provided with strong jaws with which the tiny worms bite through the delicate skin between their host's joints to suck its blood and devour the softer tissues. The branchiobdellids have no chaetae, like the leeches, and also creep in a looping fashion in the same way as they do. In the juvenile stages they feed in the same way as their relatives, on detritus and decaying organic matter.

The last group of oligochaetes is characterized by the opening for the male spermatic duct being several segments behind the ones in which the testes are situated. These are thus termed OPISTHOPORES (suborder Opisthopora; from the Greek ὄπισθεν = behind). The roughly 2500 species included in this group comprise about four-fifths of the total number of oligochaetes. By far the majority are "earthworms," the only exception being the family HAPLOTAXIDAE, which contains only a few species. *Haplotaxis* (*Phreoryctes*) *gordioides* is an inhabitant of European groundwaters and occasionally enters springs or even tapwater. It is almost 30 cm in length and possesses almost 500 segments although it is only 1 mm in diameter. As in the case of fresh-water amphipods and other groundwater animals, it is also found in the depths of large lakes. *Criodrilus lacuum* (family Criodrilidae) also lives in water, is the same length as *Haplotaxis*, and consists of about 450 segments. It is, however, much thicker (diameter 5 mm) and more similar in form to an earthworm.

The EARTHWORMS (see Color plate, p. 352) comprise several families. They live in the ground and vary in length, depending on the species (in Europe, L 2–30 cm), the GIANT EARTHWORM (*Megascolides australis*), found in the Australian bush, reaching a length of up to 3 m and a diameter of 3 cm.

The family MEGASCOLECIDAE contains many species, but is restricted in its distribution to tropical climates, while Germany's indigenous earthworms are all members of the family LUMBRICIDAE. This includes about 170 species, about thirty-five of which occur in Germany. The reader may wonder how so many species of this relatively uniform animal can be distinguished. Characteristics aiding in identification are on which segment of the body the sexual organs have their openings, how many segments the clitellum consists of, and, especially, the distribution of the chaetae which are mounted on wartlike protruberances, together with the presence of glandular thickenings of the skin in the region of the saddle. In addition, there are variations in color, pattern, body shimmer, the shape of the prostomium, and certain other features. The commonest species is the COMMON EARTHWORM (*Lumbricus terrestris*; L up to 30 cm), which inhabits chalky soil in preference to all others. In loams, on the other hand, the species most commonly encountered is *Lumbricus rubellus* (L up to 15 cm) which is red-brown to violet on the upper surface, or the smaller, chestnut-brown to brownish-violet *Lumbricus castaneus* (L up to 5 cm). In earth near waterstands or waterlogged soil, the most common species is *Eiseniella tetraedra* (L up to 5 cm), easily recognizable by its square-shaped posterior segments. Another species found in moist soil or damp leaf mold is the greenish *Allolobophora chlorotica* (L up to 7 cm). The reddish *Allolobophora rosea* (L up to 8 cm) is much less choosy in its selection of ground to inhabit. The most brightly colored European earthworm is *Eisenia foetida* (L up to 13 cm), which is yellowish, with a red or brown horizontal line on each segment. Its preference for dung and compost resulted in its German name *Mistwurm* (= dungworm), and its odor resulted in it being given the Latin name *foetida* (= stinking).

Earthworms

Fig. 12-11. Earthworms during copulation. The slime casing formed by the clitellum (twenty-sixth to thirty-first body segments) can be seen.

It may seem a waste of time to the layman to pay much attention to such inconspicuous and rarely seen animals but they are, in fact, extremely important. "It is a matter of doubt whether there are many other animals which have played such an important role in Earth's history as these lowly creatures," stated Charles Darwin in conclusion to his study, "The formation of arable ground by worms together with observations on their mode of life," a paper which appeared in 1881. Earthworms swallow earth and digest out of it organic substances which have been broken down by other animals and bacteria, excreting the inorganic parts as feces. This devouring of earth is not random, however; earthworm feces are avoided; earthworms also drag fallen leaves from the surface down into their burrows. They can also do this to young plants, and are then said to cause "damage." In comparison to this, however, there is no doubt that the good they do is far greater: the organic material is thoroughly mixed with the earth to form loam, and the earth is continually tilled by being carried from below to the surface and loosened to relatively great depths by the worms' burrows. Owing to the aeration of the soil, soil bacteria increase in population; these accelerate the rate of organic decomposition and release inorganic substances which are then taken up by plants. The earthworms' activity is therefore a basic necessity for fertile land; this can be proved by experiment, worm-free earth producing a far lower yield than earth with a worm population. As an example of how a little activity can produce big results, the piles of feces from earthworms in 1 m² of meadow were weighed over a period of a year. It was found that the amount of earth brought up varied from 4.4 to 8 kg. From this it could be calculated that earthworms turn over 11 to 20 t of earth per acre per year.

The behavior of earthworms is difficult to observe, since most of it occurs below ground. With patience, one can observe the formation of "worm piles" in the field, or the disappearance of a leaf underground (or even a bird feather as a mistake!), and the enormous power exerted by the worm on the other end. Many earthworms reproduce underground, but the common earthworm does so on the surface. After warm spring rains or during the early morning or late evening hours, these worms crawl half out of their burrows and lie together in pairs so that their heads point away from each other and the chaetae at the anterior end of the body and the region of the clitellum hook deep into the underside of the partner. Slime secreted from the clitellum of both partners surrounds them in a mucous tube covering the foreparts of their bodies. The hardening slime also closes the groove running ventrally from the male sexual openings to form a tube through which the sperm are carried to the spermatheca of the partner. This whole process takes up to two or three hours, before the worms separate again (Fig. 12-11).

The clitellum also forms a slime envelope during the fertilizing and laying of the eggs, hardening to form an egg capsule. In the common earthworm, the capsule contains only a single egg, in the dung worm, two to five. The young worms leave the capsule after three or four weeks. The number of fascinating things a patient study of this group would reveal can be drawn from Darwin's previously mentioned study. One sentence from this valuable work will be cited here: "Archeologists probably do not know to what extent they should thank the worms for the preservation of many antique objects."

The main enemies of worms below the ground are the mole, carabid larvae, the luminous centipedes (*Geophilus*; see Chapter 17) and, above all, a high water level in the soil, which causes oxygen deficiency. This forces the worms to leave their burrows; once outside, they are threatened by sunlight and, above all, hordes of worm-eating birds. Even plant-protecting sprays can penetrate the soil and be a source of danger to the worm population. Dry summer weather and winter cold, however, are not dangerous, since the worms can burrow deeper to escape them. Here they dig chambers, which they line with their feces, where they remain, rolled into a ball, until the danger period is past.

The LEECHES (Hirudinea; see Color plate, p. 352) comprise the second order of Clitellata. They have arisen from the oligochaetes. The acanthobdellids, which have chaetae on their first five body segments, as do the oligochaetes, are a bridging group between the two orders. The number of segments between the prostomium and the posterior sucker is always twenty-six. The sucker consists of four segments in the acanthobdellids, of seven in all other leeches; their total number of segments, therefore, are thirty and thirty-three, respectively. Externally, the number appears to be much greater, but several of the external rings are present to each true segment; the medicinal leech, for example, has five. With the exception of the acanthobdellids, all leeches possess a sucker surrounding the oral opening. The anus opens dorsally in front of the posterior sucker. The midgut, especially in blood-sucking species, is provided with numerous lateral pockets; these are capable of containing large quantities of blood—an adaptation to the fact that the opportunity to feed is relatively rare and long periods of time elapse between feedings. The body cavity and blood system are similar to those of the oligochaetes in the case of the acanthobdellids, but in all other leeches the body cavity is more or less occluded with connective tissue and serves to distribute the blood in place of the degenerated blood system. The ventral nerve cord is surrounded by such a blood-filled body cavity remnant, the anterior and posterior ganglia being joined to form a nerve mass supplying the anterior and posterior suckers. The funnels of the excretory organs (metanephridia) also open into these body cavity remnants and open to the exterior in pairs, by ventral pores in many segments. All leeches are hermaphroditic, with a pair of ovaries and several pairs of testes.

Copulation in the leech differs from that in oligochaetes: the gnathobdellids lie with their ventral sides toward one another but, in contrast to

▷
Leeches (order Hirudinea). Proboscid leeches (suborder Rhynchobdellae): 1. Duck leech (*Theromyzon tessulatum*), 2. *Helobdella stagnalis*, 3. *Hemiclepsis marginata*, 4. Greater snail leech (*Glossiphonia complanata*), 5. Lesser snail leech (*Glossiphonia heteroclita*), 6. Common fish leech (*Piscicola geometra*), Jawed leeches (suborder Gnathobdellae): 7. Medicinal leech (*Hirudo medicinalis medicinalis*), 7a. Medicinal leech, (section of the ventral surface), 8. Hungarian leech (*Hirudo medicinalis officinalis*), 8a. Hungarian leech (section of the ventral surface), 9. Horse leech (*Haemopis sanguisuga*), 10. Dog leech (*Erpobdella octoculata*).

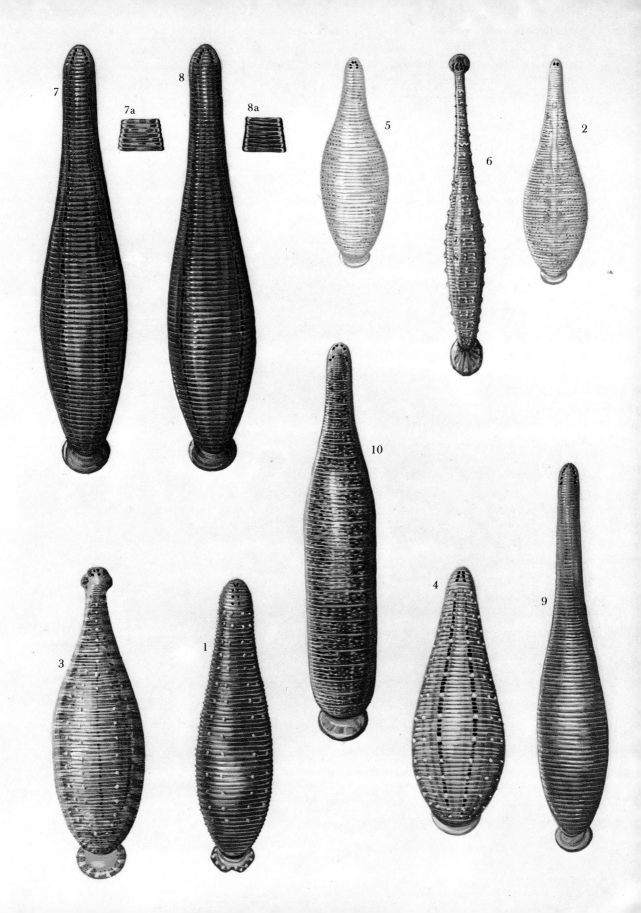

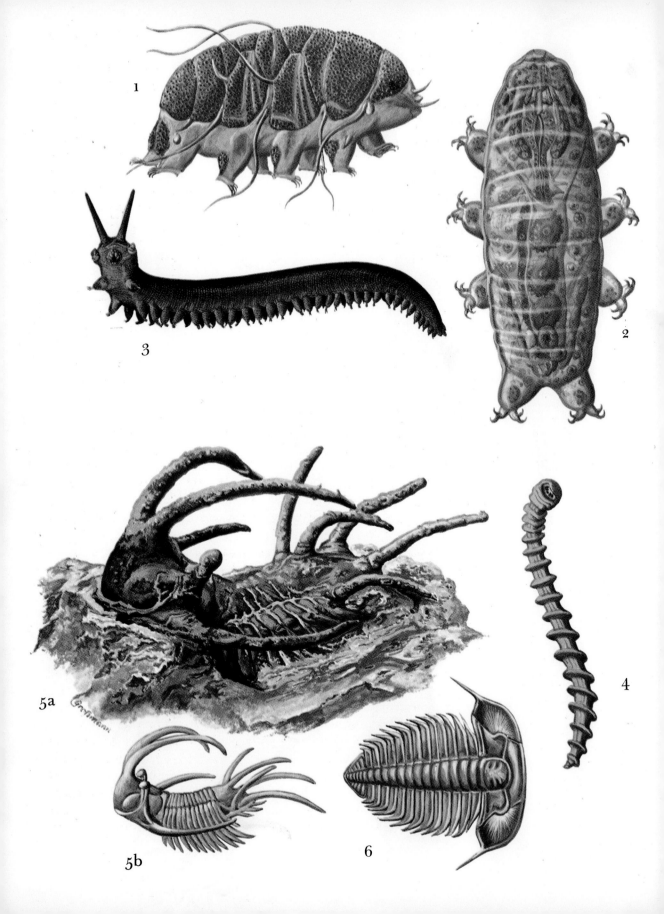

the oligochaetes, head-to-head. The partner's sperm is then injected into the vagina by means of an extrusible copulatory organ. Medicinal and horse leeches crawl on land before copulating. The majority of rhynchobdellids and epicarids reproduce in a different manner: they form spermatophores which become filled with sperm during copulation and are stuck to the partner's skin. These penetrate the epidermis and release a cell-dissolving substance which enables the sperm to enter through the musculature, into the body cavity, and finally to the ovaries. This type of copulation is reminiscent of that present in the most primitive turbellarians and also in the bedbugs (see Vol. II). The eggs are, as in the oligochaetes, laid in a mucous cocoon produced by the clitellum. Leeches "loop" when they travel, similar to looper caterpillars, alternately attaching themselves to the substrate with the anterior and posterior suckers. The acanthobdellids use their chaetae on the anterior segments to attach themselves to the ground. The majority of leeches are good swimmers. They flatten the body and perform vertical, wavelike movements which propel them forward.

Only a single species of the previously mentioned suborder ACAN-THOBDELLAE, which is placed between the oligochaetes and true leeches, has survived to the present, from the time that the leeches rose from the oligochaetes. This, *Acanthobdella peledina* (L up to 37 mm), is a parasite on fresh-water fish, especially relatives of the salmon, in northern Eurasia from Scandinavia to the Lena River area.

The suborder RHYNCHOBDELLIDS (Rhynchobdellae) is characterized by the fact that the anterior intestine can be extruded from the mouth sucker as a stabbing proboscis. The DUCK LEECH (*Theromyzon tessulatum*; L up to 5 cm; see Color plate, p. 381) penetrates the mouth and nasal passages of waterfowl, sucking blood from their nasal and esophageal epithelium; birds sometimes die of suffocation in the case of heavy infestations. This leech is present in four size classes, separated in time by a blood meal. The GLOSSIPHONIIDS (family Glossiphoniidae, e.g., *Helobdella stagnalis*; L up to 1 cm; see Color plate, p. 381) suck blood from mosquito larvae, aquatic woodlice, water fleas, and snails. *Hemiclepsis marginata* (L up to 3 cm; see Color plate, p. 381), conspicuous by its beautiful patterning and activity, feeds on small organisms and on fish and newts. The two *Glossiphonia* species common in both running and stagnant water in Europe, *Glossiphonia complanata* (L up to 3 cm; see Color plate, p. 381) and *Glossiphonia heteroclita* (L up to 1 cm; see Color plate, p. 381), suck blood from snails. Within the family Glossiphoniidae are species which also attack humans, such as the eastern European *Haementeria costata* (L up to 7 cm) and especially *Haementeria officinalis* (L up to 8 cm), which is found in Central and South America. The latter are used in Mexico as medicinal leeches to relieve blood pressure. The haementerians which suck human blood have been accused of transmitting diseases, but there is no proof of this; the rhynchobdellids, however, transmit dangerous

Suborder: Rhynchobdellae

◁

Water bears (phylum Tardigrada, greatly magnified): 1. *Echiniscus blumi*, 2. *Macrobiotus hufelandi* (female) Peripatid (phylum Onychophora): 3. *Heteroperipatus engelhardi* (female); Pentastomid (phylum Pentastomida): 4. *Armillifer armillatus*, Trilobites (class Trilobita; extinct): 5. *Ceratarges armatus*: a) Original preparation in the Senckenberg Museum, Frankfurt (magnified 2½ times), b) Reconstruction, by E. Richter (normal size); 6. *Olenus* species.

trypanosomes to fish and frogs. One unique characteristic of the glossiphoniids is their care of the young: the cocoons, which are attached to the parent's ventral surface, are covered by the body and continually supplied with fresh water by its wavelike movements. Other species also attach the cocoons to their ventral surfaces, and the young leeches remain there for an appreciable time. *Glossiphonia heteroclita* and *Helobdella stagnalis* even have a special ventral area for the attachment of the young, and the South African species *Marsupiobdella africana* has even developed a brood pouch.

The most common species of the family ICHTHYOBDELLIDAE is the COMMON FISH LEECH (*Piscicola geometra*; L up to 5 cm; see Color plates, pp. 376 and 381). It often attaches itself to plants or submerged stalks for hours or even days by means of its large posterior sucker. When a fish approaches, the leech, stimulated either by the fish's shadow or the water currents it produces, starts to wave the anterior part of its body about until it contacts the fish. It then attaches itself immediately to its prey by the anterior sucker, releasing its hold with the posterior one. It remains on the fish for several days, sometimes as long as a month. As a result of heavy infestation and its accompanying blood loss, the fish loses an appreciable amount of weight; the leech also serves as a carrier of the organism causing abdominal dropsy. Certain species of ichthyobdellids are marine and parasitize rays, teleosts, and turtles.

The suborder GNATHOBDELLAE possesses three jaw-plates in the esophagus, furnished with teeth with which they penetrate their prey's skin. The most commonly known species is the MEDICINAL LEECH (*Hirudo medicinalis medicinalis*; L up to 15 cm; see Color plate, p. 381, and compare Color plate, p. 352) which has six red or brownish longitudinal stripes dorsally and a speckled ventral surface. As it was once widely used for medicinal bleeding, and still is to a certain extent, it has become relatively rare in Europe as a result of the demand. It is replaced nowadays by the introduced HUNGARIAN MEDICINAL LEECH (*Hirudo medicinalis officinalis*; see Color plate, p. 381 and compare Color plate, p. 352), a subspecies with only four longitudinal stripes and a non-speckled ventral surface.

During feeding, the leech presses its jaws against the skin and cuts through it like a circular saw; the wound is shaped like a three-rayed star. Simultaneously, the salivary glands, whose openings are found between the jaws, secrete hirudin, which prevents the blood from clotting. The leech sucks up between 10 and 15 ml of blood; after it has dropped off, a further 20 to 50 ml continue to flow from the wound. The blood is dehydrated and thickened within the leech's intestine prior to digestion, which takes up to six months. Certain intestinal bacteria (*Pseudomonas hirudinus*) prevent the decay of the stored blood; they are also responsible for its remarkably long digestion period. This extremely restricted food availability brought about by the bacteria allows the leech to survive for half a year on one meal, and even after this time, it does not have to

Family:
Ichthyobdellidae

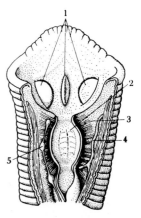

Fig. 12-12. Anterior end of a medicinal leech opened ventrally. Anterior, the three jaws furnished with teeth (1); behind these, the head of the pharynx (2) with radial sucking musculature attached (3); right, the ventral blood vessel (4); left, the ventral nerve cord (5).

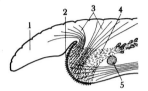

Fig. 12-13. Prostomium (1) and central jaw (2) of a medicinal leech in longitudinal section showing the muscles (3) operating the " circular saw." The openings of the salivary glands (4) are hatched, the brain (5) is dotted.

starve in cases where no new prey are immediately available. Break-down of the animal's own body tissues permits continued, although meager, survival. According to R. Lotz, the leech, which is the only animal in the whole animal kingdom which appears not to suffer from hunger, is the best astronaut for long-term space flights.

The medicinal leech has become rare in Europe, but a related species which is often confused with it, the HORSE LEECH (*Haemopis sanguisuga*; L up to 15 cm; see Color plate, p. 381), is still common in stagnant and running water. Although its Greek generic name and Latins pecies name mean "blood sucker," it actually is not one. Its jaws are too small and too few in number to cut through skin. It feeds by swallowing worms, insect larvae, and, in fact, anything which it can swallow whole. The presence of only small lateral pockets in the intestine and the large size of the anus are adaptations to this mode of feeding. Its anything–but–apt names, "horse leech" and "blood sucker," are the result of its confusion with a blood-sucking relative, the southern European *Limnotis nilotica* (L up to 10 cm). This lives in springs and puddles used by cattle and horses as drinking places. While the animals are drinking, the leeches enter their nostrils in large numbers, moving downward into the throat, larynx, and esophagus; by causing swelling of the tissues, they can even lead to the animal's death by asphyxiation. Humans are also attacked if they drink such water, especially by young leeches, which attach to the throat, tonsils, respiratory passages, or vocal cords of their victim for days, or even weeks. The French troops in Egypt suffered severely from this leech species.

Family: Haemadipsidae

The worst human pests, however, are the LAND LEECHES (family Haemadipsidae) which inhabit southern and southeastern Asia, Oceania, Madagascar, and South America, consisting of about fifty species. The CEYLON LEECH (*Haemadipsa zeylanica*; L up to 3 cm) has become well known because of its enormous population on Ceylon. It is able to creep through the smallest holes in protective clothing and cut through the skin of the person attacked without the victim feeling it. The large numbers of this leech and the extreme bleeding following its detachment lead to severe loss of blood. The leeches not only recognize their prey through vibrations but also by smell; they move upwind toward stationary men or flocks of animals. The only land leech indigenous to Europe is *Xerobdella lecontei* (L up to 4 cm), found in the Alps, which feeds on the blood of the Alpine salamander, as well as eating all kinds of small creatures. It emerges from its hiding place in the ground only during rain or on nights with high air humidity.

Suborder: Pharyngobdellae

The PHARYNGOBDELLIDS (suborder Pharyngobdellae) have no proboscis or toothed jaws; the esophagus, however, is extremely long and muscular. They are therefore pure swallowers of their prey. The common DOG LEECH (*Erpobdella octoculata*; L up to 6 cm; see Color plate, p. 381), which inhabits stagnant and slowly running water, however, is able to attach itself to sections of its prey by means of its jaws; it sucks water snails out of their shells with the aid of its esophageal muscles. The few European pharyngobdellids are all fresh-water inhabitants. Related species living in South America, which belong to the same family (Erpobdellidae), are, strangely enough, terrestrial forms ("ground leeches").

13 Onychophorids, Water Bears, and Linguatulids

The following three animal phyla, the ONYCHOPHORIDS, WATER BEARS, and LINGUATULIDS, indicate their membership among the true segmented animals by many characteristics; from these, it can be assumed that they have arisen from the "segmented worms" independently of the arthropods. They are definitely not precursors of the arthropods and thus not "missing links" between them and the segmented worms. Each of these three small phyla has developed along its own lines during evolutionary prehistory. Although they are combined to form the PARARTHROPODA (which means "animals next to the arthropods" systematically) or ONCOPODA (claw-footed animals), they are no more closely associated with each other than they are with the arthropods. There are therefore no grounds for these three phyla to be considered as having the same evolutionary origins.

Evolution; by
E. Thenius

It is known that pararthropods have existed since the Middle Cambrian, as discovery of the fossilized genus *Aysheaia*, an onychophorid, in rocks 540 million years old from British Columbia has shown; this indicates that these forms were not terrestrial but marine. This origin shows that the onychophorids must have become terrestrial through the development of transitional species which lived in moist environments (water-vapor saturated biotopes).

Phylum:
Onychophora, by
O. Kraus

The present-day ONYCHOPHORIDS (phylum Onychophora; BL up to 15 cm) are all photonegative terrestrial animals restricted to the tropics and neighboring southern continents. Apart from one white cave-dwelling species and other more gray and brownish representatives, reddish, orange, green, blue-green, or even blue species are known, some of which are even spotted or striped. They occur in environments with relatively high air humidity, especially under rotting wood and fallen leaves.

The onychophoran body is covered by only a very thin layer of chitin (barely 0.001 mm thick). The skin bears numerous papillae and has fine, close-set ring markings, which, however, have no relationship to the true body segmentation. Segmentation is externally expressed in the arrangement of the fourteen to twenty-three pairs of walking legs which, because of their form, have resulted in the entire phylum being given the German name of *Stummelfüsser* (= stub feet). The limbs appear as ringed protrusions of the body, with a type of "sole" on the underside, the tip of which has a double-hooked claw (onychium), from which they have received their scientific name of Onychophora (claw bearer). These structures compare favorably with similar claws present in the water bears and linguatulids. Two conspicuous antennae, modified walking legs, are present at the anterior end. Between these and the first pair of true walking legs are the oral papillae. They are clawless, but as a substitute they contain a well-developed gland used in defense and useful in prey capture. The animals are able to squirt a slimy, sticky secretion from these glands, projecting it for 30 to 50 cm. It falls over the attacker or prey like a net, rendering it helpless.

The oral papillae can be considered the organs of the third segment, since those of the second are modified to an oral opening and are used in feeding. Only the sicklelike blades of modified claws and the tip of the limb are retained in the second appendages. The onychophorid cuts an opening in its prey (e.g., termites, woodlice, snails, worms) with these and injects a digestive saliva. This dissolves all the soft parts of the prey's body so that the fluid can then be sucked up. The cushionlike thickening around the mouth provides the necessary air insulation. The midgut runs the whole length of the body. Substances taken up are enclosed in a permeable membrane and digested simultaneously. The indigestable portions remain enclosed in the membrane and are deposited as feces from the anus at the end of the body. Segmental excretory organs (nephridia) also serve to remove the metabolic wastes. These open ventrally in every pair of walking limbs, and thus are close to the ground.

The onychophorids' respiratory organs are especially worthy of mention, since Moseley showed that they were trachea arising from numerous tiny respiratory openings scattered over the whole body. This finding naturally resulted in suggestions that these were a primitive form of the much more advanced respiratory organs (tracheal system) of arthropods. This is, however, unlikely, since there are several points of view which suggest that respiration by means of trachea had been "discovered" independently by the onychophorids. Gas exchange is thus independent of the blood–circulatory system. It should therefore be no surprise that only a relatively simple circulatory organ is present. This is a pulsating dorsal vessel, which, in structure and function, parallels that of the arthropod system almost completely.

The onychophorids, being nocturnal, possess only two simple lensed eyes at the base of the antennae; these eyes scarcely reach a diameter of 0.3 mm. Numerous epidermal cells, however, are present. The true nervous system consists of a brain (supraesophageal ganglion) which

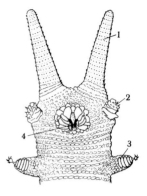

Fig. 13-1. Onychophorid: 1. Antenna, 2. Oral papilla, 3. Walking leg, 4. Jaws.

Nocturnal animals

leads into a paired ventral nerve cord; this, however, is not divided into segmental thickenings (ganglia)—in contrast to the "ladderlike" nervous system found in arthropods.

Reproduction takes place in an extremely curious manner. The male deposits its packets of sperm (spermatophores) at some point on the female's body, for example, on the back or sides. Tissue collapse occurs at this point, enabling the sperm to migrate into the female's body, where they penetrate to the ovaries and fertilize the mature ova.

The males of certain peripatids do not seem to "trouble" to look for females on which to deposit their spermatophores. If they encounter another male or a juvenile, they also attach their sperm packets to this. A shortage of females can lead, in addition to this, to all the males in one area reproducing with a single female. A biologist once investigated such a female and found that its body was covered with no less than 180 spermatophores. All transitional types have been observed, from forms which lay their large, yolk-filled eggs "normally," to other representatives in which large, yolk-filled eggs complete their development within the female, resulting in the birth of advanced young, to onychophorids in which the fertilized eggs obtain their nourishment from the mother's body. In the latter case, the eggs have little yolk and therefore are small; nourishment is provided for the embryo through a special organ (placenta). In this case, too, well-developed young are born.

To date, about seventy species of onychophorids are known. These fall into two groups, the PERIPATIDS (Peripatidae; English, "strollers") and the PERIPATOPSIDS (Peripatopsidae). The peripatids, inhabitants of tropical America (e.g., *Heteroperipatus engelhardi*; see Color plate, p. 382), have also been found in tropical western Africa and southeastern Asia. One point is certain: the present-day distribution, separated by wide oceans, must at some time in the past have been connected by some means. It is very probable that the disjointed distribution of this ancient group of animals is associated with the previous existence of a connected southern continent (Gondwanaland) which, in the meantime—according to the continental drift theory—has become destroyed and drifted apart. The present-day distribution of the peripatopsids is hardly less widely scattered: they occur in Chile, South Africa, and the Australian region, including Tasmania and New Zealand. This suggests that Antarctica was also once part of the previously mentioned southern continent of bygone eras, and the present-day distribution of the peripatopsids is considered a weighty animal-geographical argument for the hypothesis that antarctic connections with the southern continents were once present.

It is an attractive proposition, to consider the onychophorids as a type of "missing link" between the annelids and arthropods, since their body structure combines characteristics of both groups. The presence of various specially developed characters, such as the respiratory organs

Fig. 13-2. Distribution of the peripatids, which are restricted to tropical zones, and the more southern peripatopsids.

(trachea), and the origins of the jaws, refutes this. Owing to this, the onychophorids are probably a dead end, a unique, extremely ancient group of animals which—according to Kaestner—probably represents an early lateral branch of the evolutionary line terminating in the arthropods.

The WATER BEARS (phylum Tardigrada; BL usually between 0.1 and 1.0 mm, maximum 1.2 mm; see Color plate, pp. 376 and 382) are extremely small segmented animals. The body is cylindrical, usually flattened on the ventral surface, and has four stumpy pairs of legs furnished distally with conspicuously large claws. Using these, the water bears are able to move about slowly, hence their scientific name ("slow walkers"). The arrangement of the segments indicates externally that the body is comprised of a head section and four following segments which bear the legs. The body covering (cuticle) is comprised not of chitin but of a protein compound in the form of a water-permeable albuminoid which is capable of swelling; in many terrestrial species, this becomes an extremely thickened, plated armor. The color is partly dependent on the color of the visible internal organs; in brightly colored species, the external body covering contains brown, olive-green, light red, or violet substances.

The anterior section of the intestine consists of an oral tube possessing a piercing apparatus, and a muscular gizzard (sucking apparatus) connected to a short pharynx; this is followed by a wider midgut and finally the rectum. These structures indicate the mode of feeding, since the animals pierce a plant cell (e.g., a moss cell) with their stylets and then suck out the contents. They are, however, not entirely vegetarian, since they have been observed feeding on the bodies of conspecifics, nematodes, and rotifers, and have even been seen capturing the latter.

The water bears live in coastal waters—in part within the sand space system (mesopsammon)—and in ditches and lakes, and especially in the transient film of water covering moss. In the moss clumps on rocks, walls, and roofs, they must be able to withstand not only tremendous temperature changes but also periods of complete desiccation. An unusual characteristic of their water metabolism is associated with this: the inhabitants of such a "microaquarium" are able to contract their bodies into a resistant form, termed a "barrel," within a relatively short period of time. In this condition, the majority of their water content is lost and the remaining water is retained under high pressure so that it is unable either to freeze or to evaporate. Thus, the animals are able to withstand the most amazing humidity and temperature changes. Resistant stages of the genus *Macrobiotus* revived after twenty months in liquid oxygen (temperature −190 to −200°C), and could also be revived after an hour of exposure to +92°C.

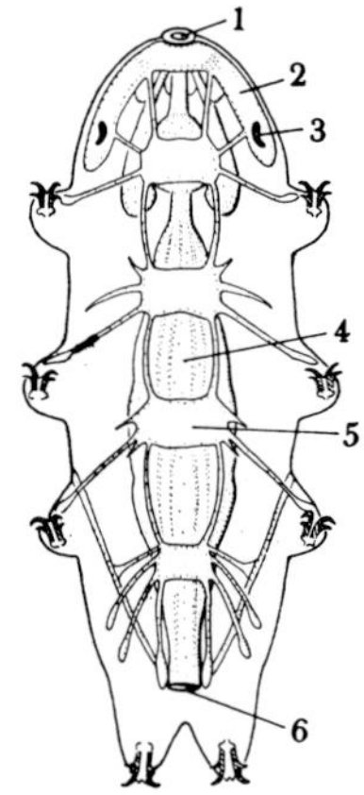

Because of their minute size, water bears have no respiratory or circulatory organs. Their nervous system is similar to that of the arth-

Phylum: Tardigrada, by O. Kraus

Fig. 13-3. Resistant stage of *Hypsibius*, ventral surface, L 0.2 mm.

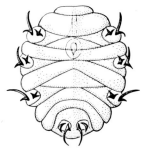

Fig. 13-4. *Macrobiotus hufelandi*. Ventral surface showing the nervous system: 1. Mouth, 2. Brain, 3. Eye, 4. Intestine, 5. Ventral ganglion, 6. Anus.

Fig. 13-5. *Echiniscoides sigismundi* climbing on the filament of the alga *Enteromorpha*.

Phylum:
Linguatulida, by
O. Kraus

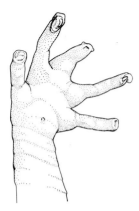

Fig. 13-6. Anterior section of a female *Cephalobaena tetrapoda*.

ropods. The supraesophageal ganglion is divided into two lobes, each containing an eye consisting of a single visual cell; following on from the subesophageal ganglion—corresponding to the body segmentation—are five ventral ganglia. Reproduction can be associated with ecdysis in the female, the ova being fertilized and protected within the shed skin. The extremely small size of the ova and also the tiny resistant form allow distribution to occur by wind currents. It is owing to this that many water bear species are distributed worldwide, having been found from the arctic to antarctic zones.

The relationships of this group to others have not been completely determined. Although their body form is most similar to that of the onychophorids, they are better considered as an independent phylum. There are at least 180 different species, divided into several orders. These cannot be discussed in more detail here. The armored *Echiniscus scrofa* (compare Color plate, p. 382), whose surface is covered with long cuticular processes, inhabits moss and lichen patches; this is also true of *Macrobiotus hufelandi* (see Color plate, p. 382), the largest water bear, the female of which reaches 1.2 mm in length. The armorless ("naked") species *Echiniscoides sigismundi* lives in algae, and inhabits European and Chinese coastal regions.

All LINGUATULIDS (phylum Linguatulida or Pentastomida) are parasites of vertebrates. This explains many unique features of their body structure. The body is wormlike, partly cylindrical and partly flattened, and varies in length between a few millimeters and, in extreme cases, 14 cm. A long posterior section follows a short anterior one with an oral opening and four hooklike claws; both sections pass into one another without any obvious point of separation. The posterior section has clear rings which are caused by tiny superficial constrictions of the body; only in rare cases are true tirelike swellings present (*Amillifer*; compare Color plate, p. 382). This external segmentation of the body, which corresponds to a similar segmentation of the underlying musculature, probably is an expression of the true segmentation. The body covering (cuticle) and the claws consist of chitin. Some species are completely transparent; many others, although colorless, are whitish; in brightly colored linguatulids, yellowish, pinkish, or reddish tones are present.

The foregut, attached to the mouth, is covered with attached muscular structures; this is a type of sucking pump which serves in the taking up of blood and mucus, on which these parasites feed. Attached to this is the midgut, which is usually a straight tube connecting with the hindgut; the anus is posterior. The body cavity, which is capacious, especially at the anterior end of the body (myxocoel), contains no excretory, respiratory, or circulatory organs. As would be expected, there are no complicated sense organs, and the nervous system of the more primitive sub-group CEPHALOBAENIDA is comparable to the general pattern of the segmented animals in that it consists of four serially arranged ventral ganglia in the anterior of the body. In all other representatives—the POROCEPHALIDA—only a uniform subesophageal mass is present, and long nerve fibers stretch from this to the posterior end of the body.

The linguatulid sexual organs have developed so that they almost fill the body cavities of mature animals. In the males, they always open between the anterior and posterior body sections; in female porocephalides, however, the sexual openings can lie far behind the center of the body. In accordance with their parasitic mode of life, linguatulids produce enormous numbers of eggs; mature females of *Linguatula serrata* in wolves, dogs, and foxes produce several million during the course of their lives.

The development of the eggs begins within the female's body. Their extremely small size (barely 0.1 mm) and, especially, the thick shell surrounding them, make observation of the developmental process extremely difficult. It is therefore understandable that it was not until 1963 that Ochse was able to report in detail on the embryonal development of at least one *Reighardia* species. According to him, rudiments of four segments are recognizable; these are retained by the water bears for their whole lives. It appears that the two anterior ones degenerate later; therefore, the four hooklike structures in the following developmental phases and in the adult animals are equivalent to the posterior segmental rudiments of the embryo. A supporting factor is the form taken by the anterior end of the primitive species *Cephalobaena tetrapoda*. Here the paired claws, which are usually arranged next to one another, still lie at the tips of two successive, cylindrical "legs"; apart from this, a pair of so-called "frontal papillae" precedes them, probably the rudiments of the first pair of segmental processes present in the embryo (Fig. 13-6).

Mature linguatulids suck blood and occur exclusively in carnivorous terrestrial vertebrates. They have been found in the lungs of reptiles, the air sacs of birds, and the respiratory organs of various mammals. Only *Linguatula serrata* and a few of its relatives attach themselves to the sinuses and nasal openings of certain mammals—for example, the wolf and dog, and very rarely in man—where they live on mucus, tissue secretions, and similar substances.

The developmental process, in which the larval form is also involved, is commonly associated with an intermediary host; details are known for only a few of the roughly sixty recognized species. That of *Linguatula serrata* is relatively complicated; an infected dog sneezes the eggs out of the nasal passages. From here, they become attached to plants and are unintentionally eaten by herbivorous mammals—rabbits and hares, for example. Within the small intestine of these intermediary hosts, the eggs hatch into tiny larvae which rapidly bore through the walls of the small intestine and are carried farther by the lymph or blood streams. A few remain within the lung alveoli, while others prefer to infest the liver. Whether or not the host surrounds the larva with a connective tissue

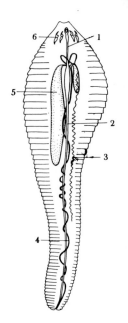

Fig. 13-7. *Linguatula serrata*, female, ventral surface. 1. Esophagus, 2. Intestine, 3. Ovary (ovarium), 4. Vagina, 5. Spermatheca (receptaculum seminis), 6. Hooks.

▷

Sea-scorpion (order Gigantostraca, extinct; see Chapter 15):
1. *Megalograptus ohioensis*;
Scorpions (order Scorpiones; see Chapter 15):
2. Giant African scorpion (*Pandinus imperator*; see Color plate, p. 407);
3. *Orthochirus innesi*;
4. Italian scorpion (*Euscorpius italicus*);
Horseshoe crabs (order Xiphosura; see Chapter 15):
5. Atlantic horseshoe crab (*Limulus polyphemus*). The scorpions are shown natural size, sea-scorpion and horseshoe crab greatly reduced.

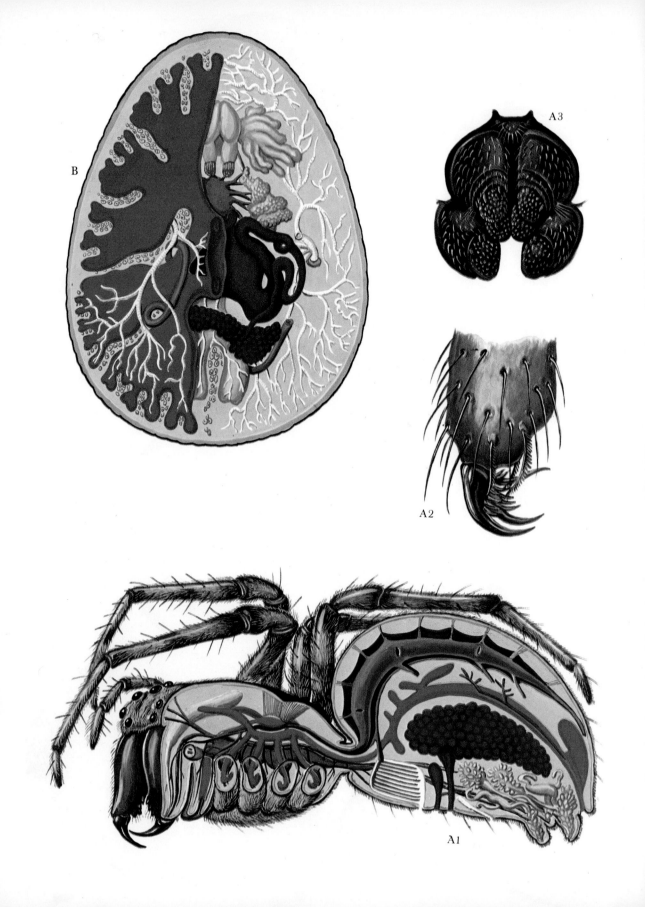

ARACHNID BODY PLAN (SPIDERS AND MITES)

A. Body structure of a spider (order Araneae, see Chapter 15):

A$_1$ Anterior and posterior ends (prosoma and opisthosoma) opened on the left side. Left, the large chelicerae, the left one opened to show the poison glands; above this the eyes with the visual nerves. The pumping pharynx arising from the mouth, horizontally lying intestine with sucking stomach, expanded by the muscles arising dorsally; lateral pockets extending to the constriction. Below this, the nervous system with all ganglia compressed into a single large mass. In the posterior part of the body (opisthosoma), the dorsal heart; below this, the intestine, here also with lateral pockets; another sac near the anus is termed the cloaca, since the excretory organs (malphigian bodies) also open into it. In contrast to those of the insects, they arise from the endoderm, and, despite their name, are not homologous structures. Anterior on the ventral surface of the abdomen, "book lungs"; behind these, the opening of the spermatheca and ovaries, followed by the respiratory tubes (trachea) and finally the silk glands (gray) and spinneret.

A$_2$ shows the leg tip of a spider with the comb claws and bristles; these allow rapid gripping and just as rapid release of the web threads.

A$_3$ shows the spinnerets ventrally from above.

B. Structure of a mite, the pigeon mite *Argas reflexus* (order Acari, see Chapter 15).

No division of the body into anterior and posterior sections, intestine with broad lateral pockets, as in many other blood-sucking animals (leech, myzostome, fish lice). The large glandular organs (gray) at the anterior end are the Genésch organs, the sticky secretion of which covers the eggs when they are laid, preventing them from desiccating. Numerous breathing tubes (white) with paired openings (visible right). Posteriorly the lateral pockets of the hind-gut (gray).

Diagrammatic coloration in A and B as in the other Color plates on body structures (e.g., Color plate, p. 352): gray = ectoderm, yellow = endoderm, carmine red = muscles, cinnibar red = female reproductive organs, violet = male reproductive organs, blue = nervous system, green = excretory organs, red-brown = blood vessels, white = trachea.

capsule, the larva continues to draw nourishment from the host, and grows, shedding its skin several times. In the case of severe infestations, the intermediary host becomes so affected that it easily falls prey to predators. The circle is then closed; within the final host, the larvae wander from the throat or stomach outward to the nasal openings. Mature *Linguatula* usually affect dogs only sightly, since they are not blood suckers.

14 The Arthropods

Phylum: Arthropoda,
by P. Rietschel

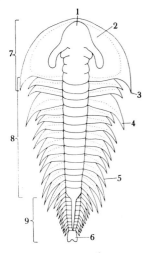

Fig. 14-1. Nomenclature for the exoskeleton of a trilobite, *Olenellus vermontanus*. 1. Glabella, 2. Cheek (gena), 3. Cheek spine, 4. Macropleura, 5. Pleura, 6. Pygidium, 7. Cephalon, 8. Prothorax, 9. Opisthothorax.

The phylum which contains by far the majority of species in the animal kingdom is the ARTHROPODS (Arthropoda). This is distinguished from the four other phyla of articulated animals by the presence of jointed appendages which are movable at the point of junction with the body and also by means of joints. A prerequisite for the formation of such joints is a characteristic also shared with the onychophorids and linguatulids: the cuticle, formed by the uppermost epidermal cell layer. It contains, in addition to protein compounds, the carbohydrate chitin, which is closely related to cellulose. The protein compound (arthropodin) and chitin are so similar in their molecular structure that both components intermesh tightly with one another. The cuticle is thus both flexible and tough. In cases where hardness is a particular requirement, the arthropodin is tanned by means of phenol. It is then termed sclerotin.

The hard parts of the exoskeleton (sclerites) thus differ from the softer joint cuticle not in their chitin content but owing to their sclerotin. When their toughness was first ascribed to chitin, they were termed heavily chitinized. This mistake in description exists in the literature even today, and it is more correct to say heavily sclerotinized. An extremely unusual change in the protein component of the cuticle can also occur: it becomes an extremely elastic substance consisting of very long chains of molecules (resilin) which, by accumulating energy, can be expanded to more than twice its original length and, on contraction, releases the energy so stored within a thousandth of a second. To date, no substance has been manufactured which can emulate this. Resilin is found in the cuticle of the wing joints in insects and is also present in the tendons of both insects and crustaceans. When these contract, they amplify muscular action.

The hardness of the cuticle has other advantages for the arthropods: not only the leg joints but also the body segments are movable, because of the presence of the softer joint cuticle. Rarely, soft organs are exposed to damage on the surface of the body, but usually they are enclosed within it and thus protected by the hard exoskeleton. The articular cavities, processes, and guide articulations must withstand greater forces, and are thus more heavily sclerotinized. The form of the joints is extremely variable, and their mechanical properties offer a source of interest for technically minded investigators. Since an arthropod is enclosed in an armor, it is capable of growing only within a very restricted space. It must, therefore, shed its original skin from time to time after it has formed a new soft and elastic one underneath. The animal expands the new cuticle by taking in air or water, and then allows it to harden. It then has time to grow within the new exoskeleton until the next ecdysis. Many arthropods undergo ecdysis at definite time intervals throughout their lives; others stop growing when they attain sexual maturity, and no longer shed their skins. The lower arthropods hatch from the egg with an incomplete number of body segments, and reach the full number after the first ecdysis. Such "anomorphosis" is found in all crustaceans, many millipedes, and apterygotes. The majority of the millipedes, the insects, and arachnids, however, hatch with their full complement of body segments.

The structure of the segments corresponds, evolutionarily speaking, to that of the annelids; the segments of a fly maggot, bee larva, or caterpillar are thus comparable to those of an earthworm. This is not true, however, for the arthropods with armored segments. In these, just as in the earthworm, the longitudinal musculature stretches from the anterior edge of one segment to its posterior edge; here, however, on the boundary between two segments, the exoskeleton forms a hard ridge (antecosta) internally, to which the muscles are attached. To enable the exoskeleton to remain flexible, it must be separated between the segments by a soft joint cuticle. There is not enough room for this at the edge of the segment itself, since this is occupied by the antecosta and the attached muscles. The joint cuticle is therefore attached anterior to this point, or sometimes behind it. The hard, sclerotinized body rings do not, therefore, correspond to the true segments, and the joint cuticle between them is not homologous to the segment boundaries of the soft-skinned arthropods or annelids (homologous = with the same evolutionary origin). An investigator who wishes to determine the true boundaries of the segments among the great variety of types present in the armored arthropods cannot depend on the visible segmentation (secondary segmentation), but must determine the sites of the antecosta inside the exoskeleton and the points of attachment of the muscles. These, and not the joint cuticle, are the boundaries between the original (primary) segments.

Since the exoskeleton determines the body shape of the arthropods, a periodic change in body form can be present in these animals during the course of their development; no other animal phylum shows this characteristic. A tremendous wealth of sensory organs is also present, formed by the cuticle and the cell layer immediately beneath it. In conjunction with

▷
True spiders (order Araneae; see Chapter 15): 1. Garden spider (*Araneus diadematus*; female; see Color plates, pp. 407 and 410). 2. Jumping spider (*Salticus scenicus*). 3. Wasp spider (*Argiope bruennichi*; see Color plate, p. 407). 4. European black widow (*Latrodectus mactans tredecimguttatus*), 5. House spider (*Tegenaria domestica*; natural size), 6. Cellar spider (*Segestria senoculata*; natural size), 7. Wolf spider (*Pisaura mirabilis*; see Color plate, p. 407), 8. Bird-eating spider (*Eurypelma soemanni*; natural size). The garden spider, jumping spider, and wolf spider are shown natural size (b) and also greatly enlarged (a).

1a

2b

2a

3

4

5

6

7b

7a

8

Groningen

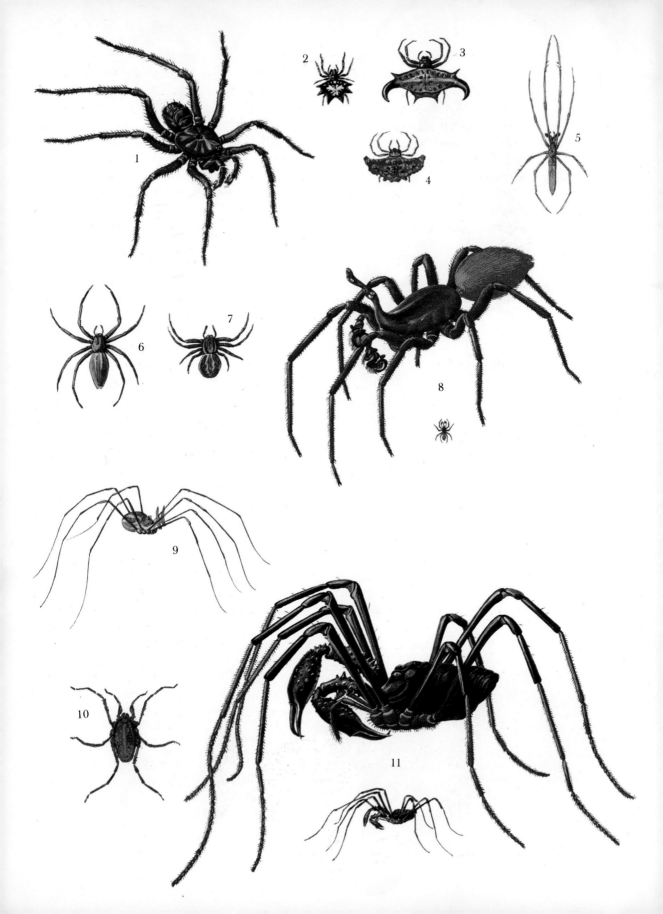

these, the nervous system and, especially, the brain, have reached a peak in development among the arthropods. The sensory experiences and multitude of behavioral patterns shown by the higher crustaceans, spiders, and higher insects offer the experimenter an inexhaustible source of new knowledge.

The exoskeleton, which allows changes in body form to occur, acts as a protective and supportive skeleton, permits the most varied movements and actions of the body, and, last but not least, facilitates the high degree of development in sensory modalities which has resulted in the arthropods being not only the phylum with the largest number of species but also the most successful in water, on land, or in the air. Representatives were present in the oceans during the Cambrian, 600 to 500 million years ago, and their development reaches back to Precambrian times. Their ancestors were, without doubt, segmented worms, but it remains an open question whether the subphyla arose independently from these or that they have a common origin.

The arthropods are divided into four or five subphyla: 1. Trilobites (Trilobita), 2. Arachnids (Chelicerata; see Chapter 15), 3. Biantennals or Crustacea (Diantennata; see Chapter 16), 4. Tracheates (Tracheata; see Chapter 17). The Pantopods (Pantopoda; see Chapter 15), included here in the Chelicerata, are often considered as a fifth phylum. With the exception of the trilobites, which were already extinct in prehistoric times, the subphyla of the arthropods are all represented in present-day fauna; some can be considered to have reached the pinnacle of their development. Other subphyla may also have occurred in ages past and long since become extinct: beautifully preserved arthropods from the Mid-Cambrian era (530 million years ago) have been found in the Burgess schists in British Columbia, but these cannot be assigned to any of the above-named subphyla.

The TRILOBITES (✠ subphylum Trilobita; see Color plate, p. 382), an extinct group of arthropods, cannot be included under the Crustacea or Chelicerata on the basis of their characteristics: BL 0.5 to 70 cm, usually between 3 and 8 cm; dorsal exoskeleton (carapace) rounded ventrally, consisting of chitin with deposits of chalk and calcium phosphate, usually patterned in colors in living animals; body divided into three sections: cephalic shield (cephalon), thorax, and pygidium; lateral subdivision of the thorax and pygidium into a central area (rachis) and lateral areas (genae). Number of free pygidial sections varies with the species; in *Agnostus*, for example, reduction to two. Size of the pygidium varies with the number of incorporated sections. Appendages biramous with a proximal endopodite (walking leg) and distal expodite (bristle-covered ramus); this, however, is not comparable to the biramous appendage of the Crustacea. Appendages of the head not modified to form mouthparts. Seven orders with over 1400 genera.

These exclusively marine animals probably fed on minute organisms.

◁
True spiders (order Araneae; see Chapter 15): 1. *Liphistius desultor* (male), 2. *Gasteracantha kuhlii* (female), 3. *Gasteracantha thorelli* (female), 4. *Pasilobus bufoninus*, 5. *Tetragnatha caudicula* (male), 6. *Micromata rosea* (female), 7. *Xysticus erraticus* (female), 8. *Walckenaera acuminata* (male, below in life size). Phalangids (order Phalangida): 9. Common phalangid (*Phalangium opilio*; male), 10. *Trogulus nepaeformis*, 11. *Ischyropsalis hellwegi* (female, below in life size).

The stomach, provided with lateral pockets, is present in the region of the cephalon; respiration was by means of gills. A pair of antennae and eyes served as sense organs. Occasionally a median tubercle is found on the cephalic shield (glabella), and was a dorsal sense organ (so-called "median eye"); bristles or similar structures are also present on the body surface as mechanoreceptors.

The trilobites developed through three different larval stages (protaspis, meraspis, and holaspis). In the prehistoric seas (Lower Paleozoic), they were distributed worldwide and were mostly benthic. Some forms lived as burrowers in sediment and some were swimmers.

Characteristics of the mode of life of the trilobites are drawn not only from their body structure but also from fossilized tracks, such as walking spoors and resting impressions. Various species were capable of rolling into a ball like woodlice. Their temporal distribution extended from the Precambrian to the Permian.

Trilobites can be divided into the following orders on the basis of the cephalic-seam development, the pygidium, and the hypostome present on the ventral surface of the cephalon: 1. Agnostida (with the genus *Agnostus* from the Upper Cambrian), 2. Redlichiida (with the genus *Paradoxides* from the Middle Cambrian), 3. Corynexochida (with the genus *Olenoides* from the Middle Cambrian), 4. Ptychopariida (with the genera *Asaphus* and *Illaenus*, both from the Ordovician), 5. Phacopida (with the genus *Phacops* from the Silurian and Devonian), 6. Lichida (with the genus *Lichas* from the Ordovician and Silurian), Odontopleurida (with the genus *Acidaspis* from the Ordovician until the Devonian).

The trilobites were once considered as the basic Crustacea, but today they are considered as a discrete subphylum on an equal footing with the other arthropod subphyla. During their evolutionary development there was a tendency for the pygidium to increase in size at the expense of the free posterior segments. In certain trilobites (e.g., *Cyclopyge*) the eyes were very large, while in others (e.g., *Pteroparia* and *Trimerocephalus*) they were reduced to rudiments or were lacking entirely. In addition, spines were developed in the region of the cephalon, thorax, and pygidium.

A series of other prehistoric arthropods from the Cambrian seas have, owing to various special characteristics, been termed the Trilobitoidea and put together with the trilobites in the TRILOBOMORPHA (Trilobitelike animals). Externally, they resemble in part arachnidlike and crustaceanlike animals, and indicate the previous variety of forms present in this animal group.

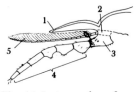

Fig. 14-2. Appendage from *Triarthus eatoni*. 1. Pleura, 2. Precoxa, 3. Coxa, 4. Telopodite, 5. Preepipodite.

15   The Arachnids and Their Relatives

Subphylum:
Chelicerata, by
O. Kraus

One of the major lines of development within the arthropods was the CHELICERATE ANIMALS (subphylum Chelicerata). This subphylum includes the spiders and their relatives. Kaestner termed them "antennaless animals," and Thenius, "shear-footed animals." Their most characteristic feature is that the first pair of appendages are not antennae but chelicerae.

The evolutionary development of the chelicerates can be followed back to the earliest known era in earth's history. The group also includes the largest segmented animal ever known, the giant sea scorpion *Pterygotus rhenanus*, which lived several hundred million years ago in the Devonian period and reached a length of over 180 cm. A large number of dwarfs also belong to this group, since many mites never reach 1 mm in size. The Chelicerata, of which, apart from the arachnids, the horseshoe crabs are also members, inhabit practically every terrestrial ecological niche and also the sea, from which they once rose. The only habitat which they have not been able to conquer is the air, since the group has never produced actively flying forms. The majority have retained their primitive predatory mode of life, while others have become parasites, and numerous mites are even vegetarian.

Their characteristics present, at first glance, an almost confusing multitude of forms: one need only compare the appearance of a scorpion with that of a garden spider. However, all chelicerates can be reduced to a basic body plan.

In every case, the body can be divided into two main sections. The anterior one is termed the prosoma, the posterior, the opisthosoma. In the anterior section, six appendage-bearing segments follow the mouth, the first being the chelicerae-bearing segment, with the chilicerae so typical for the group—a pair of appendages used in feeding. Primitively, they consist of three sections, the central one of which is so unilaterally extended that the terminal section appears to be attached laterally. In this way, a pincerlike structure is formed, which serves as a grasping organ.

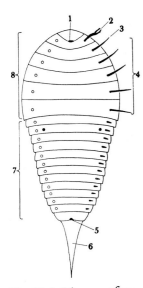

Fig. 15-1. Diagram of an arachnid (from Kraus):
1. Mouth, 2. Chelicerae, 3. Pedipalp, 4. Walking legs, 5. Anus, 6. Caudal spine, 7. Posterior of the body, 8. Anterior of the body.

Progressing from anterior to posterior, five body segments are present, the appendages of which serve in locomotion in the most primitive state. Usually, however, only the last four pairs can be termed walking legs in the true sense of the phrase. The anterior pair of legs, which lie behind the chelicerae, are usually distinguished as pedipalps, since they are usually strongly modified in various ways. For example, both the pincers of the scorpion and the copulatory organ of male spiders are developed from this structure.

When this type of body segmentation is compared with that of Crustacea or Insecta, two features are noticeable: the chelicerates have no jaws (mandibles), so they must be classed as "amandibulates" in contrast to the remaining "mandibulates." In addition, all chelicerates lack organs of touch in the form of antennae. It is, however, possible that unknown progenitors of the chelicerates once possessed antennae; this would mean, however, that these organs degenerated at an extremely early point in evolution. Fossil remains of related groups—the extremely ancient, long-extinct trilobites—possessed antennae, and findings in the embryology of the chelicerates suggest that degeneration was most probably the case.

While the anterior of the body is mainly adapted to locomotion and feeding, the main part of the posterior section is concerned with digestion, reproduction, and respiration. It is therefore to be expected that the associated appendages have degenerated or become modified in various ways. In the most primitive cases, at least thirteen posterior segments are present, as can be seen today in the scorpions; in less primitive members, however, this number has been decreased in various ways. Attached to the end of the body in certain ancient chelicerates is a structure termed the "caudal spine," which is not a true segment and has not been retained by all groups.

The chelicerates are usually divided into three main groups (classes): 1. MEROSTOMATES (class Merostomata) with only a single living order, the horseshoe crabs (Xiphosura). 2. ARACHNIDS (class Arachnida) with at least nine orders, which, besides the scorpions, includes, true spiders, phalangids, and mites. 3. PANTOPODS (class Pantopoda) with only a single order, six families, and about 500 species.

Apart from the pantopods, only the arachnids can be said to show great speciation in the present-day fauna; the merostomates are represented by only five species of horseshoe crabs which inhabit relatively isolated areas. The history of the horseshoe crabs can be followed back to the Cambrian (560 million years ago). The representatives of this group living today—similar to the trilobites—are marine benthonic creatures with a flattened body consisting of two unsegmented sections (prosoma and opisthosoma) together with a "caudal spine" (telson). Apart from the fact that trilobites and horseshoe crabs differ in body structure, they cannot be considered closely related, because the boundaries between the segments are not comparable.

Evolutionary development, by E. Thenius

In the most ancient horseshoe crabs from the Cambrian (genus *Aglaspis*), the anterior of the body (prosoma) was small and the posterior section (opisthosoma) consisted of twelve free segments. During the course of evolution, there has been a decrease in number and a uniting of these segments. The horseshoe crabs from the Carboniferous (genus *Belinurus* or *Euproops*), for example, have only nine posterior segments. In the genus *Palaeolimulus*, of more recent geological times, the posterior segments are united into a single, uniform posterior, the present-day "limulus type." This form has changed only slightly as of this point, as the genera *Psammolimulus* from the Triassic and *Mesolimulus* from the Jurassic indicate.

In the geological past, the merostomates were quite variable in form and speciation, and were worldwide in distribution; it is only in the recent geological past that they have regressed. During the Jurassic, horseshoe crabs were present in Europe; now their distribution area comprises the Atlantic coast of North America and the eastern Indian Ocean to the western Pacific. The fossil horseshoe crabs were, as are their present-day relatives, inhabitants of shallow waters, where they lived in sandy and muddy bottoms. During its development, the American horseshoe crab (genus *Limulus*) goes through a stage which, owing to its similarity to a trilobite, is termed the "trilobite larva." This larva is, however, very similar to the original primitive horseshoe crabs, and this stage would be better termed the "*Euproops* stage."

Family: giant sea-scorpions

Although the horseshoe crabs are still represented by a few "living fossils," the GIANT SEA-SCORPIONS (⊬ family Eurypteridae, genera *Pterygotus* and *Eurypterus*; BL almost up to 2 m) have become completely extinct. They resembled the scorpions in external features, but were—apart from structural variations—entirely water dwellers. They were originally purely marine; among the geologically speaking more recent forms were also representatives from brackish and fresh water. The most ancient and primitive species are therefore marine. Only the last pair of legs is modified to form a flattened swimming leg, all the other appendages serving as walking legs. The posterior part of the body consisted of twelve free segments ending in a caudal spine. This was either spinelike or leaflike. The sea-scorpions occurred from the Ordovician to the Permian (500 to about 225 million years ago). They are related evolutionarily to the ancient horseshoe crabs (*Paleomerus*) and similar forms from the Cambrian.

The present-day arachnids, which comprise so many species and are, in general, terrestrial, are little known with respect to fossil remains. It is interesting that the geologically oldest spiderlike animals are members of the scorpion group (Scorpionida); whether these ancient scorpions (genera *Palaeophonus*, *Proscorpius*) were aquatic or terrestrial was discovered only recently. Størmer was able to show that they were aquatic. True terrestrial scorpions (genus *Eoscorpius*) have been found in the Carboniferous. In more recent geological deposits, mites (genus *Protacarus*), phalangids (genus *Eotrogulus*), solpugids (genus *Protosolpuga*), and true spiders (genus *Arthrolycosa*) have been discovered. *Arthrolycosa* still possessed a segmented abdomen, a feature which now is found only in the most primitive living spiders, the Mesothelae from southeastern Asia. Although fossil spiders are not plentiful, those found indicate that the most highly developed forms, for example, the orb-web spinners, are the most recent evolutionary developments. In comparison to the horseshoe crabs, whose evolutionary peak was reached in prehistoric times, the spiders have attained theirs only recently.

All aquatic chelicerates with a large, well-developed abdomen have been put together in a single group, the MEROSTOMATES (class Merostomata). Those forms which were previously terrestrial and have returned to the aquatic environment are not included here. Merostomates are now found only in the marine habitat; extinct representatives (sea-scorpions; see Color plate, p. 393), in contrast, often inhabited brackish or fresh water. All merostomates have the following characteristics in common: a large body with exceptional segmentation. The anterior was covered with a special dorsal plate, the posterior carried a large caudal spine. Only one order is still represented today, the HORSESHOE CRABS (order Xiphosura).

The horseshoe crabs are the only living representatives of a group with a long geological history. Only a few species have survived until the present (BL including caudal spine up to 60 cm). In those marine habitats where they occur, however, they are present in extremely large numbers. Their body structure has many primitive characteristics, and they show many similarities with the basic chelicerate type.

In these large chelicerate animals of the present day, the anterior and posterior sections of the body each are covered with a uniform dorsal plate, and the caudal spine is long and powerful. The appendages of the anterior section are the same as the chelicerate basal type, the pedipalps and first to third walking legs bearing pincers. The posterior appendages, however, are modified, extremely compressed, and platelike; the first pair of appendages in this series (actually the eighth), which bears the openings of the reproductive organs dorsally, covers the remaining pairs, bearing well-developed gills, like a type of lid. Each of these gills consists of about 150 gill plates which lie one on top of the other like the pages of a book.

Horseshoe crabs live partly buried in sand or mud (Fig. 15-2). Digging into the substrate is done by the rounded anterior dorsal plate being driven forward and downward. The animals can also walk with their appendages, and even swim. These types of locomotion are commonly found in the young animals who swim, by rhythmic beating of the abdominal appendages, in an inverted position. All types of small marine animals serve as food (e.g., small fish, worms, and thin-shelled crustaceans), which are captured with the pincers and carried to the mouth. The mouth, owing to

▷
Above:
A pair of scorpions of the genus *Buthus* during courtship.
Below:
A female of the giant African scorpion (*Pandinus imperator*; see Color plate, p. 393) carries her young on her back.

▷▷
(Left from top to bottom):
The zebra spider (*Argiope lobata*) on her web with the "zig-zag" radial band. Species of the genus *Peucetia*.
The garden spider *Araneus diadematus* (see Color plates, pp. 399 and 410), a female on her web.
Right (from top to bottom):
Wasp spider (*Argiope bruennichi*; see Color plate, p. 399), a female in its preferred spot in the center of the web, which it rarely leaves.
Araneid from Mombasa.
The wolf spider (*Pisaura mirabilis*; see Color plate, p. 399) carries her egg cocoon in the chelicerae; before the eggs hatch, she spins a web around it, within which she guards the young.

▷▷▷
Araneus diadematus (see Color plates, p. 399 and 407), a garden spider from Tenerife.

▷▷▷▷
Left (from top to bottom):
Araneus quadratus lives in reeds in marshy areas where it inhabits a zone about knee-high from the ground (Greenland and Iceland, to Japan). Adapted in color to the background, the hunting
(Continued on p. 411)

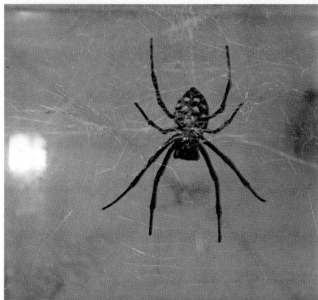

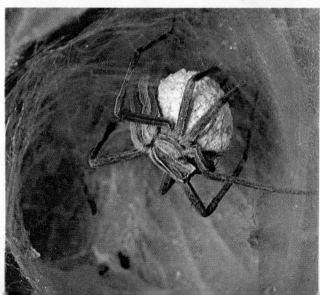

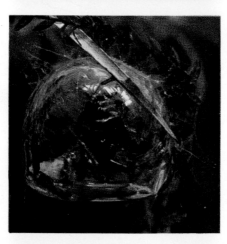

(Continued from p. 406)
spider *Micrommata rosea* (see Color plate, p. 400) lies in wait for its prey. The air bubble trapped by the hairs on the abdomen of the water spider (*Argyroneta aquatica*) must be refilled at the surface. The cocoon of *Agrocea brunnea*, which is attached to twigs, is called a "fairy lantern."
Center (from top to bottom):
Thomisus onustus captures insects on flowers. Because of the pointed processes on their abdomens, orb spiders of the tropical genus *Gasteracantha* have been called spined or thorny spiders. The water spider (*Argyroneta aquatica*) builds an underwater bell web and fills it with air, using this as its home. Young orb spiders often sit in masses on stalks and twigs.
Right (from above to below):
The male of *Eresus cinnaberinus* (see Color plate, p. 415) searching for a female (black) on the ground. The oak-leaf spider (*Araneus ceropegius*). The lipistiids, like the trap-door spiders, live in holes in the ground, the entrances of which are covered with a well-camouflaged lid. Trip-lines extending from this alert the spider. To prevent the lid closing as the spider dashes out, it always holds one leg underneath it. *Tetragnatha extensa* in cryptic position.

a curious mode of growth, is situated between the point of attachment of the anterior pair of legs.

The dorsal plate of the anterior part of the body has, on each side, a large compound eye, consisting of numerous simple eyes compressed together. Apart from this, a pair of simple median eyes is also present, lying close together medially at the anterior of the back; they are very small and easily overlooked. Little is known regarding other sensory organs.

For copulation and oviposition, the horseshoe crabs migrate in large numbers to the shallow coastal waters, resulting in massive numbers collecting in one place. The male clambers onto the female's back and holds fast by clutching the sides of the dorsal plate. Within the tidal zone, the female digs holes up to 15 cm deep, within which it lays up to 1000 eggs; these are fertilized by the male's sperm which is released at the same time. The larvae hatch after about six weeks, the mature animal developing from them. In the eastern American species *Limulus polyphemus*, it has been shown that at least sixteen ecdyses occur before sexual maturity, which comes between the ninth and twelfth year.

The present-day distribution of the horseshoe crabs appears disjointed. The genus of KING CRABS (*Limulus*), with the single species *Limulus polyphemus*, lives on the eastern coast of North America. All other representatives (genera *Trachypheus* and *Carcinoscorpinus*, which contain four species), in contrast, are restricted to the marine environment of southeastern Asia.

The SPIDERLIKE ANIMALS (class Arachnida) are terrestrial chelicerates which have gone over to air-breathing. Their respiratory organs indicate evolutionary origins in the "book gills" of the merostomates, which consist of numerous plates. By invagination, paired "lung-books" have been developed, exposed to the external environment only by paired slits on the ventral surface of the abdomen. The well-developed abdominal appendages present in the horseshoe crabs are extremely reduced in the arachnids; they are recognizable only in the case of the scorpions and spiders, although greatly modified.

Within the arachnids, at least nine orders must be recognized: 1. Scorpions, 2. Whip scorpions, 3. Palpigrades, 4. Spiders, 5. Ricinulids, 6. False scorpions, 7. Solipugids, 8. Phalangids, and 9. Mites. The number of recognized species lies in the range of about 50,000.

From a certain point of view, especially with respect to the body segmentation, the SCORPIONS (order Scorpiones) can be considered as relatively primitive arachnids. They are therefore the first to be considered here. The anterior section of the body is covered with a uniform dorsal plate. The attached posterior section consists of free segments, the last five being reduced in size to form a "tail." At the end of this, comparable to the caudal spine of the merostomates, is a bladderlike appendage extended at its tip to form a stinging spine (telson). The paired first appendages (chelicerae) are trisegmental but short; the pedipalps are strongly developed with pincers at their tips; the walking legs show no special modifications. The appendages of the rump on the ninth segment are modified to form paired, white attachments ("combs"). These are covered with numerous sensory cells which respond to mechanical stimuli; their biological significance is still not clear. The next four segments (the tenth to thirteenth) bear paired respiratory slits leading to the lung-books. Eyes are present: the anterior dorsal shield bears large paired median eyes, and at the point of junction of the dorsal and lateral surfaces, a few (from two to five) smaller lateral eyes.

Scorpions, as their flattened body indicates, are inhabitants of all types of cracks. They live partly in natural spaces, such as cracks in the rocks or leaf axils, but are also capable of digging their own resting places, using the first to third walking legs. It is understandable, when their inconspicuous, nocturnal lives are taken into account, that they are now and again found in human habitations.

Their prey consists of beetles, cockroaches, and other arthropods, which are grasped by the pincers close to the body and often are carried straight to the chelicerae. The poison spine is usually brought into use only when the prey is relatively large or struggles too violently; the scorpion then curls the "tail" over the back and the sting is inserted one or more times. The chelicerae then rip the prey apart and carry the food to the oral cavity. The poison spine also plays a part in defense, but scorpions are by no means aggressive. The tail strikes out forward (see Color plate, p. 393) only when they are cornered. The poison varies in its effect. The majority of species, including the small members of the European genus *Euscorpius* (see Color plate, p. 393), are more or less harmless to humans; others produce painful stings, swelling, and sometimes fever as well; such symptoms usually disappear after one or two days. Other species, in comparison, are extremely dangerous; their sting can lead to death, especially for children. Strangely enough, these are nearly all scorpions inhabiting tropical zones, especially those of the north African and American desert regions. The poison from the SAHARA SCORPION (*Androctonus australis*) is comparable in strength to that of the cobra, and can kill a dog within seconds.

Copulation is preceded by courtship; as the female stands opposite the male, he grasps her pedipalps with his pincers, and walks around with her; this is termed the scorpion's "dance." After the performance of certain behavior patterns, which vary from species to species, the male deposits a packet of sperm (spermatophore) on the substrate. He then moves backward, pulling his partner with him until her sexual opening is above the spermatophore and takes it up. Scorpions are in part livebearers; the egg shell usually breaks immediately after "birth," releasing the fully developed young. The females carry the young on their backs for some time.

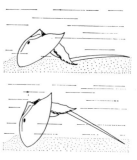

Fig. 15-2. *Limulus polyphemus.* Two possible body positions during digging into mud.

The scorpion's sting

"Maternal care" in
scorpions

The great entomologist, Jean-Henri Fabre (1823-1915) gave an especially beautiful description of maternal care in scorpions: one after the other they climb onto the mother's back. At first, they stroll leisurely along the pincers which the scorpion has turned sideways, making their climb easier. Then they sit crammed together on the mother's back. They can hold on relatively tightly with their tiny claws, and it is difficult to move them with a brush without damaging the delicate creatures. When Fabre approached the female's family too closely with a straw, she raised both pincers, ready to push the intruder away.

If the young are brushed off the mother's back, she then starts searching for them, and as soon as they are contacted, she allows them to climb back on her back again. Fabre observed that even "strange young were just as readily accepted as her own. One could almost say that the scorpion adopts them."

About 600 different species of scorpion are known. These are divided into four families and numerous genera. Only examples can be quoted here. The IMPERIAL SCORPION (*Pandinus imperator*; BL up to 8 cm; see Color plates, pp. 393 and 407), found in tropical western Africa, is one of the most impressive representatives. Other members of the genus *Pandinus* also occur in western Africa. Southern and southeastern Europe has several species of the genus *Euscorpius* (see Color plate, p. 393) which are difficult to distinguish from one another; their most northerly point of distribution is Krems, in Austria. These are all relatively small animals with very well-developed pedipalps and a conspicuously small, delicate tail.

Another group of similarly flattened crevice inhabitors is the WHIP SCORPIONS (order Pedipalpi). This means that the name of one of the groups of spiderlike animals and the scientific name for the second pairs of appendage (pedipalps) are identical, a fact which could sometimes cause confusion.

The following common features can be distinguished: the anterior and posterior parts of the body are separated from one another, in the uropygids by a constriction (reduced first abdominal ring); two—more rarely, one—pairs of lung-books. Pedipalps large and with a distal pincer or developed to form a "basket trap" (see Color plate, p. 416). The first pair of walking legs modified to form an antennalike organ of touch. Two main groups: suborder UROPYGI (BL up to 4.5 cm), 2. AMBLYPYGI (BL up to 4.5 cm).

The UROPYGIDS owe their name to a unique caudal attachment formed from the last three abdominal segments and a following segmented lash; the lash corresponds to the caudal spine of the horseshoe crabs and the last section of the scorpion "tail"—the poison gland and its spine. These animals are completely restricted to the tropics and subtropics. Copulation occurs in a similar way to that of the scorpions. Uropygids are able to defend themselves in two ways: they are able to pinch hard with the

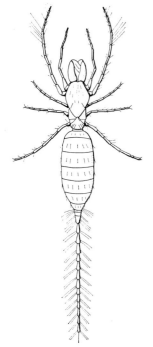

Fig. 15-3. The palpigrade *Koenenia mirabilis*. Division of the body anterior into proterosoma and two free segments, constriction of the last three metameres (segments 14–17) into rings.

pedipalps and also possess a means of "chemical warfare." Large, paired glands open on each side of the anus; uropygids are able to aim their secretion in the attacker's direction. In a carefully documented case, this substance was found to contain 84% acetic acid, 5% caprylic acid, 11% water. About 130 species are known, an example of which is *Typopeltis crucifer* (see Color plate, p. 416), from Formosa.

A completely different type of body form is found in the AMBLYPYGIDS. These, in contrast to the more elongate uropygids, have a rounded anterior and an elongated oval abdomen without a tail and poison glands. Their extremely long pedipalps are furnished with long, pointed spines (see Color plate, p. 416), forming a perfect "basket trap." The first pair of legs is modified to form an extremely long, delicately jointed antennal whip. These nocturnal crevice-inhabitors occur mainly in the tropics and subtropics; they are, however, also found in Crete and Israel. As in the uropygids, sperm transfer takes place by means of the male's spermatophore. They stimulate the female with the long, whiplike legs, and finally lead her toward the deposited sperm. About sixty different species have been described to date: *Damon medius*, from tropical western Africa, is used here as a typical representative.

The PALPIGRADES (order Palpigradi; L 2.8 mm) can follow on directly from the whip scorpions. These are minute, pale arachnids; the anterior section of the body is not covered with a uniform dorsal plate; the abdomen bears an elongated, multi-segmented lash. These delicate animals, distributed worldwide in warmer lands, live under sunken stones and in cracks with high air humidity. The first representative was discovered by Grassi in 1885 in Sicily; forty-six species have been described since then, one of which was even found in Innsbruck (Fig. 15-3).

The TRUE SPIDERS or WEB SPIDERS (order Araneae) are, without doubt, the best-known group of arachnids. With over 30,000 known species, they contain the second greatest variety of forms of all the chelicerates (next to the mites). Body length ranges from 9 cm (bird-eating spider) to 0.7 mm (smallest spider species). Anterior and posterior sections of the body separated by a deep constriction, the "pedicel," which corresponds to the first abdominal segment. Anterior section (cephalothorax) covered by uniform dorsal and ventral plates, with a group of eyes near the anterior border; the original eight eyes retained in part, sometimes reduced; true cave inhabitants completely eyeless. Abdomen more or less baglike, the segments no longer recognizable externally in mature individuals (exception: suborder Mesothelae). Chelicerae with only two joints, a basal segment, and opposable claw, with the openings of the poison glands slightly before the tips. Pedipalps in the female and immature male similar in structure to the walking legs, but shorter and more palplike. The four pairs of walking legs are supplements, especially in the web-spinning spiders, with special structures (e.g., the empodium found in the garden spider) associated with web spinning. The abdominal appendages

▷
True spiders (order Araneae):
1. *Hyptiotes paradoxus* (female above, natural size),
2. Ant spider (*Myrmarachne formicaria*; male below, natural size),
3. *Eresus cinnaberinus* (natural size to the right, female above, male below),
4. *Pholcus phalangioides* (female with eggs),
5. *Dysdera erythrina* (male; right, natural size).

Order: true spiders

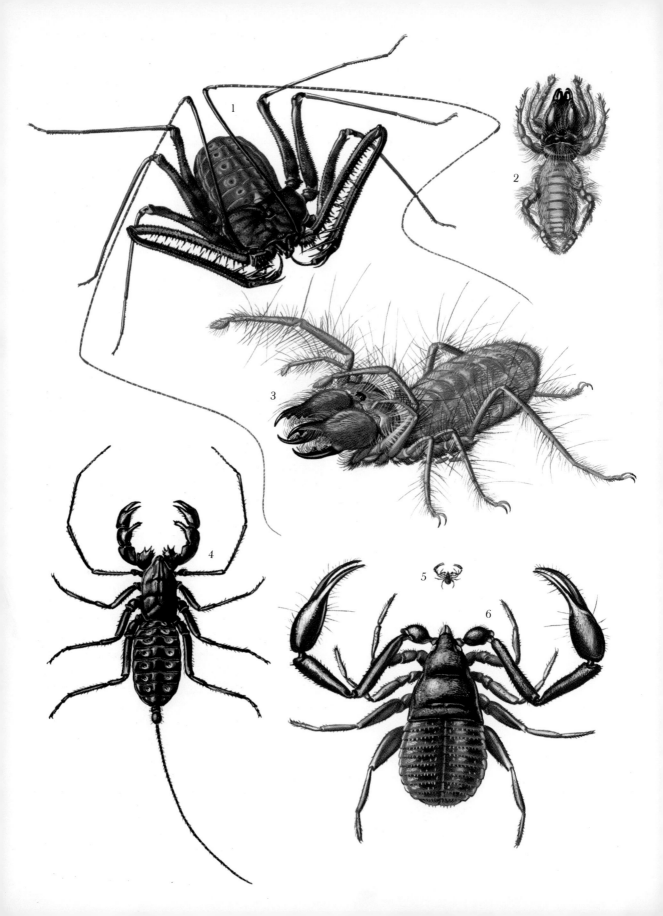

almost completely rudimentary; those of the second and third abdominal segments still in the form of lung-books (developed from the basal structure of the legs). The appendages of the fourth and fifth segments modified to form spinnerets and, due to change of position during embryology, displaced to the end of the body. All other appendages missing completely. The original segmentation of the abdomen is represented in many spiders by patterns on the dorsal surface, for example, by rows of spots.

All spiders are carnivorous. They almost always capture arthropods, especially insects; the large bird-eating spiders can, however, overcome lizards and other small vertebrates, as their name suggests. Only the representatives of certain groups build webs, that is, set traps. Numerous other forms wander about hunting or lie in wait for their prey at definite spots, for example on flowers. The wandering hunters often lay a line of silk behind them, extending it as they move and occasionally fastening it to the substrate. The formation of such a "lifeline" can easily be observed in wolf spiders. The true web-spinners, in contrast, always remain in one spot. In the simplest cases, they inhabit holes in the earth or cracks in walls, spinning threads which radiate from the opening of their tubes. If a prey animal stumbles across one of these "trip lines," the resultant vibration is transmitted to the silk-lined tube in such a way that the spider is able to determine the direction in which it must search for the prey (Fig. 15-4).

The "sheet net" is, among others, a development from this basic type of web. It is found, for example, among house spiders (e.g., genus *Tegenaria*; see Color plate, p. 399). This web, which is a source of irritation to all housewives, has a baglike living area in its farthest corner, corresponding to the previously mentioned living tube, in which the spider usually remains. This is attached to a sheet of spun silk which is so covered with trip lines that an insect can only stagger about on it.

Such types of web appear primitive when compared with those of spiders capable of producing sticky threads. In contrast to the previously mentioned trip lines, true "lime twigs" are produced in this case. There are two basically different types: the so-called cribellate spiders, the threads of whose webs produced an extremely fine "wool" which traps the prey, and acribellate spiders, which produce true adhesive threads. The cribellate spiders brush the wool from the previously spun threads by means of a special bristle structure on the fourth pair of legs. Electronmicroscopical investigations have shown that these threads contain a mass of extremely thin, loose, single threads, each with a thickness of about 0.000015 mm. They adhere—probably by surface friction—to the prey animal which, by its struggles, ensnares itself more and more in these insidious fibers. All other spiders capture their prey by means of sticky threads which—similar to a string of pearls—bear minute drops of adhesive fluid along their length.

It is impossible to describe here the great variety of webs. The web

◁
Arachnids (class Arachnida):
1. Whip scorpion (*Damon medius johnstoni*),
2. *Chelypus macronyx*,
3. Solpugid (*Solpuga lethalis*),
4. Long-tailed uropygid (*Typopeltis crucifer*),
5 and 6. Book scorpion (*Chelifer cancroides*; 5 natural size, 6 enlarged).

of the SHEET-WEB SPIDERS (family Linyphiidae) should be mentioned. This consists of strands extended in various directions, and usually hangs horizontally in bushes—for example, in fir woods in fall—and between shorter plants. The webs of the DOMED-WEB SPIDERS (family Theridiidae), to which the black widow (see Color plate, p. 399) belongs, are generally very loose and highly complex; usually only the underlying threads are adhesive. The peak of web development is found in the orb web. This type is spun by both cribellate and acribellate spiders. It is still a point of conjecture whether the orb web was "discovered" twice by the spiders, —developed independently each time—or whether its appearance in both groups is a sign of their development from a common ancestor. A complete orb web, as spun by the garden or orb spider (see Color plate, p. 399), originates firstly from a few basal threads arranged radially at more or less equal intervals. After the center has been strengthened by a thick mass of threads, the spider carries another thread in a wide spiral around the basal threads. This is only a temporary "lifeline," since a much tighter spiral, consisting of a sticky thread, is produced in the next step. The spider breaks the lifeline away at that time, and it is thus not part of the final web. The spiders often produce such webs anew each night. The material from the old web is eaten beforehand, so it is not lost to the animal.

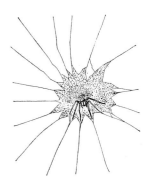

Fig. 15-4. *Segestria florentina.* The opening of the living tube is expanded and terminates in radial trip lines. The other end of the tube is used for escape. During the day, the spider remains deep within the tube, and at night it lies in wait near the opening.

Many people have an irrational fear of spiders. The reasons for this are unfounded, since only a very few species can be considered dangerous or even unpleasant to humans. By far the majority of spiders are not capable of even penetrating human skin. Even among the large bird-eating spiders, only a few forms are really dangerous; their poison is usually sufficient to kill only small vertebrates, up to about the size of a pigeon: usually the danger of blood poisoning is far more real than the effect of the spider's bite. Among the really dangerous spiders, apart from one single South American comb spider (genus *Phoneutria*), is the American genus *Loxosceles*, especially the black widow, since their poison glands contain a nerve poison. The males, owing to their small size, are harmless; the bite of the female, which is about the size of a pea, is extremely painful and, in a few cases, even deadly.

Copulation in the spiders is extremely unusual and even unique. The sexual openings in both sexes lie anteriorly in the ventral surface of the abdomen. In the male, the highly modified terminal joint of the first pair of legs (pedipalps) serves as an intromittent organ, first having to be filled with sperm. For this purpose, the male first spins a small "sperm net" and deposits a drop of sperm upon it so that this can be taken up by the copulatory organ. When the pedipalp has been loaded, the animal is ready to copulate. In the large tropical bird-eating spiders, this precursor of copulation lasts from three to four hours. Only afterward does the animal approach a female, which involves a great deal of difficulty. The prey-capture drive of the female sexual partner, which is usually the

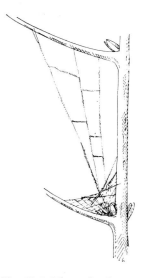

Fig. 15-5. The web of *Dictyna arundinacea.* The radial threads are 5–6 cm long.

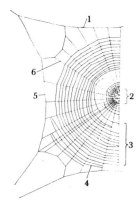

Fig. 15-6. An orb web (slightly diagrammatic), diameter 30 cm. 1. Upper edge thread (bridge), 2. Center, 3. Prey-capturing threads, 4. Sticky threads, 5. Edge threads, 6. Radial threads.

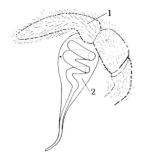

Fig. 15-7. End of the male pedipalp in *Segestria senoculata* with an especially simple copulatory apparatus (2) and its point of insertion on the tarsus (1).

larger of the two sexes, must be blocked and the animal "switched" to a copulatory mood. The male orb spider accomplishes this by beating a characteristic rhythm with its legs on the female's web. Other male spiders use visual signals; this, naturally, occurs only in species with relatively good vision. Waving and shivering movements, often associated with the presentation of a particular patterning, play a major role in these cases.

"Male wolf spiders wave at the female vigorously with both pedipalps, raise them, and at the same time perform a few dancing steps," states Wendt. True dances are performed by many male jumping spiders. These are very small, usually brightly colored animals whose mode of hunting is almost catlike. They stalk their prey carefully, spring onto its back, and kill it with a quick bite. When the male spider sees a female, however, it does not stalk it but stands high, with the legs extended. Usually it waves with the first pair of legs and moves in front of the female in this position, darting from side to side, until it finally stands face to face in front of her. If the female has remained standing quietly up to this point, the male can then "dare" to jump on her back and insert its sperm-filled pedipalp into her sexual opening (Fig. 15-8).

In a few species of spiders it is usual for the partners to grasp each other during copulation (Fig. 15-9). In the central European genus *Tetragnatha* (see Color plate, p. 410), the male places his large chelicerae at right angles to those of the female, whose chelicerae are thereby held wide apart. Because of this, the male, which maintains this hold throughout the whole copulation process, is protected against being bitten. The relationships are a little different in certain dwarf spiders (Erigonidae). Here the male bears bizarre processes in the optical region of the cephalothorax (e.g. *Walckenaera*; see Color plate, p. 400). In a few species, it has been observed that during copulation the female holds fast to these processes with the chelicerae. In one of the orb spiders of southern France, Grasshof has proved that the male, in comparison to those of other members of the group, is uncommonly small. During copulation, it jumps onto the female hanging in the middle of the web, similar to a trapeze artist. If the jump has been correctly aimed and is thus successful, it dashes its sperm-filled pedipalp into the complicated processes of the female's sexual opening. At the same time, the female buries her claws in the abdomen of her "suitor." Sperm transfer does not appear to be affected by the killing bite. Grasshof observed that, after about nine minutes, the process is completed and the female drags the male from her back to directly before the oral opening and proceeds to feed on it.

One of the hunting spiders, *Pisaura listeri* (see Color plate, p. 399), goes one step further. The male captures a fly, spins a cocoon around it, trots up to a female with it, and offers it to her as a "nuptial gift." This offering can last some time and is often repeated. The male, during this process, is extremely excited and takes up a grotesque position. *Pisaura* females which are not ready to copulate are not in the least impressed with this and even threaten their suitor and chase him away. When the fly is accepted—that is, the female has buried her claws in it—the male completes the copulation. The English arachnologist Bristowe observed that the male, after completing the copulation, often takes the nuptial gift back from the female.

On later oviposition, the female produces a silk cocoon which is usually suspended from or attached to some substrate. The wolf spider females (see Color plate, p. 399) actually carry the cocoon around attached to their abdomen. In spring, especially, these spiders with their white or greenish-blue cocoons can be seen on every stroll through the woods and fields, where they are found running along the ground or on plants. Shortly afterward, the newly hatched young can be seen crowded onto the mother's abdomen as they are carried about by her.

The suborder MESOTHELAE (BL 1–3.5 cm) includes the most primitive living spiders. *Liphistius* (see Color plate, p. 400) is the best-known genus. The abdomen always has free dorsal plates; the spinnerets have not yet reached the tip of the abdomen, instead lying centrally on its ventral surface. All nine species inhabit southeastern Asia.

The suborder of BIRD-EATING SPIDERS (Orthognatha; BL up to 9.5 cm) are large, thickly haired, brown to black spiders. Their scientific name alludes to the characteristic position of the chelicerae: as in the previously mentioned order, the basal joints are parallel to the longitudinal axis of the body and pointed forward. The claws lie next to one another and work independently. The abdomen is a simple sac; in the TRUE BIRD-EATING SPIDERS (family Aviculariidae) there is often a patch where hair is lacking. This "bald spot" comes about by the animal stroking itself with the hind legs; the hair breaks off and is thrown in the direction of an attacker. It has an extremely irritating effect on mucous membranes, especially those of the respiratory tract. Bird-eating spiders are found in all warm and hot countries.

By far the majority of spider species belong to the LABIDOGNATHIDS (suborder Labidognatha). In these, the basal joints of the chelicerae are directed downward; the claws work in opposition, and are thus arranged touching one another. Members of the family DYSDERIDAE (see Color plate, p. 415), especially species of the genus *Dysdera*, usually live under stones. Their chelicerae are elongated and the claws developed into long daggers. This is an adaptation to their mode of feeding: *Dysdera* captures and eats woodlice.

In central Europe, the family PHOLCIDAE (genus *Pholcus*) is represented by two species, one of them, *Pholcus phalangioides* (BL up to 11 mm; see Color plate, p. 415), illustrated here. These extremely long-legged animals usually hang upside-down on a delicate web. On disturbance or danger, the spider starts the web vibrating with its body and becomes difficult to see clearly. It is usually found in houses.

Fig. 15-8. The male jumping spider (*Attulus saltator*) performs a pirouette on the tips of its walking legs during courtship.

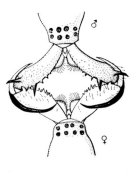

Fig. 15-9. Two *Tetragnatha* clasping one another during copulation.

The black widow

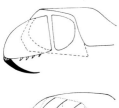

Fig. 15-10. Comparison between orthognathid (above) and labidognathid (below) chelicerae.

Family: dwarf spiders

Family: funnel spiders

Over 1300 species of the family THERIDIIDAE (BL 1–15 mm) are known. The famous BLACK WIDOW (*Latrodectus mactans*; see Color plate, p. 399) is a member of this group; this species, together with its close relatives, is very widely distributed and is also found in the Mediterranean area. At present, it is absent from central Europe and northern Eurasia. The true black widow (BL, ♀♀ up to 10 mm) is deep black, usually with red sickle-shaped patches on its abdomen; a red hour-glass figure is always present on the ventral surface of the abdomen, which allows the species to be easily recognized.

The ORB SPIDERS, in the true sense of the word (family Araneidae), consist of hundreds of species. The most important genus is *Araneus*, which includes the GARDEN SPIDER (*Araneus diademata*; BL, ♀♀ 17 mm; see Color plates, pp. 399, 408, and 409). During the day it remains in the center of its web. Other species build their hiding place near the web, maintaining contact with it by means of several "signal threads." The southern European WASP SPIDER (*Argiope bruennichi*; BL, ♀♀ up to 25 mm; see Color plates, pp. 399 and 408) is also a member of this group. It is immediately recognizable not only because of its characteristic coloration, but also by its web. It has a curious zig-zag ribbon of white silk. In warmer climes, and especially in the tropics, there are curiously horned orb spiders which have a bizarre kind of beauty (see Color plate, p. 410); the most important genus is *Gasteracantha*. Perhaps the formation of spines serves as a protection against birds. In the majority of spider species, the male is much smaller than the female, and the gasteracanthid males are especially minute, resulting in only the relatively conspicuous female being known for certain species; the corresponding male has yet to be discovered.

Walckenaera acuminata (BL up to 3.7 mm; see Color plate, p. 400) belongs to the DWARF SPIDERS (family Erigonidae), of which numerous species are known. The male's curious cephalothoracic process is used by the female to grapple onto during copulation. Many representatives of this group are able to "fly" for long distances by means of a silk-thread "kite"; they have even been known to cross oceans. These species are typical of the Indian summer since, during this time, the air is full of their gossammer, that is, with flying spiders. The family Tetragnathidae can be mentioned here, the most important genus of which is *Tetragnatha* (BL, ♀♀ up to 12 mm). They build horizontal orb webs, usually near water. The male's long chelicerae play a role during copulation (see Color plate, p. 400).

The HOUSE SPIDERS (genus *Tegenaria*) are members of the FUNNEL SPIDER group (family Agelenidae), the well-known house spider (*Tegenaria domestica*; BL, ♀♀ 14 mm; see Color plate, p. 399) being a typical representative. These are related to the WOLF SPIDERS (family Lycosidae), which consist of numerous species which live as wandering hunters, such as the genus *Pardosa* (BL up to about 8 mm). The southern European TARANTULA (genus *Lycosa*) is also a member of this group. There is a popular belief that the bite of this spider causes a "dancing frenzy"; this spider is, however, no more dangerous than other spiders of the same size.

The CRAB SPIDERS (family Thomisidae; BL 5–7 mm) do not build a web. They usually lie in wait for their prey on plants, especially flowers. The most important genera are *Xysticus* (see Color plate, p. 400) and *Ozyptila*. The prey is not chewed, but only bitten and the contents sucked out. The green *Micrommata* (see Color plates, pp. 400 and 410) belongs to the crab spiders, in the broader sense of the term. The JUMPING SPIDERS (family Salticidae) are extremely visual animals: they wander around hunting for their prey and leap on it. This family, which contains 2800 species with the widest variety of forms imaginable, is represented in Europe by the common *Salticus scenicus* (BL up to 6 mm; see Color plate, p. 399) in addition to many other species. The slender ANT SPIDER (*Myrmarachne formicaria*; see Color plate, p. 415), which is about the same size, occurs together with ants, whose shape it copies with amazing likeness.

All spiders which produce the extremely fine "wool" covering to the web, as mentioned previously, are put together in a further suborder, the CRIBELLATES (Cribellata). *Eresus cinnaberinus* (BL, ♀♀ up to 16 mm; see Color plates, pp. 410 and 415) will be taken here as an example. The blackish females live in central Europe in tubes in the earth in areas where the climate is favorable. The smaller males, which wander about, are conspicuous because of their bright coloration; when threatened, they make lashing movements with their abdomens. Some cribellates build orb webs. *Hyptiotes* (BL, ♀♀ 5 mm), however, produces only a sector of such a web; the means by which this web is secured is shown in the color plate on page 415. If prey flies into the web, *Hyptiotes* releases her hold and the trap collapses on the victim.

The rare, tropical RICINULIDS (order Ricinulei; BL up to 1 cm) owe their name to a broadening of the anterior edge of the cephalothoracic plate. A plate which is capable of being clapped forward and downward is present, covering the chelicerae. These strongly armored animals are also modified in other ways and appear to be unrelated to any of the other orders. Probably the best-known genus, *Cryptocellus*, whose name refers to its eyeless state, is an inhabitant of tropical America. The order contains about fifteen species.

The FALSE SCORPIONS (order Pseudoscorpiones or Chelonethi) are sometimes confused with the scorpions, but their similarity is only very superficial. Basically, it is only because the second pair of limbs (pedipalps) is, like that of the scorpions, furnished with a pair of pincers at the tips. As far as the rest is concerned, there are more dissimilarities than similarities between false scorpions and scorpions.

In these tiny, compressed animals (BL up to 7 mm, usually 2–4 mm) the cephalothorax with its uniform dorsal plate is attached to a definitely segmented, elongate oval abdomen; posteriorly, this is rounded and lacks

Family: crab spiders

Fig. 15-11. The black widow (*Latrodectus mactans tredecimguttatus*; female) of the Mediterranean area. The animals usually hang upside-down on their webs.

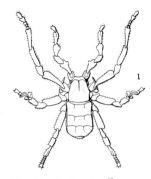

Fig. 15-12. *Cryptocellus simonis*. Male with copulatory organs on the third pair of walking legs (1).

a "tail" and sting. The third and fourth pairs of appendages are developed as normal walking legs. The respiratory organs are highly modified to form trachea (see Chapter 17), with paired openings (stigmata) on the posterior edges of the third and fourth abdominal segments. Eyes poorly developed or lacking entirely, in contrast to mechanoreceptors (simple sensory hairs, pit hairs).

The false scorpions also are, as their body structure indicates, inhabitants of cracks. They are found in all climates, usually under loose bark and stones, in leaf mold, in the dens of small mammals, bee hives, and even in the tidal region of the coast. A few species are occasionally found in houses; for example, the well-known and harmless BOOK SCORPION (*Chelifer cancroides*; see Color plate, p. 416) lives between old books and in similar places, feeding on booklice and other tiny creatures. *Lamprochernes nodosus* is also frequently found in human habitations: sexually mature females attach themselves to the legs of flies, which they do not eat but use as "helicopters." In this way, they ensure distribution of the species. Copulation occurs in a similar way as in the scorpions; the male deposits a stalked spermatophore which is taken up by the female. The actual transfer of sperm is preceded, in the various groups, by an extremely varied courtship: sometimes the male grasps its partner with the pincers so it looks as if the animals are "dancing"; in other cases, the animals only stand facing one another and the male attracts the female forward by means of special behavior patterns. To date about 1300 species are known, divided into three suborders, a number of families, and numerous genera.

Within the arachnids, the SOLIPUGIDS (order Solipuga; BL up to 7 cm) are in a curious position, since, apart from a few specializations, they show definitely primitive characteristics. The cephalothorax is not covered by a uniform plate; this is present only in the region of the first to fourth limbs (proterosoma); the two last segments of the cephalothorax have their own special dorsal plates (corresponding to the two posterior pairs of limbs). The abdomen is elongated and cylindrical, rounded posteriorly, and has separate segments. The chelicerae have two segments, in the form of enormously developed pincers (see Color plate, p. 416). The pedipalps are like walking legs, without claws. The third to sixth pairs of appendages (first to fourth pair of legs) are normally developed, with large, paired terminal claws. Respiration takes place by means of a well-developed tracheal system; no lung-books are present. The body, especially the limbs, is covered by numerous tactile hairs so that, as Kaestner says, the animal seems to "perceive with its whole body." A pair of large central eyes; lateral eyes, in contrast, represented only by rudiments.

Solipugids live in the deserts and plains of all warm countries with the exception of Australia, where, strangely enough, they are completely lacking. In Europe they are represented by the genus *Gluvia* (two species), found in Spain. Some solipugids hunt their prey by day (day solipugids);

The book scorpion

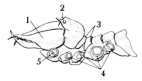

Fig. 15-13. The anterior of the body of *Galeodes graecus* (lateral view). 1. Chelicerae, 2. Eye, 3. Free cephalothoracic dorsal plate, 4. Point of attachment of the walking legs, 5. Point of attachment of the pedipalp.

the majority of species, however, are nocturnal (night solipogids). The females, especially, are voracious, capturing all kinds of arthropods, for example large grasshoppers, bettles, spiders, especially termites, and, in extreme cases, even small lizards. The powerful jaws chop, squash, and chew the victim, which finally looks like a formless lump. They are able to defend themselves vigorously, and can even give a man a bad bite; they do not, however, possess any poison.

Copulation, which has been investigated in only a few species, is especially curious. According to observations made by the Russian zoologist Birula, the male of *Galeodes*, when copulating with the female, really attacks her, jumping on her and holding her fast. In response to the male's hold, the female remains perfectly motionless. Without showing the least sign of resistance, she allows the male to drag her to a point which appears suitable to him and then turn her on her back. The male then widens the female's sexual openings with his oral appendages and releases a drop of spermal fluid which is immediately taken up in the chelicerae and literally pressed into the female's sexual openings. The widely spread edges of the opening are then pinched together by the male. Immediately afterward, he dashes off as the female "awakens" from the previous immobile state. This is because now she would only consider the male as prey, not a sexual object. Oviposition occurs later, either in natural cavities in the ground or in those dug by the female.

The systematics of the solipugids are extremely difficult and not fully explained in the majority of cases. This is especially true of generic differences and the distinguishing of species. About 840 "species" have been described. Only two especially important families can be considered here. The long-legged SOLPUGIDS (Solpugidae, with the main genus *Solpuga*; see Color plate, p. 416), mainly light brown, are distributed throughout northern Africa and Asia Minor. The RHAGODIDS (family Rhagodidae, most important genus *Rhagodes*) are found in approximately the same area. Their legs are much shorter and the last pair of walking legs has special digging processes; the anterior portion of the conspicuously cylindrical body is usually darker.

Of all the PHALANGIDS (order Phalangida), the naturalist probably knows the long-legged types (see Color plate, p. 400), which are found everywhere on bushes, plants, walls, and paths through the fields in late summer and fall. It is probably because of their long, thin legs that these animals have received the popular name of "tailors" in Germany. The group, which includes about 3200 species, also contains very different, more or less cryptic forms (see Color plate, p. 400), which are usually known only to experts in this field.

In the phalangids (BL up to 22 mm), the anterior and posterior sections of the body are joined together without any constriction between them. Cephalothorax with uniform dorsal plate which—except in a few cases—has a prominent lump bearing two eyes. The abdomen, in contrast, shows

▷
Above:
A tropical bird-eating spider of the genus *Lasiodora* shows, as do certain other genera, a "bald patch" on the normally thickly haired abdomen. On approaching an enemy, the spider strokes its hind legs across its abdomen, rubbing off the hair. These hairs can cause the enemy hours of violent skin irritation, especially if they reach the mucous membranes.
Below:
The "face" of the European tarantula (*Lycosa tarentula*). Contrary to superstition, the bite of this spider is completely harmless to humans.

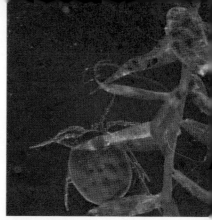

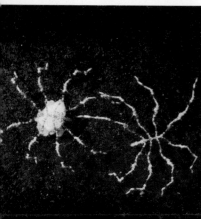

the usual segmentation with, in the obvious cases, dorsal and ventral plates. The oral appendages have three joints: the two joints above the basal one form a pincer which points downward. The pedipalps of almost all the European representatives are similar in structure to a short leg; certain cave-dwelling species and many tropical groups, however, have heavily spiked ones. The first to the fourth pairs of walking legs (third to sixth segmented appendages) are, in some cases, typical in length (the primitive condition); often, however, they are extremely long. In these cases foot joints (tarsi) are divided into numerous subsegments allowing them to grip grass stalks and stems.

Phalangids have evil-smelling odor glands, the openings of which are on the anterior corners of the dorsal plate covering the anterior of the body. In many long-legged species a type of "self-mutilation" is common: if an enemy, such as a bird, picks up a phalangid by one of its legs, the animal discards the limb by means of a special mechanism. While the phalangid escapes, the sacrificed leg remains behind and keeps wriggling for a while, drawing the enemy's attention.

All phalangids are carnivorous. Slugs and snails are the preferred food of some species (for example *Trogulus, Ischyropsalis*); other representatives capture mites, primitive insects, and similar creatures. Members of the mid-European genus *Phalangium* also feed on wind-fallen fruit such as apples and plums. Phalangids always take up their food in their oral appendages and pluck it to pieces there.

Reproduction occurs in a way completely different than that previously described for other arachnids. Instead of indirect transport of the sperm, a true copulation takes place. Usually the partners stand head to head. As a result, the sexual openings found on the ventral surface of each animal are relatively far away from one another; simple contact is not possible. It is therefore understandable that the openings leading from the male's sexual organs have been extended to form a long copulatory process; with this, the male is able to reach the female's sexual openings. Oviposition occurs by means of a tubelike ovipositor. With this, females of the *Phalangium* species, for example, are able to lay their eggs in cavities within the earth; other species use empty snail shells or attach actual clutches of eggs to the underside of stones.

Two of the three suborders are barely represented in central Europe. This is the case with respect to the relatively small, mitelike CYPHOPHTHALMIDS (Cyphophthalmi; BL up to 2 mm) which consists of about twenty-five species. These are blind, light-brown animals whose distribution areas include southern France, Corsica, and the Balkans, extending northward to the Polish Carpathians. The mainly tropical LANIATORIDS (suborder Laniotores), with about 1500 species, include extremely curious forms. The relatively rare European representatives are all small and, with a few exceptions, found only in the caves of the younger mountain ranges (Pyrenees, Alps, etc.).

◁
Left (from top to bottom):
Phalangid (Phalangida) feeding on a dead beetle.
Mite larvae using a phalangid as a means of transportation.
Mass infection of a red-spider mite (Tetranychidae). Pantopods appear to consist entirely of legs, hence their scientific name. Left, a male *Endeis panciporosa* with a ball of eggs.
Center, above:
The velvet spider mite (*Trombidium holosericeum*) lives in the upper layers of the earth and feeds mainly on insect eggs.
Right (above):
Fresh-water mite with a clutch of eggs.
Center (below): *Pycnogonum littorale* (see Color plate, p. 437) from the North Sea.
Below right: The tick *Aponomma gervaisi*, from the Indo-Australian region, usually attacks tortoises.

The most important suborder of the temperate zones is the PHALANGIDS (Palpatores), with about 1000 species. Probably the most common species in central Europe is the COMMON PHALANGID (*Phalangium opilio*; BL 3.5–7 mm), which is usually found along field paths in fall. The male (see Color plate, p. 400) is easily recognized because of its clublike processes to the oral appendages, directed upward. Members of the genus *Trogulus* (BL up to 22 mm; in European species, however, usually only up to 10 mm; see Color plate, p. 400), in contrast, are found in leaf mold and under stones. These are nocturnal and feed on small snails. In environments with a high air humidity, for example under fallen trees in untended forests, the relatively rare, heavily armored members of the genus *Ischyropsalis* (BL 5–7.5 mm; see Color plate, p. 400) can be found. Some species are able, with the aid of their large and strong chelicerae, to break open snail shells and pull the retreating animal out in bits.

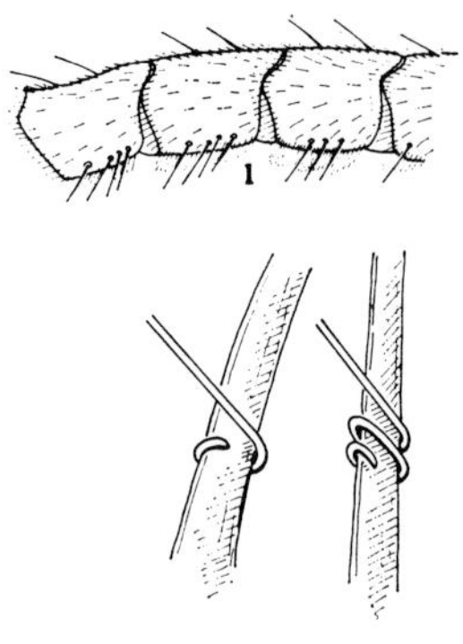

The MITES (order Acari or Acarina; BL 0.1–almost 30 mm, the latter applying to fully fed ticks), with about 10,000 species, belong to the arachnids with the most variable forms. The enormous variety of environments and types of life that these animals have adapted to has necessarily resulted in variations in form which no other group of arachnids has even partially managed to achieve. Mites alone have, during the course of their evolutionary development, become parasites on both plant and animal hosts—and this not only once but several times in different evolutionary lines. The mites have on many occasions relinquished a predatory life, which is so typical of arachnids; many feed on dead animals, plants, or moldering compounds; the meal mite (see Color plate, p. 437) is even capable of surviving on extremely dehydrated plant substances.

Body segmentation in the mites is basically different from that of other arachnids; three sections follow one another. The most anterior section (gnathosoma) is formed from the bases of the pedipalps which, together with a dorsal and ventral plate, form a uniform structure. The second (propodosma) is uniform, bearing the gnathosoma (with chelicerae and pedipalps) and the first and second pairs of legs. Attached to this at its broadest point is the posterior section of the body (hysterosoma); this consists of the body segments to which the third and fourth pairs of legs belong, together with the completely fused true abdomen; this is very much reduced, sacklike, and almost always unsegmented. The boundary between the middle and posterior sections of the body is clearly delineated (see Color plate, p. 437). Shape, structure of the appendages, etc., are so variable in the different groups that they must be referred to by examples. Only a few of the suborders can be described here, together with certain of their most important species as type specimens.

As in the true spiders, a colossal variation in types of reproductive behavior is also present in the mites. In certain of the water mites, the

Fig. 15-14. *Leiobunum rotundum*. The middle joints of the tarsus (above). 1. Section of the proximal tarsus. The multisegmented tarsi curl around a grass stalk like the tail of a New World monkey.

Order: mites

Reproductive behavior

male not only holds his partner by the tarsal claws during copulation, but actually glues her to him. Some male mites possess a copulatory organ, others transfer the sperm in the same way as the solpugids with the aid of their chelicerae, while yet others use one of the pairs of legs, in the same way as the crustaceans. The remainder use the simplest method: the animal deposits a spermatophore in the vicinity of the female, which she then finds and transports to her sexual openings.

Male ticks approach the female, which is attached to some warm-blooded animal, and creep under her body; the proboscis is then introduced into the female's sexual opening to enlarge it. The male then turns around, deposits its spermatophore, and pushes it, with the help of its proboscis and legs, into the female's sexual openings, as the male has no copulatory apparatus. The moth mite *Pyemotes herfsi* behaves even more curiously, from the human point of view: "Male moth mites play the role of 'midwives' so that they can copulate with the female as soon as she appears on life's stage," states Wendt. "This behavior is extremely practical. Female moth mites parasitize the caterpillars of small moths. They give birth to fully developed young, as do many parasitic arachnids and insects, the larval and pupal stages having taken place within the mother's body. If males are born, many of them remain around the mother's body opening, feeding on juices obtained by piercing her body with their mouthparts, and waiting for their sisters to be born."

The whole process is completed in the following way, as Kaestner describes: "If the anterior of a young female appears in the mother's birth canal, one of the males turns around, grasps his sister with his pincerlike hind legs, and pulls and levers her by releasing her body and then grasping it again, until he finally pulls her out. Such 'midwifery' is only shown with respect to the females, and certainly is not necessary, since many of the first born are females. The young female which has been drawn out is not released but is copulated with immediately."

**Suborder:
Trombidiformes**

The well-known red-spider mite belongs to the suborder TROMBIDIFORMES; many mite species belonging to the TETRANYCHIDS (family Tetranychidae; BL 0.26–0.5 mm) are also included in this broad category. These animals are parasitic, sometimes in large colonies, on plants which are often covered with their fine webs. The FRUIT-TREE RED-SPIDER MITE (*Panonychus ulmi*; BL 0.26–0.47 mm) can be taken as an example; this species attacks fruit trees and causes great damage, mainly by sucking out the contents of leaf cells. The TROMBIDIFORMS (family Trombidiidae) belong to the same suborder, *Trombidium holosericum* (BL up to 4 mm; see Color plate, p. 426) being the best-known example. These brilliant scarlet-colored animals are attracted out of the earth by the first rays of the spring sunshine, and wander about over the surface. *Trombidium* should not be confused with the equally red HARVEST MITE (*Trombicula autumnalis*; BL about 2 mm; see Color plate, p. 437), which is subterranean. The young individuals which appear in summer, sometimes in masses, are larvae and have only six legs; in contrast to the adults, they collect in moist areas on the surface of the earth and on grass. They mainly attack mammals, even humans, by attaching themselves to thin-skinned areas, boring in, and producing an extremely unpleasant skin irritation.

Various groups of trombidiformes have adapted to a life under water. This not only applies to the MARINE MITES (family Halacaridae), but also to the over 2400 species of fresh-water mites. The latter are not an evolutionarily uniform group and are grouped here only with respect to their similar mode of life. These animals are often brightly colored; for example, the THICK-LEGGED WATER MITE (*Unionicola crassipes*; BL 1–2 mm; see Color plate, p. 437). The legs of this greenish, predatory mite have, as do those of many other water mites, conspicuous swimming hairs.

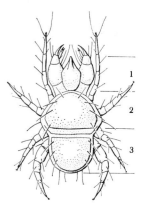

Fig. 15-15. *Cheyletus eruditus* (dorsal view). 1. Gnathosoma with chelicerae and the thick pedipalps of a predator, 2. Prepodosoma with walking legs I and II, 3. Hysterosoma with walking legs III and IV.

A further suborder, the SARCOPTIFORMES (Sarcoptiformes), includes a series of pests and nuisances worthy of mention: these include the ACARIDS (family Acaridae). The CHEESE MITE (*Tyrophagus casei*; BL 0.45–0.7 mm; see Color plate, p.437) lives on cheese but also feeds on smoked substances (ham, bacon, sausage, etc.). In the MEAL MITE (*Acarus siro*; BL 0.4–0.6 mm; see Color plate, p. 437), the male can be recognized by its first pair of legs, which is thickened at the base. This species can occur in great numbers in granaries, mills, and pantries, and is able to reproduce extremely rapidly under optimal conditions. With a decrease of the air humidity or other detrimental changes in environmental conditions, resting stages occur which are able to survive for two years or more. The HOUSE MITE (*Glyciphagus domesticus*; BL 0.3–0.75 mm; see Color plate, p. 437), in contrast, can be described only as a nuisance; a few other species lead similar lives. The animals develop in damp houses, especially in newly built ones which have not dried out properly. Groceries, wallpaper paste, and furniture stuffing such as mattress filling serve as food. These house mites cover furniture, plates, groceries, and other surfaces like a thick, crawling dust; their appearance in masses has, understandably, a very disgusting effect. The best means of combatting them is to dry out the affected rooms as quickly as possible, usually by means of heating.

The SARCOPTIDS (family Sarcoptidae) are also allied to this group. These roundish animals live as parasites in the skin of warm-blooded animals. If they attack hairless areas, the typical scabies symptoms appear; the same infection on hair-covered areas is termed mange. The HUMAN ITCH MITE (*Sarcoptes scabiei*; BL 0.18–0.4 mm; see Color plate, p. 437) bores perpendicularly into the skin and then makes tunnellike holes in the uppermost layer of living tissue, parallel to the surface. This produces a violent itching; lumps, scabs, and blisters appear on the infected areas, especially on the fingers, the folds of skin between them, the undersides of the wrists, the underarms, and other areas of soft skin.

The human itch mite

The TICKS (family Ixodidae) are probably the best-known representatives of the PARASITIC MITES (suborder Parasitiformes), often reaching a size which is unusual among the mites. All ticks feed on reptilian, avian,

Suborder: parasitic mites

and mammalian blood. Their oral appendages are adapted as boring and attachment organs. Often, the tick releases its hold only after days, when it is finally full of blood. Especially in tropical countries, ticks can be transmitters of dangerous diseases when they bite humans, sheep, cattle, and other domestic animals. The best-known central European tick, also distributed worldwide, is *Ixodes ricinus* (BL in the fully expanded ♀♀ up to 11 mm; see Color plate, p. 437). It is especially common in the damp underbrush of European woods. Only the adult female attacks various domestic and wild animals, including dogs and humans; adult males, in contrast, do not feed. The attacked skin area produces a burning irritation similar to the stings of some insects.

The PIGEON MITE (*Argas reflexus*: BL up to 4 mm; see Color plate, p. 437) belongs to the ARGASIDS (family Argasidae) and can occur in large numbers, especially in dovecotes. Only starving animals sometimes attack humans; human blood, however, is literally poison to them, and they die after ingesting it, surviving barely nine days afterward. A person, after being bitten, usually has a skin infection, which takes a long time to heal, and general signs of illness.

The extremely curiously formed PANTOPODS (class Pantopoda) are exclusively marine. They are found in the most varied habitats from the coastal zones to the ocean's depths. Their external structure is so strange that they can hardly be considered relatives of the spiders; only detailed examination of their appendages, especially the structure of the first pair, termed chelicerae, indicates their true relationships.

The anterior of the body (prosoma) has a pronounced, proboscislike process in front, sometimes rod-shaped. The number of appendages varies—a unique case among the arthropods—some forms having four walking legs (see Color plate, p. 437), others having up to nine pairs. The legs are always thin, and in many cases extremely elongated and conspicuously narrow. If the first to third pairs of appendages are present, the first pair are characteristic chelicerae consisting of three or four segments. The second pair of appendages are pedipalplike and thus termed pedipalpi. The third pair are also palplike and serve the male as ovipositors. The fourth to seventh pairs of appendages are always present and correspond to the four pairs of walking legs in other arachnids, also being developed as walking legs; their number can be increased so that six pairs of walking legs can be present. The posterior portion of the body (opisthosoma) is reduced, and is recognizable only as a stumplike appendage.

The curious external structure corresponds to an equally unusual distribution of the internal organs: the minute abdomen offers practically no space, and even the anterior region of the body, owing to its sticklike form, can offer only an extremely restricted one. It should therefore come as no surprise that the gut sends paired baglike projections into the legs, and the reproductive organs also have similar distensions.

Class: Pantopoda

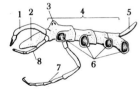

Fig. 15-16. Lateral view of *Heteronymphon kempi*, a pantopod. 1. Chelicerae, 2. Proboscis, 3. Eyes, 4. Prosoma, 5. Opisthosoma, 6. Point of attachment of the walking legs, 7. Ovigerous leg, 8. Pedipalp.

The relationships of the pantopods are—with the exception of the fact that they belong to the arachnids in general—not yet clear. An explanation of their curious arrangement of appendages must first be found. The anterior section of the body in all other arachnids bears only six pairs of appendages. When pantopods often have a larger number of pairs, this could mean that two pairs develop from a single embryological structure; on the other hand, it could mean that the posterior pairs of walking legs have developed from abdominal appendages. Apart from mechanoreceptors, pantopods usually possess four lensed eyes on the anterior section of the body.

About 500 species are known, the largest of which (BL 60 mm) reaches a diameter of about 50 mm; six families are distinguished. The COASTAL PANTOPODS (family Pycnogonidae) will be the only ones mentioned here. The SHORE PYCNOGONID (*Pycnogonum littorale*; BL up to 18 mm; see Color plates, pp. 426 and 437) is common in the North Sea coastal area. The females have only four pairs of legs; in the males, extra ovigerous legs are developed. The animals suck the juices of anemones (actinids). The BEAKED PANTOPOD (*Colossendeis proboscidea*; BL up to 50 mm; see Color plate, p. 437), which belongs to the family of DEEP-SEA PANTOPODS (Collossendeidae), lives in arctic waters; the animal illustrated here originates from the Bear Islands. In these forms, as in the shore pycnogonid, the oral appendages are reduced; the pedipalps and ovigerous legs are, in contrast, easily recognizable. Especially long-legged forms inhabit the deep oceans.

16   The Crustacea

Subphylum:
Diantennata, by
P. Rietschel

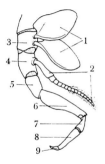

Fig. 16–1. Biramous
crustacean limb: 1. Gill
attachments (epipodite),
2. Exterior branch
(exopodite), 3. Proximal
joint (coxopodite), 4. Basal
joint (basipodite), 3 and 4:
Basal segments of the
biramous limb,
5. Ischiopodite,
6. Meropodite,
7. Carpopodite,
8. Propodite,
9. Dactylopodite,
5 to 7: Interior branch
(endopodite).

The evolution of the
Crustacea, by
E. Thenius

The Crustacea is the only class comprising the arthropod subphylum, the Diantennata. As the subphylum's name indicates, the class contains animals with two pairs of antennae, while in the subphylum Chelicerata, the first pair, and in the subphylum Tracheata, the second pair is missing. In addition, all Crustacea have biramous appendages; the basic plan is a proximal endopodite and a distal exopodite. Originally, all crustacean limbs, with the exception of the first antenna, were biramous; many investigators do not consider the first antenna as a true appendage but as a structure derived from the palpi of the annelidian prostomal ring. The crustacean body is divisible into head, thorax, and abdomen. The head consists of the antennal segments and, behind these, the first and second maxillary segments; there may also be one or two thoracic segments with maxillipeds attached to these. Biramous appendages used in locomotion are present on the thorax, and in some cases also on the abdomen. Another characteristic of the crustaceans is the presence of a larval stage with only three appendages, the two antennal pairs and a maxilla (nauplius larva). It therefore consists of part of the head; the two maxillae-bearing rings, the thorax, and the abdomen develop first during the course of further molts (metanauplius and further stages).

The Crustacea have been known from early in the earth's history and, even today, are a highly successful group. With their approximately 35,000 species, they contain four times as many species as the present-day birds. They originated in the sea, and the majority are still marine, but large numbers have invaded the fresh-water habitat and a few have even become terrestrial.

The plentiful fossil remains left by certain groups have enabled scientists to determine their fate accurately. These groups are the Conchostraca, Ostracoda, and Cirripedia among the "lower" Crustacea, and the Decapoda among the "higher" ones.

The Conchostraca, which were almost entirely fresh-water inhabitants, have been known since the Lower Paleozoic. Their carapaces, resembling bivalve molluscan shells, are especially common in the fresh-water deposits of the Carboniferous era: the genera *Leaia* and *Isaura* are commonly found in these deposits. The carapace of the Conchostraca has not changed since prehistoric times; therefore, present-day forms can correctly be termed "living fossils."

In contrast to these, the Ostracoda, which also have a bivalved dorsal carapace, show a tremendous variation in species and a true evolution extending from the Cambrian to the present. They are represented in the Cambrian by the genus *Bradorina*; they reached their evolutionary peak in the Lower Paleozoic with the LEPERDITIIDA (genus *Leperditia*) and in the Lower and Upper Paleozoic with the BEYRICHIIDA (genus *Beyrichia*). During these periods they comprised a large number of species, had an extremely sculptured carapace, and reached a length of almost 6 cm. These ostracods of prehistoric times were, until quite recently, considered to be completely extinct. Then, from New Zealand and the South Pacific Ocean regions surrounding it, living forms from the Tertiary and Quartenary periods (genus *Puncia*) were described; these are considered as the last living remnants of prehistoric Beyrichiida. At the start of the Mesozoic, the "modern" ostracods (Podocopida) replaced the Leperditiida and Beyrichiida; these are still common in fresh water.

Of the Cirripedia, only the carapaced forms have been handed down as fossils. Some lived as parasites; of these, only traces of the ACROTHORACICA and ASCOTHORACICA have been fossilized, in such forms as the remains of borings in mollusk and echinoderm shells or cysts in octocorallians (see Chapter 6). The geologically oldest member of the Thoracica is *Cyprilepas*, from the Silurian, a form from which, in certain respects, both the primitive LEPADOMORPHS as well as the more specialized BALANOPHORPHS could have arisen. The stalked forms of the Cirripedia arose much earlier than the stalkless ones. The inclusion of the above-mentioned Ascothoracica among the Cirripedia is not completely agreed upon, but they can be followed, together with the Acrothoracica and Thoracica, back to a common ancestor.

The "higher" Crustacea now contains the larger number of species. There are also many fossil remains of this group. The most primitive of the "higher" Crustacea are the PHYLLOCARIDA, which now are represented only by the LEPTOSTRACA (e.g., genus *Nebalia*). There are fossil remains of the HYMENOSTRACA, with the genus *Hymenocaris*, from the Cambrian and Ordovician, and the CERATIOCARINA, with the genus *Ceratiocaris*, from the Ordovician to the Permian.

The EUMALACOSTRACA, apart from fossil groups which extended over long geological periods (e.g., SYNCARIDA, with the genus *Paleocaris* from the Carboniferous and Permian, together with the genera *Anaspides* and *Bathynella* from the most recent geological times), also includes highly specialized crustaceans, such as the EUCARIDA, the true crabs (BRACHYURA), which first appeared during the Jurassic. The DECAPODS—

also members of the Eucarida, the most highly developed forms, with the most species of this group—can be traced back to a common ancestor in the Permian; their oldest fossilized true representatives (the genera *Pemphix* and *Aeger*) originate in the Triassic. The crabs are modern (Brachyura), and are the group of Crustacea which contains the largest number of species; they first arose in the Jurassic. In contrast to the views of certain zoologists, they are considered here as arising from the PALINURIDS from the Triassic era. While crabs were rare during the Jurassic (genera *Eocarcinus*, *Pithonoton*), they were one of the most common crustaceans in the Tertiary deposits (e.g., the genera *Dromia*, *Calappa*, *Ranina*, *Cancer*, *Portunus*, and *Lobocarcinus*). The crabs are therefore the youngest, geologically speaking, of all the crustacean groups, and they are still in full evolutionary development.

The Lower Crustacea, by P. Rietschel

At a very early stage, those crustaceans with a uniform number of body segments were grouped together and termed the "Higher Crustacea" (Malacostraca). Even Aristotle so termed them, separating them from the hard-shelled mollusks, snails, and sea urchins. All the other crustaceans, whose body segments were not so uniform in number—about half of the 35,000 species—were termed the "Lower Crustacea" (Entomostraca = "those with the indented shells"). It was soon clear that the "Lower Crustacea," in contrast to the "Higher Crustacea," were not a related group among themselves; this is why nine subclasses of the Lower Crustacea are distinguished, each of which is placed at the same level of classification as the tenth subclass, the Higher Crustacea.

The subclass CEPHALOCARIDA, with only four known species (L up to 3 mm), has survived to the present from the gray mists of antiquity. Its nearest relative is *Lepidocaris rhyniensis* (L 3 mm), known from the freshwater deposits of the Devonian period and placed in its own subclass, the LIPOSTRACA. The present-day cephalocarids inhabit the sea bottom in coastal regions. They have very primitive characteristics: the abdomen is very uniform and is divided into nineteen segments, of which the first seven bear the biramous appendages characteristic of the crustaceans. The next two pairs of appendages are much reduced, while the last ten posterior body segments bear no appendages at all. The biramous appendages are characterized by a leaflike exopodite held stiff by blood pressure, and a segmented, cylindrical endopodite which bears a claw at its tip. These legs therefore consolidate characteristics of the Anostraca and Phyllopoda with those of the other crustaceans. Even the appendage of the last head segment (the second maxilla) is similar to these abdominal appendages. A nauplius larva, typical for the crustaceans, hatches from the egg. As do the adults, the larvae live on the sea bottom and waft food particles toward the mouth with the aid of the outer branches of the second antennae. With the endites of the same appendages they pass the food under the upper lip. During successive molts, the larva develops additional body segments until their number is complete after the tenth molt: only after the eighteenth molt, however, does the animal attain the full number of appendages, becoming sexually mature at the same time.

Another extremely ancient subclass of the Crustacea is ANOSTRACA (= shell-less). As in the cephalocarids, but in contrast to the other Crustacea, the carapace, formed by the posterior head segments and anterior thoracic ones, is absent. Therefore, all the appendages lie free laterally. The abdomen is also divided into a large number of identical segments, and the thoracic legs are all very similar to one another. In addition, the nine or ten abdominal segments bear no appendages. The legs are all of the phyllopod structure and have only one segment that aids in movement. The external surface bears a leaflike gill attachment, while the internal surface is covered with bristles. These bristles meet to form a filter network in the central spaces on the underside. The leg movements produce a water current within this central space, which flows mouthward, carrying with it the plankton trapped in the filter network. At the same time, the leg movements ensure the continual presence of fresh water for breathing. The phyllopodia of the Anostraca thus do not seem to give the impression of being a primitive characteristic; they are far more highly developed, multifunctional structures. In contrast to the blind cephalocarids, the anostracids have well-developed, stalked, lateral, compound eyes.

The Anostraca swim upside down through the water, enabling the observer to see the impressive, wavelike movements of their phyllopodia. They look extremely strange and are undoubtedly survivors of prehistoric times which seem not to "want" to adapt to the present-day world. These slowly cruising and defenseless animals would long ago have been wiped out by bony fish and insect larvae had they not found a retreat into which these "modern" hunters could not penetrate. This is swiftly running water arising in springtime with the melting of snow on the mountains. Of course, the length of life of the Anostraca is, under such conditions, very short, for the ground absorbs the water soon after the snow melts. The animals have developed three ways of combating these conditions: their eggs are able to survive in dried mud for years without damage; in wet mud, they start developing at very low temperatures; and at higher temperatures the time needed for development is extremely short. If one wishes to see the slowly swimming, beautifully orange-and-turquoise-colored *Chirocephalus grubei* (L up to 28 mm) in woodland pools, one must therefore correctly judge the time of their appearance; even then, one may often look for them in vain. *Branchipus stagnalis* (L up to 23 mm; see Color plate, p. 451) can be found in short-term waterstands in open land until September.

In salty lakes and saltpans, the BRINE SHRIMP (*Artemia salina*; L up to 15 mm) has also found a safe retreat from enemies. This habitat is even more astounding when one realizes that this species has arisen not from marine ancestors but from fresh-water ones; its relatives, which have

Subclass: Anostraca

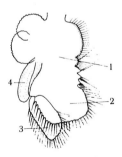

Fig. 16–2. Anostracan phyllopodium (*Chirocephalus*): 1. Base of the biramous limb, 2. Endopodite, 3. Exopodite, 4. Gills.

▷
Mites: order Acari:
1. *Panonychus ulmi*,
2. *Trombicula autumnalis*,
3. Harvest mites (*Sarcoptes scabiei*), 4. Cheese mite (*Tyrophagus casei*),
5. House mite (*Glyciphagus domesticus*), 6. Thick-legged water mite (*Unionicola crassipes*), 7. Meak mite (*Acarus siro*), 8. *Ixodes ricinus*, on the left after feeding, 9. Pigeon mite (*Argas reflexus*); Class Pantopoda: 10. *Pycnogonum littorale*, 11. *Collosendeis proboscidea*.

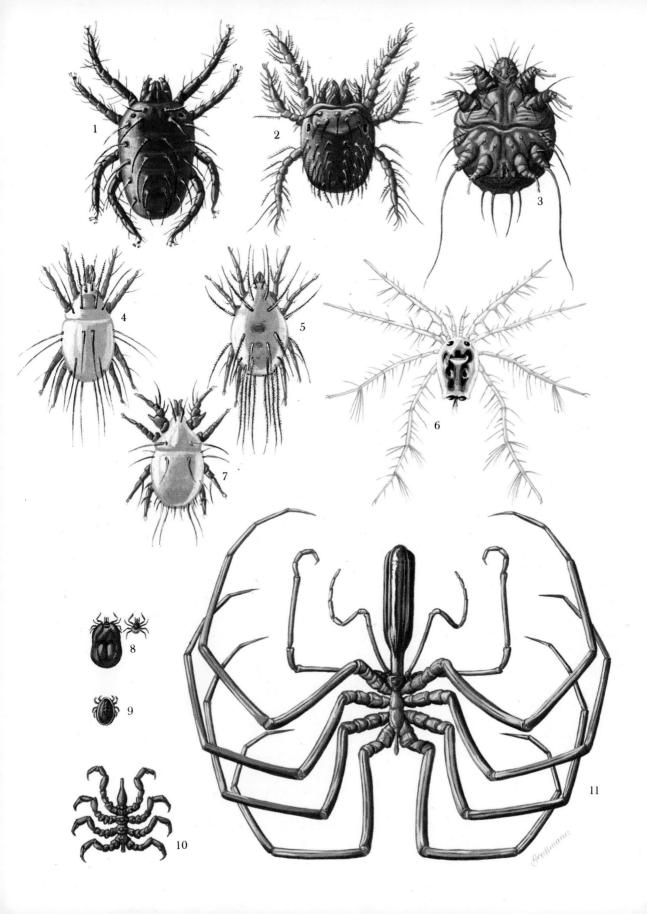

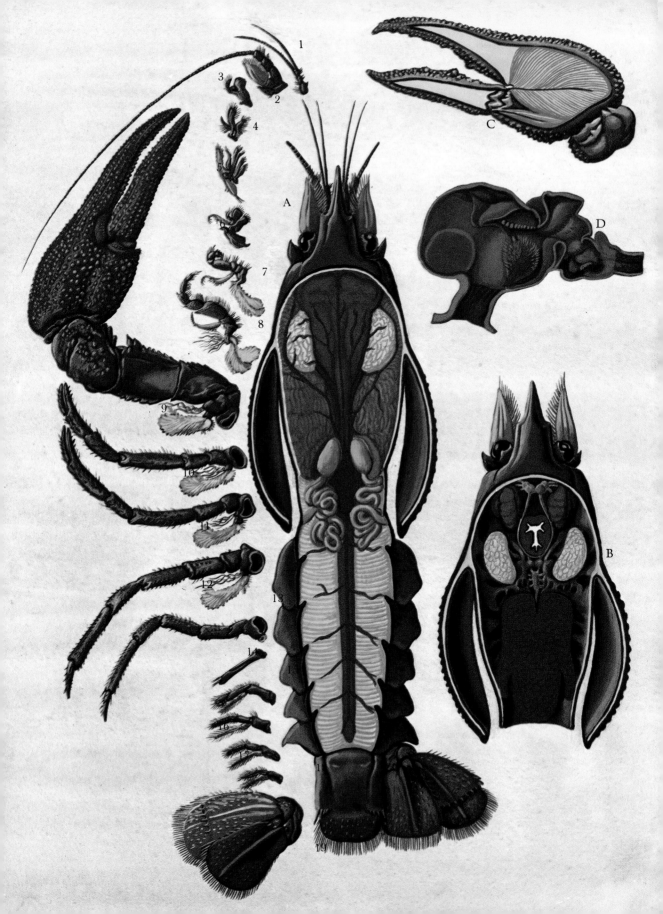

BODY STRUCTURE OF A DECAPOD (CRAYFISH)

A. Male crayfish, opened dorsally: appendages (1-19) present only on the left side and shown here detached from the body (basal joints and endopodite, dark; expodite, lighter in color). In counting from one to nineteen, the first antenna is considered as the first segmental appendage; in the text, the part of the head from which this arises is considered a true segment. The appendages and segmental nature of this part of the body is still a debatable point: if it is considered a development from the prostomial segment and tentacles of annelid ancestors, the number of true segments both here and in the text must be eighteen. The series of appendages: 1. First antenna (antennula), 2. Second antenna (antenna), endopodite with a long, whip-like process, exopodite a "scale," 3. Mandible with a chewing surface and palp, 4. First maxilla, 5. Second maxilla with respiratory plate (scaphognathite, exopodite) which, by continual scooping movements, draws the water within the gill chamber from back to front. 6-8: First to third maxillipeds, the second and third (7, 8) with gills, all with palps (exopodites). 9-13: The five walking legs, the first four with gills. 14-19: The six abdominal appendages (pleopods); in the male, the first two (14, 15) are modified to form a copulatory organ; in the female, the first pair of abdominal limbs is reduced, the following four (15-18) have long exopodites and endopodites and carry the eggs. The last (19) in both sexes is broadened to form a swimming fan (uropod) together with the telson. Section illustrations: Anterior of the head (protocephalon) with stalked eyes, both pairs of antennae (1, 2), and the base of the second antenna; ventral head area with the base of the mandibles (3-5), the maxillipeds (6-8), and walking legs (9-13), abdomen with the segments 14 to 19 and the telson. Internal organs (from anterior to posterior): in the middle, the gizzard, lateral to this, the strong muscles to the mandibles; next to these and behind, the many-branched mid-gut diverticulum; at the bottom of this, the paired lobes of the testes (the unpaired section covered by the heart); behind these, the coiled vas deferens, which opens in the basal segment of the last pair of walking legs; between the testes and vas deferens, the heart; on its upper surface, two ostia for incoming blood outside the vessel system, the arteries leaving the heart anteriorly and posteriorly. The posterior one, lengthwise along the gut, branches off to the tail muscles. Anus ventral (thus not visible) on the telson (the central lobe of the tail fan).

B. Female crayfish: Head and thoracic region (gnathothorax) opened from above, stomach, digestive diverticulum, and heart removed. The brain most anterior (supraesophageal ganglion), the two longitudinal nerve cords passing from this around the esophagus to the jaw ganglion (subesophageal ganglion), esophagus in horizontal section; on each side of this and above it, the "green glands," excretory organs (antennal glands, open on the second antenna); behind these, as in A, the strong muscles of the mandibles. Instead of the testes shown in A, here paired ovaries anteriorly, unpaired ones posteriorly, with two oviducts which open at the base of the third pair of walking legs.

C. Crayfish claw opened to show the strong closing muscles and the weaker opening muscles with their tendons.

D. Digestive canal from the mouth to the beginning of the hindgut opened longitudinally from the left: the esophagus arising from the mouth, stomach left with lenslike swellings of the lateral walls (gastroliths), a calcium store for the new carapace after each molt; right of this, the gastral teeth. Only the short, yellow section to the right (as the digestive glands which open into this) is formed from endoderm: esophagus, stomach, and hindgut are ectodermal, have a chitinous epidermis, and are molted together with the external skeleton.

Coloration in general the same as in the other illustrations; however, brown (the chitinous armor) = structures formed from the ectoderm; the exopodites of the appendages, lighter in color.

remained in fresh water, are extremely sensitive to changes in the salt content in their habitat. The brine shrimp's carapace is impervious to salt, but the shrimp cannot avoid taking up salt with its food; this continually penetrates the gut walls and enters the bloodstream. It is, however, continually excreted through the appendages of the first ten pairs of legs, and thereby rendered harmless. Calcium salts, however, do not enter the shrimp.

In contrast to the Anostraca, the brine shrimps are easily available to naturalists: in tropical-fish shops, their eggs can be bought by the thousand for very little money; many tropical-fish fanciers use them to produce live food for their animals. For this, a jam jar containing a solution of three to eight percent salt is prepared, to which a small quantity of eggs is added. The development from the hatching metanauplius larva to maturity and reproduction can then be followed. The larvae are fed with a solution of baker's yeast, and the growing and mature animals are fed with dry, powdered fishfood. A multitude of things can be observed: the hatching of the larvae, the increase in the number of their body segments and legs from molt to molt, copulation among the adults—in which the male, with its bizarrely formed second antenna, attaches itself to the female—the filtering action of the legs, and the animal's position in space. Gravity does not determine the animal's swimming position; instead, light does ("negative dorsal light response"). In addition, the brine shrimp is the master of variability. Depending on the salt content of the solution in which the animal is raised, a tremendous variety of forms can be produced, which were once considered to be separate species. Their characteristic differences, however, are not inheritable. The eggs used to raise these types must be obtained from Old World sources, for the American brine shrimps do not show this variability, and are often considered to be a separate species, *Artemia gracilis*.

In the subclass PHYLLOPODA, the range from primitive to highly developed forms is much greater than in the two previously mentioned subclasses. They obtained the name "Phyllopoda" because the majority of species obtain their food with the aid of their leaflike filtering legs; a minority, which live as hunters, have multisegmented, sticklike legs. In contrast to the members of the previously mentioned subclasses, the filterers among the Phyllopoda enclose their filter apparatus laterally by large folds in the skin behind the back of the head, the "carapace" or "shell." This can cover the whole body; in the hunters, however, it is reduced to a dorsal brood pouch.

The order NOTOSTRACA comprises primitive forms; their well-developed dorsal shell is formed from the extensive carapace and lateral folds of the head. The number of their body segments (annuli) and legs exceeds that of all their crustacean relatives: the forty or more annuli bear seventy pairs of legs. One pair of legs—as usual—is present on the first up to twelve pairs of annuli; the following annuli have an increase in leg

Subclass: Phyllopoda

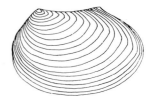

Fig. 16–3. Shell of a conchostracan.

number, and can have up to six pairs of legs. Correspondingly, the ventral muscles and nerve ganglia of the ventral nerve cord can increase in number in the posterior segments, but the dorsal muscles do not. This variation from the rules can be explained by the fact that, during development, incomplete division of the posterior of the abdomen into annuli occurs. Only a few of the last body segments lack appendages. Of what could the "Father of Systematics," Carl von Linné, have been thinking when he gave these animals the Greek name "apus," which means "footless"?

The Notostraca live mainly on the bottom of waterstands, where they crawl about with their backs toward the surface or bury themselves in the mud. Their body position is also regulated by the light received by the eyes, but in the opposite direction of the Anostraca ("positive dorsal light response"). They also feed with the help of their filtering phyllopodia, but in addition hunt insect larvae, annelids, frogs eggs, and even young tadpoles. In areas where they live together with anostracans, these shrimps form their main source of food. In detecting their prey, they use the first pair of legs, which bear long whiplike processes, and not the two pairs of small antennae hidden beneath the carapace.

Remains from prehistoric times are proof of the extreme age of the Notostraca. Slightly later fossils from the Frankish Keuper (Upper Triassic) are so exceptionally well preserved that they can even be identified as the present-day species *Triops cancriformis*, which has therefore survived unchanged for 200 million years. What sorts of changes has this species seen and survived in its environment during this timespan? As have the anostracans, they have survived everything in the "asylum" of short-term water spates and long periods of dormancy within the egg buried in dry mud. These veterans of earth's prehistory are often found together: *Triops cancriformis* (L up to 10 cm) with *Branchipus stagnalis*, *Lepidurus apus* (L up to 5 cm; see Color plate, p. 451) with *Chirocephalus grubei*. Even Goethe was so entranced by *Triops cancriformis* after he found it in the Jena area that he offered a taler—a silver coin—for the next individual to be discovered and a gulden for the second. Despite this attractive offer, Goethe was not able to obtain another of these fascinating animals.

While the posterior end of the Notostraca terminates in two long threads, that of the order Onychura terminates in a pair of strong claws. Its carapace covers the body not only as a posterior shield but also as a pair of lateral folds. Primitive forms from this order—members of the suborder Conchostraca—have also survived to the present in the asylum of the short-term water spates. Their name derives from their molluscan, shell-like carapace, the upper layer (cuticle) of which is not lost during successive molts (Fig. 16–3). Therefore, in the adult animal, the juvenile skins are arranged above one another and cover the largest and last skin in such a way that their free edges resemble the growth rings of a mollusk shell. As in the mollusks, the bivalved shell is attached by an adductor (closing) muscle. The large number of abdominal segments (thirty-six to forty) and the large number of leg pairs (ten to thirty-two), with which the animals filter out their food, are primitive characteristics.

To discover Conchostraca in the wild, a zoologist must have a great deal of luck; for geologists, on the other hand, their fossilized carapaces are commonplace, especially in Triassic and Jurassic deposits. The genus *Isaura*, earlier "*Estheria*," has been worldwide in distribution from the Devonian to the present. Of course, the bivalved carapaces of these animals, usually all that remains as fossils, do not offer many distinguishing characteristics. *Isaura* may therefore have passed through many changes of the internal organs during this time, even changes worthy of generic status. Prehistorically, the *Isaura* species, like the oldest known Conchostracan— *Rhabdostichus* from the Devonian era—were marine, while the modern representatives all live in fresh water. The laterally compressed *Limnadia lenticularis* (L 17 mm) tumbles about on the bottoms of shallow summer puddles, while *Lynceus brachyurus* (L up to 6.5 mm) swims upside down in the water. In *Limnadia*, only parthenogenetic females are known; in *Lynceus*, however, males are also present.

Following on from these honorable survivors of prehistory, the cephalocarids, the Anostraca, the Notostraca, and the Conchostraca, comes the army of "modern" waterfleas, with its numerous species, forming the suborder Cladocera of the order Onychura. Nearly all are smaller than the previously described remnants of earth's prehistory, and their exposed eyes are easily visible. Their jerky means of swimming resulted in their common name; they are well known to aquarists as a common source of live food for fish. Especially the large females of the genus *Daphnia* are used for this purpose; consequently, another common name, "daphnia," has come into use to describe all the waterfleas. Approximately 420 species have been described.

The head of the waterflea bears a cephalic shield and small first antennae with numerous sense organs. The second antennae, in contrast, are extremely large and, like all biramous appendages, consist of two branches beset with long swimming hairs. They serve as rowing organs; their beat causes the jerky swimming movements characteristic of the waterfleas. In addition, in the adults the head also bears both the compound eyes present in the larvae, and the original, paired, compressed, lateral eyes which join to form a single large eye. This process can be observed in young within the brood pouch, whose eyes, in the beginning, are still separate.

As highly developed phyllopods, the waterfleas have only a few abdominal segments with four to six pairs of legs. The posterior section, which lacks appendages and which bears two claws, is enclosed in the carapace and protrudes only occasionally, when the animal extends it. With the microscope one can see, within this structure, the hindgut and the anus, which is closed by circular muscles. The majority of waterfleas have phyllopodia which are furnished with gill processes and filter combs

▷
Copepods (subclass Copepoda) and fish lice (subclass Branchiura). Copepods:
1. *Calocalanus pavo* (order Calanoida; purely planktonic), 2 to 4: Copepods of inland waters:
2. *Diaptomus* sp. (order Calanoida; planktonic),
3. *Cyclops* sp. (order Podoplea; swimmer; see also Color plates, pp. 305/6 and 451), 4. *Canthocamptus* sp. (order Podoplea; crawler), 5 to 8: Parasitic copepods (females):
5. Sapphire shrimp (*Sapphirina fulgens*; order Podoplea), 6. Gill shrimp (*Ergasilus sieboldi*; order Podoplea), 7. Pike shrimp (*Lernaea esconia*; order Lernaeida), 8. Perch shrimp (*Achtheres percarum*; order Lernaeopodida), Fish lice:
9. Carp louse (*Argulus foliaceus*).

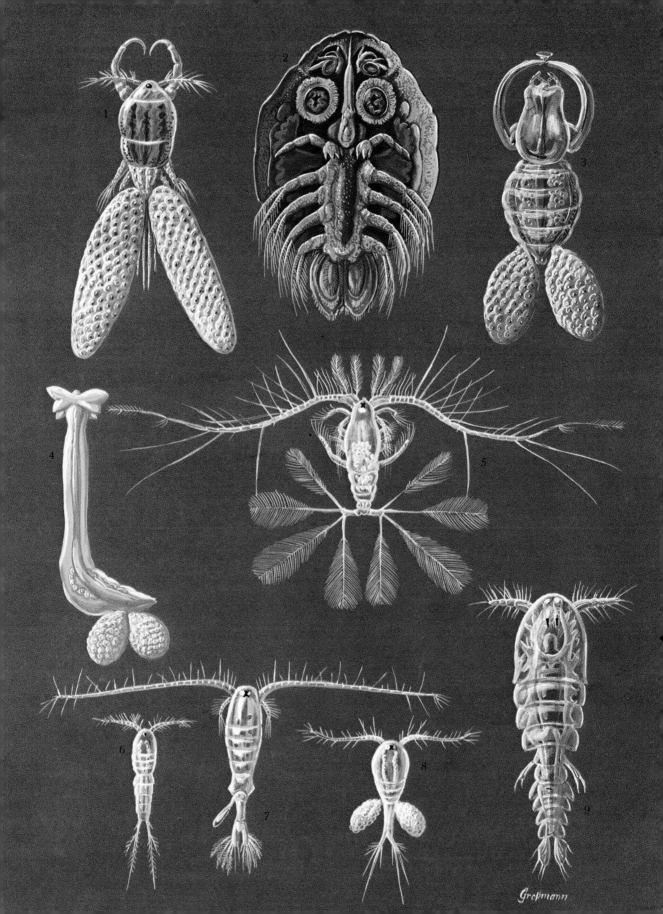

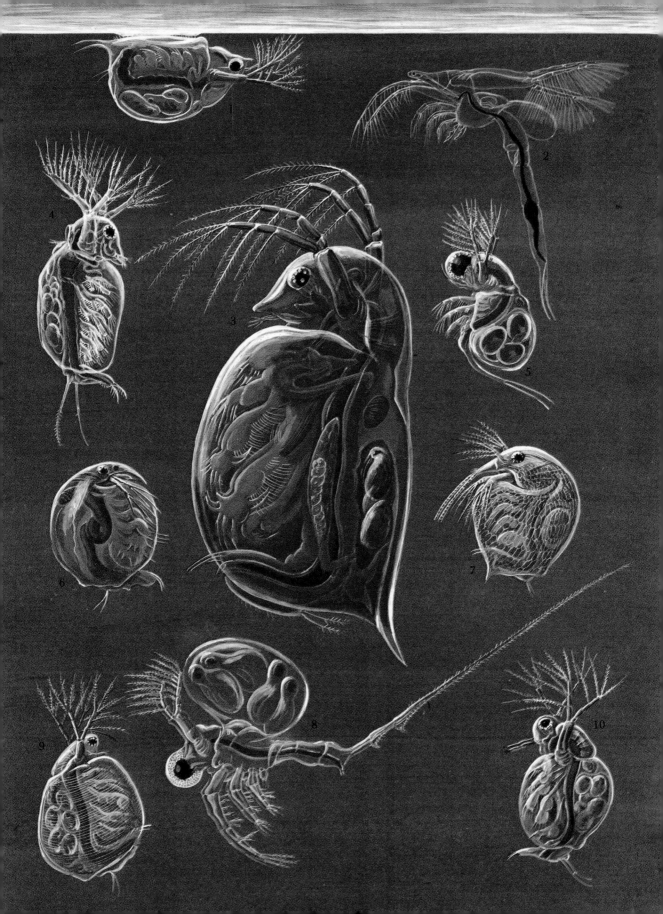

and with which they drive the water through the small space within the carapace. In this way, waterfleas filter algae out of the water and carry them toward the mouth; these give a green color to the gut and its attached hepatic diverticulum. In the anterior section of the carapace, the observer can see, with the help of the microscope, folds and an S-shaped, coiled "shell gland" which opens on the last head segment (the second maxillary segment) and is therefore also termed the "maxillary gland." This is the waterflea's excretory organ.

Seeing the beating heart through the microscope is always an impressive experience for the observer; how the heart's beat drives the colorless blood corpuscles through the spaces in the connective tissue of the shell folds and appendages is clearly visible. On the waterflea's dorsal surface, the carapace forms a "brood pouch"; the eggs and unhatched larvae are carried within this, and their development can easily be followed. In fall, however, the winter eggs, covered by a dark skin (ephippium), are found here. Under the microscope, the laterally compressed waterflea is usually visible only from the side. One gets to know the organism, however, only when it is visible from all sides. A large daphnia female which is allowed to swim freely in a small jar provides such a view when examined under a magnifying glass.

Anyone who takes water samples from the same place at different times of the year will be surprised by the waterfleas: there are changes not only the number of species represented, but also in the body form within a single species. These seasonal changes are termed "temporal variations"; many theories have been put forward regarding their cause and significance—especially since very similar examples can be observed in the totally different rotifers (see Chapter 10). Water temperature, light, and food play important roles in these variations. Another conspicuous characteristic in many species is that during spring and continuing into the summer only females are found; these reproduce parthenogenetically. Only later in the year do the small males appear, recognizable by their relatively large first antennae and lack of a brood pouch. At this time, normal reproduction and copulation take place; these eggs are winter eggs which, together with the protective dark skin of the brood pouch (ephippium = "saddle") are laid and first start their development in the following spring. Some species also produce such "resting eggs" during the spring; in the tropics these carry through the summer drought. In the short summers of the far north, however, where there is only time enough for one or, at the most, a few, reproductive cycles, the females hatching from the resting eggs can lay resting eggs without any previous copulation. Waterfleas of the genus *Moina* (see Color plate, p. 444) which inhabit temporary puddles in Europe produce males in the first or second generation, so that resting eggs are formed as quickly as possible.

Of the over eighty waterflea species endemic to Europe, only five are marine; all the others live in stagnant fresh water. A few examples of the

Annual metamorphosis in waterfleas

◁
Waterfleas:
1. *Scapholeberis mucronata*,
2. *Leptodora hyalina*,
3. *Daphnia pulex*, 4. *Sida crystallina*, 5. *Polyphemus pediculus*, 6. *Chydorus sphaericus*, 7. *Bosmina longirostris*, 8. *Bythotrephes longimanus*, 9. *Simocephalus vetulus*, 10. *Moina rectirostris*.

especially common or conspicuous species will serve to illustrate the variability of the organisms within this group of animals. The members of three families, with their six pairs of abdominal appendages and naked resting eggs (without a protective brood-pouch skin), exhibit primitive characteristics. These SIDIDAE, HOLOPEDIIDAE, and LEPTODORIDAE are put together in the family group CTENOPODS (superfamily Ctenopoda). The sidids are represented in Europe by three genera, each containing one species. *Sida crystallina* (L, ♀♀ up to 4 mm, ♂♂ 2.5 mm; see Color plate, p. 444) is a bank-living form which bears an attachment organ at the neck, with which it attaches itself to algae and other water plants. The transparent, yellow *Latona setifer* (L, ♀♀ up to 3 mm, ♂♂ 2 mm), with its beautiful red, blue, and brown patterning, lives on the bottom in mud and mold. It is easily recognizable by the apparently three-branched antennae and the bristled edges of its shell. Germany's third sidid, *Diaphanosoma brachyurum* (L, ♀♀ 1 mm, ♂♂ even smaller) lives free-swimming in clear water and has the capability, rare among animals, to make its weight the same as that of the surrounding water so that it floats completely without motion.

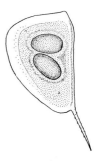

Fig. 16–4. Ephippium of the waterflea *Daphnia pulex*, with winter eggs.

Germany's largest waterflea, another member of the same superfamily, is the completely transparent *Leptodora hyalina* (L, ♀♀ up to 14 mm, ♂♂ 7–9 mm; see Color plate, p. 444), the only member of the family LEPTO-DORIDAE. Despite its size, the animal is difficult to identify while it is swimming wildly about in a jar containing other fresh-water planktonic animals; only the water current which it produces among the other organisms indicates that this species is present. Under the magnifying glass one can also see a fine black line—the gut—and a black spot—the eye. It is therefore not suprising that the large and beautiful *Leptodora* is only rarely discovered, although it is present in all types of inland waters such as the barren waters of Lake Constance, in the rich, brown waters of the Federsee in Upper Schwabia, and even in the waters of the River Main in Frankfurt's East Harbor. *Leptodora* is not a filterer but a hunter; therefore its thoracic legs are not phyllopodia but stenopodia or stick legs, and its carapace is reduced to a small brood pouch on the dorsal surface, leaving the long body and legs free. Another curious characteristic about this waterflea is that its larva hatches from the egg as a metanauplius.

Germany's largest waterflea

The third family of the ctenopods, the Holopediidae, consists of only one genus containing a single species: *Holopedium gibberum* (L, ♀♀ 2.5 mm, ♂♂ 1.5 mm). This strange waterflea, which lives in acidic moorish water, does not lose its old cuticles when it molts. Instead, these swell to form a jellylike mass which gives the originally laterally compressed animal an almost spherical form.

The following family includes the vast majority of waterfleas; they all have five pairs of thoracic legs and are placed together in the family group ANOMOPODA (superfamily Anomopoda = "those with legs departing from the norm"). Their legs do not actually do this, but, together with the carapace, they form a dense, complicated food filter. The family DAPH-

NIDAE contains all the large conspicuous species. *Daphnia magna* (L, ♀♀ up to 5 mm) is the largest; it is found in small, warm ponds and in water-storage tanks of plant nurseries. *Daphnia pulex* (L, ♀♀ up to 3.5 mm; see Color plate, p. 444) can be found among the plants in small ponds. In the same place, and quite often in company with the aforementioned species, but also in open water of the larger lakes, one can find the transparent *Daphnia longispina* (L, ♀♀ up to 2.5 mm). The even more transparent and smaller *Daphnia cucullata* (L, ♀♀ up to 2 mm), which is usually found together with the largest ctenopods of the genus *Diaptomus*, is a species restricted to lakes. This waterflea changes its form depending on the water in which it lives, but can be always distinguished from the other *Daphnia* species by the absence of a secondary eye.

The beautiful *Scapholebris mucronata* (L, ♀♀ up to 1 mm; see Color plate, p. 444) also belongs to the daphnia group. Its shell edge is free ventrally, and forms a straight line projecting backward into an equally straight spine covered with straight hairs. With this, the animal suspends itself from the underside of the water surface, along which it skims like a boat. Not to be confused with any other species, because of its pronounced forehead and tiny snout, is *Simocephalus vetulus* (*simocephalus* = "monkey-head"; L, ♀♀ up to 3 mm; see Color plate, p. 444). It is found in various inland water types and is conspicuous because of its quiet, non-jerky swimming movements. The previously mentioned genus *Moina* also belongs to this family.

Of the family Bosminidae, two species which are quite variable in form are found in Germany: *Bosmina coregoni* (L, ♀♀ up to 1 mm) and *Bosmina longirostris* (L, ♀♀ up to 0.6 mm; see Color plate, p. 444). These have only short rowing antennae, but their immovable first antennae are extremely long and situated in the head like a downward-pointing trunk. Waterfleas of the family MACROTHRICIDAE are less commonly encountered; these are characterized by an equally large but movable first antennal pair. The commonest species is *Ilyocryptus sordidus* (L, ♀♀ up to 1 mm), which is unable to swim and lives in mud. As in the case of the Conchostraca, this animal does not discard its old skins on molting, but retains them in layers over the new skin. The edges of these layers are recognizable by the delicate rows of hairs which once fringed the edges of the old carapaces. Finally, the family CHYDORIDAE, with its numerous species, should be mentioned. Its members carry the first antennae protected under a beak, and their gut is looped. The tiny, almost spherical *Chydorus sphaericus* (L, ♀♀ up to 0.5 mm; see Color plate, p. 444) is by far Germany's most common waterflea.

A "robber daphnia"

The "robber daphnia," *Leptodorus*, shows to what extent the entire body structure of a waterflea can change in the transition from a filtering to a hunting mode of life: the filtering phyllopodia become multisegmental stenopodia; the carapace which encloses the filtration system laterally is reduced to a dorsal brood pouch; the compound eye, which must detect the prey, increases in size and efficiency. The same changes can be found in the family group of the ONYCHOPODA (superfamily Onychopoda), whose original filtering ancestors are still unknown. Four pairs of completely exposed thoracic legs, which end in claws and bear no gill processes, are common to both. The family POLYPHEMIDAE has obtained its name owing to its large eyes: it is named after the Cyclopean giant, Polyphemus, of Greek mythology. These "robber daphnia" are most commonly found in European seas, where species of the genera *Podon* and *Evadne* live in the plankton. In inland waters, *Polyphemus pediculus* (L, ♀♀ up to 1.6 mm; see Color plate, p. 444) hunts in larger lakes, usually near the shores. Only in the free water of the larger and deeper lakes can one find *Bythotrephes longimanus* (L, ♀♀ without caudal spine 3 mm; see Color plate, p. 444), which has a caudal spine longer than its entire body. In Lake Constance, it reaches economic value occasionally as a main source of food for coregonids, endemic food fish.

Within the food chain, filtering waterfleas are important members of the so-called "first trophic level"; the second trophic level, especially fish larvae, feed on these. This role is especially important in stagnant inland waters. The resting eggs of waterfleas are easily transported by water birds, and they are able to populate any suitable environment rapidly. Storage tanks in plant nurseries, drainage pools in woods, bomb craters, and other small waterstands are ideal locations, but are becoming rarer because of the bad habit of using any place in which water collects as a dump for rubbish.

Subclass: Ostracoda

The subclass OSTRACODA's development extends far beyond the reduction of body segments and number of pairs of legs, as we have seen in the waterfleas. Their heads and thoraxes form such a uniform structure that for a long time the second pair of maxillipeds of the head was considered the first pair of thoracic legs. There is no visible division into segments even in the abdomen, which bears only two pairs of legs. Even these are absent in the genus *Polycope*, which lives in the interstitial spaces between sand grains on the sea's bottom. The whole animal is enclosed in two lateral shell valves to which it is attached by the posterior part of the head. At this point, the two valves can be closed by an adductor muscle working in opposition to an elastic tendon. The opening and closing of the two shells therefore occurs as in bivalve mollusks, and, as in these, the tendon can appear in conjunction with a closing joint with a longitudinal comb and longitudinal groove on the opposite side or with teeth and pits to receive them on the other side.

The deposition of calcium salts in the shell is also reminiscent of mollusks; therefore, the shells of ostracods, like those of the mollusks, have remained as fossils in deposits from previous eras. They can be found even as far back as the Upper Cambrian (about half a billion years ago). The ostracods achieved their first developmental peak in prehistoric times; during the Jurassic, they showed a second period of bloom. At the start of the Tertiary, this development became more precipitous, and extends

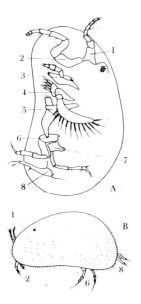

Fig. 16–5. Ostracod *Candona candida*: A. Left shell removed and only the left one of the paired appendages illustrated; B. External view: 1. First antenna, 2. Second antenna, 3. Mandible, 4. First maxilla, 5. Second maxilla, 6. First thoracic legs (crawling leg), 7. Second thoracic leg (cleaning leg), 8. Forked "tail" (furca).

to the present. The ostracods arose in the sea, and the majority of their species are still found there; in early prehistoric times, however, they invaded fresh water. The tiny, fossilized shell remains have recently become very important to geologists: since the ostracods speciated very rapidly during their period of "bloom," and since these species are typical of various epochs, they serve as excellent "type fossils" for the temporal ordering of prehistoric deposits. Earlier, the calcified hard parts of mollusks served the same purpose, but in bores and other rock samples, the shells of ostracods are present in much greater numbers. Only the equally tiny capsules of the unicellular Foraminifera and coccoliths are of such great aid in orientation to the geologist.

The number of species living today is estimated at about 12,000; of these, about 100 inhabit European inland waters. They are all small (L 0.5–2.6 mm), and even the largest living species, the marine *Gigantocypris agassizi*, is, at the most, 23 mm in length. In contrast, during prehistoric times one species, *Leperditia titanica*, reached a length of 58 mm.

The present-day ostracods are divided into two orders: the exclusively marine MYODOCOPIDA and the partly marine, partly fresh-water PODOCOPIDA. The myodocopids are characterized by a ventrally rounded shell, frequently with an anterior bulge from which the second antenna, used in locomotion, is the only appendage to protrude. Usually, they have compound eyes and a heart. In contrast, the shells of the podocopids have a straight or inwardly bulging edge and never have an opening for the second antenna. Compound eyes and a heart are lacking. They move and progress with the aid of both pairs of antennae.

This means of locomotion is unique among the animal kingdom, but, owing to the animal's extremely small size and the speed of movement of the antennae, it is extremely difficult to observe in detail. The beat of the two pairs of antennae is not uniform. The first antennae beat in a half-circle backward and move the animal in a curved path with a ventral mid-point. The second antennae beat in a half-circle ventrally, driving the animal in a curved path with a dorsal mid-point. Only when both beat together can a straight path be held. By stronger beats of the first or second pair, the animal can turn backward or forward. Ostracods live mainly on the bottom of waterstands, swimming only for short distances. They feed mainly on rotting plant material and dead animals. All that they leave of leaves which have sunk to the bottom of pools is a delicate skeletal structure of veins. In feeding, they usually employ their powerful mandibles.

The first maxillae are biramous, with a well-developed exopodite that forms a flattened gill surface. This moves continually, bringing the animal fresh water for respiration. The second maxillae can also bear such a flattened plate, but usually they are walking legs similar to the first pair of thoracic legs. With their aid, the animal can creep along the ground. The males, however, have powerful grasping teeth on their second maxillae, which they use to hold the female during copulation. It is rare to observe copulation in these animals, since it usually lasts only a very short time; in addition, in many European fresh-water ostracods, the male is rare or lacking entirely, so that the females reproduce mainly or entirely partheno-genetically. The last pair of legs in the family CYTHERIDAE is used for crawling. In the family CYPRIDIDAE, which includes most of Europe's endemic ostracods, it is used instead to clean the breathing plates and the interior of the shell, being modified for this purpose and therefore extremely mobile.

Within the folds of the ostracod shell are the paired hepatic caeca, the excretory organ, and the reproductive organs. Of all these organs, the male testes, when seen under the microscope, present a very unusual sight: the sperm cells are by far the largest of the whole animal kingdom; this, of course, is to the detriment of their number. Since they are several times longer than the entire animal, they lie between the two leaves of each shell in folds and can be pressed out by gentle pressure. A special pumplike organ serves to extrude these enormous cells. The female lays the fertilized eggs either singly or as a clutch on waterplants. From the egg, an already shelled nauplius larva, with uniramous appendages, hatches. In German species, this animal usually molts seven times. In contrast to the phyllopods, the sexually mature adult animal no longer molts.

It is very difficult to get to know the ostracods in detail, since they are so small. Apart from this, one must have a great deal of patience and delicacy to determine the various species under the microscope and to identify the various body segments. Therefore, these tiny animals have found few adherents among zoologists. Of the great number of European species, one is conspicuous because of its mode of life: Notodromas monacha (L 1.1 mm), the males of which are frequently found. While other ostracods in the aquarium creep along the bottom or swim close to it, this species swims at the water surface with its dorsal side uppermost, turning on its back, like the waterflea Scapholeberis, and grazing on the sheet of bacteria underneath the water surface. One point which makes working with ostracods easier is that they are quite simple to keep. One must only exclude their enemies—the larvae of demoiselle dragonflies, stoneflies, and also some mayflies—whose main source of food they are. In nature, the ostracods are of only secondary importance as a source of fish food; on the other hand, as devourers of rotting plant and animal tissue, they play a major role in the purification of stagnant waters.

Ostracods show no close relationship with any of the other crustacean subclasses. They form an independent branch of the Crustacea which has been in existence since prehistoric times. Nevertheless, this branch has survived successfully to the present.

In contrast to the uniform and isolated group of the ostracods, the following five subclasses of the Lower Crustacea are all closely related. During the course of evolution, their members achieved success in many

▷
Left, top to bottom: The anacostracan *Branchipus stagnalis* lives in fresh-water ponds. It swims with the ventral side up. Waterflea with an ephippium, a brood pouch separated from the rest of the carapace, in which the resistant eggs are laid. The large stomatopod *Squilla mantis* (see also Color plate, p. 459). The slipper lobster (*Scyllarus arctus*; see also Color plate, p. 459) with typically flattened segments of the second antenna.

Center, top to bottom: *Lepidurus apus*, a phyllopod, appears in short-term pools caused by melting snows in spring. Female of a copepod (*Cyclops*; see also Color plate, p. 443) bearing full egg sacs on both sides of the abdomen. A tropical stomatopod. The Norway or king lobster (*Nephrops norvegicus*; see also Color plates, pp. 459 and 460) is considered a delicacy.

Right, top to bottom: Waterflea (*Daphnia pulex*; see also Color plate, p. 444), with the brood pouch filled with eggs (left); a carnivorous waterflea (*Polyphemus pediculus*; see also Color plate, p. 444), with a compound eye (right). Ostracods (Ostracoda): The shrimp *Parapandalus narval* is often found in caves in the Mediterranean. *Stenopus hispidus*, a shrimp distributed throughout the Indo-Pacific, frequently acts as a cleaner.

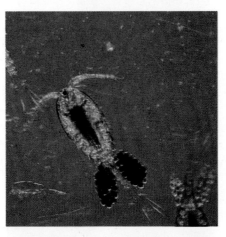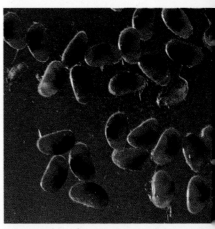
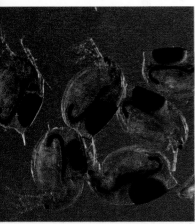

different ways, as swimmers and planktonic animals in free water, as wrigglers on the bottom, in the interstitial sand spaces, as parasites on and in animals of various classes, and as "growths" on other creatures or rocks. Mrs. Dahl grouped the Lower Crustacea under the name MAXILLOPODA.

Subclass: Copepoda

The center of interest of the Maxillopoda is the COPEPODS (subclass Copepoda). With their approximately 4000 species, they comprise the bulk of the Maxillopoda; their population density, however, is much more impressive. The marine biologist Sir Alister Hardy reckoned that, in number, they top that of all the animals on earth, including the trillions of insects. When the protozoans are excluded from this count, this scientist can be believed. It has been calculated that there are 4 million t of these minute and almost weightless creatures in the Gulf of Maine on the eastern coast of North America alone. From these numbers, one can imagine what an important role the copepods play in nature. They are the main converters of the plant material which reproduces and grows near the water surface under the influence of light, into animal protein. On the other hand, they form the main source of food for the large shoals of fish and are therefore the first link in the food chain, one whose absence is unthinkable. Copepods form the main component of plankton of inland waters; a single sweep of a plankton net through the water proves this. Here they fulfill the same task, as a link in a ramifying food chain.

Free-living copepods have a characteristic form: the externally unsegmented head does not terminate with the segment of the second maxilla but with the first thoracic segment, which bears a pair of uniramous maxillipeds. The remaining five thoracic segments each bear a pair of biramous appendages, and, together, form the externally segmented thorax. Only the fifth thoracic segment has a special role, since it grows together with the fourth abdominal segment (in the order Calanoida) or with the first (in the order Podoplea); in addition, its appendages, in the order Calanoida, differ in structure between the two sexes or, in the order Podoplea, are more or less degenerated. The trilobed eye of the nauplius larva is usually retained throughout life. The abdomen consists of five or fewer segments, lacks appendages, and ends in two forked, separated attachments (furca).

It is impossible to find a characteristic common not only to the free-living but to all the copepods, including the parasitic ones, since many of the parasites are so deviant from the norm and, in some cases, really bizarre in form. A relatively constant characteristic, however, is the "egg sacs" of the females: glands at the end of the oviduct secrete a sticky substance when the eggs are laid, surrounding them and thus gluing them together. The egg sacs formed in this way are, depending on the distance apart of the oviduct openings on the first abdominal segment (genital segment), either paired or joined together to form a single sac. Even the formless parasites have retained this characteristic of the copepods. Another relatively stable characteristic shared with the parasites is in the

◁
The European spiny lobster (*Palinurus vulgaris*; see also Color plate, p. 460) is, like the true lobster, a favorite food. In contrast to the latter, however, it has no pincers and bears numerous thorny processes on the carapace and the base of the second antennae. This is why it was given the name "spiny" lobster.
◁◁
Crayfish female carrying numerous young about on the ventral side of the abdomen.
Australian crayfish (*Cherax* species).
◁◁◁
The lobster (*Homarus gammarus*; see also Color plate. p. 460) comes out of its hiding place. The left one of the two large pincers is used to break open mollusk shells.
Australian crayfish (*Cherax* species).

course of development: following the nauplius and metanauplius stages is a "copepodid" stage. The copepodid is a copepod larva in which the separation of the abdomen from the thorax is already visible. In the monstrillids, which are parasitic as larvae, this stage is extremely modified; in this case, however, the non-parasitic adults reveal their relationship to the copepods.

The bulk of the copepods play the important role of "primary consumers" in the marine environment; the shoals of herring and mackerel follow on from them as "secondary consumers." These important shrimps are grouped together in the order CALANOIDA (Calanoida or Gymnoplea). Their small abdomens are clearly set off from the broad thoraxes; from this alone, they are easily identifiable from other copepods. A further characteristic is the presence of a heart. Their most conspicuous characteristic is, however, their long first antennae, which are spread laterally; the animals are suspended from these as from a parachute, and thus sink slowly: the calanoids are "planktonic." After having sunk for a short distance, the rowing legs of the thoracic segments start to beat, while the projecting antennae lose their stiffness. With the antennae close to the body, the animals climb upward in hops. The antennae then spread again, and the whole slow sinking process begins again. Because of their greater surface area, the feathery attachments to the antennae, legs, or abdomen increase the amount of friction generated by the body, and thus help to brake the rate of sinking.

The copepods' importance in the marine food chain

Among these planktonic organisms, *Calanus finmarchicus* (L up to 4.5 mm) is distributed worldwide, especially in the northern oceans, but also in the Mediterranean Sea and Indian Ocean. The marine calanoids drift over wide areas, followed by schools of fish which live on them; these, in turn, are followed by the fishing fleets. The planktonic organisms also drift perpendicularly: as dusk falls, they rise from the depths, and at dawn, they descend once more. *C. finmarchicus* is reputed to rise regularly 100 to 200 m to the surface and descend again into the depths, with the schools of fish following them.

Copepods of the order Calanoida are just as much at home in fresh water. *Diaptomus* species (see Color plate, p. 443) live in free water even in small ponds; *Heterocope weismanni* is found in Lake Constance and Starnbergersee, and, together with the waterflea *Bythotrephes*, is a main source of food for coregonid fish. The beautiful DIAPTOMIDS of European inland waters are easily recognizable by the way they float; the females are further characterized by the single egg sacs and the males by their right first antenna, which is bent and used as an attachment organ during copulation. The brilliant colors of these normally transparent animals can be seen through the microscope, shining red, orange-yellow, and blue. The colors are produced by oil droplets in the interior of the body. These droplets determine the animal's specific weight and thereby control its ability to float.

The PODOPLEANS (order Podoplea) are different in form: in these copepods, the wide head and thorax merge gradually into the abdomen so that they appear pear-shaped or cylindrical. Their first antennae are shorter than half the body length and are not used in floating, since these animals are not planktonic but partly swimmers and partly crawlers. In contrast to the calanoids, they have no heart. They also have numerous marine representatives, but the majority of the fresh-water copepods are members of this group. The swimmers almost all belong to the suborder CYCLOPOIDA, the crawlers, in contrast, to the suborder Harpacticoida.

While the free-swimming marine cyclopods belong to several families, our fresh-water forms all belong to the family CYCLOPIDAE and are members of the supergenus *Cyclops* (see Color plates, pp. 305/6 and 443), with several subgenera and numerous species. The diaptomids, often found together with these cyclopoids, are easily recognized by their completely different means of locomotion: their first antennae are not used as a parachute. They also progress with hopping movements of their rowing legs, but the slow sinking with antennae spread is not present between bouts of jerky swimming. While floaters obtain at least part of their food by filtering the fine plankton, members of the *Cyclops* group mainly eat decaying remains of plants and animals. Some species are also hunters, such as *Megacyclops viridis* (L up to 5 mm), one of Europe's largest cyclopids.

In the plankton of stagnant ponds, these small crustaceans are the most common of all animals, except the rotifers; the females are easily distinguishable by their paired egg sacs, the males, by their two bent first antennae. Apart from this, the naturalist also encounters large numbers of nauplius, metanauplius, and copepodid larvae—usually a rare occurrence: an animal species present in every one of its life stages at the same time and in the same place.

The bottom-dwelling HARPACTICOIDS (suborder Harpacticoida) are less conspicuous, and are often overlooked. Only a few marine species are planktonic as swimmers. The majority of marine species and all fresh-water inhabitants live on the bottom and move about mainly by crawling. Their bodies are therefore elongated and cylindrical; their abdomens are scarcely smaller than their cephalothoraxes. Among the harpacticoids as well, the marine species are divided into four families, while the freshwater species all belong to a single family, the CANTHOCAMPTIDAE. Many canthocamptids are also marine; the numerous fresh-water canthocamptids, which were once included in the single genus *Canthocamptus* (see Color plate, p. 443), are now separated into several genera.

With the parasitic copepods, we return to the suborder Cyclopoida; they have taken this path of parasitism several times and in various directions. The shrimp *Nicothoe astaci*, which lives on the gills of lobsters and spiny lobsters, still looks like a cyclopid. This also holds true for the

SAPPHIRINES (genus *Sapphirina*; see Color plate, p. 443); their females live in salps, while the males are still free-living. The color-play of light on these animals' external layer of cells gave them their name. The zoologist Karl Gegenbaur (1826–1903) called them "a sea-light in the daytime," since the individual flashes of color are sometimes sapphire blue, then gold-green, and then purple.

The parasitic copepod *Ergasilus sieboldi* (L 1–1.5 mm; see Color plate, p. 443), which is also recognizable as a cyclopid, lives in fresh water and very brackish water as well. The males are not parasitic, and the females become so only after copulation. From this point on, they swim against the current and in this way enter under the gill covers of fish and attach to their gills. Here they use their second antennae, transformed into large, three-jointed grasping claws, as hooks to hold fast. From this time on-ward, they feed on the epidermal cells of the gills. The fish is damaged not only by the wounds caused by the parasite, but also by the fungus which infests them later. Only fifty parasites on each side of the fish can reduce its oxygen uptake by two-thirds, but commonly, many more than this attack a single animal. The female copepodid seeks its host close to the bottom, so bottom-dwelling fish species are usually the ones to suffer the most severely from "gill shrimps," the slow-moving tench being the most common victim. Therefore, the disease is called, in marine fish, "tench disease." Pike are the second most commonly attacked, and then bream.

Myticola intestinalis (L up to 8 mm) has taken a bigger step into para-sitism. Originally it attacked only the Mediterranean mussel *Mytilus galloprovincialis*; but in 1939 it was found in the true mussel of the southern North Sea, probably carried there on the hulls of ships. This infestation caused severe damage to the mussel banks. The Dutch mussel harvest, as a result, sank to a tenth of its normal level between 1949 and 1950. The shrimp, which lives in the intestine of the mussel, speeds up its host's protein digestion, and at the same time increases the host's oxygen require-ments while it decreases feeding and respiration. Infected mussels are small and brown, in contrast to the yellow or orangish healthy animals. Since infection cannot be determined when the mussel is closed, infected animals get mixed with healthy ones during subsequent sale. Fresh mussels are sold even in winter, far from the sea, so a naturalist living inland also has the chance of encountering these parasitic copepods. Only its first larval stage is free-living in the sea; therefore it is possible to find seven further larval stages, as well as the adult, within the host mussel.

A completely different path to parasitism was taken by the MON-STRILLIDS (family Monstrillidae), also relatives of the *Cyclops*. Even the nauplius larva hatching from the egg is modified: it lacks a mouth and intestine; its maxilla, usually a biramous swimming appendage, is modified to form strong hooks. The nauplius is unable to swim with these; con-sequently, the French scientist Malaquin believes that the female *Haemocera*

▷
Crustacean larvae:
1. Metazoea larva of a porcelain crab (*Porcellana*), 2. Mysis stage of the Norway or king lobster (*Nephrops norvegicus*; see also Color plates, pp. 451 and 460), 3. Nauplius larva of an ostracod (*Cypris*), 4. "Pupa" of a goose barnacle (*Lepas*), 5. Metanauplius larva of an encrusting barnacle (*Balanus*), 6. Zoea larva of the heart crab (*Thia*), 7. Pseudozoea larva of a stomatopod (*Squilla*), 8. Magalopa larva of a true swimming crab (*Portunus*), 9. Zoea larva of a shrimp (*Hippolyte*), 10. Phyllosoma larva of a slipper lobster (*Scyllarus*), 11. Nauplius larva of a copepod (*Cyclops*).

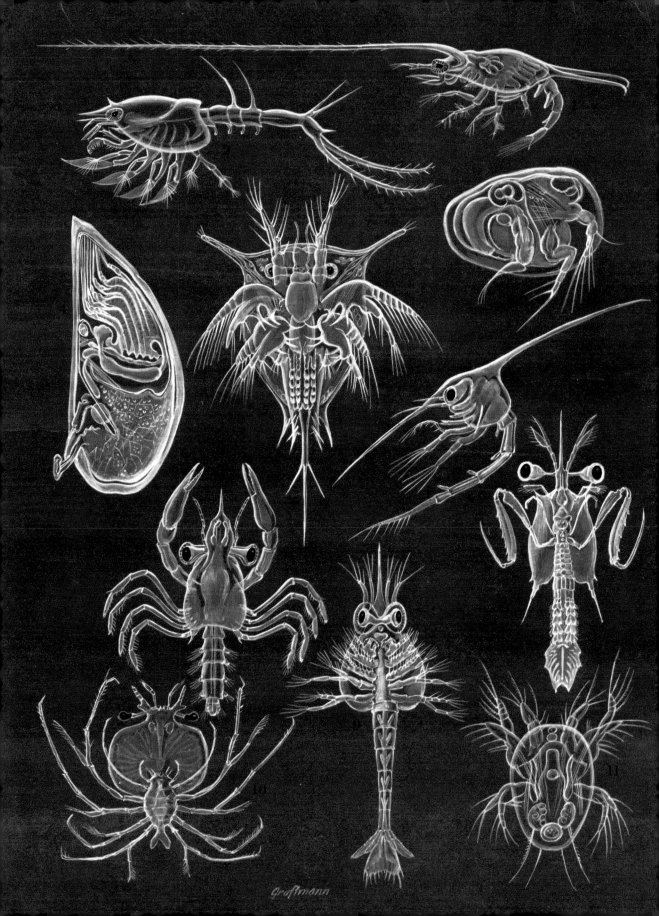

danae (also called *Cymbasoma rigidum*) purposely lays her eggs in a colony of the future host, the tubeworm *Salmacina dysteri*. The nauplius attaches itself with its hooks to the worm's skin, then molts its outer skin, together with the appendages; only its internal organs slip into the interior of its host through a hole in the worm's skin. Here it shows retrogressive development into a mass of cells which penetrates the blood system through the body cavity. It develops into a larva once more, and obtains its food from the surrounding blood with the aid of two long processes. After several molts, a male or female copepod develops from this spindle-shaped larva; these copepods hardly differ from free-living forms. When the last larval skin breaks, the hatching adult bores its way out of the worm and takes up a free-living copepod existence.

While the monstrillids show their relationship to the copepods through their adult form, despite the extremely modified larval development, the parasitic mode of life of the following copepods is so strange that a special order had to be created for them, within which only parasites are included. The members of the CALIGOIDA have either a shieldlike, flattened head and thorax or a completely spherical one. Their mouth is modified to form a sucking tube with spinelike maxillae. Two sucking discs at the base of the first antennae, together with a completely disc-shaped anterior of the body, serve the adults of *Caligus rapax* (L up to 7 mm) as the means of attachment to several marine-fish types. The nauplius and metanauplius larvae are free-swimming; from the third larval stage onward, the juveniles are attached to fish by means of a thread secreted by a gland on the head. The closely related *Caligus lacustris* (L up to 6 mm) parasitizes fresh-water fish, especially perch, pike, and bream.

The order LERNAEIDA has probably arisen from the caligoids, but here only the females are parasites, the males being free-living or attaching themselves to the females in the first copepodid stage. The females are impregnated when they are in the copepodid larval stage; after this, they lose their segmentation, their abdomen swells to a wormlike, S-shaped structure while the thorax becomes a long, thin neck, and the head produces three long, branched horns. In *Lernaeocera branchialis* (L up to 17 mm), common in the North Sea, development up to copulation takes place on flatfish; the fantastic metamorphosis of the female, however, occurs in shellfish and their relatives. The females bore into the respiratory funnel of these animals, penetrating with their head processes to the base of the gills at the point of fusion of the heart with its aorta. They do not enter the cavity of the aorta, since this could kill the host. For all highly developed parasites, there is a rule that one does not saw off the branch on which one sits. Despite this, the infected animal suffers severe damage, since the wounded aortal walls produce thickenings which narrow the blood vessel at the point of attack; this, however, takes place at exactly the point at which the oxygen-poor blood leaves the heart to travel to the gills where its oxygen content will be increased. Several *Lernaea* species

◁
Decapods (order Decapoda): 1. Giant blood-red shrimp (*Aristeomorpha foliacea*), 2. Norway or king lobster (*Nephrops norvegicus*; see also Color plates, pp. 451 and 459), 3. Large slipper lobster (*Scyllarides latus*), 4. Rock crab (*Galathea strigosa*), 5. European spiny lobster (*Palinurus vulgaris*; see also Color plate, p.454), 6. European lobster (*Homarus gammarus*; see also Color plate, p. 452).

parasitize the skin of fresh-water fish; the longest of all the copepods, *Penella balaenoptera* (L up to 32 cm), lives 5–7 cm deep in the blubber of the whale, its juvenile stages being parasites of certain marine snails.

In the LERNAEOPODIDS (order Lernaeopodida), the females are also extremely modified. Their whole body is covered with a soft skin and is divided into two sections by a constriction of the anterior region of the thorax. The second maxillae are the organs of attachment in this case. They consist of two anterior projections; these form, at their tips, a hard swelling (bulla) which attaches them to each other and to the host. The males remain small (dwarf males), retain their segmentation, and lead a parasitic mode of life only until they become sexually mature. Then they leave their host to seek a female. *Achtheres percarum* (L up to 5 mm; see Color plate, p. 443) lives in the gills of perch and pike-perch, usually covered with a thick layer of mucus secreted by the host.

The genus *Xenocoeloma* is the most extremely modified from the original copepod body form of all the parasitic copepods. These animals form a cone-shaped growth on the marine bristleworms they infect. Such a structure appears simple, but its internal form is puzzling: it is covered by the worm's skin and filled by a blind saclike projection from the worm's body cavity; between these, however, lies the muscle layer of the parasite. The male and female reproductive organs of the copepod, which is hermaphroditic, are present at the tip of the cone. At this point, the egg sacs are also developed, from the eggs of which the nauplius larvae hatch. If the egg sacs and nauplius larvae were not present, the crustaceanlike nature of *Xenocoeloma* would be completely lost.

The following subclass, the MYSTACOCARIDA, returns us to the original crustacean form. At present, only three species are included within this group. They are minute animals, at the most 0.5 mm long, which live partly in coastal groundwater and partly in the spaces between large grains of sand near the coast, at depths of about 25 m. Because of the number of their body segments, which is the same as that of the copepoda, they are included with them in the same group. Their first thoracic legs are, like those of the copepods, maxillipeds. They are thus Maxillipoda, but are more primitive than copepods in certain characteristics: their first thoracic segment, which bears the maxillipeds, is not joined to the head but is freely movable in opposition to it. Even the oral appendages are more primitively constructed than those of the copepods; of these, the mandible is a well-developed biramous appendage, an example of the original, basic form of this structure. Finally, the ventral nerve cord is still double, as in the phyllopods but not the Copepoda. Certain other characteristics, however, are more highly developed than those of other copepods: since the animal uses its appendages to push through the spaces between the grains of sand, the thoracic legs no longer have the task of rowing. They are therefore small and uniramous. The presence of four separate eyes, at the point where the nauplius eye originally was, is

Subclass: Mystacocarida

extremely curious. The nauplius eye in copepods consists of three eyes joined together; and the third, single eye, during its development, also consisted of two separate eyes. This curiosity of the mystacocarids is therefore simply a case of separation of the components of the original nauplius eye, something which also occurs in certain copepods. As far as is known, the mystacocarids hatch from the egg as metanauplius larvae. The sparsity of the known species will, no doubt, change as the fauna of the sand-space system is more fully investigated.

Subclass: Branchiura

In the FISH LICE (subclass Branchiura) we are again dealing exclusively with parasites, but in this case parasites which attach themselves to their hosts only occasionally, to feed. The hosts, as the parasites' common name indicates, are fish, both marine and fresh-water species. The fish lice, like the caligoids, are flattened and shieldlike; in contrast to the copepods, however, the shield is formed from the head cuticle and the lateral folds of the first thoracic segment. Therefore, the first thoracic appendages are included in the head, as they are in the copepods, but are not modified to form maxillipeds. The two short pairs of antennae bear hooks with which the parasite attaches itself to its host. In some species, the first maxillipeds also have attaching hooks; in the European carp louse, however, each bears a powerful sucking disc. The mandible, which serves as a penetrating tool, is difficult to distinguish. The four pairs of thoracic legs are biramous and adapted for rowing. The abdomen is more or less degenerate. It ends in two lobes which have a good blood supply and which were earlier considered to be gills; the fish lice were therefore called "gill-tails"; hence the scientific name "Branchiura," taken from the Greek. Actually, the movements of these structures aid in blood circulation. Their muscles pump the blood, which collects toward the rear of the belly side, up to the back and from there, forward to the heart.

Examination under a magnifying glass shows the conspicuous internal organs of a carp louse, especially the midgut which is provided on both sides with multibranched diverticula. As in the medicinal leech and ticks, these diverticula serve as stores for food during the long periods between each blood meal; chances of feeding do not occur too frequently and are therefore fully utilized. Other crustaceans which continually parasitize their hosts (permanent parasites) are not adapted for such storage. Apart from this, the walls of these diverticula also take up digested blood. Of the approximately seventy-five species, *Argulus scutiformis* (L up to 30 mm) is the largest. The EUROPEAN CARP LOUSE (*Argulus foliaceus*; L up to 6 mm; see Color plate, p. 443) is commonly found in fish pools, so it is often captured by naturalists in plankton hauls; it is also well known to the angler. Apart from numerous fish species, it also attacks tadpoles and newts. In fish, it causes damage not only by sucking blood but also through the wounds it makes while doing so, which later become infected with fungus.

The ASCOTHORACIDS (Ascothoracida) form a subclass which contains only about twenty-five species, exclusively parasites. Although the fossil remains of these animals are not known, the extreme age of their hosts allows extrapolation to the ancient origins of this parasite, for they infest corals and echinoderms. They have changed little from the basic body plan of the crustaceans. A bilobed shell covers the head and thorax or the entire body. In adaptation to a parasitic mode of life, the head appendages have been modified: the first antenna is a gripping, tweezerlike structure, the second is reduced, and the three oral appendages are stabbing structures. The thoracic legs are still biramous rowing organs in some cases; in internal parasites (endoparasites) they are reduced and the body segmentation has started to disappear. Eyes and a heart are lacking.

The copepods, the fish lice, and the ascothoracids have shown us how the basic crustacean form can be completely changed in adaptation to a parasitic mode of life. The subclass CIRRIPEDIA will illustrate the extent to which sessile crustaceans are also able to modify their body structure. They are so changed that most people do not realize that they are crustaceans at all. In this subclass as well, one order has taken up a parasitic mode of life and, in doing so, has become so changed that even a relationship with the arthropods is lacking. But even these show their relationship by their copepod nauplius larva.

This nauplius larva is characterized by the lateral projections on the anterior part of its body. Nauplius and metanauplius, owing to their mobility, ensure distribution of the species; the adults, which attach themselves to driftwood and the hulls of ships, also aid in this. After these stages is a larval stage which, owing to its shell, is similar to an ostracod and is called "cypris larva," after the ostracod genus *Cypris* (see Color plate, p. 459). On closer examination, this similarity is found to be purely superficial. The shell of the cypris larva is not bivalved, but covers the back completely, without a hinge. The legs are also completely different in structure: all six pairs are biramous and similar in form. The abdomen and the second antennae are reduced in size. The first pair of antennae of the future sessile cirripede, however, has a strong sucker, with which the cypris larva attaches itself as soon as it has found a suitable substrate. After it has become sessile, it stops feeding and enters a resting phase during which basic internal changes take place. One can compare this situation with that of the pupal stage of higher insects, and it is therefore termed the "cypris pupa." At the beginning, the shell is still soft; now it begins to deposit calcareous material within it, forming complicated and specially arranged plates. The six pairs of thoracic legs grow to form multisegmental "cirripodia" which ensure continual renewal of the respiratory current, which also carries plankton on which the animal feeds. On attachment, the cypris larva turns the ventral edge of the shell and the legs toward the substrate. During metamorphosis, the shell turns through a right angle and the body of the animal through a semicircle so that, finally, the animal's back is toward the substrate, and its ventral surface,

Subclass:
Ascothoracida

Sessile crustaceans:
the barnacles

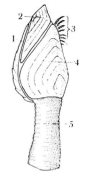

Fig. 16–6. Common goose barnacle, from the left: 1. Carina, 2. Tergum, 3. Cirripede, 4. Scutum, 5. Stalk.

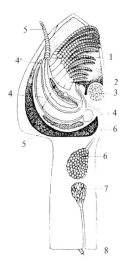

Fig. 16–7. Goose barnacle opened from the left side: 1. Last cirripede, 2. First cirripede, 3. Closing muscle cut through, 4. Gut, 4′. Gut diverticulum, 4″. Anus, 5. Testes, 5′. Copulation organ, 6. Ovary, 6′. Oviduct, 7. Cement gland, 8. First antenna.

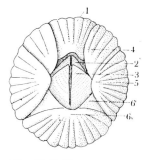

Fig. 16–8. Common barnacle seen from above: 1. Carina, 2. Tergum, 3. Scutum, 4. Latum carinale, 5. Latum, 6. Rostrolaterale with wings (6′).

with the cirripodia, is pointed away from it. Such inversion during the course of becoming sessile is known in other animal groups, such as kamptozoans and crinoids.

Cirripedes with calcareous plates are put together in the order THOR-ACICA. Of the roughly 600 known species, half of them belong to the suborder GOOSE BARNACLES (Lepadomorpha). The foremost part of the head forms a stalk attached to the substrate. The "capitulum" is attached to this, consisting of the rest of the head, the thorax with the six pairs of thoracic legs, the remains of the abdomen, and the shell lobes with their calcareous plates. The number of these varies from family to family; there are five in the COMMON GOOSE BARNACLE (*Lepas anatifera*; capitulum length up to 5 cm, stalk usually up to 10 cm; see Color plate, p. 459). Their scientific specific name means "the goose carrier"; their English name, "goose barnacle," has been derived from this. The animals really look like little geese, and, according to old myths, the barnacle goose is formed out of these creatures, which grow like plants on driftwood. In the *Topographia Hibernia* of Giraldus Cambrensis, from the late 12th Century, it is written that "certain bishops and other holy men in certain parts of Ireland are not afraid of eating geese during fasts since they are not meat and not of meat origin"—an explanation which was not accepted by Pope Innocent III who, in 1215, forbade the practice.

In the suborder BALANOMORPHA, the attachment stalk, which cor-responds to the stalk of the goose barnacles, is not visible externally. It is extremely shortened and is found within the ring of shells, the capitulum, here called the "carina." This surrounds the same parts of the body as the capitulum. The carina consists of four to eight separate plates and is frequently attached to the substrate by a similarly calcified basal plate. All these calcareous plates are firmly attached to one another, with only an opening from which the thoracic limbs protrude dorsally. A pair of shield and dorsal plates (scuta and terga), which, in contrast to the carina plates, retain their mobility, make it possible to close this opening. This occurs mainly when the barnacles must remain dry for a long period of time, such as at ebb tide. Not all types of barnacles are able to stand desiccation to the same extent: the STAR BARNACLE (*Chthamalus stellatus*; diameter 5–10 mm) attaches itself to rocky shores at the level of the medium spring tides, where it becomes submerged only every fourteen days; the COMMON BARNACLE (*Balanus balanoides*; diameter 5–15 mm), which covers every stone like a carpet in the tidal regions of the North Sea, can live in the spray zone with only occasional moisture. The RIDGED BARNACLE (*Balanus perforatus*; diameter 1.5–3 cm) and the IVORY BARNACLE (*Balanus eburneus*; diameter 2–3 cm), on the other hand, do not inhabit the tidal area; the barnacle *Balanus balanus* (see Color plate, p. 459) is to be found in large numbers from below the tidal regions to great depths, and thus avoids being left dry. The TORTOISE BARNACLE (*Chelonibia testudinaria*; diameter 1.5–2.5 cm) attaches itself to the shells of turtles and bores through the shell with rootlike processes from the carina. These penetrate not only the shell but also the bone beneath it. The basal attachment area remains skin-covered and does not form a basal plate. This is also true of the WHALE BARNACLES (family Coronunlidae), whose strongly striated carina plates bore deeply into the whale's skin as an anchor.

The suborder VERRUCOMORPHA, which has few species, in several respects is at a midpoint between the goose barnacles and the encrusting barnacles. The shortened stalk lies within the carina, but in this case it is closed by only one scutum and tergum, the two other plates being fixed to the carina. The Verrucomorpha are therefore asymmetrical. From over fifty species, only one, *Verruca stroemia* (diameter up to 5 mm) is found in the North Sea, on empty oyster and snail shells; all the others are present in the Atlantic and Mediterranean.

Almost all the Thoracica are hermaphroditic, an adaptation to their sessile mode of life. The testes surround the intestine in the thoracic region; curiously, however, their ovaries are transposed to the upper head region. In goose barnacles, therefore, they are found in the stalk, in the encrusting barnacles and Verrucomorpha, at the base of the carina. The ripe eggs enter the mantle cavity, are fertilized there, and develop into nauplius larvae which swarm out into the water. They can therefore occur in the plankton in enormous numbers; the cypris larvae then colonize the substrate far too thickly, and the majority die because of overcrowding. The lack of sufficient suitable substrates results in those barnacles which live in the middle of a group growing very slowly. The Thoracica colonize not only rocky coasts and harbor mouths, but also ships' keels. Because of the friction which they produce with the water, the ship's speed is reduced; at the same time, they destroy the ship's rust-preventive paint. Their regular removal and the repainting of the hull are expensive; barnacles therefore also play a role in economics.

The order ACROTHORACICA includes a few cirripedes which bore into the soft shells of mollusks and the calcareous skeletons of corals. They therefore require no protection by their own calcareous plates. As in the Thoracica, the anterior of the head contains the ovaries; they do not form a stalk, but an irregularly shaped attachment disc. The few thoracic limbs are uniramous. Most Acrothoracica are of separate sexes, and the males are very small (dwarf males). *Trypetesa lampas* (L, ♀♀ 7–15 mm, ♂♂ 1.2 mm) bores into the spindle of the shell in whelk species on the English coast.

The third order of cirripedes includes about 200 species of RHIZO-CEPHALA. They have progressed from a sessile mode of life to a high grade of parasitism. Their early development resembles that of the Thoracica: a nauplius larva hatches from the egg; it has the anterior horns typical of the subclass, but lacks an intestine. The tropical fresh-water rhizocephalids release a cypris larva, also without an intestine, from the egg. The nauplius larvae also metamorphose rapidly into this stage; they attach themselves by their first antennal pair to a hair on their host, a hermit crab or a crab.

Suborder:
Verrucomorpha

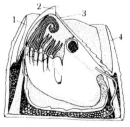

Fig. 16–9. Barnacle, opened from the left side:
1. Carina, 2. Tergum, 3. Scutum, 4. Rostrolaterale.

Now something extraordinary occurs: the whole body, including the legs, is discarded; all that remains is the head, which takes the form of a sac and bears a long spine (kentron) anteriorly. A uniform mass of cells fills the saclike head. The spine bores through the thin skin at the base of the hair, and the cell mass glides through this into the interior of the crab. Here—probably by means of the bloodstream—it passes to the thorax and grows to form a plate surrounding all the organs. Finally, it produces a tube to the body wall and penetrates this. Through this tube, the parasite now appears in a saclike form covered with a chitinous skin. Within this sac are the ovaries and spermatotheca. The sperm from the male are retained here. The male's development is known from the Japanese genus *Peltogasterella*, and is no less strange than that of the female. The males are produced from larger eggs than the females, and their nauplius and cypris larvae are also larger. They attach themselves to the female's sac, and their internal organs also regress to a cell mass. This wanders into the female's spermatotheca, where the cells become sperm cells.

In European seas, *Sacculina carcini* attacks mainly the shore crab (*Carcinus maenus*), but also other crabs. This rhizocephalid breaks out of the crab's abdomen, ventrally. It damages the crab internally, especially the hormone-secreting glands, thus preventing its victim from molting further. The rhizocephalid also frequently prevents its host's testes from developing; the abdomen of a striken male crab can therefore appear to have the same shape as that of a female.

Subclass: Higher Crustacea, by R. Altevogt

The typical characteristic of the HIGHER CRUSTACEA (subclass Malacostraca) is, with only a few exceptions, the uniform number of body segments—nineteen—which are all that have remained, during the course of evolution, of the original forty or more possessed by the primitive crustaceans. Therefore, a general rule applies to these animals: the serial and similar organs which were once present have, during the course of this higher development, been reduced in number or modified for other functions. This is illustrated, for example, by the oral appendages of crabs; these were previously walking legs and thus have been given the name "maxillipeds." This is also true of the original swimming legs (pleopods) of the middle body segments which, in certain of the Decapoda, have been adapted for a function in reproduction, acting as gonopodia and transferring the sperm to the female sexual openings. The structure of these gonopodia is a very useful characteristic in the systematics of several of the Higher Crustacea, of which ten orders with about 18,000 species are known.

Because of this large number of species, the Higher Crustacea comprise almost half the known crustaceans; they are superior to the Lower Crustacea in their great variety of forms, the enormous variety of their ecological niches, and especially in their highly developed behavior patterns. Their superior position is also emphasized by their size: the largest living arthropods are members of the Decapoda. The shrimp

Jasus huegeli of the antarctic oceans reaches 60 cm in length, the Indo-Australian crab *Pseudocarcinus gigas* has a carapace width of 40 cm, and the giant Japanese crab *Macrocheira kaempferi*, found at great depths, has a leg span of about 3 m.

Body segmentation

The body segments of the Higher Crustacea form a series of body sections (tagmata, singular tagma); however, by the attachment or fusion of one segment to the one preceding or following it, a variable number of these segments appear in different groups. The protocephalon, which bears the two pairs of antennae, can be considered the most primitive segment. Whether the first pair originally belonged to a segment or was associated with a prostomium is here, as in the insects, still a matter of discussion. In contrast, the second antennae are, without doubt, segmental appendages. The protocephalon therefore, depending on the point of view taken regarding the first antennae, consists of the prostomium and one segment, or two segments. Following this are three segments whose appendages are oral appendages (the mandible and the first and second maxillae); this section is termed the gnathocephalon. The attached thorax consists of eight segments, each of which bears a pair of legs. The first pair or pairs can be modified and used in feeding, and are thus termed maxillipeds. Their body segments are therefore called the gnathothorax. Six abdominal segments follow the eight thoracic segments; together these form the abdomen or pleon. This is terminated by a telson, corresponding to the pygidium of annelid worms; in the case of both, the anus opens here.

The section called the "head" comprises very different structures. In the leptostracans and *Bathynella* it consists of the protocephalon and gnathocephalon, as it does in the insects. Thoracic segments are attached to these, creating a "head-thorax" (cephalothorax). Usually only one segment is attached, as in amphipods and isopods; in the lobsters, however, all eight are. For a long period of time, all decapods were considered to have such a cephalothorax, but it has now been proved that, in the majority, the protocephalon is segmentally and movably separated from the gnathocephalon. This, therefore, forms a gnathocephalothorax.

As in many of the Lower Crustacea, many of the Higher ones have a skin fold that is produced backward from the last gnathocephalic segment, covering the back and sides to a greater or lesser extent. By the development and modification of dorsal plates (epimeres), this "carapace" can form a highly involved structure.

Dorsal plates are flat extensions of the segments which arise dorsally and often extend laterally down the sides. These can fuse with those of the following thoracic segments and form the basis of the crustacean's armor. This armor not only serves to protect the animal, to which end its deposits of chitin and calcareous salts can reach a high degree of rigidity, but also provides the respiratory organ with a space in which it is protected, since this remains thin-skinned and has no calcareous deposits. The blood

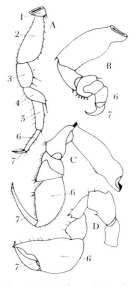

Fig. 16–10. Thoracic limbs of isopods: A. Walking leg from an aquatic isopod, B. Grasping leg from *Aega psora*, a parasite on fish, C. Grasping leg (subchela) of *Saduria entomon*, D. Pincer (chela) of a tanaidid (*Apseudes*):
1. Coxopodite,
2. Basal joint (Basipodite),
3. Ischiopodite,
4. Meropodite,
5. Carpopodite,
6. Propodite,
7. Dactylopodite.

solution (hemolymph) can thus perform its task of gas exchange in an area of multibranching tissues with a large surface area.

In certain Higher Crustacea, abdominal segments also grow together; the legless telson, for example, on which the anus opens, unites with the abdominal segments in front of it. Apart from this, other changes in form, with respect to the abdomen, took place during the course of evolution. In the crabs, or Brachyura, it is short and clapped forward against the thorax. The fact that in certain parasitic species—isopods among them—even more drastic changes in body form have come about can be explained by their specialized mode of life.

The majority of the Higher Crustacea typically have pincers developed on the thoracic legs. These serve in feeding, grasping the sexual partner, and defense. The two basic types of such pincers (chela and subchela) are shown in the color plate on page 438.

All Higher Crustacea have a common position for the openings of the sexual organs. In females, these open ventrally on the sixth thoracic segment; in males, the vas deferens opens on the eighth thoracic segment. The body covering of the Higher Crustacea is a sclerotized exoskeleton, frequently strengthened with calcium salts; it consists of an epidermis, producing a multilayered chitinous overlay which is covered externally by an epicuticula. The calcium salts are deposited by the epidermis. In larger crustaceans, the exoskeleton can also protrude internally at certain points, forming a partial endoskeleton. Numerous swellings, sensory hairs, swimming hairs, and wartlike and spinelike structures are typical of the external body covering; in some species, these may be arranged in rows and even act as sound-producing organs as are found in grasshoppers.

Coloration

Deposited or disseminated color agents often provide these crustacea with a permanent reddish, bluish, or grayish-brown color; this can, however, change rapidly in those cases where star-shaped chromatophores of the subepidermal tissue are contracted or expanded by nervous or hormonal influence. For example, the well-known North Sea prawn is blue-brown when living; its reddish color is produced only after boiling, during which the protein color-agent astaxanthin breaks down into its component proteins and a coloring agent (a carotinoid). On the other hand, certain fiddler crabs are able to change their color to fit their background—be it sandy or muddy—from almost black to pure white at every ebb tide; in this case, certain hormones produced by glands in the eyestalk contract the chromatophores. The Japanese samurai crab is of special interest, since its carapace is so patterned that it looks like a face. The ringed pattern around the legs of certain shore crabs is a good example of how brightly colored crabs can be. In smaller species, such as planktonic shrimps, the red or gold color is produced by fat drops in the interior of the body; the green or blue temporary contents of the gut also provide partial body coloration.

Crustaceans move either by swimming, climbing, striding, or digging in the loose bottom sediments. They can also float passively in the plankton. The thoracic walking legs (peraeopods) are the main means of locomotion in the Higher Crustacea; here the primitive crustacean biramous limb is so modified that the base and endopodite of these legs now resemble the legs of a spider or insect. The first pair of these walking legs—following the three pairs of maxillipeds; that is, the fourth pair of thoracic legs— bears the pincers, which arose from a later evolutionary development and have nothing in common with the original biramous limb.

Although this pair of pincers is used mainly in feeding, one can still prove that the pincers originally were walking legs. When a decapod has all its walking legs removed, or even only its most important ones, or when it discards its legs itself (autotomy), it attempts to walk on its pincers. In the Lower Crustacea, the primitive biramous limb, by flatten- ing its exopodite, forms a swimming leg. Such swimming legs (pleopods), into which the abdominal appendages are modified, are found in the shrimps and amphipods, among others. Because of their flattened form, they can also aid in respiration; their movement causes respiratory water currents and draws planktonic particles to the animal. The thin-skinned structures may also become points of attachment for gills. The last, termi- nal pair of swimming legs (the uropods) are leaf-shaped, and, together with the telson, form an effective tail fan; in co-operation with the violent contraction of the curved abdomen, this allows the animals to shoot backward—an extremely effective flight response.

The food of the Higher Crustacea varies from all vegetable and animal planktonic particles wafted past by the respiratory current and equally tiny food particles found in the bottom sediments, to algae, fruit, and larger types of food which the crab either eats as carrion or catches as living prey. The previously mentioned maxillipeds play an important role which must be considered in the light of the combined function of the oral appendages. The most primitive oral appendages, the mandibles and the first and second maxillae, have, during the course of evolution, been displaced more and more toward the esophagus; the following thoracic leg pairs follow this tendency. Thus, after the second maxillae, from inside to outside, come the first, second, and third maxillipeds; the most external part of this extremely complicated apparatus, whose structure can only be understood by examination of its evolutionary development, is therefore the third maxilliped, whose broad basal segments (the original biramous limb of the Lower Crustacea) are extremely well adapted for this function. Only in such a discrete chamber would it be possible, for example, for a shore crab to sort out the edible particles from the mud it takes up.

The food must also be sorted again before it comes to the point of actual digestion. All Higher Crustacea have a gastric mill at the end of the foregut, a characteristic structure with chitinous teeth and highly developed trapping and filtering systems. This structure is necessary, since without it the numerous lateral diverticula of the foregut and, in part, the midgut

Methods of locomotion

would otherwise become clogged with large food particles. These diverticula take up and digest the tiny food particles. Because of the filters, the large and indigestible food particles never enter the sieve system of numerous bristles and hairs, but are led directly into the main gut—the mid and end gut—and excreted. The midgut is simply the point of opening of numerous mid-gut diverticula; these secrete digestive enzymes but also aid in absorption.

The waste products of metabolism are excreted by means of nephridial tubes; these, however, do not lie in the abdomen, but in the head region, where they arise as a funnel in the remains of the coelom and then, as a twisted tube, open at the base of the second antennal pair (in the decapods) or the second pair of maxillae (in the isopods). In fresh-water shrimps, these canals are much longer than in marine ones, since the antennal or maxillary glands also regulate the salt content and water content of the body. A longer excretory canal therefore means better regulation. In fresh-water shrimps, a chambered bladder is also built into the terminal section of this canal, providing an even larger surface area and better regulation. Of course, the excretory products are mainly nitrogenous substances derived from protein breakdown, especially ammonia, ammonia salts, and the like. Some of these substances are excreted through the gills; recently, the body covering has also been discovered to act as an excretory organ.

Closely associated with feeding and metabolism is respiration, which is taken over by the interior of the carapace walls (gill chambers), thin-skinned leg attachments (epipodites), and also sections of the carapace itself. The continual wafting of fresh water past these structures is ensured by the movements of the walking legs, and by the second pair of maxillipeds in decapods, and by the abdominal legs in isopods.

There are, however, shore crabs which are capable of a temporarily terrestrial life. These crabs are characterized by a gill chamber which is protected against desiccation and serves as a fully sufficient respiratory organ, requiring only oxygen-rich water. The animal provides this by continually stirring the water in the gill chamber by means of its scaphagnothites, appendages of the second pair of maxillae; the water emerges at the mouth opening, runs ventrally along the body, has its oxygen replenished there, and re-enters the gill chamber at the point of attachment of the legs. In species of very small crabs, the short ventral surface is not sufficient to allow complete gas exchange. This is why fly-sized members of the tropical genus *Dotilla* deflect the respiratory water stream over their backs. To aid in slowing down the flow and increasing the surface area over which the water flows, there are numerous wartlike and lumplike extensions of the dorsal carapace, playing an important role in the gas exchange.

Certain isopods have gone over to an entirely terrestrial mode of life and have, in addition to the retained true gills, the so-called "white bodies." These are extensions of the abdominal legs which are penetrated by tiny breathing tubes (trachea). Other Higher Crustacea, which could almost be termed "air breathers," are able to remain out of water for several hours; these include members of the groups *Gecarcinus* and *Grapsus*. Conversely, the robber crab or coconut crab (genus *Birgus*) will drown if submerged in water for more than five hours, since their breathing chambers do not contain gills and can be considered as true lungs.

The blood pigment of the Higher Crustacea is mainly hemocyanin; this is not, however, contained in special blood cells, but is dissolved instead in the watery lymph. The crab hemolymph does, however, contain cells (hemocytes) which partly act like the white blood corpuscles of the higher animals, engulfing foreign particles (phagocytes), and are partly involved in clotting processes. An originally long, dorsal heart drives the blood through the blood-vessel remnants into open chambers (open blood system). The blood flowing through the open tissue system is drawn into the heart's interior by ostia during the heart's expansion phase, and then driven back into the anteriorly and posteriorly running blood vessels. In certain crustaceans (e.g., in the genus *Mysis*), the blood vessels to the brain region are provided with a second heart; this ensures better blood flow to the "brain" (subesophageal ganglion). The heart-beat frequency is dependent on the size of the body and the temperature. In the crayfish, it reaches 100 beats per minute.

The heavy deposition of calcium salts in the already rigid chitinous skeleton makes gradual growth of the animal impossible. Growth therefore occurs in definite spurts which take place during the time the exoskeleton is soft after each molt. These molts are a good example of hormonal processes, which are especially well developed in the crustaceans. Glands which excrete their compounds internally (endocrine glands), controlling the process, are partial developments of the nervous system and lie as neurosecretory cell islands at the base of the eyestalk in decapods (X-organ and sinus gland); their secretions control molting, chitin formation, and color change (chromatophorotropine). The non-neurosecretory gland, the Y-organ, found near the epidermis of the second maxillae or second antennae, also plays a part in the hormonal activity involved in molting, its secretions initiating the molt.

Finally, in 1954, endocrine glands which secrete a male hormone (androgen) were discovered in amphipods; afterward, they were also found in other Higher Crustacea. They lie at the end of the vas deferens; when they are transplanted into a female, it changes into a male; they therefore contain true sexual hormones. Too little is known of other hormonal secretions in the Higher Crustacea, but numerous other life processes (e.g., the carbohydrate level in the blood, calcium uptake and water uptake of the tissues) are at least partially controlled by internal secretions.

The variable body form and the different types of life which are found

▷
Above from left to right:
Panulirus ornatus, an Indo-Pacific relative of the European spiny lobster.
The crab *Ethusa mascarone* disguises itself by holding various camouflaging objects above its back. Here it carries a piece of coral branch.
The dromid crab (*Dromia vulgaris*) is a master of camouflage. Its brown "fur" is covered with greenish mud, and a few red sea anemones (*Dendrodoa grossularia*) have attached themselves to its back.
Above, center, all three photographs:
Here the dromid crab *Dromia vulgaris* drags a piece of sponge onto its back, beneath which it finally almost disappears.
Below:
The rock crab (*Cancer pagurus*) can reach a diameter of 30 cm and a weight of 6 kg. Its flesh is considered very tasty.

Growth

Endocrine organs

Sense organs

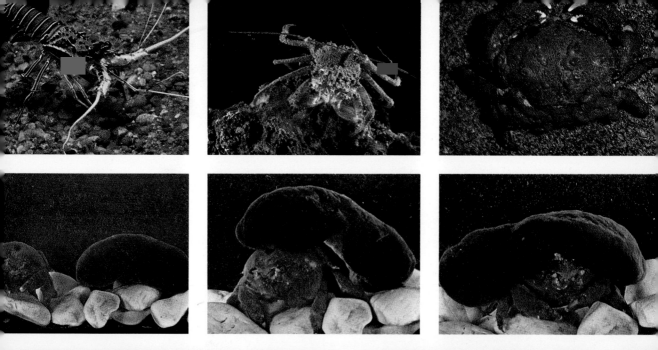

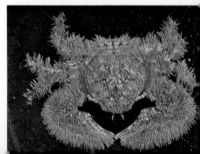

◁
Above: A ghost crab
(*Ocypode*).
Middle, from left to
right:
Male fiddler crab (*Uca
tangeri*). With the greatly
enlarged pincer on one side
of the body, extremely
effective courtship
movements are performed.
Land crabs, like
Gecarcinus lagostoma, are
practically independent of
the sea, returning to it only
to reproduce. The Chinese
crab (*Eriocheir sinensis*) has a
thick layer of "fur" on the
male's chelae. It was
originally Chinese in
distribution, but was
inadvertently brought to
European waters.
Bottom from left to right:
The spider crab (*Pugettia
gracilis*), from the Pacific.
Stone crabs like this
Lopholithodes species are not
true crabs, but are related
to the hermit crabs.
Even the gray porcelain
crab (*Porcellana platycheles*)
is not a true crab, despite
its crablike form. It is a
member of the rock crab
family (Galatheidae).
◁◁
Some encrusting barnacle
species settle on solid
substrates in the tidal zones
in tremendous numbers,
closely packed together.
Even harbor moles and
posts are covered with
them, as in this picture.
◁◁◁
The tiny, beautifully
marked painted shrimps
(*Hymenocera picta*) feed
exclusively on starfish, into
which they bore with their
pincers, ripping off pieces
of tissue. Here the shrimp
is about to turn a starfish
over by propping it up
with its body.

in the Higher Crustacea are reflected in the extremely highly developed sense organs they possess and the behavior patterns dependent on these organs. This is especially true of the sense organs concentrated in the head region. Any person who has had something to do with crabs immediately thinks here about the stalked eyes. The ghost crab *Ocypode ceratophthalmus* has up to 30,000 individual ommatidia on a single eyestalk. In contrast, however, there are much more "modest" visual organs. The compound eye of the woodlouse (genus *Porcellio*), in comparison, consists of twenty to thirty ommatidia, that of the shore crab *Carcinus maenus*, about 7000. In the single central eye of the nauplius larva, however, there are only a few pigmented cups, each with about twenty visual cells; in a few species (Notostraca) their number can reach 200 per pigment cup. As a result of this simple structure and low number, the nauplius eye can only serve as an organ to detect the presence of and direction of light. Even these capabilities, however, play an important role in the planktonic mode of life of these larvae, since, in their environment, brightness and angle of incidence of the light mean "above" and are therefore essential factors in orientation.

The compound eyes of adult crustaceans not only perceive form and color but also aid the animal in orientation toward light. Many higher crustaceans are capable, depending on the height and movement of the sun—and perhaps the moon as well—of using these signals to orient within their environments. Movement toward feeding grounds or, in land crabs, movement to release the larvae into their original habitat, the sea, are therefore controlled by such orientations. This is the case in fiddler crabs and other forms. In this, the animals—as was first proved in the case of the honeybee—use the movements of the polarized light rays to find their way about, even when the sun is covered by clouds. Landmarks also play an important role in direction finding.

Usually, however, the ommatidia of the compound eye are used in the perception of brightness, color, and form in the surroundings. They have a visual field of about 360 degrees on the horizontal plane and can therefore see "all round." Numerous crustaceans belonging to various levels of development are even able to see colors. This has been proved in an opto-kinetic cylinder, an apparatus in which colored and gray stripes of the same brightness are rotated around the animal. These cause a reflexive (optokinetic) turning of the eyestalk or the whole body when the animal sees the colored stripes. In the same apparatus it has been proved that the animals are also capable of distinguishing different levels of brightness. The Mediterranean crab *Potamon*, for example, can distinguish slight differences in brightness of only two percent.

In the same way, a fantastic ability to distinguish forms, "visual acuity," can be determined; crabs respond especially well to moving objects. Various shore crabs of the genus *Potamon* are extremely efficient at this; they respond to objects which appear within one minute of arc.

In comparison, human visual acuity can resolve half this angle; that is, it is twice as good. Certain Higher Crustacea can also perceive motionless objects; hermit crabs, fiddler crabs, and others direct their movement and orientation with respect to stationary objects in their surroundings.

All these visual abilities tend to make the decapods extremely conscious of their surroundings, resulting in their flight distance generally being extremely great. Egrets flying off 8 m away cause Indian fiddler crabs, which have a carapace diameter of about 4 cm, to run off; the European fiddler crab *Uca tangeri* can see a 1.8-m-tall man at 19 m distance. Recognition of the sexual partner by certain land crabs is also brought about by this extremely good vision.

Deep-water crabs or cave-dwelling species have less acute eyesight; in these, the sensations of touch and vibration play the most important role in the perception of the surroundings. The sense of touch is mainly associated with the first and second antennae, but hairs on the oral appendages, the claws, the walking legs, and the abdomen are also extremely sensitive to touch. These stimuli, in part, can emanate from the animal's own body; they therefore send messages to the central nervous system regarding the position of the claws and legs with respect to the rest of the body. Such organs (proprioceptors) are found as bundles of hairs at the numerous joints or as internal stretch receptors within the joints themselves. These strands of expandable sense cells are probably also the basis for vibration reception, which is highly developed in the shore crab. In the fiddler crabs, a true communications system—a "tapping code"—has evolved as a result of this ability to perceive vibrations.

Among the mechanical senses we must also include the organs of gravity reception, which inform the animal of its position in space and of turning movements. Invaginations of the carapace form bladderlike cavities at various parts of the body. In the decapods they are found at the base of the first antenna; in certain amphipods, in the head; in the *Mysis* relations, which are often open-ocean or deep-sea dwellers, on the pair of swimming legs forming the tail fan; and in certain isopods, on the telson itself. These gravity or body position sensors (statocysts) possess a statolith, a calcareous stone attached to a cushion of sense cells; depending on the position of the animal in space, different sense cells of this cushion are stimulated to different extents, indicating the body position to the animal. The interior of this organ is filled with water from the surroundings. At each molt the animal discards it, together with its statolith. The statolith either can be constructed by the crab itself and must therefore be formed anew after each molt, or it can consist of sand grains, pieces of shell, or other objects which the animal inserts into the opening of the gravity receptor. If, for example, one puts a freshly molted shrimp on iron filings, it will put these into its statocyst as statoliths. With the aid of a magnet one can therefore confuse the animal with regard to its position in

Large flight distance

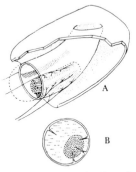

Fig. 16–11. Basal joint of a lobster's first antenna: A. Opened, B. Sensory cavity in horizontal section. The three rows of sensory hairs with sand grains act as gravity receptors; the single row of sensory hairs surrounded by water acts in perception of turning movements.

space, since the iron filings respond to the magnetic field and not to the field of gravity.

Apart from the hooklike sensory hairs which are connected to the statolith, most statocysts also contain threadlike hairs which register the turning movement of the animal, through movements of the water mass within the statocyst—as do those in our aural labyrinths. Apart from this, the degree of muscle tension is also regulated by the statocysts.

With regard to the chemical senses—especially the senses of smell and taste—in the Higher Crustacea there are large numbers of tubelike sense cells (aesthetascs) on the first antenna; they respond to substances dissolved in the water, and probably to odors in the air as well. Apart from these, organs of taste are also found at the ends of the walking legs and on the oral appendages.

Nervous system

The functioning of these extremely efficient sense organs is coordinated by the nervous system; in the Higher Crustacea this is reminiscent of the primitive type of nervous system found in the Segmentaria, to which the annelids also belong. For example, the nervous system of the crayfish indicates clearly that it has arisen from a segmented body form, for lateral branches are given off by the longitudinal nerve cords at absolutely regular intervals. The true crabs are, in this instance, the most greatly modified forms. Here several such segmental sections join to form a uniform nerve knot (ganglion), especially in the case of the subesophageal ganglion which provides innervation to the oral appendages. A smaller ganglion above the mouth sends nerves to the compound eyes and antennae; in the Higher Crustacea it has reached the same level of development as in the insects. The cephalic sense organs are thus made extremely efficient, since they are innervated from three "brain" centers (lamina ganglionaris, medulla externa, and medulla interna). An "olfactory center" is also well developed; the efferent nerves of the chemical sensory receptors from the first antennae terminate in this center. In addition to these centers, the crabs also have a nerve network beneath the epidermis, which receives stimuli from the mechanoreceptor hairs and elsewhere.

The high level of efficiency of this nervous system explains certain extremely highly developed behavior patterns, such as orientation, prey capture, and courtship. This efficiency is illustrated by certain measurements made on the speed of transmission of stimuli in the American crayfish *Orconectes*. In the two pairs of giant fibers which this animal has, excitation can progress at a rate of 20 m per second from anterior to posterior, thus making possible the rapid backward beat of the abdomen which is usually necessary for successful escape. Such an efficient nervous system also makes possible the performance of certain behavior patterns which are reminiscent not only of the insects, but of birds as well; these are illustrated in the courtship behavior of certain crabs.

The reproductive organs of the Higher Crustacea usually consist of paired ovaries and testes which lie above the intestine in the thoracic region. They often have glandular outlets which secrete the adhesive substance for cementing the eggs to the substrate or holding the sperm together in a spermatophore. In copulation, the males hold the females with their pincers, and specially modified swimming legs function as copulatory organs. The fusion of eggs and sperm usually occurs some time after copulation, when the jellylike covering of the spermatophore dissolves in either the ovary, the ovarian duct, or special reproductive cavities, thus releasing the sperm for fertilization. Courtship and copulation are often restricted to certain times of the year and phases of the moon —in the last cases, also to certain tidal states. In this way a better distribution of the larvae, which hatch later, or the larva's actual release, is brought about at high tide. The sperm often have an extremely curious structure and a penetration mechanism which, in lobsters and hermit crabs, includes an explosive, spearlike process; with the aid of this, the contents of the sperm capsule are shot into the egg.

Reproductive organs

On completion of fertilization, the eggs are extruded from the opening of the oviduct. The majority of Higher Crustacea have some form of maternal care; they attach the eggs to their own bodies until the larval development has progressed to a relatively advanced level. The Decapoda attach their egg packets to the abdominal legs, the Stomatopoda carry them between the maxillipeds, and the oceanic euphasids carry them along on their thoracic legs.

Embryological development

The development of the embryo starts with a type of cleavage reminiscent of the annelids. The series in which the separate parts of the embryo develop is also illustrative of this relationship. As in the annelids, the first segments of the body are formed simultaneously, the others serially in the Crustacea. Another characteristic of embryology is also typical of the annelids; during the time the posterior segments develop in a form of budding from the anterior ones, the most anterior body sections have already started to form organs. In the Crustacea, however, a secondary body cavity lined with cell layers (coelom) is not found in a continual form, only small sections of this being present, distributed among the body tissues (myxocoel).

The early development consists of several larval forms which become more and more variable from stage to stage by the addition of segments (anamery). The first free-swimming larva is the nauplius, as is found in the euphasids and certain primitive decapods; it has the first three pairs of appendages—the first and second antennae and the mandibles—but further body segments are lacking. This larval form is well adapted to a planktonic mode of life and swimming; it ensures the distribution of its species. As further segments are developed, the larva becomes a metanauplius, then a protozoea, and then a zoea, from which the crab develops, complete with all its organs.

Development in crustaceans with brood care

One form of development which deviates markedly from that described above is found in crustaceans with maternal care; an almost complete young adult hatches from the egg. On hatching, it just increases in size and the sexual organs develop (epimery). The more advanced the development within the egg, the better are the chances of survival, since the planktonic stage, with its accompanying dangers, is left out. The crayfish, for instance, which show advanced epimery, therefore lay only 250 eggs, while the shrimps, whose eggs hatch larvae (anameric development), produce a few hundred thousand eggs. To facilitate floating in the plankton, the zoea larvae have various processes, dorsal horns, and other such structures which are of use in species determination. Up to five zoeal stages can be identified in a single species; they finally produce the megalopoa stage—a larva which is no longer planktonic but lives on the bottom or near the shore. The first tiny crab arises from this, becoming an adult crab after numerous molts; in the majority of species, it continues growing for the rest of its life.

Molting

As already mentioned, the molts necessary for this continual mode of growth and which also take place between each larval stage, are under hormonal control. This also includes the formation of a new carapace beneath the old one. Large quantities of calcareous salts and chitin are reabsorbed from the old carapace, these being, in part, taken up by the blood (hemolymph). Under the influence of a hormone (ecdyson), which also controls molting in other arthropods, the crab discards the old carapace (exuvia), expands by taking in water, thus stretching the usually wrinkled "new skin" to a size larger than necessary, and starts growing into it later. Even lost appendages can be replaced, to a certain extent, during this process. Certain pincers and walking legs have a true "autotomy point"; the animals are able to discard their appendages almost at will and leave these to an attacking enemy. On termination of the growth period, the soft-skinned crab becomes armored again by the deposition of chitin and its hardening (sclerotization), and the deposition of calcium salts, which occur during two to three weeks. They look exactly as before, only larger. The calcareous salts required are mostly obtained through the gills from the external water; the pieces of calcium carbonate in the stomach (stomach stones or gastroliths) provide only a very small amount of this material.

The age of Higher Crustacea

The Higher Crustacea can reach extreme ages. This can be illustrated by the well-known lobster, whose pincers can be as long as a man's hand and whose body can measure up to 50 cm. Lobsters have lived from twenty-five to thirty years in aquaria; even the crayfish can reach twenty or more, although it becomes sexually mature at two or three. The European crayfish can reach 20 cm in length and weigh 135 g, while its American cousin *Orconectes*, introduced into Europe, reaches only 13 cm in length and, as a result, remains lighter.

The most primitive group of the Higher Crustacea is the LEPTOSTRA-

CIDS (order Leptostraca; BL 0.6–4 cm). The shell-like carapace around their head and abdomen made ordering them systematically difficult for a long time. The forked tail (furca) is also a characteristic which is no longer found in the Higher Crustacea. With the exception of *Nebaliopsis typica*, which lives at depths of about 150 m, where it takes up a swimming mode of life, the leptostracids inhabit the sea bottom from the tidal regions to depths of about 400 m; here they sort edible fragments from the mud on the bottom by means of their filter bristles. The NEBALIDS (genus *Nebalia*) live in the Atlantic, Mediterranean, and northern oceans, while *Nebaliopsis* is found in equatorial regions. The nebalids release their eggs into the water and have no maternal care.

Because of their curious appearance, the STOMATOPODS (order Stomatopoda) are also called mantis shrimps. The body structure of the mantis shrimp *Squilla mantis*, which can be found in certain fish markets in Spain and Italy, is extremely reminiscent of that of a praying mantis (see Vol. II). The well-developed abdomen with its terminal fan allows these shrimps to shoot backward rapidly should danger threaten; usually, however, they swim slowly and evenly about, using the flattened swimming legs. It is probable that the eyelike patterning of the tail fan is of value in holding the group together; this "tail-light" serves as a recognition signal. The stomatopods are also efficient walkers, since they live in cracks or holes which they dig. Their food consists of shrimps, fish, and mollusks, which they grasp with lightning swiftness in their forelegs (the second maxillae), which function like the blades of a folding pocketknife. No prey can escape once caught, for the leg joints are provided with sharp spines. The next three pairs of legs—the thoracic legs—have similar but smaller clawed knives. The movable head, like that of the praying mantis, has compound eyes on short stalks, and the adult animals retain the nauplius eye—a rarity among the Higher Crustacea. The mantis shrimp *Squilla mantis* (BL about 20 cm; see Color plates, pp. 451 and 459) lays about 5000 eggs, which the female cements together and carries between the mantislike feeding legs for about ten weeks. In this way, a supply of fresh water is ensured; however, during the time it carries the eggs, the female is unable to feed. *Squilla raphidea*, found in the Indo-Pacific, is a much larger species (BL 33 cm). A number of smaller mantis shrimp species (*Gonodactylus*) live on coral reefs, where they are believed to feed, in part, on sea anemones.

All stomatopods have a primitive heart which extends along the thorax and abdomen as a long tube. Segmentally arranged ostia are present along its length.

The superorder SYNCARIDS (Syncarida; BL 5 cm maximum) includes, apart from certain fossil groups and the "living fossils" of the order Anaspidacea, the smallest of all the Higher Crustacea, the BATHYNELLACIDS (order Bathynellacea). They are also primitive in the structuring of the body segments, since the head is usually not joined to the following

Order: Stomatopoda

The smallest Higher Crustacea

segments. Although the animals are extremely small, they are very important in unraveling the evolutionary development of the Higher Crustacea. As early as the middle of the 19th Century, paleontologists discovered syncaridlike shrimps in the fresh-water deposits of the Permian and Carboniferous eras. When similar living forms were discovered in 1892 in fresh-water lakes in Tasmania, this was hailed with as much excitement among experts as were the famous "living fossil" snail *Neopilina* and the coelacanth *Latimeria* (see Vol. V). After these living fossils had been discovered, examination of mudflats and interstitial sand-grain systems all over the world turned up further syncarids, and more are still being discovered.

The syncarids have no dorsal armoring (carapace)—a secondary development as an adaptation to their cryptic mode of life on or in the bottom sediments. A few special characteristics of this inconspicuous but, evolutionarily speaking, extremely important superorder have already been mentioned. In the order ANASPIDACEA, members of the genus *Anaspides* attach their large eggs individually to objects in the surroundings and leave them to develop themselves. The young animal hatches, looking just like the adult, bypassing the larval stages.

The BATHYNELLACIDS (order Bathynellacea; BL only 0.5–2 mm), the most minute of the Higher Crustacea, live in the interstitial spaces of sand grains in groundwater streams, which appear to be widely connected throughout the whole hemisphere. They are also occasionally found in the sandy floors of springs; two species are present in the deep reaches of Lake Baikal. The smallest representatives are present in the genera *Leptobathynella* and *Thermobathynella*. Of special interest is the differentiation into the numerous forms present at each point of discovery— geographical subspecies and species—which is understandable because of the different ecological conditions pertaining there. *Thermobathynella adami*, for example, can even exist in thermal springs at a temperature of 55°C in central Africa.

The large number of representatives making up the EUCARIDS (superorder Eucarida) are among the best-known of all crustaceans; their most important orders are the EUPHAUSIDS (Euphausiacea) and the DECAPODS (Decapoda). They usually have a large carapace, and the sense organs are well developed. The compound eyes, for example, are always stalked.

Order: Euphausiacea

Although the EUPHAUSIDS (order Euphausiacea; BL 4–6 mm) seem almost shrimplike, they can be distinguished immediately from the true shrimps, which belong to the Decapoda, by the end of the tail, which is covered with bristles. The lateral gills are not covered by the carapace, as is the case in the Decapoda. In addition, the maxillipeds, which are very well developed in the decapods, are lacking. The majority of euphausids are oceanic, inhabiting certain zones in enormous numbers, almost 100 million individuals per shoal. The inhabited areas are characterized by uniform temperatures and salinity, and each species of euphausid is so closely adapted to the environmental conditions in which it is found that it dies, or at least is unable to reproduce, if it drifts into a water zone with a different temperature or salinity.

Light signals aid in keeping the enormous shoals together; these are emitted from light organs, usually ten of them. These organs lie at the base of the eyestalk, on the second and seventh thoracic legs, and on the first four abdominal segments. They are lighted for a few seconds by nervous impulses. A central striped body acts as a type of "light bulb"; its phosphorescent substance, like that of the glowworms (see Vol. II), consists of two components, whose cooperative function makes it possible for the light to be emitted. Reflective cell layers behind the "light bulb" and a lens in front of it aid in making the light production more efficient. Their roles, however, have not yet been investigated in detail.

The enormous swarms of euphausids are, of course, an immense source of readily available food for many inhabitants of the oceanic regions, such as herrings, sardines, and other fish. Even the giant blue whale and baleen whale sometimes feed exclusively on these tiny animals. Commercial whalers take the appearance of euphausid swarms to be an indication of a good whaling area.

The majority of euphausids feed by filtering plant and animal plankton; they make daily migrations up and down with their food source, their movements being dependent on light intensity. At night, therefore, they are found near the surface, and during the day, at depths of several hundred meters. This can be demonstrated by echolocation, since the euphausids form a scatter-layer which returns the echo. Although traces of sunlight are still present at these depths, the ability to phosphoresce is still effective there. As in the glowworms, the light is "cold," that is, a short-wave green or blue. Whether this light also gives protection against enemies, by confusing the animal's outline, has not been proved.

The most important members of the euphausids belong to the "krill," as it is called by whalers. *Euphausia superba* inhabits cold seas; *Meganyctiphanes norvegica* lives on the edge of the Gulf Stream in Norway; in addition, members of the genus *Nematoscelis* and the carnivorous genus *Stylocheiron* should also be mentioned. *Stylocheiron* has complicated eyes, divided into sections. One section is for sight above and the other to the side; each is clearly divided from the other.

The number of species present in the order DECAPODA is similar to that of birds, for about 8500 species are known. The long-tailed forms, such as the crayfish, are distinguished from the short-tailed ones, such as the crabs; between the two is an intermediary group in which the abdomen is provided with swimming legs but is usually held curled under the thorax (e.g., the tropical mole crab *Thalassina* and the hermit crabs). Five suborders are distinguished: 1. Shrimplike macrurans (Natantia; long-tailed crabs), 2. Crawfish (Reptantia), 3. True macrurans (Astacura), 4. Anomura (intermediate forms), 5. True crabs (Brachyura).

Order: Decapoda

Suborder: Natantia

The most primitive decapods are the SHRIMPLIKE MACRURANS (suborder Natantia), with about 2000 species which inhabit all possible environments between the deep sea and fresh water, but which are mostly marine. The body is usually laterally compressed, and a tail fan is present on the relatively long abdomen. Despite this, the shrimps spend most of their time on the bottom, walking abound on their five pairs of walking legs or using them to dig in the sediment. The long first antennae, which are held close together, form a tube through which fresh water is brought to the body for respiration, as in the genus *Solenocera* from the Mediterranean. A similar structure is used by the crabs *Emerita* and *Albunea*, while *Corystes* uses its second antennae for the same purpose.

A shrimp which is caught commercially along North America's eastern coast and in the Caribbean Sea is *Penaeus setifer* (BL up to 20 cm). Its abdomen is sold as "crab tails" in preserved form all over the world. The "crab tails" of South America, Australia, and the Mediterranean region usually are also *Penaeus* species. In Italy, for example, it is *Penaeus trisulcatus*. Many genera of shrimps live almost exclusively in the deep sea, such as *Pandalus borealis* in the North Atlantic; others, conversely, are bound to the coastal zones, among them the genus *Hippolyte* with almost 200 species. The species name of *Hippolyte varians* ("changeable") honors this creature, since it takes up the background color of the green or brown alga on which it happens to be sitting. This takes place through the action of various color cells in the epidermis and is, in part, controlled through the eyes. This can be shown when the animal is kept in dim light conditions, for the red and green tones change to a pale blue. *Hippolyte varians* is also pale blue during the night. Many species live communally with sea anemones, and in this case are completely transparent, apart from a few points of color.

Family: Alpheidae

The SNAPPING SHRIMPS (family Alpheidae), live in coral heads or the shore regions of tropical seas. Their first pair of walking legs is especially strong, and is provided with a pincer. Unilaterally enlarged pincers—that is, larger left- or right-hand pincers—are differentiated from the smaller one on the other side. Members of the genus *Alpheus* are able to shoot a sharp stream of water with the snapping pincer, accompanying this with a loud report. This is used as a defensive mechanism, but also aids in prey capture, since it has a shock effect. In the snapping shrimps, which often live as pairs in holes they dig, beautiful examples of closely cooperating communities can be demonstrated. Such a symbiosis can be found between the shrimps and fish. Here the fish, which has a greater flight distance, indicates the presence of danger which the shrimp is not yet able to perceive, while the shrimp provides the fish with a refuge in its hole.

The common shrimp

The COMMON SHRIMP (*Crangon crangon*; family Crangonidae) is also a member of the Natantia. This species inhabits the Wattenzee in large numbers, despite that body of water's great variations in temperature and salt content. Such a variable environment can be inhabited only by specially adapted forms. The common shrimp is especially well equipped for this. The total annual tonnage of this shrimp captured comprises up to five percent of the Wattenzee's total fisheries catch. It is used as food for animals or humans, often as a delicacy. Only small shrimps, under 5 cm in length, are used as animal food. The common shrimp feeds at night on other small crustaceans, worms, and mollusks; during the day, it hides in the soft mud. Its life expectancy is three years; during this time, a single female can produce up to 20,000 offspring. Although the common shrimp is also found in the Baltic, the common BALTIC SHRIMP (*Palaemon squilla*) belongs to another family, Palaemonidea. This species, however, occurs in much smaller numbers and much less regularly than the common shrimp.

Many palaemonids inhabit fresh water, especially in the tropics; some even inhabit submerged caves. Such cave dwelling shrimps are present in the Carso lakes in the Balkans. In these, vision is reduced to the point of total absence of the eyes, and the sense of touch has developed into the main means of determining changes in the surroundings, something which can be surmised from the exceedingly long antennae.

The richness of their genetic background enabled the palaemonids to evolve in several directions; they thus provide good examples of special developments in behavior and adaptation to the environment. Numerous species have become symbiotic in sponges, sea anemones, corals, and even fish. This can lead to a very delicate "cleaner symbiosis"; in this, the Caribbean shrimp *Periclimenes petersoni* attracts fish by waving its antennae from "its" sea anemone. It is then allowed to remove external parasites from the fish's scales, opercula, and other parts of the body which are difficult to reach, and may even enter the fish's mouth to clean up the remains of its meal.

The CRAWFISH (suborder Reptantia: "the crawlers") can easily be distinguished from the shrimps (Natantia: "the swimmers") by their dorso-ventrally flattened bodies and the usually powerfully developed pincers on the first pair of walking legs. The abdominal legs are no longer adapted for swimming, but are used by the female as the place to attach her eggs. This group includes the SPINY LOBSTERS (family Palinuridae), recognizable by their antennae, which are longer than the body, and the spiny processes of the carapace. Large gripping pincers are lacking. The commercial use of spiny lobsters in all parts of the world is attributable to their meaty abdomens.

Suborder: Reptantia

The EUROPEAN SPINY LOBSTER (*Palinurus vulgaris*; BL up to 45 cm, weight up to 8 kg; see Color plate, p. 460) lives on the rocky coasts of the Atlantic and the Mediterranean. It is a valuable food animal. It feeds at night on snails, mollusks, and dead animals; it must therefore be caught with pots or bait. The size and weight given above are reached by lobsters in good fishing areas, such as the coast of Brittany; they are then about ten to fifteen years old. The CAPE SPINY LOBSTER (*Jasus lalandei*) is a popular

The European spiny lobster

item on the menu in South Africa and Australia; the same is true of other *Palinurus* species on the coasts of North and South America. Not only gourmets but also scientists are interested in these animals: the American species *Palinurus argus* has seasonal migrations; in this, hundreds of animals walk one after the other for more than 100 km along the bottom of the sea, and are thus able to discover suitable areas with sufficient food. It is possible that the sounds the animal produces, which were discovered only a few years ago, aid in regulating these migrations.

In the related SCYLLARIDS (family Scyllaridae), the SLIPPER LOBSTERS, the second antennae are very much compressed and covered with spine-like processes which aid in defense; these give the animals a body form which cannot be confused with any other creature. Members of the genera *Scyllarus* (see Color plates, pp. 451 and 459) and *Scyllarides* (see Color plate, p. 460) live on rocky shores and frequently on coral reefs.

The TRUE MACRURANS (suborder Astacura) are best characterized by the EUROPEAN LOBSTER (*Homarus gammarus*; see Color plate, p. 460), and the AMERICAN LOBSTER (*Homarus americanus*) typical representatives of the LOBSTERS (family Homaridae), and by the EUROPEAN CRAYFISH (*Astacus astacus*), a member of the crayfish family (Astacidae).

The lobster

The LOBSTERS inhabit rocky coasts, feeding on mollusks and dead animals during their nightly excursions, after which they always return to their home holes. This fidelity to one area is a result of their excellent ability to orient, something which they have in common with nearly all the Higher Crustacea, and of their learning ability. Seasonal wanderings have also been shown to occur in lobsters. The famous Heligoland lobster became reduced in number and size because too many were caught, and special protective measures had to be taken to save it. The females first become sexually mature at six years of age, when they are about 25 cm in length, and lay about 8000 eggs; females 40 cm in length can produce 30,000 eggs. Old lobsters molt only every second year.

Some lobsters are deep-sea inhabitants; this indicates that these are, evolutionarily speaking, extremely old. In fact, the lobsterlike crustaceans have been well represented in fossil form since the Jurassic. One lobster species inhabiting deep water is the KING LOBSTER or NORWAY LOBSTER (*Nephrops norvegicus*; see Color plates, pp. 451, 459 and 460), which is usually found along the coasts of Brittany and Norway, but has also been caught in the Mediterranean. It is recognizable by its narrow pincers.

Family: Astacidae

The CRAYFISH (family Astacidae) are restricted in distribution to the northern hemisphere. In the southern hemisphere the ecological niche they occupy is taken over by the PARASTICIDS (family Parasticidae) in South America and Madagascar (but not in Africa), and in Australia by the AUSTROASTACIDS (family Austroastacidae).

The EUROPEAN CRAYFISH (*Astacus astacus*; L, ♂♂ up to 25 cm, ♀♀ up to 18 cm) also hunts for its food at night, feeding on both plant and animal material. Male crayfish can live twenty years. The crayfish lives only in very clean water, where it hides in a hole in the bank; therefore its population has decreased drastically because of increasing water pollution. In addition, a fungus disease termed "crab plague" (infection by the fungus *Aphanomyces astaci*) almost wiped out the whole European crayfish population in 1870, and the present-day population is still threatened by it. The AMERICAN CRAYFISH (*Orconectes limosus*; BL about 10 cm), introduced into Europe in 1890, is immune to the disease and can also survive in dirty water. It does not dig holes, and can also be seen feeding during the day. Although it cannot be compared in size with the European crayfish, it is now considered the most common inland crustacean used for human consumption.

The INTERMEDIATE FORMS (suborder Anomura) combine, in their body structure, the macrurans with the brachyurans. Members of this group are the Indo-Pacific mole crabs and the marine and terrestrial hermit crabs, together with the lithodid crabs which are not true brachyurans. In addition, the western Atlantic molecrabs, curious creatures which are completely pincerless and spend their lives buried in the sand, are also included.

Intermediate forms

The MOLE CRAB (*Thalassina anomala*) lives, buried like a mole, in the mangrove swamps of the Indo-Pacific. Here it sieves food particles out of the mud, piling the mud it has sorted in huge hills which can reach a height 2 m and a width of 10 m. Since it feeds completely underground, it must continually enlarge its tunnels. As a result, it can cause great damage to rice paddies near the shore.

Many visitors to the coast know the crabs which hide their soft abdomens in snail shells and more rarely in the shells of other mollusks. These are the MARINE HERMIT CRABS (family Paguridae; see Color plates, pp. 494/5). Since a snail shell is usually spirally wound in a right-hand direction, the abdomen of the crab is usually also bent from the last larval stage (glaucothöe) onward. The last pair of walking legs, together with an area of wart-covered skin, hold the shell tight. The soft abdomen takes part in respiration and thus serves as a type of "gill." The asymmetrical body form of the hermit crab is also recognizable from the pincers, one of which is much larger than the other; the large pincer acts as a type of lid, closing the entrance to the snail shell. Since the crab grows with each molt, the snail shell it inhabits soon becomes too small. To obtain a new home, it must therefore investigate objects in the surroundings to determine whether they would be suitable and the right size. The senses of sight and touch determine the weight of the object as well as its size and "fit."

Marine hermit crabs

An interesting feature of the marine hermit crabs is the great variety of associations they have with coelenterates which are attached to the shell in which the hermit crab lives. Sponges can also grow over the hermit crab's "house." There are a large number of mollusks whose shells are used by the hermit crab; from the whelk of the North Sea coasts to to the straight tusk shell, which is not spirally wound, and the flat shells of the heart cockle. Heart cockles are also used as a refuge by the Indo-Pacific PORCELAIN CRABS (genus *Porcellanopagurus*); understandably, the abdomens of these crabs are straight and no longer have a spiral twist.

Land hermit crabs

All members of the tropical genera *Coenobita* and *Birgus* are termed TERRESTRIAL (LAND) HERMIT CRABS. *Coenobita* hides its soft abdomen in the shells of marine or terrestrial snails; the soft skin of the abdomen, with its numerous blood vessels, takes over the main role of respiration. Such a land hermit crab drowns in water, since the true gills are greatly reduced. The animal returns to the sea only to release its larvae.

This is also true of the ROBBER or COCONUT CRAB (*Birgus latro*; BL up to 32 cm; see Color plate, p. 493) from the Indo-Pacific coasts. The pincers of this crab are so powerful that it can cut coconuts from palm trees, afterward opening the nuts on the ground. It is extremely agile in climbing such palm trees. Land hermit crabs of the genus *Coenobita* have also been observed slowly climbing up trees. Travelers in the tropics have frequently reported that the coconut crab is extremely sensitive to ground vibrations. During the juvenile stages, it still uses a shell to protect its abdomen; later the abdomen is curled forward under the body.

The STONE CRABS (family Lithodidae; carapace diameter over 20 cm; see Color plate, p. 476) have the same habit; these, however, live in cold waters. Their fifth pair of legs penetrates the gill cavities and serves as a gill brush. The species *Paralithodes camtschatica* is also a member of these large crabs. The smaller *Lithodes maja* inhabits the North Atlantic. The larger species form the basis of a growing fishing and canning industry.

An unusual method of feeding is found in the Atlantic MOLE CRABS (family Hippidea). The flagellae of the antennae protrude from the sand as bows. They are furnished with comblike bristles which filter floating particles from the water as the tide retreats. When the antenna is covered with food particles, the crab pulls it in beneath the broad base of the third maxilliped and, in the true sense of the word, licks off the collected food. *Emerita talpoida* inhabits the Atlantic coast of North and Central America. As the water level falls, these curious crabs hurry in groups out of the sand and into the sea, rapidly bury themselves again with their backs to the land, and start "fishing" once more at the "right" depth.

Suborder: Brachyura

The TRUE CRABS (suborder BRACHYURA) are, without doubt, the most highly developed of all crustaceans. Their most primitive representatives are the DROMID CRABS (family Dromiidae), known from the Jurassic. Their carapace is usually covered with sponges or sea moss which the crabs cut to the size of their carapaces and plant on their backs. The SPONGE CRAB (genus *Dromia*; see Color plate, p. 473) does this with the sponge *Suberites*. The genus *Hypoconcha* (whose scientific name means "beneath the snail"), instead of this, simply holds a flat mussel shell over itself.

There is a tendency in many crabs to plant some kind of dead or living object on their backs and so disguise themselves. This is the case in the SPIDER CRABS (genus *Maja*) which belong to the SPIDER CRAB family (Majidae). Old spider crabs no longer disguise themselves. The CALAPPAS (genus *Calappa*), from the Mediterranean and warmer oceans, use their sandy-red-spotted coloration and their rounded form as a means of camouflage, making them look like a piece of shingle. Calappas are a delicacy in Spain and Italy; the German ROCK CRAB (*Cancer pagurus*; diameter about 20 cm; see Color plate, p. 473) is also a favorite item on European menus.

The form of the carapace is similar in the SWIMMING CRABS (family Portunidae). The SHORE CRAB (*Carcinus maenus*) also belongs to this family, although it lacks the characteristic feature of the swimming crabs, the flattened paddle on the fifth pair of walking legs. It is only a moderate swimmer, poor in comparison to the TRUE SWIMMING CRABS (genus *Portunus*; see Color plates, pp. 459 and 493). The true swimming crabs can swim rapidly sideways. In this way they can deftly capture their animal prey by dashing at it from their resting position on the bottom in the coastal zones. The related genus of BLUE CRABS (genus *Callinectes*) lives in the warm seas of Central America, where they are an important source of food for the inhabitants of those countries. The Indo-Pacific crab *Podopthalmus vigil* is provided with enormous eyestalks.

Many genera of the previously mentioned groups of crabs can survive, at least temporarily, in brackish water or even pure fresh water. The FRESH-WATER CRABS (family Potamidae) are permanent inhabitants of fresh water. As in the case of the fresh-water crayfish, they have dispensed with a larval development; fully developed tiny crabs hatch from the eggs. In the lands of the eastern Mediterranean and Asia Minor there are many species of the genus *Potamon*, inhabiting rivers and ditches; they are even found at altitudes of 2100 m.

The Indo-Pacific crab *Lybia tesselata* is an example of the curious body form and amazing behavioral ability found among the crabs. It lives on coral reefs, and when danger threatens it picks up a stinging sea anemone in each pincer and stretches these toward the enemy. German shore crabs also raise the pincers toward the enemy when frightened. Although the behavior of *Lybia tesselata* seems so astounding, it must be emphasized that this defense posture is really a modification of a very basic and primitive threat gesture.

Even Aristotle (384–322 B.C.) appears to have been interested in the variety and strange modes of life of the crabs. He was the first to describe the OYSTER CRAB (*Pinnoteres pisum*; diameter 1–2 cm; see Color plate, p. 493), a soft-skinned crab which frequently lives in Mediterranean bivalves. Such types of commensalism are found in many members of the PINNOTERIDS (family Pinnoteridae), to which over 120 species belong. The females usually remain "true" to their mussel and feed as true commensals from the animal and plant plankton their host wafts by. During the breeding season, they leave the mussel and meet the males, which appear to be free-living. After copulation, they return to the protection of the mussel.

The level of development of the GHOST and FIDDLER CRABS (family Ocypodidae) is especially impressive from all points of view. The GHOST CRABS (genus Ocypode; see Color plate, p. 476) were described by Linné as "very fast runners-away"; hence their name. The ghost crabs inhabit hot tropical shores, building burrows 1–2 m deep—often with two exits— sometimes even building sand pyramids as indicators of their presence. They have extremely well-developed eyes on stalks which can be folded laterally. In certain species, the actual eye hemisphere is unilaterally or bilaterally extended in "supra-optical horns," whose function is not yet clear. Some ghost crabs climb mangroves, turn over their leaves, and, with a rapid snap of their pincers, capture flies resting there in the shade— a behavior pattern which seems extremely birdlike.

Another member of this group is the FIDDLER CRABS (genus *Uca*), well known for their conspicuous courtship behavior. The STALKED FIDDLER CRAB (*Uca stylifera*) of the Indo-Pacific holds the record for the length of the supra-optical horns; because of these, it has great difficulty entering its burrow. As a result, the horns in this fiddler crab are often broken off, which seems to cause no visible detriment to the animal. As in the majority of ghost crabs, and in the fiddler crabs as well, the females have pincers of equal size, with which they pick up the mud in "handfuls" and carry it to the oral appendages to be sorted; the respiratory water store plays a major role in this sorting process. The inedible particles collect at the bases of the oral appendages; from there, the fiddler crab picks them off with its pincers and lays them aside in geometrically arranged patterns as it slowly walks forward, eating. This gives such a pleasant "human" effect that the American naturalist William Beebe once said, "it is always a little difficult to consider the fiddler crab with the objectivity which all naturalists of worth are supposed to use. The fact that they have ten legs instead of two and a hard shell instead of our assortment of internal bones becomes less and less important the more we come to know these various personalities. ..."

The males have only one feeding pincer, the other having been developed into an enormous structure used in courtship; its weight can reach half that of the entire crab. With this pincer, the males perform courtship movements which vary from species to species. In *Uca annulipes* they stretch the pincer sideways and then bring it, still stretched, in front of the body; in *Uca insignis* they wave the pincer in circles over the body, the crab at the same time standing on tiptoe, these signals indicating its readiness to mate. In successful cases, a female waved at in this way approaches the excitedly waving male and then follows him into his burrow, where copulation takes place. This courtship can take place only when the tide is low; the fiddler crabs therefore have to adapt other types of behavior—feeding, fighting, courtship, and body coloration—to the tidal rhythm.

William Beebe and his co-worker Jocelyn Crane gave a classical description of such courtship from the shores of Bahia Honda (Panama) in 1942:

"Sitting at the edge of the mudflat, we could see a dozen species of fiddler crab at the same time and, through the binoculars, observe every detail of their inexhaustible drive without disturbing the animals in the least. The pincer-waving of the male long-legged fiddler crab (*Uca stenodactyla*) was the most entertaining of all. The females and the juveniles, with two tiny pincers, trotted in their brown-spotted coats in the background, but the large males displayed in their most vivid colors: the back white and shimmering blue, the eight legs flaming red, and the huge, waving pincer a beautiful pink. ... As our eyes were drawn to a few shiny white, tiny jitterbug-fiddler crabs (*Uca saltitanta*), whose colony lay further out on the mudflat, we could see that the rhythm of their waving was very different from that of the long-legged fiddler crabs. Apart from this, they never followed females, but continued interminably their jumping on the spot, waving the large pincer backward and forward enthusiastically as soon as they were able to determine the slightest sign of female interest; they were definitely the inventors of the jitterbug. A third group of dancers, dancing fiddler crabs, dressed in ruby red and gray, would have done honor to the Russian ballet, since they stretched both pincers forward with slow, graceful gestures, at the same time raising their bodies on tiptoe and dancing in figures with a featherlike tread. ... Finally we realized the truth: each species had its own courtship dance which differed greatly from that of other species as much as the rhumba differs from the waltz, and each of these dances was just as characteristic and as much a part of courtship as the robin's song, the peacock's fan, and the spring serenade of the frogs. ...

"We observed an especially handsome emerald fiddler crab (*Uca beebei*). He strutted in his finest colors and had tried for at least an hour to attract the attention of a little gray female about 10 cm away, but without success. He stopped his show only for extremely short feeding breaks, but she seemed to be ignoring him, improving her burrow, feeding greedily, and not deigning him a glance. Finally, she stopped eating and appeared to be noticing him for the first time. This gave him a little encouragement, and his dancing became more lively; he swung his huge purple pincer up and down like a madman, and, while raising it proudly, performed curious dragging movements of his eight green legs. The female raised herself and approached a few centimeters nearer, stopping along the way for a little refreshment. The admirer danced faster and faster, and finally brought her completely under his spell. Almost hypnotized, she watched him from about 2.5 cm away.

"Now the character of his dance changed and he slowly turned round

Courtship behavior in the fiddler crab

▷
Left, from top to bottom:
The rock crab (*Galathea strigosa*) can, like the shrimp or lobsterlike crabs, jerk back rapidly with a beat of its powerful tail.
Caprellid (*Stenorhynchus seticornis*).
The portunid (*Portunus puber*) has a dorsal carapace covered thickly with fine hairs. It can reach 7 cm in length, and is found in North Sea bays.
Right, top to bottom:
Birgus latro, the coconut or robber crab, is a land hermit crab. The common swimming crab (*Portunus holsatus*) has oarlike flattenings of the last pair of legs: a characteristic of the swimming crabs.
Pinnoteres pisum, the oyster crab, lives in the mantle cavity of bivalve mollusks.
▷▷
This hermit crab (*Paguristes oculatus*) lives within a snail shell which has been overgrown by the sponge *Suberites domuncula*.
▷▷▷
Pagurus calidus, a hermit crab of the Mediterranean and Atlantic regions. On its shell it carries two sea anemones (*Calliactis parasitica*), which have grown together, and have extruded their acontia (lila).

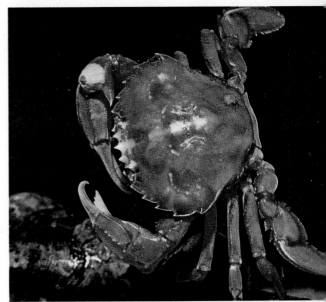

 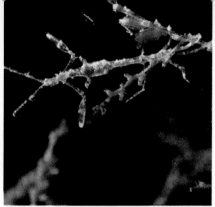 

The Chinese crab

like a mannequin under her gaze so that she could first admire his shiny green back and then his purple pincers. Finally, she approached within reach and he stroked her legs gently with his own; she did the same. They separated for a moment, and he danced a happy jig before dashing into his burrow. His shining pincer was the last to disappear, with a final irresistible wave. The female followed without hesitation."

At night, and in thick vegetation, where the previously described waving gestures cannot be seen, the perception of ground vibrations takes the place of vision. With his large pincer, the male taps signals on the ground in a rhythmic, species-specific series; these tell the female "here is a male ready to mate" or another male "here is the courtship territory of a rival." Since it is possible to copy the tapping code of each species successfully, the investigator can carry out a real "conversation" with the animals. Even in highly populated crab areas, these signals ensure that members of the same species come together. Some species are even able to produce sound by rubbing parts of their bodies together, the same being true of certain ghost crabs; certain *Ocypode* species from Africa and Indonesia, according to observers, are able to sing like birds. In Central America, one species of fiddler crab carries the descriptive name SINGING FIDDLER CRAB (*Uca musica*); another, because of its dancing and waving movements, has been named, after the Greek Muse of the Dance, the TERPSICHORE FIDDLER CRAB (*Uca terpsichore*). In Europe, there is only one crab from this group, found in the Gulf of Cadiz, in Spain, the TANGIERS FIDDLER CRAB (*Uca tangeri*; see Color plate, p. 476).

Related to the ghost and fiddler crabs are the species of the genus *Dotilla*. They form closely packed feeding communities of up to 104 individuals per 1 m², building towers similar to Eskimos igloos; they sit on these until the next flood tide. Even "sport clubs" have been observed: fiddler crabs and ghost crabs run at high speed in pairs or in groups along the shore, without any recognizable reason for doing so. Group movements of this type are also found in the MICTYRIDS or GRENADIER CRABS (Mictyridae), which belong to another family. They inhabit the Indo-Pacific, living in the tidal areas and sorting through the mud for food.

The ROCK CRABS (family Grapsidae), on the other hand, have a definite preference for rocky shores. For example, the RED ROCK CRAB (*Grapsus grapsus*), a member of this group, lives on the rocky cliffs of the Galapagos Islands as well as in other areas. The MANGROVE CRAB (*Aratus pisoni*), of the Caribbean region, climbs mangroves to above the water level, where it feeds, in part, on their leaves. The beautiful *Goniopsis pulchra*, of the eastern Pacific, also belongs to this group. The COLUMBUS or SARGASSUM CRAB (*Planes minutus*), from the Sargasso Sea, can be found crawling over the thick knots of sargassum weed.

The CHINESE CRAB (*Eriocher sinensis*; diameter, ♂♂ 7–8 cm; see Color plate, p. 476), a relative of the rock crabs, has become one of Europe's endemic animals. It was carried to harbors on the North Sea coast in 1912, and has invaded Germany's rivers and canals, reproducing there rapidly. Today it is found in Finland, Sweden, Denmark, Holland, Belgium, and parts of France. Its German name, meaning "wool crab," comes from the wooly hair covering the male's pincers. Every year, all crabs over five years old wander to the sea during or after June; there the males build thick, barred dams in the lower reaches of the rivers, which must be crossed by the females which arrive later. In this way, they ensure copulation with every female. During this migration to the sea, the animals travel between 8 and 12 km a day. The young hatch in the following May in the rills on the shore of the Wattenzee; after caring for their brood, which they attach to their legs, most of the females die. The young develop near the shore, and start wandering upstream by the time they are two years old—even traveling as far as Dresden and Prague. They disperse in the inland waters. Their huge population has made them a pest in fisheries.

A similar migration to the sea is shown by the females of the huge CARIBBEAN LAND CRAB (*Cardisoma guanhumi*; carapace diameter up to 15 cm), which, with their large defensive pincers, are extremely impressive creatures. They become quite tame when cared for by people, and take leaves and fruit from their owner's hands. They often live in the wild, far away from the sea, and dig holes up to 2 m deep, in which copulation probably takes place. In the native markets they are considered a delicacy. They are, of course, extremely difficult to catch, because of their excellent vision and response to ground vibration. The females wander to the sea in large numbers at the time of the full moon, and deposit their well-developed larvae there at the time of the spring tides.

The smaller COMMON LAND CRAB (*Gecarcinus ruricola*; carapace diameter 9 cm), which also belongs to the LAND CRAB family (family Gecarcinidae; see Color plate, p. 476), likewise undertakes migrations to the sea, although it lives many miles inland. In contrast, the common, hole-dwelling genus *Ucides* is found in mangrove swamps not far from salt water. The RED LAND CRAB (*Gigantinus lateralis*), of the Caribbean area, has, to date, been studied almost exclusively in the laboratory; this is because it is very difficult to observe this creature, which is active only at dusk and during the night, and, on top of this, normally lives under the blanket of vegetation. The red land crab deserves the name "land crab" more than its relatives, since it lives exclusively on land and is not tied to a water source. It is able to draw the water it requires from damp substrates. In this animal, too, the reproductive cycle is dependent on the phases of the moon. One interesting fact is that when the water in the tropical streams rises—which it can do at 1 m every five minutes during the rainy season—this crab climbs up the trees to escape, and, if it cannot reach a safe place in time, drowns.

Between the superorder Eucarida, with its about 8500 species, and the Peracarida is a small superorder, the PANCARIDA, which contains a single order (Thermosbaenacea) with six species (BL up to 4 mm)! They are similar to the Peracarida in most of their characteristics, but differ from them by the presence of a dorsal brood chamber. The few known species live in the sand-space system of thermal springs, in warm coastal groundwater, and in cave pools. Their distribution around the Mediterranean and far from the coast in Texas indicate that long ago their ancestors wandered from the sea into the coastal groundwater. The first species discovered (1923), *Thermosbaena mirabilis* (in English, "the wonderful"), lives in Tunisia in the sand-space system of a hot spring in the Oasis El Hama, and is continually washed out into a collecting basin where it crawls about although the water temperature is about 45°C. The crab "freezes" when the water temperature cools to only 30°C!

The superorder PERACARIDA, with its many species, is characterized by a curious brood structure: the female has a brood pouch (marsupium) on the ventral side of the body. It consists of brood plates (oostegites) which grow from the base of the thoracic legs before or after spawning, over-lapping each other to such an extent that a closed chamber is formed between the thoracic legs. The male injects its sperm into this, and the eggs are fertilized within it. Here they undergo their development. Two lines of development are found within the Peracarida, which arose from the mysids. All of the order Amphipoda belong to one evolutionary line, and the cumacids, isopods, and tanaiids belong to the other.

The evolutionary roots of the peracarids branched off into the MYSIDS at a very early stage. These shrimps are the most primitive of the group and are so similar to the most primitive eucarids, the euphasids, that both were once included in a single order. The first pair of thoracic legs is modified to form a maxilliped; the remaining seven, on the other hand, are slender biramous limbs, the interior branches of which serve in walking, the exterior ones, in swimming. The most primitive and largest mysid (BL up to 35 cm!) belongs to the genus *Gnathophausia* and lives in the deep sea. The members of the family Mysidae, which contains numerous species, bear a gravity receptor on the interior branch of the uropod. They are also able to perceive water vibrations with this organ. Its weight (statolith) consists of an organic nucleus around which layers of calcium fluoride are deposited. Since both statoliths are lost at each molt, the shrimp always has to replace the fluoride content. Species of the genus *Praunus* live in shoals in the North Sea and the Baltic. *Mysis relicta* is a species which is valuable from an animal-geographic as well as an evolutionary point of view, since it inhabits fresh water and remained as a relic species in the lakes of Scandinavia, the Mecklenberg lake region, and North America. The ancestor of this species, which inhabited Siberian lakes in the Middle Tertiary (Oligocene), became more widely distributed as the various lakes were divided and separated. In these smaller waters, specific and subspecific development continued. Since they live only

Fig. 16–12. Gravity receptor of a mysid: A. Last pair of legs of *Neomysis integer*: base of the biramous limb, 2. Endopodite with statoliths in the statocyst, 3. Exopodite. B. Statocyst in longitudinal section: the statolith borne on sensory hairs with calcium fluoride externally, internally (dotted) an organic mass.

in cold water layers, and in Mecklenberg, for example, reproduce only during the winter, it is highly likely that they originated in the Arctic.

In the order AMPHIPODS (Amphipoda) the carapace is completely degenerated, so the thoracic segments are exposed. The first thoracic segment is fused with the head; as a result, the thorax consists of only seven segments. On the abdomen (pleon), the first three pairs of legs are swimming appendages which are pointed forward. The three posterior ones (here the fourth and fifth pairs of pleopods, which, like the sixth, are termed "uropods") are jumping legs with only a few segments; the common name of this group, the WATERFLEAS, is due to this habit of springing. In most amphipods, the body is laterally compressed.

Order: Amphipoda

The family of GAMMARIDS (Gammaridae) includes marine and fresh-water representatives; the COMMON WATERFLEA (*Rivulogammarus pulex*) is found everywhere in European streams. Other waterfleas inhabit the deep oceans down to 10,000 m. In springs arising from groundwater, tiny, colorless waterfleas are sometimes brought to light: these are members of the genus *Niphargus*. They are not close relatives of the common waterflea which have wandered into the groundwater; their nearest relatives are partly blind, partly still sighted members of the genera *Eriopisa* and *Eriopisella* which live in the gravel of the sea shore. From here, the ancestors of *Niphargus* wandered into the groundwater. This, however, is not a uniform environment throughout the *Niphargus'* range of distribution. The genus *Niphargus* therefore divided into numerous species and subspecies, as did the genera *Niphargellus*, *Niphargopsis*, and *Microniphargus*. Their distribution gives information about the courses of rivers within Europe from before the start of the ice ages. *Niphargus* also lives at the bottom of the deep Alpine lakes, from 50 m down in Lake Constance, for example.

Members of the SAND FLEAS (family Talitridae) are found not only in the sea and in brackish and fresh water, but also in wet sand. The COMMON SAND FLEA (*Talitrus saltator*), for example, lives above the waterline in heaps of washed-up algae. The lives of both the BEACH FLEA (*Orchestia gammarellus*) and the sand flea depend on returning to their small ecological niche on the shore when they are blown away by the wind. The sun, the moon, and the direction of the polarized light act as their "compass" in this. If brought into the laboratory, the shrimps always move in the direction in which the sea used to be for them; that is, animals from the north side of a bay move south and those from the south side move north. Even young raised entirely in the laboratory will do this; the sea direction appears to be indelibly imprinted in a particular population, the result of a natural-selection mechanism which sentenced to death any shrimp which wandered in the wrong direction.

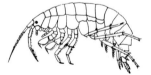

Fig. 16–13. Sand flea *Rivulogammarus lacustris*.

A very much more common inhabitant of Europe's coasts is the family COROPHIIDAE. This includes *Corophium volutator*, up to 4000 of

Fig. 16–14. Ghost shrimp lying in wait for prey.

Order: Cumacea

which can inhabit 1 m² of the shallows. It emerges from its burrow with circling antennae, to seek its food; as a result, an obvious hummock is formed at the entrance to the burrow at each ebb tide, something which cannot be missed by a wanderer over the sandbanks. Members of the GHOST SHRIMP family (Caprellidae) are found in bunches of algae and at the bases of colonial polyps throughout Germany's waters. These are unusual creatures, with sticklike thoracic segments and a reduced abdomen. The last three thoracic limbs are modified to form grasping hooks with which the animal attaches itself to the vegetable or animal "scrub"; the third and fourth pairs of legs are more or less reduced in females in the majority of species; the brood plates are borne on these segments. The two anterior pairs of legs are armed with powerful pincers (subchelae) and serve as prey-capturing organs for the shrimps, which lay in wait for their food and capture it in much the same way as a praying mantis does. They move through the underbrush in the same way as a looper caterpillar, and their bodies are wonderfully adapted to blend in with it. Germany's most common species are the GHOST SHRIMP (*Phtisica marina*; BL, ♂♂ up to 20 mm, ♀♀ up to 16 mm), which has subchelae on the third and fourth pairs of legs, and *Caprella linearis* (BL, ♂♂ up to 20 mm, ♀♀ up to 14 mm), in which the third and fourth pairs of legs are completely degenerated. With reduction of these legs and the abdomen, the family of WHALE LICE (Cyamidae) carries on from the ghost shrimps, but in their adaptation to a life attached to the skin of a whale, their body form is completely different. As in other ectoparasites (lice, feather lice, and ticks), their bodies are wide and flat and their remaining five pairs of walking legs bear powerful pincers with which the whale louse attaches itself to its host. They produce holes in the whale's skin which are as big as the palm of a man's hand and penetrate deep into the blubber. An infestation of thousands of these parasites can greatly damage even the largest whale.

The second line of development from the mysid shrimps is characterized by a complete reduction of the carapace. The respiratory function of its thin-walled interior is therefore also lacking, and the abdominal limbs (pleopods) take over this task. The last abdominal segments fuse with the end one (telson) to form a uniform section. Even the development of the excretory organs went along different lines in this group. Although antennal and maxillary glands are present in the primitive mysids, only the antennal gland has been retained in the more highly developed sand fleas; in this series, it is the maxillary gland which is retained. A gradually increasing level of development can be followed from the cumacids, through the tanaidids, to the isopods, which represent the pinnacle of development in the Crustacea.

The cumacids (order Cumacea) have a curiously shaped body, with a thoracic region which looks swollen and a thin, narrow abdomen. The shrimp *Diastylis rathkei*, of the European coasts, gives an example of a high population density: over 1200 of these animals have been counted in 1 m².

The order TANAIDACEA derives its German name, which means scissor isopod, from the flattened form of its first pair of walking legs, which bears powerful pincers. The carapace is much smaller than that of the cumacids, but still encloses a small respiration chamber on each side. Like the cumacids, the tanaidacids are marine animals; *Heterotanais oerstedi* (BL 2 mm) inhabits the German coasts; *Tanais cavolinii* (BL 5 mm) is much more common, being found in the algal growths of the Mediterranean area.

The last group of the Higher Crustaceans is the order ISOPODA. In the number of its species (about 4000) and its ability to adapt to its environment, it is at the same peak of development as the crabs. From the marine environment, they have conquered not only fresh water, but also the land. They no longer have a carapace. Their thoracic limbs are powerful walking legs, while the abdominal legs, on the other hand, are leaflike and usually so broad that the two branches can no longer lie side by side and the external branch therefore covers the interior one. Since a carapace is lacking, these appendages have taken over the task of respiration.

Members of the suborder FLABELLIFERA are provided with a tail fan. This suborder includes FISH ISOPODS of various families—which live partly on dead fish, partly on living ones held helpless in gill nets, and partly as parasites on their fish hosts—together with the family LIMNORIIDAE, which live on wood. Species of the genus *Limnoria* (BL 5 mm) —in Europe, *Limnoria lignorum*—destroy wooden ship hulls, harbor jetties, bridge piles, and other structures, despite their small size, by making their burrows close together. With the aid of a cellulose-splitting enzyme in their intestines, they can digest the wood; the fungus which invades their burrows is also used as food. Members of the suborder ASELLOTA are found in marine habitats from the coast to the deep sea (down to 9790 m). All the members of the family ASELLIDAE are fresh-water inhabitants, of which the COMMON FRESH-WATER ISOPOD (*Asellus aquaticus*) is found only in stagnant or gently flowing water. It lives on decaying vegetable material and reproduces twice a year, during spring and again in fall. It survives the winter even in shallow waters, since it is able to endure being frozen in the ice. The fresh-water isopods are an important link in the food chain of fresh-water areas, as primary consumers. The closely related CAVE ISOPOD (*Asellus cavaticus*) has no eyes, like many cave-dwelling animals, and is almost transparent. It is interesting that the species distributed in cave waters throughout southern, western, and central Germany, Switzerland, northern Austria, Belgium, France, and southern England, does not belong to the same subgenus as the common fresh-water isopod, but to a subgenus (*Proasellus*), which is mainly present in the Mediterranean area, and which is no longer found within the distribution area of the cave isopod. "This indicates that in earlier eras, even before the ice ages, the subgenus had a wider distribution area even above ground, from which it was driven out, and present-day *Asellus cavaticus* has only remained as a relic" (Stammer 1932).

Order: Tanaidacea

Suborder: Flabellifera

Suborder: Oniscoidea

One of the most successful groups of the Higher Crustacea is, without doubt, that of the LAND ISOPODS (suborder Oniscoidea), since it is found in all parts of the earth, in damp cellars and in hot, dry deserts, from the coastal areas to high in the mountains, from the equator to the northern tundra and the southern steppes. Despite this, these animals require moisture and even a real drop of water, although in small amounts. Since they are gills, the interior branches of the abdominal limbs must always be covered by a thin film of water, which is protected from evaporation by the exterior limb branches. This water is renewed by the last abdominal limb, the uropod, which forms a sucking tube through which water is brought to the gills. Even in the thoracic segments of many land isopods, there is a special water-conduction system on the ventral surface, which also carries water to the gills. They replace evaporated water with water from their food; in addition, they drink through both the mouth and the anus. They also seek out damp, protected places during the day, leaving these only during the moist night. Cellar woodlice and pill woodlice have, in addition to the gills, true air-breathing organs in the "white bodies" found on the side of all or certain of the exterior branches of the abdominal limbs. These are branched breathing tubes (trachea) similar to those of the spiders, millipedes, and insects, but developed independently of these. Almost all the species feed on plant debris, but also on living plants if these are soft and fleshy enough.

The most primitive of all the land isopods, present both in water and on land, is the BEACH WOODLOUSE (*Ligia oceanica*; BL 3 cm; see Color plate, p. 496), which lives on rocky beaches and on harbor constructions. During the day, it remains in cracks in the rocks, slightly above the water level, and flees extremely rapidly from one crack to the next if disturbed. Its escape route often ends in the water, within which it walks rather than swims. Held continually under water in an experiment, it lasted for eighty-three days, and could even feed and molt there. Of all the land isopods, this species is the most amphibious. It does not possess trachea. This is also true of the blackish SOWBUG (*Oniscus asellus*; BL up to 18 mm; see Color plate, p. 496), which is a purely terrestrial animal. The COMMON WOODLOUSE (*Porcellio scaber*), which is often found together with this species, has tracheae on the first pair of abdominal limbs.

The PILL WOODLOUSE (*Armadillidium vulgare*; see Color plate, p. 496) has tracheal organs on all five pairs of abdominal legs except the last, the uropod. This species is the most advanced, in its adaptation to a terrestrial life, of all the land isopods. It even inhabits warm walls and emerges during the day. It is able to roll into a ball; its body, of all body shapes, has the smallest relationship between surface area and volume, and, as a result, the least amount of water evaporation. In addition, when the animal rolls up, the thin skin between the armored plates and the abdominal area is enclosed, thus reducing water loss as well as protecting the limbs from attack by enemies. These land isopods have thus trod the

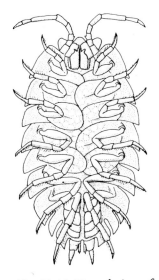

Fig. 16–15. Ventral view of common woodlouse.

same evolutionary path as the pill millipedes, which are found in the same habitat and are easily confused with the isopod when both are rolled up. The two have arrived at these adaptations independently, however.

The tiny, eyeless, white ANT ISOPOD (*Platyarthrus hoffmannseggii*; L 2–4 mm) is at home in special, moist, dark areas; it is found only in the ground nests of ant species. About two dozen kinds of ants can harbor this commensal, but it prefers those hosts which secrete large quantities of formic acid. The isopods seek out an area of the nest which is the most thickly populated with ants, where they find their main source of food, the ants' feces. They can be kept experimentally without ants just as ants can survive without their "street sweepers." While the ants usually take no notice of this inhabitant of their underground burrows, sometimes a newly arrived isopod is attacked. When this happens, it lies as if paralyzed, with the end of its abdomen directed toward its attacker. A glandular secretion then emerges from the exterior branches of its last pair of legs, smearing the ant's mouth. If the ants move to another nest, the isopods usually remain in the old one for some time, but will then follow them. They find their way by sensing the traces of formic acid left by the ants. Experiments have shown that they are able to follow continuous spoors by themselves. Apart from ant feces, they also eat fungal spores, unicellulates, and decayed plant material, preferring sugar above all; this is offered them as nectar from the root aphids found within the ants' nest (see Vol. II). As a result, the isopods can, to a certain extent, be food competitors with the ants. Usually female ant isopods are far more common than males. In Europe, they usually reproduce twice, at the most three times, per year. The low number of young produced is an adaptation to life in the ant colony, where there are few dangers and few losses. Ant isopods are often found beneath stones where ants nests are built. (After looking, please replace these carefully !) Within the milling crowd of ants, the white isopods are very conspicuous, but they soon flee to the depths of the nest, proof that they can still perceive light although they have no eyes. The ant isopods are commonest in the Mediterranean region, their main area of distribution.

We will close the varied series of the Higher Crustacea with the suborder PARASITIC ISOPODS (Epicaridea), which have still been little studied. They are parasitic mainly on other crustaceans, sucking their blood. The young animals still retain the isopodal shape, and the males keep this to maturity. The females, on the other hand, change their body form completely: the oral appendages become sucking pegs with piercing organs, and the remaining appendages disappear. Their "crustacean" nature is then no longer recognizable.

The ant isopod

Suborder: Epicaridea

17 The Tracheates

Subphylum:
Tracheata, by
P. Rietschel

The TRACHEATES (subphylum Tracheata) are air-breathing arthropods like the chelicerates, and are highly adapted to a terrestrial mode of life; even the aquatic members of this group have arisen from terrestrial forms. They obtain their name from the organs they use to breathe air, the "trachea"—tubelike invaginations of the epidermis and its covering layer (cuticle)—which branch within the body and form a network, the finest endings of which terminate in all the organs, tissues, and cells of the animal. The tips can even enter cells and provide them with oxygen from the air, a task fulfilled in most other animals by the blood. In the tracheates, their body fluid has been freed of this task: as a result, the body liquid termed "blood" in tracheates only rarely contains blood pigments which bind oxygen during its transport. In the most primitive state, the trachea open in each segment with a pair of breathing pores (stigma). Trachea can be joined between segments by longitudinal branches, and within a single segment by a lateral branch. Other terrestrial arthropods have also developed trachea, such as the onychophorids, many chelicerates, and certain crustaceans; but these air-breathing organs arose independently of those of the tracheates. They are "analogous" to them, but not "homologous."

The tracheates are also uniform in the structure of the cephalic appendages. They are more closely related to the Crustacea than the Chelicera. Originally, tracheates possessed two pairs of antennae like the Crustacea: during early development, they are formed temporarily, but then lost.

The tracheates are usually divided into two classes: A. Millipedes (Myriapoda), B. Insects (Hexapoda or Insecta; see Vol. II). Of the two, the millipedes, with respect to their body segmentation, the structure of their tracheal system, nervous system, sense organs, and their ecological adaptation, are the more primitive.

Class: Myriapoda,
by O. Kraus

Several groups of tracheates are included as millipedes (class Myriapoda). Externally, they all have one thing in common: their body consists of a large number of relatively similar segments. As a result, they have a relatively large number of legs, varying from eight to about 240. Their name "millipedes" ("thousand feet") is therefore an overstatement.

This great difference in the number of segments could indicate that the millipedes are not an evolutionary uniform group. Other characteristics of their body structure, especially the head capsule and mouthparts, point in the same direction. It is probable that the true millipedes, pauropods, centipedes, and symphylids form a class of their own, although the symphylids show definite similarities in structure to the insects. Since the relationships between these groups have not yet been determined, the general term "millipedes" will be still used for them here. From an evolutionary point of view, they are on a par ("sister group") with the insects, and at the very least are associated with them through the symphylids. All millipedes are terrestrial; in rare cases (*Hydroschendyla*) they have penetrated as far as the tidal zones of sea coasts.

Since only a few fossil remains of millipedes exist, no evolutionary series can be determined. They are known to have been present in the Upper Silurian, represented by the Archipolypoda (genus *Euphoberia*). Diplopods are known from the Upper Carboniferous (genera *Anthracoiulus*, *Glomeropsis*), centipedes since the Upper Cretaceous (genus *Calciphilus*). It is interesting that of the giant forms found in the Carboniferous, the genus *Acantherpestes*, for example, some, such as *Arthropleura*, have a very primitive body structure.

The subclass DIPLOPODA (BL up to 28 cm, usually much smaller) are also termed "true millipedes." These form a very old group of animals, since fossil remains have been obtained in large numbers from Paleozoic rocks. Bohemia's coal deposits, for example, have supplied many well-preserved remains.

As a result of their extreme age, the diplopods have experienced many changes of the earth's surface. For example, it has been proved without doubt that South America and Africa were once connected, forming a single continent. The distribution of the julid family Spirostreptidae reflects this even today. Colloquially speaking, one could say that they have "just noticed" that their distribution area has been severed by the Atlantic Ocean—a process which occurred during the Triassic and was completed during the Cretaceous era (about 80 million years ago).

In the diplopods, the abdominal segments are joined in pairs so that a double segment is formed; in more highly developed species this is a closed cylinder, somewhat tapered anteriorly. In this way, anterior segments can be telescoped into posterior ones. As a result of the doubled segments, there are also two pairs of appendages per segment (exception: the first segment lacks all appendages, second to fourth rings have only one pair). The head capsule has segmented antennae and two lateral eyes which consist of aggregations of simple eyes (ocelli). Some forms are completely blind. The most anterior oral appendages are developed into powerful mandibles. Following these is another pair of segmental append-

Subclasses: Diplopoda, Pauropoda, Chilopoda, and Symphyla, by O. Kraus

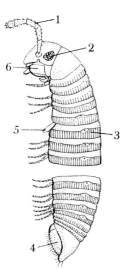

Fig. 17-1. Julid (only anterior and posterior ends shown). 1. Antenna, 2. Eyes, 3. Opening of gland secreting defensive substance, 4. Anal lobe, 5. Reproductive organ, 6. Mandibles.

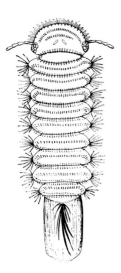

Fig. 17-2. Pselaphognath.

ages belonging to the head, which have grown together in the center, closing the oral cavity from below. Since very similar relationships are also found in the pauropods, the two groups can be joined together to form a higher category—the "Dignatha" (two jaws). The posterior of the body consists of an end piece bearing three lobelike processes, a ventral unpaired and two lateral paired plates, which open like wing doors when the animal defecates.

The majority of diplopods inhabit moist environments, especially the mulch and loam deposits in woodlands. Apart from a few carnivorous cave-dwelling forms (genus *Apfelbeckia*), they live on rotting vegetable matter, but also feed on algae, fruit, and, occasionally, dead animals. They are protected from their natural enemies in most cases by their extremely hard, armorlike exoskeletons. Apart from this, chemical defense substances are found throughout the group. The POLYDESMIDS, usually easily recognizable by their typical body form, produce hydrocyanic acid.

A lady zoologist collected large living examples of this group in Africa and put them in a plastic bag. She told me that diplopods jammed together in such numbers kill themselves in their own cyanide. Apart from this, she was also a little careless or even ignorant of the effects, for as she opened the plastic bag to introduce newly captured animals, she must have breathed in some of the gas, which made her feel ill and dizzy.

Other diplopods, the JULIDS, produce an even more dangerous poison: a mixture of two quinones. This is not only a bacteriacide but also a virulent skin and membrane irritant; it has an extremely burning taste. Since it is very similar chemically to primine (the poisonous agent in primulas), it is no wonder that the natives fear these "quinone animals" as producers of allergies. A third group of defensive substances was recently discovered in the glomerids: Schildknecht discovered an alkaloid whose bitter taste exceeded even that of quinine. A mouse which has picked up such a glomerulid in its mouth would certainly lose all interest in the animal afterward!

The reproductive organs of diplopods do not open at the end of the body but at about the point in which the second pair of legs joins the trunk. The male picks up the sperm with modified legs (usually those of the seventh body segment); the actual act of sperm transfer is brought about with the aid of these "copulatory feet." Larvae with very few body segments—and, as a result, very few pairs of legs—hatch from the eggs, which are laid some time after fertilization. The number of segments is increased from molt to molt until, finally, the correct number is reached. The budding of these body segments is reminiscent of the "teleoblastic" addition of body segments found in the annelids; perhaps this ancient process was continued and "carried on" into the evolution of the diplopods.

An extremely unusual process has been observed in the JULIDS (family Julidae): the male does not die after copulation, but molts instead; only budlike, regressed copulatory limbs are present after this time. In this form, it is similar to an animal before the molt into the sexually mature individual. After another molt, which is basically similar to the molt preceding sexual maturity, a male capable of copulating is formed again. This process can be repeated several times. If the biological meaning of this process is questioned, one can only state that this assures a surplus of sexually mature males, increasing proportionally the chance of all females to be copulated with. Speaking anthropomorphically, one could state that each male repeats its "second childhood" several times.

The number of known species is about 8000, but the number actually present is likely to be much larger. Tropical areas, especially, have not yet been seriously searched. This is driven home by the fact that collections of these animals brought back from remote areas contain far more unknown than known species. Here we can only illustrate the most important groups by means of a few examples.

All PSELAPHOGNATHS (order Pselaphognatha) are relatively small (BL up to 2–5 mm). Their external body covering is—in contrast to that of all the other diplopods—soft and non-calcareous. *Polyxenus lagurus* (see Color plate, p. 509) is common and widely distributed throughout Europe. These animals, consisting of a head and eleven body segments, have numerous delicate bristles arranged in a definite pattern on their upper surfaces. Two thick brushes of fine, glassy hairs are present posteriorly. Pselaphognaths are found everywhere under bark; I have found them regularly under the bark of apple trees in neglected orchards, especially in moist areas. They feed on algae and decaying bark. There is only one species in central Europe which reproduces entirely parthenogenically; another form in which both sexes are present is found on Sylt, a German island.

The Pselaphognatha are separated from the other diplopods, which are termed the CHILOGNATHATES (order Chilognatha). They are all more or less hard-bodied, since calcium salts are deposited in their exoskeletons.

The previously mentioned glomerids belong to the first suborder (Opisthandria), their most important genus being *Glomeris*. The most common species in Europe is *Glomeris marginata* (BL 7–20 mm), which is regularly found in woodland leaf mold. The color plate on page 509 shows this smooth, shiny animal rolled up. The GIANT MILLIPEDES (family Sphaerotheriidae) of the South African and Indo-Australian regions are much larger. The color plate on page 509 shows an African member of this group (a female) in ventral view; it can be seen clearly that these two animals have not yet developed a "telescopic cylinder" (something common to all Opisthandria). The dorsal plates cover their bodies like shields above, and the ventral surface consists of an appendage-bearing ventral plate to which a longitudinal row of lateral plates is attached on each side.

The POLYDESMIDS (Polydesmida) are the next members of the second suborder (Proterandria) to be mentioned here. In these animals, all of

▷
1. Giant millipede from tropical Africa (ventral view, natural size), 2 and 3. *Glomeris marginata*: 2 Rolled up, lateral view, 3 Extended, dorsal view, natural size. 4 and 5. Julid *Cylindroiulus londinensis*: 4 Rolled up, 5 Extended, natural size. 6. Julid *Schizophyllum sabulosum* (natural size; see Color plate, p. 496). 7. Giant julid *Graphidostreptus* from Africa (natural size). 8 and 9. Polydesmids (*Polydesmus complanatus*): 8 Natural size, 9 Enlarged 2½ times. 10. Giant polydesmid *Platyrrhacus* (natural size) from the Amazon. 11. Pselaphognath (*Polyxenus lagurus*) enlarged 30 times.

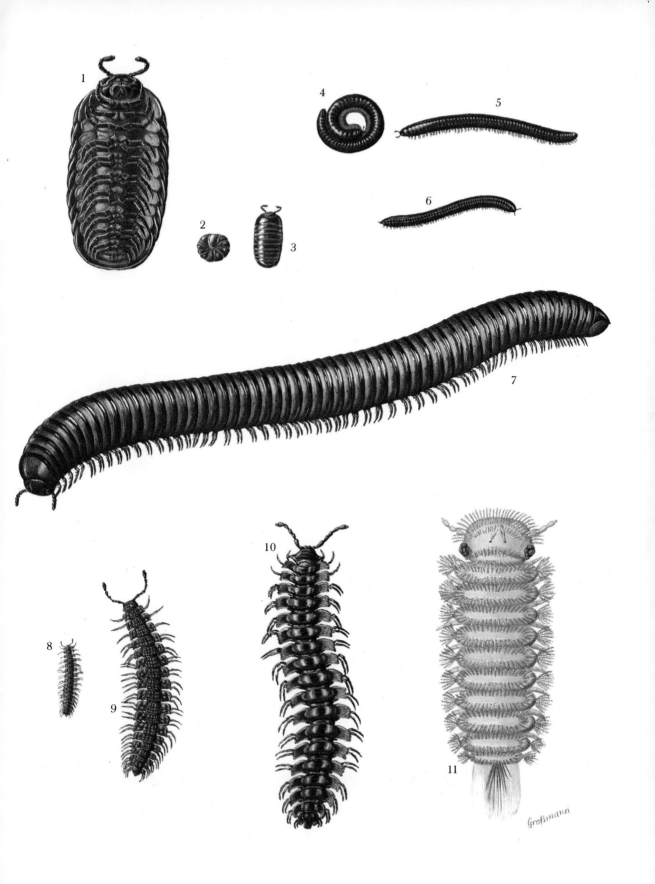

Großmann

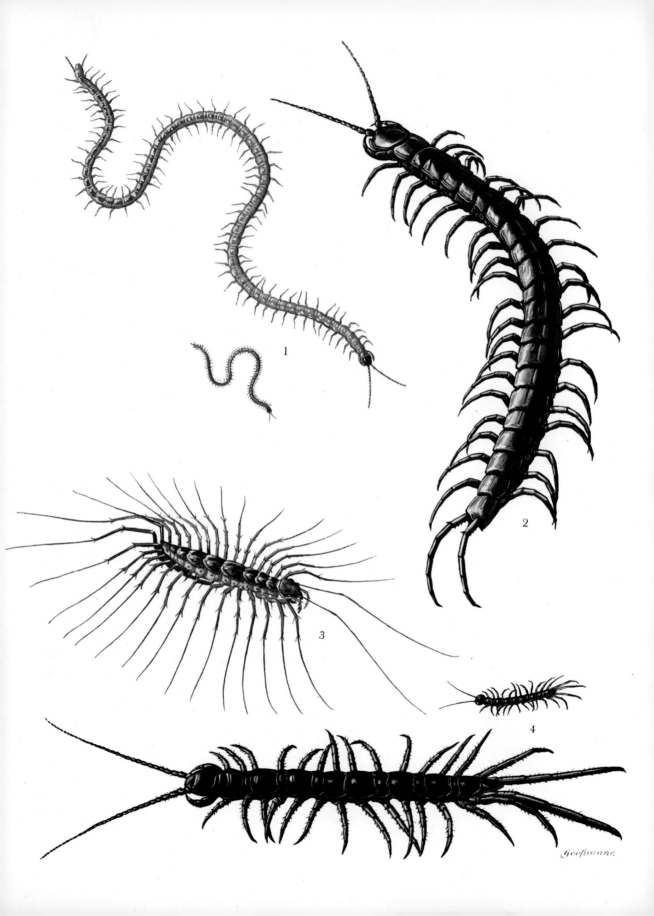

1

2

3

4

Großmann

Chilopods (subclass Chilopoda): 1. Luminous centipede (*Geophilus electricus*) below, natural size, 2. Centipede *Scolopendra subspinipes*, 3. Scutigeromorph *Scutigera coleoptrata*; 4. The so-called "millipede"—the brown centipede (*Lithobius forficatus*), above in natural size.

which are blind, the anterior half of each double segment is a simple cylinder; the posterior half, in contrast, nearly always bears a winglike lateral process on each side, from which the defensive gland pores open. The COMPRESSED POLYDESMID (*Polydesmus complanatus*; see Color plate, p. 509), widely distributed in central Europe, is frequently found under stones and wood. There is a whole series of similar species which look alike. Tropical forms sometimes reach a large size, such as *Platyrrhacus*, which lives in the rainforest of the Upper Amazon.

In contrast to these, the members of the JULIDAE have an absolutely cylindrical body form, and the number of body segments is much larger (up to 121). *Cylindroiulus londinensis* (= *C. teutonicus*; BL 18–37 mm: see Color plate, p. 509) lives mainly in the open plains areas of Europe, and can sometimes appear in such numbers as to be a pest in cultivation. *Schizophyllum sabulosum* (BL 15–30 mm; see Color plate, p. 509) is a brightly colored and relatively narrow animal with two longitudinal yellow stripes on its back. It lives in central Europe in high plant growth and thickets; its conspicuous coloration is a warning coloration (quinone animal). Very large, almost finger-thick julids live in tropical areas. A species of the tropical African genus *Graphidostreptus* is shown as an example (see Color plate, p. 509). There are numerous genera which are similar in appearance.

Subclass: Pauropoda

The PAUROPODS (subclass Pauropoda) have a head-capsule structure which is very similar to that of the diplopods: the two groups, which are probably closely related to one another, have only two pairs of oral appendages. Occasionally, a few very primitive forms with twelve dorsal plates and eleven pairs of limbs are found: usually there are six dorsal plates, however, and nine pairs of legs (number of dorsal plates lower than number of body segments; and these are usually only obvious from the ventral side).

Pauropods are delicate, sometimes colorless (white) or sometimes darkly colored ground animals which occur in leaf mold, under stones, and in cracks in the ground. They are dependent on moist surroundings; it is known that, in the case of a few species at least, death follows within a few minutes if they are exposed to normal room temperature and humidity, as a result of desiccation. As far as is known, pauropods feed mainly on fungoid mycelium which are bitten and sucked dry. To date, 370 species have been described, of which the central European *Pauropus silvaticus* is illustrated here as an example (Fig. 17-3).

Subclass: Symphyla

The SYMPHYLIDS (sublcass Symphyla; BL up to 8 mm, usually smaller) differ basically from the diplopods and pauropods in the structure of their head capsule; the symphylids have three pairs of cephalic appendages. This means that another body segment has been included in the head, separating them from the two subclasses previously described. The symphylids have this characteristic in common with the insects, and this, together with other structures, indicates that the symphylids are close relatives of the insects.

There are twelve pairs of walking legs; at the end of the body there are two styli (modified pairs of legs). The number of dorsal plates—in contrast to the relationship present in the pauropods—is greater than the number of pairs of walking legs (this has arisen by sub-segmentation of originally unified dorsal plates during development).

Fig. 17-3. *Pauropus sylvaticus.*

The pale symphylids—as their delicacy indicates—also inhabit moist environments near the ground; they are found in litter and leaf mold, and under stones. They feed on soft plant parts, including roots, and can become a pest in plant nurseries and greenhouses; they also feed on small members of the Segmentaria. About 120 species are known. The DWARF CENTIPEDE (*Scutigerella immaculata*; BL up to 8 mm) will be cited here as an example; this species is distributed throughout North America, Europe, and northern Africa, and is also found in Hawaii, often becoming a pest, as described above.

Fig. 17-4. Dwarf centipede.

There are also three body segments included in the head capsules of the CENTIPEDES (subclass Chilopoda; BL up to 26.5 cm). The basic difference between the two groups, however, is that only the two anterior pairs of limbs serve as oral appendages in the true sense of the word; the third pair is palplike in form; it can, as Kaestner so graphically described, "hold pieces of food but not break them up." Centipedes can be recognized immediately by the characteristic structure of the appendages present on the first trunk segment. These are modified to form powerful "jaws." The anterior border of the basal plate almost always bears rough teeth, while the tip of the appendage consists of a powerful, pointed claw containing poison glands.

Subclass: Chilopoda

Each of the body segments in centipedes bears a dorsal plate; the number of pairs of walking legs is always uneven, varying from 15 to more than 171; the eyes may be completely lacking, or there may be only a few simple eyes (ocelli); in the scutigeromorphs, large, complex eyes may be developed. They are all ground-living animals which feed mainly on other creatures.

The centipedes can be grouped into four orders: luminous centipedes; true centipedes; lithobiomorphs; and scutigeromorphs. The first two groups have the following in common: on leaving the egg, the larva already has the complete number of body segments. As a result, they are often put together in a group termed the "Epimorpha." The members of the other two orders, however, hatch from the egg with only four to seven pairs of walking legs, and the complete number of body segments is attained only during the molts preceding maturity; they are termed "Anamorpha," in contrast to the "Epimorpha."

The LUMINOUS CENTIPEDES (order Geophilomorpha; BL up to 22.5 cm, usually much smaller) are light yellowish-brown, delicate in structure, and easily recognizable by the large number of body segments (always more than thirty, sometimes up to 171). They move relatively slowly. They live under stones and in ground litter, especially that of the

Order: luminous centipedes

upper ground levels, where they live in cracks or bury themselves. *Pachymerium ferrugineum*, a species common in Europe and elsewhere, secretes a defensive substance, bitter to the taste and smelling like almonds, from pores in the ventral plates: this, as was shown in 1968, is hydrocyanic acid. *Geophilus electricus* (BL 3.5–4.5 cm; see Color plate, p. 510) is also a common European member of this group. The animal owes its name to a luminous substance which is also secreted through pores in the ventral plates. There are at least 1000 known species of luminous centipedes.

Order: true centipedes

The TRUE CENTIPEDES (order Scolopendromorpha; BL up to 22.5 cm) are usually powerful, reddish-brown to dark brown and greenish animals. A few have eyes; smaller forms are often blind. Centipedes are recognizable by the number of pairs of legs: they always have 21 or 23. In contrast to the luminous centipedes, they are rapid runners; they also live under stones, beneath which many species build a chamber or tunnel system. The larger representatives of warmer lands, even those of the Mediterranean area, are extremely dangerous creatures, and their bite is very painful. This is true in the case of *Scolopendra subspinipes* (see Color plate, p. 510), which is distributed throughout tropical areas and is also found in subtropical regions.

Order: Lithobiomorpha

In the LITHOBIOMORPHS (order Lithobiomorpha) the number of pairs of legs and body segments is smaller: the last two pairs of the fifteen pairs of legs present in fully grown animals are always somewhat elongated and are termed "dragging legs." The body length is about 4.5 cm, but is usually smaller (2–3 cm). These are extremely rapid runners, especially under stones. The prey is found with the organs of touch, especially the antennae; if the antennae are removed, the animal starves to death. *Lithobius forficatus* (BL up to 3.2 cm; see Color plate, p. 510) is one of the commonest larger forms of central Europe. There are about 1000 "species" described, but the actual number of true species is probably less than this.

Order: Scutigeromorpha

The extremely long-legged SCUTIGEROMORPHS (order Scutigeromorpha) are similar to the lithobiomorphs in many aspects; they also have fifteen pairs of walking legs, the narrow end segments of which are divided into numerous subsegments. The respiratory system is especially curious. Instead of the usual tracheal system found in the myriapods, single respiratory slits are found in the middle of the posterior border of the dorsal plates; these are a special structure found only in this group ("tracheal lungs").

Scutigeromorphs are very fleet hunters which, at full speed, can hardly be followed by the eye. The largest variety of forms is found in the tropics. The scutigeromorph of the Mediterranean area, *Scutigera coleoptrata* (BL up to 2.6 cm; see Color plate, p. 510), is found as far as central Europe, but is restricted to climatically appropriate areas there, such as the Kaiserstuhl area of southern West Germany. It is most commonly found in vineyards.

Systematic Classification

Owing to the large number of species included among the Lower Animals, it is impossible to include all genera and species in this systematical survey. The majority of phyla and classes included in this volume include far more animals than could be mentioned in the text. The page numbers with ★ next to them refer to color plates and marginal diagrams, page numbers without ★ to the main article; a dash instead of a page number indicates that this genus or species has not been mentioned in the text. † indicates fossils.

SUBKINGDOM UNICELLULATES (PROTOZOA)

CLASS FLAGELLATA

Order Chrysomonadina

Family Chromulinidae	—	**Family Ochromonadidae**	—
Chromulina rosanoffii	96	*Dinobryon sertularia*	96
Family Rhizochrysidae	—		
Rhizochrysis, Chrysarachnion	89	**Family Coccolithophoridae**	96

Order Cryptomonadina

Chilomonas paramaecium	97

Order Phytomonadina

Family Chlamydomonadidae	—		—
Haematococcus	107★	**Family Volvooidae**	
Chlamydomonas nivalis	—	*Volvox*	81

Order Euglenoidina

Family Euglenidae	—	*Phacus*	95
Euglena viridis	92	**Family Peranemidae**	—
E. gracilis	92	*Peranema trichophorum*	95

Order Dinoflagellata

Family Ceratidae	—	*Goniaulax*	98
Ceratium hirundinella	97	*Gryodinium*	—
		Ornithocercus magnificus	88★
Family Noctilucidae	—	*Peridinium divergens*	88★
Noctiluca miliaris	97	*Histioneis remora*	88★
		Ceratocoris horrida	88★
Family Gymnodinidae	—	*Erythropsis pavillardi*	97★
Gymnodinium catenella	98	*Merodinium*	115
G. pascheri	98		

Order Protomonadina

Family Eumonadidae	—	*T. brucei*	99
		T. rhodesiense	99
Family Craspedomonadidae (Choanoflagellates)	98	*T. melophagium*	100
Haeckel's protosponge,	99	*T. evansi, T. equinum*	100
Protospongia haeckeli	99	*T. equiperdum*	100
Salpingoeca amphoroideum	173★	*Leishmania tropica*	100
Family Trypanosomatidae (Trypanosomes)	99	*L. donovani*	100
Trypanosoma gambiense	99	*Costia necatrix*	98

Order Diplomonadina

Octomitus muris	101	G. muris	101
Giardia (Lamblia) intestinalis	101		

Order Polymastigina

Family Trichomonadidae	102	**Family Prysonymphidae**	—
Trichomonas ardindelteili	—	**Family Hypermastigidae**	102
T. hominis, T. tenax (= elongata)	102	Joenia annectus	88★
T. vaginalis	102	Trichonympha turkestanica	102★
T. fecalis	—	Microspironympha porferi	102★
		Kofoidia loriculata	102★
Family Calonymphidae	102	Hoplonympha, Rhynchonympha	102
Calonympha grassii	88★	Urinympha, Barbulanympha	102

Order Opalinia

Protopalina, Zelleriella, Cepedea	103	O. dimidiata	103
Opalina ranarum	103		

CLASS RHIZOPODA

Order Amoebina

Amoeba proteus	104★	Naegleria, Pelomyxa palustris	105
A. polypodia	104★	P. binucleata	104★
Thecamoeba verrucosa	—	Entamoeba coli, E. gingivalis	106
Hartmannella	104	Dysentery amoeba, E. histolytica	106
Astramoeba radiosa	—	Malpiphiella mellificae	109

Order Testacea

Euglypha alveolata, Arcella	110	Difflugia pyriformis	110
Lequereusia spiralis	110	Centropyxis aculeata	110

Order Foraminifera

Globigerina	110	P. pertusus	107★
G. hulloides	93★	Nummulites orbiculatus	93★
Heterostegina	110	Bulimina inflata, Nodosaria spinacosta	93★
Lagena interrupta, L. spiralis	93★	Bolivina alata	93★
Miliola striolata, M. reticulata	93★	Polystomella aculeata	93★
Peneroplis planata	93★	Elphidium (=Polystomella) crispum	110★

Order Heliozoa

Suborder Actinophrydia

Actinophrys sol, Actinosphaerium eichhorni	112

Suborder Centrohelidia

Order Radiolaria

Suborder Acantharia

Acantholoncha flavosa, Stauracantha quadrifurca	94★

Suborder Spumellaria

Hexacontium asteracanthion	94★

Suborder Nassellaria

Triceraspyris gazella,	94★	*Spiralis*	—
Crytophormia	—	*Clathrocanium reginae*	94★
Pterocorys rhinoceros	94★	*Calocyclas monumentum*	94★

Suborder Phaeodaria

CLASS SPOROZOA

Order Gregarinida

Suborder Schizogregarinida

Suborder True Gregarines (Eugregarinida)

Monocystis	116	*G. steini*	117
Gregarina blattarum, G. cuneata	117	*Corycella armata*	116★
G. polymorpha	117		

Order Coccidia

Suborder Schizococcidia

Family Eimeridae	118	**Family Haemosporidae**	119
Eimeria stiedae	118	*Plasmodium*	52
E. tenella, E. zurnii	118	Malaria parasite, *P. vivax*	120
Isopora, Toxoplasma gondii	118	*P. ovale, P. malariae*	120
		P. falciparum	120

CLASS CILIATA

Order Holotricha

Suborder Gymnostomata

Water bear, *Didinium nasutum*	128	*Dileptus anser*	128
Coleps hirtus, Lacrymaria olor	128		

Suborder Trichostomata

Colpoda cucullus	128

Suborder Hymenostomata

Paramecium	122	*P. caudatum*	124
P. aurelia	124	*Colpidium colpoda*	128
Green paramecium, *P. burseria*	124	*Ichthyophthirius multifiliis*	128

Suborder Astomata

Order Peritricha

Suborder Sessilia

Vorticella	129	*Ophrydium versatile*	129
V. nebulitera	113★	*Carchesium*	129
Epistylis, Opercularia, Zoothamnium	129		

Suborder Mobilia

Trichodina domerguei	129	*T. pediculus*	129

Order Spirotricha
Suborder Heterotricha

Stentor, *Stentor*	129	*Bursaria truncatella*	108★
Gray stentor, *S. roeseli*	130	*Balantidium coli*	130
Blue stentor, *S. coeruleus*	130	*Spirostomum ambiguum*	130
Green Stentor, *S. polymorphus*	130	*Nyctotherus cordiformis*	130

Suborder Hypotricha

Stylonychia mytilus	130

Suborder Oligotricha

Halteria grandinella	130	**Family Tintinnidae**	131

Suborder Entodiniomorpha

Entodinium caudatum	113★	*Diplodinium denticulatum*	113★

Order Chonotricha

Spirochona gemmipara	131

Order Suctoria

Choanophrya infundibulifera	132	*Ephelota gemmipara*	132
Tachyblaston ephelotensis	132	*Dendrocometes paradoxus*	132

CLASS CNIDOSPORIDA
Order Myxosporidia

Family Myxobolidae	—	*M. pfeifferi*	133
Myxobolus	133	*Henneguya*	107★

Order Actinomyxidia
Order Microsporidia

Family Nosematidae	—	*Nosema bombycis, N. apis*	134

CLASS HAPLOSPORIDA
CLASS SARCOSPORIDIA

Sarcocystis miescheriana,	135	*S. tenella*	135
S. muris	135	*S. lindemanni*	136
S. fusiformis	136		

CLASS PIROPLASMIDA

Family Theileridae	136	**Family Babesidae**	136
Theileria parya	136	*Babesia bigemina, B. canis*	137

SUBKINGDOM METAZOA
SUBKINGDOM and PHYLUM MESOZOA
Order Orthonectida

Rhopalura 138★

Order Dicyermida

Dicyema 139★   *Trichoplax adhaerens* 138

SUBKINGDOM PARAZOA
PHYLUM SPONGES (SPONGIA)
CLASS PLEOSPONGIA († ARCHAEOCYATHA)

†*Archaeocyathus* 169★

CLASS CALCAEROUS SPONGES (CALCAREA)
Order Homocoela

Family Homocoelidae	157	L. complicata, L. variabilis	155
Leucosolenia	150/151★	L. botryoides	158
L. coriacea	142		

Order Heterocoela

Family Sycettidae	158	Family Leuconidae	158
Sycetta primitiva	—	Leuconia (=Leucandra) nivea	158
Sycon	153	L. aspera	158
S. raphanus, S. setosum	153	Leucandria aspera	143
S. ciliatum	158		

Order Pharetronida

Petrobiona massiliana 158

CLASS GLASS SPONGES (HEXACTINELLIDAE)
Order Hexasterophora

| Venus' flower-basket, Euplectella aspergillum | 159 | Lophocalyx philippensis | 155★ |
| Farrea occa | 159 | | |

Order Amphidiscophora

| Hyalonema | 173★ | Pheronema raphanus | 160★ |
| H. thompsoni | 160 | Monoraphis chuni | 160 |

CLASS COMMON SPONGES (DEMOSPONGIAE)
Order Tetraxonida
Suborder Homosclerophora

| Family Oscarellidae | — | Oscarella lobularis | 161 |

Suborder Sigmatophora

| Family Tetillidae | — | Tetilla cranium | — |

Suborder Astrophora

Family Pachastrellidae	—	*Ancorina cerebrum*	—
Poecillastra compressa	—		
Family Stelletidae	—	**Family Geodiidae**	161
Stelleta grubii, Penares helleri	—	*Geodia cydonium*	161

Suborder Astromonaxonellina

Family Donatiidae	161	*Suberites domuncula*	162
Sea orange, *Tethya aurantium*	161	*S. ficus* (=*Ficulina ficus*)	162
Family Chondrosiidae	161	**Family Spirastrellidae**	—
Kidney sponge, *Chondrosia reniformis*	161	*Spirastrella cunctatrix*	146★
Family Polymastiidae	162	**Family Clionidae (Boring sponges)**	162
Polymastia mamillaris	162	*Cliona viridis, C. lobata*	—
Family Suberitidae (Cork sponges)	162	*C. celata*	163

Order Cornacuspongia

Suborder Protorhabdina

Family Mycalidae	—	*Mycale massa*	—

Suborder Poikilorhabdina

Family Myxillidae	—	*Crella rosea*	—
Myxilla rosacea	148★	**Family Microcionidae**	—
Family Tedaniidae	—	*Microciona armata*	—
Tedenia anhelanus	—	**Family Clathriidae**	—
Family Crellidae	—	*Clathria coralloides*	—

Suborder Phthinorhabdina

Family Axinellidae	—	*E. muelleri*	148★
Axinella verrucosa		*Heteromeyenia baileyi* (=*H. repens*)	153
A. polypoides	—	*H. stepanowii*	—
A. cannabina	145★	*Trochospongilla horrida*	164
Phakellia ventillabrum	166★	*Ochridaspongia rotunda*	164
Family Gelliidae	—	*Lubomirskia baicalensis*	163
Gellius angulatus	—	*Drulia* (=*Parmula*) *browni*	164
Adocia grossa, A. cinerea, A. rosea	—	**Family Chalinidae**	—
Family (fresh-water sponges) Spongillidae	163	*Haliclona oculata, H. cratera*	—
Spongilla lacustris	164		
S. fragilis	153	**Family Ciocalyptidae**	166
S. igloviformia	156	Bread crust sponge, *Halichondria panicea*	166
Fresh-water sponge, *Ephydatia fluviatilis*	143	*Hymeniacidon sanguinea*	—

Suborder Aporhabdina

Family Dysideidae	—	Fine levantine, *S. I. mollissima*	164
Dysidea fragilis, D. topha	—	Horse sponge, *Hippospongia communis*	164
D. spinifera	155	Velvet sponge, *H. c. meandriformis*	164
Family Spongidae (Bath sponges)	164	Grass sponge, *H. c. cerebriformis*	164
Dalmatian sponge, *Spongia officinalis*	164	Wool sponge, *H. canaliculata*	164
Yellow sponge, *S. irregularis*	164	**Family Aplysinidae**	166
Elephant ear sponge, *S. lamella*	164	*Verongia* (=*Aplysina*) *aerophoba*	166
Zimocca sponge, *Sp. I. zimocca*	164	*V. fistularis*	156

Order Dendroceratida

Family Halisarcidae	—	*Aplysilla rosea*	166
Jelly sponge, *Halisarca dujardini*	166	*A. sulfurea*	—
Family Darwinellidae	—	*Dendrilla rosea*	165★

SUBKINGDOM EUMETAZOA

SUBSECTION COELENTERATA

PHYLUM CNIDARIA

CLASS HYDROZOA

Order Athecata-Anthomedusae

Family Corynidae	188	*Hydractinia echinata*	190
Coryne sarsi, Sarsia tubulosa, S. gemmifera	188	**Family Rathkeidae**	—
Family Pennariidae	—	**Family Tubulariidae**	—
Family Milleporidae	237★	*Tubularia*	187
Fire Coral, *M. platyphyllus*	189	*Euphysa aurata*	184★
Family Branchiocerianthidae	—	*Corymorpha nutans*	188
Branchiocerianthus imperator	188	**Family Margelopsidae**	—
Family Eleutheriidae	—	**Family Cladonemidae**	—
Eleutheria dichotoma	184★	*Cladonema radiatum*	188
Family Clavidae	—	**Family Bougainvilliidae**	—
Cordylophora caspia	189	**Family Stylasteridae**	189
Family Hydractiniidae	—	*Stylaster*	249
Podocoryne carnea	189	**Family Eudendridae**	188

Order Limnohydrina-Limnomedusae

Craspedacusta sowerby	190	*Olindias phosphorica*	191
Gonionemus vertens	191		

Order Hydrina

Green Hydra, *Chlorohydra viridissima*	185	Grey Hydra, *H. oligactis*	187
Hydra	187	*Protohydra leuckarti*	187
Brown Hydra, *H. vulgaris*	187		

Order Halammohydrina

Halammohydra octopodides	192

Order Colonial Jellyfish (Siphonophora)

Suborder Siphonanthae

Family group Physophorae

Family Rhizophysidae	—	*Agalma elegans*	194
Family Physalidae	—	*Halistemma rubra*	271★
Portuguese Man-of-War, *Physalia physalis*	193	**Family Physophoridae**	—
Family Apolemiidae	—	*Physophora hydrostatica*	194
Family Forskalidae	194	**Family Rhodaliidae**	—
Forskalia contorta	194		

Family group Calycophorae

Family Sphaeronectidae	—	**Family Diphyidae**	195
Family Prauidae	—	*Chelophyes appendicularia*	195
Praya diphyes	271★	*Muggiaea kochii*	195
Stephanophyes superba	195	**Family Abylidae**	—
Family Hippopodiidae	—	*Abylopsis*	193★
Hippopodius hippopus	195	*A. tetragona*	—

Suborder Disconanthae

Sailors-before-the-wind, *Vellella spirans*	195	*Porpita porpita*	196
V. velella	—		

Order Thecaphora-Leptomedusae

Family Plumularidae	206	*Octorchis gegenbauri*	207
Litocarpia myriophyllum	206	**Family Campanulariidae**	—
Aglaophenia pluma	206	*Laomedea* (=*Obelia*)	205
Family Campanulinidae	206	**Family Sertulariidae**	—
Campanulina	206	Cypress moss, *Sertularia cupressina*	205
Aequorea forskalea	206	Coral moss, *Hydrallmania falcata*	206
Laodicea undulata	206	*Sertularella*	198★

Order Trachymedusae

Rhopalonema velatum, Aglantha digitalis	207	*Pantachogon rubrum*	207
Liriope tetraphylla	207	*Haliscera papillosum*	207
Trunked jellyfish, *Geryonia proboscidalis*	207	*Aglaura hemistoma*	181★

Order Narcomedusae

Pegantha, Polypodium, Cunina octonaria	208	*Solmundella bitentaculata*	208
Solmissus albescens	208		

†Order Stromatoporidea

†*Stromatopora*	177	†*Trupetostroma*	169★

†Order Sphaeractinoidea

†*Sphaeractinia*	177	

CLASS CONULATA†

CLASS TRUE JELLYFISH (SCYPHOZOA)

Order Stauromedusae

Lucernaria bathophila, L. quadricornis, Hali-clystus octoradiatus	212	*Craterolophus tethys*	212

Order Cubomedusae

Sea wasp, *Charybdea marsupialis*	212	*Chironex fleckeri*	213
Sea wasp, *Chiropsalmus quadrigatus*	213		

Order Coronatea

Nausithoe punctata	214	*Periphylla regina*	214
N. rubra	183★		

Order Semaestomae

Family Pelagiidae | — | | *C. lamarcki* | 216
Chrysaora hyoscella | 214 | | Phosphorescent jellyfish, *Pelagia noctiluca* | 214
Family Cyaneidae | 215 | | **Family Ulmariidae** | —
Lion's mane jellyfish, *Cyanea capillata* | 210 | | Common jellyfish, *Aurelia aurita* | 216
Arctic lion's mane, *C. arctica* | 215 | | *Stygiomedusa, Dactylometra quinquecirrah* | 217
Blue lion's mane or cornflower jellyfish,

Order Rhizostomae

Rhizostoma pulmo | 218 | | *C. andromeda* | 198★
R. octopus | 218 | | *Cotylorhiza tuberculata* | 218
Cassiopeia xamanchana | 218 | | *Rhopilema esculenta* | 219

CLASS ANTHOZOA

SUBCLASS † RUGOSA

†*Streptelasma,* †*Calceola* | 177 | | †*Lithostrotion* | 177
†*Cyanthophyllum* (=*Hexagonaria*) | — | | Rugosa coral, †*Dalmanophyllum* | —

SUBCLASS HEXACORALLIA

Order Sea Anemones (Actinaria)

Suborder Protanthea

Gonactinia | 222 | | *G. prolifera* | 224

Suborder Nunantheae

Family series Boloceroidaria

Bunodopsis | 223

Family series Abasilaria

Edwardsia, Peachia | 224 | | *Halcampa* | 222

Family series Endomyaria

Actinia | 221 | | *Anemonia* | 222
Sea anemone, *A. equina* | 224 | | *A. sulcata* | 224
Tealia | 222 | | *Stoichactis* | 223
T. felina | — | | *Bunodactis verrucosa* | 224
Sea dahlia, *T. crassicornis* | 200★ | | *Corynactis viridis* | 271★

Family series Mesomyaria

Genus group Inermia | — | | *Adamsia palliata* | 223
Stomphia carneola | 221 | | *Sagartia* | 222
Genus group Acontiaria | 221 | | *Cereus pedunculatus* | 224
Metridium senile | 224 | | *Aiptasia* | 221★
Hermit crab anemone, *Calliactis parasitica* | 202★ | | *A. mutabulis* | 221

Order Stony Corals (Madreporaria)
Solitary corals

Caryophylla Clavus	230	*C. clavus*	—
C. smithii	230	*Leptosammia pruvoti*	230
Flabellum	231	*Balanophyllia italica*	231
F. angulare	231	*B. elegans*	231
F. goodei	231	Star coral, *B. regia*	231
F. anthophyllum	231	*Fungia*	224
Cladocora cespitosa	231	Fungus coral, *Fungis fungites*	230

Colonial corals

Branching corals		Star coral, *Astroides calycularis*	232
Stag's-horn coral, *Acropora*	232	*Porites, Gonastrea*	232
Symphyllia, Madreporaria alcidornis	227	*Favia*	232
Madrepora oculata, Dendrophyllia ramea,	231	Brain corals	
Astrangia danae	232	Brain coral, *Diploria*	232
Lophelia pertusa	231	*Lobophyllia*	232
Massive corals		Stony coral, *Meandrina*	170★

Order Antipatheria

Cirripathes rumphii, Paranthipathes larix	233	*Anthipathes subpinnata*	233

Order Ceriantheria

Cerianthus membranaceus	234	*C. llovdii*	234
C. m. fuscus, C. m. violaceus	234	Sand anemone, *Cerianthopsis americanus*	234

Order Zoantheria

Parazoanthus axinellae	234	*E. vatovai*	234
Epizoanthus arenaceus	234	*E. incrustatus*	235

SUBCLASS TABULATA

†*Favosites*	170★

SUBCLASS OCTOCORALLIA
Order Alcyonaria

Family Haimeidae	—	*A. palmatum*	245
Haimea	245	*A. brioniense*	245
Family Cornularidae	245	*Sarcophyton ehrenbergi*	240★
Cornularia cornucopeia	245	*S. trocheliophorum*	246
Clavularia	245	*Gersimea, Anthomatus grandiflorus*	245
Family Tubiporidae (Organ pipe Corals)	245	*Dendronephthya*	241★
Tubipora purpurea	271★	*Eunephthya rosea*	271★
Family Alcyonidae	—	*Paralcyonium elegans*	263★
Dead man's hand or finger, *Alcyonium digitatum*	235	*Xenia*	240★

Order Gorgonaria
Suborder Seleraxonia

Precious corals, *Corallium*	247	*G. abyssorum. Paragorgia arborea*	249
Red precious coral, *C. rubrum*	247	False coral, *Parerythropodium coralloides*	249

Suborder Holaxonia

Eunicella (Sea fans)	250	Violet horny coral, Paramuricea chamaeleon	250
Warty coral, E. verrucosa	250	Venus's fan, Rhipidogorgia flabellum	251
White horny coral, E. stricta	250	Eugorgia rubens, Lophogorgia chilensis	251
Yellow horny coral, E. cavolini	250		

Order Blue Corals (Helioporidae)

Heliopora coerulea	251

Order Sea Pens (Pennatularia)

Family Pteroididae	252	Veretillum cynomorium	254
Seylatopsis djibutiensis	—	**Family Renillidae (Sea pansies)**	255
Grey sea pen, Pteroides griseum	253	Renilla amethystina	271★
Phosphorescent sea pen, Pennatula phosphorea	254	Sea Whip, Funicula quadrangularis	255
Acanthoptilum	254	**Family Umbellulidae**	255
Family Virgulariidae		Umbellula antarctica	255
Virgularia mirabilis, Stylatula	254	U. lindahlii	255
Family Veretillidae	254		

PHYLUM ACNIDARIA

CLASS CTENOPHORA

Subclass Tentaculifera

Sea Gooseberry, Pleurobranchia pileus	259	Eulampetia panserina (Gastrodes	
Sea nut, Mertensia ovum	259	parasiticum = larva of Eu. p.)	259
Hormiphora plumosa	259	Callianira bialata	259

Order Lobata

Leucothea multicornis	260	Mnemiopsis leidvi	260
Bolinopsis infundibulum	260	Bolina hydatina	265
B. vitrea	—	Eucharis multicornis	271★

Order Venus's Girdle (Cestidea)

Venus's Girdle, Cestus veneris	265

Order Platyctenidae

Ctenoplana	266	C. gonoctena	271★
Coeloplana	257★		

Order Sessile Ctenophores (Tjallfiellidae)

Tjallfiella tristoma	266

SUBCLASS NUDA

Order Beroidea

Melon jellyfish, Beroe	258	Sea mitre, B. cucumis	268
B. ovata	268		

SUBSECTION BILATERALIA
Main Branch Protostomia
PHYLUM PLATHELMINTHES
CLASS TURBELLARIA
Superorder Archoophora

Xenoturbella	278★	*X. bocki*	282

Order Acoela

Convoluta roscoffensis	285

Order Macrostomida

Macrostomum appendiculatum	285	*M. lineare*	285
Microstomum	282		

Order Catenulida

Catenula lemnae, Stenostomum leucops	285

Order Polycladida

Stylochus zebra	285	*Discocoelis tigrina*	281★
S. pilidium, S. frontalis	285	*Prostheceraeus roseus, P. moseley*	296★
Graffizoon lobatum	285	*Thysanozoon bronchii*	296★
Notoplana longestyletta	296★	*Yungia aurantiaca*	295★

Superorder Neophora
Order Prolecithophora
Order Lecithoepitheliata

Prorhynchus stagnalis	286

Order Seriata
Suborder Proseriata
Suborder Triclads (Triclada) or Planarians

Marine planarians (Maricola)	286	*Planaria torva*	280★
Procerodes lobata	286	*Fonticola vitta*	280★
Bdelloura	286	*Dugesia gonocephala*	286
Fresh-water planarians (Paludicola)	286	*D. lugubris*	280★
Crenobia alpina	280★	American stream planarians, *D. dorotocephala*	286
C. septentrionalis, C. meridionalis	286	Terrestrial planarians (Terricola)	286
Polycelis cornuta	286	*Rhynchodemus terrestris*	287
P. nigra	280★	Greenhouse planarians, *Bipalium kewense*	287
Bdellocephala punctata, Dendrocoelum lacteum,			

Order Neorhabdocoela
Suborder Typhloplanoida

Mesostoma ehrenbergi	287

Suborder Dalyelloida

Dalyellia viridis   288

Suborder Temnocephalida

Scutariella didactyla   288

Suborder Kalyptorhyncha

Gyratrix hermaphrodita   288

CLASS TREMATODA

SUBCLASS MONOGENA

Gyrodactylus elegans	291	Bladder fluke, *Polystoma integerrimum*	291
Diplozoon, *Diplozoon paradoxum*	291		

SUBCLASS DIGENA

Family Fasciolidae	—	*Metagonimus yokagawei*	294
Giant or Common liver fluke, *Fasciola hepatica*	292	*Heterophyes heterophyes*	294
F. gigantica	—	**Family Troglotrematidae**	—
American giant liver fluke (*Fasciolopsis magna*)	293	East Asian lung fluke, *Paragonimus westermani*	294
Giant intestinal fluke (*F. buski*)	294	North American lung fluke, *P. kellicotti*	297
Family Dicricoeliidae	—	**Family Schistosomidae (Blood flukes)**	297
Small liver fluke, *Dicrocoelium dendriticum*	293	Bladder blood fluke, *Schistosoma haematobium*	297
Family Opisthorchidae	—	Intestinal blood fluke, *S. mansoni*	297
Cat liver fluke, *Opisthorchis felineus*	294	Japanese blood fluke, *S. japonicum*	297
Upper Indian liver fluke, *O. viverrini*	294	*Trichobilharzia szidati*	297
Chinese liver fluke, *Clonorchis sinensis*	294	**Family Brachylaemidae**	—
Family Echinostomatidae	—	*Leucochloridium paradoxum*	295★
Echinostoma ilocanum	294	*L. fuscum*	298
Family Heterophyidae	—		

CLASS TAPEWORMS (CESTODA)

Order Haplobothrioidea

Order Pseudophyllidea

Fish tapeworm, *Diabothriocephalus latus*	308	*Caryophyllaeus laticeps*	309
Ligula intestinalis	309	*Biacetabulum sieboldii*	309

Order Tetrarhynchidea

Order Tetraphyllidea

Order Diphyllidea

Order Cyclophyllidea

Family Taeniidae	—	Cat tapeworm, *Hydatigena taeniseformis*	307
Beef tapeworm, *Taeniarhunchus saginatus*	299	**Family Hymenolepididae**	—
Pork tapeworm, *Taenia solium*	300	*Hymenolepis diminuta*	308
T. hydatigena	307	*H. nana*	308
T. pisiformis	307	*H. n. nana*	308
Dog tapeworm or bladder tapeworm, *Echinococcus granulosus*	303	*H. n. fraterna*	308
Multiceps multiceps	303	**Family Anoplocephalidae**	307
Diplidium caninum	—	**Family Davaineidae**	307

CLASS GNATHOSTOMULIDAE

Gnathostomula paradoxa	309	*Pterognathia grandis*	310
Gnathostomaria lutheri	310	*P. simplex*	310★
Nanognathia exigua	310	*Austrognathia*	311

PHYLUM KAMPTOZOA (=ENTROPROCTA)

Family Loxosomatidae	313	**Family Urnatellidae (Fresh-water kamptozoans)**	313
Family Pedicellinidae	313	*Urnatella gracilis*	313

CLASS NEMERTINI (BOOTLACE WORMS)
SUBCLASS ANOPLA
Order Palaeonemertini

Tubulanus annulatus	301★	*Carinia, Carinoma*	—
Cephalothrix	—		

Order Heteronemertini

Atlantic horseshoe crab, *Lineus geniculatus*	301★	*Baseodiscus*	—
Cerebratulus marginatus	302★	*Micrura alaskensis*	312★

SUBCLASS ENOPLA
Order Hoplonemertini

Amphiporus exilis	318★	*Prostoma graecense*	322
Oerstedia, Emplectonema, Ototyphlonemertes	—	*Tetrastemma quadrilineatum*	313★
Drepanophorus	—	*Geonemertes chalicophora*	322
D. crasus	—		

Order Bdellomorpha

Malacobdella	318

PHYLUM ROUNDWORMS
(ASCHELMINTHES = NEMATHELMINTHES)
CLASS GASTROTRICHA
Order Macrodasyoidea
Order Chaetonotoidea

Neodasys, Xenotrichula	327	**Family Neogosseidae**	—
Gastrotrichan, *Chaetonotus maximus*	305/306★	**Family Dasydytidae**	328

CLASS ROTIFERS (ROTATORIA)
SUBCLASS SEISONIDEA
Order Seisonida

Seison nebaliae	331

SUBCLASS EUROTATORIA

Order Digonota

Suborder Leechlike Rotifers (Bdelloida)

Family Habrotrochidae	—	*Rotaria*	326★
Habrotrocha flaviformis	331	*R. neptunia*	331
H. pusilla textrix brevilabris	329★	*R. macroceros*	305/306★
H. constricta	330	*R. rotatoria*	305/306★
Family Philodinidae	—	*Macrotrachela insulana*	331
Mniobia incrassata	331	*Philodina citrina*	305/306★

Order Monogononta

Suborder Ploima

Family Brachionidae	—	*Itura myersi*	305/306★
Epiphanes senta	332	*Notommata allantois*	325★
Macrochaetus subquadratus	325★	*N. copeus*	326★
Trichotria pocillum	326★	*Resticula nyssa*	
Platyias quadricornis	—	**Family Trichocercidae**	—
Brachionus	332	*Trichocera*	331★
B. quadridentatus	326★	*T. taurocephala*	331★
B. calyciflorus	305/306★	*Ascomorphella volvocicola*	325★
B. rubens	327★	**Family Gastropodidae**	—
Mytilina mucronata	305/306★	*Gastropus*	—
Euchlanis deflexa	305/306★	*Ascomorpha ecaudis*	325★
Keratella	332	*Chromogaster*	—
K. cochlearis	330★	**Family Dicranophoridae**	—
K. quadrata	305/306★	*Albertia*	332
Kellikottia longispina	331★	*Encentrum oxyodon*	332
Squatinella	325★	*E. incisum*	332
S. lamellaris	331★	*Dicranophorus*	326★
Lepadella	326★	*D. forcipatus*	332
L. patella	330★	**Family Asplanchnidae**	—
Colurella	325★	*Asplanchna*	328
Eudactylota eudactylota	325★	*A. priodonta*	330★
Family Lecanidae	—	*A. sieboldi*	—
Proales werneckii	332	*Asplanchnopus multiceps*	325★
P. fallaciosa	305/306★	**Family Synchaetidae**	—
Family Notommatidae	—	*Polyarthra*	332
Cephalodella	332	*Synchaeta*	326★
C. gibba	332	*S. pectinata*	305/306★
C. forficula	305/306★	**Family Microcodinidae**	—
Monommata	325★	*Microcodon clavus*	325★

Suborder Flesculariacea

Family Testudinellidae	—	*L. ceratophylli*	305/306★
Flina	—	*Floscularia ringens*	332
Trochosphaera solstitialis	325★	*Octotrocha speciosa*	325★
Horaella brehmi	330★	*Ptygura pilula*	332
Family Flosculariidae	—	**Family Conochilidae**	—
Limnias	326★	*Conochilus*	326★
L. melicerta	332	*C. unicornis*	326★

Suborder Collothecacea

Family Collothecidae — C. o. cornuta 332★
Collotheca campanulata 325★ C. gracilipes 305/306★
C. hoodii 325★ Stephanoceros fimbriatus 333
C. ornata 326★

CLASS THREADWORMS (NEMATODA)
SUBCLASS ADENOPOREA (=APHASMIDA)
Order Araeolaimida

Areaolimus, Cylindrolaimus, Plectus —

Order Monohysteridae

Monohystera, Desmolaimus, Siphonolaimus —

Order Desmodorida

Ceramonema, Desmodora, Draconema —

Order Chromatodorida

Chromatodora, Comesoma, Choanolaimus —

Order Desmoscolecida

Desmoscolex, Tricoma, Greeffiella —

Order Enoplida

Superfamilies Tripyloidea, Enoploidea and Oncholaimoidea

Tripyla, Halslaimus — Trichosomoides crassicauda 338
Dilaimus denticulatus 336

Superfamily Mermithoidea

Mermis subnigrescens 336 Hexamermis, Agamermis —

Superfamilies Trichuroidea and Dioctophymatoidea

Whipworm, Trichuris trichiura 334 Trichinas Trichinella spiralis 345
Capillaris —

Order Dorylaimida

Mononchus, Nygolaimus, — Longidorus, Xiphinema, Trichodorus 340
Eudorylaimus, Actinolaimus —

SUBCLASS SECERNENTEA (PHASMIDIA)
Order Rhabditida

Rhabditis, Diplogaster, — Neoaplectana —
Cephalobus, Scoleophilus — Dwarf threadworms, Strongyloides 343

Order Tylenchida

Ditylenchus dipsaci	340	Wheet eelworm. *H. avenae*	340
Leaf nematodes, *Aphelenchoides*	340	Root gall eelworm, *Meloidogyne*	342
Radopholus	340	*Allantonema, Entaphelenchus*	—
Rotylenchus	341	*Parasitylenchus, Heterotylenchus*	342
Tylenchorhynchus	340	*Sphaerularia bombi*	337
Pratylenchus	—	*Rotylenchulus, Nacobbus*	340
Paratylenchus, Helicotylenchus	340	*Tylenchulus*	341
Heterodera	338	*Tylenchus polyhypus*	347
Potato eelworm, *H. rostochiensis*	341	*Sphaeronema minutissimus*	334
Beet eelworm, *H. schachtii*	341	Coconut nematode, *Rhadinaphelenchus cocophilus*	340

Order Strongylia

Strongulus, Metastrongulus	—	*S. trachea*	344
Syngamus	338	*Anclyostoma duodenale*	343

Order Ascaridida

Oxuris equi	—	Roundworm, *Ascaris lumbricoides*	344
Threadworm, *Enterobius vermicularis*	344	*Ascarididia galli*	301★
Toxocara, Heterakis, Stomachus	—		

Order Spirurida

Medina worm, *Dracunculus medinensis*	345	*Dipetalonema streptocerca*	348
Spirura, Onchocerca	—	*Plancentonema gigantissimum*	—
Hair worm, *Wuchereria bancrofti*	345	*Loa loa*	—

CLASS NEMATOMORPHA (HORSEHAIR WORMS)

Superfamily Gordioidea

Gordius	349★	*G. dectici*	349
G. aquaticus	349	*Parachordodes tolosarus*	354

Superfamily Nectonematoidea

Nectonema	354

CLASS KINORHYNCHA

CLASS ACANTHOCEPHALA

Order Eoacanthocephala

Order Palaeoacanthocephala

Polymorphus, P. boschadia	355	*Acanthocephalus lucii*	355★
Fillicollis anatis	355	**Family Pomphorhynchidae**	—
Family Echinorhynchidae	—	*Pomphorynchus*	355
Echinorhynchus truttae	355		

Order Archiacanthocephala

Family Oligacanthorhynchidae	—	*hirudinaceus*	355
Giant spiny-headed worm, *Macracanthorhynchus*		*Moniliformis moniliformis*	355

PHYLUM PRIAPULIDS (PRIAPULIDA)

Unicaudate priapulid, *Priapulus caudatus* — 357   *Halicryptus spinulosus* — 357★

PHYLUM SIPUNCULIDS (SIPUNCULIDA)

Sipunculus nudus	358	*P. granulatum*	357★
Phascolosoma	—	*Golfingia, Siphonomecus multicinctus*	358
P. lurco	358	*Physcosoma granulatum*	302★

PHYLUM ECHIUROIDS (ECHIURIDA)

Echiurus echiurus	358	Green bonellia, *B. viridis*	358
Bonellia	358	*Ikeda taenioides*	358

Order Sactosomatinea

SUPERPHYLUM ARTICULATA

PHYLUM ANNELIDA

Order Free-living Polychaetes (Errantia)

Suborder Amphinomorpha

Family Amphinomidae   —   *Hermiodice carunculata*   363

Suborder Nereimorpha

Family Aphroditidae (Sea caterpillars)	363	*N. diversicolor*	365
Sea mouse, *Aphrodite aculeata*	363	*Lycastopsis raunensis*	365
Lepidonotus squamatus	363	*L. amboinensis*	365
Polynoe scolopendrina	366	*Lycastis vitabunda*	365
Lagisca extenuta	373★	*L. terrestris*	365
Family Phyllodocidae	365	**Family Nephthydidae**	—
Phyllodoce	365	**Family Glyceridae**	—
Eulalia viridis	373★	**Family Eunicidae**	363
Family Alciopidae	365	Palolo worms, *Eunice*	363
Family Tomopteridae	365	*E. gigantea*	363
Tomopteris	365★	Samoan Palolo worm, *E. viridis*	364
Family Hesionidae	—	Atlantic palolo worm, *E. fucata*	364
Family Syllidae	364	**Family Histriobdellidae**	366
Syllis cornuta	373★	*Histriobdella homari*	366
Odontosyllis	364	*Stratiodrilus tasmanicus*	366
Family Nereidae (=Lycoridae)	364	**Family Ichthyotomidae**	366
Nereis	—	*Ichthyotomus sanguineus*	366
N. virens	365		

Order Sessile Polychates (Sedentaria)

Suborder Spiomorpha

Family Spionidae	—	**Family Chaetopteridae**	—
Family Disomidae	—	*Chaetopterus variopedatus*	367
Family Orbiniidae	—		

Suborder Drilomorpha

Family Cirratulidae	—	*Arenicola*	367
Family Chorhaemidae (=Flabelligeridae)	—	Lugworm, *A. marina*	367
Family Opheliidae	—	Family Maldanidae (=Clymenidae)	—
Family Capitellidae	—	Family Sternaspididae	—
Family Arenicolidae	—	Family Oweniidae	—

Suborder Terebellomorpha

Family Pectinariidae (−Amphictenidae)	—	Family Terebellidae	
Pectinaria koreni	367	"Shell collector", *Lanice conchilega*	368
Family Ampharedidae	—		

Suborder Hermellimorpha

Family Sabellariidae (=Hermellidae)	—	Family Serpulidae	—
Sabellidae	—	Post horn worm (*Spirorbis*)	368
Spirographis spallazanii	386	*Protula tubularia*,	376★
Peacock feather worm (*Sabella pavonina*)	376★	Chalk-tube worm, *Serpula vermicularis*	376★
Fabricia (=*Amphicora*) *sabella*	369		

Order Archiannelida

Family Polygordiidae	—	Ground water polychaete, *Troglochaetus*	369
Family Protodrilidae	—	*T. beranecki*	369
Family Saccocirridae	—	Family Dinophilidae	—
Family Nerillidae	—		

CLASS MYZOSTOMIDA
Order Proboscifera

Family Myzostomidae	—	*M. cysticolum*	370
Myzostoma	370		

Order Pharyngidea

Family Cystomyzostomidae	—	Family Protomyzostomidae	—
Family Pulvinomyzostomidae	—	Family Asteriomyzostomidae	—
Family Mesomyzostomidae	—		

CLASS CLITELLATA
Order Oligochaeta
Suborder Plesiopora

Family Aeolosomatidae	371	Family Tubificidae (Tubifex worms)	372
Aelosoma hemprichi	371	*Tubifex*	309
Family Naididae	371	Common tubifex, *T. tubifex*	372
Chaetogaster, C. limnaei	371	*Branchiura sowerbyi*	377
Stylaria lacustris	371	Family Phreodrilidae	—
Ripestes parasita	371	Family Enchytraeidae	377
Aulophorus, Dero	372	*Enchytraeus albidus*	377

Suborder Prospora

Family Lumbriculidae	377	Family Branchiobdellidae	377
Lumbriculus variegatus	377	*Branchiobdella*	377

Suborder Opisthopora

Family Haplotaxidae	378	*Allolobophora chlorotica, A. rosea*	378
Haplotaxis (=Phreoryctes) gordioides	378	Common earthworm, *Lumbricus terrestris*	378
Family Criodrilidae	378	*L. rubellus, L. castaneus*	378
Criodrilus lacuum	378	**Family Acanthodrilidae**	—
Family Lumbricidae	378	**Family Megascolecidae**	378
Eiseniella tetraedra	378	*Megascolides australis*	378
Eisenia foetida	378		

Order Leeches (Hirudinea)
Suborder Acanthobdellae

Acanthobdella peleinda	383

Suborder Rhynchobdellae

Family Glossiphoniidae	383	*Hemiclepsis marginata*	381★
Glossiphonia	383	*Haementeria costata, H. officinalis*	383
Greater snail leech, *G. complanata*	383	*Marsupiobdella africana*	384
Lesser snail leech, *G. heteroclita*	383	**Family Ichthyobdellidae (=Piscicolidae) (fish**	
Helobdella stagnalis	383	**leeches)**	384
Duck leech, *Theromyzon tessulatum*	383	Common fish leech, *Piscicola geometra*	384

Suborder Gnathobdelliae

Family Hirudinidae	—	*Limnatis nilotica*	—
Medicinal leech, *Hirundo medicinalis medicinalis*	384	**Family Haemadipsidae (Land leeches)**	385
Hungarian medicinal leech, *H. m. officinalis*	384	Ceylon leech, *Haemadipsa zeylanica*	385
Horse leech (*Haemopis sanguisuga*)	384	*Xerobdella lecomtei*	—

Suborder Pharyngobdellae

Family Erpobdellidae (Free-living polychaetes)	363	Dog leech, *Erpobdella octoculata*	385

PHYLUM ONYCHOPHORA

Family Peripatidae (" Strollers ")	389	**Family Peripatopsidae**	389
Heteroperipatus engelhardi	389	*Peripatopsis moseleyi*	376★

PHYLUM WATERBEARS (TARDIGRADA)
Order Heterotardigrada
Suborder Arthrotardigrada
Suborder Echiniscoidea

Echiniscoides sigismundi	391	*E. blumi*	382★
Echiniscus scrofa	391		

Order Eutardigrada

Macrobiotus	390	*Hypsibius*	390★
M. hufelandi	391		

Order Mesotardigrada
PHYLUM LINGUATULIDA
Order Cephalobaenida

Riehardia	—	*Cephalobaena tetrapoda*	392

Order Porocephalida

Armillifer	391	*Linguatula serrata*	392
A. armillatus	382★		

PHYLUM ARTHROPODA
SUBPHYLUM TRILOBITES (†TRILOBITA)
Order †Agnostida

†*Agnostus*	401

Order †Redlichiida

†*Olenellus vermontanus*	397★	†*Paradoxides*	402

Order †Corynexochida

†*Olenoides*	402

Order †Ptychopariida

†*Olenus*	382★	†*Illaenus*	402
†*Osaphus*	—		

Order †Phacopida

†*Phacops*	402

Order †Lichida

†*Lichas*	402

Order †Odontopleurida

†*Acidapsis*	402	†*Ceratarges armatus*	382★

SUBPHYLUM CHELICERATA
CLASS MEROSTOMATA
Order †Aglaspida

†*Aglaspis*	405

Order Giant Sea Scorpions (†Eurypteridae=Gigantostraca)

Giant sea scorpion, *Pterygotus rhenanus*	405	†*Megalograptus ohioensis*	393★
†*Eurypterus*	405		

Order Horseshoe Crabs (Xiphosura)

Family Limulidae	—	†*Psammolimulus*, †*Mesolimulus*	405
†*Palaeomerus*	—	†King Crabs, *Limulus*	411
†*Belinurus*	405	Atlantic horseshoe crab, *L. polyphemus*	411
†*Euproops*, †*Palaeolimulus*	405	*Trachypheus, Carcinoscorpinus*	411
†*Weinbergina*	—		

CLASS ARACHNIDA

Order Scorpions (Scorpiones)

†*Palaeophonus*, †*Proscorpius* — 405
†*Eoscorpius* — 405
Family Buthidae — —
  Bathus — 407★
  Sahara scorpion, *Androctonus australis* — 412
Family Scorpionidae — —
  Pandinus — 413

Giant African or imperial scorpion, *Pandinus imperator* — 413
Family Chactodae — —
  Euscorpius — 412
  Italian scorpion (*E. italicus*) — 412
Family Vejovidae — —

Order Pedipalpi

Suborder Uropygi

Long-tailed uropygid, *Typopeltis crucifer* — 414

Suborder Amblypygi

Damon medius — 414

Whip scorpion, *D. m. johnstoni* — 416★

Order Palpigradi

Koenenia mirabilis — 413★

Order True Spiders (Araneae)

True spiders, †*Arthrolycosa* — 406

Suborder Mesothelae

Family Lipistiidae — —
  Liphistius — 420

L. desultor — 400★

Suborder Bird-Eating Spiders (Orthognatha)

Family Ctenizidae — —
Family Dipluridae — —
Family Aviculariidae (True bird-eating spiders) — 420

Bird-eating spiders, *Eurypelma soemanni* — 399★
Lasiodora — 425★
Family Atypidae — —

Suborder Labidognatha

Superfamily Dyseriformia

Family Dysderidae — 420
  Dysdera — 420
  D. erythrina — 415

Cellar spider, *Segestria senoculata* — 399★
S. florentina — 418★

Superfamily Scytodiformia

Family Caponiidae — —
Family Sicariidae — —

Loxosceles — 418

Superfamily Pholciformia

Family Pholcidae — —

Pholcus phalangioides — 420

Superfamily Hersiliiformia

Family Hersiliidae — —

Superfamily Web-Spinners (Argiopiformia)

Family Theridiidae (Domed-web spiders)	421	Spined or thorny spiders, *Gasteracantha*	421
Black widow, *Latrodectus mactans*	421	*G. kuhlii*	—
European black widow, *L. m. tredecimguttatus*	399★	*G. thorelli*	400★
Family Linyphiidae (Sheet-web spiders)	418	**Family Erigonidae (Dwarf spiders)**	421
Family Araneidae (Orb spiders)	421	*Walckenaera*	419
Garden spider, *Araneus diadematus*	421	*W. acuminata*	421
Oak-leaf spider, *A. ceropegius*	410★	**Family Tetragnathidae**	421
A. quadratus	410★	Trap-door spiders, *Tetragnatha*	421
Wasp spider, *Argiope bruennichi*	421	*T. extensa*	410★
Zebra spider, *A. lobata*	408★	*T. caudicula*	400★
Epeira	—	**Family Mimetidae**	—

Superfamily Lycosaeformia

Family Agelenidae (Funnel spiders)	421	*P. mirabilis*	419
House spider group, *Tegenaria*	421	*P. listeris*	420
House spider, *T. domestica*	421	**Family Lycosidae (Wolf spiders)**	421
Water spider, *Argyroneta aquatica*	410★	Tarantula, *Lycosa tarentula*	421
Family Pisauridae	—	**Family Ctenidae**	—
Pisaura	419	Comb spiders, *Phoneutria*	—

Superfamily Gnaphosiformia

Family Gnaphosidae (=Drassodidae)	—

Superfamily Clubioniformia

Family Clubionidae	—	*Micrommata*	422
"Fairy lantern", *Agroeca brunnea*	410★	*M. rosea*	422
Family Eusparassidae	—		

Superfamily Thomisiformia

Family Thomisidae (Crab spiders)	422	**Family Philodromidae**	—
Xysticus	422	**Family Salticidae (Jumping spiders)**	422
X. erraticus	422	*Salticus scenicus*	422
Thomisus onustus	410★	Ant spider, *Myrmarachne formicaria*	422
Oxyptila	—	Jumping spider, *Attulus saltator*	420★

Suborder Cribellata

Family Hypochilidae	—	*Dictyna arundinacea*	418★
Family Filistatidae	—	**Family Uloboridae**	—
Family Eresidae	—	*Hyptiotes*	422
Eresus cinnaberinus	422	*H. paradoxus*	415★
Family Acanthoctenidae	—	**Family Deinopidae**	—
Family Dictynidae	—	**Family Amaurobiidae**	—

Order Ricinulei

Cryptocellus	422	*C. simonis*	422

Order Pseudoscorpions (Pseudoscorpiones or Cheloneti)
Suborder Cthoniinea
Suborder Neobisiinea
Suborder Cheliferinea

Lamprochernes nodosus	423	Book scorpion, *Chelifer cancroides*	423

Order Solifuga

†*Protosolpuga*	406	*Solpuga*	424
Gluvia	423	*S. lethalis*	416★
Galeodes Caspius	424	**Family Rhagodidae**	424
G. graecus	423★	*Rhagodes*	424
Family Solpugidae	—		

Order Phalangids (Phalangidae)

†*Eotrogulus*	—		

Suborder Cyphophthalmi
Suborder Laniatores
Suborder True Phalangids (Palpatores)

Family Trogulidae	—	*I. hellwegi*	428
Trogulus	427	**Family Phalangiidae**	—
T. nepaeformis	428	*Phalangium*	427
Family Nemastomatidae	—	Common phalangid, *Phalangium opilio*	428
Family Ischyrosalididae	—	*Leiobunum rotundum*	428★
Ischyropsalis	427		

Order Mites (Acari or Acarina)

†*Protacarus*	405		

Suborder Holothyroidea
Suborder Parasitic Mites (Parasitiformes)

Family Parasitidae	—	**Family Trachytidae**	—
Family Poecilochiridae	—	**Family Ixodidae (Ticks)**	430
Family Laelaptidae	—	*Ixodes ricinus*	431
Family Spinturnicidae	—	**Family Argasidae**	431
Family Antennophoridae	—	Pigeon mite, *Argas reflexus*	431
Family Uropodidae	—		

Suborder Trombidiformes

Moth mite, *Pyemotes herfsi*	429	*T. autumnalis*	429
Family Eupodidae	—	**Family Erythraeidae**	—
Family Tetranychidae (Spider mites)	429	**Family Hydrachnidae**	—
Fruit-tree red-spider mites, *Panonychus ulmi*	429	**Family Limnocharidae**	—
Family Cheyletidae	—	**Family Eylaidea**	—
Family Demodicidae	—	**Family Thyasidae**	—
Family Bdellidae	—	**Family Hydraphantidae**	—
Family Halacaridae (Marine mites)	430	**Family Sperchonidae**	—
Family Limnohalacaridae	—	**Family Pontarachnidae**	—
Family Cunaxidae	—	Shell mite (*Unionicola aculeata*)	—
Family Trombidiidae	429	Thicked-legged water mite (*U. crassipes*)	430
Velvet spider mite (*Trombidium holosericeum*)	429	**Family Arrenuridae**	—

Suborder Sarcoptiformes

Family Acarida	430	**Family Analgesidae**	—
Cheese mite, *Tyrophagus casei*	430	**Family Falculiferidae**	—
Meal mite, *Acarus siro*	430	**Family Dermoglyphidae**	—
House mite, *Glyciphagus domesticus*	430	**Family Cytoditidae**	—
Family Sarcoptidae	430	**Family Listrophoridae**	—
Human itch mite (*Sarcoptes scabei*)	430	**Family Oribatidae**	—
Family Psoroptidae	—	**Family Phthiracaridae**	—
Family Pterolichidae	—		

Suborder Gall Mites (Tetrapodili)

Family Eriophyidae	—	**Family Phyllocoptidae**	—

CLASS PANTOPODA

Family Nymphonidae	—	**Family Collossendeidae (Deep-sea pantopods)**	432
Heteronymphon kempi	431★	Beaked pantopods, *Collossendeis proboscidea*	432
Family Pallenidae	—	**Family Pycnogonidae (Coastal pantopods)**	432
Family Phoxichilidiidae	—	Shore pycnogonid, *Pycnogonum littorale*	432
Family Ammothridae	—		

SUBPHYLUM DIANTENNATA

CLASS CRUSTACEA

SUBCLASS CEPHALOCARIDA

SUBCLASS LIPOSTRACA

†*Lepidocaris rhyniensis*	435

SUBCLASS ANOSTRACA

Chirocephalus	436★	Brine shrimp, *Artemia salina*	436
C. grubei	436	American brine shrimp, *A. gracilis*	440
Branchipus stagnalis	436		

SUBCLASS PHYLLOPODA

Order Notostraca

Triops cancriformis	441	*Lepidurus apus*	441

Order Onychura

Suborder Conchostraca

†*Leaia*	434	*Limnadia lenticularis*	442
†*Isaura* (=*Estheria*)	442	*Lunceus brachyurus*	—
†*Rhabdostichus*	442		

Suborder Waterfleas (Cladocera)

Superfamily Ctenopoda

Family Sididae	446	*Holopedium gibberum*	446
Sida crystallina	446	**Family Leptodoridae**	446
Latona setifer, Diaphanosoma brachyurum	446	*Leptodora*	446
Family Holopediidae	446	*L. hyalina*	446

Superfamily Anomopoda

Family Daphniidae	446	**Family Bosminidae**	447
Waterfleas, *Daphnia*	442	*Bosmina coregoni*	447
D. magna	447	*B. longirostris*	447
D. pulex	447	**Family Macrothrieidae**	447
D. longispina, D. cucullata	447	*Ilyocryptus sordidus*	447
Scapholeberis mucronata, Simocephalus vetulus	447	**Family Chydoridae**	447
Moina	445	*Chydorus sphaericus*	447
M. rectrirostris	447		

Superfamily Onychopoda

Family Polyphemidae	448	*Polyphemus pediculus*	448
Podon, Evadne	448	*Bythotrephes longimanus*	448

SUBCLASS OSTRACOSA
Order †Leperditiida

†Leperditia	434	*†L. titanica*	449

Order †Beyrichiida

†Beyrichia	434

Order Myodocopida
Suborder Myodocopa

Gigantocypris agassizi	449

Suborder Halocypriformes
Suborder Cladocopa

Polycope	448

Order Podocopida
Suborder Platycopa
Suborder Podocopa

Family Bairdiidae	—	*Notodromas monacha*	450
Family Darwinulidae	—	*Cypris*	464
Family Cyprididae	450	*Candona candida*	449

SUBCLASS COPEPODA
Order Progymnoplea
Order Calanoida (Gymnoplea)

Calanus finmarchieus	456	*Diaptomus*	447
Heterocope weismanni	—	*Calcocalanus pavo*	—

Order Propodoplea

Order Podoplea

Suborder Cyclopoida

Family Cyclopidae 457   **Family Sapphiriniidae** —
   Cyclops 457.   *Sapphirina* 458
   Megacyclops viridis 457   *S. fulgens* 443★
Family Monstrillidae 458   **Family Ergasilidae** —
   Hacmocera danae (=*Cymbasoma rigidum*) 458   Gill shrimp, *Ergasilus sieboldi* 458

Suborder Harpacticoida

Family Canthocamptidae 457   *Canthocamptus* 457

Order Caligoida

Caligus rapax, *C. lacustris* 461

Order Lernaeida

Lernaeocera branchialis 461   Pike shrimp, *L. esconia* 443★
Lernaea 461   *Penella balaenoptera* 462

Order Lernaeopodida

Perch shrimp, *Achtheres percarum* 462   *Xenocoeloma* 462

SUBCLASS MYSTACOCARIDA

Derocheilocaris —

SUBCLASS FISH LICE (BRANCHIURA)

Family Argulidae —   *A. scutiformis* 463
   Carp louse, *Argulus foliaceus* 463

SUBCLASS ASCOTHORACIDA

Family Synagogidae —   **Family Dendrogasteridae** —
Family Lauridae —

SUBCLASS CIRRIPEDIA (BARNACLES)

Order Thoracica

†*Cyprilepas* 434

Suborder Goose Barnacles (Lepadomorpha)

Family Scalpellidae —   *Lepas* 465
Family Iblidae —   Common goose barnacle, *L. anatifera* 465
Family Lepadidae —

Suborder Verrucomorpha

Verruca stroemia 466

Suborder Barnacles (Balanomorpha)

Family Chthamalidae — Common barnacle, *B. balanoides* 465
Star barnacle, *Chthamalus stellatus* 465 Ridged barnacle, *B. perforatus* 465
Family Balanidae — Ivory barnacle, *B. eburneus* 465
Encrusting barnacles, *Balanus* 465 Tortoise barnacle, *Chelonibia testudinaria* 465
B. balanus 465 **Family Coronulidae (Whale barnacles)** 466

Order Acrothoracica

Trypetesa lampas 466

Order Rhizocephala

Peltogasterella 467 *Sacculina carcini* 466

SUBCLASS HIGHER CRUSTACEANS (MALACOSTRACA)

Order †Hymenostraca

†*Hymenocaris* 434

Order †Ceratiocarina

†*Ceratiocaris* —

Order †Eumalacostraca

†*Paleocaris* 434

Superorder Phyllocarida

Leptostraca

Nebaliopsis typica 482 *Nebalia* 482

Superorder Hoplocarida

Order Stomatopoda

Family Squillidae — Mantis shrimp, *Squilla mantis* 482
Squilla 482 *S. raphidea, Gonodactylus* 482

Superorder Syncarida

Order Anaspidacea

Family Anaspididae — **Family Koonungidae** —
Anaspides 483

Order Stygocaridacea

Order Bathynellacea

Bathynella 434 *Leptobathynella, Thermobathynella adami* 483

Superorder Eucarida

Order Euphausiacea

Family Bentheuphasiidae — *Nematoscelis* 484
Family Thysanopodidae — *Stylocheiron* 484
Euphasia superba, Meganyctiphanes norvegica 484

Order Decapoda

Suborder Shrimps (Natantia)

Family Penaeidae	—	**Family Sergestidae**	—
Penaeus setifer, P. trisulcatus	485		

Superfamily Pasiphaeoida

Family Pasiphaeidae	—

Superfamily Hoplophoroida

Family Atyidae (Fresh water shrimps)	—	**Family Nematocarcinidae**	—
Family Hoplophoridae	—		

Superfamily Pandaloida

Family Pandalidae	—	*Pandalus borealis*	485

Superfamily Palaemonoida

Family Alpheidae (Snapping shrimps)	485	*H. varians*	485
Alpheus	485	**Family Palaemonidae**	—
Family Hippolytidae	—	Baltic shrimp, *Palaemon squilla*	486
Hippolyte	459★	*Periclimenes petersoni*	486

Superfamily Crangonoida

Family Processidae	—	**Family Stenopodidae**	—
Family Crangonidae	485	*Stenopus hispidus*	451★
Common Shrimp, *Crangon crangon*	485		

Suborder Crawfishes (Reptantia)

Family Polychelidae	—	*J. huegeli*	486
Family Palinuridae	435	**Family Scyllaridae (Slipper lobsters)**	487
European spiny lobster, *Palinurus vulgaris*	486	*Scyllarus*	487
P. argus	487	Slipper lobster, *S. arctus*	487
Panulirus ornatus	473★	*Scyllarides*	487
Cape spiny lobster, *Jasus lalandei*	486	Large slipper lobster, *S. latus*	487

Suborder Astacura

Family Homaridae (Lobsters)	487	European crayfish, *Astacus astacus*	487
European lobsters, *Homarus gammarus*	487	American crayfish, *Orconectes limosus*	488
American lobster, *H. americanus*	487	*Cherax*	453★
Norway lobster, *H. norvegicus*	—	**Family Parastacidae**	487
Family Crayfish (Astacidae)	487	**Family Austroastacidae**	—

Suborder Anomura

Family Axiidae	—	Family Lomisidae	—
Family Thalassinidae	—	Tropical mole crab, *Thalassina anomala*	488
Family Paguridae (Marine hermit crabs)	494★	Family Pylochelidae	—
Pagurus arrosor	222	Family Galatheidae	476★
P. bernhardus	222	Rock crab, *Galathea strigosa*	460★
P. calidus	495★	Family Porcellanidae (Porcelain crabs)	—
Porcelain crab, *Porcellanopagurus*	489	Porcellana	495★
Family Coenobitidae (Land hermit crabs)	—	Gray porcelain crab, *P. platycheles*	476★
Coenobita	489	Family Sand Crabs (Hippidae)	—
Coconut crab, *Birgus latro*	489	Emerita talpoida	489
Family Lithodidae (Stone crabs)	489	Family Albuneidae	—
Paralithodes camtschatica, *Lithodes maja*	489		

Suborder True Crabs (Brachyura)

Family Homolodromiidae	—	Family Dorippidae	—
Family Dromiidae (Dromid crabs)	489	Ethusa mascarone	473★
Sponge crabs *Dromia*	489	Family Clappidae	—
D. vulgaris	473★	Calappa	490
Hypoconch	489	Family Leucosiidae	—
Family Homolidae	—		

Superfamily Oxyrhyncha

Family Majiidae (Spider crabs)	490	Spider crabs, *Maja*	489
Spider crab, *Puggetia gracilis*	476★	Family Parthenopidae	—

Superfamily Brachyrhyncha

Family Corystidae	—	U. annulipes	491
Family Atelecyclidae	—	U. insignis	491
Heart crab, *Thia*	459★	Long-legged fiddler crab, *U. stenodactyla*	492
Family Cancridae	—	Emerald fiddler crab, *U. beebei*	492
Rock crab, *Cancer pagurus*	490	Jitterbug fiddler crab, *U. saltitanta*	492
Family Portunidae (Swimming crabs)	490	Singing fiddler crab, *U. musica*	497
Shore crab, *Carcinus maenus*	490	Terpsichore fiddler crab, *U. terpsichore*	497
True swimming crabs, *Portunus*	490	Tangier fiddler crab, *U. tangeri*	497
Common swimming crab, *P. holsatus*	493★	Dotilla	497
P. ruber	493★	Family Mictyridae (Grenadier crabs)	497
Blue crabs, *Callinectes*	490	Family Grapsidae (Rock crabs)	497
Podophthalmus vigil	490	Rock crab, *Grapsus grapsus*	497
Family Potamonidae (Fresh water crabs)	—	Mangrove crab, *Aratus pisoni*	497
Potamon	477	Goniopsis pulchra	497
Family Xanthidae	—	Columbus or Sargassum crab, *Planes minutus*	497
Lybia tesselata	490	Chinese crab, *Eriocheir sinensis*	497
Family Geryonidae	—	Family Gecarcinidae (Land crabs)	498
Family Pinnoteridae	490	Carribean land crab, *Cardisoma guanhumi*	498
Oyster crab, *Pinnoteres pisum*	490	Common land crab, *Gecarcinus ruricola*	498
Family Ocypodidae (Ghost and fiddler crabs)	491	G. lagostoma	476★
Ghost crab, *Ocypode ceratophthalmus*	491	Ucides	498
Fiddler Crabs, *Uca*	—	Red land crab, *Gigantinus lateralis*	498
Stalked fiddler crab, *U. stylifera*	491	Family Hapalocarcinidae	—

Superorder Pancarida

Order Thermosbaenacea

Thermosbaena mirabilis	499

Suborder Percarida

Order Mysidacea

Suborder Lophogastrida

Family Lophogastridae	—	**Family Eucopiidae**	—
Gnathophausia	499		

Suborder Mysida

Family Petalophthalmidae	—	*Mysis relicta, Praunus*	499
Family Mysidae	499	*Neomysis integer*	499★

Order Amphipods (Amphipoda)

Suborder Gammaridea

Family Gammariidae	496★	Beach flea, *Orchestia gammarellus*	500
Common waterflea, *Rivulogammarus pulex*	500	**Family Ampeliscidae**	—
R. lacustris	500★	**Family Haustoriidae**	—
Niphargus, Niphargellus	—	**Family Phliasidae**	—
Niphargopsis, Microniphargus	—	**Family Aoridae**	—
Eriopisa, Eriopisella	500	**Family Photidae**	—
Family Calliopiidae	—	**Family Amphithoidae**	—
Family Lysianassidae	—	**Family Corophiidae**	500
Family Leucothoidae	156	*Corophium volutator*	500
Family Talitridae (Sand fleas)	500	**Family Cheluridae**	—
Common sand flea, *Talitrus saltator*	500	**Family Podoceridae**	—

Suborder Ingolfiellidea

Suborder Laemodipodea

Family Caprellidae (Ghost shrimps)	501	Ghost shrimp, *Pseudoprotella phasma*	496★
Ghost shrimp, *Phtisica marina*	501	**Family Cyamidae (Whale lice)**	501
Caprella linearis	501		

Suborder Hyperiidea

Family Lanceolidae	—	**Family Phromimidae**	—
Family Eumimonectidae	—	**Family Lycaeidae**	—
Family Scinidae	—	**Family Oxycephalidae**	—
Family Thaumatopsidae	—	**Family Typhidae (= Platyscelidae)**	—
Family Hyperiidae	—		

Order Cumacea

Family Bodotriidae	—	**Family Pseudocumidae**	—
Family Leuconidae	158	**Family Diastylidae**	—
Family Nannastacidae	—	*Diastylis rathkei*	501
Family Lampropidae	—		

Order Spelaeopriphacez

Order Tanaidacea (=Anisopoda)

Suborder Monokonophora

Family Apseudidae	—	**Family Kalliapseudidae**	—
Apseydes	—		

Suborder Kilonophora

Family Neotanaidae	—	*Tanais cavolinii*	502
Family Paratanaidae	—	Family Tanaidae	—
Heterotanaeus oerstedi	502		

Order Isopoda

Suborder Gnathiidea

Suborder Microcerberidea

Suborder Anthuridea

Suborder Flabellifera

Family Cirolanidae	—	Family Sphaeromatidae	—
Family Aegidae	—	Subfamily Limnoriinae	—
Aega psora	—	*Limnoria lignorum*	502
Family Cymothoidae	—	Subfamily Sphaeromatinae	—
Amilocra physodes	—	Family Serolidae	—

Suborder Valvifera

Family Idoteidae	—	Family Arcturidae	—
Saduria (= Mesidotae) entomon	—		

Suborder Asellota

Family Asellidae (Fresh water isopods)	502	*Proasellus*	—
Common fresh-water isopod, *Asellus aquatica*	502	Family Parasellidae	—
Cave isopod, *A. cavaticus*	502	Family Microparasellidae	—

Suborder Phreatoicidea

Suborder Land Isopods (Oniscoidea)

Family Ligiidae	—	Family Cylisticidae	—
Beach woodlouse, *Ligia oceanica*	503	Family Porcellionidae	—
Family Trichoniscidae	—	Common woodlouse, *Porcellio scaber*	503
Family Platyarthridae	—	Family Armadillidiidae	—
Ant isopod, *Platyarthrus hoffmannseggii*	504	Pill woodlouse, *Armadillidium vulgare*	503
Family Oniscidae	—	Family Armadillidae	—
Sowbug, *Oniscus asellus*	503	Family Tylidae	—

Suborder Parasitic Isopods (Epicaridea)

Family Bopyridae	—	Family Entoniscidae	—
Family Dajidae	—		

SUBPHYLUM TRACHEATES (TRACHEATA)

CLASS MYRIAPODA

SUBCLASS PILOPODA

Order Pselaphognatha

Polyxenus lagurus 508

Order Chilognatha
Suborder Opisthandria

Family Glomeridae — *G. marginata* 508
Glomeris 508 **Family Sphaerotheriidae (Giant millipedes)** 508

Suborder Proterandria

Family Group Polydesmida

Compressed polydesmid, *Polydesmus complanatus* 511 *Platyrrhacus* 511
Brachydesmus superus 496★ *Apfelbeckia* 507

Family Group Nematophora

Family Group Julida

Family Julidae 507 *Schizophullum sabulosum* 511
Cylindroiulus londoninensis (= *C. teutonicus*) 511 *Graphidostreptus* 511

SUBCLASS PAUROPODA

Paruopus silvaticus —

SUBCLASS SYMPHYLA

Dwarf centipede, *Scutigerella immaculata* 512

SUBCLASS CENTIPEDES (CHILOPODA)

Superorder Epimorpha

Order Luminous Centipedes (Geophilomorpha)

Pachymerium ferrugineum 513 Luminous centipede, *Geophilus electricus* 513

Order Giant Centipedes (Scolopendromorpha)

Scolopendra subspinipes 513

Superorder Anamorpha

Order Lithobiomorpha

Lithobius forficatus 513

Order Scutigeromorpha

Scutigera coleoptrata 513

On the Zoological Classification and Names

For many years, zoologists and botanists have tried to classify animals and plants into a system which would be a survey of the abundance of forms in fauna and flora. Such a system, of course, may be established under very different aspects. Since Charles Darwin, his predecessors, and his successors have found that all creatures have evolved out of common ancestors, species of animals and plants have been classified according to their natural relationships. Our knowledge about the phylogeny, and thus the relationship of each living being to the other, is augmented every year by new discoveries and insights. Old ideas are replaced with more recent and more appropriate ones. Therefore, the natural classification of the animal kingdom (and the plant kingdom) is subject to changes. Furthermore, the opinions of zoologists, who are working on the classification of animals into the various groups, are anything but uniform. These differences and changes are usually insignificant. The classification of vertebrates into the classes of fish, amphibia, reptiles, birds, and mammals has been fixed for many decades. Only the Cyclostomata were recently separated from the fish and all other classes of vertebrates as the "jawless" Agnatha (comp. Vol. IV).

The animal kingdom has been split into several sub-kingdoms and these were again divided into further sections, subsections, and so on. The scale of the most important systematic categories follows in a descending rank order:

Kingdom
Sub-kingdom
Phylum
Subphylum
Class
Subclass
Superorder
Order
Suborder
Infraorder
Family
Subfamily
Tribe
Genus
Subgenus
Species
Subspecies

The scientific names of the animals and their spelling follow the international rules for the zoological nomenclature as agreed upon by the XV International Congress for Zoology and are obligatory for all zoological publications. The name of the genus, which is a Latin or Latinized noun, is singular and capitalized. After the name of the genus follows the name of the species and of the subspecies. The names of the species and subspecies may be nouns or adjectives, and they are spelled in the lower case. The name of a subgenus, which is formed in the same manner as a genus, may be added in brackets following the name of the genus. The names of the tribes, subfamilies, families, and superfamilies are plural capitalized nouns. They are formed from the name of a given genus by adding to the principal word the endings -ini for the tribe, -inae for the subfamily, -idae for the family, and -oidea for the superfamily. The names of the authors who were the first to describe and to name a species, subspecies, or group of animals should be cited with the year of this naming at least once in each scientific publication. The name of the author and year are not enclosed in brackets when the species or subspecies is classified as belonging to the same genus with which the author had originally classified it. They are in brackets when another genus name is used in the present publication. The scientific names of the genus, subgenus, species, and subspecies are supposed to be printed with different letters, usually italics.

Animal Dictionary

I. English—German—French—Russian

For scientific names of species see the German-English-French-Russian section of this dictionary or the index.

ENGLISH NAME	GERMAN NAME	FRENCH NAME	RUSSIAN NAME
Acarids	Vorratsmilben	Acaridés	Амбарные клещи
Acorn barnacle	Gemeine Seepocke	Balane commun	Обыкновенный морской жолудь
Acroporids	Baumförmige Korallen	Acroporidés	
Alcyonarian corals	Lederkorallen	Alcyonaires	Пробковые полипы
Alpheids	Knallkrebschen	Alphéidés	Креветки
American lobster	Amerikanischer Hummer	Homard d'Amérique	Американский омар
Amphipods	Flohkrebse	Amphipodes	Бокоплавы
Anthipathids	Schwarze Korallen	Anthipathidés	Антипаты
Anthozoans	Blumentiere	Anthozoaires	Коралловые полипы
Aphroditids	Seeraupen	Aphroditidés	Афродитовые
Arachnides	Spinnentiere	Arachnides	
Araneids	Radnetzspinnen i. e. S.	Aranéidés	Настоящие пауки
Argasids	Lederzecken	Argasidés	Аргазиды
Arthropods	Gliederfüßer	Arthropodes	Членистоногие
Astomates	Mundlose	Astomes	
Aviculariids	Echte Vogelspinnen	Aviculariidés	
Barnacle	Gemeine Entenmuschel	Anatife	Обыкновенная морская уточка
Barnacles	Rankenfüßer, Seepocken	Cirripèdes, Balanomorphes	Усоногие ракообразные
Beach flea	Küstenhüpfer	Orchestie des rivages	Береговой скакун
— fleas	Strandflöhe	Talitridés	Песочные скакуны
Beef tapeworm	Rinderbandwurm	Ténia du bœuf	Невооруженный цепень
Black widow spider	Schwarze Witwe	Veuve noire	
Blue corals	Blaukorallen	Hélioporidés	
— crabs	Blaukrabben	Callinectes	
Bonellia	Grüne Bonellia	Bonellie verte	Зеленая бонеллия
Bosminids	Rüsselkrebse	Bosminidés	
Brain corals	Neptunsgehirn	Diploriens	
Branchiobdellids	Kiemenegel	Branchiobdellidés	
By-the-wind-sailors	Segelqualle	Velelle	Парусница
Calcareous sponges	Kalkschwämme	Calcariens	Известковые губки
Caprellids	Gespenstkrebse	Capréllidés	Капрелловые
Centipedes	Hundertfüßer	Chilopodes	Губоногие многоножки
Chelicerates	Chelicerentiere	Chélicèriens	Хелицеровые
Chigger	Erntemilbe	Lepte automnal	Полевая краснотелка
Chonotriches	Trichterwimperlinge	Chonotriches	
Ciliates	Wimpertiere	Ciliés	Реснячные
Cirripeds	Rankenfüßer	Cirripèdes	Усоногие ракообразные
Clionids	Bohrschwämme	Clionidés	Клионовые
Collossendeids	Tiefsee-Asselspinnen	Collossendéidés	Глубоководные морские пауки
Comb jelly	Seestachelbeere	Pleurobrachie velue	
— jellyfishes	Rippenquallen	Cténophores	Гребневики
Common barnacle	Gemeine Seepocke	Balane commun	Обыкновенный морской жолудь
— jellyfish	Ohrenqualle	Aurélie	
— shrimp	Nordseegarnele	Crangon commun	
Copepods	Ruderfußkrebse	Copépodes	Веслоногие рачки
Coronates	Tiefseequallen	Coronates	
Coronulids	Wal-Seepocken	Coronulidés	Китовые морские жолуди
Crab spiders	Krabbenspinnen	Thomisidés	Пауки-бокоходы
Crabs	Echte Krabben	Brachyures	Крабы
Craspedomonadids	Kragengeißler	Craspedomonadidés	Воротничковые жгутиконосцы
Crawfishes	Ritterkrebse	Reptantides	
Crayfish	Europäischer Flußkrebs	Écrevisse commune	Широкопалый речной рак
Crayfishes	Flußkrebse	Astacidés	Речные раки
Crustaceans	Krebstiere	Crustacés	Ракообразные
Ctenopods	Kammfüßer	Cténopodes	
Cyaneids	Haarquallen	Cyanéidés	Цианеевые

ENGLISH NAME	GERMAN NAME	FRENCH NAME	RUSSIAN NAME
Cyclopids	Hüpferlinge	Cyclopidés	Циклопиды
Dead men's finger	Meerhand	Main de mer	Алциона
Decapods	Zehnfüßer	Décapodes	Десятиногие ракообразные
Dinoflagellates	Panzergeißler	Dinoflagellés	Панцирные жгутиконосцы
Diplozoon	Doppeltier	Diplozoon paradoxal	Странный спайник
Dog tapeworm	Hülsenwurm	Ténia échinocoque	Эхинококк
Dromid crabs	Wollkrabben, Wollkrebse	Dromies velues, Dromidés	Волосатые крабы
Dysderids	Sechsaugen	Dysdéridés	
Earthworm	Gemeiner Regenwurm	Ver de terre	Обыкновенный дождевой червь
Earthworms	Wenigborster	Oligochaetes	Малощетинковые черви
Echiurids	Igelwürmer	Echiures	Эхиуриды
English crab	Taschenkrebs	Crabe tourteau	Большой сухопутный краб
Epicarids	Schmarotzerasseln	Épicaridés	
Erigonids	Zwergspinnen	Erignidés	
Eurypterids	Riesenskorpione	Euryptéridés	Эвриптериды
False scorpions	Afterskorpione	Pseudoscorpions	Ложноскорпионы
Fiddler crabs	Winkerkrabben	Crabes à signaux	Манящие крабы
Fish lice	Fischläuse	Branchiures	Рыбьи вши
— louse	Karpfenlaus	Argule foliacé	Карповая вошь
— tapeworm	Fischbandwurm	Ténia du poisson	
Flagellates	Geißeltiere	Flagellés	Жгутиковые
Flatworms	Strudelwürmer	Turbellaires	Ресничные черви
Flukes	Saugwürmer	Trématodes	Сосальщики
Foraminifera	Lochträger	Foraminifères	Фораминиферы
Fresh water amphipods	Brunnenkrebse	Gammares de fontaines	
— — crabs	Flußkrebse	Astacidés	Речные раки
— — isopod	Gemeine Wasserassel	Aselle aquatique	Водяной ослик
— — isopods	Wasserasseln	Asellidés	Водные мокрицы
— — sponges	Süßwasserschwämme	Spongillidés	Бадяги
Fungus corals	Pilzkorallen	Coraux-champignons	Груздевики
Funnel-web spiders	Trichterspinnen	Agélénidés	Мешковые пауки
Garden spider	Kreuzspinne	Epéire diadème	Паук крестовик
Gasteracanthes	Stachelspinnen	Gastéracanthes	
Gastrotriches	Bauchhaarlinge	Gastrotriches	Брюхоресничные черви
Ghost crabs	Sandkrabben	Ocypodes	Песчаные крабы
Glomerids	Saftkugler	Gloméridés	Клубовидковые
Glossiphoniids	Knorpelegel	Glossiphónidés	Плоские пиявки
Grass spiders	Trichterspinnen	Agélénidés	Мешковые пауки
Green crab	Strandkrabbe	Crabe enragé	Европейский краб
Gribbles	Holzbohrasseln	Limnoridés	Сверлящие мокрицы
Gymnostomates	Nacktmünder	Gymnostomes	
Haemadipsids	Landegel	Hémadipsidés	Наземные пиявки
Halacarids	Meereswassermilben	Halacaridés	Морские клещи
Haplotaxids	Brunnenwürmer	Haplotaxidés	
Harvestman	Gemeiner Weberknecht	Faucheur	Обыкновенный сенокосец
Harvestmen	Weberknechte	Phalangides	Сенокосцы
Hermit crabs	Meeres-Einsiedlerkrebse	Paguridés	Морские раки отшельники
Heterotrichous ciliates	Verschiedenbewimperte	Hétérotriches	
Holotrichous ciliates	Ganzbewimperte	Holotriches	
Horse sponge	Pferdeschwamm	Éponge de toilette	
Horsehair worms	Saitenwürmer	Nématomorphes	Волосатиковые черви
Horseshoe crabs	Schwertschwänze, Königskrabben	Xiphosures, Limules	Мечехвосты
House scorpion	Bücherskorpion	Chélifère cancroide	Книжный ложноскорпион
— spider	Hausspinne	Araignée domestique	Домовый паук
— spiders	Hauswinkelspinnen	Araignées domestiques	Домовые пауки
Hunting spiders	Wolfsspinnen	Lycosidés	Тарантуловые
Hydrocorallids	Hydrokorallen	Hydrocoralliaires	Гидрокораллы
Hydrozoans	Hydrozoen	Hydrozoaires	Гидроидные
Ichthyobdellids	Fischegel	Ichthyobdéllidés	
Isopods	Asseln	Isopodes	Равноногие ракообразные
Itch mite	Krätzemilbe des Menschen	Gale de l'homme	Чесоточный зудень
Jellyfishes, anemones and corals	Hohltiere	Cœlentérés	Кишечнополостные
Jointed worms	Gliederwürmer	Annélides	
Julides	Schnurfüßer i. w. S.	Julides	Кивсяки
Julids	— i. e. S.	Julidés	Кивсяки
Jumping spiders	Springspinnen	Salticidés	Пауки-скакуны
Kamptozoans	Kelchwürmer	Kamptozoaires	Камптозои
King crabs	Königskrabben	Limules	Обыкновенный мечехвост
Kinorhynches	Hakenrüßler	Kinorhynches	Киноринхи
Land crab	Gemeine Landkrabbe	Crabe terrestre	Обыкновенный сухопутный краб
— crabs	Landkrabben i. e. S.	Gécarcinidés	Сухопутные крабы
Large jellyfishes	Echte Quallen	Scyphozoaires	Сцифоидные
Leech	Medizinischer Blutegel	Sangsue médicale	Лечебная пиявка
Leeches	Egel	Hirudinés	Пиявки

ENGLISH NAME	GERMAN NAME	FRENCH NAME	RUSSIAN NAME
Leuconids	Knollenkalkschwämme	Leuconidés	
Linguatulida	Zungenwürmer	Linguatulides	Пятиустки
Lions mane	Gelbe Haarqualle		Волосатая цианея
Lithistides	Steinschwämme	Lithistides	
Lithodid crabs	Steinkrabben	Lithodidés	Каменные крабы
Lobates	Lappenrippenquallen	Lobés	
Lobster	Europäischer Hummer	Homard d'Europe	Европейский омар
Lobsters	Hummer	Homaridés	Омары
Lugworm	Sandwurm	Arénicole des pêcheurs	Пескожил
Luminous centipede	Leuchtender Erdläufer	Géophile luisant	Светлянка
— centipedes	Erdläufer	Géophilomorphes	Светлянковые
Madreporian corals	Steinkorallen	Madrépores	Мадрепоровые кораллы
Many-celled animals	Echte Vielzeller	Eumétazoaires	Настоящие много- клеточные
Mawworm	Spulwurm	Ascaride lombricoïde	Человеческая аскарида
Maxillipods	Kieferfüßer	Maxillipodes	
Merostomates	Hüftmünder	Merostomates	Меростомовые
Mictyrids	Armeekrabben	Mictyridés	
Millipeds	Tausendfüßer, Doppelfüßer	Myriapodes, Diplopodes	Кивсяковые
Mites	Milben	Acariens	Клещи
Mitre jellyfish	Gurkenqualle	Béroé allongé	
Mole crabs	Sandkrebse	Hippidés	
Money spiders	Baldachinspinnen	Linyphidés	
Moon jellyfish	Ohrenqualle	Aurélie	
Mushroom corals	Pilzkorallen	Coraux-champignon	Груздевики
Myzostomes	Saugmünder	Myzostomides	Мизостомиды
Naidids	Wasserschlängler	Naididés	Вьюнки
Nature's ploughman	Gemeiner Regenwurm	Ver de terre	Обыкновенный дождевой червь
Northern comb jelly	Gurkenqualle	Béroé allongé	
Norway lobster	Kaiserhummer	Néphrops	Норвежский омар
Ocypodids	Renn- und Winkerkrabben	Ocypodidés	Песчаные крабы
Oligotrichous ciliates	Wenigbewimperte	Oligotriches	
Oncopods	Krallenfüßer	Oncopodes	
Onychopods	—	Onychopodes	
Orb-web spiders	Haubennetzspinnen	Théridiidés	Пауки-ткачи
Organ-pipe corals	Orgelkorallen	Tubiporidés	Органчиковые
Ostracods	Muschelkrebse	Ostracodes	Ракушковые рачки
Ox tapeworm	Rinderbandwurm	Ténia du bœuf	Невооруженный цепень
Oyster crab	Muschelwächter	Crabe de moule	Гороховый краб
Palpatores	Weberknechte i. e. S.	Palpateurs	
Pantopods	Asselspinnen	Pantopodes	Многоколенчатые
Pauropods	Wenigfüßer	Pauropodes	Пауроподы
Pedipalps	Skorpionspinnen	Pédipalpes	Жгутоногие пауки
Peritrichous ciliates	Glockentierchen	Péritriches	Кругоресничные инфузории
Pholcids	Zitterspinnen	Pholcidés	
Phyllopods	Blattfußkrebse	Phyllopodes	Жаброногие ракообразные
Pillbug	Rollassel	Armadille vulgaire	Шаровидка
Plathelminthes	Plattwürmer	Plathelminthes	Плоские черви
Poritids	Lochkorallen	Poritidés	Поритовые
Portuguese man-of-war	Seeblase	Physalie	Сифонофора физалия
Potamids	Süßwasserkrabben	Potamidés	Пресноводные крабы
Prawns	Garnelenartige Langschwanzkrebse	Natantides	Креветковые
Precious corals	Edelkorallen	Coraux	Благородные кораллы
Protozoans	Einzeller	Protozoaires	Простейшие животные
Pycnogonids	Ufer-Asselspinnen	Pycnogonidés	Береговые морские пауки
Radiolaria	Radiolarien	Radiolaires	Лучевики
Rat tapeworm	Zwergbandwurm der Maus		Мышиный карликовый цепень
Red precious coral	Rote Edelkoralle	Corail rouge	Благородный коралл
Renillids	Seestiefmütterchen	Rénillidés	
Rhizopods	Wurzelfüßer	Rhizopodes	Корненожки
River crayfish	Amerikanischer Flußkrebs	Écrevisse américaine	Американский речной рак
Robber crab	Palmendieb	Crabe des cocotiers	Пальмовый вор
Rock barnacle	Gemeine Seepocke	Balane commun	Обыкновенный морской жолудь
— crab	Felsenkrabbe	Grapse des rochers	
— crabs	Springkrabben	Grapsidés	
— -dwelling crab	Taschenkrebs	Crabe tourteau	Большой сухопутный краб
— lobster	Europäische Languste	Langouste européenne	Европейский лангуст
Roundworm	Spulwurm	Ascaride lombricoïde	Человеческая аскарида
Roundworms	Schlauchwürmer	Aschelminthes	
Sand anemone	Amerikanische Zylinderrose		Американский цериант
Schistosomids	Pärchenegel	Schistosomidés	Кровяные двуустки

ENGLISH NAME	GERMAN NAME	FRENCH NAME	RUSSIAN NAME
Scorpions	Skorpione	Scorpions	Скорпионы
Scyllarids	Bärenkrebse	Scyllaridés	Раки-медведи
Sea anemones	Purpurrose, Aktinien	Actinies	Актиниевые
– blubber	Gelbe Haarqualle		Волосатая цианея
– finger	Meerhand	Main de mer	Алциона
– mitre	Gurkenqualle	Béroé allongé	
– -pens	Seefedern	Pennatulaires	Морские перья
– walnut	Seestachelbeere	Pleurobrachie velue	
Seaworms	Vielborster	Polychaetes	Многощетинковые черви
Shore crab	Strandkrabbe	Crabe enragé	Европейский краб
Siphonophores	Staatsquallen	Siphonophores	Сифонофоры
Slipper animalcules	Pantoffeltierchen	Paramécie	Туфельки
Soft coral	Meerhand	Main de mer	Алциона
Sowbug	Mauerassel	Cloporte des murs	Стенная мокрица
Sphaerotheriids	Riesenkugler	Sphérothéridés	
Spider crabs	Seespinnen	Araignées de mer	Майи
– –	Dreieckskrabben	Majidés	Майи
Spiders	Echte Spinnen	Araniens	Пауки
Spiny-headed worms	Kratzer	Acanthocéphales	Колючеголовые черви
Spiny lobster	Europäische Languste	Langouste européenne	Европейский лангуст
– lobsters	Langusten	Palinuridés	Лангусты
Spirotriches	Spiralwimperlinge	Spirotriches	
Sponges	Schwammtiere	Spongiaires	Губки
Spongids	Badeschwämme	Spongidés	Туалетные губки
Sporozoans	Sporentierchen	Sporozoaires	Споровики
Star coral	Sternkoralle		Бокальчатый коралл
Stentor	Trompetentierchen	Stentor	Трубачи
Stomatopods	Mundfüßer	Stomatopodes	Ротоногие ракообразные
Stony corals	Steinkorallen	Madrépores	Мадрепоровые кораллы
Strapworms	Schnurwürmer	Némertines	Немертины
Suberitids	Korkschwämme	Subéritidés	Туалетные губки
Sun animalcules	Sonnentierchen	Heliozoaires	
– jelly	Gelbe Haarqualle		Волосатая цианея
Swimming crabs	Schwimmkrabben	Portunidés	Крабы-пловунцы
Swine tapeworm	Schweinebandwurm	Ténia du porc	Вооруженный цепень
Symphylids	Zwergfüßer	Symphyles	Сколопендреллы
Tapeworms	Bandwürmer	Cestodes	Ленточные черви
Tarantulas	Taranteln	Tarentules	Тарантулы
Tardigrades	Bärtierchen	Tardigrades	Тихоходки
Tetragnathids	Streckkiefer	Tétragnathidés	
Theridiids	Kugelspinnen	Therididés	
Thousend-legged worms	Tausendfüßer	Myriapodes	Многоножки
Threadworms	Fadenwürmer	Nématodes	Круглые черви
Tick	Holzbock	Tique	Собачий клещ
Ticks	Zecken	Ixodidés	Иксодовые клещи
Tracheates	Tracheentiere	Trachéates	Трахейные
Trematodes	Saugwürmer	Trématodes	Сосальщики
Trichina	Trichine	Trichine	Спиральная трихина
Trichostomates	Wimpermünder	Trichostomes	
Trilobites	Dreilapper	Trilobites	Трилобиты
Trombidiis	Laufmilben i. e. S.	Trombidiidés	Краснотелки
True scorpions	Skorpione	Scorpions	Скорпионы
– shrimps	Garnelenartige Langschwanzkrebse	Natantides	Креветковые
– swimming crabs	Echte Schwimmkrabben	Portunes	Крабы-пловунцы
Trypanosomatids	Trypanosomen	Trypanosomatidés	Трипанозомы
Tubifex	Gemeiner Schlammröhrenwurm	Tubifex	Обыкновенный трубочник
Tubificids	Schlammröhrenwürmer	Tubificidés	Трубочники
Unorganized animals	Mitteltiere	Mésozoaires	Мезозои
Urnatellids	Süßwasser-Kelchwürmer	Urnatellidés	
Venus's-flower-basket	Gießkannenschwamm	Euplectelle	
– girdle	Venusgürtel	Ceste de Vénus	
– girdles	–	Cestides	
Water bears	Bärtierchen	Tardigrades	Тихоходки
– fleas	Wasserflöhe	Cladocères	Ветвистоусые ракообразные
Whale lice	Walläuse	Cyamidés	Китовые вши
Wheel animalcules	Rädertiere	Rotatores	Коловратки
Whip scorpions	Skorpionspinnen	Pédipalpes	Жгутоногие пауки
White crab	Karibische Landkrabbe		Карибский сухопутный краб
Wolf spiders	Wolfsspinnen	Lycosidés	Тарантуловые

II. German—English—French—Russian

GERMAN NAME	ENGLISH NAME	FRENCH NAME	RUSSIAN NAME
Abgeplatteter Bandfüßer		Polydesme à lamelles	Плоский многосвяз
Acanthobdellae		Acanthobdellés	Акантобделлы
Acanthocephala	Spiny-headed worms	Acanthocéphales	Колючеголовые черви
Acari	Mites	Acariens	Клещи
Acaridae	Acarids	Acaridés	Амбарные клещи
Acarus siro		Acare de farine	Чесоточный клещ
Achtstrahlige Korallen		Octocoraux	Восьмилучевые коралловые полипы
Acnidaria		Acnidaires	Нестрекающие кишечнополостные
Acroporidae	Acroporids	Acroporidés	
Actinaria	Sea anemones	Actinies	Актиниевые
Actinia	– –	–	Актинии
– equina		Actinie pourpre	Конская актиния
Afterskorpione	False scorpions	Pseudoscorpions	Ложноскорпионы
Agelenidae	Funnel-web spiders	Agélénidés	Мешковые пауки
Aktinien	Sea anemones	Actinies	Актиниевые
Älchen	Threadworms	Nématodes	Круглые черви
Alcyonaria	Alcyonarian corals	Alcyonaires	Пробковые полипы
Alcyonium digitatum	Sea finger	Main de mer	Алциона
Alpheidae	Alpheids	Alphéidés	Креветки
Amerikanische Flußplanarie		Dugésie de rivières	Американская речная планария
– Zylinderrose	Sand anemone		Американский цериант
Amerikanischer Flußkrebs	River crayfish	Écrevisse américaine	Американский речной рак
– Hummer	American lobster	Homard d'Amérique	Американский омар
Amoebina		Amoebines	Амебы
Amphipoda	Amphipods	Amphipodes	Бокоплавы
Ancylostoma duodenale		Ankylostome duodénal	Анкилостома
Androctonus australis		Androcton d'Afrique du Nord	Сахарский скорпион
Anemonia sulcata		Anémone de mer	Бороздчатая анемона
Annelida	Jointed worms	Annélides	Кольчатые черви
Anomura		Anomures	Неполнохвостые раки
Anthipatharia		Anthipathariens	Антипатовые
Anthipathidae	Anthipathids	Anthipathidés	Антипаты
Anthozoa	Anthozoans	Anthozoaires	Коралловые полипы
Aphrodite aculeata		Aphrodite hérissée	Колючая афродита
Aphroditidae	Aphroditids	Aphroditidés	Афродитовые
Arachnida	Arachnids	Arachnides	Паукообразные
Araneae	Spiders	Araniens	Пауки
Araneidae	Araneids	Aranéidés	Настоящие пауки
Araneus diadematus	Garden spider	Epéire diadème	Паук крестовик
Arenicola marina	Lugworm	Arénicole des pêcheurs	Пескожил
Argasidae	Argasids	Argasidés	Аргазиды
Argulus foliaceus	Fish louse	Argule foliacé	Карповая вошь
Arktische Riesenqualle		Cyanée artique	Арктическая цианея
Armadillidium vulgare	Pillbug	Armadille vulgaire	Шаровидка
Artemia salina		Artémie	Соляный рачек
Arthropoda	Arthropods	Arthropodes	Членистоногие
Ascaris lumbricoides	Roundworm	Ascaride lombricoïde	Человеческая аскарида
Asc-helminthes	Roundworms	Aschelminthes	
Asellidae	Fresh water isopods	Asellidés	Водные мокрицы
Asellus aquaticus	– – isopod	Aselle aquatique	Водяной ослик
Asseln	Isopods	Isopodes	Равноногие ракообразные
Asselspinnen	Pantopods	Pantopodes	Многоколенчатые
Astacidae	Freshwater crabs	Astacidés	Речные раки
Astacura		Astacures	Длиннохвостые раки
Astacus astacus	Crayfish	Écreuisse commune	Широкопалый речной рак
Astroides calycularis	Star coral		Бокальчатый коралл
Atlantischer Palolo		Eunice atlantique	Атлантический палоло
Aufgußtierchen	Ciliates	Ciliés	Реснничные
Aurelia		Aurélies	Аурелии
– aurita	Common jellyfish	Aurélie	
Aviculariidae	Aviculariids	Avicularidés	
Bachflohkrebs		Gammare de ruisseaux	Бокоплав-блоха
Badeschwämme	Spongids	Spongidés	Туалетные губки
Balanomorpha	Barnacles	Balanomorphes	Морские жолуди
Balanus balanoides	Common barnacle	Balane commun	Обыкновенный морской жолудь
Baldachinspinnen	Money spiders	Linyphidés	
Bandfüßer		Polydesmides	Многосвязки

GERMAN NAME	ENGLISH NAME	FRENCH NAME	RUSSIAN NAME
Bandwürmer	Tapeworms	Cestodes	Ленточные черви
Bärenkrebse	Scyllarids	Scyllaridés	Раки-медведи
Bärtierchen	Waterbears	Tardigrades	Тихоходки
Bauchhaarlinge	Gastrotrichans	Gastrotriches	Брюхоресничные черви
Bauchwimperlinge	Hypotrichans	Hypotriches	
Baumförmige Korallen	Acroporids	Acroporidés	
Bdelloida	Bdelloidans	Bdelloides	
Beroe cucumis	Sea mitre	Béroé allongé	
Bewaffneter Bandwurm	Swine tapeworm	Ténia du porc	Вооруженный цепень
Birgus latro	Robber crab	Crabe des cocotiers	Пальмовый вор
Blasenwurm	Dog tapeworm	Ténia échinocoque	Эхинококк
Blattfußkrebse	Phyllopods	Phyllopodes	Жаброногие ракообразные
Blaues Trompetentierchen	Blue stentor	Stentor bleu	Синий трубач
Blaukorallen	Blue corals	Hélioporidés	
Blaukrabben	— crabs	Callinectes	
Blumentiere	Anthozoans	Anthozoaires	Коралловые полипы
Blutsauger		Sangsue du cheval	Ложноконская пиявка
Bohrschwämme	Clionids	Clionidés	Сверлящие губки
Bonellia viridis	Bonellia	Bonellie verte	Зеленая бонеллия
Borstenegel		Acanthobdellés	Акантобделлы
Bosminidae	Bosminids	Bosminidés	
Brachyura	Crabs	Brachyures	Крабы
Branchiobdellidae	Branchiobdellids	Branchiobdellidés	
Branchipus stagnalis		Branchipe commun	Жаброног
Branchiura	Fish lice	Branchiures	Рыбьи вши
Braune Hydra		Hydre commune	Обыкновенная гидра
Brauner Steinläufer		Mille-pattes commun	Обыкновенная много-ножка
Brunnenkrebse	Fresh water amphipods	Gammares de fontaines	
Brunnenwürmer	Haplotaxids	Haplotaxidés	
Bücherskorpion	House scorpion	Chélifère cancroide	Книжный ложноскорпион
Calappa		Calappe	Стыдливые крабы
Calcarea	Calcareous sponges	Calcariens	Известковые губки
Callinectes	Blue crabs	Callinectes	
Cancer pagurus	Rock-dwelling crab	Crabe tourteau	Большой сухопутный краб
Caprella linearis		Caprelle-chèvre	Морская козочка
Caprellidae	Caprellids	Capréllidés	Капрелловые
Carcinus maenas	Shore crab	Crabe enragé	Европейский краб
Cardisoma guanhumi	White crab		Карибский сухопутный краб
Ceriantharia		Cérianthaires	Цериантовые
Cerianthopsis americanus	Sand anemone		Американский цериант
Cerianthus lloydii		Cérianthe nordique	Северный цериант
— *membranaceus*		Cérianthe de Méditerranée	Перепончатый цериант
Cestidea	Venus's girdles	Cestides	
Cestoda	Tapeworms	Cestodes	Ленточные черви
Cestus veneris	Venus's girdle	Ceste de Vénus	
Ceylonegel		Sangsue de Ceylan	Цейлонская пиявка
Chaetopterus variopedatus		Chétoptère luisant	Пергаментный трубкожил
Chelicerata		Chélicériens	
Chelicerentiere		—	Хелицеровые
Chelifer cancroides	House scorpion	Chélifère cancroide	Книжный ложноскорпион
Chelonethi	False scorpions	Pseudoscorpions	Ложноскорпионы
Chilopoda	Centipedes	Chilopodes	Губоногие многоножки
Chinesischer Leberegel		Clonorchis chinois	Китайская двуустка
Chlorohydra viridissima		Hydre verte	Зеленая гидра
Chrysaora		Chrysaores	Хризаоры
Ciliata	Ciliates	Ciliés	Ресничные
Cirripedia	Barnacles	Cirripèdes	Усоногие ракообразные
Cladocera	Water fleas	Cladocères	Ветвистоусые рако-образные
Clionidae	Clionids	Clionidés	Сверлящие губки
Clonorchis sinensis		Clonorchis chinois	Китайская двуустка
Cnidaria	Cnidarians	Cnidaires	Стрекающие кишечно-полостные
Coccidia	Coccidians	Coccidies	Кокцидии
Coelenterata	Jellyfishes, anemones and corals	Cœlentérés	Кишечнополостные
Collossendeidae	Collossendeids	Collossendéidés	Глубоководные морские пауки
Collossendeis proboscidea		Pantopode à trompe	Хоботковый морской паук
Colpoda cucullus		Colpode	Колпода
Conchostraca		Conchostraques	Раковинные жаброноги
Copepoda	Copepods	Copépodes	Веслоногие рачки
Corallium	Precious corals	Coraux	Благородные кораллы
— *rubrum*	Red precious coral	Corail rouge	Благородный коралл
Coronulidae	Coronulids	Coronulidés	Китовые морские жолуди
Crangon crangon	Common shrimp	Crangon commun	Обыкновенная креветка

GERMAN NAME	ENGLISH NAME	FRENCH NAME	RUSSIAN NAME
Craspedomonadidae	Craspedomonadids	Craspedomonadidés	Воротничковые жгутико- носцы
Crustacea	Crustaceans	Crustacés	Ракообразные
Ctenophora	Comb jellyfishes	Cténophores	Гребневики
Ctenopoda	Ctenopods	Cténopodes	
Cumacea	Cumaceans	Cumacés	Кумовые ракообразные
Cyamidae	Whale lice	Cyamidés	Китовые вши
Cyanea		Cyanées	Цианеи
– arctica		Cyanée artique	Арктическая цианея
– capillata	Lions mane		Волосатая цианея
Cyaneidae	Cyaneids	Cyanéidés	Цианеевые
Cyclopidae	Cyclopids	Cyclopidés	Циклопиды
Dalmatiner Schwamm		Éponge officinal	Адриатическая губка
Daphnien	Water fleas	Cladocères	Ветвистоусые рако- образные
Darm-Pärchenegel		Bilharzie de l'intestin	Кишечная кровяная двуустка
Decapoda	Decapods	Décapodes	Десятиногие ракообразные
Deckennetzspinnen	Money spiders	Linyphidés	
Deuterostomia		Deutérostomes	Вторичноротые
Dibothriocephalus latus	Fish tapeworm	Ténia du poisson	
Dickbeinige Wassermilbe		Unionicole à grosses pattes	Толстоногий водный клещ
Dicrocoelium dendriticum		Dicrocélium fer-de-lance	Ланцетовидная двуустка
Didinium nasutum	Water bear	Didinium	
Digene Saugwürmer		Trématodes digènes	Дигенетические сосаль- щики
Dinoflagellata	Dinoflagellates	Dinoflagellés	Панцирные жгутиконосцы
Diplopoda	Millipeds	Diplopodes	Кивсяковые
Diploria	Brain corals	Diploriens	
Diplozoon paradoxum	Diplozoon	Diplozoon paradoxal	Странный спайник
Doppelfüßer	Millipeds	Diplopodes	Кивсяковые
Doppeltier	Diplozoon	Diplozoon paradoxal	Странный спайник
Dörnchenkorallen		Anthipathariens	Антипатовые
Dracunculus medinensis		Filaire de Médine	Ришта
Dreieckskrabben	Spider crabs	Majidés	Майи
Dreilapper	Trilobites	Trilobites	Трилобиты
Dromia	Dromid crabs	Dromies velues	Волосатые крабы
Dromiidae	– –	Dromidés	Волосатые крабы
Dugesia dorotocephala		Dugésie de rivières	Американская речная планария
Dysderidae	Dysderids	Dysdéridés	
Echinococcus granulosus	Dog tapeworm	Ténia échinocoque	Эхинококк
Echiurida	Echiurids	Echiures	Эхиуриды
Echiurus echiurus		Echiure	Эхиур
Echte Krabben	Crabs	Brachyures	Крабы
– Quallen	Large jellyfishes	Scyphozoaires	Сцифоидные
– Schwimmkrabben	True swimming crabs	Portunes	Крабы-пловунцы
– Spinnen	Spiders	Araniens	Пауки
– Vielzeller	Many-celled animals	Eumétazoaires	Настоящие многоклеточ- ные
– Vogelspinnen	Aviculariids	Avicularidés	
Edelkorallen	Precious corals	Coraux	Благородные кораллы
Edelkrebs	Crayfish	Écrevisse commune	Широкопалый речной рак
Egel	Leeches	Hirudinés	Пиявки
Eigentliche Langschwanzkrebse		Astacures	Длиннохвостые раки
Einkammerige		Foraminifères à coquille mono- thalame	Однокамерные форамини- феры
Einschwänziger Priapswurm		Priapule	Однохвостый приапул
Einzeller	Protozoans	Protozoaires	Простейшие животные
Enchytraeus albidus		Enchytrée	Энхитрея
Entamoeba histolytica		Amibe dysentérique	Дизентерийная амеба
Entenmuscheln	Barnacles	Cirripèdes	
–		Lepadomorphes	Морские уточки
Enterobius vermicularis		Ver d'intestin	Обыкновенная острица
Entomostraca	Entomostraca	Entomostraques	Низшие ракообразные
Entoprocta	Kamptozoans	Kamptozoaires	Камптозои
Epicaridae	Epicarids	Épicaridés	
Erdläufer	Luminous centipedes	Géophilomorphes	Светляковые
Erigonidae	Erigonids	Erigonidés	
Eriocheir sinensis		Crabe chinois	Китайский мохнаторукий краб
Erntemilbe	Chigger	Lepte automnal	Полевая краснотелка
Errantia		Polychaetes errants	Свободноползающие полихеты
Eumetazoa	Many-celled animals	Eumétazoaires	Настоящие многоклеточ- ные
Eunice fucata		Eunice atlantique	Атлантический палоло

GERMAN NAME	ENGLISH NAME	FRENCH NAME	RUSSIAN NAME
Euphausiacea	Euphausiaceans	Euphausiacés	Эфраузиевые рако-образные
Euplectella aspergillum	Venus's-flower-basket	Euplectelle	
Europäische Languste	Spiny lobster	Langouste européenne	Европейский лангуст
Europäischer Flußkrebs	Crayfish	Écrevisse commune	Широкопалый речной рак
– Hummer	Lobster	Homard d'Europe	Европейский омар
Eurypteridae	Eurypterids	Euryptéridés	
Fadenwürmer	Threadworms	Nématodes	Круглые черви
Fangschreckenkrebse	Stomatopods	Stomatopodes	Ротоногие ракообразные
Fasciola gigantica		Douve du foie géante	Гигантская печеночная двуустка
– hepatica		Grande douve du foie	Печеночная двуустка
Felsenkrabbe	Rock crab	Grapse des rochers	
Festsitzende Vielborster		Polychaetes sédentaires	Сидячие полихеты
Filzwurm		Aphrodite hérissée	Колючая афродита
Fischasseln		Cymothoïdés	Рыбные мокрицы
Fischbandwurm	Fish tapeworm	Ténia du poisson	
Fischegel	Ichthyobdellids	Ichthyobdéllidés	Рыбьи пиявки
Fischläuse	Fish lice	Branchiures	
Flagellata	Flagellates	Flagellés	Жгутиковые
Flohkrebse	Amphipods	Amphipodes	Бокоплавы
Flußkrebse	Fresh water crabs	Astacidés	Речные раки
Foraminiferen	Foraminifera	Foraminifères	Фораминиферы
Freilebende Vielborster		Polychaetes errants	Свободноползающие полихеты
Fühlerlose	Chelicerates	Chélicèrieus	Хелицеровые
Fungia	Fungus corals	Coraux-champignons	Груздевики
Ganzbewimperte	Holotrichous ciliates	Holotriches	
Garnelenartige Langschwanz-krebse	Trues shrimps	Natantides	Креветковые
Gastrotricha		Gastrotriches	Брюхоресничные черви
Gecarcinidae	Land crabs	Gécarcinidés	Сухопутные крабы
Gecarcinus ruricola	– crab	Crabe terrestre	Обыкновенный сухопутный краб
Geißelskorpione		Uropyges	Скорпионопауки
Geißeltiere			
Geisterkrabben	Flagellates	Flagellés	Жгутиковые
Gelbe Haarqualle	Ghost crabs	Ocypodes	Песчаные крабы
Gemeine Entenmuschel	Lions mane		Волосатая цианея
	Barnacle	Anatife	Обыкновенная морская уточка
– Landkrabbe	Land crab	Crabe terrestre	Обыкновенный сухопутный краб
– Seepocke	Common barnacle	Balane commun	Обыкновенный морской жолудь
– Wasserassel	Fresh water isopod	Aselle aquatique	Водяной ослик
Gemeiner Fischegel		Sangsue piscicole	Обыкновенная рыбья пиявка
– Leberegel		Grande douve du foie	Печеночная двуустка
– Regenwurm	Earthworm	Ver de terre	Обыкновенный дождевой червь
– Schlammröhrenwurm		Tubifex	Обыкновенный трубочник
– Strandfloh		Talitre sauteur	Песочный скакун
– Weberknecht	Harvestman	Faucheur	Обыкновенный сенокосец
Geophilomorpha	Luminous centipedes	Géophilomorphes	Светляковые
Geophilus electricus	– centipede	Géophile luisant	Светлянка
Gespenstkrebse	Caprellids	Capréllidés	Капрелловые
Gießkannenschwamm	Venus's-flower-basket	Euplectelle	
Glasschwämme		Hexactinellides	Стеклянные губки
Gliederfüßer	Arthropods	Arthropodes	Членистоногие
Gliedertiere	Articulates	Articulates	
Gliederwürmer	Jointed worms	Annélides	Кольчатые черви
Glockentierchen	Peritrichous ciliates	Péritriches	Кругоресничные инфузории
Glomeridae	Glomerids	Gloméridés	Клубовидковые
Glossiphoniidae	Glossiphoniids	Glossiphonidés	Плоские пиявки
Gordius aquaticus		Gordius aquatique	Водяной волосатик
Gorgonaria		Gorgonaires	Горгонарии
Granat	Common shrimp	Crangon commun	Обыкновенная креветка
Grapsidae	Rock crabs	Grapsidés	
Grapsus grapsus	– crab	Grapse des rochers	
Graue Hydra		Hydre grise	Серая гидра
– Seefeder		Ptéroide gris	Серое морское перо
Gregarinen		Grégarines	Грегарины
Grenadierkrabben	Mictyrids	Mictyridés	
Großer Leberegel		Grande douve du foie	Печеночная двуустка
– Rückenschaler		Triops	Большой щитень

GERMAN NAME	ENGLISH NAME	FRENCH NAME	RUSSIAN NAME
Grüne Bonellia	Bonellia	Bonellie verte	Зеленая бонеллия
– Hydra		Hydre verte	Зеленая гидра
Grubenwurm		Ankylostome duodénal	Анкилостома
Gurkenqualle	Sea mitre	Béroé allongé	
Haarquallen	Cyaneids	Cyanéidés	Цианеевые
Haarwurm		Wuchérérie	Вухерия
Haemadipsa zeylanica		Sangsue de Ceylan	Цейлонская пиявка
Haemadipsidae	Haemadipsids	Hémadipsidés	Наземные пиявки
Haemopis sanguisuga		Sangsue du cheval	Ложноконская пиявка
Hakenrüßler	Kinorhynches	Kinorhynches	Киноринхи
Halacaridae	Halacarids	Halacaridés	Морские клещи
Haplotaxidae	Haplotaxids	Haplotaxidés	
Harnblasen-Pärchenegel		Bilharzie de la vessie	Мочеводная кровяная двуустка
Harnblasen-Saugwurm		Polystomum de vessie	Лягушечья многоустка
Haubennetzspinnen	Orb-web spiders	Thériididés	Пауки-ткачи
Hausspinne	House spider	Araignée domestique	Домовый паук
Hauswinkelspinnen	– spiders	Araignées domestiques	Домовые пауки
Helioporidae	Blue corals	Hélioporidés	
Heliozoa	Sun animalcules	Héliozoaires	
Heterocoela		Hétérocoeles	Сотовидные губки
Heterotricha	Heterotrichous ciliates	Hétérotriches	
Heuschreckenkrebse	Stomatopods	Stomatopodes	Ротоногие ракообразные
Hexacorallia		Héxacoraux	Шестилучевые коралловые полипы
Hexactinellida		Hexactinellides	Стеклянные губки
Hippidae	Mole crabs	Hippidés	
Hippospongia communis	Horse sponge	Éponge de toilette	
Hirudinea	Leeches	Hirudinés	Пиявки
Hirudo medicinalis medicinalis	Leech	Sangsue médicale	Лечебная пиявка
Höhere Krebse		Malacostraques	Высшие ракообразные
Hohltiere	Jellyfishes, anemones and corals	Cœlentérés	Кишечнополостные
Holotricha	Holotrichous ciliates	Holotriches	
Holzbock	Tick	Tique	Собачий клещ
Holzbohrasseln	Gribbles	Limnoridés	Сверлящие мокрицы
Homaridae	Lobsters	Homaridés	Омары
Homarus americanus	American lobster	Homard d'Amérique	Американский омар
– *gammarus*	Lobster	– d'Europe	Европейский омар
Hüftmünder	Merostomates	Merostomates	Меростомовые
Hülsenwurm	Dog tapeworm	Ténia échinocoque	Эхинококк
Hummer	Lobsters	Homaridés	Омары
Hundertfüßer	Centipedes	Chilopodes	Губоногие многоножки
Hüpferlinge	Cyclopids	Cyclopidés	Циклопиды
Hydra oligactis		Hydre grise	Серая гидра
– *vulgaris*		– commune	Обыкновенная гидра
Hydrokorallen	Hydrocorallids	Hydrocoralliaires	Гидрокораллы
Hydropolypen	Hydrozoans	Hydrozoaires	Гидроидные
Hydrozoen	–	–	Гидроидные
Hymenolepis nana		Ténia nain	Карликовый цепень
– – *fraterna*	Rat tapeworm		Мышиный карликовый цепень
– – *nana*		– – de l'homme	Человеческий карликовый цепень
Ichthyobdellidae	Ichthyobdellids	Ichthyobdéllidés	Рыбьи пиявки
Igelwürmer	Echiurids	Echiures	Эхиуриды
Infusorien	Ciliates	Ciliés	Ресничные
Isopoda	Isopods	Isopodes	Равноногие ракообразые
Ixodes ricinus	Tick	Tique	Собачий клещ
Ixodidae	Ticks	Ixodidés	Иксодовые клещи
Japanischer Pärchenegel		Bilharzie d'extrème-Orient	Японская кровяная двуустка
Julida	Julides	Julides	Кивсяки
Julidae	Julids	Julidés	Кивсяки
Kaiserhummer	Norway lobster	Néphrops	Норвежский омар
Kalkschwämme	Calcareous sponges	Calcariens	Известковые губки
Kammfüßer	Ctenopods	Cténopodes	
Kammquallen	Comb jellyfishes	Cténophores	Гребневики
Kamptozoa	Kamptozoans	Kamptozoaires	Камптозои
Kanker	Harvestmen	Phalangides	Сенокосцы
Kappentierchen		Colpode	Колпода
Karibische Landkrabbe	White crab		Карибский сухопутный краб
Karpfenlaus	Fish louse	Argule foliacé	Карповая вошь
Käsemilbe		Acare du fromage	Сырный акар
Katzenleberegel		Douve du chat	Кошачья печеночная двуустка

GERMAN NAME	ENGLISH NAME	FRENCH NAME	RUSSIAN NAME
Kelchwürmer	Kamptozoans	Kamptozoaires	Камптозои
Kellerassel		Cloporte des caves	Погребная мокрица
Kieferfüßer	Maxillipods	Maxillipodes	
Kiemenegel	Branchiobdellids	Branchiobdellidés	
Kiemenschwänze	Fish lice	Branchiures	Рыбьи вши
Kinorhyncha	Kinorhynches	Kinorhynches	Киноринхи
Kleiner Leberegel		Dicrocélium fer-de-lance	Ланцетовидная двуустка
Knallkrebschen	Alpheids	Alphéidés	Креветки
Knollenkalkschwämme	Leuconids	Leuconidés	
Knorpelegel	Glossiphoniids	Glossiphonidés	Плоские пиявки
Köderwurm	Lugworm	Arénicole des pêcheurs	Пескожил
Kokzidien		Coccidies	Кокцидии
Kompaßquallen		Chrysaores	Хризаоры
Königskrabben	King crabs	Limules	Обыкновенный мечехвост
Korkschwämme	Suberitids	Subéritidés	
Krabbenspinnen	Crab spiders	Thomisidés	Пауки-бокоходы
Kragengeißler	Craspedomonadids	Craspedomonadidés	Воротничковые жгутико-носцы
Krallenfüßer	Oncopods	Oncopodes	
—	Onychopods	Onychopodes	
Krätzemilbe des Menschen	Itch mite	Gale de l'homme	Чесоточный зудень
Kratzer	Spiny-headed worms	Acanthocéphales	Колючеголовые черви
Krebstiere	Crustaceans	Crustacés	Ракообразные
Kreuzspinne	Garden spider	Epéire diadème	Паук крестовик
Kronenquallen	Coronates	Coronates	
Krustenanemonen		Zooanthaires	Зоантарии
Kugelspinnen	Theridiids	Therididés	
Kumazeen		Cumacés	Кумовые ракообразные
Küstenhüpfer	Beach flea	Orchestie des rivages	Береговой скакун
Landasseln		Oniscoides	Мокрицевые
Landegel	Haemadipsids	Hémadipsidés	Наземные пиявки
Landkrabben i. e. S.	Land crabs	Gécarinidés	Сухопутные крабы
Landplanarien		Terricoles	Наземные планарии
Lanzettegel		Dicrocélium fer-de-lance	Ланцетовидная двуустка
Langusten	Spiny lobsters	Palinuridés	Лангусты
Lappenrippenquallen	Lobates	Lobés	
Latrodectus mactans	Black widow spider	Veuve noire	
Laufmilben		Trombidiformes	Краснотелковые клещи
— i. e. S.	Trombidiids	Trombidiidés	Краснотелки
Lederkorallen	Alcyonarian corals	Alcyonaires	Пробковые полипы
Lederzecken	Argasids	Argasidés	Аргазиды
Lepadomorpha		Lepadomorphes	Морские уточки
Lepas anatifera	Barnacle	Anatife	Обыкновенная морская уточка
Leptostraken		Leptostraques	Тонкопанцирные
Leuchtender Erdläufer	Luminous centipede	Géophile luisant	Светлянка
Leuchtkrebse		Euphausiacés	Эфраузиевые рако-образные
Leuconidae	Leuconids	Leuconidés	
Ligia oceanica		Ligie des rivages	Береговая мокрица
Ligula intestinalis		Ligule des intestins	Ремнец
Limnoridae	Gribbles	Limnoridés	Сверлящие мокрицы
Limulus	King crabs	Limules	Обыкновенный мечехвост
Linguatula serrata		Linguatule nasale	Носовая пятиустка
Linguatulida	Linguatulides	Linguatulides	Пятиустки
Linyphiidae	Money spiders	Linyphidés	
Lithistida	Lithistides	Lithistides	
Lithobiomorpha		Lithobiomorphes	Костянковые
Lithobius forficatus		Mille-pattes commun	Обыкновенная много-ножка
Lithodidae	Lithodid crabs	Lithodidés	Каменные крабы
Loa loa		Filaire loa	Червь-лоа
Lobata	Lobates	Lobés	
Lochkorallen	Poritids	Poritidés	Поритовые
Lochträger	Foraminifera	Foraminifères	Фораминиферы
Lumbricus terrestris	Earthworm	Ver de terre	Обыкновенный дождевой червь
Lycosa	Tarantulas	Tarentules	Тарантулы
Lycosidae	Wolf spiders	Lycosidés	Тарантуловые
Macracanthorhynchus hirudinaceus		Acanthocéphale géant	Гигантский скребень
Madenwurm		Ver d'intestin	Обыкновенная острица
Madrepora		Madrépores	Мадрепоры
Madreporaria	Stony corals	Madrépores	Мадрепоровые кораллы
Maja	Spider crabs	Araignées de mer	Майи
Majidae	— —	Majidés	Майи

GERMAN NAME	ENGLISH NAME	FRENCH NAME	RUSSIAN NAME
Malacostraca		Malacostraques	Высшие ракообразные
Malmignatte	Black widow spider	Veuve noire	
Mandibellose	Amandibulates	Amandibulés	
Mandibeltiere	Mandibulates	Mandibulés	
Maricola		Maricoles	Морские планарии
Mauerassel	Sowbug	Cloporte des murs	Стенная мокрица
Maxillipoda	Maxillipods	Maxillipodes	
Medinawurm		Filaire de Médine	Ришта
Medizinischer Blutegel	Leech	Sangsue médicale	Лечебная пиявка
Meeres-Einsiedlerkrebse	Hermit crabs	Paguridés	Морские раки отшельники
Meeresplanarien		Maricoles	Морские планарии
Meereswassermilben	Halacarids	Halacaridés	Морские клещи
Meerhand	Sea finger	Main de mer	Алциона
Meerquappe		Echiure	Эхиур
Meerschwert	Venus's girdle	Ceste de Vénus	
Megascolides australis		Lombric géant	Гигантский дождевой червь
Mehlmilbe		Acare de farine	Чесоточный клещ
Merostomata	Merostomates	Merostomates	Меростомовые
Mesozoa	Unorganized animals	Mésozoaires	Мезозои
Mictyridae	Mictyrids	Mictyridés	
Milben	Mites	Acariens	Клещи
Mittelkrebse		Anomures	Неполнохвостые раки
Mittelmeer-Zylinderrose		Cérianthe de Méditerranée	Перепончатый цериант
Mitteltiere	Unorganized animals	Mésozoaires	Мезозои
Monogene Saugwürmer		Trématodes monogènes	Моногенетические сосальщики
Monothalamia		Foraminifères à coquille mono-thalame	Однокамерные фораминиферы
Mundfüßer	Stomatopods	Stomatopodes	
Mundlose	Astomates	Astomes	
Muschelkrebse	Ostracods	Ostracodes	Ракушковые рачки
Muschelschaler		Conchostraques	Раковинные жаброноги
Muschelwächter	Oyster crab	Crabe de moule	Гороховый краб
Myriapoda	Thousand-legged worms	Myriapodes	Многоножки
Mysidacea	Mysidaceans	Mysidacés	Мизиды
Myzostomida	Myzostomes	Myzostomides	Мизостомиды
Nacktamöben		Amoebines	Амебы
Nacktmünder	Gymnostomates	Gymnostomes	
Naiden	Naidids	Naididés	Вьюнки
Naididae	Naidids	—	Вьюнки
Nasentierchen	Water bear	Didinium	
Nasenwurm		Linguatule nasale	Носовая пятиустка
Natantia	Trues shrimps	Natantides	Креветковые
Nemat-helminthes	Roundworms	Némathelminthes	
Nematoda	Threadworms	Nématodes	Круглые черви
Nematomorpha		Nématomorphes	Волосатиковые черви
Nemertini	Strapworms	Némertines	Немертины
Nephrops norvegicus	Norway lobster	Néphrops	Норвежский омар
Neptunsgehirn	Brain corals	Diploriens	
Nessellose Hohltiere		Acnidaires	Нестрекающие кишечно-полостные
Nesselquallen		Cyanées	Цианеи
Nesseltiere		Cnidaires	Стрекающие кишечно-полостные
Neumünder		Deutérostomes	Вторичноротые
Neumundtiere		Deutérostomes	Вторичноротые
Niedere Krebse		Entomostraces	Низшие ракообразные
Niphargus	Fresh water amphipods	Gammares de fontaines	
Nordische Zylinderrose		Cérianthe nordique	Северный цериант
Nordseegarnele	Common shrimp	Crangon commun	Обыкновенная креветка
Norwegischer Hummer	Norway lobster	Néphrops	Норвежский омар
Notostraca		Notostraques	Щитни
Octocorallia		Octocoraux	Восьмилучевые коралловые полипы
Ocypode	Ghost crabs	Ocypodes	Песчаные крабы
Ocypodidae	Ocypodids	Ocypodidés	Песчаные крабы
Ohrenqualle	Common jellyfish	Aurélie	
Ohrenquallen		Aurélies	Аурелии
Oligochaeta	Earthworms	Oligochaetes	Малощетинковые черви
Oligotricha	Oligotrichous ciliates	Oligotriches	
Oncopoda	Oncopods	Oncopodes	
Oniscoidea		Oniscoides	Мокрицевые
Oniscus asellus	Sowbug	Cloporte des murs	Стенная мокрица
Onychophora		Onychophores	Коготные черви
Onychopoda	Onychopods	Onychopodes	

GERMAN NAME	ENGLISH NAME	FRENCH NAME	RUSSIAN NAME
Onychura	Onychures	Onychures	
Opisthorchis felineus		Douve du chat	Кошачья печеночная двуустка
Orchestia gammarellus	Beach flea	Orchestie des rivages	Береговой скакун
Orgelkorallen	Organ-pipe corals	Tubiporidés	Органчиковые
Oronectes limosus		Écrevisse américaine	Американский речной рак
Orthognatha		Orthognathes	Пауки-птицееды
Ostracoda	Ostracods	Ostracodes	Ракушковые рачки
Ostseegarnele		Palémon de la Baltique	Балтийская креветка
Oxyuris equi		Oxyure chevalin	Лошадиная острица
Paguridae	Hermit crabs	Paguridés	Морские раки отшельники
Palaemon squilla		Palémon de la Baltique	Балтийская креветка
Palinuridae	Spiny lobsters	Palinuridés	Лангусты
Palinurus vulgaris	– lobster	Langouste européenne	Европейский лангуст
Palmendieb	Robber crab	Crabe des cocotiers	Пальмовый вор
Palpatores	Palpatores	Palpateurs	
Paludicola	Paludicoles	Paludicoles	Пресноводные планарии
Pantoffeltierchen	Slipper animalcules	Paramécie	Туфельки
Pantopoda	Pantopods	Pantopodes	Многоколенчатые
Panzergeißler	Dinoflagellates	Dinoflagellés	Панцирные жгутиконосцы
Panzerkrebse		Astacures	Длиннохвостые раки
Paramecium	Slipper animalcules	Paramécie	Туфельки
Parasitiformes		Parasitiformes	Гамазоидные клещи
Pärchenegel	Schistosomids	Schistosomidés	Кровяные двуустки
Pauropoda	Pauropods	Pauropodes	Пауроподы
Pedipalpi	Whip scorpions	Pédipalpes	Жгутоногие пауки
Pennatularia	Sea-pens	Pennatulaires	Морские перья
Pentastomida	Linguatulides	Linguatulides	Пятиустки
Pergamentwurm		Chétoptère luisant	Пергаментный трубкожил
Peritricha	Peritrichous ciliates	Péritriches	Кругоресничные инфузории
Pferdeaktinie		Actinie pourpre	Конская актиния
Pferde-Madenwurm		Oxyure chevalin	Лошадиная острица
Pferdeschwamm	Horse sponge	Éponge de toilette	
Phalangida	Harvestmen	Phalangides	
Phalangium opilio	Harvestman	Faucheur	Обыкновенный сенокосец
Pharyngobdellae		Pharyngobdellés	Глоточные пиявки
Pholcidae	Pholcids	Pholcidés	
Phyllopoda	Phyllopods	Phyllopodes	Жаброногие ракообразные
Physalia physalis	Portuguese man-of-war	Physalie	Сифонофора физалия
Pilzkorallen	Fungus corals	Coraux-champignons	Груздевики
Pinnoteres pisum	Oyster crab	Crabe de moule	Гороховый краб
Piscicola geometra		Sangsue piscicole	Обыкновенная рыбья пиявка
Pistolenkrebse	Alpheids	Alphéidés	Креветки
Plathelminthes	Plathelminthes	Plathelminthes	Плоские черви
Plattenegel	Glossiphoniids	Glossiphonidés	Плоские пиявки
Plattwürmer	Plathelminthes	Plathelminthes	Плоские черви
Pleurobrachia pileus	Sea walnut	Pleurobrachie velue	
Polychaeta	Seaworms	Polychaetes	Многощетинковые черви
Polydesmida		Polydesmides	Многосвязки
Polydesmus complanatus		Polydesme à lamelles	Плоский многосвяз
Polystoma integerrimum		Polystomum de vessie	Лягушечья многоустка
Polythalamia		Foraminifères à coquille po-lythalame	Многокамерные фораминиферы
Porcellio scaber		Cloporte des caves	Погребная мокрица
Porentierchen	Foraminifera	Foraminifères	Фораминиферы
Poritidae	Poritids	Poritidés	Поритовые
Portugiesische Galeere	Portuguese man-of-war	Physalie	Сифонофора физалия
Portunidae	Swimming crabs	Portunidés	Крабы-пловунцы
Portunus	True swimming crabs	Portunes	Крабы-пловунцы
Posthörnchenwürmer		Spirorbes	Спирорбис
Potamidae	Potamids	Potamidés	Пресноводные крабы
Priapswürmer	Priapulids	Priapulides	Приапулиды
Priapulus caudatus		Priapule	Однохвостый приапул
Protostomia		Protostomes	Первичноротые
Protozoa	Protozoans	Protozoaires	Простейшие животные
Pseudoscorpiones	False scorpions	Pseudoscorpions	Ложноскорпионы
Pteroides griseum		Ptéroide gris	Серое морское перо
Purpurrose	Sea anemones	Actinies	Актинии
Purpurseerose		Actinie pourpre	Конская актиния
Pycnogonidae	Pycnogonids	Pycnogonidés	Береговые морские пауки
Pycnogonum littorale		Pycnogonon littoral	Береговой морской паук
Quappwurm		Echiure	Эхиур
Rädertiere	Wheel animalcules	Rotatores	Коловратки
Radiata	Jellyfishes, anemones and corals	Cœlentérés	Кишечнополостные

GERMAN NAME	ENGLISH NAME	FRENCH NAME	RUSSIAN NAME
Radiolarien	Radiolaria	Radiolaires	Лучевики
Radnetzspinnen i. e. S.	Araneids	Aranéidés	
Rankenfüßer	Barnacles	Cirripèdes	
Reiterkrabben	Ghost crabs	Ocypodes	Песчаные крабы
Renillidae	Renillids	Rénillidés	
Renn- und Winkerkrabben	Ocypodids	Ocypodidés	Песчаные крабы
Reptantia	Crawfishes	Reptantides	
Rhizocephala		Rhizocéphales	Корнеголовые
Rhizopoda	Rhizopods	Rhizopodes	Корненожки
Rhizostomae		Rhizostomes	Ризостомы
Rhynchobdellae		Rhynchobdellés	Хоботные пиявки
Riemenbandwurm		Ligule des intestins	Ремнец
Riesenkratzer		Acanthocéphale géant	Гигантский скребень
Riesenkugler	Sphaerotheriids	Sphérothéridés	
Riesenläufer		Scolopendromorphes	Сколопендры
Riesen-Leberegel		Douve du foie géante	Гигантская печеночная двуустка
Riesenregenwurm		Lombric géant	Гигантский дождевой червь
Riesenskorpione	Eurypterids	Euryptéridés	Эвриптериды
Rindenkorallen		Gorgonaires	Горгонарии
Rinderbandwurm	Beef tapeworm	Ténia du bœuf	Невооруженный цепень
Ringelwürmer	Jointed worms	Annélides	Кольчатые черви
Rippenquallen	Comb jellyfishes	Cténophores	Гребневики
Ritterkrebse	Crawfishes	Reptantides	
Rivulogammarus pulex		Gammare de ruisseaux	Бокоплав-блоха
Rollassel	Pillbug	Armadille vulgaire	Шаровидка
Rotatoria	Wheel animalcules	Rotatores	Коловратки
Rote Edelkoralle	Red precious coral	Corail rouge	Благородный коралл
Rückenschaler		Notostraques	Щитни
Ruderfußkrebse	Copepods	Copépodes	Веслоногие рачки
Ruhramöbe		Amibe dysentérique	Дизентерийная амеба
Rüssel-Asselspinne		Pantopode à trompe	Хоботковый морской паук
Rüsselegel		Rhynchobdellés	Хоботные пиявки
Rüsselkrebse	Bosminids	Bosminidés	
Saftkugler	Glomerids	Gloméridés	Клубовидковые
Sahara-Skorpion		Androcton d'Afrique du Nord	Сахарский скорпион
Saitenwürmer	Horsehair worms	Nématomorphes	Волосатиковые черви
Salticidae	Jumping spiders	Salticidés	Пауки-скакуны
Salzkrebschen		Artémie	Соляный рачек
Sammetmilbe		Trombidion	Шелковая краснотелка
Sandkrabben	Ghost crabs	Ocypodes	Песчаные крабы
Sandkrebse	Mole crabs	Hippidés	
Sandwurm	Lugworm	Arénicole des pêcheurs	Пескожил
Saphirkrebse		Sapphirines	Сафирные рачки
Sapphirina		—	Сафирные рачки
Sarcoptes scabiei	Itch mite	Gale de l'homme	Чесоточный зудень
Saugmünder	Myzostomes	Myzostomides	Мизостомиды
Saugtierchen		Sucteurs	Сосущие инфузории
Saugwürmer	Flukes	Trématodes	Сосальщики
Schalamöben		Testacés	Раковинные амебы
Schamkrabben		Calappe	Стыдливые крабы
Scheibenquallen	Large jellyfishes	Scyphozoaires	Сцифоидные
Scherenfüßer	Chelicerates	Chélicériens	
Schistosoma haematobium		Bilharzie de la vessie	Мочеводная кровяная двуустка
— japonicum		— d'extrême-Orient	Японская кровяная двуустка
— mansoni		— de l'intestin	Кишечная кровяная двуустка
Schistosomidae	Schistosomids	Schistosomidés	Кровяные двуустки
Schlammröhrenwürmer	Tubificids	Tubificidés	Трубочники
Schlauchwürmer	Roundworms	Aschelminthes	
Schlundegel		Pharyngobdellés	Глоточные пиявки
Schmarotzerasseln	Epicarids	Épicaridés	
Schmarotzermilben		Parasitiformes	Гамазоидные клещи
Schneider	Harvestmen	Phalangides	
Schnurfüßer i. e. S.	Julids	Julidés	Кивсяки
Schnurfüßer i. w. S.	Julides	Julides	Кивсяки
Schnurwürmer	Strapworms	Némertines	Немертины
Schwammtiere	Sponges	Spongiaires	Губки
Schwarze Korallen	Anthipathids	Anthipathidés	Антипаты
Schwarze Witwe	Black widow spider	Veuve noire	
Schweinebandwurm	Swine tapeworm	Ténia du porc	Вооруженный цепень
Schwertschwänze	Horseshoe crabs	Xiphosures	Мечехвосты
Schwimmkrabben	Swimming crabs	Portunidés	Крабы-пловунцы

GERMAN NAME	ENGLISH NAME	FRENCH NAME	RUSSIAN NAME
Scolopendromorpha		Scolopendromorphes	Сколопендры
Scorpiones	Scorpions	Scorpions	Скорпионы
Scutigeromorpha		Scutigéromorphes	Мухоловковые
Scyllaridae	Scyllarids	Scyllaridés	Раки-медведи
Scyphozoa	Large jellyfishes	Scyphozoaires	Сцифоидные
Sechssaugen	Dysderids	Dysdéridés	
Sechsstrahlige Korallen		Héxacoraux	Шестилучевые коралло-вые полипы
Sedentaria		Polychaetes sédentaires	Сидячие полихеты
Seeanemonen	Sea anemones	Actinies	Актиниевые
Seeblase	Portuguese man-of-war	Physalie	Сифонофора физалия
Seefedern	Sea-pens	Pennatulaires	Морские перья
Seemannshand	Sea finger	Main de mer	Алциона
Seemaus		Aphrodite hérissée	Колючая афродита
Seepocken	Barnacles	Balanomorphes	Морские жолуди
–	--	Cirripèdes	Усоногие ракообразные
Seeraupen	Aphroditids	Aphroditidés	Афродитовые
Seerosen	Sea anemones	Actinies	Актиниевые
Seeskorpione	Eurypterids	Euryptéridés	Эвриптериды
Seespinnen	Spider crabs	Araignées de mer	Майи
Seestachelbeere	Sea walnut	Pleurobrachie velue	
Seestiefmütterchen	Renillids	Rénillidés	
Segelqualle	By-the-wind-sailors	Velelle	Парусница
Siphonophora		Siphonophores	Сифонофоры
Sipunculida		Sipunculides	Сипункулиды
Skolopender		Scolopendromorphes	Сколопендры
Skorpione	Scorpions	Scorpions	Скорпионы
Skorpionspinnen	Whip scorpions	Pédipalpes	Жгутоногие пауки
Solifuga		Solifuges	Сольпуги
Sommerkiemenfuß		Branchipe commun	Жаброног
Sonnentierchen	Sun animalcules	Heliozoaires	
Spaltfüßer		Mysidacés	Мизиды
Sphaerotheriidae	Sphaerotheriids	Sphérothéridés	
Spinnenasseln		Scutigéromorphes	Мухоловковые
Spinnentiere	Arachnides	Arachnides	Паукообразные
Spinnenverwandte	Chelicerates	Chélicèriens	
Spiralwimperlinge	Spirotriches	Spirotriches	
Spirorbis		Spirorbes	Спирорбис
Spirotricha	Spirotriches	Spirotriches	
Spongia	Sponges	Spongiaires	Губки
Spongia officinalis		Éponge officinal	Адриатическая губка
Spongidae	Spongids	Spongidés	Туалетные губки
Spongillidae	Fresh water sponges	Spongillidés	Бадяги
Sporentierchen	Sporozoans	Sporozoaires	Споровики
Sporozoa	–	–	Споровики
Springkrabben	Rock crabs	Grapsidés	
Springspinnen	Jumping spiders	Salticidés	Пауки-скакуны
Spritzwürmer		Sipunculides	Сипункулиды
Spulwurm	Roundworm	Ascaride lombricoïde	Человеческая аскарида
Staatsquallen	Siphonophores	Siphonophores	Сифонофоры
Stachelspinnen	Gasteracanthes	Gastéracanthes	
Steinkorallen	Stony corals	Madrépores	Мадрепоровые кораллы
Steinkrabben	Lithodid crabs	Lithodidés	Каменные крабы
Steinläufer		Lithobiomorphes	Костянковые
Steinschwämme	Lithistides	Lithistides	
Stentor	Stentor	Stentor	Трубачи
– *coeruleus*		– bleu	Синий трубач
Sternkoralle	Star coral		Бокальчатый коралл
Stomatopoda	Stomatopods	Stomatopodes	Ротоногие ракообразные
Strahlentierchen	Radiolaria	Radiolaires	Лучевики
Strandassel		Ligie des rivages	Берёговая мокрица
Strandflöhe	Beach fleas	Talitridés	Песочные скакуны
Strandkrabbe	Shore crab	Crabe enragé	Европейский краб
Streckkiefer	Tetragnathids	Tétragnathidés	
Strudelwürmer	Flatworms	Turbellaires	Ресничные черви
Stummelfüßer		Onychophores	Коготные черви
Suberitidae	Suberitids	Subéritidés	
Suctoria		Sucteurs	Сосущие инфузории
Süßwasser-Kelchwürmer	Urnatellids	Urnatellidés	
Süßwasserkrabben	Potamids	Potamidés	Пресноводные крабы
Süßwasserplanarien		Paludicoles	Пресноводные планарии
Süßwasserschwämme	Freshwater sponges	Spongillidés	Бадяги
Symphyla	Symphylids	Symphyles	Сколопендреллы
Taenia solium	Swine tapeworm	Ténia du porc	Вооруженный цепень
Taeniarhynchus saginatus	Beef tapeworm	– – bœuf	Невооруженный цепень
Talitridae	Beach fleas	Talitridés	Песочные скакуны
Talitrus saltator		Talitre sauteur	Песочный скакун

GERMAN NAME	ENGLISH NAME	FRENCH NAME	RUSSIAN NAME
Taranteln	Tarantulas	Tarentules	Тарантулы
Tardigrada	Waterbears	Tardigrades	Тихоходки
Taschenkrebs	Rock-dwelling crab	Crabe tourteau	Большой сухопутный краб
Taschenquallen	Coronates	Coronates	
Tausendfüßer	Thousend-legged worms	Myriapodes	Многоножки
– i. e. S.	Millipeds	Diplopodes	Кивсяковые
Tauwurm	Earthworm	Ver de terre	Обыкновенный дождевой червь
Tegenaria	House spiders	Araignées domestiques	Домовые пауки
– domestica	– spider	Araignée domestique	Домовый паук
Terricola		Terricoles	Наземные планарии
Testacea	Testaceans	Testacés	Раковинные амёбы
Tetragnathidae	Tetragnathids	Tétragnathidés	
Theridiidae	Orb-web spiders	Thérididés	Пауки-ткачи
–	Theridiids	Thérididés	
Thomisidae	Crab spiders	Thomisidés	Пауки-бокоходы
Tiefsee-Asselspinnen	Collossendeids	Collossendéidés	Глубоководные морские пауки
Tiefseequallen	Coronates	Coronates	
Topfwurm		Enchytrée	Энхитрея
Tracheata	Tracheates	Trachéates	Трахейные
Tracheentiere	–	–	Трахейные
Trematodes	Flukes	Trématodes	Сосальщики
Trichine	Trichina	Trichine	Спиральная трихина
Trichinella spiralis	–	–	Спиральная трихина
Trichostomata	Trichostomates	Trichostomes	
Trichterspinnen	Funnel-web spiders	Agélénidés	Мешковые пауки
Trichterwimperlinge	Chonotriches	Chonotriches	
Trilobita	Trilobites	Trilobites	Трилобиты
Trilobitenverwandte		Trilobitomorphes	Трилобитовые
Trilobitomorpha		–	Трилобитовые
Triops cancriformis		Triops	Большой щитень
Trombicula autumnalis	Chigger	Lepte automnal	Полевая краснотелка
Trombidiformes		Trombidiformes	Краснотелковые клещи
Trombidiidae	Trombidiids	Trombidiidés	Краснотелки
Trombidium holosericum		Trombidion	Шёлковая краснотелка
Trompetentierchen	Stentor	Stentor	Трубачи
Trypanosomatidae	Trypanosomatids	Trypanosomatidés	Трипанозомы
Trypanosomen	–	–	Трипанозомы
Tubifex tubifex	Tubifex	Tubifex	Обыкновенный трубочник
Tubificidae	Tubificids	Tubificidés	Трубочники
Tubiporidae	Organ-pipe corals	Tubiporidés	Органчиковые
Turbellaria	Flatworms	Turbellaires	Ресничные черви
Tyrophagus casei		Acare du fromage	Сырный акар
Uca	Fiddler crabs	Crabes à signaux	Манящие крабы
Ufer-Asselspinne		Pycnogonon littoral	Береговой морской паук
Ufer-Asselspinnen	Pycnogonids	Pycnogonidés	Береговые морские пауки
Unbewaffneter Bandwurm	Beef tapeworm	Ténia du bœuf	Невооружённый цепень
Unionicola crassipes		Unionicole à grosses pattes	Толстоногий водный клещ
Urmünder		Protostomes	Первичноротые
Urmundtiere		–	Первичноротые
Urnatellidae	Urnatellids	Urnatellidés	
Uropygi		Uropyges	Скорпионопауки
Velella spirans	By-the-wind-sailors	Velelle	Парусница
Venusgürtel	Venus's girdles	Cestides	
–	– girdle	Ceste de Vénus	
Verschiedenbewimperte	Heterotrichous ciliates	Hétérotriches	
Vielborster	Seaworms	Polychaetes	Многощетинковые черви
Vielkammerige		Foraminifères à coquille polythalame	Многокамерные фораминиферы
Vogelspinnen i. w. S.		Orthognathes	Пауки-птицееды
Vorratsmilben	Acarids	Acaridés	Амбарные клещи
Wabenkalkschwämme		Hétérocoeles	Сотовидные губки
Wachsrose		Anémone de mer	Бороздчатая анемона
Walläuse	Whale lice	Cyamidés	Китовые вши
Wal-Seepocken	Coronulids	Coronulidés	Китовые морские жёлуди
Walzenspinnen		Solifuges	Сольпуги
Wanderfilarie		Filaire loa	Червь-лоа
Wasserasseln	Fresh water isopods	Asellidés	Водные мокрицы
Wasserflöhe	Water fleas	Cladocères	Ветвистоусые ракообразные
Wasserkalb		Gordius aquatique	Водяной волосатик
Wasserkälber		Nèmatomorphes	Волосатиковые черви
Wasserschlängler	Naidids	Naididés	Вьюнки
Weberknechte	Harvestmen	Phalangides	Сенокосцы
– i. e. S.	–	Palpateurs	
Webespinnen	Spiders	Araniens	Пауки

GERMAN NAME	ENGLISH NAME	FRENCH NAME	RUSSIAN NAME
Wechseltierchen		Amoebines	Амебы
Weichschaler		Malacostragues	Высшие ракообразные
Weiße Korallen		Madrépores	Мадрепоры
Wenigbewimperte	Oligotrichous ciliates	Oligotriches	
Wenigborster	Earthworms	Oligochaetes	Малощетинковые черви
Wenigfüßer	Pauropods	Pauropodes	ПаУроподЫ
Widderkrebs		Caprelle-chèvre	Морская козочка
Wimpermünder	Trichostomates	Trichostomes	
Wimpertiere	Ciliates	Ciliés	Ресничные
Winkerkrabben	Fiddler crabs	Crabes à signaux	Манящие крабы
Wolfsspinnen	Wolf spiders	Lycosidés	Тарантуловые
Wollhandkrabbe		Crabe chinois	Китайский мохнаторукий краб
Wollkrabben	Dromid crabs	Dromies velues	Волосатые крабы
Wollkrebse	– –	Dromidés	Волосатые крабы
Wuchereria bancrofti		Wuchérérie	Вухерия
Wurzelfüßer	Rhizopods	Rhizopodes	Корненожки
Wurzelkrebse		Rhizocéphales	Корнеголовые
Wurzelmundquallen		Rhizostomes	Ризостомы
Xiphosura	Horseshoe crabs	Xiphosures	Мечехвосты
Zecken	Ticks	Ixodidés	Иксодовые клещи
Zehnfüßer	Decapods	Décapodes	Десятиногие ракообразные
Zitterspinnen	Pholcids	Pholcidés	
Zoantharia		Zooanthaires	Зоантарии
Zungenwürmer	Linguatulides	Linguatulides	Пятиустки
Zwergbandwurm		Ténia nain	Карликовый цепень
– der Maus	Rat tapeworm		Мышиный карликовый цепень
– des Menschen		Ténia nain de l'homme	Человеческий карликовый цепень
Zwergfüßer	Symphylids	Symphyles	Сколопендреллы
Zwergspinnen	Erigonids	Erigonidés	
Zylinderrosen		Cérianthaires	Цериантовые

III. French—German—English—Russian

FRENCH NAME	GERMAN NAME	ENGLISH NAME	RUSSIAN NAME
Acanthobdellés	Borstenegel		Акантобделлы
Acanthocéphale géant	Riesenkratzer		Гигантский скребень
Acanthocéphales	Kratzer	Spiny-headed worms	Колючеголовые черви
Acare de farine	Mehlmilbe		Чесоточный клещ
– du fromage	Käsemilbe		Сырный акар
Acaridés	Vorratsmilben	Acarids	Амбарные клещи
Acariens	Milben	Mites	Клещи
Acnidaires	Nessellose Hohltiere		Нестрекающие кишечнополостные
Acroporidés	Baumförmige Korallen	Acroporids	
Actinie pourpre	Purpurseerose		Конская актиния
Actinies	Aktinien, Purpurrose	Sea anemones	Актиниевые
Agélénidés	Trichterspinnen	Funnel-web spiders	Мешковые пауки
Alcyonaires	Lederkorallen	Alcyonarian corals	Пробковые полипы
Alphéidés	Knallkrebschen	Alpheids	Креветки
Amibe dysentérique	Ruhramöbe		Дизентерийная амеба
Amibes à coquilles	Schalamöben		Раковинные амебы
Amoebines	Nacktamöben		Амебы
Amphipodes	Flohkrebse	Amphipods	Бокоплавы
Anatife	Gemeine Entenmuschel	Barnacle	Обыкновенная морская уточка
Androcton d'Afrique du Nord	Sahara-Skorpion		Сахарский скорпион
Anémone de mer	Wachsrose		Бороздчатая анемона
Ankylostome duodénal	Grubenwurm		Анкилостома
Annélides	Gliederwürmer	Jointed worms	Кольчатые черви
Anomures	Mittelkrebse		Неполнохвостые раки
Anthipathariens	Dörnchenkorallen		Антипатовые
Anthipathidés	Schwarze Korallen	Anthipathids	Антипаты
Anthozoaires	Blumentiere	Anthozoans	Коралловые полипы
Aphrodite hérissée	Seemaus		Колючая афродита
Aphroditidés	Seeraupen	Aphroditids	Афродитовые
Arachnides	Spinnentiere	Arachnides	Паукообразные
Araignée domestique	Hausspinne	House spider	Домовый паук

FRENCH NAME	GERMAN NAME	ENGLISH NAME	RUSSIAN NAME
Araignées de mer	Seespinnen	Spider crabs	Майи
— domestiques	Hauswinkelspinnen	House spiders	Домовые пауки
Aranéidés	Radnetzspinnen i. e. S.	Araneids	Настоящие пауки
Araniens	Echte Spinnen	Spiders	Пауки
Arénicole des pêcheurs	Sandwurm	Lugworm	Пескожил
Argasidés	Lederzecken	Argasids	Аргазиды
Argule foliacé	Karpfenlaus	Fish louse	
Armadille vulgaire	Rollassel	Pillbug	
Artémie	Salzkrebschen		Шаровидка
Arthropodes	Gliederfüßer		Соляный рачек
Ascaride lombricoïde	Spulwurm	Arthropods	Членистоногие
Aschelminthes	Schlauchwürmer	Roundworm	Человеческая аскарида
Aselle aquatique	Gemeine Wasserassel	Roundworms	
Asellidés	Wasserasseln	Fresh water isopod	Водяной ослик
Astacidés	Flußkrebse	— — isopods	Водные мокрицы
Astacures	Eigentliche Langschwanzkrebse	— — crabs	Речные раки
Astomes	Mundlose		Длиннохвостые раки
Aurélie	Ohrenqualle	Astomates	
Aurélies	Ohrenquallen	Common jellyfish	
Avicularidés	Echte Vogelspinnen		Аурелии
Balane commun	Gemeine Seepocke	Aviculariids	
		Common barnacle	Обыкновенный морской жолудь
Balanomorphes	Seepocken	Barnacles	Морские жолуди
Béroé allongé	Gurkenqualle	Sea mitre	
Bilharzie de la vessie	Harnblasen-Pärchenegel		Мочеводная кровяная двуустка
— — l'intestin	Darm-Pärchenegel		Кишечная кровяная двуустка
— d'extrème-Orient	Japanischer Pärchenegel		Японская кровяная двуустка
Bonellie verte	Grüne Bonellia	Bonellia	Зеленая бонеллия
Bosminidés	Rüsselkrebse	Bosminids	
Brachyures	Echte Krabben	Crabs	Крабы
Branchiobdellidés	Kiemenegel	Branchiobdellids	
Branchipe commun	Sommerkiemenfuß		Жаброног
Branchiures	Fischläuse	Fish lice	Рыбьи вши
Calappe	Schamkrabben		Стыдливые крабы
Calcariens	Kalkschwämme	Calcareous sponges	Известковые губки
Callinectes	Blaukrabben	Blue crabs	
Caprelle-chèvre	Widderkrebs		Морская козочка
Caprellidés	Gespenstkrebse	Caprellids	Капрелловые
Cérianthaires	Zylinderrosen		Цериантовые
Cérianthe de Méditerranée	Mittelmeer-Zylinderrose		Перепончатый цериант
— nordique	Nordische Zylinderrose		Северный цериант
Ceste de Vénus	Venusgürtel	Venus's girdle	
Cestides	—	— girdles	
Cestodes	Bandwürmer	Tapeworms	Ленточные черви
Chélicèriens	Chelicerentiere	Chelicerates	Хелицеровые
Chélifère cancroide	Bücherskorpion	House scorpion	Книжный ложноскорпион
Chétoptère luisant	Pergamentwurm		Пергаментный трубкожил
Chilopodes	Hundertfüßer	Centipedes	Губоногие многоножки
Chrysaores	Kompaßquallen		Хризаоры
Ciliés	Wimpertiere	Ciliates	Ресничные
Cirripèdes	Rankenfüßer	Barnacles	Усоногие ракообразные
Cladocères	Wasserflöhe	Water fleas	Ветвистоусые ракообразные
Clionidés	Bohrschwämme	Clionids	Сверлящие губки
Clonorchis chinois	Chinesischer Leberegel		Китайская двуустка
Cloporte des caves	Kellerassel		Погребная мокрица
— — murs	Mauerassel	Sowbug	Стенная мокрица
Cnidaires	Nesseltiere		Стрекающие кишечнополостные
Coccidies	Kokzidien		Кокцидии
Cœlentérés	Hohltiere	Jellyfishes, anemones and corals	Кишечнополостные
Collossendéidés	Tiefsee-Asselspinnen	Collossendeids	Глубоководные морские пауки
Colpode	Kappentierchen		Колпода
Conchostraques	Muschelschaler		Раковинные жаброноги
Copépodes	Ruderfußkrebse	Copepods	Веслоногие рачки
Corail rouge	Rote Edelkoralle	Red precious coral	Благородный коралл
Coraux	Edelkorallen	Precious corals	Благородные кораллы
— -champignons	Pilzkorallen	Fungus corals	Груздевики
Coronates	Tiefseequallen	Coronates	
Coronulidés	Wal-Seepocken	Coronulids	Китовые морские жолуди
Crabe chinois	Wollhandkrabbe		Китайский мохнаторукий краб

FRENCH NAME	GERMAN NAME	ENGLISH NAME	RUSSIAN NAME
— de moule	Muschelwächter	Oyster crab	Гороховый краб
— des cocotiers	Palmendieb	Robber crab	Пальмовый вор
— enragé	Strandkrabbe	Shore crab	Европейский краб
— terrestre	Gemeine Landkrabbe	Land crab	Обыкновенный сухопутный краб
— tourteau	Taschenkrebs	Rock-dwelling crab	Большой сухопутный краб
Crabes	Echte Krabben	Crabs	Крабы
— à signaux	Winkerkrabben	Fiddler crabs	Манящие крабы
Crangon commun	Nordseegarnele	Common shrimp	Обыкновенная креветка
Craspedomonadidés	Kragengeißler	Craspedomonadids	Воротничковые жгутиконосцы
Crevette grise	Nordseegarnele	Common shrimp	Обыкновенная креветка
Crustacés	Krebstiere	Crustaceans	Ракообразные
Cténophores	Rippenquallen	Comb jellyfishes	Гребневики
Cténopodes	Kammfüßer	Ctenopods	
Cumacés	Kumazeen		Кумовые ракообразные
Cyamidés	Walläuse	Whale lice	Китовые вши
Cyanée artique	Arktische Riesenqualle		Арктическая цианея
Cyanées	Nesselquallen		Цианеи
Cyanéidés	Haarquallen	Cyaneids	Цианеевые
Cyclopidés	Hüpferlinge	Cyclopids	Циклопиды
Cymothoïdés	Fischasseln		Рыбные мокрицы
Décapodes	Zehnfüßer	Decapods	Десятиногие ракообразные
Deutérostomes	Neumundtiere, Neumünder		Вторичноротые
Dicrocélium fer-de-lance	Lanzettegel		Ланцетовидная двуустка
Didinium	Nasentierchen	Water bear	
Dinoflagellés	Panzergeißler	Dinoflagellates	Панцирные жгутиконосцы
Diplopodes	Doppelfüßer	Millipeds	Кивсяковые
Diploriens	Neptunsgehirn	Brain corals	
Diplozoon paradoxal	Doppeltier		Странный спайник
Douve du chat	Katzenleberegel		Кошачья печеночная двуустка
— — foie géante	Riesen-Leberegel		Гигантская печеночная двуустка
Dromidés	Wollkrebse	Dromid crabs	Волосатые крабы
Dromies velues	Wollkrabben	— crabs	Волосатые крабы
Dugésie de rivières	Amerikanische Flußplanarie		Американская речная планария
Dysdéridés	Sechsaugen	Dysderids	
Echiure	Meerquappe		Эхиур
Echiures	Igelwürmer	Echiurids	Эхиуриды
Écrevisse américaine	Amerikanischer Flußkrebs	River crayfish	Американский речной рак
— commune	Europäischer Flußkrebs	Crayfish	Широкопалый речной рак
Enchytrée	Topfwurm		Энхитрея
Entomostraques	Niedere Krebse		Низшие ракообразные
Epéire diadème	Kreuzspinne	Garden spider	Паук крестовик
Épicaridés	Schmarotzerasseln	Epicarids	
Éponge de toilette	Pferdeschwamm	Horse sponge	
— officinale	Dalmatiner Schwamm		Адриатическая губка
Ergionidés	Zwergspinnen	Erigonids	
Eumétazoaires	Echte Vielzeller	Many-celled animals	Настоящие многоклеточные
Eunice atlantique	Atlantischer Palolo		Атлантический палоло
Euphausiacés	Leuchtkrebse		Эвфаузиевые ракообразные
Euplectelle	Gießkannenschwamm	Venus's-flower-basket	
Euryptéridés	Riesenskorpione	Eurypterids	Эвриптериды
Faucheur	Gemeiner Weberknecht	Harvestman	Обыкновенный сенокосец
Filaire de Médine	Medinawurm		Ришта
— loa	Wanderfilarie		Червь-лоа
Flabellifères	Fächerschwanzasseln	Flabellifera	
Flagellés	Geißeltiere	Flagellates	Жгутиковые
Foraminifères	Lochträger	Foraminifera	Фораминиферы
— à coquille monothalame	Einkammerige Lochträger		Однокамерные фораминиферы
— — — polythalame	Vielkammerige Lochträger		Многокамерные фораминиферы
Gale de l'homme	Krätzemilbe des Menschen	Itch mite	Чесоточный зудень
Gammare de ruisseaux	Bachflohkrebs		Бокоплав-блоха
Gammares de fontaines	Brunnenkrebse	Fresh water amphipods	
Gastéracanthes	Stachelspinnen	Gasteracanthes	
Gastrotriches	Bauchhaarlinge	Gastrotriches	Брюхоресничные черви
Gécarcinidés	Landkrabben i. e. S.	Land crabs	Сухопутные крабы
Géophile luisant	Leuchtender Erdläufer	Luminous centipede	Светлянка
Géophilomorphes	Erdläufer	— centipedes	Светлянковые
Gloméridés	Saftkugler	Glomerids	Клубовидковые
Glossiphonidés	Knorpelegel	Glossiphoniids	Плоские пиявки

FRENCH NAME	GERMAN NAME	ENGLISH NAME	RUSSIAN NAME
Gordius aquatique	Wasserkalb		Водяной волосатик
Gorgonaires	Rindenkorallen		Горгонарии
Grande douve du foie	Großer Leberegel		Печеночная двуустка
Grapse des rochers	Felsenkrabbe	Rock crab	
Grapsidés	Springkrabben	— crabs	
Grégarines	Gregarinen		Грегарины
Gymnostomes	Nacktmünder	Gymnostomates	
Halacaridés	Meereswassermilben	Halacarids	Морские клещи
Haplotaxidés	Brunnenwürmer	Haplotaxids	
Hélioporidés	Blaukorallen	Blue corals	
Heliozoaires	Sonnentierchen	Sun animalcules	
Hémadipsidés	Landegel	Haemadipsids	Наземные пиявки
Hétérocoeles	Wabenkalkschwämme		Сотовидные губки
Hétérotriches	Verschiedenbewimperte	Heterotrichous ciliates	
Héxacoraux	Sechsstrahlige Korallen		Шестилучевые кораловые полипы
Hexactinellides	Glasschwämme		Стеклянные губки
Hippidés	Sandkrebse	Mole crabs	
Hirudinés	Egel	Leeches	Пиявки
Holotriches	Ganzbewimperte	Holotrichous ciliates	
Homard d'Amérique	Amerikanischer Hummer	American lobster	Американский омар
— d'Europe	Europäischer Hummer	Lobster	Европейский омар
Homaridés	Hummer	Lobsters	Омары
Hydre commune	Braune Hydra		Обыкновенная гидра
— grise	Graue Hydra		Серая гидра
— verte	Grüne Hydra		Зеленая гидра
Hydrocoralliaires	Hydrokorallen	Hydrocorallids	Гидрокораллы
Hydrozoaires	Hydrozoen	Hydrozoans	Гидроидные
Hymenostomes	Hautmünder	Hymenostomates	
Hypotriches	Bauchwimperlinge	Hypotridos	
Ichthyobdéllidés	Fischegel	Ichthyobdellids	Рыбьи пиявки
Isopodes	Asseln	Isopods	Равноногие ракообразные
Ixodidés	Zecken	Ticks	Иксодовые клещи
Julidés	Schnurfüßer i. e. S.	Julids	Кивсяки
Kamptozoaires	Kelchwürmer	Kamtozoans	Камптозои
Kinorhynches	Hakenrüßler	Kinorhynches	Киноринхи
Langouste européenne	Europäische Languste	Spiny lobster	Европейский лангуст
Lepadomorphes	Entenmuscheln		Морские уточки
Lepte automnal	Erntemilbe	Chigger	Полевая краснотелка
Leptostraques	Leptostraken		Тонкопанцирные
Leuconidés	Knollenkalkschwämme	Leuconids	
Ligie des rivages	Strandassel		Береговая мокрица
Ligule des intestins	Riemenbandwurm		Ремнец
Limnoridés	Holzbohrasseln	Gribbles	Сверлящие мокрицы
Limules	Königskrabben	King crabs	Обыкновенный мечехвост
Linguatule nasale	Nasenwurm		Носовая пятиустка
Linguatulides	Zungenwürmer	Linguatulida	Пятиустки
Linyphidés	Baldachinspinnen	Money spiders	
Lithistides	Steinschwämme	Lithistides	
Lithobiomorphes	Steinläufer		Костянковые
Lithodidés	Steinkrabben	Lithodid crabs	Каменные крабы
Lobés	Lappenrippenquallen	Lobates	
Lombric	Gemeiner Regenwurm	Earthworm	Обыкновенный дождевой червь
— géant	Riesenregenwurm		Гигантский дождевой червь
Lycosidés	Wolfsspinnen	Wolf spiders	Тарантуловые
Madrépores	Weiße Koralle		Мадрепоры
—	Steinkorallen	Stony corals	Мадрепоровые кораллы
Main de mer	Meerhand	Sea finger	Алциона
Majidés	Dreieckskrabben	Spider crabs	Майи
Malacostraques	Höhere Krebse		Высшие ракообразные
Mandibulés	Mandibeltiere	Mandibulates	
Maricoles	Meeresplanarien		Морские планарии
Maxillipodes	Kieferfüßer	Maxillipods	
Merostomates	Hüftmünder	Merostomates	Меростомовые
Mésozoaires	Mitteltiere	Unorganized animals	Мезозои
Mictyridés	Armeekrabben	Mictyrids	
Mille-pattes commun	Brauner Steinläufer		Обыкновенная многоножка
Myriapodes	Tausendfüßer	Thousend-legged worms	Многоножки
Mysidacés	Spaltfüßer		Мизиды
Myzostomides	Saugmünder	Myzostomes	Мизостомиды
Naididés	Wasserschlängler	Naidids	Вьюнки
Natantides	Garnelenartige Langschwanz-krebse	Trues shrimps	Креветковые
Nématodes	Fadenwürmer	Threadworms	Круглые черви

FRENCH NAME	GERMAN NAME	ENGLISH NAME	RUSSIAN NAME
Nématomorphes	Saitenwürmer	Horsehair worms	Волосатиковые черви
Némertines	Schnurwürmer	Strapworms	Немертины
Néphrops	Kaiserhummer	Norway lobster	Норвежский омар
Notostraques	Rückenschaler		Щитни
Oxyure chevalin	Pferde-Madenwurm		Лошадиная острица
Octocoraux	Achtstrahlige Korallen		Восьмилучевые коралловые полипы
Ocypodes	Sandkrabben	Ghost crabs	Песчаные крабы
Ocypodidés	Renn- und Winkerkrabben	Ocypodids	Песчаные крабы
Oligochaetes	Wenigborster	Earthworms	Малощетинковые черви
Oligotriches	Wenigbewimperte	Oligotrichous ciliates	
Oncopodes	Krallenfüßer	Oncopods	
Oniscoides	Landasseln		Мокрицевые
Onychophores	Stummelfüßer		Коготные черви
Onychopodes	Krallenfüßer	Onychopods	
Orchestie des rivages	Küstenhüpfer	Beach flea	Береговой скакун
Orthognathes	Vogelspinnen i. w. S.	Orthognathes	Пауки-птицееды
Ostracodes	Muschelkrebse	Ostracods	Ракушковые рачки
Paguridés	Meeres-Einsiedlerkrebse	Hermit crabs	Морские раки отшельники
Palémon de la Baltique	Ostseegarnele		Балтийская креветка
Palinuridés	Langusten	Spiny lobsters	Лангусты
Palpateurs	Weberknechte i. e. S.	Palpatores	
Paludicoles	Süßwasserplanarien		Пресноводные планарии
Pantopode à trompe	Rüssel-Asselspinne		Хоботковый морской паук
Pantopodes	Asselspinnen	Pantopods	Многоколенчатые
Paramécie	Pantoffeltierchen	Slipper animalcules	Туфельки
Parasitiformes	Schmarotzermilben		Гамазоидные клещи
Pauropodes	Wenigfüßer	Pauropods	Пауроподы
Pédipalpes	Skorpionspinnen	Whip scorpions	Жгутоногие пауки
Pennatulaires	Seefedern	Sea-pens	Морские перья
Péritriches	Glockentierchen	Peritrichous ciliates	Кругоресничные инфузории
Phalangides	Weberknechte	Harvestmen	Сенокосцы
Pharyngobdellés	Schlundegel		Глоточные пиявки
Pholcidés	Zitterspinnen	Pholcids	
Phyllopodes	Blattfußkrebse	Phyllopods	Жаброногие ракообразные
Physalie	Seeblase	Portuguese man-of-war	Сифонофора физалия
Plathelminthes	Plattwürmer	Plathelminthes	Плоские черви
Pleurobrachie velue	Seestachelbeere	Sea walnut	
Polychaetes	Vielborster	Seaworms	Многощетинковые черви
— errants	Freilebende Vielborster		Свободноползающие полихеты
— sédentaires	Festsitzende Vielborster		Сидячие полихеты
Polydesme à lamelles	Abgeplatteter Bandfüßer		Плоский многосвяз
Polydesmides	Bandfüßer		Многосвязки
Polystomum de vessie	Harnblasen-Saugwurm		Лягушечья многоустка
Poritidés	Lochkorallen	Poritids	Поритовые
Portunes	Echte Schwimmkrabben	True swimming crabs	Крабы-пловунцы
Portunidés	Schwimmkrabben	Swimming crabs	Крабы-пловунцы
Potamidés	Süßwasserkrabben	Potamids	Пресноводные крабы
Pou des poissons	Karpfenlaus	Fish louse	Карповая вошь
Priapule	Einschwänziger Priapswurm		Однохвостый приапул
Priapulides	Priapswürmer		Приапулиды
Protostomes	Urmundtiere, Urmünder		Первичноротые
Protozoaires	Einzeller	Protozoans	Простейшие животные
Pseudoscorpions	Afterskorpione	False scorpions	Ложноскорпионы
Ptéroide gris	Graue Seefeder		Серое морское перо
Pycnogonides	Ufer-Asselspinnen	Pycnogonids	Береговые морские пауки
Pycnogonon littoral	Ufer-Asselspinne		Береговой морской паук
Radiolaires	Radiolarien	Radiolaria	Лучевики
Rénillidés	Seestiefmütterchen	Renillids	
Reptantides	Ritterkrebse	Crawfishes	
Rhizocéphales	Wurzelkrebse		Корнеголовые
Rhizopodes	Wurzelfüßer	Rhizopods	Корненожки
Rhizostomes	Wurzelmundquallen		Ризостомы
Rhynchobdellés	Rüsselegel		Хоботные пиявки
Rotatores	Rädertiere	Wheel animalcules	Коловратки
Salticidés	Springspinnen	Jumping spiders	Пауки-скакуны
Sangsue de Ceylan	Ceylonegel		Цейлонская пиявка
— du cheval	Pferdeegel		Ложноконская пиявка
— médicale	Medizinischer Blutegel	Leech	Лечебная пиявка
— piscicole	Gemeiner Fischegel		Обыкновенная рыбья пиявка
Sapphirines	Saphirkrebse		Сафирные рачки
Schistosomidés	Pärchenegel	Schistosomids	Кровяные двуустки
Scolopendromorphes	Riesenläufer		Сколопендры

FRENCH NAME	GERMAN NAME	ENGLISH NAME	RUSSIAN NAME
Scorpion des livres	Bücherskorpion	House scorpion	Книжный ложноскорпион
Scorpions	Skorpione	Scorpions	Скорпионы
Scutigéromorphes	Spinnenasseln		Мухоловковые
Scyllaridés	Bärenkrebse	Scyllarids	Раки-медведи
Scyphozoaires	Echte Quallen	Large jellyfishes	Сцифоидные
Siphonophores	Staatsquallen		Сифонофоры
Sipunculides	Spritzwürmer		Сипункулиды
Solifuges	Walzenspinnen		Сольпуги
Sphérothéridés	Riesenkugler	Sphaerotheriids	
Spirorbes	Posthörnchenwürmer		Спирорбис
Spirotriches	Spiralwimperlinge	Spirotriches	
Spongiaires	Schwammtiere	Sponges	Губки
Spongidés	Badeschwämme	Spongids	Туалетные губки
Spongillidés	Süßwasserschwämme	Fresh water sponges	Бадяги
Sporozoaires	Sporentierchen	Sporozoans	Споровики
Stentor	Trompetentierchen	Stentor	Трубачи
– bleu	Blaues Trompetentierchen		Синий трубач
Stomatopodes	Mundfüßer	Stomatopods	Ротоногие ракообразные
Subéritidés	Korkschwämme	Suberitids	
Sucteurs	Saugtierchen		Сосущие инфузории
Symphyles	Zwergfüßer	Symphylids	Сколопендреллы
Talitre sauteur	Gemeiner Strandfloh		Песочный скакун
Talitridés	Strandflöhe	Beach fleas	Песочные скакуны
Tardigrades	Bärtierchen	Waterbears	Тихоходки
Tarentules	Taranteln	Tarantulas	Тарантулы
Ténia du bœuf	Rinderbandwurm	Beef tapeworm	Невооруженный цепень
– – poisson	Fischbandwurm	Fish tapeworm	
– – porc	Schweinebandwurm	Swine tapeworm	Вооруженный цепень
– échinocoque	Hülsenwurm	Dog tapeworm	Эхинококк
– nain	Zwergbandwurm		Карликовый цепень
– – de l'homme	– des Menschen		Человеческий карликовый цепень
Terricoles	Landplanarien		Наземные планарии
Testacés	Schalamöben		Раковинные амебы
Tétragnathidés	Streckkiefer	Tetragnathids	
Thérididés	Haubennetzspinnen, Kugelspinnen	Orb-web spiders	Пауки-ткачи
Thomisidés	Krabbenspinnen	Crab spiders	Пауки-бокоходы
Tique	Holzbock	Tick	Собачий клещ
Trachéates	Tracheentiere	Tracheates	Трахейные
Trématodes	Saugwürmer	Flukes	Сосальщики
– digènes	Digene Saugwürmer		Дигенетические сосальщики
– monogènes	Monogene Saugwürmer		Моногенетические сосальщики
Trichine	Trichine	Trichina	Спиральная трихина
Trichostomes	Wimpermünder	Trichostomates	
Trilobites	Dreilapper	Trilobites	Трилобиты
Trilobitomorphes	Trilobitenverwandte		Трилобитовые
Triops	Großer Rückenschaler		Большой щитень
Trombidiformes	Laufmilben		Краснотелковые клещи
Trombidiidés	Laufmilben i. e. S.	Trombidiids	Краснотелки
Trombidion	Sammetmilbe		Шелковая краснотелка
Trypanosomatidés	Trypanosomen		Трипанозомы
Tubifex	Gemeiner Schlammröhrenwurm	Tubifex	Обыкновенный трубочник
Tubificidés	Schlammröhrenwürmer	Tubificids	Трубочники
Tubiporidés	Orgelkorallen	Organ-pipe corals	Органчиковые
Turbellaires	Strudelwürmer	Flatworms	Ресничные черви
Unionicole à grosses pattes	Dickbeinige Wassermilbe		Толстоногий водный клещ
Urnatellidés	Süßwasser-Kelchwürmer	Urnatellids	
Uropyges	Geißelskorpione		Скорпионопауки
Velelle	Segelqualle	By-the-wind-sailors	Парусница
Ver de Guinée	Medinawurm		Ришта
– – terre	Gemeiner Regenwurm	Earthworm	Обыкновенный дождевой червь
– d'intestin	Madenwurm		Обыкновенная острица
– solitaire du bœuf	Rinderbandwurm	Beef tapeworm	Невооруженный цепень
– – – porc	Schweinebandwurm	Swine tapeworm	Вооруженный цепень
Veuve noire	Schwarze Witwe	Black widow spider	
Wuchérérie	Haarwurm		Вухерия
Xiphosures	Schwertschwänze	Horseshoe crabs	Мечехвосты
Zooanthaires	Krustenanemonen		Зоантарии

IV. Russian—German—English—French

RUSSIAN NAME	GERMAN NAME	ENGLISH NAME	FRENCH NAME
Адриатическая губка	Dalmatiner Schwamm		Éponge officinale
Акантобделлы	Borstenegel		Acanthobdellés
Актиниевые	Aktinien	Sea anemones	Actinies
Актинии	Purpurrosen	– –	–
Алциона	Meerhand	– finger	Main de mer
Амбарные клещи	Vorratsmilben	Acarids	Acaridés
Амебы	Nacktamöben		Amoebines
Американский омар	Amerikanischer Hummer	American lobster	Homard d'Amérique
Американский речной рак	– Flußkrebs		Écrevisse américaine
Американский цериант	Amerikanische Zylinderrose	Sand anemone	
Анкилостома	Grubenwurm		Ankylostome duodénal
Антипатовые	Dörnchenkorallen		Anthipathariens
Антипаты	Schwarze Korallen	Anthipathids	Anthipathidés
Аргазиды	Lederzecken	Argasids	Argasidés
Арктическая цианея	Arktische Riesenqualle		Cyanée artique
Атлантический палоло	Atlantischer Palolo		Eunice atlantique
Аурелии	Ohrenquallen		Aurélies
Афродитовые	Seeraupen	Aphroditids	Aphroditidés
Бадяги	Süßwasserschwämme	Freshwater sponges	Spongillidés
Балтийская креветка	Ostseegarnele		Palémon de la Baltique
Береговая мокрица	Strandassel		Ligie des rivages
Береговой морской паук	Ufer-Asselspinne		Pycnogonon littoral
Береговой скакун	Küstenhüpfer	Beach flea	Orchestie des rivages
Береговые морские пауки	Ufer-Asselspinnen	Pycnogonids	Pycnogonidés
Бесстебельчатые усоногие	Seepocken	Barbacles	Balanomorphes
Бихорхи	Walzenspinnen		Solifuges
Биченосцы	Geißeltiere	Flagellates	Flagellés
Благородные кораллы	Edelkorallen	Precious corals	Coraux
Благородный коралл	Rote Edelkoralle	Red precious coral	Corail rouge
Бокальчатый коралл	Sternkoralle	Star coral	
Бокоплав-блоха	Bachflohkrebs		Gammare de ruisseaux
Бокоплавы	Flohkrebse	Amphipods	Amphipodes
Большой сухопутный краб	Taschenkrebs	Rock-dwelling crab	Crabe tourteau
Большой щитень	Großer Rückenschaler		Triops
Бороздчатая анемона	Wachsrose		Anémone de mer
Брюхоресничные черви	Bauchhaarlinge	Gastrotriches	Gastrotriches
Бычий солитер	Unbewaffneter Bandwurm	Beef tapeworm	Ténia du boeuf
Веслоногие рачки	Ruderfußkrebse	Copepods	Copépodes
Ветвистоусые рако-образные	Wasserflöhe	Water fleas	Cladocères
Водные мокрицы	Wasserasseln	Fresh water isopods	Asellidés
Водяной волосатик	Wasserkalb		Gordius aquatique
Водяной ослик	Gemeine Wasserassel	Fresh water isopod	Aselle aquatique
Волосатая цианея	Gelbe Haarqualle	Lions mane	
Волосатиковые черви	Saitenwürmer	Horsehair worms	Nématomorphes
Волосатые крабы	Wollkrebse	Dromid crabs	Dromidés
Волосатые крабы	Wollkrabben	– –	Dromies velues
Вооруженный цепень	Schweinebandwurm	Swine tapeworm	Ténia du porc
Воротничковые жгутиконосцы	Kragengeißler	Craspedomonadids	Craspedomonadidés
Восьмилучевые коралловые полипы	Achtstrahlige Korallen		Octocoraux
Вторичноротые	Neumundtiere, Neumünder		Deutérostomes
Вухерия	Haarwurm		Wuchérérie
Высшие ракообразные	Höhere Krebse		Malacostraques
Вьюнки	Wasserschlängler	Naidids	Naididés
Гамазоидные клещи	Schmarotzermilben		Parasitiformes
Гастротрихи	Bauchhaarlinge	Gastrotriches	Gastrotriches
Гигантостраки	Riesenskorpione	Eurypterids	Euryptéridés
Гигантская печеночная двуустка	Riesen-Leberegel		Douve du foie géante
Гигантские ракоскорпионы	Riesenskorpione	–	Euryptéridés
Гигантский дождевой червь	Riesenregenwurm		Lombric géant
Гигантский скребень	Riesenkratzer		Acanthocéphale géant
Гидроидные	Hydrozoen	Hydrozoans	Hydrozoaires
Гидрокораллы	Hydrokorallen	Hydrocorallids	Hydrocoralliaires
Глоточные пиявки	Schlundegel		Pharyngobdellés
Глубоководные морские пауки	Tiefsee-Asselspinnen	Collossendeids	Collossendéidés
Горгонарии	Rindenkorallen		Gorgonaires

RUSSIAN NAME	GERMAN NAME	ENGLISH NAME	FRENCH NAME
Гороховый краб	Muschelwächter	Oyster crab	Crabe de moule
Гребневики	Rippenquallen	Comb jellyfishes	Cténophores
Грегарины	Gregarinen		Grégarines
Груздевики	Pilzkorallen	Fungus corals	Coraux-champignos
Губки	Schwammtiere	Sponges	Spongiaires
Губоногие многоножки	Hundertfüßer	Centipedes	Chilopodes
Дафнии	Wasserflöhe	Water fleas	Cladocères
Десятиногие ракообразные	Zehnfüßer	Decapods	Décapodes
Дигенетические сосальщики	Digene Saugwürmer		Trématodes digènes
Дизентерийная амеба	Ruhramöbe		Amibe dysentérique
Длиннохвостые раки	Eigentliche Langschwanzkrebse		Astacures
Домовые пауки	Hauswinkelspinnen	House spiders	Araignées domestiques
Домовый паук	Hausspinne	– spider	Araignée domestique
Древоточцы	Wasserasseln	Fresh water isopods	Asellidés
Европейский краб	Strandkrabbe	Shore crab	Crabe enragé
Европейский лангуст	Europäische Languste	Spiny lobster	Langouste européenne
Европейский омар	Europäischer Hummer	Lobster	Homard d'Europe
Жаброног	Sommerkiemenfuß		Branchipe commun
Жаброногие ракообразные	Blattfußkrebse	Phyllopods	Phyllopodes
Жгутиковые	Geißeltiere	Flagellates	Flagellés
Жгутиконосцы	–		
Жгутоногие пауки	Skorpionspinnen	Whip scorpions	Pédipalpes
Жгутоусы	Wenigfüßer	Pauropods	Pauropodes
Зеленая бонеллия	Grüne Bonellia	Bonellia	Bonellie verte
Зеленая гидра	– Hydra		Hydre verte
Зоантарии	Krustenanemonen		Zooanthaires
Известковые губки	Kalkschwämme	Calcareous sponges	Calcariens
Иксодовые клещи	Zecken	Ticks	Ixodidés
Исполинские щитни	Riesenskorpione	Eurypterids	Euryptéridés
Каменные крабы	Steinkrabben	Lithodid crabs	Lithodidés
Камптозои	Kelchwürmer	Kamptozoans	Kamptozoaires
Капрелла	Widderkrebs		Caprelle-chèvre
Капрелловые	Gespenstkrebse	Caprellids	Capréllidés
Карибский сухопутный краб	Karibische Landkrabbe	White crab	
Карликовый цепень	Zwergbandwurm		Ténia nain
Карповая вошь	Karpfenlaus	Fish louse	Argule foliacé
Кивсяки	Schnurfüßer i. e. S.	Julids	Julidés
Кивсяки	– i. w. S.	Julides	Julides
Кивсяковые	Doppelfüßer	Millipeds	Diplopodes
Киноринхи	Hakenrüßler	Kinorhynches	Kinorhynches
Китайская двуустка	Chinesischer Leberegel		Clonorchis chinois
Китайский мохнаторукий краб	Wollhandkrabbe		Crabe chinois
Китовые вши	Walläuse	Whale lice	Cyamidés
Китовые морские жолуди	Wal-Seepocken	Coronulids	Coronulidés
Кишечная кровяная двуустка	Darm-Pärchenegel		Bilharzie de l'intestin
Кишечнополостные	Hohltiere	Jellyfishes, anemones and corals	Cœlentérés
Клещи	Milben	Mites	Acariens
Клионовые	Bohrschwämme	Chionids	Clionidés
Клубовидковые	Saftkugler	Glomerids	Gloméridés
Книжный ложноскорпион	Bücherskorpion	House scorpion	Chélifère cancroide
Коготные черви	Stummelfüßer		Onychophores
Коленчатоногие	Asselspinnen	Pantopods	Pantopodes
Коловратки	Rädertiere	Wheel animalcules	Rotatores
Колпода	Kappentierchen		Colpode
Кольчатые черви	Gliederwürmer	Jointed worms	Annélides
Кольчецы	–	– –	
Колючая афродита	Seemaus		Aphrodite hérissée
Колючеголовые черви	Kratzer	Spiny-headed worms	Acanthocéphales
Кокцидии	Kokzidien		Coccidies
Конская актиния	Purpurseerose		Actinie pourpre
Коралловые полипы	Blumentiere	Anthozoans	Anthozoaires
Корнеголовые	Wurzelkrebse		Rhizocéphales
Корненожки	Wurzelfüßer	Rhizopods	Rhizopodes
Короткохвостые раки	Echte Krabben	Crabs	Brachyures
Костянка	Brauner Steinläufer		Mille-pattes commun
Костянковые	Steinläufer		Lithobiomorphes
Кошачья печеночная двуустка	Katzenleberegel		Douve du chat
Крабовые пауки	Krabbenspinnen	Crab spiders	Thomisidés
Крабы	Echte Krabben	Crabs	Brachyures
Крабы-пловунцы	Schwimmkrabben	Swimming crabs	Portunidés
Крабы-пловунцы	Echte Schwimmkrabben	True swimming crabs	Portunes

RUSSIAN NAME	GERMAN NAME	ENGLISH NAME	FRENCH NAME
Краснотелковые клещи	Laufmilben		Trombidiformes
Краснотелки	– i. e. S.	Trombidiids	Trombidiidés
Красный коралл	Rote Edelkoralle	Red precious coral	Corail rouge
Креветки	Knallkrebschen	Alpheids	Alphéidés
Креветковые	Garnelenartige Langschwanzkrebse	Trues shrimps	Natantides
Кровяные двуустки	Pärchenegel	Schistosomids	Schistosomidés
Круглые черви	Fadenwürmer	Threadworms	Nématodes
Кругоресничные инфузории	Glockentierchen	Peritrichous cilistes	Péritriches
Кумовые ракообразные	Kumazeen		Cumacés
Лангусты	Langusten	Spiny lobsters	Palinuridés
Ланцетовидная двуустка	Lanzettegel		Dicrocélium fer-de-lance
Лентецы	Bandwürmer	Tapeworms	Cestodes
Ленточные черви	–		
Лечебная пиявка	Medizinischer Blutegel	Leech	Sangsue médicale
Ложноконская пиявка	Pferdeegel		– du cheval
Ложноскорпионы	Afterskorpione	False scorpions	Pseudoscorpions
Лошадиная острица	Pferde-Madenwurm		Oxyure chevalin
Лучевики	Radiolarien	Radiolaria	Radiolaires
Лягушечья многоустка	Harnblasen-Saugwurm		Polystomum de vessie
Мадрепоровые кораллы	Steinkorallen	Stony corals	Madrépores
Мадрепоры	Weiße Korallen		
Майи	Seespinnen	Spider crabs	Araignées de mer
Майи	Dreieckskrabben	– –	Majidés
Малоногие	Wenigfüßer	Pauropods	Pauropodes
Малощетинковые черви	Wenigborster	Earthworms	Oligochaetes
Малый солитер	Zwergbandwurm		Ténia nain
Манящие крабы	Winkerkrabben	Fiddier crabs	Cŕabes à signaux
Медицинская пиявка	Medizinischer Blutegel	Leech	Sangsue médicale
Мезозои	Mitteltiere	Unorganized animals	Mésozoaires
Меростомовые	Hüftmünder	Merostomates	Merostomates
Мечехвосты	Schwertschwänze	Horseshoe crabs	Xiphosures
Мешковые пауки	Trichterspinnen	Funnel-web spiders	Agélénidés
Мизиды	Spaltfüßer		Mysidacés
Мизостомиды	Saugmünder	Myzostomes	Myzostomides
Многокамерные фораминиферы	Vielkammerige		Foraminifères à coquille polythalame
Многоколенчатые	Asselspinnen	Pantopods	Pantopodes
Многоногие	Tausendfüßer	Thousand-legged worms	Myriapodes
Многоножки	–	– –	
Многосвязки	Bandfüßer		Polydesmides
Многощетинковые черви	Vielborster	Seaworms	Polychaetes
Мокрицевые	Landasseln		Oniscoides
Моногенетические сосальщики	Monogene Saugwürmer		Trématodes monogènes
Морская козочка	Widderkrebs		Caprelle-chèvre
Морские анемоны	Purpurrose	Sea anemones	Actinies
Морские жолуди	Seepocken	Barnacles	Balanomorphes
Морские клещи	Meereswassermilben	Halacarids	Halacaridés
Морские кольчецы	Vielborster	Seaworms	Polychaetes
Морские пауки	Asselspinnen	Pantopods	Pantopodes
Морские перья	Seefedern	Sea-pens	Pennatulaires
Морские планарии	Meeresplanarien		Maricoles
Морские раки отшельники	Meeres-Einsiedlerkrebse	Hermit crabs	Paguridés
Морские уточки	Entenmuscheln		Lepadomorphes
Мочеводная кровяная двуустка	Harnblasen-Pärchenegel		Bilharzie de la vessie
Мухоловковые	Spinnenasseln		Scutigéromorphes
Мышиный карликовый цепень	Zwergbandwurm der Maus	Rat tapeworm	
Наземные пиявки	Landegel	Haemadipsids	Hémadipsidés
Наземные планарии	Landplanarien		Terricoles
Настоящие много-клеточные	Echte Vielzeller	Many-celled animals	Eumétazoaires
Настоящие пауки	Radnetzspinnen	Araneids	Aranéidés
Невооруженный цепень	Rinderbandwurm	Beef tapeworm	Ténia du bœuf
Нематоды	Fadenwürmer	Threadworms	Nématodes
Немертины	Schnurwürmer	Strapworms	Némertines
Неполнохвостые раки	Mittelkrebse		Anomures
Нестрекающие кишечно-полостные	Nessellose Hohltiere		Acnidaires
Низшие ракообразные	Niedere Krebse		Entomostraques
Нитчатые черви	Fadenwürmer	Threadworms	Nématodes
Норвежский омар	Kaiserhummer	Norway lobster	Néphrops
Носовая пятиустка	Nasenwurm		Linguatule nasale
Обыкновенная гидра	Braune Hydra		Hydre commune

RUSSIAN NAME	GERMAN NAME	ENGLISH NAME	FRENCH NAME
Обыкновенная креветка	Nordseegarnele	Common shrimp	Crangon commun
Обыкновенная многоножка	Brauner Steinläufer		Mille-pattes commun
Обыкновенная морская уточка	Gemeine Entenmuschel	Barnaclo	Anatife
Обыкновенная острица	Madenwurm		Ver d'intestin
Обыкновенная рыбья пиявка	Gemeiner Fischegel		Sangsue piscicole
Обыкновенный дождевой червь	— Regenwurm	Earthworm	Ver de terre
Обыкновенный мечехвост	Königskrabben	King crabs	Limules
Обыкновенный морской жолудь	Gemeine Seepocke	Common barnacle	Balane commun
Обыкновенный речной рак	Europäischer Flußkrebs	Crayfish	Écrevisse commune
Обыкновенный сенокосец	Gemeiner Weberknecht	Harvestman	Faucheur
Обыкновенный сухопутный краб	Gemeine Landkrabbe	Land crab	Crabe terrestre
Обыкновенный трубочник	Gemeiner Schlammröhrenwurm	Tubifex	Tubifex
Однокамерные фораминиферы	Einkammerige		Foraminifères à coquille monothalame
Однохвостый приапул	Einschwänziger Priapswurm		Priapule
Олигохеты	Wenigborster	Earthworms	Oligochaetes
Омары	Hummer	Lobsters	Homaridés
Органчиковые	Orgelkorallen	Organ-pipe corals	Tubiporidés
Пальмовый вор	Palmendieb	Robber crab	Crabe des cocotiers
Панцирные жгутиконосцы	Panzergeißler	Dinoflagellates	Dinoflagellés
Парамеции	Pantoffeltierchen	Slipper animalcules	Paramécies
Парусница	Segelqualle	By-the-wind-sailors	Velelle
Пастбищные клещи	Zecken	Ticks	Ixodidés
Паук крестовик	Kreuzspinne	Garden spider	Epéire diadème
Пауки	Echte Spinnen	Spiders	Araniens
Пауки-бокоходы	Krabbenspinnen	Crab spiders	Thomisidés
Пауки-птицееды	Vogelspinnen i. w. S.		Orthognathes
Пауки-скакуны	Springspinnen	Jumping spiders	Salticidés
Пауки-ткачи	Haubennetzspinnen	Orb-web spiders	Thérididés
Паукообразные	Spinnentiere	Arachnids	Arachnides
Пауроподы	Wenigfüßer	Pauropods	Pauropodes
Первичноротые	Urmundtiere		Protostomes
Пергаментный трубкожил	Pergamentwurm		Chétoptère luisant
Пескожил	Sandwurm	Lugworm	Arénicole des pêcheurs
Песочные скакуны	Strandflöhe	Beach fleas	Talitridés
Песочный скакун	Gemeiner Strandfloh		Talitre sauteur
Песчаные крабы	Renn- und Winkerkrabben	Ocypodids	Ocypodidés
Песчаные крабы	Sandkrabben	Ghost crabs	Ocypodes
Печеночная двуустка	Großer Leberegel		Grande douve du foie
Пиявки	Egel	Leeches	Hirudinés
Плоские пиявки	Knorpelegel	Glossiphoniids	Glossiphonidés
Плоские черви	Plattwürmer	Plathelminthes	Plathelminthes
Плоский многосвяз	Abgeplatteter Bandfüßer		Polydesme à lamelles
Погребная мокрица	Kellerassel		Cloporte des caves
Полевая краснотелка	Erntemilbe	Chigger	Lepte automnal
Полихеты	Vielborster	Seaworms	Polychaetes
Поритовые	Lochkorallen	Poritids	Poritidés
Пресноводные крабы	Süßwasserkrabben	Potamids	Potamidés
Пресноводные планарии	Süßwasserplanarien		Paludicoles
Приапулиды	Priapswürmer	Priapulides	Priapulides
Пробковые полипы	Lederkorallen	Alcyonarian corals	Alcyonaires
Простейшие животные	Einzeller	Protozoans	Protozoaires
Пятиустки	Zungenwürmer	Linguatulida	Linguatulides
Равноногие ракообразные	Asseln	Isopods	Isopodes
Радиолярии	Radiolarien	Radiolaria	Radiolaires
Раки-медведи	Bärenkrebse	Scyllarids	Scyllaridés
Раковинные амебы	Schalamöben		Testacés
Раковинные жаброноги	Muschelschaler		Conchostraques
Ракообразные	Krebstiere	Crustaceans	Crustacés
Ракушковые рачки	Muschelkrebse	Ostracods	Ostracodes
Ремнец	Riemenbandwurm		Ligule des intestins
Ресничные	Wimpertiere	Ciliates	Ciliés
Ресничные черви	Strudelwürmer	Flatworms	Turbellaires
Речные раки	Flußkrebse	Freshwater crabs	Astacidés
Ризостомы	Wurzelmundquallen		Rhizostomes
Ришта	Medinawurm		Filaire de Médine
Роговые кораллы	Rindenkorallen		Gorgonaires
Ротоногие ракообразные	Mundfüßer	Stomatopods	Stomatopodes
Рыбные мокрицы	Fischasseln		Cymothoidés
Рыбий вши	Fischläuse	Fish lice	Branchiures

RUSSIAN NAME	GERMAN NAME	ENGLISH NAME	FRENCH NAME
Рыбьи пиявки	Fischegel		Ichthyobdéllidés
Сафирные рачки	Saphirkrebse		Sapphirines
Сахарский скорпион	Sahara-Skorpion		Androcton d'Afrique du Nord
Сверлящие губки	Bohrschwämme	Clionids	Clionidés
Сверлящие мокрицы	Holzbohrasseln	Gribbles	Limnoridés
Светлянка	Leuchtender Erdläufer	Luminous centipede	Géophile luisant
Светлянковые	Erdläufer	— centipedes	Géophilomorphes
Свиной солитер	Schweinebandwurm	Swine tapeworm	Ténia du porc
Свободнополозающие полихеты	Freilebende Vielborster		Polychaetes errants
Северный цериант	Nordische Zylinderrose		Cérianthe nordique
Сенокосцы	Weberknechte	Harvestmen	Phalangides
Серая гидра	Graue Hydra		Hydre grise
Серое морское перо	— Seefeder		Ptéroide gris
Сидячие полихеты	Festsitzende Vielborster		Polychaetes sédentaires
Синий трубач	Blaues Trompetentierchen		Stentor bleu
Сипункулиды	Spritzwürmer		Sipunculides
Сифонофора физалия	Seeblase	Portuguese man-of-war	Physalie
Сифонофоры	Staatsquallen	Siphonophores	Siphonophores
Сколопендреллы	Zwergfüßer	Symphylids	Symphyles
Сколопендры	Riesenläufer		Scolopendromorphes
Скорпионопауки	Geißelskorpione		Uropyges
Скорпионы	Skorpione	Scorpions	Scorpions
Собачий клещ	Holzbock	Tick	Tique
Сольпуги	Walzenspinnen		Solifuges
Соляный рачек	Salzkrebschen		Artémie
Сосальщики	Saugwürmer	Flukes	Trématodes
Сосущие инфузории	Saugtierchen		Sucteurs
Сотовидные губки	Wabenkalkschwämme		Hétérocoeles
Спиральная трихина	Trichine	Trichina	Trichine
Спирорбис	Posthörnchenwürmer		Spirorbes
Споровики	Sporentierchen	Sporozoans	Sporozoaires
Стебельчатые усоногие	Entenmuscheln		Lepadomorphes
Стеклянные губки	Glasschwämme		Hexactinellides
Стенная мокрица	Mauerassel	Sowbug	Cloporte des murs
Стенторы	Trompetentierchen	Stentor	Stentor
Странный спайник	Doppeltier	Diplozoon	Diplozoon paradoxal
Стрекающие кишечнополостные	Nesseltiere		Cnidaires
Стыдливые крабы	Schamkrabben		Calappe
Сухопутные крабы	Landkrabben i. e. S.	Land crabs	Gécarcinidés
Сцифоидные	Echte Quallen	Large jellyfishes	Scyphozoaires
Сырный акар	Käsemilbe		Acare du fromage
Тарантуловые	Wolfsspinnen	Wolf spiders	Tarentules Lycosidés
Тарантулы	Taranteln	Tarantulas	—
Тихоходки	Bärtierchen	Waterbears	Tardigrades
Толстоногий водный клещ	Dickbeinige Wassermilbe		Unionicole à grosses pattes
Тонкопанцирные	Leptostraken		Leptostraques
Трахейные	Tracheentiere	Tracheates	Trachéates
Трилобитовые	Trilobitenverwandte		Trilobitomorphes
Трилобиты	Dreilapper	Trilobites	Trilobites
Трипанозомы	Trypanosomen	Trypanosomatids	Trypanosomatidés
Трубачи	Trompetentierchen	Stentor	Stentor
Трубочники	Schlammröhrenwürmer	Tubificids	Tubificidés
Туалетные губки	Badeschwämme	Spongids	Spongidés
Турбеллярии	Strudelwürmer	Flatworms	Turbellaires
Туфельки	Pantoffeltierchen	Slipper animalcules	Paramécies
Тысяченожки	Tausendfüßer	Thousand-legged worms	Myriapodes
Усоногие ракообразные	Rankenfüßer	Barnacles	Cirripèdes
Фаланги	Walzenspinnen		Solifuges
Фораминиферы	Lochträger	Foraminifera	Foraminifères
Хелицеровые	Chelicerentiere	Chelicerates	Chélicériens
Хоботковый морской паук	Rüssel-Asselspinne		Pantopode à trompe
Хоботные пиявки	Rüsselegel		Rhynchobdellés
Хризаоры	Kompaßquallen		Chrysaores
Цейлонская пиявка	Ceylonegel		Sangsue de Ceylan
Цериантовые	Zylinderrosen		Cérianthaires
Цестоды	Bandwürmer	Tapeworms	Cestodes
Цианеевые	Haarquallen	Cyaneids	Cyanéidés
Цианеи	Nesselquallen		Cyanées
Циклопиды	Hüpferlinge	Cyclopids	Cyclopidés
Человеческая аскарида	Spulwurm	Roundworm	Ascaride lombricoïde
Человеческий карликовый цепень	Zwergbandwurm des Menschen		Ténia nain de l'homme
Червь-лоа	Wanderfilarie		Filaire loa
Чесоточный зудень	Krätzemilbe des Menschen	Itch mite	Gale de l'homme

RUSSIAN NAME	GERMAN NAME	ENGLISH NAME	FRENCH NAME
Чесоточный клещ	Mehlmilbe		Acare de farine
Членистоногие	Gliederfüßer	Arthropods	Arthropodes
Шаровидка	Rollassel	Pillbug	Armadille vulgaire
Шелковая краснотелка	Sammetmilbe		Trombidion
Шестилучевые коралловые полипы	Sechsstrahlige Korallen		Héxacoraux
Широкопалый речной рак	Europäischer Flußkrebs	Crayfish	Écreuisse commune
Щетинковые пиявки	Borstenegel		Acanthobdellés
Щитни	Rückenschaler		Notostraques
Эвриптериды	Riesenskorpione	Eurypterids	Euryptéridés
Эьфаузиевые ракообразные	Leuchtkrebse		Euphausiacés
Эпхитрея	Topfwurm		Enchytrée
Эхинококк	Hülsenwurm	Dog tapeworm	Ténia échinocoque
Эхиур	Meerquappe		Echiure
Эхиуриды	Igelwürmer	Echiurids	Echiures
Язычковые	Zungenwürmer	Linguatulida	Linguatulides
Японская кровяная двуустка	Japanischer Pärchenegel		Bilharzie d'extrème-Orient

Conversion Tables of Metric to U.S. and British Systems

U.S. Customary to Metric		Metric to U.S. Customary	
To convert	Multiply by	To convert	Multiply by
—— Length ——			
in. to mm.	25.4	mm. to in.	0.039
in. to cm.	2.54	cm. to in.	0.394
ft. to m.	0.305	m. to ft.	3.281
yd. to m.	0.914	m. to yd.	1.094
mi. to km.	1.609	km. to mi.	0.621
—— Area ——			
sq. in. to sq. cm.	6.452	sq. cm. to sq. in.	0.155
sq. ft. to sq. mi.	0.093	sq. m. to sq. ft.	10.764
sq. yd. to sq. m.	0.836	sq. m. to sq. yd.	1.196
sq. mi. to ha.	258.999	ha. to sq. mi.	0.004
—— Volume ——			
cu. in. to cc.	16.387	cc. to cu. in.	0.061
cu. ft. to cu. m.	0.028	cu. m. to cu. ft.	35.315
cu. yd. to cu. m.	0.765	cu. m. to cu. yd.	1.308
—— Capacity (liquid) ——			
fl. oz. to liter	0.03	liter to fl. oz.	33.815
qt. to liter	0.946	liter to qt.	1.057
gal. to liter	3.785	liter to gal.	0.264
—— Mass (weight) ——			
oz. avdp. to g.	28.35	g. to oz. avdp.	0.035
lb. avdp. to kg.	0.454	kg. to lb. avdp.	2.205
ton to t.	0.907	t. to ton	1.102
l. t. to t.	1.016	t. to l. t.	0.984

Abbreviations

U.S. Customary	Metric
avdp.—avoirdupois	cc.—cubic centimeter(s)
ft.—foot, feet	cm.—centimeter(s)
gal.—gallon(s)	cu.—cubic
in.—inch(es)	g.—gram(s)
lb.—pound(s)	ha.—hectare(s)
l. t.—long ton(s)	kg.—kilogram(s)
mi.—mile(s)	m.—meter(s)
oz.—ounce(s)	mm.—millimeter(s)
qt.—quart(s)	t.—metric ton(s)
sq.—square	
yd.—yard(s)	

TEMPERATURE

AREA

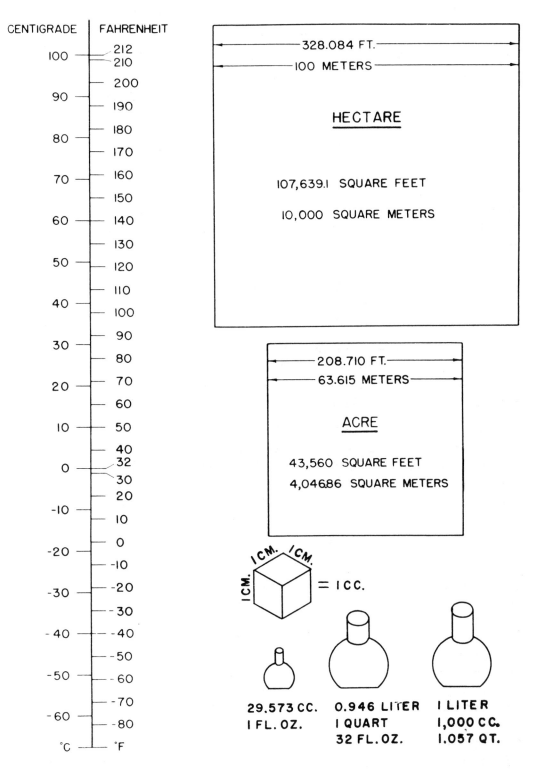

CENTIGRADE | FAHRENHEIT

100 — 212
210
200
90 — 190
180
80 — 170
160
70 — 150
140
60 — 130
120
50 — 110
100
40 — 90
80
30 — 70
60
20 — 50
40
32
0 — 30
20
10
-10
-20 — 0
-10
-20
-30 — -30
-40 — -40
-50
-50 — -60
-70
-60 — -80
°C — °F

HECTARE

328.084 FT.
100 METERS

107,639.1 SQUARE FEET

10,000 SQUARE METERS

ACRE

208.710 FT.
63.615 METERS

43,560 SQUARE FEET
4,046.86 SQUARE METERS

1 CM. 1 CM. 1 CM. = 1 CC.

29.573 CC.
1 FL. OZ.

0.946 LITER
1 QUART
32 FL. OZ.

1 LITER
1,000 CC.
1.057 QT.

WEIGHT

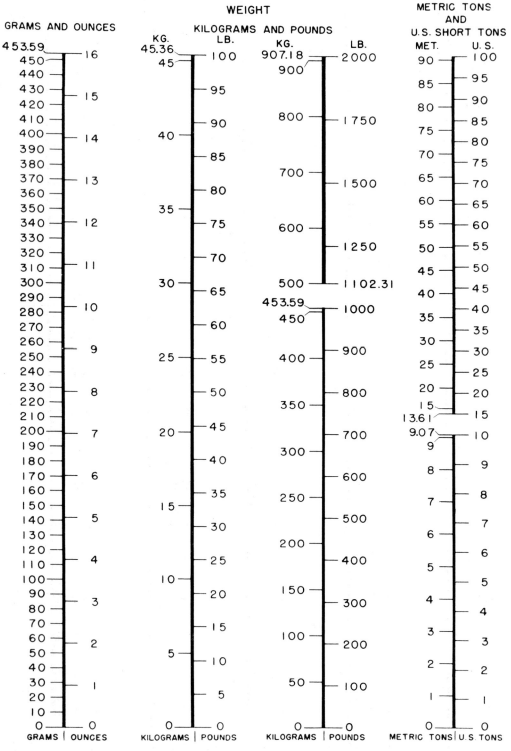

GRAMS AND OUNCES — KILOGRAMS AND POUNDS — METRIC TONS AND U.S. SHORT TONS

GRAMS | OUNCES
KILOGRAMS | POUNDS
KILOGRAMS | POUNDS
METRIC TONS | U.S. TONS

LENGTH: MILLIMETERS AND INCHES

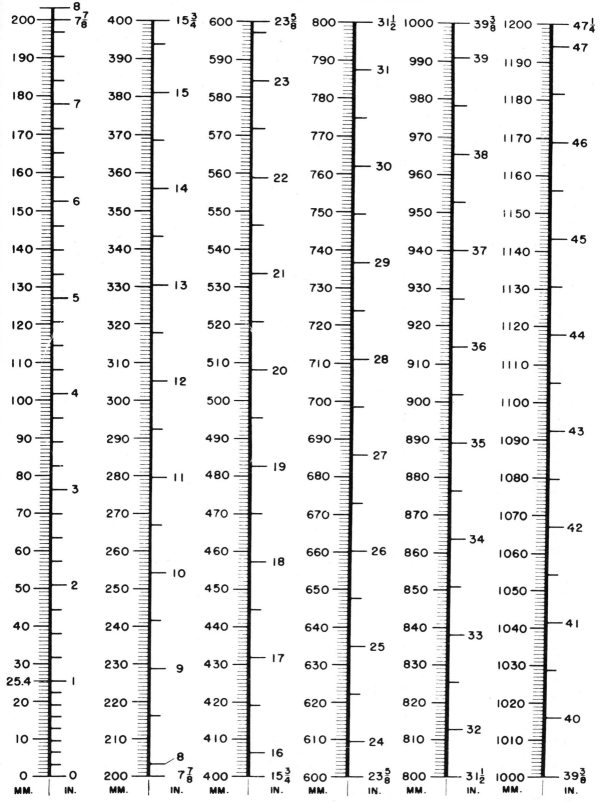

LENGTH

METERS AND FEET

KILOMETERS AND MILES

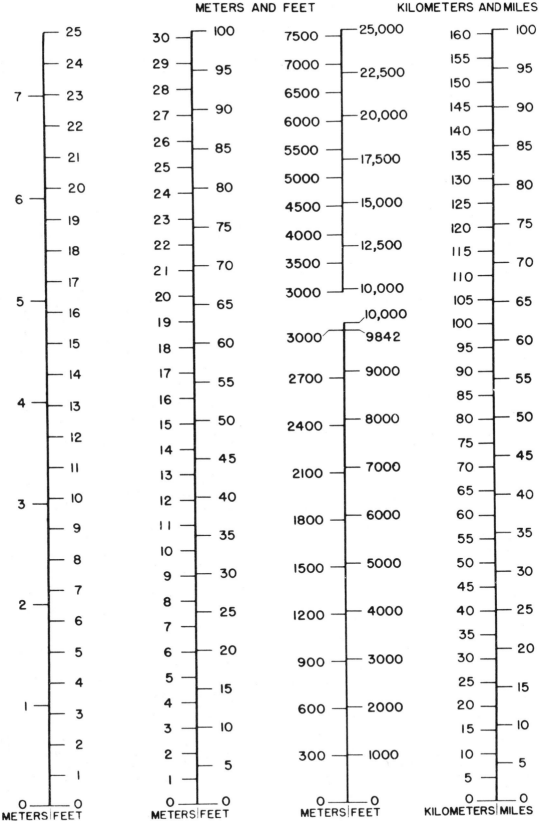

Supplementary Readings

These references of books and articles published in scientific journals deal with animals and topics that are covered in this volume. Some of these were the original sources on which the content of this book is based. These titles are intended as an aid to readers who are interested in additional information and more detailed coverage of the subjects contained in this book.★

★Supplementary Readings prepared by John B. Brown.

Books and Monographs

THE ANIMAL KINGDOM

Allee, W. C. and others. 1949. *Principles of Animal Ecology*. W. B. Saunders and Company, Philadelphia.

Bates, M. 1960. *The Forest and the Sea*. Random House, New York.

Bonner, J. T. 1952. *Morphogenesis*. Princeton University Press, Princeton, N.J.

Browning, T. O. 1963. *Animal Populations*. Harper and Row, New York.

Burnett, A. L. and T. Eisner. 1964. *Animal Adaptation*. Holt, Rinehart and Winston, New York.

Colbert, E. H. 1955. *Evolution of the Vertebrates*. John Wiley and Sons, New York.

Darling, F. E. and J. P. Milton, Eds. 1966. *Future Environments of North America*. Natural History Press, New York.

Darwin, C. 1859. *The Origin of Species*. London.

Fox, S. W. and K. Dose. 1972. *Molecular Evolution and the Origin of Life*. Freeman, San Francisco.

Giese, A. C. 1968. *Cell Physiology*. W. B. Saunders Company, Philadelphia.

Hamburgh, M. and E. J. W. Barrington, Eds. 1971. *Hormones in Development*. Appleton-Century-Crofts, New York.

Hughes, G. M. 1965. *Comparative Physiology of Vertebrate Respiration*. Harvard University Press, Cambridge, Mass.

Huxley, J. S. 1943. *Evolution, The Modern Synthesis*. Harper, New York.

Kormondy, E. J. 1969. *Concepts of Ecology*. Prentice-Hall, Inc., Englewood Cliffs, N.J.

Kummel, B. 1961. *History of the Earth*. W. H. Freeman and Company, San Francisco.

Lockwood, A. P. M. 1963. *Animal Body Fluids and Their Regulation*. Harvard University Press, Cambridge, Mass.

Mayr, E. 1963. *Animal Species and Evolution*. Harvard University Press, Cambridge, Mass.
—, E. G. Linsley and R. L. Usinger. 1953. *Methods and Principles of Systematic Zoology*. McGraw-Hill Book Co., Inc., New York.

Odum, E. P. 1963. *Ecology*. Holt, Rinehart and Winston, New York.

Prosser, C. L. and F. A. Brown, Jr. 1961. *Comparative Animal Physiology*. Saunders, Philadelphia.

Romer, A. S. 1970. *The Vertebrate Body*. W. B. Saunders Co., Philadelphia.
—. 1972. *The Procession of Life*. Doubleday, Garden City, New York.

Schenk, E. T. and J. H. McMasters. 1956. *Procedure in Taxonomy*, 3rd ed. Stanford University Press, Stanford, California.

Schmidt-Nielsen, K. 1960. *Animal Physiology*. Prentice-Hall, Englewood Cliffs, N.J.

—. 1972. *How Animals Work*. Cambridge University Press, New York.

Simpson, G. G. 1949. *The Meaning of Evolution*. Yale University Press, New Haven, Conn.
—. *The Major Features of Evolution*. Columbia University Press, New York.
—. 1951. *Horses*. Oxford University Press, New York.
—. 1961. *Principles of Animal Taxonomy*. Columbia University Press, New York.

Slobodkin, L. B. 1961. *Growth and Regulation of Animal Populations*. Holt, Rinehart and Winston, New York.

Srb, A. M., R. D. Owen and R. S. Edgar, 1965. *General Genetics*, 2nd ed. W. H. Freeman and Co., San Francisco.

Stear, E. B. and A. H. Kadish, Eds. 1969. *Hormonal Control Systems*. American Elsevier Publishing Company, Inc., New York.

Stebbins, G. L. 1971. *Processes of Organic Evolution*. Prentice-Hall, Inc., Englewood Cliffs, N.J.

Sturtevant, A. H. and G. W. Beadle. 1939. *An Introduction to Genetics*. W. B. Saunders Co., Philadelphia. (Reprinted in paperback by Dover.)

Swanson, C. P. *The Cell*. Prentice-Hall, Inc., Englewood Cliffs, N.J.

Thompson, D. W. 1942. *On Growth and Form*, 2nd ed. Cambridge University Press, Cambridge.

Walker, M. I. 1971. *Amateur Photomicrography*. Focal Press, New York and London.

Wallace, B. and A. M. Srb. 1961. *Adaptation*. Prentice-Hall, Englewood Cliffs, N.J.

Watson, J. D. 1965. *The Molecular Biology of the Gene*. W. A. Benjamin, New York.

Williams, G. C. 1966. *Adaptation and Natural Selection*. A Critique of Some Current Evolutionary Thought. Princeton University Press, Princeton, N.J.

BEHAVIOR

Aronson, L. and others. 1970. *Development and Evolution of Behavior*. W. H. Freeman and Co., San Francisco.

Bastock, M. 1967. *Courtship: An Ethological Study*. Heinemann Educational Books, Ltd., London.

Clegg, P. C. and A. G. Clegg. 1959. *Hormones, Cells and Organisms*. The Role of Hormones in Mammals. Stanford University Press, Stanford, Calif.

Darwin, C. 1872. *The Expression of the Emotions in Man and Animals*. Phoenix Science Series, University of Chicago Press, Chicago.

Eibl-Eibesfeldt, I. 1970. *Ethology. The Biology of Behavior*. Holt, Rinehart and Winston, New York.

Etkin, W. 1964. *Social Behavior and Organization among Vertebrates*. The University of Chicago Press, Chicago.

Ewer, R. F. 1968. *Ethology of Mammals*. Plenum Press, New York.

Fraenkel, G. S. and D. L. Gunn. 1940. *The Orientation of Animals*. Kineses, Taxes and Compass Reactions. Oxford University Press, New York. (Reprinted in paperback by Dover.)

Hediger, H. 1950. *Wild Animals in Captivity*. Butterworths Scientific

Publications, London. (Reprinted in paperback by Dover.)

—. 1955. *Studies of the Psychology and Behaviour of Captive Animals in Zoos and Circuses*. Butterworths Scientific Publications, London. (Reprinted in paperback by Dover.)

—. 1969. *Man and Animal in the Zoo*. Zoo Biology. Delacorte Press, New York.

Hinde, R. A. 1966. *Animal Behaviour*: A Synthesis of Ethology and Comparative Psychology. McGraw-Hill Book Company, New York.

—. 1972. *Non-Verbal Communication*. Cambridge University Press, New York.

Klopfer, P. H. and J. Hailman. 1967. *An Introduction to Animal Behavior*: Ethology's First Century. Prentice-Hall, Inc., Englewood Cliffs, N.J.

—, —. 1972. *Control and Development of Behavior*: An Historical Sample from the Pens of Ethologists. Addison-Wesley Publishing Co., Reading, Mass.

—, —. 1972. *Function and Evolution of Behavior*: An Historical Sample from the Pens of Ethologists. Addison-Wesley Publishing Co., Reading, Mass.

Lorenz, K. 1952. *King Solomon's Ring*. Crowell, New York.

—. 1965. *Evolution and Modification of Behavior*. The University of Chicago Press, Chicago.

—. 1970, 1971. *Studies in Animal and Human Behavior*, 2 vols. Harvard University Press, Cambridge, Mass.

Manning, A. 1967. *An Introduction to Animal Behavior*. Edward Arnold, Ltd., London.

Marler, P. and W. J. Hamilton III. 1966. *Mechanisms of Animal Behavior*. John Wiley & Sons, Inc., New York.

McGill, T. E., Ed. 1965. *Readings in Animal Behavior*. John Wiley and Sons, Inc., New York.

Pribram, K. H., Ed. 1969. *On the Biology of Learning*. Harcourt, Brace and World, Inc., New York.

Roeder, K. D. 1963. *Nerve Cells and Insect Behavior*. Harvard University Press, Cambridge, Mass.

Schiller, C. H. 1957. *Instinctive Behavior*. International University Press, New York.

Thorpe, W. H. 1963. *Learning and Instinct in Animals*, 2nd ed. Harvard University Press, Cambridge, Mass.

Tinbergen, N. 1951. *The Study of Instinct*. Oxford University Press, New York and Oxford.

—. 1960. *The Herring Gull's World*. Basic Books, New York.

—. 1965. *Animal Behavior*. Time, Inc., New York.

—. 1968. *Curious Naturalists*. Basic Books, New York.

Von Frisch, K. 1955. *The Dancing Bees*. Harcourt, Brace and Co., New York.

Wickler, W. 1968. *Mimicry in Plants and Animals*. McGraw-Hill Book Co., New York.

—. 1972. *The Sexual Code*. The Social Behavior of Animals and Men. Doubleday, Garden City.

ANIMALS COVERED IN THIS VOLUME

Barnes, R. D. 1968. *Invertebrate Zoology*. W. B. Saunders Co., Philadelphia.

Barrington, E. J. W. 1965. *The Biology of Hemichordata and Protochordata*. Oliver & Boyd, London.

—. 1967. *Invertebrate Structure and Function*. Houghton Mifflin Co. Boston.

Bayer, F. M. and H. B. Owre. 1968. *The Free-Living Lower Invertebrates*. Macmillan Co., New York.

Baylis, H. A. and R. Daubney. 1926. *A Synopsis of the Families and Genera of Nematoda*. Richard Clay and Sons, London.

Beck, D. E. and L. F. Braithwaite. 1962. *Invertebrate Zoology*: Laboratory Workbook. Burgess Publishing Co., Minneapolis, Minn.

Berrill, J. N. 1950. *The Tunicata*. Johnson Reprint Corp., New York.

Bick, H. 1972. *Ciliated Protozoa*. An Illustrated Guide to the Species Used as Biological Indicators in Freshwater Biology. World Health Organization, Geneva.

Bird, A. F. *The Structure of Nematodes*. 1971. Academic Press, New York.

Borradaile, L. A., F. A. Potts, L. E. S. Eastham and J. T. Saunders. 1935. *The Invertebrata*. Macmillan Co., New York.

Brinkhurst, R. O., B. G. M. Jamieson and others. 1972. *Aquatic Oligochaeta of the World*. University of Toronto Press, Buffalo, N.Y.

Brown, F. A., Jr. 1950. *Selected Invertebrate Types*. John Wiley and Sons, Inc., New York.

Brusca, R. C. 1973. *A Handbook to the Common Intertidal Invertebrates of the Gulf of California*. The University of Arizona Press, Tuscon, Arizona.

Buchsbaum, R. M. 1948. *Animals Without Backbones*. University of Chicago Press, Chicago.

—, and L. J. Milne. 1960. *The Lower Animals*: Living Invertebrates of the World. Doubleday and Co., New York.

Bullough, W. S. 1950. *Practical Invertebrate Anatomy*. Macmillan Co., New York.

Cameron, T. W. 1934. *The Internal Parasites of Domestic Animals*. A. and C. Black, London.

Carter, G. S. 1951. *A General Zoology of the Invertebrates*. Sedgwick and Jackson, London.

Chitwood, B. G. and M. B. Chitwood. 1950. *An Introduction to Nematology*. Chitwood, Baltimore.

Clark, R. B. 1964. *Dynamics of Metazoan Evolution*: The Origin of the Coelom and Segments. Clarendon Press, Oxford.

Cloudsley-Thompson, J. L. 1961. *Land Invertebrates*. Methuen, London.

Coatney, C. R., W. E. Collins, Mc. W. Warren and P. G. Contacos. 1971. *The Primate Malarias*. National Institutes of Health, Bethesda, Maryland.

Corliss, J. O. 1961. *The Ciliated Protozoa*. Pergamon Press, New York.

Croll, N. A. 1971. *The Behavior of Nematodes*. Their Activity, Senses and Responses. Saint Martinis, New York.

Dales, R. P. 1963. *Annelids*. Hutchinson & Co., Ltd., London.

Darwin, C. 1896. *The Structure and Distribution of Coral Reefs*. John Murray, London.

Davis, C. C. 1955. *The Marine and Fresh-water Plankton*. Michigan State University Press, East Lansing.

Dawes, B. 1946. *The Trematoda*. Cambridge University Press, New York.

Donner, J. 1966. *Rotifers*. Translated and adapted by H. G. S. Wright. Frederick, Warne & Co., Ltd., New York.

Dougherty, E. C., Ed. 1963. *The Lower Metazoa*: Comparative Biology and Phylogeny. University of California Press, Berkeley.

Drew, G. A. 1920. *A Laboratory Manual of Invertebrate Zoology*. W. B. Saunders, Co., Philadelphia.

Edmonson, W. T. 1959. *The Rotifera*. In: Freshwater Biology. Ed. by H. B. Ward and G. P. Whipple, Revised Edition Edited by W. T. Edmonson. Wiley, New York.

Erasmus, D. A. 1973. *The Biology of Trematodes*. Crane Russak, New York.

Gardiner, M. S. 1972. *Biology of the Invertebrates*. McGraw-Hill Book Co., New York.

Gibson, R. 1972. *Nemeteans*. Humanities-Hillary House, New York.

Goodey, J., Ed. 1963. *Soil and Freshwater Nematodes*. Rev. Ed., Methuen, London.

Gosner, K. L. 1971. *Guide to Identification of Marine and Estuarine Invertebrates*. Wiley-Interscience, New York.

Gould, M. C. and others. 1972. *Invertebrate Oogenesis II*. MSS Information Corp., New York.

Green, J. 1963. *A Biology of Crustacea*. Quadrangle Books, Chicago.

Griffiths, M. 1968. *Echidnas*. International Series of Monographs in Pure and Applied Biology, Zoology Division, Vol. 38. Pergamon Press, Ltd., Oxford, England.

Haozi, J. 1963. *The Evolution of Metazoa*. Macmillan Co., New York.

Harvey, E. B. 1956. *The American Arbacia and Other Sea Urchins*. Princeton University Press, Princeton, N. J.

Hegner, R. W. and J. G. Engemann. 1968. *Invertebrate Zoology*. Macmillan Co., New York.

Hickman, C. P. 1967. *Biology of the Invertebrates*. C. Y. Mosby Co., St. Louis.

Hill, D. L. 1972. *The Biochemistry and Physiology of Tetrahymena*. Academic Press, New York.

Hoare, C. A. 1972. *The Trypanosomes of Mammals*. A Zoological Monograph. Blackwell, Oxford, England.

Hyman, L. H. 1940. *The Invertebrates*: Vol. I. *Protozoa Through Ctenophora*. McGraw-Hill Book Co., New York.

—. 1951. *The Invertebrates*: Vol. II. *Platyhelminthes and Rhynchocoela: The Acoelomate Bilateria*. McGraw-Hill Book Co., New York.

—. 1951. *The Invertebrates*: Vol. III. *Acanthcephala, Aschelminthes, and Entoprocta*: The Pseudocoelomate Bilateria. McGraw-Hill Book Co., New York.

—. 1955. *The Invertebrates*: Vol. IV. *Echinodermata: The Coelomate Bilateria*. McGraw-Hill Book Co., New York.

—. 1959. *The Invertebrates*: Vol. V. *Smaller Coelomate Groups: Chaetognatha, Hemichordata, Pogonophora, Horonida, Ectoprocta, Brachiopoda, Siponculida; The Coelomate Bilateria*. McGraw-Hill Book Co., New York.

Jeon, K. W., Ed. 1973. *The Biology of Amoeba*. Academic Press, New York.

Kaestner, A. 1967. *Invertebrate Zoology*: Vol. I. *Porifera, Cnidaria, Platyhelminthes, Aschelminthes, Mollusca, Annelida, and Related Phyla*. Translated by H. W. Levi and L. R. Levi. John Wiley and Sons, Inc., Interscience Publishers, New York.

Kheysin, Y. M. 1972. *Life Cycles of Coccidia of Domestic Animals*. University Park Press, Baltimore.

Lapage, G. 1957. *Animals Parasitic in Man*. Penguin Books, Baltimore.

Laverack, M. S. 1963. *The Physiology of Earthworms*. Pergamon Press Elmsford, N. J.

Lenhoff, H. M. and W. F. Loomis, Eds. 1961. *The Biology of Hydra and Some Other Coelenterates*. University of Miami Press, Coral Gables, Florida.

Levine, N. D. 1973. *Protozoan Parasites of Domestic Animals and Man*. Burgess, Minneapolis.

Mackinnon, D. L. and J. Hawes. 1961. *Protozoa*. Oxford University Press, London.

Mann, K. H. 1961. *The Leeches*. Pergamon Press, Elmsford, N. J.

Margulis, L. 1972. *The Origin of Eukaryotic Cells*. Yale University Press, New Haven, Conn.

Marshall, A. J. and W. D. Williams, Eds. 1972. *Textbook of Zoology: Invertebrates*. American Elsevier, New York.

Mayer, A. G. 1910. *Medusae of the World*. Vol. I–III. Carnegie Institution, Washington, D.C.

Meglitsch, P. A. 1967. *Invertebrate Zoology*. Oxford University Press, New York.

Moore, R. C. 1967. *Treatise of Invertebrate Paleontology*, Vols. I & II. The Geological Society of America, Inc., and The University of Kansas, Lawrence.

Mortenson, T. H. 1927. *Handbook of Echinoderms of the British Isles*. Oxford University Press.

Mulligan, H. W., Ed. 1970. *The African Trypanosomiases*. Wiley-Interscience, New York.

Murray, J. W. 1972. *An Atlas of British Recent Foraminiferids*. American Elsevier, New York.

Nichols, D. 1962. *The Echinoderms*. Hutchinson & Co., Ltd., London.

Pennak, R. W. 1953. *Fresh-water Invertebrates of the United States*. Ronald Press Co., New York.

Pérez-Miravette, A., Ed. 1972. *Behaviour of Microorganisms*. Proceedings of a Congress, Mexico City, Aug. 1970. Plenum, New York.

Pratt, H. S. 1935. *A Manual of the Common Invertebrate Animals*. Blakiston Co., Philadelphia.

Robson, G. C. 1932. *A Monograph of Recent Cephalopoda*. British Museum, London.

Rogers, W. P. 1962. *The Nature of Parasitism*. Academic Press, New York.

Rounds, H. D. 1968. *Invertebrates*. Reinhold Book Corp., New York.

Russell, F. S. 1955. *Medusae of the British Isles*. Cambridge University Press.

Russell-Hunter, W. D. 1968. *A Biology of Lower Invertebrates*. Macmillan Co., New York.

Sawyer, R. T. 1972. *North American Freshwater Leeches, Exclusive of the Piscicolidae, with a Key to all Species*. University of Illinois Press, Urbana, Ill.

Schechter, V. 1959. *Invertebrate Zoology*. Prentice-Hall, Englewood Cliffs, N.J.

Schmitt, W. L. 1965. *Crustaceans*. University of Michigan Press, Ann Arbor.

Shrock, R. and W. Twenhofel. 1953. *Principles of Invertebrate Paleontology*. McGraw-Hill, New York.

Smith, F. G. W. 1971. *Atlantic Reef Corals*. A Handbook of the Common Reef and Shallow-water Corals of Bermuda, the Bahamas, Florida, the West Indies and Brazil. University of Miami Press, Coral Gables, Florida.

Stephen, A. C. and S. J. Edmonds. 1972. *The Phyla Sipunccula and Echinura*. British Museum of Natural History, London.

Stephenson, J. 1930. *The Oligochaeta*. Oxford University Press, London.

Stoddart, D. R. and M. Yonge, Eds. 1971. *Regional Variation in Indian Ocean Coral Reefs*. Zoological Society of London Symposium No. 28, London, May 1970. Academic Press, New York.

Tartar, V. 1961. *The Biology of Stentor*. Pergamon Press, Elmsford, N.J.

Thompson, J. A. 1927. *Brachiopod Morphology and Genera*. New Zealand Board of Science and Art Manual 7, Dominion Museum, Wellington, New Zealand.

Thorne, G. 1961. *Principles of Nematology*. McGraw-Hill Book Co., New York.

Van Cleave, H. J. 1931. *Invertebrate Zoology*. McGraw-Hill Book Co., New York.

Vernberg, W. B. and F. J. Vernberg. 1972. *Environmental Physiology of Marine Animals*. Springer-Verlag, New York.

Wardle, R. A. and J. A. Mcleod. 1952. *Zoology of Tapeworms*. University of Minnesota Press, St. Paul.

Webster, J. M., Ed. 1972. *Economic Nematology*. Academic Press, London.

Wells, M. J. 1968. *Lower Animals*. McGraw-Hill, New York.

Whittington, H. B. and W. D. I. Rolfe, Eds. 1963. *Phylogeny and Evolution of Crustacea*. Cambridge Museum of Comparative Zoology, Cambridge, Mass.

Wiens, H. J. 1962. *Atoll Environment and Ecology*. Yale University Press, New Haven, Conn.

Wilmoth, J. H. 1967. *Biology of Invertebrata*. Prentice-Hall, Englewood Cliffs, New Jersey.

Winn, H. E. and B. L. Olla, Eds. 1972. *Behavior of Marine Animals*. Vol. I, *Invertebrates*. Plenum, New York.

Yamaguti, S. 1958. *Systema Helminthum*: Vol. I. *The Digenetic Trematodes of Vertebrates*. John Wiley and Sons, Inc., Interscience Publishers, New York.

—. *Systema Helminthum*: Vol. II. *The Cestodes of Vertebrates*. John Wiley and Sons, Interscience Publishers, New York.

—. 1962. *Systema Helminthum*: Vol. III. *The Hematodes of Vertebrates*. John Wiley and Sons, Inc., Interscience Publishers, New York.

—. 1963. *Systema Helminthum*: Vol. IV. *Monogena and Aspidocotylea*. John Wiley and Sons, Inc., Interscience Publishers, New York.

—. 1963. *Systema Helminthum*: Vol. V. *Acanthocephala*. John Wiley and Sons, Inc., Interscience Publishers, New York.

German Books and Monographs

(Some have been published in English.)

VOLUMES I–XIII

A. Encyclopedias

(1) *Knaurs Tierreich in Farben.* Beginning 1956. 7 Vols. Droemersche Verlagsanstalt, Munich/Zurich.
(2) *Urania Tierreich.* Beginning 1967. 6 Vols. Urania Verlag, Leipzig/Jena/Berlin; Harri Deutsch, Frankfurt M./Zürich.
(3) Zur Strassen, O., Ed.: *Brehms Tierleben.* 1930. 4th ed. 13 Vols. Bibliographisches Institut, Leipzig.
(4) Müller, A. H. 1963–1970. *Lehrbuch der Paläozoologie.* 3 Vols. in 7 Parts, VEB Fischer, Jena.

B. General Reference Works

(1) *ABC Biologie.* Harri Deutsch, Frankfurt M.
(2) Ax, P. 1960. *Die Entdeckung neuer Organisationstypen im Tierreich.* NB 258.
(3) Geiler, H., W. Hennig, and H. Dathe. *Taschenbuch der Zoologie.* 4 Vols. VEB Thieme, Leipzig.
(4) Kaestner, A. Beginning 1967. *Lehrbuch der Speziellen Zoologie.* Fischer, Jena/Stuttgart.
(5) Kühn, A. 1964. *Grundriss der allgemeinen Zoologie.* 15th ed. Thieme, Stuttgart.
(6) Rensch, B., and G. Dücker. 1963. *Biologie II (Zoologie).* Fischer-Lexikon, Fischer, Frankfurt M.
(7) *Das Tierreich.* Beginning 1953. 7 Vols. in 14 Parts. SG.
(8) Vogel, G., and H. Angermann, 1968. *dtv-Atlas zur Biologie.* 2 Vols. 2nd ed.
(9) Wurmbach, H. 1968, 1970. *Lehrbuch der Zoologie.* 2 Vols. 2nd ed. Fischer, Stuttgart.

C. Books on Specific Habitats

(1) De Haas, W., and F. Knorr. 1965. *Was lebt im Meer?* Franckh, Stuttgart.
(2) Brauer, A., Ed. *Die Süsswasserfauna Deutschlands.* Fischer, Jena.
(3) Brohmer, P. *Deutschlands Pflanzen- und Tierwelt.* 7 Vols. Quelle und Meyer, Heidelberg.
(4) Engelhardt, W. 1962. *Was lebt im Tümpel, Bach und Weiher?* Franckh Stuttgart.
(5) Grimpe, G., and E. Wagler, Eds. 1925–1955. *Die Tierwelt der Nord- und Ostsee.* Akademische Verlagsgesellschaft, Leipzig.
(6) Luther, W., and K. Fiedler. 1967. *Die Unterwasserfauna der Meeresküsten.* Parey, Hamburg/Berlin.
(7) Riedl, R. 1963. *Fauna und Flora der Adria.* Parey, Hamburg/Berlin.
(8) Wesenberg-Lund, C. 1939. *Biologie der Süsswassertiere.* Springer, Berlin/Vienna.

D. Parasites

(1) Brumpt, E., M. Neveu-Lemaire, and A. Ehrhardt. *Praktischer Leitfaden der Parasitologie des Menschen.* Springer, Berlin/Göttingen/Heidelberg.
(2) Buchner, O. 1960. *Tiere als Mikrobenzüchter.* VW 75. Springer, Berlin/Heidelberg.
(3) Kotlan, A. 1960. *Helminthologie.* UAW.
(4) Martini, E. *Wege der Seuchen.* Enke, Stuttgart (3rd ed.).
(5) Osche, G. 1966. *Die Welt der Parasiten.* VW 87. Springer, Berlin/Heidelberg.
(6) Piekarski, G. 1954. *Lehrbuch der Parasitologie.* Springer, Berlin/Heidelberg.

E. Handbooks

(1) Brohmer, P. 1969. *Fauna von Deutschland.* Quelle und Meyer, Heidelberg.
(2) —, P. Ehrmann, and G. Ulmer, Eds. 1929–1936. *Die Tierwelt Mitteleuropas.* Quelle und Meyer, Leipzig.

(3) Dahl, F. u. D., and P. Peus. Eds. *Die Tierwelt Deutschlands und der angrenzenden Meeresteile.* VEB Fischer, Jena.
(4) Döberlein, L. 1955. *Bestimmungsbuch* (3 Vols. with Insects, Mollusks and Vertebrates). Oldenbourg, Munich/Berlin.
(5) Stresemann, E. 1955, 1967, 1969. *Exkursionsfauna.* 3 Vols. in 4 Parts. Volk und Wissen, Berlin.

F. Paleontology

(1) Kuhn, O. 1949. *Lehrbuch der Paläozoologie.* Schweizerbarth, Stuttgart.
(2) Müller, A. H. 1963–1970. *Lehrbuch der Paläozoologie.* 3 Vols. in 7 Parts. VEB Fischer, Jena.
(3) —, and H. Zimmermann. *Aus Jahrmillionen, Tiere der Vorzeit.* VEB Fischer, Jena.
(4) Thenius, E. 1970. *Paläontologie.* Franckh, Stuttgart.
(5) —. 1963. *Versteinerte Urkunden.* Springer, Berlin/Vienna.

G. Behavior

(1) Baeumer, E. 1964. *Das Dumme Huhn. Verhalten des Haushuhns.* Kosmos-Bändchen 242. Franckh, Stuttgart.
(2) V. Buddenbrock, W. 1956. *Wie orientieren sich die Tiere?* Kosmos-Bändchen 212. Franckh, Stuttgart.
(3) Dethier, V. G., and E. Stellar. 1964. *Das Verhalten der Tiere.* Franckh, Stuttgart.
(4) Eibl-Eiblesfedlt. I. 1969. *Grundriss der vergleichenden Verhaltensforschung.* Piper, Munich.
(5) Koehler, O. 1968. *Die Aufgabe der Tierpsychologie.* Wissenschaftliche Buchgesellschaft, Darmstadt.
(6) Köhler, W. 1921. *Intelligenzprüfungen an Menschenaffen.* Springer, Berlin.
(7) Lorenz, K. 1963. *Das sogenannte Böse.* Borotha-Schöler, Vienna.
(8) —. 1949. *Er redete mit dem Vieh, den Vögeln und den Fischen.* Borotha-Schöler, Vienna.
(9) —. 1950. *So kam der Mensch auf den Hund.* Borotha-Schöler, Vienna.
(10) —. 1965. *Über tierisches und menschliches Verhalten.* Piper, Munich.
(11) Remane, A. 1960. *Das soziale Leben der Tiere.* Rowohlts deutsche Enzyklopädie, Vol. 97.
(12) Tinbergen, N. 1956. *Instinktlehre.* 2nd ed. Parey, Berlin.
(13) —. 1967. *Tiere untereinander.* 2nd ed. Parey, Berlin.

H. External Anatomy

(1) Portmann, A. *Die Tiergestalt.* Herder-Bücherei 240.
(2) Wickler, W. 1971. *Mimikry.* 2nd ed. Kindler, Munich.

I. Physiology

(1) Hanke, W. 1969. *Hormone.* SG 1141/1141a.
(2) Schmidt-Nielsen, K. 1965. *Physiologie der Tiere.* Franckh, Stuttgart.
(3) Wells, M. 1968. *Wunder primitiven Lebens.* KUB.

J. The Cell

(1) Loewy, A. G., and P. Siekevitz. 1967. *Die Zelle.* BLV.
(2) McElroy, W. 1964. *Biochemie und Physiologie der Zelle.* Franckh, Stuttgart.
(3) Swanson, C. P. 1964. *Die Zelle.* Franckh, Stuttgart.

K. Ontogeny

(1) Ebert, J. D. 1967. *Entwicklungsphysiologie.* BLV.
(2) Pflugfelder, O. 1969. *Lehrbuch der Entwicklungsgeschichte und Entwicklungsphysiologie der Tiere.* 2nd ed. VEB Fischer, Jena.
(3) Seidel, H. 1953. *Entwicklungsphysiologie der Tiere.* SG 1162/1163.

L. Genetics

(1) Botsch, W. 1969. *Morsealphabet des Lebens.* Kosmos-Bändchen 245. Franckh, Stuttgart.
(2) Beadle, G. and M. 1969. *Sprache des Lebens.* Eine Einführung in die Genetik. Fischer, Frankfurt M.

(3) Kühn, A. 1965. *Grundriss der Vererbungslehre*. 4th ed. Quelle und Meyer, Heidelberg.
(4) Levine, R. P. 1966. *Genetik*. BL.

M. Evolution

(1) De Beer, G. 1966. *Atlas der Evolution*. BLV.
(2) Dobzhansky, Th. 1962. *Die Entwicklung des Menschen*. Parey, Berlin.
(3) Heberer, G. 1949. *Allgemeine Abstammungslehre*. Musterschmidt, Göttingen.
(4) Lange, E. 1971. *Mechanismen der Evolution*. NB 433.
(5) Mertens, R. 1947. *Die "Typostrophen-Lehre" im Lichte des Darwinismus*. Kramer, Frankfurt M.
(6) Rensch, B. 1954. *Neuere Probleme der Abstammungslehre*. 2nd ed. Enke, Stuttgart.
(7) Savage, J. M. 1966. *Evolution*, BL.
(8) Schindewolf, O. H. 1947. *Fragen der Abstammungslehre*. Kramer, Frankfurt M.
(9) Simpson, G. G. 1951. *Zeitmasse und Ablaufformen der Evolution*. German translation by G. Heberer. Musterschmidt, Göttingen.

N. Zoogeography

(1) De Latin, G. 1967. *Grundriss der Zoogeographie*. VEB Fischer, Jena.
(2) Schilder, F. A. 1956. *Lehrbuch der allgemeinen Zoogeographie*. VEB Fischer, Jena.

O. Philosophy

(1) Hartmann, M. 1948. *Die philosophischen Grundlagen der Naturwissenschaften*. Fischer, Jena.
(2) Rensch, B. 1968. *Biophilosophie*. Fischer, Stuttgart.

P. Zoos and Nature Reserves

(1) Engelhardt, W. 1968. *Die letzten Oasen der Tierwelt*. 6th ed. Umschau-Verlag, Frankfurt M.; Pinguin-Verlag, Innsbruck.
(2) Kirchshofer, R. 1966. *Zoologische Gärten der Welt. Die Welt des Zoo*. Umschau-Verlag, Frankfurt M.; Pinguin-Verlag, Innsbruck.

Q. Animal photography

(1) Bode, F. *Mikrophotographie für Jedermann*. Franckh, Stuttgart.
(2) Olberg, G. 1960. *Wissenschaftliche Tierphotographie*. VEB Deutscher Verlag der Wissenschaften, Berlin.
(3) Paysan, K. 1963. *Naturfotografie für Jedermann*. Franckh, Stuttgart.

R. Magazines

(1) *Das Tier*. Internationale Tierillustrierte. Hallwag, Bern and Stuttgart.
(2) *Kosmos*. Bild unserer Welt. Franckh, Stuttgart.
(3) *Mikrokosmos*. Zeitschrift für angewandte Mikroskopie, Mikrobiologie, Mikrochemie und mikroskopische Technik. Franckh, Stuttgart.

VOLUME I

Protozoa

Doflein, F. and E. Reichenow. 1953. *Lehrbuch der Protozoenkunde*. 6th ed. VEB Fischer, Jena.
Matthes, D. and F. Wenzel. 1966. *Wimpertiere (Ciliata)*. Franckh, Stuttgart.
Mayer, M. 1966. *Kultur und Präparation der Protozoen*. Franckh, Stuttgart.
Göke, G. 1963. *Meeresprotozoen*. Franckh, Stuttgart.
Grell, K. G. 1956. *Protozoologie*. 1956. 2nd ed. Springer, Berlin/Heidelberg/New York.
Grospietsch, T. 1965. *Wechseltierchen (Rhizopoden)*. Franckh, Stuttgart.
Schönborn, W. 1966. *Beschalte Amöben (Testacea)*. NB 357.
Other References: D 1, D 6, E 3, and F 2.

Sponges

Other References: C 1, C 3, C 5, C 6, C 7, C 8, E 3, and F 2.

Coelenterates

Wenzel, B. 1958. *Glastiere des Meeres—Rippenquallen*. NB 213.
Other References: C 1, C 3, C 5, C 6, C 7, C 8, E 3, and F 2.

Flatworms

Henke, G. 1962. *Die Strudelwürmer des Süsswassers*. NB 299.
Löliger-Müller, B. 1957. *Parasitische Würmer*. NB 192.
Other References: C 2, C 8, D 1, D 3, and D 6.

Rotifers

Donner, J. 1956. *Rädertiere (Rotatorien)* Franckh, Stuttgart.
—. 1965. *Ordnung Bdelloidea (Rotatoria, Rädertiere)*. Akademie-Verlag, Berlin.
Voigt, M. 1957. *Rotatoria. Die Rädertiere Mitteleuropas*. Bornträger, Berlin.
Wulfert, K. 1969. *Rädertiere*. NB 416.

Nematodes (Threadworms)

Decker, H. 1969. *Phytonematologie*. VEB Deutscher Landwirtschaftsverlag, Berlin.
Goffart, H. 1951. *Nematoden der Kulturpflanzen Europas*. Parey, Berlin.
Kämpfe, L. 1952. *Rüben- und Kartoffelälchen*. NB 80.
Meyl, A. H. 1961. *Fadenwürmer (Nematoden)*. Franckh, Stuttgart.
Müller, B. 1957. *Parasitische Würmer*. NB 192.
Other References: D 1, D 3, D 6, and E 3.

Annelids

Herter, K. 1967. *Der Medizinische Blutegel und seine Verwandten*. NB 381.
Darwin, C. 1882. *Die Bildung der Ackererde durch die Thätigkeit der Würmer mit Beobachtung über deren Lebensweise*. C. Darwin's collected works translated by von J. V. Carus, Vol. XIV/1. Schweizerbarth, Stuttgart.
Other References: C 1, C 2, C 3, C 5, C 6, C 7, C 8, and E 3.

Arachnids

Crome, W. 1951. *Die Wasserspinne*. NB 44.
—. 1956. *Taranteln, Skorpione und Schwarze Witwen*. NB 167.
Hirschmann, W. 1966. *Milben*. Franckh, Stuttgart.
Karg, W. 1962. *Räuberische Milben im Boden*. NB 296.
Müller, E. W. 1960. *Milben an Kulturpflanzen*. NB 270.
Pötzsch, J. 1963. *Von der Brutpflege heimischer Spinnen*. NB 324.
Weygold, P. 1966. *Moos- und Bücherskorpione*. NB 356.
Wiehle, H. 1949. *Vom Fanggewebe einheimischer Spinnen*. NB 12.
—. 1964. *Aus dem Spinnenleben Wärmerer Länder*. NB 138.
Other References: C 3 and E 3.

Crustaceans

Herbst, H. V. 1962. *Blattfusskrebse (Phyllopoden: Echte Blattfüsser und Wasserflöhe)*. Franckh, Stuttgart.
Kiefer, F. F. 1960. *Ruderfusskrebse (Copepoden)*. Franckh, Stuttgart.
Lüling, K. H. 1953. *Schmarotzende Ruderfusskrebse*. NB 99.
Müller, H. *Die Flusskrebse*. NB 121.
Panning, A. 1952. *Die Chinesische Wollhandkrabbe*. NB 70.
Vollmer, C. 1952. *Wasserflöhe*. NB 45.
—. 1952. *Diemenfuss, Hüpferling und Muschelkrebs*. NB 57.
Other References: C 1, C 2, C 3, C 5, C 6, C 7, C 8, E 3, and F 2.

Tracheates

Dobroruka, L. J. 1961. *Hundertfüsser*. NB 281.
Seifert, G. 1961. *Die Tausendfüsser*. NB 273.
Other References: E 3.

Picture Credits

Artists: S. Bousani-Baur (pp. 183, 184, 271). K. Grossmann (pp. 87, 173, 174, 277, 278, 315, 316, 351, 352, 382, 393, 394, 399, 400, 415, 416, 437, 438, 443, 444, 459, 470, 509, 510). S. Milla (pp. 88, 93, 94, 113, 114, 283, 284, 289, 290, 305/306, 381). K. Wulfert (pp. 325 and 326).

Scientific consultants of the artists: H. -R. Haefelfinger (Bousani-Baur). W. Hohorst (Milla). P. Rietschel (Grossmann, Milla).

Color photos: Ax (pp. 148 above middle left, 198 above middle, 296 below left, 302 middle left, below left and below right, 373 above middle right, 376 below left, 426 below middle, 452 above left, 476 below left and below middle, 496 middle above middle and above right). Böck (pp. 296 below right, 410 middle above middle, 496 above middle right). Burton/Photo Researchers (pp. 410 below middle right, 426 middle right, 453 above right, 493 below right, 496 above left and below middle left). Chaumeton/Jacana (p. 451 below left). Collignon (p. 264 below). Cropp (pp. 261 and 262 above left). Danesch (pp. 408 below right and 476 middle left). Dossenbach (p. 410 below right). Eder (p. 376 above right). Fotosub (pp. 263 above right, 302 above left, 373 below, 451 below middle right), Frickhinger (pp. 107 lower middle left, lower left, lower middle right and lower right, 108 below left and middle right, 238 above left, 374 below, 451 above left). Grave/Jacana (pp. 107 above left and above middle right and 108 above left and middle left). Gruhl (pp. 47, 238 below left, 242 below left). Haefelfinger (pp. 148 below left, above right, upper middle right and below right, 198 middle right and lower left, 202 upper right, 204 upper left, 238 middle left, 262 below left, lower middle left, lower middle right and below right, 263 below middle and below right, 295 above right, 296 above right and below middle, 301 below left and below right, 302 middle right, 373 above middle left, below middle left, 451 below middle left, 473 above middle and middle from left to right, 476 below right, 496 above middle left and above middle). Hansen (pp. 201, 408 upper left, middle left and middle right, 425 below). Hoppe (p. 198 lower middle right). Interholc (pp. 451 lower right, 473 upper left, 493 upper left, 494, 495). Jatzke (pp. 199, 200 lower left). Kilian (p. 148 upper left, lower middle left and lower middle right). Knorr (pp. 46, 48 top and bottom, 145, 149 upper right, 197, 262/263 top). Kopp (pp. 146, 149 lower right, 152, 202 lower left, 204 lower left and lower right, 237, 244, 272, 375 bottom, 493 middle left). Köster (pp. 202 upper left, 204 upper right, 243, 376 upper left). Lanceau/Jacana (p. 496 lower right). Lauckner (pp. 239 and 240 lower left). Lautenschlager (p. 302 upper right). Marcuse (p. 451 middle lower middle. Moosleitner (pp. 147 and 373 top). Okapia (p. 493 upper right).

Paysan (p. 451 lower middle). Pflteschinger (pp. 410 upper middle left, lower middle left, upper middle, middle lower middle and upper middle right, 426 upper middle left). Photo Researchers (pp. 408 upper right, 476 middle, 496 middle lower middle). Quedens (pp. 475 and 493 middle right). P. Rietschel (pp. 198 lower right and 295 middle). S. Rietschel (p. 264 top). Rozendaal (pp. 45, 149 upper left, 198 middle left, 240 upper middle left, 241, 242 upper left and middle left, 376 upper middle right, 454, 493 lower left). Sauer (pp. 107 upper middle left and upper right, 108 upper right and middle right, 295 lower left, 376 lower middle left, lower middle right and lower right, 407 top, 410 lower left, 426 upper left, lower middle left and upper right, 496 lower left). Schrempp (pp. 410 top left, lower middle and top right and 496 lower middle). H. Schumacher (pp. 240 top left and 426 lower left). Siedel (pp. 108 lower right, 198 upper left, 451 upper middle left, middle upper middle, upper right and upper middle right, 476 middle right). Sillner (p. 149 lower middle left and lower right). Six (pp. 296 middle, 301 middle right, 426 lower right, 473 bottom). Submarine (pp. 200 lower left, upper right and lower right, 263 lower left). Summ/Jacana (p. 451 upper middle). Taylor (pp. 149 upper middle left, upper middle right and middle right, 240 lower middle left, 295 lower right, 296 upper left). Thare/Bavaria (p. 203). Tomsick (p. 374 top). V-Dia Verlag (pp. 198 upper right, 295 upper left, 496 lower middle right). Vasserot/Jacana (p. 473 upper right). Visage/Jacana (p. 408 lower left). Weber (p. 426 upper middle). Weischer (p. 301 upper left, middle left and upper right). Wickler (p. 474). ZFA (pp. 150/151, 202 lower right. 407 bottom, 409, 425 top, 452/453 bottom, 476 top).

Photo layout: J. Kühn.

Line drawings: Donner (pp. 331, 332, 333 bottom, 337). Kilian (pp. 140–166). Rietschel (pp. 138, 139, 275, 299, 307, 309, 323, 354, 355, 357, 358, 359, 360, 362, 364, 366, 368, 370, 371, 379, 384, 385, 434, 440, 441, 446, 449, 466, 467, 470, 480, 501, 502, 503, 505). Line drawings after authors' sketches: Althuber (pp. 298 bottom and 304). Ax (pp. 310 and 312). Bousani-Baur (pp. 176–259). Diller (pp. 27, 28, 38, 39, 40, 43, 60, 97, 98, 99, 101, 102, 103, 104, 110, 116, 118, 123, 124, 125, 130, 132, 133, 168, 169, 170, 171, 281, 287, 292, 293, 298 top, 318, 319, 320, 328 top, 334, 335, 336, 338, 339, 340, 342, 348, 388, 390, 391, 396, 397, 402, 403, 412, 413, 418, 419, 420, 421 top, 423, 428, 430, 432, 512, 514). Kacher (pp. 67–69). Steffel (pp. 41, 42, 286, 311, 328 bottom, 329, 330, 333 top). Page 421 bottom is reproduced with the kind permission of "Natur und Museum". Distribution maps: Diller (p. 54) and Steffel (all others).

Index

Abylopsis, 193★
Acantharia, 94★, 112
Acanthaster planci ("Crown of Thorns" starfish), 278
Acantherpestes, 506
Acanthobdella peledina, **383**
Acanthobdellae, 380, **383**
Acanthocephala (Spiny-headed worms), 316★, 324, **354f**, 356f
Acanthocephalus lucii, 355★
Acantholoncha flavosa, 94★
Acanthoptilum, **254**
Acari (Mites), 394★, 411, **478**ff
Acaridae (Acarids), **430**
Acarina, see Acari
Acarus siro (Meal mite), 428, **430**, 437★
Achtheres percarum (Perch shrimp), 443★, **462**
Acidaspis, 402
Acnidaria, 176
Acoela, 280
Acontiaria, 221
Acropora (Stag's horn coral), 226, **232**
Acroporidae (Treelike corals), 232
Acrothoracica, 434, **466**
Actinaria (Sea anemones), 220f, **255f**, 277★
Actinia, 221f
— *equina* (Sea anemone), 50, **224**, 277★
Actinomyxidia, 132, **133**
Actinophrys sol, **112**, 305/306★
Actinosphaerium eichhorni, **112**, 305/306★
Actinozoa, 180
Actinula larva, 177★, 182, 221
Adamsia palliata, 223
Aelosoma hemprichi, **371**
Aelosomatidae, 371
Aeolidia papillosa (Sea slug), 221
Aequorea forskalea, 206
African scorpion, see Giant African scorpion
Agalma elegans, **194**
Agelenidae (Funnel spiders), 421f
Aglantha digitalis, **207**
Aglaophenia pluma, **206**
Aglaspis, 405
Aglaura hemistoma, 181★
Agnostida, 402
Agnostus, **401**
Agroeca brunnea ("Fairy lantern"), 410★
Aiptasia, 221★
— *mutabilis*, 221, 271★
Albertia, 332
Albunea, 485
Alciopidae, 365f
Alcyonaria (Alcyonarian corals), 178, 235
Alcyonium brioniense, **245**
— *digitatum* (Dead man's hand or finger), **235f**, 240★, 245, 271★
— *palmatum*, 45★, **245**
Aldrovandi, Ulisses, 245
Algura hemistoma, 184★
Allolobophora chlorotica, **378**
— *rosea*, **378**
Alpheidae (Snapping shrimps), 485
Alpheus, **485**
Amblypygi, 413f
American brine shrimp (*Artemia gracilis*), 440
American crayfish (*Orconectes limosus*), 479, 481, **488**
American giant liver fluke (*Fasciolopsis magna*), 293

American lobster (*Homarus americanus*), 487
American stream planarian (*Dugesia dorotocephala*), **286f**
Ammophila campestris (Digger wasp), 57
Amoeba polypodia, 104★
— *proteus*, 104★, 107★, 305/306★
Amoebae, 87★
Amoebic dysentery, 105m, **106**, 109
Amoebina (Shell-less amoebae), 104ff
Amphicora, see *Fabricia*
Amphictenidae, see Pectinariidae
Amphidiscophora, 160
Amphinomorpha, 363
Amphipoda (Waterfleas), 500
Amphiporus exilis, 318★
Anaspidacea, 483
Anaspides, 434, **483**
Anclyostoma duodenale, 343f
Androctonus australis (Sahara scorpion), **412**
Anemone fish, 223
Anemones (Anthozoa), **177**, **219**ff
Anemonia, 222
— *sulcata*, **224**, 271★
Anomopoda, 446
Anomura, 484, **488**
Anoplocephalidae, 307
Anostraca, 435f, 436★
Ant isopod (*Platyarthrus hoffmannseggii*), **504**
Ant spider (*Myrmarachne formicaria*), 415★, **422**
Anthipatheria, 220, **232f**
Anthipathes subpinnata, **233**
Anthipathidae (Black corals), 233
Anthomastus grandiflorus, **245**
Anthozoa (Anemones), **177**, **219**ff
Anthracoiulus, 506
Apfelbeckia, **507**
Aphelenchoides (Leaf nematodes), 340
Aphrodite aculeata (Sea mouse), **363**, 373★
Aphroditidae (Sea caterpillars), 363
Aplysilla rosea, **166**
Aplysina aerophoba, see *Verongia aerophoba*
Aplysinidae, 166
Aponomma gervaisi, 426★
Arachnida (Spider-like animals), 404, **411f**, 416★
Araneae (True spiders or web spiders), 394★, 399★, 400★, 411, **414**, 415★
Araneidae (Orb spiders), 419★, **421**
Araneus
— *ceropegius* (Oak-leaf spider), 410★
— *diadematus* (Garden spider), 399,★ 408★, 409★, **421**
— *quadratus*, 410★
Aratus pisoni (Mangrove crab), **497**
Arcella, **110**
— *vulgaris*, 305/306★
Archaeocyatha, 139, **168**
Archaeocyathus, 169★
Archiacanthocephala, 355
Archiannelida, 363, **369**
Archipolypoda, 506
Archoophora, **281f**
Arctic lion's mane (*Cyanea arctica*), 215
Arenicola marina (Lugworm), **367**, 367★

Argas reflexus (Pigeon mite), 394★, **431**, 437★
Argasidae, 431
Argiope
— *bruennichi* (Wasp spider), 399★, 408★, **421**
— *lobata* (Zebra spider), 408★
Argulus foliaceus (Carp louse), 443★, **463**
— *scutiformis*, **463**
Argyroneta aquatica (Water spider), 410★
Aristeomorpha foliacea (Giant blood-red shrimp), 460★
Aristotle, 82, 435, 490
Armadillidium vulgare (Pill woodlouse), 496★, **503**
Armillifer, **391**
— *armillatus*, 382★
Arrow worms (Chaetognatha), 85
Artemia
— *gracilis* (American brine shrimp), 440
— *salina* (Brine shrimp), **436f**
Arthrolycosa (True spiders), 406
Arthropleura, 506
Arthropoda (Arthropods), 82, 85, 361, **397**ff
Articulata, 85, 360ff
Asaphus, 402
Ascaridia galli, 301★
Ascaris lumbricoides (Maw-worm), 333, 335, 337, **344**
Asc-helminthes (Roundworms), 85, 316★, **323ff**
Ascomorpha ecaudis, 325★
Ascomorphella volvocicola, 325★
Ascothoracica, 434, **463f**
Asellidae (Fresh-water isopods), 502
Asellota, 502
Asellus aquaticus (Common fresh-water isopod), **502**
— *cavaticus* (Cave isopod), **502**
Aspidiophorus, 323★
Asplanchna, 328, 330f
— *priodonta*, 330★
Asplanchnopus muticeps, 325★
Astacidae (Crayfish), 487
Astacura (True macrurans), 484, **487**
Astacus astacus (European crayfish), **487f**
Astomata, **128**
Astrangia danae, 232
Astroides calycularis (Star coral), 232, 238★, 271★, 277★
Athecata-Anthomedusae, 182, **187f**
Atlantic horseshoe crab (*Limulus polyphemus*), 393★, **411**, 412★
Atlantic Palolo worm (*Eunice fucata*), **364**
Atolla, **214**
Attulus saltator (Jumping spider), 420★
Aulophorus, **372**
Aurelia, 210
— *aurita* (Common jellyfish), 183★, 210, 216f
Austalopithecines, 25
Austroastacidae, 487
Austrognathai, **311**
Aviculariidae (True bird-eating spiders), 420
Axinella cannabina, 145★, 234
Aysheaia, 387

Babesia bigemina, **137**
— *canis*, **137**
Babesidae, 136

Bag jellyfish, see Deep-sea jellyfish
Balanomorpha, 434, **465**
Balanophyllia, 230, 231★
— *elegans*, **231**
— *italica*, **231**
— *regia* (Star coral), **231**
Balantidium coli, **130**, 130★
Balanus (Encrusting barnacles), 465
— *balanoides* (Common barnacle), 465, 465★, 466★
— *balanus*, 459★, 465
— *eburneus* (Ivory barnacle), 465
— *perforatus* (Ridged barnacle), 465
Baltic shrimp (*Palaemon sequilla*), 486
Barbulanympha, 102
Barnacles (Balanomorpha), 435, 465
Barnes, H. J., 213
Basket sponges (Cornacuspongida), 139, 143, 163f
Bath sponges (Spongidae), **164**, 165★, 165m
Bathynella, 434, 465
Bathynellacea, 482f
Bdellocephala punctata, 280★
Bdelloidea (Leechlike rotifers), 331
Bdellomorpha, 318
Bdelloura, 286
Beach, F. A., 79
Beach flea (*Orchestia gammarellus*), **500**
Beach woodlouse (Ligia oceanica), 496★, **503**
Beaked pantopod (*Colossendeis proboscidea*), 432, 437★
"Beaker jellyfish" (Stauromedusae), 211f, 211★
Beebe, William, 491f
Beef tapeworm (*Taeniarhynchus saginatus*), 53, 278★, 299f, 299★
Beet eelworm (*Heterodera schachtii*), 341
Behavior, 56ff
Beklemischev, 281
Belinurus, 405
Beroe (Melon jellyfish), **258**
— *cucumis* (Sea mitre), **268**
— *ovata*, **268**, 271★
Beroidea (Melon comb jellyfish), 267f
Beyrichia, 434
Beyrichiida, 434
Biacetabulum sieboldii, **309**
Biantennals or Crustacea (Diantennata), 401, **433ff**, 433★
Biatopes, 51
Bilateralia (Bilaterally symmetrical animals), 269f, 275
Bilharz, Theodor, 297
Bipalium kewense (Greenhouse planarian), **287**
Bird-eating spiders (Orthognatha), 420, 421★
Birgus (Coconut crab or robber crab), 472
— *latro*, **489**, 493★
Birula, 424
Black corals (Anthipathidae), 233
"Black water fever", 100
Black widow (*Latrodectus mactans*), 421
Bladder blood fluke (*Schistosoma haematobium*), 297
Bladder fluke (*Polystoma integerrimum*), **291**
Bladder tapeworm, see Dog tapeworm
Blood flukes (*Schistosomidae*), 297

Heavy type indicates the main entry, an asterisk ★ indicates an illustration, and m indicates a distribution map.

Blue corals (*Helioporidae*), 246, **251**, 251*

Blue crabs (*Callinectes*), **490**

Blue lion's mane or cornflower jellyfish (*Cyanea lamarckii*), 183*, **216**

Blue stentor (*Stentor coeruleus*), **130**, 305/306*

Bolina hydatina, **265**

Bolinopsis infundibulum, **260**, 263*

Bolivina alata, 93*

Bonellia viridis (Green bonellia), 302*, 351*, **358f**

Book scorpion (*Chelifer cancroides*), 416*, **423**

Boot-lace worms or strapworms (Nemertini), 85, **313ff**, 315*

Boring sponges (Clionidae), **162f**, 163*

Bosmina coregoni, **447**
— *longirostris*, 444*, **447**

Bosminidae (Bosminids), 447

Brachydesmus superus, 496*

Brachionus, 332
— *calycifloris*, 305/306*, 330
— *quadridentatus*, 326*
— *rubens*, 327*

Brachyura (True crabs), 434f, 469, 484, **489**

Bradurina, 434

Brain coral (*Diploria*), **232**

"Brain worm", see *dicrocoelium dendriticum*

Branchiobdella, **377**

Branchiobdellidae (Branchiobdellids), 377

Branchiocerianthus imperator, **188**

Branchipus stagnalis, **436**, 441, 451*

Branchiura (Fish lice), 463

Branchiura sowerbyi, **377**

Bread crust sponge (*Halichondria panicea*), **166**

Brehm, Alfred Edmund, 19, 22, 162

Brine shrimp (*Artemia salina*), **436f**

Bristowe, 420

Brown hydra (*Hydra vulgaris*), **187**

Bryozoa, 82, 166, 312

Bulimina inflata, 93*

Bunodactis verrucosa, 201*, **224**, 271*

Bunodopsis, **223**

Bursaria truncatella, 108*

Buthus, 407*

Bythotrephes longimanus, 444*, **448**, 456

Calanoida (Gymnoplea), 443*, **455**

Calanus finmarchieus, **456**

Calappa, 435, **490**

Calcarea (Calcareous sponges), 139, 157f

Calcareous sponges (Calcerea), 139, 157f

Calceola, **177**

Calciphilus, 506

Caligoida, 461

Caligus lacustris, **461**
— *rapax*, **461**

Calliactis parasitica (Hermit crab anemone), 202*, 222, 493*

Callianira bialata, **259**, 271*

Callinectes (Blue crabs), **490**

Calocalanus pavo, 443*

Calocyclas monumentum, 94*

Calonympha grassii, 88*

Calonymphidae, 102

Calycophorae, 193, 195

Calyx nicaeensis, 149*

Cambrensis, Giraldus, 465

Cameron, 274

Campanopsis gegenbauri, 207

Campanulina, **206**, 206*

Campanulinidae, 206

Cancer pagurus (Rock crab), 435,

473*, **490**

Candona candida, 449*

Canthocamptidae, 457

Canthocamptus, 443*, **457**

Cape spiny lobster (*Jasus lalendei*), **486**

Caprella linearis, **501**

Caprellidae (Ghost shrimps), 501

Carabus, 44

Carchesium, 108*, **129**

Carcinoscorpinus, 411

Carcinus maenas (Shore crab), 467, 477, **490**

Cardisoma guanhumi (Caribbean land crab), **498**

Caribbean land crab (*Cardisoma guanhumi*), **498**

Carp louse (*Argulus foliaceus*), 443*, 463

Caryophylla
— *clavus*, **230**, 271*
— *smithii*, **230**

Caryophyllaeus laticeps, **309**

Cassiopeia
— *andromeda*, 198*
— *xamachana*, **218**

Castrodes parasiticum, see *Eulampetia pancerina*

Cat liver fluke (*Opisthorchis felineus*), **294**

Cat tapeworm (*Hydatigena taeniaformis*), **307**

Catenula lemnae, 282, **285**

Catenulida, 285

Cave isopod (*Asellus cavaticus*), **502**

Cell, animal, 27f, 27*, 28*

Centipedes (Chilopoda), 512

Centropyxis aculeata, **110**

Cepedea, **103**

Cephalobaena tetrapoda, 391*, **392**

Cephalobaenida, 391

Cephalocarida, 435

Cephalodella
— *forficula*, 305/306*
— *gibba*, **332**

Ceratarges armatus, 382*

Ceratiocarina, 434

Ceratiocaris, 434

Ceratium hirundinella, 88*, **97**

Ceratocoris horrida, 88*

Cerebratulus marginatus, 302*

Cereus pedunculatus, 202*, **224**

Cerianthia, 220, **233f**

Cerianthopsis americanus (sand anemone), **234**

Cerianthus lloydii, **234**
— *membranaceus*, 203*, **234**, 271*
— — *fuscus*, **234**
— — *violaceus*, **234**

Cestidea (Venus's girdles), 259f, **265**

Cestoda (Tapeworms), 278*, **298f**

Cestus veneris (Venus's girdle), 265f, 271*

Ceylon leech (*Haemadipsa zeylanica*), **385**

Chaetogaster limnaei, **371**

Chaetognatha (Arrow worms), 85

Chaetonotoidea, 327

Chaetonotus maximus (Gastrotrichan), 305/306*, 323*

Chaetopterus variopedatus, **367**, 367*

Chalk-tube worm (*Serpula vermicularis*), 376*

Charybdea marsupialis, 212f

Cheese mite (*Tyrophagus casei*), **430**, 437*

Chelicerata (Arachnids), 82, 401, **403ff**, 403*

Chelifer cancroides (Book scorpion), 416*, **423**

Chelonethi, see False scorpions

Chelonibia testudinaria (Tortoise barnacle), **465**

Chelophyes appendiculata, **195**, 271*

Chelypus macronyx, 416*

Cherax, 453*

Cheyletus eruditus, 430*

Chigger or Harvest mite (*Trombicula autumnalis*), **429**, 437*

Chilognatha, 508

Chilomonas paramecium, **97**

Chilopoda (Centipedes), 512

Chinese crab (*Eriocheir sinensis*), 476*, **497**

Chinese liver fluke (*Clonorchis sinensis*), **294**

Chirocephalus, 436*
— *grubei*, **436**, 441

Chironex fleckeri, 183*, **213**

Chiropsalmus quadrigatus (Sea wasp), 211, **213**

Chlorohydra viridissima (Green hydra), **185ff**

Choanoflagellates Craspedomonadidae), 85, **98**

Choanophrya infundibulifera, **132**, 132*

Chondrosia reniformis (Kidney sponge), 161f

Chondrosiidae, **161**

Chonotricha, **131**

Chordata (Back-boned animals), 85

Chromulina rosanoffii, **96**

Chrysaora, **211**
— *hyoscella*, 183*, **214f**

Chyrsarachnion, 89

Chrysomonadina (Goldmonadina), 96f

Chthamalus stellatus (Star barnacle), **465**

Chun, 195, 267

Chydoridae, 447

Chydorus sphaericus, 444*, **447**

Ciliata (Ciliates), 85, 87*, **122**, 125*

Ciocalyptidae, **166**

Cirripathes rumphii, **233**

Cirripedia (Barnacles), 433f, **464**

Cladocera (Waterfleas), **442**

Cladocora cespitosa, 231f, 238*

Cladonema radiatum, **188**, 188*

Clathrina, 154*
— *coriacea*, 149*

Clathrocanium reginae, 94*

Clavularia, **245**

Cliona celata, **163**

Clionidae (Boring sponges), 162f, 163*

Clitellata (Earthworms and leeches), 362, **370**

Clonorchis sinensis (Chinese liver fluke), **294**

Cnidaria (Cnidarians), 85, 176, **178ff**

Cnidosporidia, 132

Coastal pantopods (Pycnogonidae), 432

Cobb, N. A., 333

Coccidia (Coccidians), 115, **118**

Coccidiosis, 118

Coccolithophoridae 96f

Coconut crab, see robber crab

Coconut nematode (*Rhadinaphelenchus cocophilus*), 340f

Coelenterata (Jellyfishes, anemones and corals), 81, 85, 166, 174*, **176ff**, 176*, 277*

Coeloplana, 257*, 277*
— *gonoctena*, 271*

Coenobita (Land hermit crabs), 489

Coleps hirtus, 128, 305/306*

Collossendeidae (Deep-sea pantopods), 432

Collossendeis proboscidea (Beaked pantopod), **432**, 437*

Collotheca campanulata, 325*
— *gracilipes*, 305/306*
— *hoodii*, 325*

— *ornata*, 326*
— — *cornuta*, 332*

Collothecacea, 333

Colpidium colpoda, **128**, 305/306*

Colpoda cucullus, **128**

Columbus or Sargassum crab (*Planes minutus*), **497**

Colurella, 325*

Comb jellyfishes (Ctenophora), 85, **256ff**, 256*, 277*

Comb spiders (Phoneutria), 418

Common barnacle (*Balanus balanoides*), 465, 465*, 466*

Common earthworm (*Lumbricus terrestris*), 361*, **378f**, 378*

Common fish leech (*Piscicola geometra*), 376*, 381*, **384**

Common fresh-water isopod (*Asellus aquaticus*), **502**

Common goose barnacle (*Lepas anatifera*), 459*, **465**

Common jellyfish (*Aurelia aurita*), 183*, 210, 216f

Common land crab (*Gecarcinus ruricola*), **498**

Common liver fluke, see Giant or common liver fluke

Common phalangid (*Phalangium opilio*), 416*, **428**

Common sand flea (*Talitrus saltator*), **500**

Common shrimp (*Crangon crangon*), **485**

Common sponges (Demospongiae), 139, 149*, 152*, 158, **160f**

Common swimming crab (*Portunus holsatus*), **493**

Common tubifex (*Tubifex tubifex*), **372**

Common waterflea (*Rivulogammarus pulex*), **500**

Common woodlouse (*Porcellio scaberi*), **503**, 503*

Compressed comb jellyfish (Platyctenidea), 266

Compressed polydesmid (*Polydesmus complanatus*, 509*, **511**

Conchostraca, 433f, 440*, 441

Conick, L. de, 348

Conjugation, 125ff, 125*

Conochilus unicornis, 326*, 332

Conulata (Conularia), 177

Convoluta roscoffensis, 282, **285**

Copepoda (Copepods), 443*, **455**

Coral moss (*Hydrallmania falcata*), **206**

Coral reef, 46/47*, 48*, 227f, 239*

Corallium (Precious corals), 247ff, 247*
— *abyssorum*, **249**
— *rubrum* (Red precious coral), 242*, **247**

Cordylophora caspia, **189**

Cork sponges (Suberitidae), **162**

Cornacuspongida (Basket sponges), 139, 143, **163f**

Cornflower jellyfish, see Blue lion's mane

Cornularia cornucopeia, **245**, 271*

Cornularidae, 245

Coronata (Deep-sea jellyfish, a.k.a., bag or crown jellyfish), 211, **213f**

Coronulidae (Whale barnacles), 466

Corophiidae, 500

Corophium volutator, **500**

Corycella armata, 116*

Corymorpha nutans, 184*, **188**

Corynactis viridis, 271*

Coryne sarsi, **188**

Corynexochida, 402

Corynidae, 188

Corystes, **485**

Costia necatrix, 98
Cotylorhiza tuberculata, 197★, **218**
Crab spiders (*Thomisidae*), **422**
Crabs, see True crabs
Crane, Jocelyn, 492
Crangon crangon (Common shrimp), **485**
Crangonidae, 485
Craspedacusta sowerby, **190**f
Craspedomonadidae (Choanoflagellates), 85, **98**
Craterolophus tethys, 198★, **212**
Crawfish (Reptantia), 484, **486**
Crayfish (Astacida), **487**
Crenobia alpina, 280★
— — *meridionalis*, **286**
— — *septentrionalis*, **286**
Cribellata, 422
Criodrilidae, 378
Criodrilus lacuum, **378**
"Crithidia" form, epimastigote, **99**, 99★
Crown jellyfish, see deep-sea jellyfish
"Crown-of-Thorns"starfish (*Acanthaster planci*), 208
Crustacea (Crustaceans), 82, **433**ff, **433**★
Cryptocellus simonis, **422**. 422★
Cryptomonadina, 97
Ctenophora (Comb jellyfishes), 85, **256**ff, 256★, 277★
Ctenoplana, **266**, 271★, **277**★
Ctenopoda, 446
Cubomedusae ("Fire jellyfish"), 211, **212**f
Cumacea (Cumaceans), 501
Cunina octonaria, **208**
Cuspidella undulata, 206
Cuvier, George, 82, 360
Cyamidae (Whale lice), 501
Cyanea (Giant jellyfish), 210f, 215
— *arctica* (Arctic lion's mane), **215**
— *capillata* (Lion's mane jellyfish), **210**, 215
— *lamarckii* (Blue lion's mane or cornflower jellyfish), 183★, **216**
Cyaneidae, 215
Cyathophyllum (= *Hexagonaria*), 177
Cyclophyllidea, 303
Cyclopidae (Cyclopids), 457
Cyclopoida, 457
Cyclops, 305/306★, 443★, 451★, **457**, 459★
Cyclopyge, 402
Cydippidea, **259**
Cylindroiulus londinensis, 509★, **511**
Cymbasoma rigidum, see *Haemocera danae*
Cyphophthalmi, 427
Cypress moss (*Sertularia cupressina*), **205**
Cyprididae, 450
Cyprilepas, 434
Cypris, 459★, **464**
Cyrtophormia spiralis, 94★
Cysticercus, 300★, 303
Cystoidea, 168
Cytheridae, 450

Dactylometra quinquecirrah, **217**
Dahl, F., 455
Dalmanophyllum (Rugosa coral), 169★
Dalmatian sponge (*Spongia officinalis*), 148★, **164**
Dalyellia viridis, **288**
Dalyelloida, 288
Damon medius, **414**
— — *johnstoni* (Whip scorpion), 416★
Daphnia (Waterfleas), 442f
— *cucullata*, **447**
— *longispina*, **447**
— *magna*, **447**
— *pulex*, 444★, 446★, **447**, 451★
Daphniidae, 446
Darwin, Charles, 19, 43, 80, 82, 167, 227, 357, 379f
Dasydytidae, 328
Davaineidae, 307
Dead man's hand or finger (*Alcyonium digitatum*), 235f, 240★, 245, 271★
"Death worm"(*Necator americanus*), 343★
Decapoda (Decapods), 433, 438★, 460★, **483**f
Deep-sea jellyfish a.k.a., bag or crown jellyfish, (Coronata), 211, **213**f
Deep-sea pantopods (Collossendeidae), 432
Delap, M. L., 214
Demospongiae (Common sponges), 139, 149★, 152★, 158, **160**f
Dendrilla rosea, 165★
Dendroceratida, 139, **166**
Dendrocoelum lacteum, 280★
Dendrocometes paradoxus, **132**
Dendrodoa grossularia, 473★
Dendronephthya, 241★, 242★
Dendrophyllia ramea, **231**
Dero, 372
Derpanophorus crassus (Ribbon nemertine), 301★
Deuterostomia, 81, 269
Development of animals, 40f, 40★
Diabothriocephalus latus (Fish tapeworm), 278★, 289★, **308**f, 308★
Diantennata (Biantennals or Crustacea), 401, **433**ff, 433★
Diaphanosoma brachyurum, 446
Diaptomus, 443★, **447**, 456
Diastylis rathkei, **501**
Dicranophorus, 326★
— *forcipatus*, 332
Dicrocoelium dendriticum (Small liver fluke), 284★, **293**
Dictyna arundinacea, 418★
Dictyocaulus viviparus (Lungworm) 335
Dicyema, 139★
Dicyemida, 138
Didinium nasutum (Water bear), 123, **128**, 305/306★
Diffugia
— *pyriformis*, **110**
— *urceolata*, 305/306★
Digenea (Digenous flukes), 274f, 278★, 288, **292**
Digononta, 329, **331**
Dilaimus denticulatus, 336
Dileptus anser, **128**
Dinobryon sertularia, **96**, 305/306★
Dinoflagellata, 88★, **97**
Dipetalonema streptocerca, 348
Diphyidae, 195
Diplodinium denticulatum, 113★
Diplomonadina, **101**
Diplopoda (True millipedes), **506**
Diploria (Brain coral), **232**
Diplozoon paradoxum (Diplozoon), **291**, 292★
Dipylidium caninum, **307**
Discomedusae, 180
Disconanthae, 192, 195
Diseocoelis tigrina, 281★
Ditylenchus dipsaci, 340f
Diurella, see *Trichocera*
Dog leech (*Erpobdella octoculata*), 381★, **385**
Dog tapeworm or bladder tapeworm (*Echinococcus granulosus*), 290★, **303**

Döhl, J., 79
Domed-web spiders (Theridiidae), 418, **421**
Donatiidae, 161
Dotilla, 471, **497**
Doyle, Conan, 215
Dracunculus medinensis (Medina worm), 333, **345**
Drilomorpha, 367
Dromia (Sponge crabs), 435, **489**
— *vulgaris*, 473★
Dromid crabs (Dromiidae), 489
Dromiidae (Dromid crabs), 489
Drulia (= *Parmula*) *browni*, 156, **164**, 164★
Duck leech (*Theromyzon tessulatum*), 381★, **383**
Dugesia dorotocephala (American stream planarian), 286f
— *gonocephala*, 280★, **286**, 295★
— *lugubris*, 280★
Dwarf centipede (*Scutigerella immaculata*), 512
Dwarf spiders (Erigonidae), 419, **421**
Dwarf threadworms (*Strongyloides*), **343**
Dysdera, 420
— *erythrina*, 415★
Dysderidae, 420
Dysentery amoeba (*Entamoeba histolytica*), **106**, 109
Dysidea fragilis
— *spinifera*, 155

Earthworm, see *Lumbricus terrestris*
Earthworms and leeches (Clitellata), 362, **370**
East Asian lung fluke (*Paragonimus westermani*), 294
Echiniscoides sigismundi, 391, 391★
Echiniscus blumi, 382★
— *scrofa*, 389★
Echinococcus granulosus (Dog tapeworm or bladder tapeworm), 290★, **303**
Echinodermata, 81, 85
Echinorhynchus truttae, **355**
Echinostoma ilocanum, 294
Echiurida, 85, 302★, **358**f
Echiurus echiurus, 302★, **358**, 359★
Edwardsia, 224
"Eelworms", see Threadworms
Ehrenberg, Christian Gottfried, 324
Eibl-Eibesfeldt, 63
Eimeria bovis, 118
— *stiedae*, 118, 118★
— *tenella*, **118**
— *zurnii*, **118**
Eimeridae, 118
Eisenia foetida, **378**
Eiseniella tetraedra, **378**
Elephant ear sponge (*Spongia officinalis lamella*), **164**
Elephantiasis, 345
Eleutheria dichotoma, 184★
Elphidium (= *Polystomella*) *crispum*, 110★
Emerald fiddler crab (*Uca beebei*), **492**
Emerita talpoida, 485, **489**
Encentrum incisum, **332**
— *oxyodon*, **332**
Enchytraeidae, 377
Enchytraeus albidus, **377**
Endeis panciporosa, 426★
Endoprocta, see Kamptozoa
Entamoeba coli, **106**
— *gingivalis*, **106**
— *histolytica* (Dysentery amoeba), **106**, 109
Enterobius vermicularis (Threadworm), **344**

Entodiniomorpha, **131**
Entodinium caudatum, 113★
Entomostraca ("Lower Crustacea"), **435**ff
Eocarcinus, 435
Eoscorpius, 405
Eotrogulus, 406
Ephelota gemmipara, **132**
Ephydatia fluviatilis (Freshwater sponge), 143, 143★, 153, **164**
— *muelleri*, 148★, 153
Epicaridea (Parasitic isopods), 504
Epiphanes senta, 305/306★, **332**
Epistylis, 129
Epizoanthus
— *arenaceus*, 204★, **234**
— *incrustatus*, **235**
— *vatovai*, **234**
Eresus cinnaberinus, 410★, 415★, **422**
Ergasilus sieboldi (Gill shrimp), 443★, **458**
Erigonidae (Dwarf spiders), 419, **421**
Eriocheir sinensis (Chinese crab), 476★, **497**
Eriopisa, **500**
Eriopisella, **500**
Eropobdella octoculata (Dog leech), 381★, **385**
Erpobdellidae ("Ground leeches"), 386
Errantia (Free-living polychaetes), 363
Erythropsis pavillardi, 97★
Estheria, see *Isaura*
Ethusa mascarone, 473★
Eucarida, 434
Eucharis multicornis, 271★
Euchlanis deflexa, 305/306★
Eucoccidia, 118
Eudactylota eudactylota, 325★, 331★
Eudendridae, 188
Euglena gracilis, 90, **92**f
— *viridis*, 88★, **92**f
Euglenoidina, 92
Euglypha alveolata, 110, 305/306★
Eugorgia rubens, 251
Eugregarina (Gregarina), **116**
Eugregarinida, 116
Eulalia viridis, 373★
Eulampetia pancerina (a.k.a., *castrodes parasiticum*), **259**
Eumalacostraca, 434
Eumetazoa (True multicellulates), 85, 139
Eunephthya rosea, 271★
Eunice (Palolo worms), 363f
— *fucata* (Atlantic Palolo worm), **364**
— *gigantea*, 363
— *viridis* (Samoan Palolo worm), **364**
Eunicella (Sea fans), 250
— *cavolini* (Yellow horny coral), 243★, **250**f, 263★
— *stricta* (White horny coral), **250**, 262★
— *verrucosa* (Warty coral), 246, **250**
Eunicidae, 363
Eupagurus prideauxi, 223
Euphasia superba, **484**
Euphausiacea (Euphausiaceans), 483
Euphoberia, 506
Euphysa aurata, 184★
Euplectella aspergillum (Venus's flower-basket), **159**f
Eupolymnia nebulosa, 376★
Euproops, 405
European black widow (*Latrodectus mactans tredemcimguttatus*), 399★
European crayfish (*Astacus astacus*), **487**f

European lobster (*Homarus gammarus*), 452★, 460★, **487**
European spiny lobster (*Palinurus vulgaris*), 454★, 460★, **486**
Eurypelma soemanni (Bird-eating spider), 399★
Eurypteridae (Giant sea-scorpions), 405
Eurypterus, 405
Euscorpius, 412f
— *italicus* (Italian scorpion), 393★, **412**
Evadne, 448
Evolution, 43f
Exoskeleton, 397ff, 397★

Fabre, Jean-Henri, 57, 413
Fabricia sabella, **369**
False coral (*Parerathropodium coralloides*), **249**
False scorpions (Pseudoscorpiones), 411, **422f**
Farrea occa, **159**
Fasciola gigantica
— *hepatica* (Giant or common liver fluke), **292**, 293★, 294★, 295★, 297★
Fasciolopsis buski (Giant intestinal fluke), **294**
— *magna* (American giant liver fluke), **293**
Favia, 227★, **232**
Favosites, 170★
Ferster, 78
Fibula nolitangere, 157
Fiddler crabs (*Uca*), 491
Filaria, 345
Filaroidea, 345, 348
Filicollis anatis, **355**
Filinia, 332
Fine Levantine (*Spongia officinalis mollissima*), **164**
Fire coral (*Millepora platyphyllus*), **189**, 189★, 198★, 237★, 238★
"Fire jellyfish" (Cubomedusae), 211, **212f**
Fish leech, see Common fish leech
Fish lice (Branchiura), 463
Fish tapeworm (*Dibothriocephalus latus*), 278★, **289★**, 308f, 308★
Fixed action patterns, 60
Flabellifera, 502
Flabellum
— *angulare*, **231**
— *anthophyllum*, **231**
— *goodei*, **231**
Flagellata (Flagellates), 88★, **91f**
Flatworms (Platyhelminthes), 85, **273ff**, 274★
Floscularia ringens, 326★, **332**, 332★
Flosculariacea, 332
Flukes (Trematodes), 278★, **288f**
Fonticula vitta, 280★
Foramenifera, 81, 91, 93★, **110**
Forskalia contorta, 194
Forskalidae, 194
Franz, H., 331
Free-living polychaetes (Errantia), **363**
Fresh-water crabs (Potamidae), 368, **490**
Fresh-water isopods (Asellidae), 502
Fresh-water kamptozoans (Urnatellidae), 313
Fresh-water planarians (Paludicola), 286f
Fresh-water polyps (Hydrina), 177, **182f**
Fresh-water sponges (Spongillidae), **163**
Frondicularia alata, 93★
Fruit-tree red-spider mite (*Panony-*

chus ulmi), **429**, 437★
Fungia, 224–228
— *fungites* (Fungus coral), **230**
Fungus coral (*Fungia fungites*), **230**
Funicula quadrangularis (Sea whip), **255**, 271★
Funnel spiders (Agelenidae), 421f

Galathea strigosa (Rock crab), 460★, 493★
Galatheidae, 476★
Galba truncatula (Liver fluke snail), 292
Galeodes caspius, 424
— *graecus*, 423★
Gammaridae, 496★, 500
Garden spider (*Araneus diadematus*), 399★, 408★, 409★, 421
Gardner, 78
Gasteracantha (Spined or thorny spiders), 410★, 421
— *thorelli*, 400★
Gastrotricha (Gastrotrichans), 316★, 324f
Gecarcinidae (land crabs), 498
Gecarcinus, 472
— *lagostoma*, 476★
— *ruricola* (Common land crab), **498**
Gegenbaur, Karl, 458
Geodia cydonium, 161
Geodiidae, 161
Geonemertes chalicophora, 332
Geophilomorpha (Luminous centipedes), 512
Geophilus electricus (Luminous centipede), 380, 510★, **513**
Gephyrea or Starworms, 356
Gersemia, 245
Geryonia proboscidalis (Trunked jellyfish), 184★, **207**
Gesner, Konrad, 82
Ghost crab (*Ocypode ceratophthalmus*), 476★, 477, **491**
Ghost and fiddler crabs (Ocypodidae), 491
Ghost shrimp (*Phtisica marina*), **501**, 501★
Ghost shrimps (Caprellidae), 501
Giant African or imperial scorpion (*Pandinus imperator*), 393★, 407★, **413**
Giant intestinal fluke (*Fasciolopsis buski*), **294**
Giant jellyfish (*Cyanea*), 210f, 215
Giant millipedes (Spaerotheriidae), 508
Giant or common liver fluke (*Fasciola hepatica*), **292**, 293★, 294★, 295★, 297★
Giant sea-scorpion (*Pterygotus rhenanus*), 403, **405**
Giant sea-scorpions (Eurypteridae), 405
Giant spiny-headed worm (*Macrocanthorhynchus hirudinaceus*), **355**
Giardia (*Lamblia*) *intestinalis*, **101**, 101★
— *muris*, **101**
Gigantinus lateralis (Red land crab), **498**
Gigantocypris agassizi, **449**
Gigantostraca (Sea -scorpion), 393★
Gilbert, 79
Gill shrimp (*Ergasilus sieboldi*), 443★, **458**
Glass sponges (Hexactinellida), 139, **159**
Globigerina, **110**
— *hulloides*, 93★
Glomeris marginata, 496★, **508**, 509★
Glomeropsis, 506
Glossiphonia, 383
— *complanata* (Greater snail leech),

381★, **383**
— *heteroclita* (Lesser snail leech), 381★, **383f**
Glossiphoniidae (Glossiphoniids), 383
Gluvia, 423
Glyciphagus domesticus (House mite), **430**, 437★
Gnathobdellae (Jawed leeches), 380, 381★, **384**
Gnathophausia, 499
Gnathostomaria lutheri, 309★, **310**, 310★, 311★
Gnathostomula paradoxa, **309ff**, 311★
Gnathostomulida, 309ff, 315★
Goldmonadina (Chrysomonadina), 96f
Golfingia, **358**
Gonactinia, 222
— *prolifera*, **224**
Gonastrea, 232
Goniaulax, **98**
Gonionemus vertens, **191**
Goniopsis pulchra, **497**
Gonodactylus, **482**
Goose barnacles (Lepadomorpha), 434, 464★, **465**, 465★
Gordioidea, 349
Gordius, 349★
— *aquaticus*, **349**
— *dectici*, **349**
Gorgonaria, 235, 246f
Graffizoon lobatum, **285**
Graphidostreptus, 509★, **511**
Grapsidae (Rock crabs), 497
Grapsus grapsus (Red rock crab), 472, **497**
Grass sponge (*Hippospongia communis cerebriformis*), **164**
Grasshog, 419
Grassi, 414
Gray hydra (*Hydra oligactis*), **187**, 198★
Gray sea pen (*Pteroides griseum*), **253**, 271★
Gray stentor (*Stentor roeseli*), **130**
Greater snail leech (*Glossiphonia complanata*), 381★, **383**
Green bonellia (*Bonellia viridis*), 302★, 351★, **358f**
Green hydra (*Chlorohydra viridissima*), **185ff**
Green paramecium (*Paramecium bursaria*), **124**
Green stentor (*Stentor polymorphus*), 113★, **130**
Greenhouse planarian, (*Bipalium kewense*), **287**
Gregarina blattarum, **117**
— *cuneata*, **117**
— *polymorpha*, **117**
— *steini*, **117**
Gregarinida, 116
Grell, K. G., 115
Grenadier crabs (Mictyridae), 497
"Ground leeches" (Erpobdellidae), 386
Ground water polychaete (*Troglochaetus*), 369
Gymnodinium catenella, **98**
— *pascheri*, **98**
Gymnoplea, see Calanoida
Gymnostomata, 128
Gyratrix hermaphrodita, 288
Gyrodactylus elegans, 291, 292★
Gyrodinium, **98**

Habrotrocha constricta, **330**
— *flaviformis*, **331**
— *pusilla textrix bervilabris*, 329★
Hadzi, 80
Haeckel, Ernst, 80ff, 112, 192

Haeckel's protosponge (*Protospongia haeckeli*), 98★, **99**
Haemadipsa zeylanica (Ceylon leech), **385**
Haemadipsidae (Land leeches), 385
Haematococcus, 107★
Haemenentia costata, 383
— *officinalis*, **383**
Haemocera danae (=*Cymbasoma rigidum*), 458f
Haemopis sanguisuga (Horse leech), 381★, **385**
Haemosporidae, **119**
Haimea, 245
Hair worm (*Wuchereria bancrofti*) 345, 348
Halacaridae (Marine mites), 430
Halammohydra octopodides, **192**
Halammohydrina, 182, 191f
Halcampa, 222
Halichondria panicea (Bread crust sponge), 166
Haliclystus octoradiatus, **212**
Halicryptus spinulosus, 357★
Halisarca dujardini (Jelly sponge), 166
Haliscera papillosum, **207**
Halistemma rubra, 271★
Halteria grandinella, **130**
Haplosporidia, 135
Haplotaxidae (Haplotaxids), 378
Haplotaxis (=*Phreoryctes*) *gordioides*, 378
Hardy, Sir Alister, 455
Harpacticoida, 457
Hartmanella, 104
Hartmann, 98
Harvest mite (*Trombicula autumnalis*), 429, 437★
Harvestman, see Common phalongid (*Phalangium opilio*)
Helicotylenchus, 340
Heliopora coerulea, 251
Helioporidae (Blue corals), 246, 251, 251★
Heliozoa (Sun animalcules), 111f
Helobdella stagnalis, 381★, **383f**
Hemiclepsis marginata, 381★, **383**
Hemimycale columella, 148★
Henneguya, 107★
Hermellimorpha, 367f
Hermiodice carunculata, 363, 373★
Hermit crab anemone (*Calliactis parasitica*), 202★, 222
Hertwigella volvocicola, see *Ascomorphella volvocicola*, 332
Heterocoela, 139, **158**
Heterocope weismanni, 456
Heterodera, 338
— *avenae* (Wheat eelworm), **340**
— *rostrochiensis* (Potato eelworm) 301★, **341**
— *schactii* (Beet eelworm), 341
Heteromeyenia baileyi (=*H. repens*), 153
Heteronemertini, 313, 315★
Heteronereis, See Nereis
Heteronymphon kempi, 431★
Heteroperipatus engelhardi, 382★, **389**
Heterophyes heterophyes, **294**
Heterostegina, 110
Heterotanais oerstedi, **502**
Heterotricha (Heterotrichous ciliates), 129f
Heterotrophic feeding, 26
Heterotylenchus, 342
Hexacontium asteracanthion, 94★
Hexacorallia (Hexaradiate corals), 177, **220**
Hexactinellida (Glass sponges), 139, **159**
Hexagonaria, see "*Cyathophyllum*", 177

Hexapoda, 505
Hexaradiate corals (Hexacorallia), 177–220
Hexasterophora, 159
"Higher crustacea" (Malacostraca), 435, 467ff
Hippidea (Mole crabs), 489
Hippolyte, 459★
— *varians*, 485
Hippopodius hippopus, 195
Hippospongia canaliculata (Wool sponge), 164
— *communis* (Horses ponge), 164
— — *cerebriformis* (Grass sponge), 164
— — *meandriformis* (Velvet sponge), 164
Hirudinea (Leeches), 85, 352★, 380, 381★
Hirundo medicinalis medicinalis (Medicinal leech), 381★, 384, 384★, 385★
— — *officinalis* (Hungarian medicinal leech), 381★, 384
Histioneis remora, 88★
Histriobdella homari, 366
Histriobdellidae, 366
Holaxonia, 246, 250
Holopediidae, 446
Holopedium gibberum, 446
Holotricha, 128
Holst, Erich von, 65, 79
Homalopterygia, 85
Homaridae (Lobsters), 487
Homarus americanus (American lobster), 487
— *gammarus* (European lobster), 452★, 460★, 487
Hominidae, 25
Homiothermic animals, 32
Homo erectus, 26
— *sapiens*, 26
Homocoela, 139, 157f
Homocoelidae, 157f
Hoplonemertini, 318f
Hoplonympha, 102
Hoplonymphinae, 102
Horaella brehmi, 330★
Hormiphora plumosa, 259
Hormones, 33f
Horse leech (*Haemopis sanguisuga*), 381★, 385
Horse sponge (*Hippo spongia communis*), 164
Horse worm (*Oxyuris equi*), 338
"Horsehair worms" (Nematomorpha), 316★ 324, 348ff
Horseshoe crab, see Atlantic horseshoe crab
Horseshoe crabs (Xiphosura), 493★, 404, 406f
House mite (*Glyciphagus domesticus*) 430, 437★
House spider (*Tegenaria domestica*), 399★, 417, 421
Human itch mite (*Sarcoptes scabei*), 430, 437★
Hungarian medicinal leech (*Hirudo medicinalis officinalis*), 381★, 384
Huxley, Julian, 101
Huxley, Thomas Henry, 82
Hyalonema, 173★
— *thomsoni*, 160, 161★
Hydatigena taeniaformis (Cat tapeworm), 307
Hydatina senta, 141
Hydra
— *oligactis* (Gray hydra), 187, 198★
— *vulgaris* (Brown hydra), 187
Hydractinia echinata, 190, 192
Hydrallmania falcata (Coral moss), 206

Hydrina (Freshwater polyps), 177, 182f
Hydrocorallidae, 189
Hydromedusa, 179★, 180, 180★
Hydrorhiza or stolon net, 180
Hydroschendyla, 506
Hydrozoa, 169, 174★, 177, 180
Hyman, 80
Hymenocaris, 434
Hymenocera
— *picta* (Painted shrimp), 474★
Hymenolepis diminuta, 308
— *nana*, 308, 308★
— — *fraterna*, 308
— — *nana*, 308
Hymenostomata, 128
Hymenostraca, 434
Hypermastigidae, 102
Hypoconcha, 489
Hypotricha, 130
Hypsibius, 390★
Hyptiotes, 422
— *paradoxus*, 415★
Ichthyobdellidae, 384
Ichthyophthirius multifiliis, 128
Ichthyotomidae, 366
Ichthyotomus sanguineus, 366
Ikeda taenioides, 358
Illaenus, 402
Ilyocryptus sordidus, 447
Imperial scorpion, see Giant African scorpion
Imprinting, 67f
Inheritance, 41f, 43★
Innate releasing mechanism (IRM), 61
Intestinal blood fluke (*Schistosoma mansoni*), 283★, 297
Isaura, 434, 442
Ischyropsalis, 427
— *hellwegi*, 400★, 428
Isopoda, 468★, 502
Isospora, 118f
Italian scorpion (*Euscorpius italicus*), 393★, 412
Itura myersi, 305/306★
Ivory barnacle (*Balanus eburneus*), 465
Ixodes ricinus, 431, 437★
Ixodidae (Ticks), 430f

Japanese blood fluke (*Schistosoma japonicum*), 297
Jasus
— *huegeli*, 468
— *lalandei* (Cape spiny lobster), 486
Jawed leeches (Gnathobdellae), 380, 381★, 384
Jelly sponge (*Halisarca dujardini*), 166
Jellyfish, see Common jellyfish (*Aurelia*)
Jellyfish, anemones and corals, see Coelenterata
Jitterbug fiddler crab (*Uca saltitanta*), 492
Joenia annectens, 88★
Julidae, 507, 507★
Jumping spiders (Salticidae), 422

Kaestner, Alfred, 259, 390, 403, 423, 429, 512
Kalyptorhyncha, 288
Kamptozoa, 82, 85, 312f, 315★, 352★
Kellikottia longispina, 331★, 332
Keratella, 332
— *cochlearis*, 330★
— *quadrata*, 305/306★
Kidney sponge (*Chondrosia reniformis*), 161f

King crabs (*Limulus*), 405, 411
King lobster, see Norway lobster
Kinorhyncha, 316★, 324, 354, 356f
Kirsteuer, 310
Koenenia mirabilis, 413★
Kofoidia loriculata, 102★
Köhler, Wolfgang, 70ff, 72★, 75
Kohts, Nadie, 78
Köllikeria fasciculata, 184★
Konermann, 79
Krüger, F., 367
Kudo, 109

Labidognatha, 420, 421★
Lacaze-Duthiers, 247
Lacrymaria olor, 128, 305/306★
Lagena interrupta, 93★
— *spiralis*, 93★
Lagisca extenuta, 373★
Lamarck, Jean Baptist, 43, 82
Lamblia, see *Giardia*
Lamprochernes nodosus, 423
Land crabs (Gecarcinidae), 498
Land hermit crabs (*Coenobita, Birgus*), 498
Land isopods (Oniscoidea), 503
Land leeches (Haemadipsidae), 385
Laniatores, 427
Lanice conchilega ("Shell collector"), 368
Laodicea undulata, 206
Laomedea (= *Obelia*), 205, 206★
Large slipper lobster (*Seyllarides latus*), 460★, 487
Lasiodora, 425★
Latona setifer, 446
Latrodectus
— *mactans* (Black widow), 421
— *tredemcimguttatus* (European black widow), 399★
Leaf nematodes (*Aphelenchoides*) 340
Leaia, 434
Learning, 66
Lecithoepitheliata, 285f
Lecqueureusia spiralis, 305/306★
Leeches (Hirudinea), 85, 352★, 380, 381★
Leechlike rotifers (Bdelloidea), 331
Leeuwenhoek, Anthony van, 185
Leiobunum rotundum, 428★
Leishmania donovani, 100
— *tropica*, 100
"Leishmania" form, amastigote, 99, 99★
Lepadella, 326★
— *patella*, 330★
Lepadomorpha (Goose barnacles), 434, 464★, 465, 465★
Lepas anatifera (Common goose barnacle), 459★, 465
Leperditia, 434
— *titanica*, 449
Leperditiida, 434
Lepidocaris rhyniensis, 435
Lepidonotus squamatus, 363
Lepidurus apus, 441, 451★
Leptobathynella, 483
Leptodora hyalina, 444★, 446
Leptodoridae, 446
Leptodorus, 447
"Leptomonas" form, promastigote, 99, 99★
Leptopsammia pruvoti, 230
Leptostraca, 434, 481f
Lequeureusia spiralis, 110
Lermaeocera branchialis, 461
Lernaea, 461
— *esconia* (Pike shrimp), 443★
Lernaeida, 461
Lernaeopodida, 462
Lesser snail leech (*Glossiphonia heteroclita*), 381★, 383f
Leucandria aspera, 143

Leuckart, Rudolf, 292
Leuckartiaria nobilis, 184★
Leucochloridium fuscum, 295★, 298
— *paradoxum*, 295★
Leuconia aspera, 158
— (= *Leucandra*) *nivea*, 158
Leuconidae, 158
Leucosolenia, 150/151★, 154, 154★
— *botryoides*, 158
— *complicata*, 155
— *coriacea*, 142, 173★
— *variabilis*, 155
Leucothea multicornis, 260, 267f
Leucothoidae, 156
Lichas, 402
Lichida, 402
Ligia oceanica (Beach woodlouse), 496★, 503
Ligula intestinalis, 309
Limax amoeba, (see *Hartmannella*), 104★,
Limnadia lenticularis, 442
Limnias, 326★
— *ceratophylli*, 305/306★
— *melicerta*, 325★, 332, 332★
Limnohydrina-Limnomedusae, 182 190f
Limnoria lignorum, 502
Limnoridae, 502
Limnotis nilotica, 385
Limulus (King crabs), 405, 411
— *polyphemus* (Atlantic horseshoe crab), 393★, 411, 412★
Lineus geniculatus (Mediterranean bootlace worm), 301★
Linguatula serrata, 392, 392★
Linguatulida or Pentastomida, 85, 387
Linne, Carl von, 93f, 186, 360, 441, 491
Linyphiidae (Sheet-web spiders), 418
Lion's mane jellyfish (*Cyanea capillata*), 210, 215
Liphistius, 420
— *desultor*, 400★
Lipostraca, 435
Liriope tetraphylla, 207
Lithistida ("stone sponges"), 160, 168
Lithobiomorpha, 513
Lithobius forficatus, 510★, 513
Lithodes maja, 489
Lithodidae (Stone crabs), 489
Lithostrotion, 177
Litocarpia myriophyllum, 206, 206★
Liver fluke, 292ff
Liver fluke snail (*Galba truncatula*), 292
Loa loa, 348
Lobata, 259
Lobocarcinus, 435
Lobophyllia, 232
Lobsters (Homaridae), 487
Lohmann, H., 97
Longidorus, 340
Long-legged fiddler crab (*Uca stenodactyla*), 492
Long-tailed uropygid (*Typopeltis crucifer*), 414, 416★
Lophelia pertusa, 231, 231★, 249
Lophocalyx philippensis, 155★
Lophogorgia chilensis, 251
Lopholithodes (Stone crab), 476★
Lorenz, Konrad, 65, 67, 70
Lotz, R., 385
"Lower Crustacea" (Entomostraca), 435ff
Loxosceles, 418
Loxosomatidae, 313
Lubomirskia baicalensis, 163
Lucernaria bathophila, 212
— *quadricornis*, 212

Ludwig, H., 324
Lugworm (*Arenicola marina*), **367**, 367★
Lumbricidae, 378
Lumbriculidae, 377
Lumbriculus variegatus, 377
Lumbricus castaneus, **378**
— *rubellus*, **378**
— *terrestris* (common earthworm), 361★, **378**f, 378★
Luminous centipede (*Geophilus electricus*), 380, 510★, **513**
Luminous centipedes (Geophilomorpha), 512
Lung flukes, see North American, East Asian Lung flukes
Lungworm (*Dictyocaulus viviparus*), 335
Lybia tesselata, **490**
Lycastis terrestris, **365**
— *vitabunda*, 365
Lycastopsis amboinensis, **365**
— *raunensis*, 365
Lycoridae, see Nereidae
Lycosa tarentula (Tarantula), **421**, 425★
Lycosidae (Wolf spiders), 421
Lydekker, Richard, 19
Lynceus brachyurus, **442**

Macracanthorhynchus hirundinaceus (Giant spiny-headed worm), **355**
Macrobiotus, 390
— *hufelandi*, 382★, 390★, **391**
— *macronx*, 305/306★
Macrochaetus subquadratus, 325★
Macrocheira kaempferi, 468
Macrodasyoidae, 327
Macrostomida, 285
Macrostomum appendiculatum, **285**
Macrothrieidae, 447
Macrotrachela insulana, **331**
Madrepora oculata, **231**
Madreporaria or Scleractinia (Stony corals), 48★, 166, 169, 177, 220, **224**f, 277★
Madreporia alcidornis, 227
Magnus, D., 62, 252
Maja (Spider crabs), **489**
Majiidae (Spider crabs), 490
Malacobdella, 318
Malacostraca ("Higher Crustacea"), **435**, **467**ff
Malaguin, 458
Malaria, 119ff, 120m
Malpiphiella mellificae, **109**
Mangrove crab (*Aratus pisoni*), **497**
Mantis shrimp (*Squilla mantis*), 451★, 459★, **482**
Maricola (Marine planarians), 286
Marine hermit crabs (Paguridae), 488, 494★, 495★
Marine mites (Halacaridae), 430
Marine planarians (Maricola), 286
Marsupiobdella africana, **384**
Martini, E., 100, 121
Mastisophora, 92
Maw-worm (*Ascaris lumbricoides*), 333, 335, 337, **344**
Maxillopoda, **455**
Mayor, Dr., 227
McConnell, 286f
Meal mite (*Acarus siro*), 428, **430**, 437★
Meandrina (Stony coral), 170★
Medicinal leech (*Hirundo medicinalis medicinalis*), 381★, **384**, 384★, 385★
Medina worm (*Dracunculus medinensis*), 333, **345**
Megacyclops viridis, **457**
Megalograptus ohioensis, 393★
Megalomma vesiculosum, 374★

Meganyctiphanes norvegica, **484**
Megascolecidae, 378
Megascolides australis, **378**
Meloidogyne (Root gall nematode), 342
Melon comb jellyfish (Beroidea), 267f
Mendel, Gregor, 42
Mermis subnigrescens, **336**, 338
Mermithoidea, 342
Merodinium, 115
Merostomata, 404, **406**
Mertensia ovum (Sea nut), **259**
Mesidotea (see *Saduria*)
— *entomon*, 496★
Mesolimulus, 405
Mesonemertini, 313, 315★
Mesostoma ehrenbergi, 280, **287**
Mesothelae, 406, 414, **420**
Mesozoa, 85, **138**
Metagenesis, 181
Metagonimus yokagawai, **294**
Metazoa (Multicellulates), 85, 305/306★
Metridium senile, 199★, **224**, 271★
Microcodon clavus, 325★
Microcosmus sulcatus, 249
Micrommata rosea, 400★, 410★, **422**
Microniphargus, **500**
Microspironympha porferi, 102★
Microsporidia, 132, **133**f
Microstomum, 282, 286★
— *lineare*, **285**
Micrura alaskensis, 312★
Mictyridae (Grenadier crabs), 497
Miliola reticulata, 93★
— *striolata*, 93★
Millepora platyphyllus (Fire coral), **189**, 189★, 198★, 237★, 238★
Milleporidae, 237★
Millipedes (Myriapoda), **505**f
Mimicry, 63
Mites (Acari), 394★, 411, **428**ff
Mnemiopsis leidyi, **260**, 271★
Mniobia incrassata, **331**
Mobiila (Motile peritricha), **129**
Moina, 445
— *rectirostris*, 444★, 447
Mole crabs (Hippidea), 489
Mollusca (Mollusks), 85
Moniliformis moniliformis, 355
Monocystis, **116**
Monogenea (Monogenetic flukes), 274, **288**f
Monogononta, 329, **331**
Monommata, 325★
Monoraphis chuni, 143, **160**, 162★
Monothalamia, 110
Monstrillidae, 458
Moseley, 388
Moth mite (*Pyemotes herfsi*), **429**
Motile peritricha (Mobilia), **129**
Muggiaea kochii, **195**
Mühlens, 101
Müller, O. F., 324
Müller's larva, 285, 287★, 352★
Multicellulates (Metazoa), 85, 305/306★
Multiceps multiceps, **303**
Myodocopida, 449
Myriapoda (Millipedes), **505**f
Myrmarachne formicaria (Ant spider), 415★, **422**
Mysidae, 499
Mysis relicta, 472, **499**
Mystacocarida, 462
Mytilicola intestinalis, **458**
Mytilina mucronata, 305/306★
Myxilla rosacea, 148★
Myxobolus pfeifferi, 107★, **133**
Myxosporidia, 132, **133**, 133★, 138
Myxostoma cysticolum, **370**
Myzostomida, 362, **369**f, 369★, 371★

Nacobbus, 340
Naegleria, **105**
Naididae, 371
Nanognathia exigua, **310**
Narcomedusae, 182, **207**f
Nasselaria, 94★
Natantia (Shrimplike macrurans), 484, **485**
Nausithoe punctata, 183★, 209, **214**
— *rubra*, 183★
Nebalia, 434, **482**
Nebaliopsis typica, 482
Necator americanus ("Death worm"), 343f
Nectonema, 354
Nectonematoidea, 354
Nellmann-Trendelenburg, 71★
Nemat-helminthes, see Asc-helminthes
Nematocysts, 174★, 179f
Nematoda (Threadworms), 316★, 324, **333**f
Nematodirus battus, **339**
Nematomorpha ("Horsehair worms"), 316★, 324, **348**ff
Nematoscelis, **484**
Nemertini (Boot-lace worms or strapworms), 85, **313**ff, 315★
Neoaplectana, 343
Neodasys, 327
Neodonta, 287
Neomysis integer, 499★
Neoophora, 285, **285**f
Neorhabdocoela, 287
Nephrops norvegicus (Norway or king lobster), 451★, 459★, 460★, **487**
Nereidae, 364
Nereimorpha, 363
Neries
— *diversicolor*, **365**, 373★
— *virens*, 365
Nicolai, Jürgen, 69
Nicothoe astaci, 457
Niphargellus, 500
Niphargopsis, 500
Niphargus, 500
Noctiluca miliaris, 88★, **97**f
Nodosaria spinacosta, 93★
North American lung fluke (*Paragonimus kellicotti*), **297**
Norway or king lobster (*Nephrops norvegicus*), 451★, 459★, 460★, **487**
Nosema apis, 134f
— *bombycis*, **134**
Notodromas monacha, **450**
Notommata allantois, 325★
— *copeus*, 326★, 330★
Notoplana longestyletta, 296★
Notostraca, 440f, **477**
Nuda, 256, **267**
Nummulites orbiculatus, 93★
Nyctotherus cordiformis, **130**

Oak leaf spider (*Araneus ceropegius*), 410★
Obelia (see also *Laomedea*)
Oceania armata, 184★
Ochridaspongia rotunda, **164**
Octocorallia (Octorodiate corals), 166, 178, 220, **235**f, 235★
Octomitus muris, 101
Octopus vulgaris, 180
Octoradiate corals (Octocorallia), 166, 178, 220, **235**f, 235★
Octorchis gegenbauri, 207
Octotrocha speciosa, 325★
Ocypode, 476★, **497**
— *ceratophthalmus* (Ghost crab), 477, **491**
Ocypodidae (Ghost and fiddler crabs), 491
Odontopleurida, 402
Odontosyllis, 364

Ohridia, 348
Oken, L., 130
Olenellus vermontanus, 397★
Olenoides, 402
Olenus, 382★
Oligochaeta, 352★, 362, **370**
Oligotricha, 130
Olindias phosphorica, **191**, 198★
Oncopoda, 387
Oniscoidea (Land isopods), 503
Oniscus asellus (Sowbug), 496★, **503**
Onychophora, 85, 361, 382★, **387**ff, 388★
Onychopoda, 448
Onychura, 441f
Opalina dimidiata, **103**
— *ranarum*, 103, 103★
Opalinina (Opalins), **103**
Opercularia, **129**
Ophrydium versatile, **129**
Ophryotrocha puerilis (Marine annelid), 39
Opisthandria, 508
Opisthorchis felineus (Cat liver fluke), **294**
— *viverrini* (Upper Indian liver fluke), **294**
Orb spiders (Araneidae), 419★, **421**
Orchestia gammarellus (Beach flea), **500**
Orconectes limosus (American crayfish), 479, 481, **488**
Organ pipe corals (Tubiporidae), **245**, 271★
Organs, 29f
"Oriental sore", 100
Ornithocercus magnificus, 88★
Orthochirus innesi, 393★
Orthognatha (Bird-eating spiders), 420, 421★
Orthonectida, 138
Oscarella lobularis, 154, **161**
Osche, G., 299, 392
Ostracoda, 169, 433f, 448f
Owen, Richard, 159
Oxytricha, 130
Oxyuris equi (Horse worm), **338**
Oyster crab (*Pinnoteres pisum*), **490**, 493★
Ozyptila, 422

Pachymerium ferrugineum, 513
Paguridae (Marine hermit crabs), 488, 494★, 495★,
Pagurites oculatus, 494★
Pagurus
— *arrosor*, 222
— *bernhardus*, 222
— *calidus*, 495★
Palaemon squilla (Baltic shrimp) **486**
Palaeoacanthocephala, 355
Palaeolimulus, 405
Palaeonemertini, 313
Palaeophonus, 405
Paleocaris, 434
Paleomerus, 405
Palinuridae, 435
Palinurus argus, **487**
— *vulgaris* (European spiny lobster), 454★, 460★, **486**
Pallas, Peter Simon, 83
Palolo worms (*Eunice*), 363f
Palpigradi, 411, **414**
Palpatores, 428
Paludicola (Fresh-water planarians), 286f
Pancarida, 499
Pandalus borealis, **485**
Pandinus imperator (Giant African or imperial scorpion), 393★, 407★, **413**
Pandorina morum, 305/306★

Panonychus ulmi (Fruit-tree red-spider mite), **429**, 437★
Pantachogon rubrum, **207**
Pantapoda, 401, 404, **431**f
Panulirus ornatus, 473★
Parachordodes tolosarus, 354
Paradoxides, 402
Paragonimus kellicotti (North American lung fluke), **297**
— *westermani* (East Asian lung fluke), **294**
Paragorgia arborea, **249**
Paralcyonium elegans, 263★
Paralithodes camtschatica, **489**
Paramecium, 56, 108★, **122**f, 123▲, 124★, 128, 305/306★
— *aurelia*, 113★, **124**, 126f
— *bursaria* (Green paramecium), **124**
— *caudatum*, **124**, 126
Paramuricea chamaeleon (Violet horny coral), 243★, **250**
Paranthipathes larix, **233**
Parapandalus narval, 451★
Pararthropoda, 387
Parasites, 52, 343
Parasitic isopods (Epicaridea), 504
Parasitic mites (Parasitiformes), **430**f
Parasitiformes (Parasite mites), **430**f
Parasitylenchus, **342**
Parastacidae, 487
Paratylenchus, 340
Parazoa, 85, 139
Parazoanthus axinellae, 45★, 204★, **234**
Pardosa, 421
Parerythropodium coralloides (False coral), **249**
Parmula browni, see *Drulia browni*
Parvancorina, 81
Pasilobus bufoninus, 400★
Pasteur, Louis, 134
Pauropoda, 511
Pauropus silvaticus, **511**, 512★
Pavlov, Ivan, 66
Peachia, 224
Peacock feather worm (*Sabella pavonina*), 376★
"Peanut worms" (Sipunculida), 81, 85, 356, **357**f
Pectinaria koreni, **367**
Pedicellinidae, 313
Pedipalpi (Whip scorpions), 411, **413**
Pegantha, **208**
Pelagia noctiluca (Phosporescent jellyfish), 210, **214**
Pelomyxa binucleata, 104★
— *palustris*, **105**
Peltogasterella, **467**
Penaeus setifer, **485**
— *trisulcatus*, **485**
Penella
— *balaenoptera*, **462**
Peneroplis pertusus, 107★
— *planata*, 93★
Pennatula phosphorea (Phosphorescent sea pen), **254**, 263★
Pennatulacea (Sea pens), 178
Pennatularia (Sea pens), 246, **251**f, 253★
Pentacoela (Protochordates), 85
Pentastomida, see also Linguatulida, 361, 382★
Peracarida, 499
Peranema trichophorum, 88★, **95**
Perch shrimp (*Achtheres percarum*), 443★, **462**
Periclimenes petersoni, **486**
Peridinium
— *divergens*, 88★
Perigonimus, 184★

Peripaticae ("Strollers"), 389, 389m
Peripatopsidae, 389
Peripatopsis moseleyi, 376★
Periphylla regina, **214**
Peritricha, **128**f
Petrobiona massiliana, 158
Peucetia, 408★
Pfungat, 76
Phacopida, 402
Phacops, 402
Phacus, 95
Phakellia ventillabrum, 166★
Phalangida, 400★, 411, **424**, 426★
Phalangium, 427
— *opilio'* (Common phalangid), 400★, **428**
Pharetronida, **158**
Pharyngidea, 370
Pharyngobdellae, 385
Phascolosoma
— *granulatum*, 367★, 358★
— *lurco*, **358**
Pheronema raphanus, 160★
Philodina citrina, 305/306★
Pholcus phalangioides, 415★, **420**
Phoneutria (Comb spiders), 418
Phosphorescent jellyfish (*Pelagia noctiluca*), 210, **214**
Phosphorescent sea pen (*Pennatula phosphorea*), **254**, 263★
Phragmatopoma, **368**
Phtisica marina (Ghost shrimp), **501**, 501★
Phyllocarida, 434
Phyllodoce, **365**
Phyllodocidae, 365
Phyllopoda, 435, **440**f
Physalia physalis (Portuguese man-of-war), **193**f, 198★, 271★
Physcosoma granulatum, 302★
Physophora hydrostatica, **194**, 271★
Physophorae, 193f
Phytomonadina, **95**f
Pigeon mite (*Argas reflexus*), 394★, **431**, 437★
Pike shrimp (*Lernaea esconia*), 443★
Pilidium larva, 322, 322★
Pill woodlouse (*Armadillidium vulgare*), 496★, **503**
Pinnoteres pisum (Oyster crab), **490**, 493★
Pinnoteridae, 490
Pisaura
— *listeri*, **419**
— *mirabilis* (Wolf spider), 399★, 408★, 420
Piscicola geometra (Common fish leech), 376★, 381★, **384**
Piscicolidae, see Icthyobdellidae
Pithecanthropus, 26
Pithonaton, 435
Placentonema gigantissimum, 334
Planaria torva, 280★
Planes minutus (Columbus or Sargassum crab), **497**
Planula larva, **177**★, 182
Plasmodium, 52, 119f
— *falciparum*, 120
— *malariae*, 120
— *ovale*, 120
— *vivax* (Malaria parasite), 114★, **120**
Platyarthrus hoffmannseggii (Ant isopod), **504**
Platyctenidea (Compressed bomb jellyfish), 266
Platyhelminthes (Flatworms), 85, **273**ff, 274★
Platyrrhacus, 509★, **511**
Platzias quadricornis, 305/306★
Pleospongia, 139
Plesiopora, 371

Pleurobrachia pileus (Sea gooseberry), 256, 258, **259**, 271★, 277★
Ploima, 331
Plumularidae, 206
Podocopida, 434, **449**
Podocoryne carnea, **189**
Podon, 448
Podophthalmus vigil, **490**
Podoplea, 443★, 455, **457**
Poikilothermic animals, 32
Polyarthra, 332
Polycelis cornuta, 280★, **286**
— *nigra*, 280★
Polychaeta, 81, 352★, **362**f, 363★
Polycladida, 285, 295★, 296★
Polycope, **448**
Polydesmida, 507f
Polydesmus complanatus (Compressed polydesmid), 509★, **511**
Polymastia mamillaris, **162**
Polymastigina, 88★, **102**f
Polymastiidae, **163**
Polymorphus boschadia, **355**
Polynoe scolopendrina, 366
Polyphemidae, 448
Polyphemus pediculus, 444★, **448**, 451★
Polypodium, 208
Polystoma integerrimum (Bladder fluke), **291**
Polystomella aculeata, 93★
— *crispa*, see *Elphidium crispum*
Polythalamia, see Foramenifera
Polyxenus lagurus, **508**, 509★
Pomphorhynchus, 355
Porcelain crabs (*Porcellanopagurus*), 489
Porcellana (Porcelain crabs), 459★
— *platycheles* (Gray porcelain crab), 476★
Porcellanopagurus (Porcelain crabs), 489
Porcellio scaber (Common woodlouse), **503**, 503★
Porites, 227, **232**
Poritidae, 232
Pork tapeworm (*Taenia solium*), 298★, **300**, 303★
Porocephalid, 392
Porpita porpita, **196**
Portuguese man-of-war (*Physalia physalis*), **193**f, 198★, 271★
Portunidae (Swimming crabs), 490
Portunus (True swimming crabs), 435, 459★, **490**
— *holsatus* (Common swimming crab), 493★
— *ruber*, 493★
Potamidae (Fresh-watercrabs), 368, **490**
Potamon, 477, 490
Potato eelworm (*Heterodera rostochiensis*), 301★, **341**
Praunus, 499
Praya diphyes, 271★
Precious corals (*Corallium*), 247ff, 247★
Priapulida (Priapulid worms), 85, 324, **356**ff
Priapulus caudatus (Unicaudate priapulid), 356★, **357**, 357m
Proales fallaciosa, 305/306★
— *werneckii*, 332
Proasellus, **502**
Proboscidleeches (Rhynchobdellae), 381★, 383f
Proboscifera, 370
Procerodes lobata, **286**
Prorhynchus stagnalis, 281, **286**
Proscorpius, 405
Prosopora, 377
Prostheceraeus moseley, 296★
— *roseus*, 296★

Prostoma graecense, 302★, **322**
Protacarus, 405
Proterandria, 511
Protochordates (Pentacoela), 85
Protohydra leuckarti, **187**
Protomonadina, 98
Protonemertini, 313, 315★
Protopalina, **103**
Protosolpuga (Solpugids), 406
Protospongia haeckeli (Haeckel's protosponge), 98★, **99**
Protostomia, 81, **269**, 275
Protozoa (Unicellulates), 85, 87★, **89**ff, 305/306★
Protula tubularia, 376★
Psammolimulus, 405
Pselophognatha, 507★, 508
Pseudocarcinus gigas, 468
Pseudophyllidea, 308
Pseudopodia, 103f
Pseudoprotella phasma (Ghost shrimp), 496★
Pseudoscorpiones (False scorpions), 411, **422**f
Pterocorallia, see Rugosa
Pterocorys rhinoceros, 94★
Pterognathia grandis, **310**
— *simplex*, 310★
Pteroides griseum (Gray sea pen), 253, 271★
Pteroididae, 252
Pteroparia, 402
Pterygotus rhenanus (Giant sea scorpion), 403, **405**
Ptychopariida, 402
Ptygura pilula, 332
Pugettia gracilis (Spider crab), 476★
Puncia, 434
Pycnogonidae (Coastal pantapods), 432
Pycnogonum littorale (Shore pycnogonid), 426★, **432**, 437★
Pyemotes herfsi (Moth mite), **429**
Pyrsonymphidae, 102

Radiata, 85
Radiolaria, 81, 91, 94★, **112**f
Radiosa amoeba, 104★, **104**f
Radopholus, 340
Ragactis pulchra, 271★
Ray, John, 82
Réaumur, 186
Red land crab (*Gigantinus lateralis*), **498**
Red precious coral (*Corallium rubrum*), 242★, **247**
Red rock crab (*Grapsus grapsus*), 472, **497**
Redlichiida, 402
Reef corals, 228ff
Reefs, 166ff, 168★
Reidl, 251
Reighardia, **392**
Remane, Adolf, 19?, 324
Renilla amethystina, 271★
Renillidae (Sea pansies), 255
Rensch, Bernard, 79
Reproduction, 37ff, 38★, 39★,
Reptantia (Crawfish), 484, **486**
Respiration, 31f
Resticula nyssa, 325★
Rhabdostichus, 442
Rhadinaphelenchus cocophilus (Coconut nematode), 340f
Rhagodes, 424
Rhagodidae, 424
Rhipidogorgia flabellum (Venus's fan), **251**
Rhizocephala, 83, **466**
Rhizochrysis, 89
Rhizopoda, 85, **103**f
Rhizostoma, 217f
— *octopus*, **218**

— *pulmo*, 183★, 211, **218**, 272★
Rhizostomae (Rhizostome jelly-fish), 211, **217f**, 217★
Rhopalonema velatum, **207**
Rhopalura, **138**★
Rhopilema esculenta, **219**
Rhynchobdellae (Proboscid leeches), 381★, 383f
Rhynchocoelia 313
Rhynchodemus terrestris, **287**
Rhynchonympha, **102**
Richter, E., 382★
Richthofenia (Corallike branchio-pods), 170
Ricinulei, 422
Ridged barnacle (*Balanus perforatus*) **465**
Riedl, R., 310f
Ripestes parasita, **371f**
Rivulogammarus lacustris, 500★
— *pulex* (Common waterflea), **500**
Robber or coconut crab (*Birgus latro*), **489**, 493★
Rock crab (*Cancer pagurus*), 435, 473★, 490
Rock crabs (Grapsidae), 497
Root gall nematode (*Meloidogyne*), 342
Rotaria, 326
— *macroceros*, 305/306★
— *neptunia*, 305/306★, 325★, **331**
— *rotatoria*, 305/306★, 328★
Rotatoria (Rotifers), 316★, 324, **328ff**
Rotifers (Rotatoria), 316★, 324, **328ff**
Rotylenchulus, 341
Rotylenchus, 340
Roundworms (Aschelminthes), 85, 316★, **323ff**
Rugosa or Pterocorallia, 169, 177f, 220

Sabella pavonina (Peacock feather worm), 376★
Sabellaria spinulosa, **368**
Sacculina, 31
— *carcini*, **466**
Sagartia, 222
Sahara scorpion (*Androctonus australis*), **412**
Sailors-before-the-wind (*Velella spirans*), 195f, 198, **205**, 271★
Salmacina dysteri (Tubeworm), 461
Salpa fusiformis, 259
Salpingoeca amphoroideum, 173★
Salticidae (Jumping spiders), 422
Salticus scenicus, 399★, **422**
Samoan Palolo worm (*Eunice viridis*). 364
Sand anemone (*Cerianthopsis americanus*), **234**
Sand fleas (Talitridae), 500
Sapphirina, **458**
— *fulgens*, 443★
Sarcocystis fusiformis, **136**
— *lindemanni*, **136**
— *miescheriana*, **135**
— *muris*, **135**
— *tenella*, 135f
Sarcodina, **103**
Sarcophyton ehrenbergi, 240★
— *trocheliophorum*, 240★, 242★, **246**
Sarcoptes scabei (Human itch mite), **430**, 437★
Sarcoptidae, 430
Sarcoptiformes, 430
Sarcosporidia, **135**
Sargassum crab, see Columbus crab
Sarsia gemmifera, **188**
— *tubulosa*, **188**
Saville-Kent, 227

Scapholeberis mucronata, 444★, **447**, 450
Scatophaga stercoraria (Dungfly), 50
Schildknecht, 507
Schistosoma haematobium (Bladder blood fluke), **297**
— *japonicum* (Japanese blood fluke), **297**
— *mansoni* (Intestinal blood fluke), 283★, **297**
Schistosomidae (Blood flukes), 297
Schizococcidia, **118**
Schizogregarinida, 116
Schizophyllum sabulosum, 496★, 509★, **511**
Schmidt, Oskar, 369
Scleractinia, see Madreporaria
Scleraxonia, 246, **247**
Scolopendra subspinipes, 510★, **513**
Scolopendromorpha (True centipedes), 513
Scorpiones (Scorpions), 393★, **411f**
Scorpionida, 405
Scutariella didactyla, **288**
Scutigera coleoptrata, 496★, 510★, **513**
Scutigerella immaculata (Dwarf centipede), **512**
Scutigeromorpha, 513
Scyllaridae (Slipper lobsters), 487
Scyllarides latus (Large slipper lobster), 460★, **487**
Scyllarus arctus (Slipper lobster), 451★, 459★, **487**
Scyphostoma polyp, 209★
Scyphozoa (True jellyfish), 174★, 177, **208f**
Scytalopsis djiboutiensis, **252**
Sea anemone (*Actinia equina*), 50, **224**, 277★
Sea anemones (Actinaria), 220f, 255f, 277★
Sea caterpillars (Aphroditidae), 363
Sea dahlia (*Tealia crassicornis*), 200★
Sea fans (*Eunicella*), 250
Sea gooseberry (*Pleurobrachia pileus*), 256, 258, **259**, 271★, 277★
Sea mitre (*Beroe cucumis*), **268**
Sea mouse (*Aphrodite aculeata*), **363**, 373★
Sea nut (*Mertensia ovum*), **259**
Sea orange (*Tethya aurantium*), 148★, **161**
Sea pansies (Renillidae), 255
Sea pens (Pennatulacea and pennatularia), 178, 246, **251f**, 253★
Sea scorpions. see Giant sea-scorpions
Sea wasp (*Chiropsalmus quadrigatus*), 211, **213**
Sea whip (*Funicula quadrangularis*), **255**, 271★
Sedentaria (Sessile polychaetes), 363, **366f**
Segestria florentina, 418★
— *senoculata* (Cellar spider), 399★, 419★
Segmented worms (Annelida), 85, 352★, **360ff**
Seison nebaliae, **331**
Seisonida, 330
Semaestomae, 211, **214f**
Seriata, 286
Serpula vermicularis (Chalk-tube worm), 376★
Serpulimorpha, 367f
Sertularella, **198**
Sertularia cupressina (Cypress moss), **205**
Sessile comb jellyfish (Tjallfiel-lidea), **266f**
Sessile peritricha (Sessilia), **129**
Sessile polychaetes (Sendentaria) 363, **366f**

Sessilia (Sessile peritricha), **129**
Sheet-web spiders (Linyphiidae), 418
" Shell collector " (*Lanice conchilega*), **368**
Shelled amoebae (Testacea), **109f**
Shell-less amoebea (Amoebina), **104ff**
Shore crab (*Carcinus maenas*), 467, 477, 490
Shore pycnogonid (*Pycnogonum littorale*), 426★, **432**, 437★
Shrimplike macrurans (Natantia), 484, **485**
Sida crystallina, 444★, **446**
Sididae, 446
Silicoflagellata, 97
Simocephalus vetulus, 444★, **447**
Singing fiddler crab (*Uca musica*), **497**
Siphonanthea, 192f
Siphonomecus multicinctus, **358**
Siphophora, 182, **192**
Sipunculida (" Peanut worms "), 81, 85, 356, **357f**
Sipunculus nudus, 302★, **358**
Sisyra, 155
" Sleeping sickness ", 99, 101
Slipper lobster (*Scyllarus arctus*), 451★, 459★, **487**
Small liver fluke (*Dicrocoelium dendriticum*), 284★, **293**
Snail leeches (*Glossiphonia*), 383
Snapping shrimps (Alpheidae), 485
Solenocera, 485
Solicmann, 78★
Solipuga, 411, **423**
Solmissus albescens, 184★, **208**
Solmundella bitentaculata, 184★, **208**
Solpuga, 424
— *lethalis* (Solpugid), 416★
Solpugidae, 424
Sowbug (*Oniscus asellus*), 496★, **503**
Spemann, Hans, 40
Sphaeractinia, **177**
Sphaeractinoidea, **177**
Sphaeronema minutissimus, 334
Sphaerotheriidae (Giant millipedes), 508
Sphaerularia bombi, 337, 337★
Spider crabs (*Maja*), **489**
Spider webs 417f, 418★, 419★, 421
Spider-like animals (Arachnida), 404, **411f**, 416★
Spiders, see Araneae
Spined or thorny spiders (*Gasteracantha*), 410★, **421**
Spiny lobsters (Palinuridae), **486**
Spiny-headed worms (Acanthocephala), 316★, 324, **354f**
Spiomorpha, 367
Spiralia, 361f
Spirastrella cunctatrix, 146★, 147★, 149★
Spirobranchus giganteus, 375★
Spirochona gemmipara, **131f**
Spirographis spallanzanii, 368, 374★, 375★
Spirorbis, **368**
Spirostomum ambiguum, **130**
Spirostreptidae, 506
Spirotricha, **129**
Spirurida, 345
Sponge crabs (*Dromia*), 435, 474★, **489**
Spongia irregularis (Yellow sponge), **164**
— *officinalis* (Dalmation sponge), 148★, **164**
— — *lamella* (Elephant ear sponge), 164
— — *mollissima* (Fine levantine), **164**
— — *zimocca* (Zimocca sponge), **164**

Spongiae (Sponges), **139f**, 140★, 173★
Spongidae (Bath sponges), **164**, 165★, 165m
Spongilla
— *fragilis*, 153
— *igloviformia*, 156
— *lacustris*, 148★, 153, 156, **164**
Spongillidae (Fresh-water sponges), 163
Sporangia, 105
Sporozoa, **115f**
Spumellaria, 94★
Squatinella, 325★
— *lamellaris*, 331★
Squilla
— *mantis* (Mantis shrimp), 451★, 459★. **482**
— *raphidea*, **482**
Stag's horn coral (*Acropora*), 226, **232**
Stalked fiddler crab (*Uca stylifera*), 491
Stammer, 502
Star barnacle (*Chthamalus stellatus*), **465**
Star coral (*Astroides calycularis*), **232**, 238★, 271★, **277**★
Star coral (*Balanophyllia regia*), **231**
Stauracantha quadrifurca, 94★
Stauromedusae (" Beaker jelly-fish "), **211f**, 211★
Steenstrupia nutans, 184★, **188**
Steinböck, 80
Stenopus hispidus, 451★
Stenorhynchus seticornis, 493★
Stenostomum leucops, 282, **285**
Stentor, 108★, **129f**
— *coeruleus* (Blue stentor), **130**, 305/306★
— *polymorphus* (Green stentor), 113★, **130**
— *roeseli* (Gray stentor), **130**
Stephanoceros fimbriatus, 326★, 331★, **333**
Stephanophyes superba, **195**
Sterrer, 310
Stoichactis, 223
Stomatopoda, 482
Stomphia carneola, 221
Stone crabs (Lithodidae), 489
" Stone sponges " (Lithistida), 160, 168
Stony corals (Madreporaria), 48★, 166, 169, 177, 220, **224f**, 277★
Størmer, 405
Strapworms, see Boot-lace worms
Stratiodrilus tasmanicus, 366
Stream planarian, see American stream planarian
Streptelasma, 177
Strobilation, 209★, 209
Stromatopora, 169, **177**
Stromatoporoidea, **177**
Strongyloides (Dwarf threadworms), **343**
Stygiomedusa, 210, **217**
Stylaria lacustris, **371**
Stylaster, 249
Stylasteridae, **189**, 237★
Stylatula, **254**
Stylocheiron, 484
Stylochus frontalis, 285
— *pilidium*, **285**
— *zebra*, **285**
Stylonychia mytilus, **130**, 130★, 305/306★
Suberites domuncula, 143, 153, 156, **162**
— *ficus* (= *Ficulina ficus*), 154, 156, **162**
Suberitidae (Cork sponges), **162**
Suctoria, 122, **131f**,

Sun animalcules (Heliozoa), 111f
Sutherland, 79
Swimming crabs (Portunidae), 490
Sycettidae, 158
Sycon ciliatum, 148★, 154★, 155, 158
— raphanus, 153
— setosum, 153
Syllidae, 364
Syllis cornuta, 373★
Symbiosis, 53
Symphyla, 511f
Symphyllia, 227
Syncarida, 434, 482
Synchaeta, 326★, 331
— pectinata, 305/306★
Syngamus, 338, 338★
— trachea, 344
Synura urella, 305/306★

Tabulata, 169, 177, 220
Tachyblaston ephelotensis, 132
Taenia hydatigena, 307
— pisiformis, 307
— saginata, see Taeniarhynchus saginatus
— solium (Pork tapeworm), 298★, 300, 303★
Taeniarhynchus saginatus (Beef tapeworm), 53, 278★, 299f, 299★
Talitridae (Sand fleas), 500
Talitrus saltator (Common sand flea), 500
Tanaidacea, 502
Tanais cavolinii, 502
Tangiers fiddler crab (Uca tangeri), 476★, 478, 497
Tapeworms (Cestoda), 278★, 298f
Tarantula (Lycosa tarentula), 421, 425★
Tardigrada (Water bears), 85, 361, 376★, 382★, 387, 390f
Tealia, 222
— crassicornis (Sea dahlia), 200★
Tegenaria domestica (House spider), 399★, 417, 421
Temnocephalida, 288
Tentaculata (Comb jellies), 85
Tentaculate comb jellyfish (Tentaculifera), 256, 258f
Tentaculifera (Tentaculate comb jellyfish), 256, 258f
Terebellomorpha, 367
Terpsichore fiddler crab (Uca terpsichore), 497
Terrestrial planarians (Terricola), 286f
Terricola (Terrestrial planarians), 286f
Testacea (Shelled amoebae), 109f
Tethya aurantium (Sea orange), 148★, 161
Tetragnatha (Trap-door spiders), 419, 420★, 421
— caudicula, 400★
— extensa, 410★
Tetragnathidae, 421
Tetranychidae, 429
Tetrastemma quadrilineatum, 313★
Tetraxonida, 139, 161, 170
Thalassina anomala (Tropical mole crab), 484, 488
Thecaphora–Leptomedusae, 182, 205f
Theileria parva, 136
Theileridae, 136
Thenius, Erich, 403
Theridiidae (Domed-web spiders), 418, 421
Thermobathynella adami, 483
Thermosbaena mirabilis, 499
Thermosbaenacea, 499
Theromyzon tessulatum (Duck leech), 381★, 383

Thia (Heart crab), 459★
Thick-legged water mite (Unionicola crassipes), 430, 437★
Thienemann, A., 138
Thomas, 292
Thomisidae (Crab spiders), 422
Thomisus onustus, 410★
Thoracica, 434, 465
Threadworm (Enterobius vermicularis), 344
Threadworms (Nematoda), 316★, 324, 333f
Thysanozoon brochii, 296★
Ticks(Ixodidae), 430f
Tinbergen, Niko, 62
Tintinnidae, 131
Tissues, 29
Tjallfiella tristoma, 266f
Tjallfiellidea (Sessile comb jellyfish), 266f
Tomopteridae, 365f
Tomopteris, 365★
Tortoise barnacle (Chelonibia testudinaria), 465
Toxoplasma gondii, 118
Tracheata, 401, 505ff
Trachymedusae, 182, 207
Trachypheus, 411
Trap-door spiders (Tetragnatha), 419, 420★, 421
Treelike corals (Acroporidae), 232
Trematodes (Flukes), 278★, 288f
Trembley, Abraham, 185ff
Triarthus cafoni, 402★
Triaxonida, 170
Tribrachidium, 81
Triceraspyris gazella, 94★
Trichinas (Trichinella spiralis), 301★, 345f
Trichinella spiralis (Trichinas), 301★ 345f
Trichinosis, 346
Trichobilharzia szidati, 297
Trichocera taurocephala, 331★
Trichocerca, 325★, 326★
Trichodina domerguei, 129
— pediculus, 113★, 129
Trichodorus, 340
Trichomonadidae, 102
Trichomonas ardindelteili
— hominis, 102
— tenax (= Elongata), 102
— vaginalis, 102
Trichonympha turkestanica, 102★
Trichoplax adhaerens, 138
Trichosomoides crassicauda, 338
Trichostomata, 128
Trichotria pocillum, 326★
Trichuris trichiura (Whipworm), 334
Triclada, a.k.a., Planaria, 286
Trilobita, 82, 382★, 401f
Trilobitoidea, 402
Trilobomorpha, 402
Trimerocephalus, 402
Triops cancriformis, 441
Trochophora, 352★
Trochosphaera solstitialis, 325★
Trochospongilla horrida, 164
Troglochaetus (Ground water polychaete), 369
— beranecki, 369, 369★
Trogulus, 427
— nepaeformis, 400★, 428
Trombicula autumnalis (Harvest mite), 429, 437★
Trombidiformes, 429
Trombidiidae, 429
Trombidium holosericeum (Velvet spider mite), 426★, 429
Tropical mole crab (Thalassina anomala), 484, 488

True bird-eating spiders (Aviculariidae), 420
True centipedes (Scolopendromorpha), 513
True crabs (Brachyura), 434f, 469, 484, 489
True jellyfish (Scyphozoa), 177, 208f
True macrurans (Astacura), 484, 487
True millipedes (Diplopoda), 506
True shrimps, see Natantia
True spiders or web spiders (Araneae), 394★, 399★, 400★, 411, 414, 415★
True swimming crabs (Portunus), 459★, 490
Trunked jellyfish (Geryonia proboscidalis), 184★, 207
Trupetostroma, 169★
Trypanosoma brucei, 99f, 100★
— equinum, 100
— equiperdum, 100
— evansi, 100
— gambiense, 99, 99m, 107★
— melophagium, 100
— rhodesiense, 99, 99m
"Trypanosoma" form (Trypomastigote), 99, 99★
Trypanosomatidae (Trypanosomes), 99, 99★
Trypetesa lampas, 466
Tubifex, 309
— tubifex (Common tubifex), 372
Tubificidae (Tubifex worms), 372
Tubipora purpurea, 271★
Tubiporidae (Organ pipe corals), 245
Tubulanus annulatus, 301★
Tubularia, 187, 188★
Turbellaria, 275★, 278★, 280f
Turritopsis, 208
Tylenchorhynchus, 340
Tylenchulus, 341
Tylenchus polyhypnus, 347
Typhloplanoida, 287
Typopeltis crucifer (Long-tailed uropygid), 414, 416★
Tyrophagus casei (Cheese mite), 430 437★

Uca (Fiddler crabs), 491
— annulipes, 491
— beebei (Emerald fiddler crab), 492
— insignis, 491
— musica (Singing fiddler crab), 497
— saltitanta (Jitterbug fiddler crab), 492
— stenodactyla (Long-legged fiddler crab), 492
— stylifera (Stalked fiddler crab), 491
— tangeri (Tangiers fiddler crab), 476★, 478, 497
— terpsichore (Terpsichore fiddler crab), 497
Ucides, 498
Ulrich, Werner, 81
Umbellula antarctica, 255, 271★
— lindahlii, 266
Umbellulidae, 255
Unicaudate priapulid (Priapulus caudatus), 356★, 357, 357m
Unicellulates (Protozoa), 85, 87★, 89ff, 305/306★
Unionicola crassipes (Thick-legged water mite), 430, 437★
Upper Indian liver fluke (Opisthorchis viverrini), 294
Urinympha, 102

Urnatella gracilis, 313
Urnatellidae (Fresh-water kamptozoans), 313
Uropygi, 413f

Velella spirans (Sailors-before-the-wind), 195f, 198★, 205, 271★
Velvet spider mite (Trombidium holosericum), 426★, 429
Velvet sponge (Hippospongia communis meandriformis), 164
Ventriculita, 160
Venus's fan (Rhipidogorgia flabellum), 251
Venus's-flower-basket (Euplectella aspergillum), 159f
Venus's girdle (Cestus veneris), 265f, 271★
Venus's girdles (Cestidea), 259f, 265
Veretillidae, 254
Veretillum cynomorium, 254, 262★, 263★
Verongia (= Aplysina) aerophoba, 166
— fistularis, 156
Verruca stroemia, 466
Verrucomorpha, 466
Verrucosa amoeba, 104★, 104
Vertebrata, 85
Violet horny coral (Paramuricea chamaelon), 243★, 250
Virgularia mirabilis, 254
Virgulariidae, 254
Vogel, H., 309
Vogt, Karl, 360
Volvox, 81, 133
Vorticella, 108★, 129
— microstoma, 305/306★
— nebulitera, 113★

Walckenaera, 419
— acuminata, 400★, 421
Warty coral (Eunicella verrucosa), 246, 250
Wasp spider (Argiope bruennichi), 399★, 408★, 421
Water bear (Didinium nasutum), 123, 128, 305/306★
Water bears (Tardigrada), 85, 361, 376★, 382★, 387, 390f
Water fleas, (Amphipoda), 500
Waterfleas (Cladocera), 442
Web spiders, see True spiders
Weinland, 227
Wendt, 429
Whale barnacles (Coronulidae), 466
Whale lice (Cyamidae), 501
Wheat eelworm (Heterodera avenae), 340
Whip scorpion (Damon medius johnstoni), 416★
Whip scorpions (Pedipalpi), 411, 413
Whipworm (Trichuris trichiura), 334
White horny coral (Eunicella stricta), 250, 262★
Wierzejski, 156
Will, F., 267
Wolf spider (Pisaura mirabilis), 399★. 408★, 420
Wolf spiders (Lycosidae), 421
Woodlouse, see Common woodlouse
Wool sponge (Hippospongia canaliculata), 164
Wuchereria bancrofti (Hair worm), 345, 348
Xenia, 240★
Xenocoeloma, 462
Xenotrichula, 327

Xenoturbella, 278★, 281
— *bocki*, **282**
Xerobdella lecomtei, **385**
Xiphinema, 340
Xiphosura (Horseshoe crabs), 393★, 404, **406f**

Xysticus erraticus, 400★, 422

Yellow horny coral (*Eunicella cavolini*), 243★, **250f**, 263★
Yellow sponge (*Spongia irregularis*), **164**

Yonge, C. M., 227
Yungia aurantiaca, 295★

Zebra spider (*Argiope lobata*), 408★
Zeier, 76
Zelleriella, **103**

Zervos, Skevos, 157
Zimocca sponge (*Spongia zimocca*), **164**
Zoantharia, 220, **234**
Zoothamnium, **129**
Zooxanthellae, 97, 115

Abbreviations and Symbols

C, °C . . . . Celsius, degrees centigrade

C.S.I.R.O. . . . . Commonwealth Scientific and Industrial Res. Org. (Australia)

f . . . . following (page)

ff . . . . following (pages)

L . . . . total length (from tip of nose [bill] to end of tail)

I.R.S.A.C. . . . Institute for Scientific Res. in Central Africa, Congo

I.U.C.N. . . . . Intern. Union for Conserv. of Nature and Natural Resources

BH . . . . body height

HRL . . . . head-rump length (from nose to base of tail or end of body)

N, N- . . . . . . North, Northern, North-

NE, NE- . . . . Northeast, Northeastern, Northeast-

E, E- . . . . East, Eastern, East-

S, S- . . . . South, Southern, South-

TL . . . . tail length

SE, SE- . . . . . Southeast, Southeastern, Southeast-

SW, SW- . . . Southwest, Southwestern, Southwest-

W, W- . . . . . West, Western, West-

♂ . . . . male

♂♂ . . . . males

♀ . . . . female

♀♀ . . . . females

♂♀ . . . . pair

+ . . . . extinct

$\frac{2 \cdot 1 \cdot 2 \cdot 3}{2 \cdot 1 \cdot 2 \cdot 3}$ . . . tooth formula, explanation in Volume X

▷ . . . . following (opposite page) color plate

▷▷ . . . . Color plate or double color plate on the page following the next

▷▷▷ . . . . Third color plate or double color plate (etc.)

Endangered species and subspecies